Microbial production of food ingredients, enzymes and nutraceuticals

© Woodhead Publishing Limited, 2013

Related titles:

Novel enzyme technology for food applications (ISBN 978-1-84569-132-5)

Enzymes: Biochemistry, biotechnology, clinical chemistry (ISBN 978-1-904275-27-5)

Protective cultures, antimicrobial metabolites and bacteriophages for food and beverage biopreservation (ISBN 978-1-84569-669-6)

Details of these books and a complete list of titles from Woodhead Publishing can be obtained by:

- visiting our web site at www.woodheadpublishing.com

- contacting Customer Services (e-mail: sales@woodheadpublishing.com; fax: +44 (0) 1223 832819; tel.: +44 (0) 1223 499140 ext. 130; address: Woodhead Publishing Limited, 80, High Street, Sawston, Cambridge CB22 3HJ, UK)

- in North America, contacting our US office (e-mail: usmarketing@ woodheadpublishing.com; tel.: (215) 928 9112; address: Woodhead Publishing, 1518 Walnut Street, Suite 1100, Philadelphia, PA 19102-3406, USA

If you would like e-versions of our content, please visit our online platform: www. woodheadpublishingonline.com. Please recommend it to your librarian so that everyone in your institution can benefit from the wealth of content on the site.

We are always happy to receive suggestions for new books from potential editors. To enquire about contributing to our Food Science, Technology and Nutrition series please send your name, contact address and details of the topic/s you are interested in to nell.holden@woodheadpublishing.com. We look forward to hearing from you.

The Woodhead team responsible for publishing this book:
Commissioning Editor: Sarah Hughes
Publications Coordinator: Anneka Hess
Project Editor: Cathryn Freear
Editorial and Production Manager: Mary Campbell
Production Editor: Richard Fairclough
Copyeditor: Marilyn Grant
Proofreader: Trevor Birch
Cover Designer: Terry Callanan

© Woodhead Publishing Limited, 2013

Woodhead Publishing Series in Food Science, Technology and Nutrition: Number 246

Microbial production of food ingredients, enzymes and nutraceuticals

Edited by
Brian McNeil, David Archer, Ioannis Giavasis
and Linda Harvey

WOODHEAD
PUBLISHING

Oxford   Cambridge   Philadelphia   New Delhi

© Woodhead Publishing Limited, 2013

Published by Woodhead Publishing Limited,
80 High Street, Sawston, Cambridge CB22 3HJ, UK
www.woodheadpublishing.com
www.woodheadpublishingonline.com

Woodhead Publishing, 1518 Walnut Street, Suite 1100, Philadelphia,
PA 19102-3406, USA

Woodhead Publishing India Private Limited, G-2, Vardaan House, 7/28 Ansari Road,
Daryaganj, New Delhi – 110002, India
www.woodheadpublishingindia.com

First published 2013, Woodhead Publishing Limited

© Woodhead Publishing Limited, 2013. Note: the publisher has made every effort to ensure
that permission for copyright material has been obtained by authors wishing to use such
material. The authors and the publisher will be glad to hear from any copyright holder it
has not been possible to contact.
The authors have asserted their moral rights.

This book contains information obtained from authentic and highly regarded sources.
Reprinted material is quoted with permission, and sources are indicated. Reasonable efforts
have been made to publish reliable data and information, but the authors and the publisher
cannot assume responsibility for the validity of all materials. Neither the authors nor the
publisher, nor anyone else associated with this publication, shall be liable for any loss,
damage or liability directly or indirectly caused or alleged to be caused by this book.

Neither this book nor any part may be reproduced or transmitted in any form or by any
means, electronic or mechanical, including photocopying, microfilming and recording, or by
any information storage or retrieval system, without permission in writing from Woodhead
Publishing Limited.

The consent of Woodhead Publishing Limited does not extend to copying for general
distribution, for promotion, for creating new works, or for resale. Specific permission must
be obtained in writing from Woodhead Publishing Limited for such copying.

Trademark notice: Product or corporate names may be trademarks or registered trademarks,
and are used only for identification and explanation, without intent to infringe.

British Library Cataloguing in Publication Data
A catalogue record for this book is available from the British Library.

Library of Congress Control Number: 2013930108

ISBN 978-0-85709-343-1 (print)
ISBN 978-0-85709-354-7 (online)
ISSN 2042-8049 Woodhead Publishing Series in Food Science, Technology and Nutrition
(print)
ISSN 2042-8057 Woodhead Publishing Series in Food Science, Technology and Nutrition
(online)

The publisher's policy is to use permanent paper from mills that operate a sustainable
forestry policy, and which has been manufactured from pulp which is processed using
acid-free and elemental chlorine-free practices. Furthermore, the publisher ensures that the
text paper and cover board used have met acceptable environmental accreditation
standards.

Typeset by Toppan Best-set Premedia Limited, Hong Kong
Printed by MPG Printgroup

© Woodhead Publishing Limited, 2013

Contents

© Woodhead Publishing Limited, 2013

© Woodhead Publishing Limited, 2013

© Woodhead Publishing Limited, 2013

© Woodhead Publishing Limited, 2013

© Woodhead Publishing Limited, 2013

© Woodhead Publishing Limited, 2013

© Woodhead Publishing Limited, 2013

Contributor contact details

(* = main contact)

Editors

Professor Brian McNeil
SIPBS Hamnett Wing
University of Strathclyde
161 Cathedral St
Glasgow
G4 0RE
UK

E-mail: b.mcneil@strath.ac.uk

Professor David Archer
School of Biology
Institute of Genetics
University of Nottingham
UK

E-mail: david.archer@nottingham.
ac.uk

Dr Ioannis Giavasis, Lecturer
Technological Educational Institute
of Larissa
Greece

E-mail: igiavasis@teilar.gr

Dr Linda Harvey
SIPBS Hamnett Wing
University of Strathclyde
161 Cathedral St
Glasgow
G4 0RE
UK

Chapter 1
Dr Brian J. B. Wood
SIPBS Hamnett Wing, Arbuthnott
Building
University of Strathclyde
161 Cathedral Street
Glasgow
G4 0RE
UK

E-mail: b.j.b.wood@btinternet.com

Chapter 2
Associate Professor Wanwipa
Vongsangnak*
Center for Systems Biology
Soochow University
Shizi Street No. 1, Suzhou
Jiangsu 215006
China

E-mail: wanwipa@suda.edu.cn

© Woodhead Publishing Limited, 2013

Professor Jens Nielsen
Department of Chemical and
    Biological Engineering
Chalmers University of Technology
SE-412 96 Gothenburg
Sweden

E-mail: nielsenj@chalmers.se

Dr L. M. Harvey, Dr M. Fazenda
    and Professor Brian McNeil
SIPBS Hamnett Wing
University of Strathclyde
161 Cathedral St
Glasgow
G4 0RE
UK

E-mail: b.mcneil@strath.ac.uk

Chapter 3
Valeria Mapelli,* Carl Johan
    Franzén and Lisbeth Olsson
Department of Chemical and
    Biological Engineering
Industrial Biotechnology
Chalmers University of Technology
SE- 41296 Gothenburg
Sweden

E-mail: valeria.mapelli@
    chalmers.se; franzen@chalmers.
    se; lisbeth.olsson@chalmers.se

Chapter 4
Dr Paul A Hoskisson
Strathclyde Institute of Pharmacy
    and Biomedical Sciences
University of Strathclyde
Hamnett Wing
161 Cathedral Street
Glasgow
G4 0RE
UK

E-mail: paul.hoskisson@strath.ac.
    uk

Chapter 5
Professor Robert J. Seviour*
La Trobe University
Bendigo
Victoria 3552
Australia

E-mail: r.seviour@latrobe.edu.au

Chapter 6
Professor Brian McNeil* and Dr
    L.M. Harvey
SIPBS Hamnett Wing
University of Strathclyde
161 Cathedral St
Glasgow
G4 0RE
UK

E-mail: b.mcneil@strath.ac.uk

Professor N. J. Rowan
Athlone Institute of Technology
Ireland

Dr I. Giavasis, Lecturer
Technological Educational Institute
    of Larissa
Greece

Chapter 7

Stuart M. Stocks
Novozymes A/S
Krogshoejvej 36
2880 Bagsvaerd
Denmark

E-mail: STUS@novozymes.com

© Woodhead Publishing Limited, 2013

Chapter 8
Yves Waché
Unit BioMA (Biotechology &
 Food microbiology)
Université de Bourgogne
AgroSup Dijon,
1, esplanade Erasme 21000 Dijon
France

E-mail: yves.wache@u-bourgogne.fr

Chapter 9
Professor S. Sanchez,* Dr B. Ruiz,
 Professor R. Rodríguez-Sanoja
Departamento de Biología
 Molecular y Biotecnología,
Instituto de Investigaciones
 Biomédicas,
Universidad Nacional Autónoma
 de México
Tercer circuito Exterior s/n, Ciudad
 Universitaria
Mexico D.F. 04510
Mexico

E-mail: sersan@biomedicas.unam.
 mx

Professor L. B. Flores-Cotera
Departamento de Biotecnología y
 Bioingeniería
Cinvestav-IPN
Ave. Instituto Politécnico Nacional
 No. 2508
Col. San Pedro Zacatenco
México D.F. 07360
Mexico

Chapter 10
H. Dvora and M. A. G. Koffas*
Department of Chemical
 Engineering
Rensselaer Polytechnic Institute
110 Eight Street
Troy
New York 12180
USA

E-mail: koffam@rpi.edu

Chapter 11
Karsten Hellmuth
Chr Hansen Nienburg GmbH
Process Development
Gr. Drakenburger Strasse 93-97
D-31582 Nienburg
Germany

Johannes M. van den Brink*
Chr Hansen A/S, Innovation
Enzymes Department
Bøge Allé 10-12
2970 Hørsholm
Denmark

E-mail: dkhvb@chr-hansen.com

Chapter 12
M. Sauer,* D. Mattanovich, and
 H. Marx
Department of Biotechnology
BOKU Wien-VIBT – University of
 Natural Resources and Life
 Sciences
Muthgasse 18
1190 Vienna
Austria

E-mail: michael.sauer@boku.ac.at

© Woodhead Publishing Limited, 2013

Chapter 13
Dr F. Grattepanche and Professor
    C. Lacroix*
ETH Zürich
Laboratory of Food Biotechnology
Institute of Food, Health and
    Nutrition
Schmelzbergstrasse 7, LFV C20
8092 Zürich
Switzerland

E-mail: christophe.lacroix@ilw.agrl.
    ethz.ch

Chapter 14
D. G. Burke, P. D. Cotter* and
    R. P. Ross
Teagasc Food Research Centre
Moorepark
Fermoy
County Cork
Ireland

E-mail: paul.cotter@teagasc.ie;
    paul.ross@teagasc.ie

C. Hill
University College Cork
Ireland

Chapter 15
Hideyuki Suzuki
Division of Applied Biology
Kyoto Institute of Technology
Matsugasaki
Sakyo-ku
Kyoto 606-8585
Japan

E-mail: hideyuki@kit.ac.jp

Chapter 16
Dr Ioannis Giavasis, Lecturer
Technological Educational Institute
    of Larissa
Greece

E-mail: igiavasis@teilar.gr

Chapter 17
Tom Granström and Matti Leisola*
Department of Biotechnology and
    Chemical Technology
Aalto University
FI-00076 AALTO
Finland

E-mail: matti.leisola@aalto.fi

Chapter 18
Dr Thu-Ha Nguyen and Prof.
    Dietmar Haltrich*
Food Biotechnology Laboratory
Department of Food Sciences and
    Technology
University of Natural Resources
    and Life Sciences
(BOKU Wien)
Muthgasse 18
A-1190 Vienna
Austria

E-mail: dietmar.haltrich@boku.
    ac.at

Chapter 19
Professor Colin Ratledge
Department of Biological Sciences
University of Hull
Hull
HU6 7RX
UK

E-mail: c.ratledge@hull.ac.uk

© Woodhead Publishing Limited, 2013

Chapter 20
Dr Barbara Klein* and Professor
  Dr Rainer Buchholz
Institute of Bioprocess Engineering
Department of Chemical and
  Biological Engineering
Friedrich Alexander University of
  Erlangen-Nuremberg
Paul-Gordan-Str.3
91052 Erlangen
Germany

E-mail: Barbara.Klein@bvt.cbi.
  uni-erlangen.de; rainer.
  buchholz@bvt.cbi.uni-erlangen.
  de

Chapter 21
Professor Dr Jose Luis Revuelta
Departamento de Microbiología y
  Genética
Universidad de Salamanca
C. M. Unamuno, E. Departamental
37007 Salamanca
Spain

E-mail: revuelta@usal.es

© Woodhead Publishing Limited, 2013

Woodhead Publishing Series in Food Science, Technology and Nutrition

© Woodhead Publishing Limited, 2013

© Woodhead Publishing Limited, 2013

© Woodhead Publishing Limited, 2013

© Woodhead Publishing Limited, 2013

© Woodhead Publishing Limited, 2013

© Woodhead Publishing Limited, 2013

© Woodhead Publishing Limited, 2013

© Woodhead Publishing Limited, 2013

© Woodhead Publishing Limited, 2013

Foreword

Microbes are fantastic creatures. They provide rapid generation times, genetic flexibility, unequaled experimental scale and manageable study systems. The practice of industrial microbiology has its roots deep in antiquity. Long before their discovery, microorganisms were exploited to serve the needs and desires of humans. They preserved milk, fruits and vegetables, and enhanced the quality of life by the resultant beverages, cheeses, bread, pickled foods and vinegar.

The use of yeasts dates back earlier than 7000 BC when the conversion of sugar to alcohol by yeasts was used to make beer in Sumeria and Babylonia. Wine was made in China as early as 7000 BC and in Assyria in 3500 BC. In 4000 BC, the Egyptians discovered that carbon dioxide generated by the action of brewer's yeast could leaven bread. Ancient peoples made cheese with molds and bacteria. Reference to wine can be found in the book of Genesis, where it is noted that Noah consumed a bit too much of the beverage. According to the Talmud, 'a man without salt and vinegar is a lost man'. The Assyrians treated chronic middle ear diseases with vinegar, and Hippocrates treated patients with it in 400 BC. According to the New Testament, vinegar was offered to Jesus on the cross. For thousands of years, moldy cheese, meat and bread were employed in folk medicine to heal wounds. By 100 BC, ancient Rome had over 250 bakeries making leavened bread. As a method of preservation, milk was fermented to lactic acid to make yoghurt and also converted into Kefyr and Koumiss using *Kluyveromyces* species in Asia. The use of molds to saccharify rice in the Koji process occurred by 700 AD.

By the 14th century AD, the distillation of alcoholic spirits from fermented grain, a practice thought to have originated in China or the Middle

© Woodhead Publishing Limited, 2013

East, was common in many parts of the world. Vinegar manufacture began in Orleans, France, at the end of the 14th century and the surface technique used is known as the Orleans method.

In the 17th century, Antonie van Leeuwenhoek, in the Netherlands, turning his simple lens to the examination of water, decaying matter and scrapings from his teeth, reported the presence of tiny 'animalcules', that is, moving organisms less than a thousandth the size of a grain of sand. He was a Dutch merchant with no university training but his spare time interest was the construction of microscopes. This lack of university connection might have caused his discoveries to remain unknown had it not been for the Royal Society in England and its secretary, Henry Oldenburg, who corresponded with European science amateurs. From 1673 to 1723, Leeuwenhoek's great powers as a microscopist were communicated to the Royal Society in a series of letters.

Most scientists thought that microbes arose spontaneously from non-living matter. What followed was a 100-year argument over spontaneous generation, aptly called the 'War of the Infusions'. Proponents had previously claimed that maggots were spontaneously created from decaying meat, but this was discredited by Redi. By this time, the theory of spontaneous generation, originally postulated by Aristotle, among others, had been discredited with respect to higher forms of life, so its proponents concentrated their arguments on bacteria. The theory did seem to explain how a clear broth became cloudy via growth of large numbers of such 'spontaneously generated microorganisms' as the broth aged. However, others believed that microorganisms only came from previously existing microbes and that their ubiquitous presence in air was the reason why they would develop in organic infusions after gaining access to these rich liquids. Three independent investigators, Charles Cagniard de la Tour of France, Theodor Schwann and Friedrich Traugott Kützing of Germany, proposed that the products of fermentation, chiefly ethanol and carbon dioxide, were created by a microscopic form of life. This concept was bitterly opposed by the leading chemists of the period (such as Jöns Jakob Berzelius, Justus von Liebig and Friedrich Wöhler), who believed fermentation to be strictly a chemical reaction; they maintained that the yeast in the fermentation broth was lifeless decaying matter. Organic chemistry was flourishing at the time and the opponents of the living microbial origin were initially quite successful in putting forth their views. Interest in the mechanisms of these fermentations resulted in later investigations by Louis Pasteur, which not only advanced microbiology as a distinct discipline, but also led to the development of vaccines and concepts of hygiene which revolutionized the practice of medicine.

In 1850, Davaine detected rod-shaped objects in the blood of anthrax-infected sheep and was able to produce the disease in healthy sheep by inoculation of such blood. In the next 25 years, Pasteur of France and John Tyndall of Britain demolished the concept of spontaneous generation and

© Woodhead Publishing Limited, 2013

proved that existing microbial life came from pre-existing life. In the 1850s, Pasteur had detected two optical types of amyl alcohol, D and L, but he was not able to separate the two. For this reason, he began to study living microbes carrying out fermentation. Pasteur concluded in 1857 that fermentation was a living process of yeast. In 1861, he proved the presence of microbes in air and discredited the theory of spontaneous generation of microbes. At this point fermentation microbiology was born, but it took almost two decades, until 1876, to disprove the chemical hypothesis of Berzelius, Liebig and Wöhler, that is that fermentation was the result of contact with decaying matter.

Pasteur was called on by the distillers of Lille to find out why the contents of their fermentation vats were turning sour. He noted through his microscope that the fermentation broth contained not only yeast cells, but also bacteria that could produce lactic acid. He was able to prevent such souring by a mild heat treatment, which later became known as 'pasteurization'. One of his greatest contributions was to establish that each type of fermentation was mediated by a specific microorganism. Furthermore, in a study undertaken to determine why French beer was inferior to German beer, he demonstrated the existence of strictly anaerobic life, that is, life in the absence of air.

In 1877, Moritz Traube proposed that (i) protein-like materials catalyzed fermentation and other chemical reactions and (ii) they were not destroyed by doing these things. This was the beginning of the concept of what we call enzymes today. He also proposed that fermentation was carried out via multistage reactions in which the transfer of oxygen occurred from one part of a sugar molecule to another, finally forming some oxidized compound like carbon dioxide and a reduced compound such as alcohol. The field of biochemistry became established in 1897 when Eduard Buchner found that cell-free yeast extracts, lacking whole cells, could convert sucrose into ethanol. Thus, the views of Pasteur were modified and it became understood that fermentation could also be carried out in the absence of living cells.

The work of Louis Pasteur pointed to the importance of the activity of non-pathogenic microbes in wine and beer in producing alcohol. This realization resulted in a large number of microbial primary metabolites of commercial importance being produced by fermentation. Primary metabolism involves an interrelated series of enzyme-mediated catabolic, amphibolic and anabolic reactions which provide biosynthetic intermediates and energy, and convert biosynthetic precursors into essential macromolecules such as DNA, RNA, proteins, lipids and polysaccharides. It is finely balanced and intermediates are rarely accumulated. By deregulating the primary metabolism, overproduction of many primary metabolites was achieved in the fermentation industry. Commercially, the most important primary metabolites were amino acids, vitamins, flavor nucleotides, organic acids and alcohols.

© Woodhead Publishing Limited, 2013

With such a fantastic background, the book edited by McNeil, Archer, Giavasis and Harvey provides the reader with an excellent view of how microbes are used today to make an amazing list of food ingredients, foods, beverages, flavors, enzymes and neutraceuticals. It is divided into two sections. Part I describes the use of bioprocessing to make food ingredients and the application of modern techniques such as systems biology, metabolic engineering and synthetic biology, and with advances in fermentation technology including monitoring of fermentations, control of microbial cultures, process scaleup and scaledown to make food ingredients and industrial enzymes used in the food industry. Part II describes the use of microorganisms for the manufacture of food flavors, carotenoids, flavanoids, terpenoids, enzymes, organic acids, probiotics, bacteriocins, amino acids, biopolymers, nutraceuticals, oligosaccharides, polyunsaturated fatty acids and vitamins. It is amazing to me that so many important food-related topics could be covered in one book but indeed it is. The book will be useful to industrial microbiologists, food technologists and students interested in the application of microorganisms for useful purposes. No longer will people think of microbes only as undesirable agents of disease or sources of antibiotics to combat such diseases. They will learn of the useful application of these wonderful creatures to make their lives more enjoyable and healthy and to contribute to our ever-growing increase in human life expectancy.

Arnold L. Demain
Research Institute for Scientists Emeriti (RISE)
Drew University
Madison, New Jersey, USA

© Woodhead Publishing Limited, 2013

1

Bioprocessing as a route to food ingredients: an introduction

B. J. B. Wood, University of Strathclyde, UK

DOI: 10.1533/9780857093547.1

Abstract: The earliest bioprocessing, although not understood as such at the time, was producing foods and fermented beverages. The processes were essentially artisan in nature. They must have begun at village level and such processes are still found in developing countries. Developments in societies' complexity drove the establishment of more organised production and marketing, which in turn encouraged process standardisation and more uniform and reliable products. Although food and beverages were the principal products, the methods for producing them also resulted in technical materials such as acetic acid, ethanol and lactic acid being manufactured. The early development of microbiology was intimately linked with food and beverage industrialisation, as well as the drive to understand the nature of diseases. Today the biotechnological industries encompass many organism types, vast resources and enormous product diversity. The shift from fossil materials to renewables will both drive further innovation in biotechnology and increase the scope for its products' applications, for example, polymerised lactic acid as a replacement for petrochemical polymers. Often these 'substitutes' offer additional advantages, such as easy biodegradation. The future for bioprocessing in food ingredient production, but also in the wider industrial sphere, is very bright. Fully developing it may require an interesting fusion of modern technologies such as stirred tank reactor fermentors with reinvented and modernised versions of ancient technologies such as solid substrate (koji) fermentations. This book demonstrates the potential and actual developments across the biotechnological spectrum.

Key words: alcoholic beverage production, oriental food fermentations, prebiotics, probiotics and neutraceuticals, solid substrate fermentation, stirred tank reactors.

1.1 Food fermentation as an ancient technology: an overview

1.1.1 Fermentation for food modification and conservation

We cannot know when food fermentation began, but all known cultures use it to modify and/or conserve foods. By their very nature most food products

© Woodhead Publishing Limited, 2013

are intrinsically unstable, under attack from enzymes present in them as a result of their production, and by the many microbes and other organisms present in the foodstuff's environment, eager to feast on this food resource. Even those few that are relatively stable, such as seeds, are only relatively so, and in exchange for this enhanced storage time, these tend to be fairly unappetizing unless cooked or otherwise modified. Even hunter-gatherers, at the start of human culture, accustomed though they were to the feast-then-famine lifestyle that was the inevitable outcome of that existence, must surely have sought methods to save any excess food from the hunt for another day.

When the first human settlements began, presumably as simple agriculture developed (although recent excavations in Turkey [see Norenzayan, 2012] suggest that some form of settlement, at least for religious functions, may have preceded agricultural cultivation), the need to store harvests, plant or animal, must have become essential for settled groups to function. The first human settlements preceded developments such as writing or other means for recording information, indeed this was probably one of the driving forces in inventing recording methods for even such basic things as noting crop yields, and its inevitable consequence, calculating the portion to be grabbed by the ruling classes. Thus we cannot know how humans came to realise that some of the same agents that destroy foods can also act to conserve and improve them.

The understanding that microbes exist, obey biological laws and can be used in a controlled way, are essentially outcomes of our development of the scientific method. That is not to say that our remote ancestors did not see such things as the mycelium that developed as a mould overgrew a food item, or the yeast mass that developed as a wine or beer underwent fermentation. However, it is far less certain that they understood that these were living things, indeed the 'vitalist' and other arguments over (for example) yeast's nature and its mode of action, were still matters for heated debate until well into the 19th century.

1.1.2   Microbial production of industrial chemicals as well as food ingredients

Our primary focus here is on microbial activities in their roles connected with foods, although this is in some ways an artificial distinction. Microbes' roles in generating a range of products important in manufacturing and other activities are essentially indistinguishable from their directly food-related functions. For example, consider ethanoic (acetic) acid. This still has many technological applications in addition to its food flavouring and preserving uses. Today these industrial uses are largely met by synthetic, petrochemically derived product, although I have visited a substantial manufactory in Indonesia where sugar cane molasses are fermented to ethanol and then distilled to a high purity product and catalytically oxidised to

© Woodhead Publishing Limited, 2013

ethanoic acid, yielding three product streams (ethanol, ethanoic acid and ethyl ethanoate) for the Japanese electronics industry. In fact ethanoic acid has been produced by fermentation for millennia and its manifold applications in medicine, manufacturing, pigment production, and so on developed and have evolved over these millennia.

The same is true for other organic acids, particularly lactic acid, while yet other acids, traditionally derived exclusively from plant resources (such as citric acid), are required in such vast amounts today that it has become essential to develop fermentations to meet this ever-expanding demand, these acids being too complex for synthetic routes from petrochemicals to be economically viable up to the present time. Lactic acid is a particularly illuminating example here. Apart from its many food and beverage applications, it always had industrial uses. Traditionally these were met by fermentation production. Latterly, petrochemical manufacture became important. Now, with the development of biodegradable polymers for various applications, not least as plastic shopping bags, there is a huge revival in production by fermentation, driven in substantial part by the need for a single optical isomer to enable polymer strands to be synthesized. An example of this is the big investment in Thailand based on fermentation production from that country's extensive carbohydrate resources. Although there must be concerns that competition for these resources will increase prices paid for staple energy foods by the country's poorest inhabitants, there are certain to be positive impacts on employment and export earnings. Thus in this case there is a reversion to traditional production for good economic reasons. Such developments must, however, be considered on a case-by-case basis, and there can be no general rule, either that petrochemical production will supplant fermentation, or that the opposite will be the case.

Another interesting non-food application of microbes is in pigment production. For example, dyes from lichens played an important role in supplying the rich colours essential for dyeing the wool that was then woven into traditional Scottish tartans. Lichens grow only slowly, their being symbioses between cyanobacteria (or sometimes algae) and filamentous fungi, and were adapted to grow in rigorous conditions unsuited for most life forms. Thus, traditional dyeing was unsustainable as markets for Scottish traditional clothing items grew and these pigments have been largely supplanted by azo and other synthetic dyes. Lichens are notoriously difficult to study in the laboratory and academic study of them is generally in serious decline, but there is an active group still working with them in Thailand's Ramkhamhaeng University. One line of enquiry is determining what conditions impel the fungal partner in the lichen to activate the sites of DNA that encode the enzymes responsible for synthesising the complex pigment molecules. This should eventually supply a deeper understanding of how the symbiosis works, why the pigments are produced and, it is to be hoped, how to operate fermentation to generate the pigments on an economically viable scale.

© Woodhead Publishing Limited, 2013

The foregoing discussion is an idiosyncratic and superficial glance at a complex subject. Subsequent sections of this chapter will consider some of the issues raised in a little more detail, but a true in-depth examination of the issues will be left to the individual chapters that comprise this treatise. It is this author's commission to set the scene within which subsequent chapters will combine to generate a compelling argument that microbial options for producing food ingredients, enzymes and nutraceuticals have a major part to play in developing a food industry that is fit for purpose in the 21st century. In this connection it is essential to appreciate that, for such developments to be economically viable it is vital to think beyond narrowly food-orientated applications and see that processes originally derived from (often) traditional methods will have commercial non-food applications. It has been suggested, for example by the late John Bu'lock (personal communication) that around the late 1940s the leading industrial nations had to decide between relying on traditional fermentation to generate industrial organic feedstocks and supplanting them with petrochemical products. At the time the choice was driven by simple economics; cheap petrochemical feedstocks and highly reliable chemical engineering processes that easily scaled up to vast outputs meant that fermentation could not compete, so fermentation was confined to specialist food applications where chemical technology was not applicable. Now, as petrochemical stocks are seen to be seriously limited, while demand for products and intermediates traditionally supplied by fermentations inexorably rises, it is salutary to look back to the middle of last century and the 3rd edition of Prescott and Dunn's *Industrial Microbiology* (Prescott and Dunn, 1959) and realise what an extraordinary range of essential materials can be generated by fermentation, although we now have to factor in the fact that currently some of the feedstocks for these processes are also human and farm animal foods, posing the risk that a return to heavy reliance on fermentations will impact adversely on food and feed prices.

1.2   Solid substrate fermentations (SSF) and stirred tank reactor (STR) technology: relative industrial dominance

As inspection of the six editions of Prescott and Dunn's work will demonstrate, SSF was the preferred option for many industrial fermentations until the 1940s. Certain processes, notably ethanol and lactic acid production, but also procedures such as acetone–butanol fermentations, were liquid processes, and enzyme production and most other procedures reliant on filamentous fungi were derived from koji processes associated with traditional products such as soy sauce and rice wine. Even citric acid fermentation was effected by mould growing as a felt of mycelium floating on a thin layer of growth medium. The initial production of penicillin for the battlegrounds

© Woodhead Publishing Limited, 2013

of the Second World War was also conducted in this manner, although despite commandeering milk bottles, hospital urine bottles and bedpans, and finally the designing of 'penicillin' flasks, it was evident that the heavy demand for penicillin G could not be met in this way. This is generally regarded as the start of the drive to use filamentous fungi in STR growth. This has been so successful that even fermentation for citric acid has finally been coerced into STR, although the very special conditions needed to divert the mould's metabolism into overproducing the acid meant that this was a particularly difficult feat.

Was the early reliance on SSF just inertial, following on from ancient and successful technologies? Was it really unreasonable to expect STR methods to develop easily? Very few moulds grow suspended in a water column in nature and those that do are mostly highly specialised organisms that have no current commercial interest. Commercially important organisms are isolated from environments where they are growing on leaf litter, rotting fruit and other vegetation, in the soil, and so on. All of these are essentially SSF situations. This means that the interesting organisms are not adapted to growing suspended in liquid. Indeed from their perspective this habitat, particularly when we add in the stresses, mechanical and otherwise, required to achieve optimal mixing and aeration, must seem rather hostile, and so it may be seen as rather remarkable that STR technology is so dominant now. All experience suggests that these fungi want to grow attached to something and everyone who has operated an STR growing them is acutely aware that they will anchor themselves to any attachment points that are available, even the stirrer paddles, on occasion, despite the shear forces operating there.

1.2.1   The case of vinegar production using bacteria

It should be appreciated that these considerations, although most evident for fungal fermentations, are more generally applicable, even to bacteria. Consider vinegar production. The most ancient methodologies, culminating in the Orleans Vinegar Process, depended on a film of vinegar bacteria forming on the surface of an unstirred body of alcoholic liquid, the vinegar bacteria being bound together by a slimy cellulosic exopolysaccharide layer exuded by the bacteria. The process, although slow, inefficient in terms of productive capacity and hence expensive to run, is still preferred for the finest culinary wine vinegar because it is considered to produce the best vinegar in terms of flavour and overall organoleptic quality. However, vinegar has many uses, medicinal, as a food preservative, as a sanitising agent and antimicrobial, and as the source of ethanoic acid for industrial applications before the petrochemical era, to quote the more obvious examples. Thus the Industrial Revolution vastly increased demand for vinegar and it became evident that traditional methods were inadequate to meet that demand.

© Woodhead Publishing Limited, 2013

In response to this the 'Quick' process was developed to generate vinegar in a shorter time span and with considerably increased control over how the process operated. Essentially the apparatus for this process comprises a tower packed with a support medium (classically beech or birch twigs or shavings, these woods being chosen because they do not taint the vinegar with resins or other odour-donating or taste-changing substances). The liquid to be converted into vinegar is sprayed over the top of this packed material, collected at the bottom of the tower and continually recycled through the tower until the ethanol has all been used up. At the same time a pump blows air countercurrent to the liquid up the tower.

A fully operational tower can convert a batch of feedstock into vinegar in seven to ten days. The significance that this process has for the current discussion is that the material packed into the tower is not simply there to maximise gas exchange into the liquid phase. On the contrary, the wood quickly becomes coated with a bacterial biofilm where most of the oxidation takes place. Successive runs with fresh feedstock gradually increase the biomass layer until eventually the tower has to be cleaned out and repacked with fresh support medium. Given that this is an easily operated, reliable and forgiving (interruptions to operation because of power supply outages damage the biofilm, but when normal operation is resumed it quickly recovers and returns to full productivity) system, it seems a little surprising that it was not more generally applied, for example to fungal fermentations. However the key to the Quick Vinegar process's success lies in the powerful antimicrobial properties possessed by ethanoic acid. Put simply, very little except vinegar bacteria (various members of the genera *Acetobacter* and *Gluconobacter*; for further discussion of these organisms and their biology and nomenclature see Adams, 1998) will grow in a properly operated tower. When a newly packed tower is brought into operation the feedstock is a mixture of about 80% ethanolic solution and 20% vinegar. In subsequent recharges a proportion of the finished vinegar is always left behind in the tower. Thus the process air requires only simple filtration, rather than passage through bacteria-retaining filters, and the tower is open to the environment. The most modern production technologies use very powerful stirring and forced aeration in STR, which while highly efficient require advanced technology, as these systems are very sensitive to even slight interruptions in their operation, with the cells dying within a short time of such interruption. For a fuller discussion of the various technologies employed for industrial vinegar production see Adams (1998).

Note that the key fact for this discussion is that the 'Quick' process tower's operation depends upon the bacteria attaching to a support material. In nature there are other examples where this provision of something to which bacteria can attach also seems to be very significant. For example, when studying carbon budget in a sea-loch affected by discharges from a factory processing seaweed for alginate production (Tyler, 1983), it was observed that where finely divided residue from the process was suspended

© Woodhead Publishing Limited, 2013

in the water column the total bacterial count was much higher than in areas without this material, even though the extraction process meant that the residue was recalcitrant and difficult to metabolise, suggesting that it was a relatively inert support for the organisms and that they were utilising it principally as a 'home' from which to exploit dissolved organic material. This principle was illustrated in a particularly dramatic fashion when inspecting a molasses to ethanol fermentation factory in Thailand. The residual liquor ('pot ale') after the ethanol had been distilled off was being discharged into a holding pond which had grown over time into a small lake. We were puzzled by how little evidence we could find for anything happening to the easily biodegradable dissolved organic substances in the pot ale lake. However, we noticed that in some areas bubbling indicated that biodegradation and methane production were happening, and it was confirmed that these areas correlated with places where there were solids suspended in the water, providing a support for the methanogens. The conclusion may be drawn that even bacteria and archaea can benefit from attachment to a suitable solid substrate.

1.2.2   The case of potable water processing and wastewater treatment

Most people would guess that alcohol production is, in aggregate, the world's largest fermentation industry. In fact that title must surely go to the potable water processing and wastewater treatment industries. Both of these rely heavily on processes that are essentially SSF. The sand filter is at the heart of processing raw potable water into safe drinking water. By definition, water that is to be treated for a public water supply will be high quality, with low bacterial counts and dissolved organic matter content. However, to be suitable for the final treatments, such as chlorination or ozone treatment, the raw feedstock must have dissolved organic matter and bacterial counts reduced to a minimum. The sand filter is much more than a passive device to remove suspended solids. In fact the sand grains are the support for a complex biofilm incorporating bacteria and other organisms that utilise dissolved organic material, organisms that feed on the microbes present in the water, and so forth.

Sewage treatment also depends for its success on biofilms. This is most obvious in the traditional trickling filter, where the liquid runs over stones or other solid support that holds a complex biofilm, probably not unlike that in the potable water sand filter, and discharges the same functions such as removing bacteria and mineralising organic matter. Just as in vinegar production, this slow and land-hungry process was insufficient for the conurbations that developed in the industrial revolution, although it serves smaller rural communities well to this day. The response here was the ubiquitous 'activated sludge' process where the wastewater is violently stirred and aerated to speed up mineralisation and microbe removal. This does not sound like an SSF, but the sludge that gives the process its name is

© Woodhead Publishing Limited, 2013

composed of tiny flocs, each a complex biological aggregate, most interesting under the microscope for the protozoa that are actively feeding on microbes suspended in the water. While this is not obviously an SSF when compared with the other examples cited above, it is reasonable to interpret the complex flocs as providing a habitat substantially analogous to a biofilm attached to a solid substrate.

The recent development in wastewater treatment, sometimes described as a 'rotating biodisc contactor' takes the biofilm concept even further. Commercial installations comprise a set of vertical discs mounted sequentially on a rod that is rotated. The assemblage is contained in an enclosure into which the water to be treated is admitted to a depth just below the discs' centre. As the discs rotate, their surfaces constantly pass through the liquid and then through the air; a biofilm quickly develops on each disc and, as the liquid is replenished at one end while the treated effluent is discharged from the other, biochemical oxygen demand (BOD) and microbial burden are reduced through the action of the organisms comprising the biofilm. Excess biofilm sloughs off the discs, collects in the bottom of the containment vessel and is eventually removed. It seems that the changes in composition of the wastewater as it flows through the system may influence the organisms making up the biofilm on successive discs and reports suggest that when such a system is operating optimally, the final discs are largely populated by diatoms, even though there will be little or no light in the enclosure. This system seems particularly tolerant towards changes in hydraulic loading and the influent liquid's BOD, making it especially attractive for campsites and other resort areas where the loading changes substantially between weekdays and weekends.

A few years ago there was interest in applying laboratory-scale biodiscs to cultivate fungi and other organisms that may exhibit a preference for attached growth and can be expected to exhibit physiologically different behaviour when so attached, compared with growth in liquid suspension. Such systems also permit ready change from a medium optimised for cell growth to one designed to effect a process, such as elaboration of a desired chemical or a biotransformation, but I am not aware that such systems have been adopted industrially.

1.2.3   Current SSF industrial applications and research
Returning to more conventional SSF based on the koji systems associated with traditional oriental fungal processes, the great successes achieved by STR systems meant that SSF largely disappeared from the more developed economies such as Europe and North America, although mushroom cultivation, not normally thought of as such, but in fact an archetypal SSF, underwent considerable expansion. In its home territories (for example China, Japan, South East Asia) SSF continued to supply soy sauce, sake and many other traditional food fermentations using the essential primary

© Woodhead Publishing Limited, 2013

transformation from indigestible polymers (protein, starches, other long-chain carbohydrates) found in seeds such as rice, wheat and soybeans into simple sugars, peptides, amino acids, and so on, that are easily assimilated by the yeasts and lactic acid bacteria that complete the production process.

Although India has no traditional fermentations that depend upon fungal SSF in any obvious way analogous to the koji process, some research institutes there, notably the Central Food Technology Research Institute (CFTRI) in Mysore and the Regional Research Laboratory in Trivandrum (TRRL) have maintained active research groups investigating the technology. Among the outcomes of this continuing interest have been two useful publications edited by Karanth (1991) and Pandey (1994). Doelle *et al.* (1992) edited a multi-author work, perhaps intended more for the European, American and Australasian markets, but notable for the number of contributors from India. More recently the Agricultural University of the Netherlands, Wageningen has generated several PhD theses and their associated peer-reviewed publications; typical examples are Smits (1998), concerning modelling fungal growth and activity in SSF, and Nagel (2002) dealing with methods for controlling the temperature and the moisture content within koji-style SSF.

Applications where microbes are attached to a support material, and medium trickles over the system thus created, have been noted in relation to some processes, but it was acknowledged that the problems involved in maintaining sterility under these conditions resulted in (i) applications that are limited by what can grow in the system (vinegar) or (ii) where, for success, it is desirable that a natural consortium develop from organisms already present in the environment (water and sewage treatment). However, with the improved technologies now available, systems exist where, for example, animal cell lines that require attachment to a physical support can be grown in liquid suspension because the system incorporates strips or ribbons of a suitable support material. It would be interesting to learn about applications of such systems to microbes such as filamentous fungi, and also what would happen if a tower were packed with such material, sterilised and operated aseptically after inoculation with an organism of interest.

Thus, while STR systems are now the dominant industrial vehicles for industrial fermentations, it seems reasonable to suppose that SSF can make a useful contribution in the future. The successes for STR systems are in part attributable to the capacity for controlling parameters such as pH, temperature, oxygen levels, medium composition, and so forth, to very fine tolerances. On the other hand SSF are difficult to control in traditional systems, but the resurgence in interest in these systems in places such as Wageningen and the continuing research in centres such as Mysore and Trivandrum, allied with the traditional food fermentations in Asia, which are undergoing extensive development to meet modern markets and operating conditions, suggest that the future for SSF may be reasonably bright. Inevitably however the contributors to the present text are principally

© Woodhead Publishing Limited, 2013

focussed on STR technologies, accurately representing the current situation.

1.3   Development of bioprocessing as a route to food ingredients: the history of koji

For the modern biological scientist, the fact that these technologies were developed and became highly productive and evidently reliable industrial processes in the absence of any knowledge about microorganisms and their capabilities and requirements, or indeed their very existence, is striking and rather puzzling. However, reading Joseph Needham's *Science and Civilization in China* (2000), it becomes clear that these and numerous other biotechnology-based industrial processes were substantial contributors to China's economic activity many centuries before the Common Era started. This required not only experts able to carry out the processes and control their progress, but also the complex commercial infrastructure required to supply the required food ingredients (soybeans, rice, wheat, salt, barley and so forth), in the right amounts and at the right time and also, of course, the capacity to market the products.

To the microbiologist these processes are also impressive for their complex succession of microbes, each effecting a distinct function, and in the right order to generate the specific organoleptic properties associated with the food or condiment. Consider soy sauce; first the koji mould needs to grow for a carefully regulated time, too short and the required enzymes will not have been synthesised in sufficient quantity for efficient and sufficient hydrolysis of seed storage components in the next stage. On the other hand, if the koji is incubated for a little too long then changes associated with the mould beginning to enter the sporulation stage cause the mild and slightly sweet odour present at the beginning of mould growth to be supplanted by harsh odours that would adversely affect the finished product's organoleptic quality. Also the mould growth is a highly aerobic and exothermic process, requiring good aeration, but this carries the risk of drying the mass of beans, wheat and mould, so inhibiting mould growth. Balancing these requirements is difficult using modern technology, even with our understanding of the mould and its needs and functions.

The mould itself is remarkable and something of an enigma. *Aspergillus oryzae* has only been found in soy sauce, miso and rice wine production facilities and does not seem to occur in nature outside these breweries. It is closely related to *Aspergillus flavus*, a toxin-producing mould, although *A. oryzae* does not produce toxins and always seems to be perfectly safe as a participant in food fermentations. *A. oryzae* is of unknown origin and how it developed from its presumed ancestor is most mysterious. In sake production, different *A. oryzae* strains are employed; they produce higher levels of amylolytic enzymes and lower protease levels than are found in soy sauce

© Woodhead Publishing Limited, 2013

moulds; the moulds used for miso are closer to the sake than the soy sauce moulds in their enzymatic activities. All these technologically important matters had been empirically developed long before enzymes and their functions were known.

In the second, salt-mash or moromi stage, the product from the koji stage is mixed with brine. This stops the mould's growth and the combination of high salt level and anaerobiosis in the mixture quickly kill it, adding hydrolytic enzymes contained within the mycelium to the mixture exported from the hyphae to digest the polymers present in the food substrate. Even though the salt concentrations are not favourable to the enzymes' activity, there is substantial further hydrolysis during the long salt-mash incubation. This does not mean that the mash is sterile, on the contrary, specialised lactic acid bacteria, originally classified as *Pediococcus*, but now assigned to the genus *Tetragenococcus*, specifically *Tetragenococcus halophilus*, which, by producing lactic acid and so acidifying the brew, produce conditions favourable for yeasts to grow. These yeasts generate flavours essential to the organoleptic qualities of the finished soy sauce. In miso fermentation, the salt mash contains less free water, but the paste so generated supports a microbial succession comparable to that in soy sauce moromi. For rice wine production there is no salt added when the koji is mixed with water, so, to prevent spoilage it is essential to develop a lactic fermentation as soon as possible, followed by a yeast fermentation to generate ethanol; because the sugar level in the fermentation is low, with fermentable maltose and glucose being quickly fermented by the yeast, high alcohol levels result.

The vital role played by the initial koji process in hydrolysing complex proteins and carbohydrates to simple components available for yeasts and lactic acid bacteria to ferment, is clearly comparable to the way that malting barley produces enzymes capable of hydrolysing the complex components present in barley seeds. We do not know how either of these processes was discovered, but while it is fairly easy to imagine malting being observed as a consequence of steeping grains prior to cooking them, it is difficult to see why mould growth, normally regarded as spoilage, should be identified as a useful process if properly controlled. Industrial applications for koji moulds in the early years of industrial enzyme production have been noted above. Latterly these enzymes are generated in STR processes, but SSF thrives in soy sauce, miso and rice wine production. There has been some revival of interest in applying SSF now that advanced engineering permits greater control for koji processes. Doelle *et al.* (1992) provide a useful compilation of information current at that time and successive PhD theses from Wageningen University of Agriculture (for example Smits, 1998; Nagel, 2002) advance the possibilities of rediscovering and applying the koji process in modern industrial settings.

Finally, it must be appreciated that other SSF, analogous to kojis, are found in many traditional food production processes. A good example is traditional production of the delicious Indonesian food called tempe

© Woodhead Publishing Limited, 2013

(tempeh). Essentially, cooked, dehulled soybeans are wrapped in banana leaves and left overnight. *Rhizopus* spp. grow and form tightly meshed mycelia, binding the beans into a solid cake that can be cooked to yield a nutritious and richly flavoured food unexpectedly rich in Vitamin B12. An international symposium (Sudarmadji *et al.*, 1997) provided much information concerning this valuable food, little known outside Indonesia, although it is available in the Netherlands and is also of interest to vegetarians interested in 'alternative' lifestyles.

1.4   Conclusion: food biotechnology past, present and future

In the preceding material I have attempted to demonstrate that biotechnological applications in food production have ancient lineage and are worldwide in their scope. Most people, even some skilled in biotechnology, do not appreciate the full impact that microbial processes have on our foods, now and throughout human history. This may be better appreciated by looking at texts such as this author's *Microbiology of Fermented Foods* (1985, 1998), Knoor's *Food Biotechnology* (1987), Hutkins' *Microbiology and Technology of Fermented Foods* (2006) and the forthcoming *The Oxford Handbook of Food Fermentations*, edited by Bamforth and Ward. Finally, I must reference the now rather elderly but still unique and irreplaceable *Fermented Foods of the World: A Dictionary and Guide* compiled by Geoffrey Campbell-Platt (1987). These technologies are still important in all known human cultures and continue to benefit us through their effects on our foods.

Currently, fermented foods and biotechnology applied to food and flavouring production continue to develop in scope and importance. This applies equally to modern high-technology innovations and to traditional products. An example of the latter is the extraordinary growth shown by the Kikkoman Corporation. This company is the largest Japanese soy sauce producer; in the 1970s it decided to develop its market in the USA by building a brewery in Wisconsin. Since then it has gone on to build new production facilities in other parts of the world, for example in the USA, Europe (Groningen, The Netherlands) and South East Asia. Fifty years ago most Westerners, if they knew soy sauce at all, saw it only as a bottle of brown stuff on the tables in Chinese restaurants, but now it is as much a staple as salt and pepper in many kitchens and the flavour (umami) that it contributes to foods where it is employed is generally recognised as a true fifth taste ranking beside salt, sweet, sour and bitter. Another excellent example is yogurt. Again, 50 years ago this was 'exotic' to most consumers in the West, although an everyday food in East Europe and Central Asia. Now, changing tastes and astute marketing make it an everyday product, not only in Europe, the USA, Australia, and so on, but, most remarkably, even in a

© Woodhead Publishing Limited, 2013

country such as Thailand, where milk was not traditionally consumed in the way found in Northern Europe. It is a useful source of calcium and is tolerated even by some people with milk or lactose intolerance. However, it is doubtful if most consumers consider these matters when deciding to purchase and eat it. Therefore its place in the supermarket must be principally attributed to skills deployed by marketing experts. There is an interesting contrast with kefir, another fermented milk product from East Europe and Russia, where it is held to possess important therapeutic properties. While certainly not *un*obtainable in the United Kingdom, kefir is definitely a niche market product. On the other hand I found 'fruit-flavored kefir' in every supermarket that I visited while working in the USA nearly 20 years ago. The reasons for this notable difference elude me.

The current text surely demonstrates that the future of food biotechnology is assured. More importantly however, it shows that innovation is key to future development. We cannot predict what foods will be of increased importance in the future, although we can presumably assume that the staples will remain as important calorie and protein sources. What then can we surmise from the contents list of this present text? Different readers will draw different conclusions, depending in large part on their interests and background, so I can only present an opinion from my own perspective as an elderly and traditional microbiologist. For me the most striking thing about the contents is the way in which it is assumed (correctly) that future development must require that humans assert control over every aspect of processes. This is not to suggest that traditional processes were free from direct human intervention, indeed the examples described above make it clear that this is not the case. However, traditional processing required persuading things that were little understood by the practitioner to perform in ways only a little removed from what would happen without human intervention. Here we start off with systems biology and metabolic engineering, two concepts that would have seemed esoteric to most working in fermented food technology a couple of decades ago. However, I suggest that the topics discussed under this heading show connections with matters such as the studies in understanding and regulating solid substrate fermentations referred to above.

We will need to assert close control over processes involving both eukaryotes (yeasts and filamentous fungi) and food bacteria if we are to develop their full potential. Furthermore it will not stop with these organisms; the enormous potential possessed by algae, but so far largely unexploited, will only be realised when we can direct their activities with confidence, yet these organisms are relatively easy to grow at industrial scale.

Recently there has been a lively interest in both the scientific and popular information media concerning so-called 'pharming'; a useful and informed discussion of this issue is presented by Hodson (2012) in *New Scientist*. The neologism has been coined to describe technologies for producing

© Woodhead Publishing Limited, 2013

pharmaceuticals, antibodies and other therapeutically useful materials in standard farm organisms genetically modified to synthesise the desired product when farmed like any other crop plant (or even animal). Despite the uninformed lay objections to GM in general, where therapeutic applications are the intended outcome, the objections are muted or absent.

The scope for genetic modification using well-established methods involving organisms like *Agrobacterium* is well established and relatively easy to apply. However, plant cell cultures are slow growing and require complex media and carefully regulated conditions. Unicellular algae are robust and require simple culture media to generate high biomass yields. There is a chapter here discussing their potential in neutraceutical production; surely they also have potential for 'pharming'? The chapters in Section I show what can be done with other eukaryotes (yeasts and filamentous fungi), so why not algae?

Section II deals with fermentor and fermentation technology. I often feel that these matters are too lightly dismissed when GM experts discuss what they can do with their arts, but growing such organisms will require skill and experience in designing and operating fermentors on an industrial scale. It is essential to appreciate that organisms are very sensitive to the environment in which they are supposed to grow, for example persuading *Aspergillus* cultivars to produce citric acid in STR was a real challenge, even though the same strains produced it in reasonable quantities when allowed to grow as a film on the surface of a liquid medium. I have already noted the marked propensity of various organisms to grow attached to any available surface and it seems reasonable to think that future developments involving STR supplied with suspended material to which organisms can attach themselves will merit investigation. The chapter titles in this section use key words such as monitoring, control, design, optimising, scale-up and product recovery, suggesting that the practicalities of fermentation operation will receive their due place.

It is thus half way through the book's contents list before we encounter specific products and this is exactly as it should be. The list is too extensive for useful summary here, but I will identify one area that specially interests me, namely pre- and probiotics. It pleases me to see that these receive separate treatment and I will mention one novel product that represents a particular challenge because of its important application in infant care, particularly in less developed countries and difficult environments such as war zones. Kitaoka *et al.* (2011) discuss the structure and functions of the complex oligosaccharides found in human milk, which seem to have important roles in guiding development of a good bifidobacterial population in the human neonate gut. They describe a rather lengthy *in vitro* procedure for generating one component of this complex mixture of oligosaccharides, but surely there must be a potential for developing more efficient methods to produce representative oligosaccharides using continuous flow and immobilised enzymes or microbes genetically engineered to synthesise them?

© Woodhead Publishing Limited, 2013

Finally, although I have given an overall summary of this book, I must note that I have not seen all the chapters in full. I have only seen summaries or abstracts for some of the chapters and this leaves me both eager to read all the chapters in full, and perhaps slightly envious of the young workers whose life work it will be to address applying the material here in solving real world problems and developing exciting new industrial applications.

1.5   References

ADAMS M R (1998), 'Vinegar', in *Microbiology of Fermented Foods*, Wood, B J B (ed.), Blackie A & P, London.

BAMFORTH, C W and WARD R E (eds) (in press) *The Oxford Handbook of Food Fermentations*. Oxford University Press, New York.

CAMPBELL-PLATT G (1987), *Fermented Foods of the World: A dictionary and guide*, Butterworth, London.

DOELLE H W, MITCHELL D A and ROLZ C E (1992), *Solid Substrate Cultivation*, Elsevier, London.

HODSON H (2012), 'Field of dreams', *New Scientist*, **214** (2867), 40–3.

HUTKINS R W (2006), *Microbiology and Technology of Fermented Foods*, Blackwell, Iowa.

KARANTH N G (1991), *Short Term Course on Solid Substrate Fermentation*, Central Food Technological Research Institute, Mysore.

KITAOKA M, KATAYAMA T and YAMAMOTO K (2011), 'Metabolic pathway of human milk oligosaccharides in bifidobacteria', in *Lactic acid bacteria and Bifidobacteria: Current progress in advanced research*, Sonomoto K and Yokota A (eds), Caister, Norfolk, 53–66.

KNOOR D (1987), *Food Biotechnology*, Marcel Dekker, New York.

NAGEL F-J (2002), *Process Control of Solid-state Fermentation: Simultaneous control of temperature and moisture content*, PhD Thesis, Wageningen Universiteit.

NEEDHAM J (2000), *Science and Civilisation in China*, Volume VI *Biology and biological technology*, Part V, *Fermentations and food science*, the University Press, Cambridge.

NORENZAYAN A (2012), 'The God issue: Religion is the key to civilisation', *New Scientist*, **213** (2856), 17th March, p 42.

PANDEY A (1994), *Solid State Fermentation*, Wiley Eastern, New Delhi.

PRESCOTT S C and DUNN C G (eds) (1959) *Industrial Microbiology*, 3rd edition, McGraw Hill, New York, pp viii + 945.

SMITS J P (1998), *Solid State Fermentation: Modelling fungal growth and activity*. PhD Thesis, Wageningen Universiteit.

SUDARMADJI S SUPARMO and RAHARJO S (1997), *Reinventing the Hidden Miracle of Tempe*, Indonesian Tempe Foundation, Jakarta.

TYLER I D (1983), *A Carbon Budget for Creran, a Scottish Sea Loch*, PhD Thesis, University of Strathclyde.

WOOD B J B (1985), *Microbiology of Fermented Foods*, 1st Edition, Elsevier, London.

WOOD B J B (1998), *Microbiology of Fermented Foods*, 2nd Edition, Blackie, London.

© Woodhead Publishing Limited, 2013

Part I

Systems biology, metabolic engineering of industrial microorganisms and fermentation technology

© Woodhead Publishing Limited, 2013

2

Systems biology methods and developments of filamentous fungi in relation to the production of food ingredients

W. Vongsangnak, Soochow University, China and J. Nielsen, Chalmers University of Technology, Sweden

DOI: 10.1533/9780857093547.1.19

Abstract: Systems biology is rapidly evolving in many areas of biological sciences. In this chapter, we present a review of systems biology of filamentous fungi in relation to the production of enzymes, chemicals and food ingredients. We summarize the current status of systems biology through functional genomics (i.e. genomics, transcriptomics, proteomics, metabolomics) and bioinformatics of different food-related filamentous fungi. In addition, we present a number of case studies dealing with systems biology and functional genomics through the development of a genome-scale metabolic model of filamentous fungi, which serve as important cell factories in food biotechnology.

Key words: filamentous fungi, food biotechnology, metabolic model, systems biology.

2.1 Introduction

Systems biology is an important component of many biological studies where high throughput-omics technologies are used to obtain a quantitative description of biological systems (Bruggeman and Westerhoff, 2007; Kitano, 2002; Kirschner, 2005). The essence of systems biology is the combination of mathematical and/or computational models with experimental data to gain new insights into biological systems. Its potential benefits can be classified into two broad areas: (1) advances in our basic understanding of biological systems and (2) the development of novel and/or improved bio-based processes. Concerning the advancement in basic biology, systems biology is used to gain new insights into molecular mechanisms

© Woodhead Publishing Limited, 2013

and biological complexity occurring in living cells and to analyze the relationships between their many components (Nielsen, 2009; Stephanopoulos *et al.*, 2004). In terms of the use of systems biology for design of novel bio-based processes, the use of mathematical models for prediction of cellular function is particular important as this allows model-based designs as in other engineering fields (Nielsen, 2009).

In the area of food biotechnology, systems biology is mostly directed towards improving production of cellular enzymes and the production of chemicals, including organic acids, fatty acids, pigments, vitamins and other food ingredients (Archer, 2000; Hjort, 2006), but there is also interest in enhancing the characteristics of starter cultures, such as yeasts, molds and edible mushrooms (Hesseltine, 1983; Leisegang *et al.*, 2006; Sakaguchi *et al.*, 1992; Takahashi and Carvalho, 2010) used for food production through the use of systems biology. In food biotechnology the properties of microorganisms used for industrial fermentations were traditionally improved through random mutagenesis and further strain selection, but in recent years more directed approaches have been followed, such as genetic engineering (Kuipers, 1999). However, classical approaches are very time-consuming and may have problems caused by the side-effects of strain construction and strain selection. Moreover, not all engineering target possibilities are exploited owing to lack of knowledge of metabolic and regulatory processes in the cells (Kuipers, 1999). Furthermore, the use of random mutagenesis is not compatible with the current consumer demand for high food safety where all aspects of the development process can be documented.

As high standards of food safety and quality are expected by different organizations, such as the Food and Drug Administration (FDA) in the USA and World Health Organization (WHO), the whole production process of food ingredient by microbial cell factories should be carefully monitored and well-quantified to avoid the formation of unwanted by-products. Certainly, systems biology can be a promising tool for the design and development of microbial cells that can be used for improved production, both in terms of yield and productivity but also in terms of food safety. Thus, we propose that systems biology has the potential to have a major impact on the food industry as it can support improvement in manufacturing processes, substantially reduce the costs of production and improve the safety of the resulting food products.

2.2   Filamentous fungi as cell factories for food biotechnology

Filamentous fungi have been used in food production since ancient times. Many extracellular enzymes, chemicals, including organic acids, fatty acids,

© Woodhead Publishing Limited, 2013

lipids and other food ingredients are produced by filamentous fungi, which are characterized as important microbial cell factories (Archer *et al.*, 2008; Hjort, 2006). Currently, several different species of filamentous fungi are used in the food industry. Filamentous ascomycetes, such as *Aspergillus oryzae*, *Aspergillus niger*, *Aspergillus sojae* and *Aspergillus awamori* are commonly recognized as industrial cell factories for the production of technical enzymes and organic acids used in food products (Bennett, 2010). On the other hand, filamentous zygomycetes, such as *Mucor circinelloides* and *Mortierella alpina* are associated with their pharmacological properties and nutritional values (Ratledge, 2002). Other filamentous fungi belonging to the deuteromycetes are used as cell factories for food biotechnology, such as *Trichoderma reesei* since it is a highly efficient producer of many extracellular enzymes, for example cellulase and other enzymes for degradation of complex polysaccharides (Nakari-Setala and Penttila, 1995). Other filamentous fungi are important for food preservation, production of colour pigments, flavours, vitamins and meat substitutes, for example single-cell protein. Table 2.1 lists some food-related filamentous fungi and their applications in food biotechnology.

With regard to fermented foods, *A. oryzae* has been well-known for fermenting rice wine (sake), soy sauce (shoyu) and soybean paste (miso) in Japan and China for thousands of years. In these food fermentation processes, *A. oryzae* plays an important role in efficient in secretion of enzymes, for example amylase, which are used to break down starch to sugar which can be fermented. During a koji culture, the koji mold *A. oryzae* secretes a variety of enzymes (e.g. aminopeptidase and protease) to digest soybean protein in industrial soy sauce production. Owing to its high potential for secretory production of enzymes, *A. oryzae* has been extensively used in food biotechnology (Machida *et al.*, 2008). In addition to the production of enzymes, recently *A. oryzae* was found to be potentially useful in C4-dicarboxylic acid production (e.g. malic acid) (Brown *et al.*, 2011). Organic acids are used as ingredients in many food products, for example citric acid is used in many soft drinks, and *A. niger* is commonly used as a cell factory for citric acid production (Papagianni, 2007). Since *A. niger* has much versatility and diversity among its strains, some strains can also be used for production of the enzyme glucoamylase, which is an industrially important hydrolytic enzyme of biotechnological significance in the starch industry. Glucoamylase is generally used in the production of glucose syrup and high fructose corn syrup (James and Lee, 1997).

In the production of food ingredients, the model organism *M. circinelloides* has also been recognized for its biotechnological interest as a source of carotenoids and lipids. In particular, it is a source of long-chain polyunsaturated fatty acids (PUFAs) which are essential fatty acids, for example gamma-linoleic acid (GLA), an omega-6 acid (Kennedy *et al.*, 1993; Ratledge, 2004). For other essential PUFAs, *M. alpina* has been used commercially to

© Woodhead Publishing Limited, 2013

Table 2.1 Applications of some filamentous fungi often used in food biotechnology

Filamentous fungi	Food biotechnological importance	Enzymes and food ingredients
Aspergillus oryzae	Fermented foods, enzymes and C4-acid production	Amylase, protease, malic acid
Aspergillus sojae	Enzyme production	Xylanase, α-arabinofuranosidase
Aspergillus niger	Enzyme and organic acid production	Glucoamylase, citric acid
Aspergillus awamori	Enzyme production and organic acid production	Amylase, protease, citric acid
Aspergillus foetidus	Enzyme production	Glucoamylase, tannase
Aspergillus japonicus	Enzyme production	Glucoamylase, pectic enzymes, β-fructofuranosidase
Ashbya gossypii	Vitamin production	Riboflavin (vitamin B2)
Agaricus bisporus	Food supplement	Edible mushroom (champignon)
Agaricus blazei	Food supplement	Edible sun mushroom
Fusarium venenatum	Protein production	Single-cell protein (SCP)
Fusarium graminearum	Food formulation	Biomass
Mucor circinelloides	Lipids and fatty acids production	Gamma-linolenic acid (GLA)
Mortierella alpina	Lipids and fatty acids production	Arachidonic acid (ARA)
Myceliophthora thermophila	Enzyme production	Cellulase, hemicellulase
Penicillium camemberti	Enzyme production	Proteinase, peptidase
Penicillium roquefortii	Enzyme production	Proteinase, peptidase
Rhizopus oligosporus	Enzyme production and antioxidant	Protease, glucoamylase, phenolic compounds
Trichoderma reesei	Enzyme production	Cellulase, β-glucanase, xylanase

Data was collected from Bennett (2010), Hesseltine (1983), Hjort (2006), Pandey *et al.* (1999), Sakaguchi *et al.* (1992) and Takahashi and Carvalho (2010).

produce arachidonic acid (ARA), another omega-6 acid (Ratledge, 2004; Sakuradani and Shimizu, 2009) which is often included in fish oil supplement and infant formula (Simopoulos, 2011). In addition, *M. alpina* is also used to produce omega-3 PUFAs on a small scale by biotechnological routes (Bajpai *et al.*, 1991), such as eicosapentaenoic acid (EPA) as food and nutritional supplements (Trabal *et al.*, 2010).

© Woodhead Publishing Limited, 2013

2.3   Systems biology of food-related filamentous fungi

Systems biology is the integrative study of entire biological systems and may involve the study of how all the components in the system interact. To understand biological systems from a holistic point of view, it is beneficial to integrate multiple data from functional genomics analyses including genomics, transcriptomics, proteomics and metabolomics into a mathematical model. In the following, we describe stepwise how genomics, comparative genomics, functional genomics and bioinformatics can be useful for advancing towards systems biology of filamentous fungi in relation to the production of food ingredients. A schematic analysis of the progression of systems biology in the context of advancing food biotechnology is illustrated in Fig. 2.1.

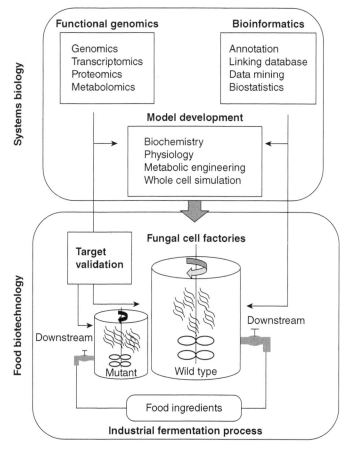

Fig. 2.1   Schematic analysis of the progression of the systems biology method and the development of food biotechnology.

© Woodhead Publishing Limited, 2013

2.3.1 Genomics and comparative genomics of food-related filamentous fungi

Genomics deals with the identification of the genomic DNA sequence of an organism. The first genome sequence for filamentous fungi was released in 2003 where The Neurospora-Fungal Genome Initiative project sequenced the genome of the model organism *Neurospora crassa* (Galagan *et al.*, 2003). This project provided the first insight into the genomic basis of the biological diversity of filamentous fungi. Soon after, the genome of the first *Aspergillus* species, namely *Aspergillus nidulans*, was sequenced by the Whitehead Institute/MIT Center for Genome Research, now called the Broad Institute (Galagan *et al.*, 2005a). Since then a large number of genome sequencing projects for filamentous fungi have been published (Galagan *et al.*, 2005b). By 2011, the sequences of more than 300 genomes of filamentous fungi had been analyzed (www.ncbi.nlm.nih.gov/genomes/leuks.cgi). These include industrially important species for the production of fermented foods, industrial enzymes and chemicals that are relevant to human health and life. Table 2.2 lists the status of genomic information for food-related filamentous fungi.

In many food-related applications, the Food and Drug Administration (FDA) in the USA and World Health Organization (WHO) have classified *A. oryzae* as generally recognized as safe (GRAS), but there were concerns about the close relation of *A. oryzae* to several other *Aspergillus* species, such as *Aspergillus flavus* which is an aflatoxin producer. This was one of

Table 2.2 Summary of genomic information currently available for food-related filamentous fungi

Filamentous fungi	Strain	Genome size (Mb)	Number of genes	References
Aspergillus oryzae	RIB 40	37.2	12 074	Machida *et al.* (2005)
Aspergillus niger	ATCC 1015	34.9	11 200	Baker (2006)
	CBS 513.88	33.9	14 165	Pel *et al.* (2007)
Aspergillus sojae	NBRC 4239	39.5	13 033	Sato *et al.* (2011)
Aspergillus awamori	RIB 2604	34.7	N/A[b]	On-going research
Ashbya gossypii	AGD	9.2	4 718	Dietrich *et al.* (2004)
Fusarium venenatum	A3/5	~38–42	~10 000	Berka *et al.* (2004)
Myceliophthora thermophila	C1	38.5	9 499	Berka *et al.* (2011); Visser *et al.* (2011)
Mucor circinelloides	CBS 277.49	36.6	11 719	JGI database[a]
Mortierella alpina	ATCC 32222	38.4	11 631	Wang *et al.* (2011)
Trichoderma reesei	QM6a	~34	9 129	Martinez *et al.* (2008)

[a]Data is available at website: genome.jgi-psf.org/Mucci2/Mucci2.home.html
[b]Not available.

© Woodhead Publishing Limited, 2013

the rationales, in addition to difficulties in studying *A. oryzae* by conventional genetic methods and its lack of a sexual cycle, for establishing a genome project (Machida *et al.*, 2005). For this project the strain RIB40 was chosen since it shares typical characteristics of morphology, growth and amylase production with many industrial strains used for saké brewing and it can also produce a high level of protease activity which is important for soy source fermentation. The complete genome was sequenced by a whole-genome shotgun (WGS) approach resulting in the release of the genome of the *A. oryzae* strain RIB40 (Machida *et al.*, 2005). The project was run as a close collaboration between the National Institute of Advanced Industrial Science and Technology (AIST) and the National Institute of Technology and Evaluation (NITE), although there were several other members of the *A. oryzae* Genome Analysis Consortium (Machida, 2002; Machida *et al.*, 2005).

Based on genome sequencing, it is possible to evaluate aspects related to food safety. Thus, a comparative functional genomics study between *A. oryzae* and *A. flavus* was performed. Interestingly, genome comparison between these two fungi showed a high sequence similarity of about 99% (Rokas *et al.*, 2007) and DNA comparison within the aflatoxin gene cluster showed a sequence similarity of more than 96% (Payne *et al.*, 2006). However, to check the expression of the aflatoxin cluster, an analysis of expressed sequence tags (ESTs) was performed. The results clearly identified that no ESTs, except for *aflJ* and *norA*, in the aflatoxin gene cluster were found in *A. oryzae* (Akao *et al.*, 2007). These may have occurred due to mutations in putative binding sites of transcription regulators controlling expression levels of genes in the aflatoxin cluster (Tominaga *et al.*, 2006). As a result of these mutations, probably introduced in several secondary metabolism gene clusters, *A. oryzae* fails to produce many of the secondary metabolites produced by *A. flavus* (Amaike and Keller, 2011; Khaldi *et al.*, 2008; Rokas *et al.*, 2007). Moreover, phylogenetic analysis based on comparative sequence of the *aflR* genes from *A. oryzae* and *A. flavus* clearly showed phylogenetic differences of this gene (Chang *et al.*, 2006; Rokas *et al.*, 2007).

These results support the industrial use of *A. oryzae* in food production as there is no expression of any unfavourable genes to human consumption (Machida *et al.*, 2008). Since *A. oryzae* is widely used for food biotechnology, comparison of the *A. oryzae* genome was performed with another two *Aspergillus* genomes, *A. nidulans* and *A. fumigatus*. Interestingly, it was revealed that *A. oryzae* has a larger specific expansion of metabolic genes than these two fungi (Galagan *et al.*, 2005a). Notably, the specific gene expansion includes genes that encode a wide variety of secretory hydrolytic enzymes involved in the metabolism of amino acids and as transporters for amino acids as well as in sugar uptake. These functions might be expected to be important for degrading and metabolizing various raw materials, such as rice, soybean and wheat in food fermentation processes. This evidence

© Woodhead Publishing Limited, 2013

supports the idea that *A. oryzae* is a model filamentous fungus for fermented food production (Hesseltine, 1983; Machida *et al.*, 2005; Rokas *et al.*, 2007).

Another example of an industrially important species is *A. niger* which exhibits a great diversity in its phenotype and its ability to produce large amounts of useful chemicals and enzymes. Therefore the genome of an industrial enzyme-producing *A. niger* strain CBS 513.88 was sequenced by Gene Alliance/DSM (Pel *et al.*, 2007) and the genome sequence was analyzed in the context of enzyme production. Moreover, the genome sequencing and analysis of an acidogenic strain ATCC 1015 was selected by the DOE Joint Genome Institute (JGI) for citric acid production (Baker, 2006). Comparative genomics analysis was performed with regard to key genes and key pathways for optimization of industrial fermentation processes for these two different *A. niger* strains, (Andersen *et al.*, 2011). The results indicated differences in metabolic genes in both strains, which could point out their capabilities in different industrial applications. For the *A. niger* strain CBS 513.88, there are many genes enriched in glucoamylase A, tRNA-synthase and transporters involved in amino acid metabolism and protein synthesis. In contrast, *A. niger* strain ATCC 1015 has a high number of genes and pathways associated with organic acid production.

Genome sequences of other filamentous fungi that are used in food ingredients production have also been sequenced, for example several genomes of the order *Mucorales*, (especially *M. circinelloides*) and the order *Mortierellales* (especially *M. alpina*). The genome sequence of *M. circinelloides* was recently announced by JGI (genome.jgi-psf.org/Mucci2/Mucci2. home.html). This has opened up an opportunity for identification of different gene targets involved in essential fatty acids production, that is polyunsaturated fatty acids (PUFAs) such as gamma-linolenic acid (GLA). More effort to sequence the whole genome of *M. alpina* was later initiated to investigate the mechanisms of lipid synthesis and triacylglycerol accumulation. The genome of *M. alpina* has been determined, and is approximately size of 38.4 megabase pairs (Mb), through close collaboration of different teams (Wang *et al.*, 2011). Currently, the *M. alpina* genome is used as a scaffold for investigating lipid accumulation in relevant filamentous fungi. This useful knowledge can further be applied in process optimization of food ingredients production, the production of PUFAs.

2.3.2 Functional genomics and bioinformatics of food-related filamentous fungi

Functional genomics is the identification of biological information represented by genomic elements, for examples, molecular function, biological process and/or cellular component. Today functional genomics is a major challenge for filamentous fungi, as for most food-related filamentous fungi, more than 50% of genes present in each genome (Table 2.2) that have not

© Woodhead Publishing Limited, 2013

yet been assigned a function. One of the key methods in functional genomics is genome comparison using bioinformatics. Here the function of an unknown protein can be inferred from its evolutionary relationship with proteins of known or putative function. Since orthologous proteins in different organisms often share the same function, several bioinformatics methods have been developed for comparative sequence analysis for functional assignment of genes (Rehm, 2001). There are several efforts to classify genes from a variety of organisms into different functional classes that can be further used for specific functional prediction, such as GO (Harris *et al.*, 2004), COGs (Tatusov *et al.*, 2000), Pfam (Finn *et al.*, 2008), MIPS (Mewes *et al.*, 1997), TIGRFAMs (Haft *et al.*, 2003), Gene3D (Lees *et al.*, 2010), SUPERFAMILY (Gough and Chothia, 2002), PROSITE (Sigrist *et al.*, 2010), PANTHER (Thomas *et al.*, 2003) and Interproscan (Quevillon *et al.*, 2005). In comparison with bacteria, where comparative genomics works very efficiently for predictive annotation, filamentous fungi are much less studied at the genome level and there are therefore many more unknown functions. However, clearly the improved annotation in one organism may lead to a general advance in the field, as this will allow use of genome comparison for improved annotation of many other organisms.

Another key challenge in functional genomics studies is when there is limited sequence similarity of a gene with any other gene identified in other organisms. This raises the question of how to identify the functional role of a gene. Traditional genetic analysis, such as gene knockout (Galli-Taliadoros *et al.*, 1995) and gene trapping approaches (Durick *et al.*, 1999), is laborious and time-intensive so it is therefore interesting to develop more high-throughput techniques. Here transcriptome analysis can be useful and there are several examples of the use of transcriptome studies for uncovering protein functions, identifying common regulatory systems (e.g. regulatory motif and transcription factor) as well as mapping signal transduction pathways and analyzing metabolic pathways. Below are given few examples of these kinds of studies for *A. niger* and *A. oryzae*. For *A. niger*, Yuan *et al.* (2008) demonstrated the use of custom-made *A. niger* Affymetrix GeneChip® Microarray (dsmM_ANIGERa_coll), provided by DSM Food Specialties (Delft, The Netherlands), for identification of a large number of new alpha-glucan acting enzymes and their physiological functions. For *A. oryzae*, transcriptome studies used a cDNA microarray of *A. oryzae* for identification of significant genes and corresponding protein functions in the central carbon metabolism when the cells were shifting from glucose-rich to glucose-poor liquid cultures (Machida, 2002; Maeda *et al.*, 2004). Furthermore, cDNA microarray analysis of *A. oryzae* could also be used to monitor how *A. oryzae* produces a wide range of hydrolytic enzymes, with a high yield, in solid state cultivations on different solid-phase media, for example wheat bran, rice bran, soy bean waste (Maeda *et al.*, 2004). Based on these studies of both *A. niger* and *A. oryzae*, several new enzymatic

© Woodhead Publishing Limited, 2013

functions that had commercial value and several for applications in the food industry, were identified and this shows how genomics and functional genomics can advance food biotechnology.

To advance the optimization of fermentation processes, uncovering the basis of transcriptional regulation of cellular metabolism is valuable. Andersen *et al.* (2008a) designed the first-generation of a tri-*Aspergillus* species Affymetrix GeneChip (3AspergDTU GeneChip) in order to use it as a tool for comparative transcriptome analysis of *A. nidulans*, *A. niger* and *A. oryzae*, and this tri-*Aspergillus* species Affymetrix GeneChip has been used in several studies to identify common regulatory systems in *Aspergillus* species. As shown by Vongsangnak *et al.* (2010), comparative transcriptome analysis can provide new insights into the global regulatory structure of carbon metabolism in *A. oryzae* and *A. niger*. When these fungi were grown on different carbon sources, for example glucose, maltose, glycerol or xylose, it was found that the responses of co-expressed genes between these two fungi were highly conserved (Andersen *et al.*, 2008a; Salazar *et al.*, 2009; Vongsangnak *et al.*, 2009). Through further extensive studies of common regulatory systems, it was found that there were simultaneous classifications of co-expressed genes with similar correlation patterns between their expression profiles. Using the concept of 'guilt by association' (Lockhart and Winzeler, 2000) for gene expression profiles, it was possible to uncover common regulatory motifs and common transcription factors between *A. oryzae* and *A. niger* regulating the utilization of carbon sources (Vongsangnak *et al.*, 2010). The information obtained can be directly applied to strain improvement in industrial fermentation processes. We can design promoter binding sites for over-expression of the gene of interest, for example one can use the transcription factor *xln*R which binds specifically to the binding site 5′-GGCTAAA-3′ (van Peij *et al.*, 1998). This can further lead to induction of xylanolytic genes involved in industrial xylose fermentation processes in *A. niger* (van Peij *et al.*, 1998) and *A. oryzae* (Marui *et al.*, 2002; Noguchi *et al.*, 2009).

In future studies, it is useful that many transcriptome datasets of filamentous fungi are available at the Gene Expression Omnibus (GEO) database (www.ncbi.nlm.nih.gov/geo/) and the Array Express database (www.ebi.ac.uk/arrayexpress). In addition to transcriptome analysis, functional genomics also involves the use of proteomics for the detection of functional and structural evolutionary relationships between proteins. Nowadays, there are different proteomics-based technologies (Banks *et al.*, 2000) and tools deposited, such as at the ExPASy proteomics Server of Swiss Institute of Bioinformatics (expasy.org/tools/), which can be used to identify proteins and/or their structures and functions (Gasteiger *et al.*, 2003; Tyers and Mann, 2003). As one of the success stories of *A. oryzae*, Oda *et al.* (2006) reported proteomics analysis that was used to determine protein secretion profiles and to identify further the difference between secreted proteins under solid-state and submerged culture conditions. The results showed that many well-known proteins, such as glucoamylase A, α-glucosidase and β-galactosidase, were identified in the submerged culture. In contrast, some hypothetical proteins with low homology to known proteins were also identified. This indicates that *A. oryzae* secreted some new proteins of unknown function in the solid-state condition (Oda *et al.*, 2006). This study led to discovery of novel proteins with limited sequence similarities. For functional annotation, proteomics analysis is often used to identify membrane proteins as reported by Ouyang *et al.* (2010), who used proteomics to annotate the functions of membrane proteins associated with the glycoconjugates and cell wall biosynthesis in the membrane proteome of *A. fumigatus*.

Metabolomics is another high-throughput experimental tool that can complement transcriptomics and proteomics. Specifically, metabolomics involves the rapid high-throughput characterization of the small molecule metabolites found in an organism and aims to increase understanding of how changes in metabolite levels affect phenotypes (Mapelli *et al.*, 2008). Therefore, metabolomics is often used to gain insight into microbial metabolism and cellular physiology. Currently, several metabolomics resources are available at general metabolomics sites and databases, for example COLMAR Metabolomics Web Portal (spinportal.magnet.fsu.edu), small molecules, for example ChEBI database (www.ebi.ac.uk/chebi) and ChemFinder (www.chemfinder.com) as well as biochemical pathways, for example KEGG (www.kegg.com), MetaCyc (metacyc.org) and ExPASy (expasy.org). Besides, there are several companies providing metabolomics services for both industrial and academic clients, for example metabonomics (www.metabometrix.com), metabolon (www.metabolon.com), and so on. They serve as key drivers for providing information on a broad range of metabolites. However, it remains hard to cover a whole metabolome and there is a need for efficient methods for interpretation of metabolomics data.

2.4   Beyond functional genomics to metabolic modelling

The actual challenge in systems biology is to develop mathematical models that can precisely predict the outcome of genetic or environmental changes and can be used as scaffolds to understand phenotypic behaviours in living organisms. With the accessibility of functional genomics derived from annotation of genome sequence data, it is possible to look into the inter-relationship of genes and their functional roles in the context of metabolism. Below we discuss how genome-scale metabolic models, which are at the core of systems biology, can be used in relation to the production of food ingredients.

© Woodhead Publishing Limited, 2013

2.4.1 Building genome-scale metabolic models and some examples

Building metabolic models involves three main processes: (1) metabolic network reconstruction, (2) mathematical formulation and (3) metabolic model validation. A very important process in generating metabolic model is the reconstruction of the metabolic network (Price *et al.*, 2003). In particular, it is important that the reconstruction process is performed so that as much information as possible is built into the model. Thus, the reconstruction process is typically based on a large number of data resources and involves several steps, as illustrated in Fig. 2.2. It usually starts with a step using functional genomics data where information about annotated genes, enzymes and biochemical reactions is used to defined gene–protein–reaction (GPR) associations. In this step each enzyme or protein is assigned to its appropriate sub-cellular localization. Following the process of the network reconstruction, the presence of various metabolic pathways in the cell is collected and the pathways are systematically curated using biochemistry information and literature. Once all pathways are reconstructed, the next step is to identify the gaps in the metabolic network and fill them using a directed bioinformatics search for the missing genes. Thereafter follows the mathematical formulation process where the identified reactions and pathways are converted into a quantitative description of the metabolic network. In this process, the set of reactions and their participating metabolites are converted to a stoichiometric matrix (S).

Under the assumption of steady state for all the metabolites, it is possible to use the mathematical metabolic model for simulation using the concept of flux balance analysis (FBA) (Varma and Palsson, 1994). This

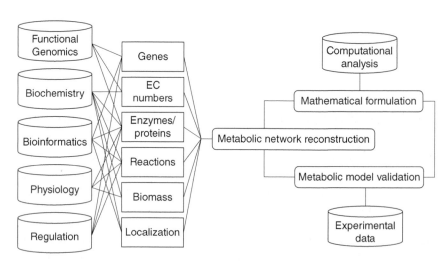

Fig. 2.2 Diagram of whole process of genome-scale metabolic model development.

© Woodhead Publishing Limited, 2013

allows prediction of cellular metabolism and phenotypic behaviour using constraint-based modelling. In constraint-based modelling an objective function needs to be formulated and here biomass formation and specific metabolite formation have been found to describe the growth of *A. oryzae* (Vongsangnak *et al.*, 2008) and citric acid production of *A. niger* very well (Andersen *et al.*, 2008b).

To gain quantitative data for cellular phenotypic behaviour, it is necessary to validate the model by comparing *in silico* simulation results with data from experimental results (e.g. growth and production formation). In this process, the model may have to be refined so that its predictions are in closer agreement with experimental results. Data from different levels of cellular processes may be incorporated for model improvement, for example addition of regulatory constraints resolves inconsistencies when regulation occurs on higher levels than the metabolic flux level (Borodina *et al.*, 2005). The whole process, including data integration, to achieve a genome-scale metabolic model is illustrated in Fig. 2.2. To date, genome-scale metabolic models have been developed for several filamentous fungi (Cvijovic *et al.*, 2010) and in relation to food biotechnology metabolic models have been developed for *Aspergillus* species. This involves a stoichiometric model for the central metabolism (David *et al.*, 2003; Melzer *et al.*, 2007) and a full genome-scale model of *A. niger* (Andersen *et al.*, 2008b), which is used as a cell factory for citric acid production, as well as a genome-scale model for *A. oryzae* (Vongsangnak *et al.*, 2008), which is used as a cell factory for enzyme production and for production of sake and soya sauce. Besides current research and development, there are on-going projects building metabolic networks of other food-related filamentous fungi such as *M. circinelloides*, which is used for production of essential PUFAs (e.g. GLA). The reconstructed *M. circinelloides* network would provide a framework for GPR associations in *M. circinelloides* and could be used for analysis of lipogenesis pathways.

2.4.2   Using systems biology to monitor enzyme production

In order to optimize industrial enzyme production, there is a need to understand the fundamental processes underlying protein synthesis and secretion by the cells, that is how is enzyme production linked to metabolism through global regulation and which genes or pathways are key players for high level enzyme production? Vongsangnak *et al.* (2011) performed a systems biology study of *A. oryzae* in relation to the production of the enzyme α-amylase; the study design is illustrated in Fig. 2.3.

The study involved comparative analysis of a strain of *A. oryzae* that produces high levels of α-amylase (CF1.1) with a reference strain (A1560) using transcriptome analysis, flux analysis and physiological studies (e.g. biomass and enzyme activities). In order to use modelling as a monitoring tool to identify fundamental metabolic processes that are important in the

© Woodhead Publishing Limited, 2013

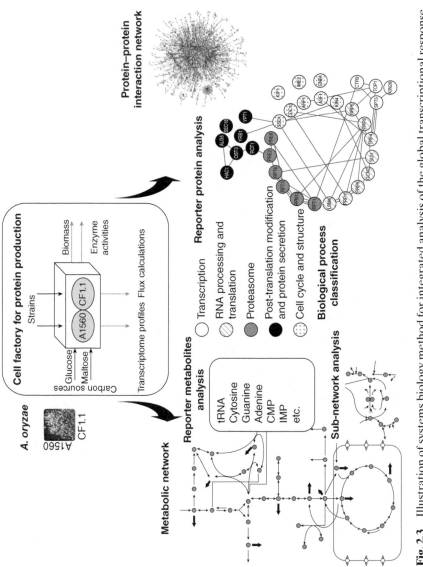

Fig. 2.3 Illustration of systems biology method for integrated analysis of the global transcriptional response to α-amylase over-production in *Aspergillus oryzae*.

© Woodhead Publishing Limited, 2013

production of α-amylase, the genome-scale metabolic model of *A. oryzae* (Vongsangnak *et al.*, 2008) was combined with transcriptome data and further applied for identification of reporter metabolites and highly correlated metabolic sub-networks using integrative data analysis (Patil and Nielsen, 2005). Interestingly, the integrative data analysis led to the identification of 25 key metabolites that were mainly distributed in purine and pyrimidine nucleotide biosynthesis. The results are biologically reasonable as formations of ribonucleic acid (RNA) and deoxyribonucleic acid (DNA) in nucleotide biosynthesis are very important in protein production. In addition, key genes encoding enzymes or transporters in response to increased protein production were identified based on metabolic sub-network analysis of *A. oryzae*. Besides identification of nucleotide metabolism (purine and pyrimidine biosynthesis) as the key sub-networks, key genes encoding enzymes involved in amino acid metabolism were significantly changed in the metabolic sub-networks from a pair-wise comparison of two *A. oryzae* strains, a wild type strain and a transformant strain over-producing amylase (Vongsangnak *et al.*, 2011).

Additionally, the study also involved reconstructing a global protein–protein interaction network of *A. oryzae* by identification of putative components through comparative genomics and interactomics (e.g. protein–protein interaction) between *A. oryzae* and *Saccharomyces cerevisiae*. Based on 140,849 protein pairs from the yeast *S. cerevisiae* BIOGRID database with 3514 orthologues genes (1 : 1 pairwise orthologues) between *A. oryzae* and *S. cerevisiae*, the reconstructed interaction network of *A. oryzae* contained 2704 individual proteins with 48,483 putative interactions of protein pairs. In order to identify key proteins using the reporter feature algorithm (Oliveira *et al.*, 2008), the gene expression data set from comparative transcriptome analysis between two strains of *A. oryzae* were combined with the reconstructed protein–protein interaction network of *A. oryzae*. Interestingly, this reconstructed protein–protein interaction network of *A. oryzae* could be used for identification of 33 key proteins and related co-regulated modules in the transcriptional response to high level α-amylase production. This study illustrates the use of systems biology for study of a complex biological system, namely protein synthesis and secretion, for identification of putative bottlenecks in enzyme production and this may lead to further improvement in industrial enzyme production (Vongsangnak *et al.*, 2011).

2.5   Systems biology perspectives on food biotechnology and food safety

From the above, it is clear that there are currently relatively few direct applications of systems biology in food biotechnology, but as indicated there are some obvious prospects for a much wider application of this advanced

© Woodhead Publishing Limited, 2013

technological approach in the food industry. Currently, many food processes rely on microbial fermentations, for example beer, wine, other alcoholic beverages, sausages, yoghurts and soya sauce, and despite the fact that many of these processes have been carried out for generations, there is still interest in improving them. Improvement can be in terms of improved industrial production, for example improved productivity and yield as discussed above, but probably more important is improvement of food quality and food safety.

Improvement of food quality can be obtained through changing the taste, texture and appearance of the food product and these improvements can come about through engineering, either directed genetic engineering or classical strain improvement of the microorganism used in the fermentation process. Systems biology is very well suited to drive this development process, as the global approach to understanding metabolism and physiology of the organism allows for identification of novel targets for strain improvement. Here we have seen much development in the field of filamentous fungi in recent years, where a number of systems biology tools have been developed, including tools for genome-wide transcription analysis and detailed genome-scale metabolic models. These tools are currently applied in the chemical and biotechnological industry for strain improvement through metabolic engineering and these techniques will find direct applications in the production of food ingredients that are being produced by fermentation, for example citric acid. These tools are, however, also very well suited to address issues in the food industry, where the problems of optimization are often more complex owing to the presence of more complex starting materials for the fermentation, for example milk, wort or meat, and often more complex objectives, for example texture, appearance and taste, which are determined not by a single metabolite and/or chemical compound, but by mixtures. Here advanced metabolic modelling can be used to identify the role of different pathways and how they can operate in a balanced fashion, to ensure proper conversion of the many different chemical compounds in the starting material to the desired spectrum of metabolites. We do not know of any such applications at present, but this is clearly an opportunity for application of systems biology in the food industry.

Also, in the area of food safety, systems biology may play an important role. Obviously, systems biology can be applied to analysis of new plant-derived food products and for evaluation of their safety in terms of the presence of possible allergens and/or toxic compounds, but also in the area of fermented foods when food safety issues arise. As there is much reluctance, in particular in Europe, to use of genetically modified organisms in food production, the food industry is constrained to rely on classical strain improvement, which involves mutagenesis and selection.

Even though classical strain improvement has been used extensively in the past, it is a technique that may result in activation of latent pathways

© Woodhead Publishing Limited, 2013

that lead to production of toxic compounds. Advanced metabolomics, but also genome sequencing and detailed metabolic modelling, may allow for advanced diagnosis of mutated strains and hereby systems biology can assist greatly in assessing the food safety of novel processes. In the future it may, however, be far more difficult to get approval for novel processes based on random mutagenesis, unless there is clear documentation provided for the changes that have occurred in the processes. Owing to the complexity, this may lead to a requirement for genome-sequencing to be applied to all new strains and further characterization of what metabolic capabilities the mutated strains have obtained. This will require systems biology to be applied as a standard technology in food companies.

Thus, we foresee an expanding role for systems biology in the food industry, in particular in relation to fermentation-based processes and to applications of filamentous fungi in the food industry, as these organisms have a complex metabolism that may be dramatically altered with the development of novel processes. However, through systems biology we are confident that filamentous fungi can find wide applications in many novel food products in the future and that these applications will have both a high quality and be safe.

2.6   Acknowledgements

Wanwipa Vongsangnak is supported by National Natural Science Foundation of China grant (No. 31200989) and starting grant (No. Q410700111) from Soochow University.

2.7   References

AKAO T, SANO M, YAMADA O, AKENO T, FUJII K, GOTO K, OHASHI-KUNIHIRO S, TAKASE K, YASUKAWA-WATANABE M, YAMAGUCHI K, KURIHARA Y, MARUYAMA J, JUVVADI P R, TANAKA A, HATA Y, KOYAMA Y, YAMAGUCHI S, KITAMOTO N, GOMI K, ABE K, TAKEUCHI M, KOBAYASHI T, HORIUCHI H, KITAMOTO K, KASHIWAGI Y, MACHIDA M and AKITA O (2007), 'Analysis of expressed sequence tags from the fungus *Aspergillus oryzae* cultured under different conditions', *DNA Res*, **14**, 47–57.

AMAIKE S and KELLER N P (2011), '*Aspergillus flavus*', *Annu Rev Phytopathol*, **49**, 10.1–10.27.

ANDERSEN M R, VONGSANGNAK W, PANAGIOTOU G, MARGARITA P S, LEHMANN L and NIELSEN J (2008a), 'A tri-species *Aspergillus* microarray–Advancing comparative transcriptomics', *Proc Nat Acad Sci*, **105**, 4387–92.

ANDERSEN M R, NIELSEN M L and NIELSEN J (2008b), 'Metabolic model integration of the bibliome, genome, metabolome and reactome of *Aspergillus niger*', *Mol Syst Biol*, **4**, 178.

ANDERSEN M R, SALAZAR M P, SCHAAP P J, VAN DE VONDERVOORT P J, CULLEY D, THYKAER J, FRISVAD J C, NIELSEN K F, ALBANG R, ALBERMANN K, BERKA R M, BRAUS G H, BRAUS-STROMEYER S A, CORROCHANO L M, DAI Z, VAN DIJCK P W, HOFMANN G, LASURE L L, MAGNUSON J K, MENKE H, MEIJER M, MEIJER S L, NIELSEN J B, NIELSEN M L, VAN OOYEN A J, PEL H J, POULSEN L, SAMSON R A, STAM H, TSANG A, VAN DEN BRINK J M, ATKINS A,

© Woodhead Publishing Limited, 2013

AERTS A, SHAPIRO H, PANGILINAN J, SALAMOV A, LOU Y, LINDQUIST E, LUCAS S, GRIMWOOD J, GRIGORIEV I V, KUBICEK C P, MARTINEZ D, VAN PEIJ N N, ROUBOS J A, NIELSEN J and BAKER S E (2011), 'Comparative genomics of citric-acid-producing *Aspergillus niger* ATCC 1015 versus enzyme-producing CBS 513.88', *Genome Res*, 21, 885–97.

ARCHER D B (2000), 'Filamentous fungi as microbial cell factories for food use', *Curr Opin Biotech*, 11, 478–83.

ARCHER D B, CONNERTON I F and MACKENZIE D A (2008), 'Filamentous fungi for production of food additives and processing aids', *Adv Biochem Eng Biotechnol*, 111, 99–147.

BAJPAI P, BAJPAI P K and WARD O P (1991), 'Eicosapentaenoic acid (EPA) production by *Mortierella alpina* ATCC 32222', *Appl Biochem Biotech*, 31, 267–72.

BAKER S E (2006), '*Aspergillus niger* genomics: past, present and into the future', *Med Mycol*, 44, S17–21.

BANKS R E, DUNN M J, HOCHSTRASSER D F, SANCHEZ J C, BLACKSTOCK W, PAPPIN D J and SELBY P J (2000), 'Proteomics: new perspectives, new biomedical opportunities', *Lancet*, 356, 1749–56.

BENNETT J W (2010), 'An Overview of the Genus *Aspergillus*', in *Aspergillus: Molecular Biology and Genomics*, Machida M and Gomi K (eds), Caister Academic Press, Norfolk, UK, 1–18.

BERKA R M, NELSON B A, ZARETSKY E J, YODER W T and REY M W (2004), 'Genomics of *Fusarium venenatum:* an alternative fungal host for making enzymes', *Appl Mycol Biotechnol*, 4, 191–203.

BERKA R M, GRIGORIEV I V, OTILLAR R, SALAMOV A, GRIMWOOD J, REID I, ISHMAEL N, JOHN T, DARMOND C, MOISAN M C, HENRISSAT B, COUTINHO P M, LOMBARD V, NATVIG D O, LINDQUIST E, SCHMUTZ J, LUCAS S, HARRIS P, POWLOWSKI J, BELLEMARE A, TAYLOR D, BUTLER G, DE VRIES R P, ALLIJN IE, VAN DEN BRINK J, USHINSKY S, STORMS R, POWELL A J, PAULSEN I T, ELBOURNE L D, BAKER S E, MAGNUSON J, LABOISSIERE S, CLUTTERBUCK A J, MARTINEZ D, WOGULIS M, DE LEON A L, REY M W and TSANG A (2011), 'Comparative genomic analysis of the thermophilic biomass-degrading fungi *Myceliophthora thermophila* and *Thielavia terrestris*', *Nat Biotechnol*, 29, 922–7.

BORODINA I, KRABBEN P and NIELSEN J (2005), 'Genome-scale analysis of *Streptomyces coelicolor* A3 (2) metabolism', *Genome Res*, 15, 820–9.

BROWN S, LUTTRINGER S, YAVER D and BERRY A (2011), *Methods for Improving Malic Acid Production in Filamentous Fungi*. United States Patent Application 20110053233. 3 March 2011.

BRUGGEMAN F J and WESTERHOFF H V (2007), 'The nature of systems biology', *Trends Microbiol*, 15, 45–50.

CHANG P K, EHRLICH K C and HUA S S T (2006), 'Cladal relatedness among *Aspergillus oryzae* isolates and *Aspergillus flavus* S and L morphotype isolates', *Int J Food Microbiol*, 108, 172–7.

CVIJOVIC M, OLIVARES R, AGREN R, DAHR N, VONGSANGNAK W, NOOKAEW I, PATIL K and NIELSEN J (2010), 'BioMet Toolbox: genome-wide analysis of metabolism', *Nucleic Acid Res*, 38, W144–9.

DAVID H, AKESSON M and NIELSEN J (2003), 'Reconstruction of the central carbon metabolism of *Aspergillus niger*', *Eur J Biochem*, 270, 4243–53.

DIETRICH F S, VOEGELI S, BRACHAT S, LERCH A, GATES K, STEINER S, MOHR C, PÖHLMANN R, LUEDI P, CHOI S, WING R A, FLAVIER A, GAFFNEY T D and PHILIPPSEN P (2004), 'The *Ashbya gossypii* genome as a tool for mapping the ancient *Saccharomyces cerevisiae* genome', *Science*, 304, 304–7.

DURICK K, MENDLEIN J and XANTHOPOULOS K G (1999), 'Hunting with traps: genome-wide strategies for gene discovery and functional analysis', *Genome Res*, 9, 1019–25.

FINN R D, TATE J, MISTRY J, COGGILL P C, SAMMUT S J, HOTZ H, CERIC G, FORSLUND K, SEAN R, EDDY S R, ERIK L L, SONNHAMMER E L and BATEMAN A (2008), 'The Pfam protein families database', *Nucleic Acids Res*, 36, D281–8.

© Woodhead Publishing Limited, 2013

GALAGAN J E, CALVO S E, BORKOVICH K A, SELKER E U, READ N D, JAFFE D, FITZHUGH W, MA L J, SMIRNOV S, PURCELL S, REHMAN B, ELKINS T, ENGELS R, WANG S, NIELSEN CB, BUTLER J, ENDRIZZI M, QUI D, IANAKIEV P, BELL-PEDERSEN D, NELSON M A, WERNER-WASHBURNE M, SELITRENNIKOFF C P, KINSEY J A, BRAUN E L, ZELTER A, SCHULTE U, KOTHE G O, JEDD G, MEWES W, STABEN C, MARCOTTE E, GREENBERG D, ROY A, FOLEY K, NAYLOR J, STANGE-THOMANN N, BARRETT R, GNERRE S, KAMAL M, KAMVYSSELIS M, MAUCELI E, BIELKE C, RUDD S, FRISHMAN D, KRYSTOFOVA S, RASMUSSEN C, METZENBERG R L, PERKINS D D, KROKEN S, COGONI C, MACINO G, CATCHESIDE D, LI W, PRATT R J, OSMANI S A, DESOUZA C P, GLASS L, ORBACH M J, BERGLUND J A, VOELKER R, YARDEN O, PLAMANN M, SEILER S, DUNLAP J, RADFORD A, ARAMAYO R, NATVIG D O, ALEX L A, MANNHAUPT G, EBBOLE D J, FREITAG M, PAULSEN I, SACHS M S, LANDER E S, NUSBAUM C and BIRREN B (2003), 'The genome sequence of the filamentous fungus *Neurospora crassa*', *Nature*, **422**, 859–68.

GALAGAN J E, CALVO S E, CUOMO C, MA L J, WORTMAN J R, BATZOGLOU S, LEE S I, BAŞTÜRKMEN M, SPEVAK C C, CLUTTERBUCK J, KAPITONOV V, JURKA J, SCAZZOCCHIO C, FARMAN M, BUTLER J, PURCELL S, HARRIS S, BRAUS G H, DRAHT O, BUSCH S, D'ENFERT C, BOUCHIER C, GOLDMAN G H, BELL-PEDERSEN D, GRIFFITHS-JONES S, DOONAN J H, YU J, VIENKEN K, PAIN A, FREITAG M, SELKER E U, ARCHER D B, PEÑALVA M A, OAKLEY B R, MOMANY M, TANAKA T, KUMAGAI T, ASAI K, MACHIDA M, NIERMAN W C, DENNING D W, CADDICK M, HYNES M, PAOLETTI M, FISCHER R, MILLER B, DYER P, SACHS M S, OSMANI S A and BIRREN B W (2005a), 'Sequencing of *Aspergillus nidulans* and comparative analysis with *A. fumigatus* and *A. oryzae*', *Nature*, **438**, 1105–15.

GALAGAN J E, HENN M R, MA L J, CUOMO C A and BIRREN B (2005b), 'Genomics of the fungal kingdom: Insights into eukaryotic biology', *Genome Res*, **15**, 1620–31.

GALLI-TALIADOROS L A, SEDGWICK J D, WOOD S A and KÖRNER H (1995), 'Gene knock-out technology: a methodological overview for the interested novice', *J Immunol Methods*, **181**, 1–15.

GASTEIGER E, GATTIKER A, HOOGLAND C, IVANYI I, APPEL R D and BAIROCH A (2003), 'ExPASy: The proteomics server for in-depth protein knowledge and analysis', *Nucleic Acids Res*, **31**, 3784–8.

GOUGH J and CHOTHIA C (2002), 'SUPERFAMILY: HMMs representing all proteins of known structure. SCOP sequence searches, alignments and genome assignments', *Nucleic Acids Res*, **30**, 268–72.

HAFT D H, SELENGUT J D and WHITE O (2003), 'The TIGRFAMs database of protein families', *Nucleic Acids Res*, **31**, 371–3.

HARRIS M A, CLARK J, IRELAND A, LOMAX J, ASHBURNER M, FOULGER R, EILBECK K, LEWIS S, MARSHALL B, MUNGALL C, RICHTER J, RUBIN G M, BLAKE J A, BULT C, DOLAN M, DRABKIN H, EPPIG J T, HILL DP, NI L, RINGWALD M, BALAKRISHNAN R, CHERRY J M, CHRISTIE K R, COSTANZO M C, DWIGHT S S, ENGEL S, FISK D G, HIRSCHMAN J E, HONG E L, NASH R S, SETHURAMAN A, THEESFELD C L, BOTSTEIN D, DOLINSKI K, FEIERBACH B, BERARDINI T, MUNDODI S, RHEE SY, APWEILER R, BARRELL D, CAMON E, DIMMER E, LEE V, CHISHOLM R, GAUDET P, KIBBE W, KISHORE R, SCHWARZ E M, STERNBERG P, GWINN M, HANNICK L, WORTMAN J, BERRIMAN M, WOOD V, DE LA CRUZ N, TONELLATO P, JAISWAL P, SEIGFRIED T, WHITE R and GENE ONTOLOGY CONSORTIUM (2004), 'The Gene Ontology (GO) database and informatics resource', *Nucleic Acids Res*, 32, D258–61.

HESSELTINE C W (1983), 'Microbiology of oriental fermented foods', *Ann Rev Microbiol*, **37**, 575–601.

HJORT C M (2006), 'Production of food additives using filamentous fungi', in *Genetically Engineered Food: Methods and Detection*, Heller K L (ed.), Wiley-VCH, Weinheim, 95–107.

JAMES J A and LEE BH (1997), 'Glucoamylases: microbial sources, industrial applications and molecular biology: A review', *J Food Biochem*, **21**, 1–52.

KENNEDY M J, READER S L and DAVIES R J (1993), 'Fatty acid production characteristics of fungi with particular emphasis on gamma linolenic acid production', *Biotechnol Bioeng*, **42**, 625–34.

© Woodhead Publishing Limited, 2013

KHALDI N, COLLEMARE J, LEBRUN M H and WOLFE K H (2008), 'Evidence for horizontal transfer of a secondary metabolite gene cluster between fungi', *Genome Biol,* **9**, R18.

KIRSCHNER M W (2005), 'The meaning of systems biology', *Cell,* **121**, 503–4.

KITANO H (2002), 'Systems biology: a brief overview', *Science,* **295**, 1662–4.

KUIPERS O P (1999), 'Genomics for food biotechnology: prospects of the use of high-throughput technologies for the improvement of food microorganisms', *Curr Opin Biotechnol,* **10**, 511–6.

LEES J, YEATS C, REDFERN O, CLEGG A and ORENGO C (2010), 'Gene3D: merging structure and function for a thousand genomes', *Nucleic Acids Res,* **38**, D296–300.

LEISEGANG R, NEVOIGT E, SPIELVOGEL A, KRISTAN G, NIEDERHAUS A and STAHL U (2006), 'Fermentation of food by means of genetically modified yeast and filamentous fungi', in *Genetically Engineered Food: Methods and Detection,* Heller K L (ed.), Wiley-VCH, Weinheim, 78–86.

LOCKHART D J and WINZELER E A (2000), 'Genomics, gene expression and DNA arrays', *Nature,* **405**, 827–36.

MACHIDA M (2002), 'Progress of *Aspergillus oryzae* genomics', *Adv Appl Microbiol,* **51**, 81–106.

MACHIDA M, ASAI K, SANO M, TANAKA T, KUMAGAI T, TERAI G, KUSUMOTO K, ARIMA T, AKITA O, KASHIWAGI Y, ABE K, GOMI K, HORIUCHI H, KITAMOTO K, KOBAYASHI T, TAKEUCHI M, DENNING D W, GALAGAN J E, NIERMAN WC, YU J, ARCHER D B, BENNETT J W, BHATNAGAR D, CLEVELAND T E, FEDOROVA N D, GOTOH O, HORIKAWA H, HOSOYAMA A, ICHINOMIYA M, IGARASHI R, IWASHITA K, JUVVADI PR, KATO M, KATO Y, KIN T, KOKUBUN A, MAEDA H, MAEYAMA N, MARUYAMA J, NAGASAKI H, NAKAJIMA T, ODA K, OKADA K, PAULSEN I, SAKAMOTO K, TOSHIHIKO SAWANO T, TAKAHASHI M, TAKASE K, TERABAYASHI Y, WORTMAN JR, YAMADA O, YAMAGATA Y, ANAZAWA H, HATA Y, KOIDE Y, KOMORI T, KOYAMA Y, MINETOKI T, SUHARNAN S, TANAKA A, ISONO K, KUHARA S, OGASAWARA N and KIKUCHI H (2005), 'Genome sequencing and analysis of *Aspergillus oryzae*', *Nature,* **438**, 1157–61.

MACHIDA M, YAMADA O and GOMI K (2008), 'Genomics of *Aspergillus oryzae*: Learning from the history of koji mold and exploration of its future', *DNA Res,* **15**, 173–83.

MAEDA H, SANO M, MARUYAMA Y, TANNO T, AKAO T, TOTSUKA Y, ENDO M, SAKURADA R, YAMAGATA Y, MACHIDA M, AKITA O, HASEGAWA F, ABE K, GOMI K, NAKAJIMA T and IGUCHI Y (2004), 'Transcriptional analysis of genes for energy catabolism and hydrolytic enzymes in the filamentous fungus *Aspergillus oryzae* using cDNA microarrays and expressed sequence tags', *Appl Microbiol Biotechnol,* **65**, 74–83.

MAPELLI V, OLSSON L and NIELSEN J (2008), 'Metabolic footprinting in microbiology: methods and applications in functional genomics and biotechnology', *Trends Biotechnol,* 26, **490**–7.

MARTINEZ D, BERKA R M, HENRISSAT B, SALOHEIMO M, ARVAS M, BAKER S E, CHAPMAN J, CHERTKOV O, COUTINHO P M, CULLEN D, DANCHIN E G, GRIGORIEV I V, HARRIS P, JACKSON M, KUBICEK C P, HAN C S, HO I, LARRONDO L F, DE LEON A L, MAGNUSON J K, MERINO S, MISRA M, NELSON B, PUTNAM N, ROBBERTSE B, SALAMOV A A, SCHMOLL M, TERRY A, THAYER N, WESTERHOLM-PARVINEN A, SCHOCH C L, YAO J, BARABOTE R, NELSON M A, DETTER C, BRUCE D, KUSKE C R, XIE G, RICHARDSON P, ROKHSAR D S, LUCAS S M, RUBIN E M, DUNN-COLEMAN N, WARD M and BRETTIN T S (2008), 'Genome sequencing and analysis of the biomass-degrading fungus *Trichoderma reesei* (syn. *Hypocrea jecorina*)', *Nat Biotechnol,* **26**, 553–60.

MARUI J, TANAKA A, MIMURA S, DE GRAAFF L H, VISSER J, KITAMOTO N, KATO M, KOBAYASHI T and TSUKAGOSHI N (2002), 'A transcriptional activator, AoXlnR, controls the expression of genes encoding xylanolytic enzymes in *Aspergillus oryzae*', *Fungal Genet Biol,* **35**, 157–69.

MELZER G, DALPIAZ A, GROTE A, KUCKLICK M, GÖCKE Y, JONAS R, DERSCH P, FRANCOLARA E, NÖRTEMANN B and HEMPEL D (2007), 'Metabolic flux analysis using stoichiometric

© Woodhead Publishing Limited, 2013

models for *Aspergillus niger*: comparison under glucoamylase-producing and nonproducing conditions', *J Biotechnol*, **132**, 405–17.

MEWES H W, ALBERMANN K, HEUMANN K, LIEBL S and PFEIFFER F (1997), 'MIPS: a database for protein sequences, homology data and yeast genome information', *Nucleic Acids Res*, **25**, 28–30.

NAKARI-SETALA T and PENTTILA M (1995), 'Production of *Trichoderma reesei* cellulases on glucose-containing media', *Appl Environ Microbiol*, **61**, 3650–5.

NIELSEN J (2009), 'Systems biology of lipid metabolism: From yeast to human', *FEBS Lett*, **583**, 3905–13.

NOGUCHI Y, SANO M, KANAMARU K, KO T, TAKEUCHI M, KATO M and KOBAYASHI T (2009), 'Genes regulated by AoXlnR, the xylanolytic and cellulolytic transcriptional regulator, in *Aspergillus oryzae*', *Appl Microbiol Biotechnol*, **85**, 141–54.

ODA K, KAKIZONO D, YAMADA O, IEFUJI H, AKITA O and IWASHITA K (2006), 'Proteomic analysis of extracellular proteins from *Aspergillus oryzae* grown under submerged and solid-state culture conditions', *Appl Environ Microbiol*, **72**, 3448–57.

OLIVEIRA A P, PATIL K R and NIELSEN J (2008), 'Architecture of transcriptional regulatory circuits is knitted over the topology of biomolecular interaction networks', *BMC Syst Biol*, **2**, 16.

OUYANG H, LUO Y, ZHANG L, LI Y and JIN C (2010), 'Proteome analysis of *Aspergillus fumigatus* total membrane proteins identifies proteins associated with the glycoconjugates and cell wall biosynthesis using 2D LC-MS/MS', *Mol Biotechnol*, **44**, 177–89.

PANDEY A, SELVAKUMAR P, SOCCOL CR and POONAM S (1999), Solid state fermentation for the production of industrial enzymes', *Curr Sci*, **77**, 149–62.

PAPAGIANNI M (2007), 'Advances in citric acid fermentation by *Aspergillus niger*: Biochemical aspects, membrane transport and modeling', *Biotechnol Adv*, **25**, 244–63.

PATIL K R and NIELSEN J (2005), 'Uncovering transcriptional regulation of metabolism by using metabolic network topology', *Proc Nat Acad Sci*, **102**, 2685–9.

PAYNE G A, NIERMAN W C, WORTMAN J R, PRITCHARD B L, BROWN D, DEAN R A, BHATNAGAR D, CLEVELAND T E, MACHIDA M and YU J (2006), 'Whole genome comparison of *Aspergillus flavus* and *A. oryzae*', *Med Mycol*, **44**, 9–11.

PEL H J, DE WINDE J H, ARCHER D B, DYER P S, HOFMANN G, SCHAAP P J, TURNER G, DE VRIES R P, ALBANG R, ALBERMANN K, ANDERSEN M R, BENDTSEN J D, BENEN J A, VAN DEN BERG M, BREESTRAAT S, CADDICK M X, CONTRERAS R, CORNELL M, COUTINHO P M, DANCHIN E G, DEBETS A J, DEKKER P, VAN DIJCK P W, VAN DIJK A, DIJKHUIZEN L, DRIESSEN A J, D'ENFERT C, GEYSENS S, GOOSEN C, GROOT G S, DE GROOT P W, GUILLEMETTE T, HENRISSAT B, HERWEIJER M, VAN DEN HOMBERGH J P, VAN DEN HONDEL CA, VAN DER HEIJDEN R T, VAN DER KAAIJ R M, KLIS F M, KOOLS H J, KUBICEK C P, VAN KUYK P A, LAUBER J, LU X, VAN DER MAAREL M J, MEULENBERG R, MENKE H, MORTIMER M A, NIELSEN J, OLIVER S G, OLSTHOORN M, PAL K, VAN PEIJ N N, RAM A F, RINAS U, ROUBOS J A, SAGT C M, SCHMOLL M, SUN J, USSERY D, VARGA J, VERVECKEN W, VAN DE VONDERVOORT P J, WEDLER H, WÖSTEN H A, ZENG A P, VAN OOYEN A J, VISSER J and STAM H (2007), 'Genome sequencing and analysis of the versatile cell factory *Aspergillus niger* CBS 513.88', *Nat Biotechnol*, **25**, 221–31.

PRICE N D, PAPIN J A, SCHILLING C H, and PALSSON B (2003), 'Genome-scale microbial *in silico* models: The constraints-based approach', *Trends Biotechnol*, **21**, 162–9.

QUEVILLON E, SILVENTOINEN V, PILLAI S, HARTE N, MULDER N, APWEILERR and LOPEZ R (2005), 'InterProScan: protein domains identifier', *Nucleic Acids Res*, **33**, W116–20.

RATLEDGE C (2002), 'The biochemistry and molecular biology of lipid accumulation in oleaginous microorganisms', *Adv Appl Microbiol*, **51**, 1–51.

© Woodhead Publishing Limited, 2013

RATLEDGE C (2004), 'Fatty acid biosynthesis in microorganisms being used for single cell oil production', *Biochimie*, **86**, 807–15.

REHM B H (2001), 'Bioinformatic tools for DNA/protein sequence analysis, functional assignment of genes and protein classification', *Appl Microbiol Biotechnol*, **57**, 579–92.

ROKAS A, PAYNE G, FEDOROVA N D, BAKER S E, MACHIDA M, YU J, GEORGIANNA D R, DEAN R A, BHATNAGAR D, CLEVELAND T E, WORTMAN J R, MAITI R, JOARDAR V, AMEDEO P, DENNING D W and NIERMAN W C (2007), 'What can comparative genomics tell us about species concepts in the genus *Aspergillus*?', *Stud Mycol*, **59**, 11–7.

SAKAGUCHI K, TAKAGI M, HORIUCHI H and GOMI K (1992), 'Fungal enzymes used in oriental food and beverage industries' in *Applied Molecular Genetics of Filamentous Fungi*, Kinghorn J R and Turner G (eds), Blackie Academic and Professional, Glasgow, UK, 54–99.

SAKURADANI E and SHIMIZU S (2009), 'Single cell oil production by *Mortierella alpina*', *J Biotechnol*, **144**, 31–6.

SALAZAR M, VONGSANGNAK W, ANDERSEN M, PANAGIOTOU G and NIELSEN J (2009), 'Uncovering transcriptional regulation of glycerol metabolism in aspergilli through genome-wide gene expression data analysis', *Mol Genet Genomics*, **282**, 571–86.

SATO A, OSHIMA K, NOGUCHI H, OGAWA M, TAKAHASHI, T, OGUMA T, KOYAMA Y, ITOH T, HATTORI M and HANYA Y (2011), 'Draft genome sequencing and comparative analysis of *Aspergillus sojae* NBRC4239', *DNA Res*, **18**, 165–76.

SIGRIST C J, CERUTTI L, DE CASTRO E, LANGENDIJK-GENEVAUX P S, BULLIARD V, BAIROCH A and HULO N (2010), 'PROSITE, a protein domain database for functional characterization and annotation', *Nucleic Acids Res*, **38**, D161–6.

SIMOPOULOS A P (2011), 'Evolutionary aspects of diet: The omega-6/omega-3 ratio and the brain', *Mol Neurobiol*, **44**, 203–15.

STEPHANOPOULOS G, ALPER H and MOXLEY J (2004), 'Exploiting biological complexity for strain improvement through systems biology', *Nat Biotechnol*, **22**, 1261–7.

TAKAHASHI J A and CARVALHO S A (2010), 'Nutritional potential of biomass and metabolites from filamentous fungi', *Current Research, Technology and Education Topics in Applied Microbiology and Microbial Biotechnology*, A. Mendez-Vilas (ed.), Formatex Research Centre, Spain, 1126–35.

TATUSOV R L, GALPERIN M Y, NATALE D A and KOONIN E V (2000), 'The COG database: a tool for genome-scale analysis of protein functions and evolution', *Nucleic Acids Res*, **28**, 33–6.

THOMAS P D, KEJARIWAL A, CAMPBELL M J, MI H, DIEMER K, GUO N, LADUNGA I, ULITSKY-LAZAREVA B, MURUGANUJAN A, RABKIN S, VANDERGRIFF J A and DOREMIEUX O (2003), 'PANTHER: a browsable database of gene products organized by biological function, using curated protein family and subfamily classification', *Nucleic Acids Res*, **31**, 334–1.

TOMINAGA M, LEE Y H, HAYASHI R, SUZUKI Y, YAMADA O, SAKAMOTO K, GOTOH K and AKITA O (2006), 'Molecular analysis of an inactive aflatoxin biosynthesis gene cluster in *Aspergillus oryzae* RIB strains', *Appl Environ Microbiol*, **72**, 484–90.

TRABAL J, LEYES P, FORGA M and MAUREL J (2010), 'Potential usefulness of an EPA-enriched nutritional supplement on chemotherapy tolerability in cancer patients without overt malnutrition', *Nutr Hosp*, **25**, 736–40.

TYERS M and MANN M (2003), 'From genomics to proteomics', *Nature*, **422**, 193–7.

VAN PEIJ N, VISSER J and DE GRAAFF L (1998), 'Isolation and analysis of *xlnR*, encoding a transcriptional activator co-ordinating xylanolytic expression in *Aspergillus niger*', *Mol Microbiol*, **27**, 131–42.

VARMA A and PALSSON B O (1994), 'Stoichiometric flux balance models quantitatively predict growth and metabolic by-product secretion in wild-type *Escherichia coli* W3110', *Appl Environ Microbiol*, **60**, 3724–31.

© Woodhead Publishing Limited, 2013

VISSER H, JOOSTEN V, PUNT P J, GUSAKOV A V, OLSON P T, JOOSTEN R, BARTELS J, VISSER J, SINITSYN A P, EMALFARB M A, VERDOES J C and WERY J (2011), 'Development of a mature fungal technology and production platform for industrial enzymes based on a *Myceliophthora thermophila* isolate, previously known as *Chrysosporium lucknowense* C1', *Ind Biotechnol*, **7**, 214–23.

VONGSANGNAK W, OLSEN P, HANSEN K, KROGSGAARD S and NIELSEN J (2008), 'Improved annotation through genome-scale metabolic modeling of *Aspergillus oryzae*', *BMC Genomics*, **9**, 245.

VONGSANGNAK W, SALAZAR M, HANSEN K and NIELSEN J (2009), 'Genome-wide analysis of maltose utilization and regulation in aspergilli', *Microbiology*, **155**, 3893–902.

VONGSANGNAK W, NOOKAEW I, SALAZAR M and NIELSEN J (2010), 'Analysis of genome-wide co-expression and co-evolution of *Aspergillus oryzae* and *Aspergillus niger*', *OMICS J Integr Biol*, **14**, 165–75.

VONGSANGNAK W, HANSEN K and NIELSEN J (2011), 'Integrated analysis of the global transcriptional response to α-amylase over-production by *Aspergillus oryzae*', *Biotechnol Bioeng*, **108**, 1130–9.

WANG L, CHEN W, FENG Y, REN Y, GU Z, CHEN H, WANG H, THOMAS M J, ZHANG B, BERQUIN I M, LI Y, WU J, ZHANG H, SONG Y, LIU X, NORRIS J S, WANG S, DU P, SHEN J, WANG N, YANG Y, WANG W, FENG L, RATLEDGE C, ZHANG H and CHEN Y Q (2011), 'Genome characterization of the oleaginous fungus *Mortierella alpina*', *PLOS One*, **6**, e28319.

YUAN X L, VAN DER KAAIJ R M, VAN DEN HONDEL C A, PUNT P J, VAN DER MAAREL M J, DIJKHUIZEN L and RAM A F (2008), '*Aspergillus niger* genome-wide analysis reveals a large number of novel alpha-glucan acting enzymes with unexpected expression profiles', *Mol Genet Genomics*, **279**, 545–61.

© Woodhead Publishing Limited, 2013

3

Systems biology methods and developments for *Saccharomyces cerevisiae* and other industrial yeasts in relation to the production of fermented food and food ingredients

V. Mapelli, C. J. Franzén and L. Olsson, Chalmers University of Technology, Sweden

DOI: 10.1533/9780857093547.1.42

Abstract: This chapter describes the use of the systems biology tool box in the production of food and food ingredients based on yeast fermentation. Challenges and possibilities of the application of systems biology are described in relation to the production of yeast fermented food and to novel production of food ingredients and nutraceuticals based on yeast fermentation. While brewer's yeast remains the main and best characterized microorganism used for food and beverage production, the chapter also describes how systems biology tools can be valuable in the implementation of novel cell factories for food ingredients using so-called non-conventional yeasts.

Key words: antioxidants, flavours, functional colours, wine and beer, yeast biotechnology.

3.1 Introduction

The use of yeast and other microorganisms for food production is probably the oldest application of microbial biotechnology. Yeast fermented food is still one of the main components of human nutrition and nowadays in depth knowledge of the physiology, genetics and metabolism of this eukaryotic microorganism allows the production of tailor-made products. A comprehensive view of the whole biological system is possible today thanks to the development of systems biology, whose tool box includes advanced experimental and computational methodologies.

In this chapter, we define the basic tools and aims of systems biology, in connection with the possibilities and challenges of a systems biology

© Woodhead Publishing Limited, 2013

approach to the production of food and food ingredients using yeast. In the production of fermented food, the application and future perspectives of systems biology are described in relation to wine and beer production, where the main challenges are connected to the use of mixed yeast cultures and industrial yeast strains. Food ingredients are also discussed, including compounds that are or can be produced by yeast fermentation and then added to food following purification. The focus is turned to flavours and colouring whose production via yeast fermentation has great potential.

Nutraceuticals are a further class of food-related products, typically defined as food supplements with bioactive and beneficial properties for human health. Here we consider the production of yeast enriched with beneficial metabolites with antioxidant activity and present recent research on the improvement of existing yeast-based nutraceuticals.

The brewer's yeast *Saccharomyces cerevisiae* is the most used yeast in food production and also the best characterized at physiological, metabolic and molecular levels and therefore most of the systems biology tools and models have been developed for this yeast. However, novel yeast-based production exploiting so-called non-conventional yeasts which have favourable properties has good potential: a few are discussed at the end of this chapter, underlying how systems biology tools can be valuable for implementation of novel cell factories for food ingredients.

3.2   History of yeast science: it all started with food

The origin of fermented food can be traced back to the origins of civilisation in places where growth of cereal crops was possible. In fact, the art of brewing and leavening bread is typically dated back to 4000–3500 BC in Egypt and in the Mesopotamian region. With no doubt, fermentation occurred spontaneously and accidentally at that time and most probably the liquor of fermented barley was even safer than water, as the fermentative activity of yeasts could prevent the growth of noxious microorganisms. In Europe, brewing was first developed in abbeys by monks who studied and improved the ancient Egyptian techniques. In this way they also bypassed the problem of undrinkable water from rivers and lakes by converting it into beer, recognized as a healthy, tasty and nutritional beverage, as reviewed by Corran (1975).

The first depictions of yeasts as part of the brewing process date back to 1680, when Antoine van Leeuwenhoek observed them under a microscope; however, at that time, yeasts were far from being considered living organisms (Barnett, 1998). The first rational description of alcoholic fermentation was made by the chemist Antoine-Laurent Lavoisier who described the transformation of cane sugar upon fermentation as the transformation of sugar into carbon dioxide, alcohol and acetic acid. However, fermentation activity was not attributed to yeasts. Studies on alcoholic fermentation continued, primarily to avoid spoilage of wine.

© Woodhead Publishing Limited, 2013

Improvements in microscopy were crucial for the definition of yeast as living organism: between 1836 and 1838 Cagniard-Latour, Kütznig and Schwann defined the seminal principles of yeast physiology (Barnett, 1998). In addition, Schwann defined yeast as a fungus and attributed to it the process of fermentation of wine sugar and, importantly, he was the first one to define 'metabolic' transformations. Later on, the chemists Wöhler, von Liebig and Berzelius opposed the theory of yeast as a living organism. Although nowadays we know that yeast is, in fact, a living organism, these chemists introduced the concept of enzymology, as they interpreted the conversion of sugar into alcohol in terms of catalysis (Barnett, 1998, 2000).

The nature of yeast as a living organism was finally established by the studies carried out by Louis Pasteur, who extensively and quantitatively analysed alcoholic fermentations and stated that 'changes of fermentation are associated with a vital activity. [...] Alcoholic fermentation never occurs without simultaneous organization, development and multiplication of cells or the continued life of cells already formed'. Pasteur also determined that substances other than ethanol and CO_2, such as glycerol and succinic acid, were derived from the transformation of sugar during fermentation (Barnett, 2000).

Success in isolating single yeast species was first achieved by Emil Christian Hansen at the Carlsberg Laboratory (Barnett and Lichtenthaler, 2001): he finally worked out that specific features of beer were mainly ascribable to the presence of specific yeast strains and he managed to isolate a single yeast strain which was able to produce good beer consistently and reproducibly. Therefore, Hansen set the basis for what nowadays is required from industrial microbial production, that is the use of pure cultures to achieve reproducible fermentation processes and thereby lead to products with well defined characteristics.

The development of novel techniques was definitely decisive in paving the way to in depth knowledge of yeasts and biological systems in general. Following the history of discoveries on the biological mechanisms ruling the life of yeast, it is fascinating to observe how small 'pieces' have been (and are still being) assembled, turning into a detailed 'picture' of entire biological systems that nowadays can be actually studied as a whole, thanks to what we describe below as 'systems biology tools'. Thanks to continuous scientific developments, yeasts can now be seen not only as means of producing food, but also as a means of improving food.

3.3   Systems biology: possibilities and challenges in relation to food

The development of systems biology tools makes it possible to look at microbial systems as a whole and at different levels, from the genome to the metabolome, passing through transcriptome and proteome. The

© Woodhead Publishing Limited, 2013

ultimate aim of systems biology is to use comprehensive data sets to build models that can be used for better understanding of the biological entities and also to predict system behaviour following certain perturbations. The majority of systems biology technology and models have been developed on laboratory strains of *Saccharomyces cerevisiae*; however, characterization of industrial strains and non-*Saccharomyces* species via application of systems biology tools is spreading and involving yeasts which are typically used for the production of food and food ingredients.

3.3.1   Systems biology: a complex tool box for model construction

The production of diverse compounds by the use of microorganisms and genetically modified microorganisms can be traced back to the 1980s, when the use of designed production organisms was attempted in order to replace the classical synthetic routes (Pass, 1981). Although recombinant DNA technologies developed and used at that time were very similar to the ones in use today, the major difference between the 1980s and now is the existence of reliable and more and more precise molecular technologies dedicated to characterization and modification of microorganisms, together with the availability of a number of annotated genome sequences.

Systemic studies in the field of biology represent a huge step forward because the function of living organisms cannot be satisfactorily addressed by looking at isolated molecules and reactions (Bruggeman and Westerhoff, 2007). The possibility of looking at whole systems enables us to elucidate how the control of a function/process within a system is distributed over several network components.

The first published microbial genome sequences were mainly driven by the medical research field (Otero and Nielsen, 2010): the genome sequence of *Saccharomyces cerevisiae* was among the first to be published, probably because of its importance as a eukaryotic model system. The entire genome sequence of *S. cerevisiae* (Goffeau *et al.*, 1996) represented the first available eukaryotic genome and a tremendous advance, even though the definition of the mere genomic source code provided only an inventory of the 6000 yeast genes. The real advantages brought by the availability of genome sequences have come once genetic sequences have been annotated. The discipline that aims to annotate each gene has been defined functional genomics. Functional genomics relies on the development and exploitation of experimental and theoretical tools to link gene sequences to their products (i.e. enzymes) and therefore to their specific functions (Bruggeman and Westerhoff, 2007). The in-depth characterization brought by functional genomics represents the crucial bridge between genome sequencing and systems biology.

The essence of systems biology is the combination of mathematical modelling and experimental biology. The final objective of systems biology is the quantitative description of the biological system under study by defining

© Woodhead Publishing Limited, 2013

a mathematical model, which, in some cases can be the final outcome of the study and might be used to predict the behaviour of the biological system under specific conditions, while in other cases, the model can be used to extract further information from experimental data (Nielsen and Jewett, 2008).

Two different systems biology approaches have been defined by the construction of mathematical models: the top-down and the bottom-up systems biology approaches (Bruggeman and Westerhoff, 2007). The top-down approach relies on data provided through the so-called *x-omics* tools (see Appendix), which are referred to as technologies able to supply sets of global data (i.e. genome wide data) at different cellular levels. The main objective of top-down systems biology is to discover novel molecular mechanisms through integration of *x-omics* data. The strength of this approach is its completeness, as it is built on genome-wide data and it can lead to the discovery of functionally related processes (Tanay *et al.*, 2004; Beyer *et al.*, 2006). The insights gained are not hypothesis driven but rather based upon induction (Kell and Oliver, 2004) and have to be tested and validated by experiments.

Bottom-up systems biology is founded on the deduction of functional properties from very detailed knowledge of a characterized sub-system; this knowledge is then translated into mathematical models used to simulate the behaviour of the system (Bruggeman and Westerhoff, 2007). Generally, bottom-up driven models are able to describe only a subset of the entire biological system, as quantitative information might not be enough to describe the whole set of interactions between all components. However, genome-scale metabolic models (GSMM) which are among the bottom-up driven models are considered fairly global in their approach (Nielsen and Jewett, 2008). GSMMs are based on collecting the stoichiometry of each biochemical reaction and associating each reaction with the related annotated gene(s). Through the application of flux balance analysis (Orth *et al.*, 2010), this would result in biochemical models that can be used to predict the relationships between genes and functions within a cellular metabolic network and to simulate growth and/or product formation (Famili *et al.*, 2003; Förster *et al.*, 2003; Price *et al.*, 2004).

Importantly, GSMMs provide a solid framework for integrating *x-omics* data. Several algorithms have been developed aiming to predict metabolic changes caused by genetic modifications (Burgard *et al.*, 2003; Patil *et al.*, 2005; Pharkya *et al.*, 2004). One of the limitations of currently available GSMMs is the lack of information about the strength of the described network interactions; hence they can predict and simulate the effects only of disruption of network connectivity (e.g. gene knock out). In order to overcome this limitation, regulatory and thermodynamic constraints will be the next step in metabolic modelling (Soh *et al.*, 2011).

Besides elucidating the fundamental principles ruling biological systems, tools developed by systems biology find many applications in industrial biotechnology. In this case, the effort put into developing new computational tools to simulate biological behaviour is primarily meant to generate better production strains and also to speed up metabolic engineering strategies for the development and optimization of cell factories for production of different chemicals.

Yeasts have long been used for food production and the yeast *S. cerevisiae* is traditionally the best characterized of the yeast species, for applications, biochemical investigations and genetics. The number of yeast species, whose genomes have been annotated, has now increased (see Table 3.1) and includes both yeasts which are relevant in the medical field (i.e. human pathogens) and yeasts that have natural properties which are considered to be of high industrial interest for food production and novel biotechnological production of different chemicals.

The reconstruction of several genome scale metabolic models has been implemented for *S. cerevisiae* (See Table 3.2), while only very recently two GSMMs for the methylotrophic yeast *Pichia pastoris,* of interest because of its efficient production of heterologous proteins, have been published (Chung *et al.*, 2010; Sohn *et al.*, 2010).

3.3.2 Yeast systems biology and production of food and food ingredients

The use of systems biology tools in food-related production might be seen from two different angles, depending on the application: production of fermented food or production of food ingredients. Specifically, production of fermented food (i.e. bread, wine and beer) typically makes use of the so called 'industrial' yeast strains, which are known to be characterized by intrinsic genetic complexity, as they can be genetically diverse, prototrophic, homothallic and often aneuploid, polyploid or alloploid, as reviewed by de Winde (2003). These traits can hinder the possibility of carrying out proper functional genomics studies and easy application of metabolic engineering tools that aim to modify robustly such strains for the development of better processes or products. Brewing, distilling and baker's yeasts are, with some exceptions, species of *Saccharomyces* genus: nevertheless, genomic exploration of non-*Saccharomyces* species of industrial interest is increasing the amount of available information (Table 3.1) (Souciet *et al.*, 2009).

Besides technical difficulties in applying molecular biology tools to industrial strains, the potential presence of genetically modified organisms (GMO) in food is subjected to very strict regulations, since in this case GMOs would be consumed by humans and then released into the environment. Engineered organisms that enter the food chain have to be approved, classified and labelled according to the principles assessed for GMO food-related safety (http://www.who.int/foodsafety/biotech/codex_taskforce/en/). Additionally, despite approval of such GMOs, the risk of failing commercialization is high, owing to no or very low acceptance by the consumers (Nevoigt, 2008; Pretorius and Bauer, 2002).

© Woodhead Publishing Limited, 2013

Table 3.1 Publicly available yeast genome sequences

Organism	Relevance	Size (kb)	Number of ORFs	Number of proteins	Publication
Candida albicans	Medical	27,599,338	14,264	14,230	Butler *et al.* (2009); Jones *et al.* (2004)
Candida dubliniensis	Medical	14,618,422	6095	5860	Jackson *et al.* (2009)
Candida glabrata	Medical	12,358,371	5536	5224	Dujon *et al.* (2004)
Candida tropicalis	Medical	14,680,443	6441	6254	Butler *et al* (2009)
Clavispora lusitaniae	Medical	12,114,892	6153	5936	Butler *et al.* (2009)
Debaryomyces hansenii	Industrial (cheese production)	12,211,410	6674	6308	Dujon *et al.* (2004)
Hansenula polymorpha	Model organisms; industrial				Eldarov *et al.* (2011); Ramezani-Rad *et al.* (2003)
Kluyveromyces lactis	Industrial	10,769,738	5445	5094	Dujon *et al.* (2004); Zivanovic *et al.* (2005)
Kluyveromyces polysporus		14,696,275	5798	5376	Scannell *et al.* (2007)
Kluyveromyces thermotolerans		10,416,446	5536	5102	Souciet *et al.* (2009)
Kluyveromyces waltii	Molecular evolution studies	10,912,112			Kellis *et al.* (2004)
Lodderomyces elongisporus	Medical	15,511,671	5908	5799	Butler *et al.* (2009)
Meyerozyma guilliermondii	Medical	10,609,954	6062	5920	Butler *et al.* (2009)
Pichia pastoris	Industrial	9,287,744	5083	5056	De Schutter *et al.* (2009)
Saccharomyces bayanus	Wine yeast	11,865,314			Kellis *et al.* (2003)
Saccharomyces castellii	Molecular evolution studies	11,219,539	5870	5592	Cliften *et al.* (2003)
Saccharomyces cerevisiae	Model organisms; industrial	12,157,105	6327	5882	Goffeau *et al.* (1996)
Saccharomyces kudriazevii	Molecular evolution studies	11,177,778			Cliften *et al.* (2003)
Saccharomyces mikatae	Molecular evolution studies	11,470,251			Kellis *et al.* (2003)
Saccharomyces paradoxus	Wine yeast	11,872,617			Kellis *et al.* (2003)
Saccharomyces pastorianus	Brewing yeast	22,382,146			Nakao *et al.* (2009)
Scheffersomyces stipitis	Industrial	15,441,179	5816	5816	Jeffries *et al.* (2007)
Schizosaccharomyces pombe	Model organism	12,591,251	5883	5020	Wood *et al.* (2002)
Spathaspora passalidarum	Industrial	13,182,099	5983	5983	Wohlbach *et al.* (2011)
Yarrowia lipolytica	Industrial	20,598,813	7415	6496	Dujon *et al.* (2004); Kerscher *et al.* (2001)
Zygosaccharomyces rouxii	Food production	9,764,635	5332	4991	de Montigny *et al.* (2000)

Data are retrieved from National Center for Biotechnology Information (NCBI) and Kyoto Encyclopedia of Genes and Genomes (KEGG). Data are updated to March 2012.

© Woodhead Publishing Limited, 2013

Table 3.2 Publicly available yeast genome scale metabolic models

Genome sequence characteristics				Metabolic network characteristics				
Genome sequence	Reference	Genome dimension (kb)	No. ORFs	Model ID	No. Reactions	No. Metabolites	Compartments	Publication
Saccharomyces cerevisiae S288C	Goffeau et al. (1996)	12,069	5860	iFF708	1175	584	Cytoplasm, mitochondria, extracellular	Förster et al. (2003)
Saccharomyces cerevisiae S288C	Goffeau et al. (1996)	12,069	5860	iND750	1489	646	Cytoplasm, mitochondria, peroxisome, nucleus, ER, golgi apparatus, vacuole, extracellular	Duarte et al. (2004)
Saccharomyces cerevisiae S288C	Goffeau et al. (1996)	12,069	5860	iLL672	1038	636	Cytoplasm, mitochondria, extracellular	Blank et al. (2005a)
Saccharomyces cerevisiae S288C	Goffeau et al. (1996)	12,069	5860	iIN800	1431	1013	Cytoplasm, mitochondria, extracellular	Nookaew et al. (2008)
Saccharomyces cerevisiae S288C	Goffeau et al. (1996)	12,069	5860	NA[b]	2342	2657[a]	Cytoplasm, mitochondria, peroxisome, nucleus, ER, golgi apparatus, vacuole, extracellular	Herrgård et al. (2008)
Pichia pastoris CBS704	De Schutter et al. (2009)	94,300	5313	PpaMBEL1254	1254	1147	Cytoplasm, mitochondria, peroxisome, nucleus, ER, golgi apparatus, vacuole, extracellular	Sohn et al. (2010)
Pichia pastoris GS115	De Schutter et al. (2009)	94,300	5313	PpaMBEL1254	1254	1147	Cytoplasm, mitochondria, peroxisome, nucleus, ER, golgi apparatus, vacuole, extracellular	Sohn et al. (2010)
Pichia pastoris X-33	De Schutter et al. (2009)	94,300	5313	iPP668	1361	1177	Cytoplasm, mitochondria, peroxisome, nucleus, ER, golgi apparatus, vacuole, extracellular	Chung et al. (2010)

Data are updated to March 2012.
[a] 1494 metabolites and 1163 proteins or complexes.
[b] http://www.comp-sys-bio.org/yeastnet/.

© Woodhead Publishing Limited, 2013

The definition of GMOs and the related legislation are no longer applicable if the genetic modification of a specific strain derives from the so-called 'self-cloning' (Nevoigt, 2008), meaning that the DNA of the specific organism derives exclusively from species that are phylogenetically related. Therefore, classical methods of development of yeast strains used in the production of food are typically employed (i.e. mutagenesis and hybridization), as reviewed by Pretorius and Bauer (2002) and de Winde (2003), both to overcome some technical difficulties in modifying industrial strains and to circumvent the use of microorganisms officially classified as GMOs.

However, in spite of no or very scarce commercialization of GM yeasts (i.e. those modified using recombinant DNA technologies), studies are ongoing and recombinant yeast strains with industrially interesting improved properties have been generated (a few examples are available: Husnik et al., 2006; Nevoigt et al., 2002; Cambon et al., 2006; Aldhous, 1990; Osinga et al., 1989) by targeting specific genetic modifications and may be commercialized when the public perception has turned more positive towards GMO-based food and food products.

Therefore, the current trend of applying systems biology tools to yeasts used for production of fermented food mostly relates to clarifying at which level environmental changes (i.e. fermentation conditions) or 'classical genetic modifications' influence yeast and process performances. Transcriptome and metabolome analyses are the most applied techniques in this field. In fact, gene expression profiles could help in identifying possible modifications of the process in order to minimize or maximize the transcription of genes responsible for specific activities that might influence the production process or the features of the final product. Metabolomics is especially important in food and food production, as the complex metabolite profile generated during fermentation is the main factor responsible for determining the final taste and aroma of the product. Metabolomics results might be integrated into transcriptome data to give important information on the complex connections of gene expression, metabolic activities and process setup. Hereby, in this context x-omics tools can be used as quality control tools throughout the fermentation process. Nonetheless, experimental data obtained through study and characterization of industrial yeasts might contribute to a certain extent to the improvement of existing genome-scale metabolic models. A schematic timeline of the development of systems biology as a tool in food production is reported in Fig. 3.1.

The production of food ingredients via establishment of yeast cell factories does not have the same strict limitations that apply to GMOs used for production of fermented food. In fact, the final product (i.e. the food ingredient) would be purified, so that no GMOs will come into contact with consumers and neither will it be released into the environment. In this case, GSMMs tend to be used to simulate the biological behaviour of the strain under study after specific perturbations, which can be both environmental and genetic. In this way, targeted metabolic engineering strategies (i.e.

© Woodhead Publishing Limited, 2013

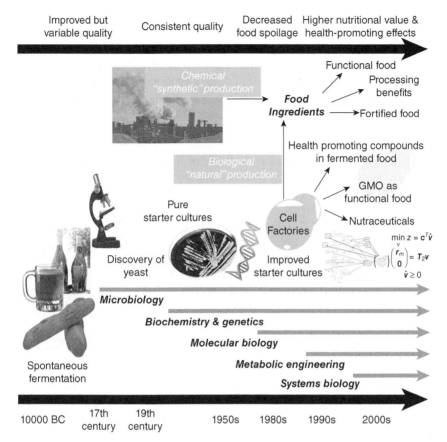

Fig. 3.1 Schematic timeline of the development of systems biology as a tool in the production of high value food ingredients. The relevant biotechnological research fields, some key milestones in the microbiology of foods, important changes in the quality and characteristics of food and potential products are indicated. Today, systems biology and metabolic engineering are used to design cell factories mainly for the production of food ingredients and nutraceuticals and to identify process improvements. In the future, GMO with desirable properties may become commercially viable food ingredients *per se*, if public opinion allows it and if all safety issues are cleared.

expression of heterologous genes or gene knock out) for the production of a specific compound can be first predicted by simulating their effects on a reconstructed GSMM. Experimental data obtained from the analysis of the engineered strain(s) have then to be used to improve and modify the model further (Nielsen, 2001).

   An important issue that should be considered when producing food ingredients is their actual functional status once they have been added to the food. There are two main angles from which this issue can be discussed:

© Woodhead Publishing Limited, 2013

one is linked to the actual level of bioavailability of the specific ingredient once it has been added to the food. This topic is of special importance when dealing with bioactive compounds. There are several examples showing that different compounds introduced in the diet are absorbed by the organism at varying levels depending on the specific food matrix (de Pascual-Teresa *et al.*, 2006; Jeanes *et al.*, 2004). The second angle that must be considered is the preservation of the specific ingredient throughout the process of food production. To address this concern, analysis of metabolite profiles of food is desirable, but it might also be hard owing to the challenges linked to the development of analytical methods able to handle food matrices that are typically very complex. This problem might also occur during analysis of fermented food, for example during determination of flavour profiles in beer and wine. Concerns about the food matrix and development of analytical methods have been addressed especially in the field of metabolomics: the United States National Institute of Standards and Technology (NIST) has developed matrix reference materials with the intention of validating analytical methods for the measurement of several kinds of compounds, e.g. vitamins, minerals, fatty acids (Sharpless *et al.*, 2004, 2007).

3.4   Systems biology tools for fermented food

Fermented foods have been traditionally used as a way to preserve the raw material (food) and yeast has been used mainly for its ability to produce ethanol and carbon dioxide. Nonetheless, organic acids produced by yeast during fermentation also contribute to the preservation. The major food products derived from yeast fermentation are beer, wine and other fermented beverages and bread.

Taste, flavour and aroma are very important features of fermented food and many variants of wine and beer are available. Consequently, it is no longer merely the yeast's ability to produce ethanol and carbon dioxide that is of major concern, but its ability to produce a range of metabolites that contribute to the product taste, aroma and texture. In this perspective, systems biology tools have a role to play as the cellular metabolic network can be mapped in detail and provide a guide to process conditions that favour the preferred metabolism.

A specific challenge is that many industrial *S. cerevisiae* strains used in these applications are poorly characterized. Furthermore, wine and beer yeasts also include species other than *S. cerevisiae*, hybrids of different *Saccharomyces* species or different genera (as reviewed by Donalies *et al.*, 2008). Although the majority of the yeasts applied belong to *S. cerevisiae sensu stricto* species, some fermented food/beverages are produced by mixed cultures, which further complicate the application of systems biology tools.

© Woodhead Publishing Limited, 2013

Even though genetically modified microorganisms are still not well accepted in consumer products, many different metabolic engineering approaches have demonstrated the possibility of improving yeast for application in beer, wine and bread making (as reviewed by Donalies *et al.*, 2008 and Saerens *et al.*, 2010). Here we will concentrate our discussion on the perspective and challenges of using systems biology tools in existing processes.

3.4.1   Systems biology tools applied to wine production

Fermentation of grape must to wine is an old process, but wine production has now become quite challenging, as wine makers have to guarantee the production of wine with specific and reproducible features and, at the same time, be cost-competitive (Pretorius and Bauer, 2002). The quality of wine is conditioned by several factors: the quality of grape cultivars (De Luca, 2011), the wine making techniques and the microorganisms present on the grapes during the fermentation and during ageing/conservation of the wine (Ciani *et al.*, 2010). In the past, spontaneous fermentation was the most common practice, with poorly predictable and sometimes undesirable outcomes. Although the use of natural yeasts for wine fermentation has not been completely abandoned, today, the use of defined starter cultures as inoculum of grape must has become common practice (Pretorius, 2000). *S. cerevisiae* is the most commonly used yeast species in starter cultures, as it plays a major role during fermentation and the use of *S. cerevisiae* starter cultures might guarantee the suppression of the growth of other yeasts which have low fermentation aptitudes and/or produce spoilage metabolites (Ciani *et al.*, 2010). However, the role of non-*Saccharomyces* yeasts which typically take part in spontaneous fermentation is now being reassessed, as they can have a positive role in the composition of the wine, thanks to the production of specific metabolites or to the enzymatic activities that they harbour (Fernandez *et al.*, 2000). Hence, the inclusion of non-*Saccharomyces* yeasts in mixed starter cultures has been proposed as a valuable tool to take advantage of the outcomes of spontaneous fermentation (Ciani *et al.*, 2006), while avoiding the risks of stuck fermentation. However, a good physiological and genetic characterization of interesting *Saccharomyces* species and other yeast genera is necessary, in order to select the best candidates.

Systems biology tools provide valuable tools to characterize the physiology of wine strains. While the limited knowledge of the genetic background of non-*Saccharomyces* yeasts hinders the use of some molecular tools, DNA microarrays have been used to characterize gene expression and genotyping of industrial wine *Saccharomyces* yeasts which often have genomic patterns not present in laboratory strains (Marks *et al.*, 2008; Pizarro *et al.*, 2008; Hauser *et al.*, 2001; Hughes *et al.*, 2000). This approach leads to the

© Woodhead Publishing Limited, 2013

identification of genes that confer advantageous features on industrial yeasts. Furthermore, comparative analysis of gene expression profiles has been used to validate the possibility of extrapolating valuable information from experimental laboratory approach to solve problems occurring in industrial fermentations, showing that laboratory-based results accurately reflect real industrial processes (Rossouw *et al.*, 2010, 2012).

If application of transcriptome analysis is hindered in fermentations where non-*Saccharomyces* genera are present, metabolomics is instead more easily applicable in these cases and provides crucial and clear information on the actual features of the wine: aroma, flavour and body of the wine. Importantly, results from metabolic footprinting analysis have been proven to meet the outcomes of human sensory analysis on wines produced using different kinds of yeast strains (Hyma *et al.*, 2011). Yeast strains can be then selected for their ability to produce favourable metabolites (Viana *et al.*, 2008). Nevertheless, great consideration should be given to the fact that metabolic footprinting of mixed fermentations can substantially vary from the bare combination of metabolic footprinting observed in fermentations carried out using single species. In fact, there is a reciprocal metabolic influence between different populations that should not be neglected (Comitini *et al.*, 2011; Howell *et al.*, 2006; Anfang *et al.*, 2009). Wine is very complex and, therefore, metabolite profiling typically results in large data sets and, consequently, statistical methods to extract valuable information are required. Chemometrics could be applied to analyse wine metabolite profiles providing quality control of the ongoing process and to monitor population dynamics within the fermentation (Howell *et al.*, 2006; Son *et al.*, 2009).

The integration of gene expression with metabolite profiles throughout fermentation has been shown to be even more informative in terms of finding specific genetic features that are responsible for a certain metabolic/ aroma profile (Rossouw *et al.*, 2008). This data integration has allowed the establishment of multivariate statistical models that can be used to predict genetic modifications potentially promoting the increase/decrease of specific metabolites and consequent aroma(s). The proteome profile of wine fermentation could also give hints on the selection of yeasts that might be included in starter cultures, in fact, certain enzymes (e.g. poly-galacturonase, β-D-xylosidase and proteases) that are often produced by non-*Saccharomyces* yeasts can enhance wine quality and simplify some wine making procedures (Fernandez *et al.*, 2000; Strauss *et al.*, 2001).

In conclusion, transcriptome, proteome and exo-metabolome analyses play an important role in understanding different aspects of wine fermentation, including strain selection, quality control of the process/product and identification of genetic targets that could enhance the development of novel strains with favourable characteristics. However, the industrial implementation of sophisticated analytics and the related chemometric analysis might not be a realistic possibility in the near future. Nevertheless, the

© Woodhead Publishing Limited, 2013

amount of validated data produced so far on laboratory-scale could result in the development of user-friendly sensors which are likely to be more applicable to industrial scale fermentation. For instance, a so-called electronic tongue has been developed to provide a user-friendly device able to reveal sensory features of wines on the basis of their chemical composition (Rudnitskaya *et al.*, 2010a, 2010b). Such a device could be used for on-line monitoring of industrial wine fermentations.

3.4.2 Systems biology tools applied to beer production

Although beer is a traditional product, constant work is performed to develop both the production process and the product, as reviewed by Piddocke and Olsson (2009). The producers aim for more efficient processes: one way to go is to develop high-gravity processes, that is beer is produced with a high substrate concentration leading to a high ethanol concentration. The final beer is diluted after the fermentation to the alcohol level that the product is intended for. This approach improves the water usage economy. However, higher demands are put on the cellular performance owing to the additional stress put on the yeast as consequence of high concentrations of substrate, ethanol and higher osmolarity. In such cases, systems biology tools can be useful to map the consequences of altered fermentation conditions. Additional developments taking place in brewing are driven by the consumers' needs, which include designed flavour profiles and avoidance of undesirable off-flavours. Low alcohol or low calorie beer are further current trends, as reviewed by Dequin (2001). These trends call for changes in secondary metabolite production, but also in the processing conditions and the raw material used for the process also has a great influence on the cellular metabolism.

The genetic complexity of Brewer's yeasts makes transcriptomics more complex, as traditionally used DNA arrays have been based on laboratory strains. Two problems arise, first non-*S. cerevisiae* genes are present in lager strains, for instance and second genes that are homologous of *S. cerevisiae* genes might have different nucleotide sequences, hence array hybridization might be weak (Smart, 2007). In spite of these limitations, several studies have been performed using DNA arrays which have helped the investigators to gain insight into physiological responses. For example, transcriptome analysis using Affymetrix® DNA microarrays has been used to map the influence of high gravity fermentation conditions and the possible influence of protease additions to the fermentations using the lager yeast Weihenstephan 34/70 (Piddocke *et al.*, 2011). Genes involved in stress responses as well as in amino acid metabolism were identified as changing significantly. The use of glucose or maltose adjuncts to achieve high gravity conditions triggered significant differences in responses at the transcriptome level. Also, cross-regulation with amino acid uptake and metabolism was detected.

© Woodhead Publishing Limited, 2013

Since nutrient limitation is among the key challenges of beer production, a few studies have been carried out on the transcriptome response following shortage of certain nutrients. The influence of zinc (Zn)-limitation on cellular metabolism and transcriptome (De Nicola *et al.*, 2007) resulted in the identification of a Zn-specific Zap1 regulon and mapping of the influence of Zn on storage carbohydrate metabolism (Hazelwood *et al.*, 2009). By performing the study on a laboratory strain, much more reliable transcriptome studies could be performed. On the other hand, the observed responses might not be directly transferable to Brewers' yeast owing to the genetic and phenotypic differences between strains. The dynamics of *Saccharomyces carlsbergensis* transcriptomic response (using Gene Filters®) were studied during the time course of lager beer production (Olesen *et al.*, 2002). Many significant changes occurred during the production process and the authors concluded that even though there are significant drawbacks in using transcriptome analysis tools designed for *S. cerevisiae*, their approach contributed to important insights that could not have been achieved using a laboratory strain.

Foam stability and haze formation are common challenges in beer production. Both issues are related to proteins produced during the brewing process. The proteins found in beer originate both from the yeast metabolism and from the raw material. A few reports have emerged on using proteome and peptidome analysis as approaches to gain more detailed information. Using two-dimensional (2D) gel electrophoresis followed by mass spectrometric identification, a detailed mapping of beer proteins has been made: proteins that can be used as beer quality markers with respect to foam formation and haze formation were identified (Iimure *et al.*, 2011). Furthermore, the presence of certain intracellular proteins indicates yeast cell damage. Such markers might also be useful as process quality indicators, as cell damage should be avoided during the fermentation. In addition, some peptides that can exist in beer might be allergenic and toxic in certain cases and proteomics/peptidomics has been demonstrated as a possible way of detecting such compounds (Picariello *et al.*, 2011).

Amino acid metabolism and the presence of amino acids is very important in brewing. First, amino acids function as a nitrogen source for the yeast. Beer fermentation (especially in high gravity conditions) is often limited by nitrogen source availability which influences both the yeast physiology and the metabolites produced. Several beer flavours originate from amino acids and peptides that mainly originate from the barley raw material, being released during the malting. There is also some practice of using proteases to release more bioavailable nitrogen. In addition to enabling a deeper insight into the process, measuring the amino acid profiles might give a measure of the beer quality and could also be used for authenticity control (Pomilio *et al.*, 2010). Such an analysis is not an *'omic'*- analysis, but rather a targeted analysis.

© Woodhead Publishing Limited, 2013

In a recent study, metabolome analysis was used to elucidate the effect of protease addition on high gravity fermentations, using glucose or maltose adjuncts for fortification. A methodology was used, which allowed determination of amino acids, non-amino acids and their derivatives. In particular, the metabolome analysis demonstrated that the effect of protease was more pronounced on maltose-rich media, as the levels of intracellular amino acids were significantly influenced in this case (Piddocke *et al.*, 2011). The use of metabolome analysis allowed a detailed mapping of the responses on the metabolite level, which also showed the complex relationship between nitrogen metabolism, central carbon metabolism and flavour production. In another study, metabolome analysis was used to compare the difference in stress responses between a laboratory strain and an industrial strain used for high gravity fuel ethanol production (Devantier *et al.*, 2005). Multivariate data analysis showed that the laboratory strain responded to the stress conditions (i.e. high substrate condition and high ethanol concentrations) by increasing metabolic fluxes around the pyruvate node, which could be considered a way for yeast to sustain energy production.

Yeast metabolites and amino acids are not the sole compounds of interest in beer; many other compounds present in beers have attracted considerable attention and might be interesting targets in metabolome analysis. In particular, vitamins, minerals and antioxidants are components that make beer a beverage with recognized health benefits as well as good nutritional values, as reviewed by Bamforth (2002).

3.5   Production of flavours from yeasts

Aroma and flavour compounds resulting from natural yeast metabolic activities include mainly higher alcohols and their esters, carbonyl- and sulphur-compounds. Yeast-produced flavours can be divided into three main categories (Querol and Fleet, 2006): (i) yeast metabolic products (e.g. xylitol and vanillin discussed below); (ii) yeast cell mass derived products (e.g. amino acids and polysaccharides); (iii) complex products arising from interaction of yeast products and components of food matrix (e.g. peptides and amino acids deriving from proteases excreted by yeast).

3.5.1   Xylitol

A relevant example of a yeast-produced flavour is the sugar alcohol xylitol, which derives from reduction of D-xylose and is typically used as sweetener in food and beverages as it has almost the same sweetness as sucrose, but a lower energy value (Granström *et al.*, 2007a). Furthermore, xylitol has been shown to prevent dental caries (Mäkinen, 1992). Industrial production of xylitol is currently performed via chemical reduction of D-xylose,

© Woodhead Publishing Limited, 2013

achieving 50–60% yields. Microbial production of xylitol represents an alternative to the chemical route and research has been focusing on the use of *S. cerevisiae* to synthesize xylitol from glucose and on the use of *Candida* species. Although the use of *Candida* spp. in production of food and food ingredients is limited, owing to their opportunistic pathogenic nature, they have some advantages in xylitol production thanks to their ability to naturally take up xylose and to their robustness in maintaining a stable intracellular redox balance (Granström *et al.*, 2007b).

Oxygen limitation has been shown to be the determining factor for xylitol over-production in *Candida guilliermondii* and *Candida tropicalis* (Granström *et al.*, 2001, 2002). Under these conditions NAD^+ is not regenerated at sufficient levels via oxidative phosphorylation and the NAD-dependent conversion of xylitol to xylulose is therefore inhibited. Only few studies on genetic engineering of *C. tropicalis* have been reported (Ko *et al.*, 2006; Ahmad *et al.*, 2012), mainly because of the scarce availability of molecular/genetic tools applicable to these yeasts. The majority of studies on the improvement of xylitol biosynthesis focused on optimization of the process and growth conditions (Granström *et al.*, 2002; Kwon *et al.*, 2006; Kim *et al.*, 1997), while further insights into the *C. tropicalis* metabolism were obtained via application of metabolic flux analysis (MFA) using a metabolic model built on *Candida milleri* (Granström *et al.*, 2000) and further modified (Granström *et al.*, 2002). Fluxes were analysed using data from oxygen limited chemostat cultivations in the presence of high xylose concentration. Results suggested a possible metabolic flux distribution wherein glycolytic and pentose phosphate fluxes increase along with the deepening of redox imbalance, resulting in higher xylose consumption and xylitol formation and also the possible use of malate and acetaldehyde as redox sinks for NADH.

The best microbial process for xylitol production has been achieved via bioprocess optimization (Kwon *et al.*, 2006), yielding a specific xylitol productivity of 12 g l^{-1} h^{-1} through cell-recycle fermentation.

3.5.2   Successful systems biology approach for vanillin biosynthesis

Vanillin is to date the global leader among aroma compounds on the market (Hansen *et al.*, 2009). The natural source of vanillin (3-methoxy-4-hydroxybenzaldehyde) is the cured seed pod of the orchid *Vanilla planifolia* (Ramachandra Rao and Ravishankar, 2000); however, of the more than 16,000 metric tonnes on the market every year, only 0.25% is from the natural source. The demand for vanillin is currently fulfilled by chemical synthesis from ferulic acid (i.e. a component of lignin and a side chain in cereal hemicellulose) or fossil hydrocarbons (Walton *et al.*, 2003). An alternative route for production of vanillin is bioconversion of ferulic acid or eugenol, exploiting the capability of several fungi and bacteria to metabolize these compounds into vanillin or vanillic acid (Ramachandra Rao and Ravishankar,

© Woodhead Publishing Limited, 2013

2000). Upon identification of the genes responsible for such bioconversion activity, 2.9 g vanillin/l were obtained from recombinant *Escherichia coli* cultures converting ferulic acid or eugenol (Overhage *et al.*, 2003; Barghini *et al.*, 2007).

However, the possibility of converting glucose via *de novo* biosynthesis of vanillin has now become attractive, as it gives the possibility of using carbon sources that are cheaper than ferulic acid. Hansen et al. (2009) have proven the potential of using the GRAS (generally recognized as safe) yeasts *Schizosaccharomyces pombe* and *S. cerevisiae* (Fig. 3.2) as cell factories for vanillin production. Production of 65 mg l^{-1} and 45 mg l^{-1} vanillin from glucose was obtained from recombinant *S. pombe* and *S. cerevisiae* cultivations, respectively. However, *S. cerevisiae* was potentially more attractive because it accumulated higher levels of vanillin intermediates. The authors speculated that the lower vanillin levels detected could be due to the ability of this yeast to reduce vanillin into the corresponding alcohol (Hansen *et al.*, 2009).

In order to avoid reduction of vanillin and limit its toxic effect, the co-expression of a glucosyl transferase from *Arabidopsis thaliana* allowed the conversion of vanillin to vanillin β-D-glucoside (VG). VG presents several advantages over vanillin: it has lower toxicity, is more soluble than vanillin and has higher value than vanillin as it can be slowly converted into vanillin by enzymes, hence augmenting the sensory event (Hansen *et al.*, 2009; Ikemoto *et al.*, 2003). Biosynthesis of VG has been demonstrated in *S. pombe*, showing 80% glucosylation of the synthesized vanillin. However, since the *S. cerevisiae* strain engineered for vanillin production showed higher production potential than *S. pombe* (Hansen *et al.*, 2009), *S. cerevisiae* was further engineered for production of VG via expression of the *A. thaliana* glucosyl transferase (Brochado *et al.*, 2010), resulting in >100 mg VG/l and only <7 mg vanillin/l in aerobic batch cultures.

The recombinant VG-producing *S. cerevisiae* strain was used as a starting point to increase the yields of VG by introducing further genetic modifications. The study represents one of the very few examples of yeast metabolic engineering driven by *in silico* design of genetic alterations using genome scale metabolic models (Förster *et al.*, 2003) and a number of bioinformatics algorithms applied to the genome scale metabolic models (See Table 3.3). In order to predict accurately potential favourable mutations for the biosynthesis of the target product, the authors set different constraints on the model, considering different possible metabolic states of *S. cerevisiae*. Full respiratory metabolism and respiro-fermentative metabolism were considered and for both flux distribution was calculated via flux balance analysis (FBA) to maximize cell biomass production. A further FBA simulation was performed to calculate the flux distribution when maximizing VG production, constraining the model with experimental data obtained from chemostat cultivations. As stoichiometric models cannot predict changes in rates, productivity cannot be optimized, but the function which can be optimized

© Woodhead Publishing Limited, 2013

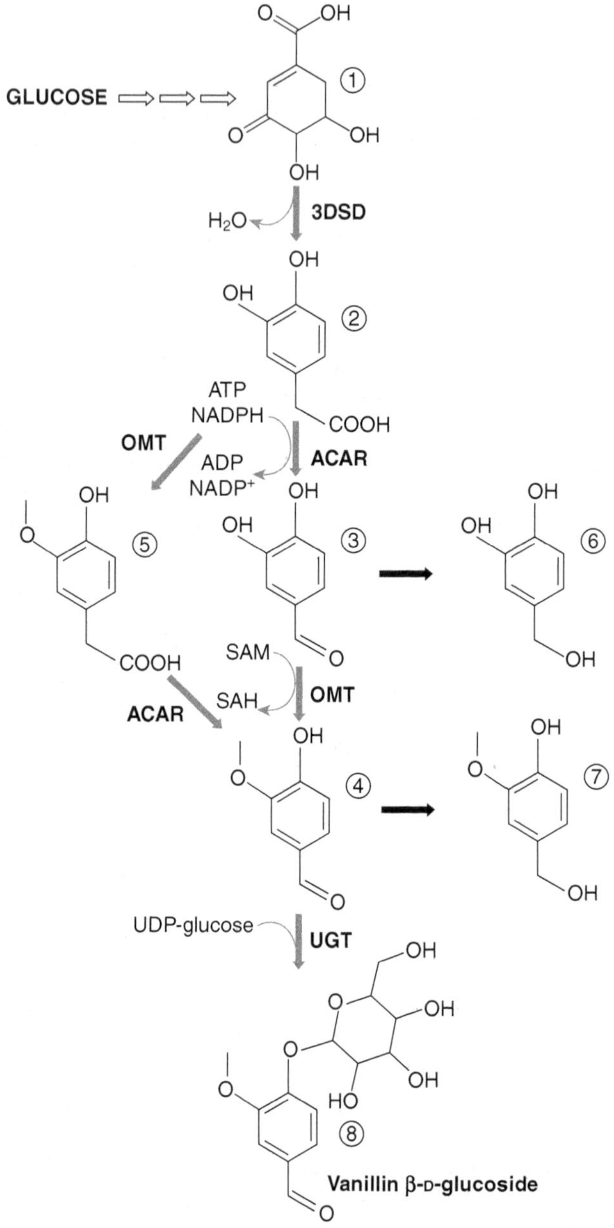

Fig. 3.2 Biosynthetic scheme of the *de novo* vanillin β-D-glucoside pathway in *S. cerevisiae* and *S. pombe*. The figure is adapted from Hansen *et al.* (2009) and Brochado *et al.* (2010) and outlines the reactions necessary for the conversion of glucose into vanillin and vanillin β-D-glucoside. White arrows are metabolic reactions naturally occurring in yeast; grey arrows are enzyme reactions introduced via metabolic engineering; black arrows are inherent yeast metabolic reactions resulting in undesirable by-products. 3DSD, 3-dehydroshikimate dehydratase; OMT, *o*-methyltransferase; ACAR, aryl carboxylic acid reductase; UGT, UDP-glucosyltransferase; SAM, *S*-adenosylmethionine; SAH, *S*-adenosylhomocysteine. Metabolites can be identified as 1, dehydroshikimic acid; 2, protocathecuic acid; 3, protocathecuic aldehyde; 4, vanillin; 5, vanillic acid; 6, protocathecuic acohol; 7, vanillyl alcohol; 8, vanillin β-D-glucoside.

© Woodhead Publishing Limited, 2013

Table 3.3   Bioinformatics tool box used for improvement of vanillin production

Tool	Description	Reference
Genome Scale Metabolic Model	Model iFF708 based on *S. cerevisiae* S288C genome sequence	Förster *et al.* (2003) Goffeau *et al.* (1996)
OptKnock	Algorithm predicting which reactions have to be removed from a metabolic network (i.e. via specific gene deletion) ensuring that the drain towards metabolites/compounds necessary for growth resources (i.e. carbons, redox potential and energy) must be accompanied, due to stoichiometry, by the production of the desired chemical.	Burgard *et al.* (2003)
OptGene	Extension of OptKnock allowing maximization of non-linear objective functions, while at the same time accounting for non-linear constraints on the metabolic network	Patil *et al.* (2005)
MOMA	Method of minimization of metabolic adjustment based on the hypothesis that knockout metabolic fluxes undergo minimal redistribution compared to wild type flux configuration. In Brochado *et al.* (2010), MOMA was used as a biological objective function with wild type flux distributions spanning three major modes of yeast metabolism.	Segré *et al.* (2002)

is (Growth × Product yield) also defined as biomass product coupled yield (BPCY) (Patil *et al.*, 2005).

The OptGene algorithm (Patil *et al.*, 2005) was used to run simulations and to predict reaction knockout targets that would have optimized BPCY and VG. Concomitantly, the MOMA (minimization of metabolic adjustment) algorithm was used to predict possible knockout mutations (Segré *et al.*, 2002). The MOMA algorithm is based on the assumption that the objective of mutants is to minimize their metabolic distance from the wild-type. The suggested targets were verified using the OptKnock algorithm (Burgard *et al.*, 2003), which maximizes the flux towards a desired product while the flux distribution optimized for growth or another objective is maintained.

The modifications suggested by the simulations were evaluated in terms of biomass and VG yields and some suggested mutations were selected for experimental validation. A further and novel parameter used to rank the suggested mutations was the Reward-Risk-Ratio (R^3), were BPCY is the reward and metabolic adjustment the risk. Depending on the reference flux

© Woodhead Publishing Limited, 2013

distribution (i.e. metabolic state) used, different targets for gene deletion were suggested. Four mutants were finally generated: pyruvate decarboxylase (*PDC1*) mutant (*pdc1Δ*); NADP$^+$-dependent glutamate dehydrogenase (*GDH1*) mutant (*gdh1Δ*), the double mutant *pdc1Δgdh1Δ* and the double mutant over-expressing the NAD$^+$-dependent glutamate dehydrogenase (*GDH2*) (*pdc1Δgdh1ΔGDH2o.e.*).

The strains *pdc1Δ* and *pdc1Δgdh1ΔGDH2o.e.* were shown to be the best performing and led to 1.5-fold increase of VG yield in batch cultivations, while a two-fold improvement of VG titre (500 mg VG/l) was observed for the *pdc1Δ* mutant in continuous cultures. Experimental data obtained were used to refine the model and the phenotypes of the mutants were simulated via FBA, using data from chemostat cultivations as model constraints. Analysis of these simulations led to the conclusion (and further hypothesis to be tested) that a better respiratory metabolism connected to a lower glucose overflow favours VG production and that metabolites playing a potential role in the engineered strains include most of the co-factors (e.g. NADPH, ATP).

This work on the biotechnological production of vanillin and vanillin β-D-glucoside represents an exhaustive example of the application of computational systems biology tools and genome scale metabolic models to increase/optimize product production using simulations as predictive tools. Although limitations of this kind of approach exist and are due to the difficulty of integrating regulatory and kinetic information into genome scale metabolic models, the actual application of the metabolic engineering cycle has been clearly shown, including a refining of the model on the basis of experimental data.

3.6   Food colouring: functional colours

Since sight is the first sense we use to get in contact with external objects, colouring plays a big role in the perception of food. In fact, people are used to associate specific colours with specific flavours; therefore, food manufacturers often add pigments to their products. Even though the food colour might be artificial, in many cases such colours have beneficial functions, since the colouring agents can be vitamins (e.g. riboflavin) or vitamin precursors with antioxidant activity (e.g. carotenoids).

3.6.1   Riboflavin and flavin mononucleotide production by yeast
Riboflavin, also known as vitamin B2, is an essential vitamin in human nutrition and animal feeding, as it is the precursor of the two coenzymes: flavin mononucleotide (FMN) and flavin adenine dinucleotide (FAD). Nowadays, most commercial riboflavin is produced biotechnologically via

© Woodhead Publishing Limited, 2013

microbial biosynthesis. The biotechnological processes are mainly based on the use of *Bacillus subtilis*, *Ashbya gosypii* or the yeast *Candida famata*, as reviewed by Abbas and Sibirny (2011) and Kato and Park (2011).

Riboflavin is used as both a food supplement and yellow colouring for beverages (i.e. colouring E-101) and the annual demand for riboflavin is about 6000 metric tonnes. The biochemical pathway leading to riboflavin synthesis has been fully elucidated by Bacher and colleagues (Bacher *et al.*, 1997a, 1997b) and mechanisms of synthesis, transport and regulation in natural producer microorganisms are exhaustively reviewed in Abbas and Sibirny (2011). Flavinogenic yeasts that are natural over-producers of riboflavin include *Pichia guillermondii* (used as model yeast), *C. famata*, *Debaryomyces hansenii*, *Schwanniomyces occidentalis* and *Candida albicans*. All these yeasts synthesize high levels of riboflavin under iron-restriction, since riboflavin biosynthesis is transcriptionally repressed by iron ions. Industrial riboflavin producers have been obtained via classical selection and mutagenesis, followed in some cases by metabolic engineering.

The *C. famata* strain dep8 (ATCC 20849) obtained via random mutagenesis is the riboflavin producing yeast that has been used on an industrial scale (Heefner, 1988; Heefner *et al.*, 1992) achieving 21 g riboflavin/l. However, the genetic instability of this strain has shifted current production towards the use of *A. gosypii* or *B. subtilis*. An additional *C. famata* strain has been recently isolated combining random mutagenesis and metabolic engineering (Dmytruk *et al.*, 2011): this strain was more stable than dep8, but its flavinogenic activity was 30% lower.

FMN, which is the phosphorylated form of riboflavin, is also used for food colouring (i.e. E-106) and fortification and in the pharmaceutical industry. FMN is currently produced via chemical phosphorylation of riboflavin; however, even the most purified preparations contain about 25% impurities. Recently, a recombinant and stable riboflavin over-producing *C. famata* strain has been engineered for over-production of FMN, by introducing the riboflavin kinase *FMN1* gene of *D. hansenii* (Yatsyshyn *et al.*, 2009). Metabolic engineering coupled to medium optimization (Yatsyshyn *et al.*, 2010) resulted in 230 mg FMN/l in batch cultivations.

Although biotechnological production of riboflavin is most practiced, there is great room for improvement, as carbon source conversion does not exceed 4%, regardless of the microorganism (Abbas and Sibirny, 2011). Possible improvement of the process might derive from further systems biology studies potentially able to give the whole picture of the biological systems. In particular, RNA sequencing technologies might help to determine the key mutations obtained via traditional mutagenesis leading to a riboflavin over-producer strain. If this kind of information is used in reverse engineering of genome scale metabolic models, reactions important for overproduction of riboflavin would be included *a priori* in the model, providing the starting point for further improvements.

© Woodhead Publishing Limited, 2013

3.6.2   Carotenoids

Carotenoids are a class of 600 related isoprenoids that give yellow to red colour to fruit, flowers and vegetables. The use of carotenoids as food colouring has long been used and chemical synthesis is the most common way to produce these compounds as pure preparations (Gordon and Bauernfeind, 1982). The mechanisms of natural biosynthesis of carotenoids have been elucidated and they are extensively reviewed by Fraser and Bramley (2004). Under the impulse of the beneficial effects elicited by these compounds on animal health (von Lintig, 2010; Fraser and Bramley, 2004), genetic engineering strategies have been developed in order to generate fortified vegetables with higher content of carotenoids, as reviewed by Fraser *et al.* (2009).

Within the carotenoid class, the two main targets for industrial production are astaxanthin (red) and β-carotene (yellow-orange). The former has very high antioxidant potential (Guerin *et al.*, 2003) and is used in cosmetic production and as food additive in nutraceuticals. β-carotene is a precursor of astaxanthin and in animal metabolism is essential for the biosynthesis of vitamin A.

Among the yeasts, *Xanthophyllomyces dendrorhous* is a natural producer of astaxanthin and a few attempts to increase astaxanthin yields or to modify the carotenoid metabolism of this yeast have been made (Verdoes *et al.*, 2003; Visser *et al.*, 2003), although they always resulted in low productivity. *X. dendrorhous* genes responsible for carotenoid biosynthesis have been cloned in *S. cerevisiae*, aiming at producing functional yeast with enriched carotenoid content and at developing a cell factory for astaxanthin (Ukibe *et al.*, 2009) or β-carotene production. While the maximum yields reported for astaxanthin were lower in *S. cerevisiae* than in *X. dendrorhous*, the β-carotene yield was up to 5.9 mg β-carotene/g dry cell weight (Verwaal *et al.*, 2007), in contrast to 0.7 mg β-carotene/g dry cell weight obtained in a *X. dendrorhous* mutant optimized for β-carotene production (Verdoes *et al.*, 2003; Visser *et al.*, 2003). Although the major industrial suppliers of β-carotene rely on chemical synthesis, *S. cerevisiae* has been shown to be a promising platform for microbial cell factory of β-carotene, but further improvements are still required.

3.7   Antioxidants

Several and diverse kinds of organic compounds can have antioxidant activity and be beneficial for animal and human health. Among the compounds with valuable nutritional and therapeutic properties are the flavonoids and stilbenes, which are typically synthesized by plants as secondary metabolites through the phenylpropanoid pathway, as reviewed by Ververidis *et al.* (2007). Attempts to establish yeast-based microbial cell factories via metabolic engineering of *S. cerevisiae* have been made and are exhaustively

© Woodhead Publishing Limited, 2013

reviewed by Siddiqui *et al.* (2012). Yeast-based biosynthesis aims to achieve the production of pure products (i.e. after downstream purification steps), or the production of fermented food using recombinant yeast that releases the compound of interest. The latter has been attempted using a resveratrol synthesizing yeast for production of white wine with similar resveratrol content as some red wines (Wang *et al.*, 2011).

A third way to deliver antioxidant compounds is in the form of nutraceuticals, that is components of nutrient food delivered in the typical form of pharmaceuticals. This is the case for selenium (Se) food supplements currently on the market which consist of tablets of Se-enriched yeast. Studies and improvements of Se-yeast production are the focus of the following section.

3.7.1 Yeast-based nutraceuticals for selenium food supplements

Selenium (Se) is an essential element for most prokaryotes and higher eukaryotes. Although excess Se causes a disease in humans and animals called selenosis, Se shortage is also the basis of a number of diseases (Koller and Exon, 1986). The essential nature of Se is determined by the presence of Se-proteins, which have Se-cysteine in their active site and are typically involved in coping with oxidizing agents, as reviewed by Birringer *et al.* (2002), thanks to the physical/chemical properties of Se which make this element prone to oxidation. In addition to Se-proteins, an important role in animal physiology is played by low molecular weight Se-compounds, whose role in preventing cell damage and cancer has been shown and primarily ascribed to methylselenol (CH_3SeH) which derives from the metabolism of inorganic and organic forms of Se (reviewed by Rayman, 2005).

It is very hard to produce food supplements that can readily deliver CH_3SeH owing to its high reactivity; therefore, food supplements typically contain precursors of methylselenol, that is Se-methylselenocysteine (SeMCys) and γ-glutamyl-Se-methylselenocysteine (γ-glu-SeMcys). These two molecules are the typical source of Se in the human diet, as they are accumulated by Se-accumulator edible plants, such as broccoli and garlic (Cai *et al.*, 1995) and have been shown to have potential anti-cancer properties (Dong *et al.*, 2001; Lee *et al.*, 2006).

Although some Se supplements on the market contain inorganic forms of Se (e.g. Na_2SeO_3), organic compounds containing Se are thought to be safer, as possible overdosage would not result into selenosis. Se-enriched yeast is the most common nutraceutical on the market (Moesgaard and Paulin, 2008) containing Se-methionine (SeMet) as the major organic Se-compound (Larsen *et al.*, 2004) which can exert beneficial effects, but might have some drawbacks, as it can be incorporated into proteins in place of methionine. Interestingly, clinical trials carried out delivering Se-enriched yeast (Clark *et al.*, 1996) or pure SeMet (Lippman *et al.*, 2009) have shown that pure SeMet had no effect as a cancer preventive agent, while

© Woodhead Publishing Limited, 2013

Se-enriched yeast resulted into 50% reduction of cancer incidence. This suggests that Se-metabolites other than SeMet, although present at lower levels, play an important role. Therefore, attention has been focused onto the development of analytical techniques able to define very detailed maps of the Se-metabolome of Se-enriched yeast (Dernovics *et al.*, 2009; Far *et al.*, 2010).

Furthermore, *S. cerevisiae* strains able to synthesize SeMCys have been generated (Mapelli *et al.*, 2011) via expression of heterologous genes from *Brassica oleracea* and *Arabidopsis thaliana*. Upon a fine-tuned fed-batch fermentation process, Mapelli *et al.* (2011) showed that the recombinant yeasts had a lower total Se content per gram of cell dry weight, but the amount of the targeted free organic Se-compounds was higher compared with the commercial Se-enriched yeast.

Complete sets of data including Se-metabolite profile in yeast and physiological parameters related to fermentation processes run in the presence of a selenium source (Mapelli *et al.*, 2012) are an important source of information for the construction of metabolic models that might contribute to the development of a Se-enriched yeast with a more beneficial Se-metabolite profile. Furthermore, a detailed description of Se metabolism in yeast would add a relevant piece of information to the general knowledge of yeast metabolism. In fact, Se is not an essential element for yeast; thereby study of yeast metabolism of Se would add further pathways to yeast metabolic maps that could then be used to design novel strategies for the production of Se-enriched yeast with added beneficial value.

3.8   Non-conventional yeasts for food and food ingredients

The long and traditional use of *S. cerevisiae* and closely related species is at the core of the definition of 'non-conventional yeasts' for species that are not closely related to *S. cerevisiae*. However, the knowledge that we have nowadays on genome evolution and metabolism tells us that *S. cerevisiae* and its relatives should be considered the 'exception' (Blank *et al.*, 2005b). Systems biology has been applied to non-conventional yeasts to a much lesser extent, compared to *S. cerevisiae*, but thanks to the potential interest in the use of these yeasts for different applications, *omics* tools have begun to be used. Among non-conventional yeasts there are a few that can potentially be used for food ingredient production.

Yarrowia lipolytica is an oleaginous yeast that has been isolated from several environments, including food. Like other oleaginous yeasts, it is able to grow on hydrophobic substrates and to accumulate lipids up to more than 20% of its dry weight (Ratledge, 1991). Thanks to the availability of the complete genome sequence, *Y. lipolytica* represents a good candidate to study the mechanisms of lipid production and accumulation. Furthermore, genetic information would also allow possible metabolic engineering

© Woodhead Publishing Limited, 2013

strategies that would lead to the production of specific kinds of lipids. Being classified as a GRAS (generally recognized as safe) organism, *Y. lipolytica* has a certain potential to be used as producer of polyunsaturated fatty acids as food additives, nutraceuticals or single cell oils.

The accumulation of lipids in *Y. lipolytica* depends on nutrient availability: excess carbohydrates and nitrogen deficiency promote lipid biosynthesis and hinder growth (Ratledge and Wynn, 2002). The regulation of the metabolism at the transcriptomic level has been studied by Morin *et al.* (2011), who concluded that at the basis of lipid biosynthesis there is a re-routing of carbon fluxes induced by nitrogen limitation, rather than enhanced lipid metabolism activity.

Further insights into the mechanisms regulating the accumulation of lipids in oleaginous yeasts throughout different growth phases have been achieved via the proteome analysis of *Rhodosporidium toruloides* (Liu *et al.*, 2009) and *Lipomyces starkey* (Liu *et al.*, 2011), the latter species being able to accumulate over 70% of its cell biomass as lipids. Although the genome of these yeasts is not fully sequenced, the information retrieved via such proteomic analyses is valuable for a deeper understanding of the biochemical events ruling microbial oleaginicity and for the rational engineering of oleaginous yeasts.

It has been shown that *Y. lipolytica* also presents a citric acid production phase coinciding with a lipid turnover phase (Makri *et al.*, 2010; Papanikolaou and Aggelis, 2009). As citric acid is widely used in the food industry both for food preservation and food flavouring and *Y. lipolytica* is able to consume several industrial and agro-industrial by-products, yeast-based production of citric acid starting from cheap raw material could also be considered as a microbial cell factory in addition to the established production by *Aspergillus niger*.

The yeast *Pichia anomala* has been isolated from several kinds of natural habitats and is known for its antimicrobial activity. It has also been shown to have an extraordinary robustness to environmental stresses. Interest in the potential use of this yeast in the food industry is mainly related to its antimicrobial activity, which can be exploited as a food preservative. In addition, the presence of *P. anomala* in feed resulted in an increased content of amino acids and a decrease in anti-nutritive phosphate-containing compounds, owing to the presence of phytase activity (Schnürer and Jonsson, 2011). In addition, the presence of *P. anomala* in the production of food and beverages can have some advantages related to the production of flavouring metabolites, such as volatile esters, as reviewed by Walker (2011). The metabolite profile of *P.anomala* growing under oxygen limitation has been examined (Fredlund *et al.*, 2004) and showed that arabitol, trehalose and glycerol accumulated in the yeast under these conditions. As these metabolites are known as stress protecting agents, their accumulation can improve yeast viability and activity upon long storage. Unfortunately, no systems biology studies have been performed on this yeast yet, but they might be

© Woodhead Publishing Limited, 2013

useful to clarify the actual mode of action involved in the anti-microbial activity and the regulation of the entire metabolism. In fact, while the presence of *P. anomala* can be advantageous for certain aspects of food production, it has been found to be the cause of 'chemical adulteration' of food and feed, possibly related to the production of ethyl acetate (Walker, 2011).

The discovery of yeast species different from *S. cerevisiae* and their physiological characterization has brought about increasing interest in their full characterization, both to exploit them for industrial fermentations or to prevent their growth, in case of pathogenic yeasts. A step forward in this direction has been made via the identification of the metabolic network topology and the quantification of fluxes of 14 hemi-ascomycetous yeasts of different relevance: from industrial to medical interest (Blank *et al.*, 2005b). This comparative study, based on the analysis of ^{13}C-labelling patterns, revealed that the major differences are at the split between glycolysis and the pentose-phosphate pathway, as well as the split between respiratory and the respiro-fermentative production of ethanol.

3.9   Conclusions

Production of food and food ingredients via microbial fermentation is characterized by an intrinsic complexity. The raw material and the final products typically have a complex composition; fermented food is often the result of the metabolic activity of mixed microbial populations, resulting in the production of a complex ensemble of primary and secondary metabolites that are responsible for the main and specific features of a certain food or beverage.

Omics methodologies have been developed in order to unravel the complexity of biological systems and deliver a picture of them as close as possible to reality. However, *omics* technologies at every level (i.e. from transcriptomics to metabolomics) have still to be improved in order to deal with the further complexity inherent in food. Mathematical models integrating biological and biochemical information are the essence of systems biology and many of the existing models have been built on the basis of *omics* data or provide a solid framework for integration of *omics*-derived data (e.g. genome scale metabolic models). Although these models can be very efficiently used in industrial biotechnology for the development of strains that would improve the production process, their use is poorly considered in the field of food production. One of the main reasons at the root of such hesitation might be related to the close connection between use of the models and the generation of genetically modified organisms which are typically avoided or not well accepted if used in food or food-related products.

However, we deem that the combination of data derived from *omics* analyses with thorough physiological characterization of the production

© Woodhead Publishing Limited, 2013

organism(s) might lead to further development of mathematical models that will be able to include physiological features, hence allowing improvement in fermentation processes without implementing any genetic modification. While the formulation of this kind of comprehensive and more detailed models might still require some time, the use of *omics* technologies could be currently implemented in the field of food and food ingredient production. In fact, the data retrieved from this comprehensive analysis will have an advantage both in terms of process control and quality assessment, but will also be important to characterize fully genetically modified organisms and the products resulting from their use in production process. This latter application may offer a solution to the safety issues related to the use of genetically modified microorganisms entering the food chain as the characteristics of the microbe/food product would be mapped in detail.

In conclusion, systems biology methods mainly developed for laboratory yeast strains and for fundamental research can find very useful applications in several branches of the food and food ingredient production processes. Furthermore, the high level of complexity inherent in food should not be seen only as a hindrance in the application of existing methodologies, but as a motivation driving further development.

3.10   References

ABBAS, C. A. and SIBIRNY, A. A. (2011). 'Genetic control of biosynthesis and transport of riboflavin and flavin nucleotides and construction of robust biotechnological producers'. *Microbiol Mol Biol Rev*, **75**, 321–60.

AHMAD, I., SHIM, W. Y., JEON, W. Y., YOON, B. H. and KIM, J. H. (2012). 'Enhancement of xylitol production in *Candida tropicalis* by co-expression of two genes involved in pentose phosphate pathway'. *Bioprocess Biosyst Eng*, **35**, 199–204.

ALDHOUS, P. (1990). 'Genetic engineering. Modified yeast fine for food'. *Nature*, **344**, 186.

ANDERSON, N. L. and ANDERSON, N. G. (1998). 'Proteome and proteomics: new technologies, new concepts, and new words'. *Electrophoresis*, **19**, 1853–61.

ANFANG, N., BRAJKOVICH, M. and GODDARD, M. R. (2009). 'Co-fermentation with *Pichia kluyveri* increases varietal thiol concentrations in Savignon Blanc'. *Aust. J Grape Wine R*, **15**, 1–8.

BACHER, A., EBERHARDT, S., FISCHER, M., MORTL, S., KIS, K., KUGELBREY, K., SCHEURING, J. and SCHOTT, K. (1997a). 'Biosynthesis of riboflavin: lumazine synthase and riboflavin synthase'. *Methods Enzymol*, **280**, 389–99.

BACHER, A., RICHTER, G., RITZ, H., EBERHARDT, S., FISCHER, M. and KRIEGER, C. (1997b). 'Biosynthesis of riboflavin: GTP cyclohydrolase II, deaminase, and reductase'. *Methods Enzymol*, **280**, 382–9.

BAMFORTH, C. W. (2002). 'Nutritional aspects of beer–a review'. *Nutr Res*, 227–37.

BARGHINI, P., DI GIOIA, D., FAVA, F. and RUZZI, M. (2007). 'Vanillin production using metabolically engineered *Escherichia coli* under non-growing conditions'. *Microb Cell Fact*, **6**, 13.

BARNETT, J. A. (1998). 'A history of research on yeasts. 1: Work by chemists and biologists 1789–1850'. *Yeast*, **14**, 1439–51.

BARNETT, J. A. (2000). 'A history of research on yeasts 2: Louis Pasteur and his contemporaries, 1850–1880'. *Yeast*, **16**, 755–71.

© Woodhead Publishing Limited, 2013

BARNETT, J. A. and LICHTENTHALER, F. W. (2001). 'A history of research on yeasts 3: Emil Fischer, Eduard Buchner and their contemporaries, 1880–1900'. *Yeast*, **18**, 363–88.

BEYER, A., WORKMAN, C., HOLLUNDER, J., RADKE, D., MOLLER, U., WILHELM, T. and IDEKER, T. (2006). 'Integrated assessment and prediction of transcription factor binding'. *PLoS Comput Biol*, **2**, e70.

BIRRINGER, M., PILAWA, S. and FLOHE, L. (2002). 'Trends in selenium biochemistry'. *Nat Prod Rep*, **19**, 693–718.

BLANK, L. M., KUEPFER, L. and SAUER, U. (2005a). 'Large-scale 13C-flux analysis reveals mechanistic principles of metabolic network robustness to null mutations in yeast'. *Genome Biol*, **6**, R49.

BLANK, L. M., LEHMBECK, F. and SAUER, U. (2005b). 'Metabolic-flux and network analysis in fourteen hemiascomycetous yeasts'. *FEMS Yeast Res*, **5**, 545–58.

BROCHADO, A. R., MATOS, C., MOLLER, B. L., HANSEN, J., MORTENSEN, U. H. and PATIL, K. R. (2010). 'Improved vanillin production in baker's yeast through *in silico* design'. *Microb Cell Fact*, **9**, 84.

BRUGGEMAN, F. J. and WESTERHOFF, H. V. (2007). 'The nature of systems biology'. *Trends Microbiol*, **15**, 45–50.

BURGARD, A. P., PHARKYA, P. and MARANAS, C. D. (2003). 'Optknock: A bilevel programming framework for identifying gene knockout strategies for microbial strain optimization'. *Biotechnol Bioeng*, **84**, 647–57.

BUTLER, G., RASMUSSEN, M. D., LIN, M. F., SANTOS, M. A. S., SAKTHIKUMAR, S., MUNRO, C. A., RHEINBAY, E., GRABHERR, M., FORCHE, A., REEDY, J. L., AGRAFIOTI, I., ARNAUD, M. B., BATES, S., BROWN, A. J. P., BRUNKE, S., COSTANZO, M. C., FITZPATRICK, D. A., DE GROOT, P. W. J., HARRIS, D., HOYER, L. L., HUBE, B., KLIS, F. M., KODIRA, C., LENNARD, N., LOGUE, M. E., MARTIN, R., NEIMAN, A. M., NIKOLAOU, E., QUAIL, M. A., QUINN, J., SANTOS, M. C., SCHMITZBERGER, F. F., SHERLOCK, G., SHAH, P., SILVERSTEIN, K. A. T., SKRZYPEK, M. S., SOLL, D., STAGGS, R., STANSFIELD, I., STUMPF, M. P. H., SUDBERY, P. E., SRIKANTHA, T., ZENG, Q., BERMAN, J., BERRIMAN, M., HEITMAN, J., GOW, N. A. R., LORENZ, M. C., BIRREN, B. W., KELLIS, M. and CUOMO, C. A. (2009). 'Evolution of pathogenicity and sexual reproduction in eight *Candida* genomes'. *Nature*, **459**, 657–62.

CAI, X. J., BLOCK, E., UDEN, P. C., ZHANG, X., B.D., Q. and SULLIVAN, J. J. (1995). 'Allium chemistry: identification of selenoamino acids in ordinary and selenium-enriched garlic, onion, and broccoli using gas-chromatography with atomic-emission detection'. *J Agric Food Chem*, **43**, 1754–7.

CAMBON, B., MONTEIL, V., REMIZE, F., CAMARASA, C. and DEQUIN, S. (2006). 'Effects of GPD1 overexpression in *Saccharomyces cerevisiae* commercial wine yeast strains lacking ALD6 genes'. *Appl Environ Microbiol*, **72**, 4688–94.

CHUNG, B. K., SELVARASU, S., CAMATTARI, A., RYU, J., LEE, H., AHN, J. and LEE, D. Y. (2010). 'Genome-scale metabolic reconstruction and *in silico* analysis of methylotrophic yeast *Pichia pastoris* for strain improvement'. *Microb Cell Fact*, **9**, 50.

CIANI, M., BECO, L. and COMITINI, F. (2006). 'Fermentation behaviour and metabolic interactions of multistarter wine yeast fermentations'. *Int J Food Microbiol*, **108**, 239–45.

CIANI, M., COMITINI, F., MANNAZZU, I. and DOMIZIO, P. (2010). 'Controlled mixed culture fermentation: a new perspective on the use of non-*Saccharomyces* yeasts in winemaking'. *FEMS Yeast Res*, **10**, 123–33.

CLARK, L. C., COMBS, G. F., JR., TURNBULL, B. W., SLATE, E. H., CHALKER, D. K., CHOW, J., DAVIS, L. S., GLOVER, R. A., GRAHAM, G. F., GROSS, E. G., KRONGRAD, A., LESHER, J. L., JR., PARK, H. K., SANDERS, B. B., JR., SMITH, C. L. and TAYLOR, J. R. (1996). 'Effects of selenium supplementation for cancer prevention in patients with carcinoma of the skin. A randomized controlled trial. Nutritional Prevention of Cancer Study Group'. *Jama*, **276**, 1957–63.

© Woodhead Publishing Limited, 2013

CLIFTEN, P., SUDARSANAM, P., DESIKAN, A., FULTON, L., FULTON, B., MAJORS, J., WATERSTON, R., COHEN, B. A. and JOHNSTON, M. (2003). 'Finding functional features in *Saccharomyces* genomes by phylogenetic footprinting'. *Science*, **301**, 71–6.

COMITINI, F., GOBBI, M., DOMIZIO, P., ROMANI, C., LENCIONI, L., MANNAZZU, I. and CIANI, M. (2011). 'Selected non-*Saccharomyces* wine yeasts in controlled multistarter fermentations with *Saccharomyces cerevisiae*'. *Food Microbiol*, **28**, 873–82.

CORRAN, H. S. (1975). *A History of Brewing*, David & Charles Publishers, Vermont, Canada.

DE LUCA, V. (2011). Wines. *Comprehensive Biotechnology*. 2nd edition, Elsevier, Amsterdam, 241–55.

DE MONTIGNY, J., STRAUB, M.-L., POTIER, S., TEKAIA, F., DUJON, B., WINCKER, P., ARTIGUENAVE, F. and SOUCIET, J.-L. (2000). 'Genomic exploration of the Hemiascomycetous yeasts: 8. *Zygosaccharomyces rouxii*'. *FEBS Lett*, **487**, 52–5.

DE NICOLA, R., HAZELWOOD, L. A., DE HULSTER, E. A., WALSH, M. C., KNIJNENBURG, T. A., REINDERS, M. J., WALKER, G. M., PRONK, J. T., DARAN, J. M. and DARAN-LAPUJADE, P. (2007). 'Physiological and transcriptional responses of *Saccharomyces cerevisiae* to zinc limitation in chemostat cultures'. *Appl Environ Microbiol*, **73**, 7680–92.

DE PASCUAL-TERESA, S., HALLUND, J., TALBOT, D., SCHROOT, J., WILLIAMS, C. M., BUGEL, S. and CASSIDY, A. (2006). 'Absorption of isoflavones in humans: effects of food matrix and processing'. *J Nutr Biochem*, **17**, 257–64.

DE SCHUTTER, K., LIN, Y.-C., TIELS, P., VAN HECKE, A., GLINKA, S., WEBER-LEHMANN, J., ROUZE, P., VAN DE PEER, Y. and CALLEWAERT, N. (2009). 'Genome sequence of the recombinant protein production host *Pichia pastoris*'. *Nat Biotech*, **27**, 561–6.

DE WINDE, J. H. (ED.) (2003). *Functional Genetics of Industrial Yeasts*, Springer-Verlag, Berlin.

DEQUIN, S. (2001). 'The potential of genetic engineering for improving brewing, wine-making and baking yeasts'. *Appl Microbiol Biotechnol*, **56**, 577–88.

DERNOVICS, M., FAR, J. and LOBINSKI, R. (2009). 'Identification of anionic selenium species in Se-rich yeast by electrospray QTOF MS/MS and hybrid linear ion trap/orbitrap MS'. *Metallomics*, **1**, 317–29.

DEVANTIER, R., SCHEITHAUER, B., VILLAS-BOAS, S. G., PEDERSEN, S. and OLSSON, L. (2005). 'Metabolite profiling for analysis of yeast stress response during very high gravity ethanol fermentations'. *Biotechnol Bioeng*, **90**, 703–14.

DMYTRUK, K. V., YATSYSHYN, V. Y., SYBIRNA, N. O., FEDOROVYCH, D. V. and SIBIRNY, A. A. (2011). 'Metabolic engineering and classic selection of the yeast *Candida famata* (*Candida flareri*) for construction of strains with enhanced riboflavin production'. *Metab Eng*, **13**, 82–8.

DONALIES, U. E., NGUYEN, H. T., STAHL, U. and NEVOIGT, E. (2008). 'Improvement of *Saccharomyces* yeast strains used in brewing, wine making and baking'. *Adv Biochem Eng Biotechnol*, **111**, 67–98.

DONG, Y., LISK, D., BLOCK, E. and IP, C. (2001). 'Characterization of the biological activity of gamma-glutamyl-Se-methylselenocysteine: a novel, naturally occurring anticancer agent from garlic'. *Cancer Res*, **61**, 2923–8.

DUARTE, N. C., HERRGARD, M. J. and PALSSON, B. O. (2004). 'Reconstruction and validation of *Saccharomyces cerevisiae* iND750, a fully compartmentalized genome-scale metabolic model'. *Genome Res*, **14**, 1298–309.

DUJON, B., SHERMAN, D., FISCHER, G., DURRENS, P., CASAREGOLA, S., LAFONTAINE, I., DE MONTIGNY, J., MARCK, C., NEUVEGLISE, C., TALLA, E., GOFFARD, N., FRANGEUL, L., AIGLE, M., ANTHOUARD, V., BABOUR, A., BARBE, V., BARNAY, S., BLANCHIN, S., BECKERICH, J.-M., BEYNE, E., BLEYKASTEN, C., BOISRAME, A., BOYER, J., CATTOLICO, L., CONFANIOLERI, F., DE DARUVAR, A., DESPONS, L., FABRE, E., FAIRHEAD, C., FERRY-DUMAZET, H., GROPPI, A., HANTRAYE, F., HENNEQUIN, C., JAUNIAUX, N., JOYET, P., KACHOURI, R., KERREST, A., KOSZUL, R., LEMAIRE, M., LESUR, I., MA, L., MULLER, H., NICAUD, J.-M., NIKOLSKI, M.,

© Woodhead Publishing Limited, 2013

OZTAS, S., OZIER-KALOGEROPOULOS, O., PELLENZ, S., POTIER, S., RICHARD, G.-F., STRAUB, M.-L., SULEAU, A., SWENNEN, D., TEKAIA, F., WESOLOWSKI-LOUVEL, M., WESTHOF, E., WIRTH, B., ZENIOU-MEYER, M., ZIVANOVIC, I., BOLOTIN-FUKUHARA, M., THIERRY, A., BOUCHIER, C., CAUDRON, B., SCARPELLI, C., GAILLARDIN, C., WEISSENBACH, J., WINCKER, P. and SOUCIET, J.-L. (2004). 'Genome evolution in yeasts'. *Nature*, **430**, 35–44.

ELDAROV, M. A., MARDANOV, A. V., BELETSKY, A. V., RAVIN, N. V. and SKRYABIN, K. G. (2011). 'Complete sequence and analysis of the mitochondrial genome of the methylotrophic yeast *Hansenula polymorpha* DL-1'. *FEMS Yeast Res*, **11**, 464–72.

FAMILI, I., FORSTER, J., NIELSEN, J. and PALSSON, B. O. (2003). '*Saccharomyces cerevisiae* phenotypes can be predicted by using constraint-based analysis of a genome-scale reconstructed metabolic network'. *Proc Natl Acad Sci U S A*, **100**, 13134–9.

FAR, J., PREUD'HOMME, H. and LOBINSKI, R. (2010). 'Detection and identification of hydrophilic selenium compounds in selenium-rich yeast by size exclusion-microbore normal-phase HPLC with the on-line ICP-MS and electrospray Q-TOF-MS detection'. *Anal Chim Acta*, **657**, 175–90.

FERNANDEZ, M., UBEDA, J. F. and BRIONES, A. I. (2000). 'Typing of non-*Saccharomyces* yeasts with enzymatic activities of interest in wine-making'. *Int J Food Microbiol*, **59**, 29–36.

FIEHN, O. (2001). 'Combining genomics, metabolome analysis, and biochemical modelling to understand metabolic networks'. *Comp Funct Genomics*, **2**, 155–68.

FÖRSTER, J., FAMILI, I., FU, P., PALSSON, B. Ø. and NIELSEN, J. (2003). 'Genome-scale reconstruction of the *Saccharomyces cerevisiae* metabolic network'. *Genome Res*, **13**, 244–53.

FRASER, P. D. and BRAMLEY, P. M. (2004). 'The biosynthesis and nutritional uses of carotenoids'. *Prog Lipid Res*, **43**, 228–65.

FRASER, P. D., ENFISSI, E. M. and BRAMLEY, P. M. (2009). 'Genetic engineering of carotenoid formation in tomato fruit and the potential application of systems and synthetic biology approaches'. *Arch Biochem Biophys*, **483**, 196–204.

FREDLUND, E., BROBERG, A., BOYSEN, M. E., KENNE, L. and SCHNURER, J. (2004). 'Metabolite profiles of the biocontrol yeast *Pichia anomala* J121 grown under oxygen limitation'. *Appl Microbiol Biotechnol*, **64**, 403–9.

GOFFEAU, A., BARRELL, B. G., BUSSEY, H., DAVIS, R. W., DUJON, B., FELDMANN, H., GALIBERT, F., HOHEISEL, J. D., JACQ, C., JOHNSTON, M., LOUIS, E. J., MEWES, H. W., MURAKAMI, Y., PHILIPPSEN, P., TETTELIN, H. and OLIVER, S. G. (1996). 'Life with 6000 Genes'. *Science*, **274**, 546–67.

GORDON, H. T. and BAUERNFEIND, J. C. (1982). 'Carotenoids as food colorants'. *Crit Rev Food Sci Nutr*, **18**, 59–97.

GRANSTRÖM, T. B., ARISTIDOU, A. A., JOKELA, J. and LEISOLA, M. (2000). 'Growth characteristics and metabolic flux analysis of *Candida milleri*'. *Biotechnol Bioeng*, **70**, 197–207.

GRANSTRÖM, T., OJAMO, H. and LEISOLA, M. (2001). 'Chemostat study of xylitol production by *Candida guilliermondii*'. *Appl Microbiol Biotechnol*, **55**, 36–42.

GRANSTRÖM, T., ARISTIDOU, A. A. and LEISOLA, M. (2002). 'Metabolic flux analysis of *Candida tropicalis* growing on xylose in an oxygen-limited chemostat'. *Metab Eng*, **4**, 248–56.

GRANSTRÖM, T. B., IZUMORI, K. and LEISOLA, M. (2007a). 'A rare sugar xylitol. Part I: the biochemistry and biosynthesis of xylitol'. *Appl Microbiol Biotechnol*, **74**, 277–81.

GRANSTRÖM, T. B., IZUMORI, K. and LEISOLA, M. (2007b). 'A rare sugar xylitol. Part II: biotechnological production and future applications of xylitol'. *Appl Microbiol Biotechnol*, **74**, 273–6.

GUERIN, M., HUNTLEY, M. E. and OLAIZOLA, M. (2003). '*Haematococcus* astaxanthin: applications for human health and nutrition'. *Trends Biotechnol*, **21**, 210–16.

HANSEN, E. H., MOLLER, B. L., KOCK, G. R., BUNNER, C. M., KRISTENSEN, C., JENSEN, O. R., OKKELS, F. T., OLSEN, C. E., MOTAWIA, M. S. and HANSEN, J. (2009). '*De novo* biosynthesis

© Woodhead Publishing Limited, 2013

of vanillin in fission yeast (*Schizosaccharomyces pombe*) and baker's yeast (*Saccharomyces cerevisiae*)'. *Appl Environ Microbiol*, **75**, 2765–74.

HAUSER, N. C., FELLENBERG, K., GIL, R., BASTUCK, S., HOHEISEL, J. D. and PEREZ-ORTIN, J. E. (2001). 'Whole genome analysis of a wine yeast strain'. *Comp Funct Genomics*, **2**, 69–79.

HAZELWOOD, L. A., WALSH, M. C., LUTTIK, M. A., DARAN-LAPUJADE, P., PRONK, J. T. and DARAN, J. M. (2009). 'Identity of the growth-limiting nutrient strongly affects storage carbohydrate accumulation in anaerobic chemostat cultures of *Saccharomyces cerevisiae*'. *Appl Environ Microbiol*, **75**, 6876–85.

HEEFNER, D. E. A. (1988). *Riboflavin Producing Strains of Microorganisms, Method for selecting, and Method for Fermentation*. World Patent: 88/09822.

HEEFNER, D. L., WEAVER, C. A., YARUS, M. J. and BURDZINSKI, L. A. (1992). *Method for Producing Riboflavin with Candida Famata*. USA Patent Application: 5/64303.

HERRGÅRD, M. J., SWAINSTON, N., DOBSON, P., DUNN, W. B., ARGA, K. Y., ARVAS, M., BLUTHGEN, N., BORGER, S., COSTENOBLE, R., HEINEMANN, M., HUCKA, M., LE NOVERE, N., LI, P., LIEBERMEISTER, W., MO, M. L., OLIVEIRA, A. P., PETRANOVIC, D., PETTIFER, S., SIMEONIDIS, E., SMALLBONE, K., SPASIC, I., WEICHART, D., BRENT, R., BROOMHEAD, D. S., WESTERHOFF, H. V., KIRDAR, B., PENTTILA, M., KLIPP, E., PALSSON, B. O., SAUER, U., OLIVER, S. G., MENDES, P., NIELSEN, J. and KELL, D. B. (2008). 'A consensus yeast metabolic network reconstruction obtained from a community approach to systems biology'. *Nat Biotechnol*, **26**, 1155–60.

HOWELL, K. S., COZZOLINO, D., BARTOWSKY, E. J., FLEET, G. H. and HENSCHKE, P. A. (2006). 'Metabolic profiling as a tool for revealing *Saccharomyces* interactions during wine fermentation'. *FEMS Yeast Res*, **6**, 91–101.

HUGHES, T. R., ROBERTS, C. J., DAI, H., JONES, A. R., MEYER, M. R., SLADE, D., BURCHARD, J., DOW, S., WARD, T. R., KIDD, M. J., FRIEND, S. H. and MARTON, M. J. (2000). 'Widespread aneuploidy revealed by DNA microarray expression profiling'. *Nat Genet*, **25**, 333–7.

HUSNIK, J. I., VOLSCHENK, H., BAUER, J., COLAVIZZA, D., LUO, Z. and VAN VUUREN, H. J. (2006). 'Metabolic engineering of malolactic wine yeast'. *Metab Eng*, **8**, 315–23.

HYMA, K. E., SAERENS, S. M., VERSTREPEN, K. J. and FAY, J. C. (2011). 'Divergence in wine characteristics produced by wild and domesticated strains of *Saccharomyces cerevisiae*'. *FEMS Yeast Res*, **11**, 540–51.

IIMURE, T., KIHARA, M., ICHIKAWA, S., ITO, K., TAKEDA, K. and SATO, K. (2011). 'Development of DNA markers associated with beer foam stability for barley breeding'. *Theor Appl Genet*, **122**, 199–210.

IKEMOTO, T., MIMURA, K. and KITAHARA, T. (2003). 'Formation of fragrant materials from odourless glycosidically-bound volatiles on skin microflora (part 2)'. *Flavour Fragrance J*, 45–7.

JACKSON, A. P., GAMBLE, J. A., YEOMANS, T., MORAN, G. P., SAUNDERS, D., HARRIS, D., ASLETT, M., BARRELL, J. F., BUTLER, G., CITIULO, F., COLEMAN, D. C., DE GROOT, P. W., GOODWIN, T. J., QUAIL, M. A., MCQUILLAN, J., MUNRO, C. A., PAIN, A., POULTER, R. T., RAJANDREAM, M. A., RENAULD, H., SPIERING, M. J., TIVEY, A., GOW, N. A., BARRELL, B., SULLIVAN, D. J. and BERRIMAN, M. (2009). 'Comparative genomics of the fungal pathogens *Candida dubliniensis* and *Candida albicans*'. *Genome Res*, **19**, 2231–44.

JEANES, Y. M., HALL, W. L., ELLARD, S., LEE, E. and LODGE, J. K. (2004). 'The absorption of vitamin E is influenced by the amount of fat in a meal and the food matrix'. *Br J Nutr*, **92**, 575–9.

JEFFRIES, T. W., GRIGORIEV, I. V., GRIMWOOD, J., LAPLAZA, J. M., AERTS, A., SALAMOV, A., SCHMUTZ, J., LINDQUIST, E., DEHAL, P., SHAPIRO, H., JIN, Y.-S., PASSOTH, V. and RICHARDSON, P. M. (2007). 'Genome sequence of the lignocellulose-bioconverting and xylose-fermenting yeast *Pichia stipitis*'. *Nat Biotech*, **25**, 319–26.

JONES, T., FEDERSPIEL, N. A., CHIBANA, H., DUNGAN, J., KALMAN, S., MAGEE, B. B., NEWPORT, G., THORSTENSON, Y. R., AGABIAN, N., MAGEE, P. T., DAVIS, R. W. and SCHERER, S. (2004).

© Woodhead Publishing Limited, 2013

'The diploid genome sequence of *Candida albicans*'. *Proc Natl Acad Sci U S A*, **101**, 7329–34.

KATO, T. and PARK, E. Y. (2011). 'Riboflavin production by *Ashbya gossypii*'. *Biotechnol Lett*, **34**, 611–8.

KELL, D. B. and OLIVER, S. G. (2004). 'Here is the evidence, now what is the hypothesis? The complementary roles of inductive and hypothesis-driven science in the post-genomic era'. *Bioessays*, **26**, 99–105.

KELLIS, M., PATTERSON, N., ENDRIZZI, M., BIRREN, B. and LANDER, E. S. (2003). 'Sequencing and comparison of yeast species to identify genes and regulatory elements'. *Nature*, **423**, 241–54.

KELLIS, M., BIRREN, B. W. and LANDER, E. S. (2004). 'Proof and evolutionary analysis of ancient genome duplication in the yeast *Saccharomyces cerevisiae*'. *Nature*, **428**, 617–24.

KERSCHER, S., DURSTEWITZ, G., CASAREGOLA, S., GAILLARDIN, C. and BRANDT, U. (2001). 'The complete mitochondrial genome of *Yarrowia lipolytica*'. *Comparative and Functional Genomics*, **2**, 80–90.

KIM, S. Y., KIM, J. H. and OH, D. K. (1997). 'Improvement of xylitol production by controlling oxygen supply in *Candida parapsilosis*'. *J Ferment Bioeng*, **83**, 267–70.

KO, B. S., KIM, J. and KIM, J. H. (2006). 'Production of xylitol from D-xylose by a xylitol dehydrogenase gene-disrupted mutant of *Candida tropicalis*'. *Appl Environ Microbiol*, **72**, 4207–13.

KOLLER, L. D. and EXON, J. H. (1986). 'The two faces of selenium–deficiency and toxicity–are similar in animals and man'. *Can J Vet Res*, **50**, 297–306.

KWON, S. G., PARK, S. W. and OH, D. K. (2006). 'Increase of xylitol productivity by cell-recycle fermentation of *Candida tropicalis* using submerged membrane bioreactor'. *J Biosci Bioeng*, **101**, 13–18.

LARSEN, E. H., HANSEN, M., PAULIN, H., MOESGAARD, S., REID, M. and RAYMAN, M. (2004). 'Speciation and bioavailability of selenium in yeast-based intervention agents used in cancer chemoprevention studies'. *J AOAC Int*, **87**, 225–32.

LEE, S. O., YEON CHUN, J., NADIMINTY, N., TRUMP, D. L., IP, C., DONG, Y. and GAO, A. C. (2006). 'Monomethylated selenium inhibits growth of LNCaP human prostate cancer xenograft accompanied by a decrease in the expression of androgen receptor and prostate-specific antigen (PSA)'. *Prostate*, **66**, 1070–5.

LIPPMAN, S. M., KLEIN, E. A., GOODMAN, P. J., LUCIA, M. S., THOMPSON, I. M., FORD, L. G., PARNES, H. L., MINASIAN, L. M., GAZIANO, J. M., HARTLINE, J. A., PARSONS, J. K., BEARDEN, J. D., 3RD, CRAWFORD, E. D., GOODMAN, G. E., CLAUDIO, J., WINQUIST, E., COOK, E. D., KARP, D. D., WALTHER, P., LIEBER, M. M., KRISTAL, A. R., DARKE, A. K., ARNOLD, K. B., GANZ, P. A., SANTELLA, R. M., ALBANES, D., TAYLOR, P. R., PROBSTFIELD, J. L., JAGPAL, T. J., CROWLEY, J. J., MEYSKENS, F. L., JR., BAKER, L. H. and COLTMAN, C. A., JR. (2009). 'Effect of selenium and vitamin E on risk of prostate cancer and other cancers: the Selenium and Vitamin E Cancer Prevention Trial (SELECT)'. *Jama*, **301**, 39–51.

LIU, H., ZHAO, X., WANG, F., LI, Y., JIANG, X., YE, M., ZHAO, Z. K. and ZOU, H. (2009). 'Comparative proteomic analysis of *Rhodosporidium toruloides* during lipid accumulation'. *Yeast*, **26**, 553–66.

LIU, H., ZHAO, X., WANG, F., JIANG, X., ZHANG, S., YE, M., ZHAO, Z. K. and ZOU, H. (2011). 'The proteome analysis of oleaginous yeast *Lipomyces starkeyi*'. *FEMS Yeast Res*, **11**, 42–51.

MÄKINEN, K. K. (1992). 'Dietary prevention of dental caries by xylitol – clinical effectiveness and safety'. *J Appl Nutr*, **44**, 16–28.

MAKRI, A., FAKAS, S. and AGGELIS, G. (2010). 'Metabolic activities of biotechnological interest in *Yarrowia lipolytica* grown on glycerol in repeated batch cultures'. *Bioresource Technol*, **101**, 2351–8.

© Woodhead Publishing Limited, 2013

MAPELLI, V., HILLESTROM, P. R., KAPOLNA, E., LARSEN, E. H. and OLSSON, L. (2011). 'Metabolic and bioprocess engineering for production of selenized yeast with increased content of seleno-methylselenocysteine'. *Metab Eng*, **13**, 282–93.

MAPELLI, V., HILLESTROM, P. R., PATIL, K., LARSEN, E. H. and OLSSON, L. (2012). 'The interplay between sulphur and selenium metabolism influences the intracellular redox balance in *Saccharomyces cerevisiae*'. *FEMS Yeast Res*, **12**, 20–32.

MARKS, V. D., HO SUI, S. J., ERASMUS, D., VAN DER MERWE, G. K., BRUMM, J., WASSERMAN, W. W., BRYAN, J. and VAN VUUREN, H. J. (2008). 'Dynamics of the yeast transcriptome during wine fermentation reveals a novel fermentation stress response'. *FEMS Yeast Res*, **8**, 35–52.

MOESGAARD, S. and PAULIN, H. S. (2008). *A Selenium Yeast Product, a Method of Preparing a Selenium Yeast Product and the Use of the Product for Preparing Food, a Dietary Supplement or a Drug.* Europe patent application: EP1478732.

MORIN, N., CESCUT, J., BEOPOULOS, A., LELANDAIS, G., LE BERRE, V., URIBELARREA, J. L., MOLINA-JOUVE, C. and NICAUD, J. M. (2011). 'Transcriptomic analyses during the transition from biomass production to lipid accumulation in the oleaginous yeast *Yarrowia lipolytica*'. *PLoS One*, **6**, e27966.

NAKAO, Y., KANAMORI, T., ITOH, T., KODAMA, Y., RAINIERI, S., NAKAMURA, N., SHIMONAGA, T., HATTORI, M. and ASHIKARI, T. (2009). 'Genome sequence of the lager brewing yeast, an interspecies hybrid'. *DNA Res*, **16**, 115–29.

NEVOIGT, E. (2008). 'Progress in metabolic engineering of *Saccharomyces cerevisiae*'. *Microbiol Mol Biol Rev*, **72**, 379–412.

NEVOIGT, E., PILGER, R., MAST-GERLACH, E., SCHMIDT, U., FREIHAMMER, S., ESCHENBRENNER, M., GARBE, L. and STAHL, U. (2002). 'Genetic engineering of brewing yeast to reduce the content of ethanol in beer'. *FEMS Yeast Res*, **2**, 225–32.

NIELSEN, J. (2001). 'Metabolic engineering'. *Appl Microbiol Biotechnol*, **55**, 263–83.

NIELSEN, J. and JEWETT, M. C. (2008). 'Impact of systems biology on metabolic engineering of *Saccharomyces cerevisiae*'. *FEMS Yeast Res*, **8**, 122–31.

NOOKAEW, I., JEWETT, M. C., MEECHAI, A., THAMMARONGTHAM, C., LAOTENG, K., CHEEVADHANARAK, S., NIELSEN, J. and BHUMIRATANA, S. (2008). 'The genome-scale metabolic model iIN800 of *Saccharomyces cerevisiae* and its validation: a scaffold to query lipid metabolism'. *BMC Syst Biol*, **2**, 71.

OLESEN, K., FELDING, T., GJERMANSEN, C. and HANSEN, J. (2002). 'The dynamics of the *Saccharomyces carlsbergensis* brewing yeast transcriptome during a production-scale lager beer fermentation'. *FEMS Yeast Res*, **2**, 563–73.

ORTH, J. D., THIELE, I. and PALSSON, B. O. (2010). 'What is flux balance analysis?' *Nat Biotechnol*, **28**, 245–8.

OSINGA, K. A., BEUDEKER, R. F., VAN DER PLAAT, J. B. and J.A., D. H. (1989). *New Yeast Strains Providing for an Enhanced Rate of the Fermentation of Sugars, a Process to Obtain Such Yeasts and the Use of These Yeasts.* Eur Patent: EP306107.

OTERO, J. M. and NIELSEN, J. (2010). 'Industrial systems biology'. *Biotechnol Bioeng*, **105**, 439–60.

OVERHAGE, J., STEINBUCHEL, A. and PRIEFERT, H. (2003). 'Highly efficient biotransformation of eugenol to ferulic acid and further conversion to vanillin in recombinant strains of *Escherichia coli*'. *Appl Environ Microbiol*, **69**, 6569–76.

PAPANIKOLAOU, S. and AGGELIS, G. (2009). 'Biotechnological valorization of biodiesel derived glycerol waste through production of single cell oil and citric acid by *Yarrowia lipolytica*'. *Lipid Technology*, **21**, 83–87.

PASS, F. (1981). 'Biotechnology, a new industrial revolution'. *J Am Acad Dermatol*, **4**, 476–7.

PATIL, K. R., ROCHA, I., FORSTER, J. and NIELSEN, J. (2005). 'Evolutionary programming as a platform for *in silico* metabolic engineering'. *BMC Bioinformatics*, **6**, 308.

© Woodhead Publishing Limited, 2013

PHARKYA, P., BURGARD, A. P. and MARANAS, C. D. (2004). 'OptStrain: a computational framework for redesign of microbial production systems'. *Genome Res*, **14**, 2367–76.

PICARIELLO, G., BONOMI, F., IAMETTI, S., RASMUSSEN, P., PEPE, C., LILLA, S. and FERRANTI, P. (2011). 'Proteomic and peptidomic characterisation of beer: Immunological and technological implications'. *Food Chem*, **124**, 1718–26.

PIDDOCKE, M. and OLSSON, L. (2009). 'Beer brewing, applications of metabolic engineering and other strategies for process improvement in beer production'. In *The Encyclopedia of Industrial Biotechnology: Bioprocess, Bioseparation and Cell Technology*. Flickinger, E. M. C. (ed.), John Wiley & Sons.

PIDDOCKE, M. P., FAZIO, A., VONGSANGNAK, W., WONG, M. L., HELDT-HANSEN, H. P., WORKMAN, C., NIELSEN, J. and OLSSON, L. (2011). 'Revealing the beneficial effect of protease supplementation to high gravity beer fermentations using "-omics" techniques'. *Microb Cell Fact*, **10**, 27.

PIZARRO, F. J., JEWETT, M. C., NIELSEN, J. and AGOSIN, E. (2008). 'Growth temperature exerts differential physiological and transcriptional responses in laboratory and wine strains of *Saccharomyces cerevisiae*'. *Appl Environ Microbiol*, **74**, 6358–68.

POMILIO, A. B., DUCHOWICZ, P. R., GIRAUDO, M. A. and CASTRO, E. A. (2010). 'Amino acid profiles and quantitative structure–property relationships for malts and beers'. *Food Res Int*, 965–71.

PRETORIUS, I. S. (2000). 'Tailoring wine yeast for the new millennium: novel approaches to the ancient art of winemaking'. *Yeast*, **16**, 675–729.

PRETORIUS, I. S. and BAUER, F. F. (2002). 'Meeting the consumer challenge through genetically customized wine-yeast strains'. *Trends Biotechnol*, **20**, 426–32.

PRICE, N. D., REED, J. L. and PALSSON, B. O. (2004). 'Genome-scale models of microbial cells: evaluating the consequences of constraints'. *Nat Rev Microbiol*, **2**, 886–97.

QUEROL, A. and FLEET, G. H. (EDS) (2006). *Yeasts in Food and Beverages*, Springer Verlag, Berlin, Heidelberg.

RAMACHANDRA RAO, S. and RAVISHANKAR, G. A. (2000). 'Vanilla flavour: production by conventional and biotechnological routes'. *J Sci Food Agric*, **80**, 298–304.

RAMEZANI-RAD, M., HOLLENBERG, C. P., LAUBER, J., WEDLER, H., GRIESS, E., WAGNER, C., ALBERMANN, K., HANI, J., PIONTEK, M., DAHLEMS, U. and GELLISSEN, G. (2003). 'The *Hansenula polymorpha* (strain CBS4732) genome sequencing and analysis'. *FEMS Yeast Res*, **4**, 207–15.

RATLEDGE, C. (1991). 'Microorganisms for lipids'. *Acta Biotechnol*, **11**, 429–38.

RATLEDGE, C. and WYNN, J. P. (2002). 'The biochemistry and molecular biology of lipid accumulation in oleaginous microorganisms'. *Adv Appl Microbiol*, **51**, 1–51.

RAYMAN, M. P. (2005). 'Selenium in cancer prevention: a review of the evidence and mechanism of action'. *Proc Nutr Soc*, **64**, 527–42.

ROSSOUW, D., NAES, T. and BAUER, F. F. (2008). 'Linking gene regulation and the exo-metabolome: a comparative transcriptomics approach to identify genes that impact on the production of volatile aroma compounds in yeast'. *BMC Genomics*, **9**, 530.

ROSSOUW, D., VAN DEN DOOL, A. H., JACOBSON, D. and BAUER, F. F. (2010). 'Comparative transcriptomic and proteomic profiling of industrial wine yeast strains'. *Appl Environ Microbiol*, **76**, 3911–23.

ROSSOUW, D., JOLLY, N., JACOBSON, D. and BAUER, F. F. (2012). 'The effect of scale on gene expression: commercial versus laboratory wine fermentations'. *Appl Microbiol Biotechnol*, **93**, 1207–19.

RUDNITSKAYA, A., NIEUWOUDT, H. H., MULLER, N., LEGIN, A., DU TOIT, M. and BAUER, F. F. (2010a). 'Instrumental measurement of bitter taste in red wine using an electronic tongue'. *Anal Bioanal Chem*, **397**, 3051–60.

RUDNITSKAYA, A., ROCHA, S. M., LEGIN, A., PEREIRA, V. and MARQUES, J. C. (2010b). 'Evaluation of the feasibility of the electronic tongue as a rapid analytical tool

© Woodhead Publishing Limited, 2013

for wine age prediction and quantification of the organic acids and phenolic compounds. The case study of Madeira wine'. *Anal Chim Acta*, **662**, 82–9.

SCANNELL, D. R., FRANK, A. C., CONANT, G. C., BYRNE, K. P., WOOLFIT, M. and WOLFE, K. H. (2007). 'Independent sorting-out of thousands of duplicated gene pairs in two yeast species descended from a whole-genome duplication'. *Proc Natl Acad Sci U S A*, **104**, 8397–402.

SAERENS, S. M., DUONG, C. T. and NEVOIGT, E. (2010). 'Genetic improvement of brewer's yeast: current state, perspectives and limits'. *Appl Microbiol Biotechnol*, **86**, 1195–212.

SCHNÜRER, J. and JONSSON, A. (2011). '*Pichia anomala* J121: a 30-year overnight near success biopreservation story'. *Antonie Van Leeuwenhoek*, **99**, 5–12.

SEGRÉ, D., VITKUP, D. and CHURCH, G. M. (2002). 'Analysis of optimality in natural and perturbed metabolic networks'. *Proc Natl Acad Sci U S A*, **99**, 15112–17.

SHARPLESS, K. E., GREENBERG, R. R., SCHANTZ, M. M., WELCH, M. J., WISE, S. A. and IHNAT, M. (2004). 'Filling the AOAC triangle with food-matrix standard reference materials'. *Anal Bioanal Chem*, **378**, 1161–7.

SHARPLESS, K. E., THOMAS, J. B., CHRISTOPHER, S. J., GREENBERG, R. R., SANDER, L. C., SCHANTZ, M. M., WELCH, M. J. and WISE, S. A. (2007). 'Standard reference materials for foods and dietary supplements'. *Anal Bioanal Chem*, **389**, 171–8.

SIDDIQUI, M. S., THODEY, K., TRENCHARD, I. and SMOLKE, C. D. (2012). 'Advancing secondary metabolite biosynthesis in yeast with synthetic biology tools'. *FEMS Yeast Res*, **12**, 144–70.

SMART, K. A. (2007). 'Brewing yeast genomes and genome-wide expression and proteome profiling during fermentation'. *Yeast*, **24**, 993–1013.

SOH, K. C., MISKOVIC, L. and HATZIMANIKATIS, V. (2011). 'From network models to network responses: integration of thermodynamic and kinetic properties of yeast genome-scale metabolic networks'. *FEMS Yeast Res*, **12**, 129–43.

SOHN, S. B., GRAF, A. B., KIM, T. Y., GASSER, B., MAURER, M., FERRER, P., MATTANOVICH, D. and LEE, S. Y. (2010). 'Genome-scale metabolic model of methylotrophic yeast *Pichia pastoris* and its use for in silico analysis of heterologous protein production'. *Biotechnol J*, **5**, 705–15.

SON, H. S., HWANG, G. S., KIM, K. M., KIM, E. Y., VAN DEN BERG, F., PARK, W. M., LEE, C. H. and HONG, Y. S. (2009). '(1)H NMR-based metabolomic approach for understanding the fermentation behaviors of wine yeast strains'. *Anal Chem*, **81**, 1137–45.

SOUCIET, J. L., DUJON, B., GAILLARDIN, C., JOHNSTON, M., BARET, P. V., CLIFTEN, P., SHERMAN, D. J., WEISSENBACH, J., WESTHOF, E., WINCKER, P., JUBIN, C., POULAIN, J., BARBE, V., SEGURENS, B., ARTIGUENAVE, F., ANTHOUARD, V., VACHERIE, B., VAL, M. E., FULTON, R. S., MINX, P., WILSON, R., DURRENS, P., JEAN, G., MARCK, C., MARTIN, T., NIKOLSKI, M., ROLLAND, T., SERET, M. L., CASAREGOLA, S., DESPONS, L., FAIRHEAD, C., FISCHER, G., LAFONTAINE, I., LEH, V., LEMAIRE, M., DE MONTIGNY, J., NEUVEGLISE, C., THIERRY, A., BLANC-LENFLE, I., BLEYKASTEN, C., DIFFELS, J., FRITSCH, E., FRANGEUL, L., GOEFFON, A., JAUNIAUX, N., KACHOURI-LAFOND, R., PAYEN, C., POTIER, S., PRIBYLOVA, L., OZANNE, C., RICHARD, G. F., SACERDOT, C., STRAUB, M. L. and TALLA, E. (2009). 'Comparative genomics of protoploid Saccharomycetaceae'. *Genome Res*, **19**, 1696–709.

STRAUSS, M. L., JOLLY, N. P., LAMBRECHTS, M. G. and VAN RENSBURG, P. (2001). 'Screening for the production of extracellular hydrolytic enzymes by non-*Saccharomyces* wine yeasts'. *J Appl Microbiol*, **91**, 182–90.

TANAY, A., SHARAN, R., KUPIEC, M. and SHAMIR, R. (2004). 'Revealing modularity and organization in the yeast molecular network by integrated analysis of highly heterogeneous genomewide data'. *Proc Natl Acad Sci U S A*, **101**, 2981–6.

UKIBE, K., HASHIDA, K., YOSHIDA, N. and TAKAGI, H. (2009). 'Metabolic engineering of *Saccharomyces cerevisiae* for astaxanthin production and oxidative stress tolerance'. *Appl Environ Microbiol*, **75**, 7205–11.

© Woodhead Publishing Limited, 2013

VERDOES, J. C., SANDMANN, G., VISSER, H., DIAZ, M., VAN MOSSEL, M. and VAN OOYEN, A. J. (2003). 'Metabolic engineering of the carotenoid biosynthetic pathway in the yeast *Xanthophyllomyces dendrorhous* (*Phaffia rhodozyma*)'. *Appl Environ Microbiol*, **69**, 3728–38.

VERVERIDIS, F., TRANTAS, E., DOUGLAS, C., VOLLMER, G., KRETZSCHMAR, G. and PANOPOULOS, N. (2007). 'Biotechnology of flavonoids and other phenylpropanoid-derived natural products. Part I: Chemical diversity, impacts on plant biology and human health'. *Biotechnol J*, **2**, 1214–34.

VERWAAL, R., WANG, J., MEIJNEN, J. P., VISSER, H., SANDMANN, G., VAN DEN BERG, J. A. and VAN OOYEN, A. J. (2007). 'High-level production of beta-carotene in *Saccharomyces cerevisiae* by successive transformation with carotenogenic genes from *Xanthophyllomyces dendrorhous*'. *Appl Environ Microbiol*, **73**, 4342–50.

VIANA, F., GIL, J. V., GENOVES, S., VALLES, S. and MANZANARES, P. (2008). 'Rational selection of non-*Saccharomyces* wine yeasts for mixed starters based on ester formation and enological traits'. *Food Microbiol*, **25**, 778–85.

VISSER, H., VAN OOYEN, A. J. and VERDOES, J. C. (2003). 'Metabolic engineering of the astaxanthin-biosynthetic pathway of *Xanthophyllomyces dendrorhous*'. *FEMS Yeast Res*, **4**, 221–31.

VON LINTIG, J. (2010). 'Colors with functions: elucidating the biochemical and molecular basis of carotenoid metabolism'. *Annu Rev Nutr*, **30**, 35–6.

WALKER, G. M. (2011). '*Pichia anomala*: cell physiology and biotechnology relative to other yeasts'. *Antonie Van Leeuwenhoek*, **99**, 25–34.

WALTON, N. J., MAYER, M. J. and NARBAD, A. (2003). 'Vanillin'. *Phytochemistry*, **63**, 505–15.

WANG, Y., HALLS, C., ZHANG, J., MATSUNO, M., ZHANG, Y. and YU, O. (2011). 'Stepwise increase of resveratrol biosynthesis in yeast *Saccharomyces cerevisiae* by metabolic engineering'. *Metab Eng*, **13**, 455–63.

WANG, Z., GERSTEIN, M. and SNYDER, M. (2009). 'RNA-Seq: a revolutionary tool for transcriptomics'. *Nat Rev Genet*, **10**, 57–63.

WILKINS, M. R., PASQUALI, C., APPEL, R. D., OU, K., GOLAZ, O., SANCHEZ, J.-C., YAN, J. X., GOOLEY, A. A., HUGHES, G., HUMPHERY-SMITH, I., WILLIAMS, K. L. and HOCHSTRASSER, D. F. (1996). 'From proteins to proteomes: Large scale protein identification by two-dimensional electrophoresis and arnino acid analysis'. *Nat Biotech*, **14**, 61–5.

WOHLBACH, D. J., KUO, A., SATO, T. K., POTTS, K. M., SALAMOV, A. A., LABUTTI, K. M., SUN, H., CLUM, A., PANGILINAN, J. L., LINDQUIST, E. A., LUCAS, S., LAPIDUS, A., JIN, M., GUNAWAN, C., BALAN, V., DALE, B. E., JEFFRIES, T. W., ZINKEL, R., BARRY, K. W., GRIGORIEV, I. V. and GASCH, A. P. (2011). 'Comparative genomics of xylose-fermenting fungi for enhanced biofuel production'. *Proc Natl Acad Sci U S A*, **108**, 13212–7.

WOOD, V., GWILLIAM, R., RAJANDREAM, M. A., LYNE, M., LYNE, R., STEWART, A., SGOUROS, J., PEAT, N., HAYLES, J., BAKER, S., BASHAM, D., BOWMAN, S., BROOKS, K., BROWN, D., BROWN, S., CHILLINGWORTH, T., CHURCHER, C., COLLINS, M., CONNOR, R., CRONIN, A., DAVIS, P., FELTWELL, T., FRASER, A., GENTLES, S., GOBLE, A., HAMLIN, N., HARRIS, D., HIDALGO, J., HODGSON, G., HOLROYD, S., HORNSBY, T., HOWARTH, S., HUCKLE, E. J., HUNT, S., JAGELS, K., JAMES, K., JONES, L., JONES, M., LEATHER, S., MCDONALD, S., MCLEAN, J., MOONEY, P., MOULE, S., MUNGALL, K., MURPHY, L., NIBLETT, D., ODELL, C., OLIVER, K., O'NEIL, S., PEARSON, D., QUAIL, M. A., RABBINOWITSCH, E., RUTHERFORD, K., RUTTER, S., SAUNDERS, D., SEEGER, K., SHARP, S., SKELTON, J., SIMMONDS, M., SQUARES, R., SQUARES, S., STEVENS, K., TAYLOR, K., TAYLOR, R. G., TIVEY, A., WALSH, S., WARREN, T., WHITEHEAD, S., WOODWARD, J., VOLCKAERT, G., AERT, R., ROBBEN, J., GRYMONPREZ, B., WELTJENS, I., VANSTREELS, E., RIEGER, M., SCHAFER, M., MULLER-AUER, S., GABEL, C., FUCHS, M., FRITZC, C., HOLZER, E., MOESTL, D., HILBERT, H., BORZYM, K., LANGER, I., BECK, A., LEHRACH, H., REINHARDT, R., POHL, T. M., EGER, P., ZIMMERMANN, W., WEDLER, H., WAMBUTT, R., PURNELLE, B.,

© Woodhead Publishing Limited, 2013

GOFFEAU, A., CADIEU, E., DREANO, S., GLOUX, S., LELAURE, V., *et al.* (2002). 'The genome sequence of *Schizosaccharomyces pombe*'. *Nature*, **415**, 871–80.
YATSYSHYN, V. Y., ISHCHUK, O. P., VORONOVSKY, A. Y., FEDOROVYCH, D. V. and SIBIRNY, A. A. (2009). 'Production of flavin mononucleotide by metabolically engineered yeast *Candida famata*'. *Metab Eng*, **11**, 163–7.
YATSYSHYN, V. Y., FEDOROVYCH, D. V. and SIBIRNY, A. A. (2010). 'Medium optimization for production of flavin mononucleotide by the recombinant strain of the yeast *Candida famata* using statistical designs'. *Biochem Eng J*, 52–60.
ZIVANOVIC, Y., WINCKER, P., VACHERIE, B., BOLOTIN-FUKUHARA, M. and FUKUHARA, H. (2005). 'Complete nucleotide sequence of the mitochondrial DNA from *Kluyveromyces lactis*'. *FEMS Yeast Research*, **5**, 315–22.

3.11   Appendix: glossary of the systems biology tool box

Systems biology can be defined as the quantitative collection, analysis and integration of whole gen*ome* scale data sets enabling the construction of biologically relevant mathematical models. Accordingly, the tools providing data for systems biology are part of what we call the *x-omics* tool box.

- Fluxomics: study of the complete set of metabolic fluxes measured or calculated in a given metabolic reaction network. A flux is defined as the rate at which input metabolites (substrates) are converted into output metabolites (products), where the rate is the concentration or the mass of reactant consumed and product formed per unit time.
- Functional genomics: discipline aiming at complete annotation of sequenced genomes via the development of theoretical and experimental tools for determining gene function by linking the gene sequences to their products (i.e. proteins) and functions. It is the fundamental step towards the development of various types of genome-scale models.
- Genomics: discipline within the genetic field aiming to determine the entire DNA sequence of organisms.
- Metabolomics: comprehensive and quantitative analysis of the complete set of metabolites (i.e. metabolome) synthesized by a cell/organism (Fiehn, 2001), wherein a metabolite is defined as any substrate or product participating in a reaction catalysed by any gene product. It includes analysis of both extracellular (exometabolome) and intracellular metabolites (endometabolome). The complete inventory and quantification of the metabolome requires the application of a number of analytical techniques, owing to the very diverse chemical/physical nature of metabolites.
- Proteomics: large scale analysis of the proteome (Wilkins *et al.*, 1996), which is the entire complement of proteins of an organism/cell. Proteomics includes identification of protein function and of protein–protein interactions. The goal of proteomics is a comprehensive, quantitative description of protein expression and its changes under the influence of specific perturbations (Anderson and Anderson, 1998).

© Woodhead Publishing Limited, 2013

- RNA-sequencing: recently developed approach to determine transcriptome profiles via high-throughput sequencing of total RNA or total mRNA in a quantitative manner. In contrast to hybridization-based techniques, RNA-sequencing is not limited to detecting transcripts that correspond to existing genomic sequences (Wang *et al.*, 2009).
- Transcriptomics: genome-wide determination of gene expression levels via analysis of mRNA levels in a specific cell population. Key targets of transcriptomics are to catalogue all species of transcripts, to determine the transcriptional structure of genes and to quantify the changes in expression levels under specific environmental or genetic perturbations. High-throughput techniques based on DNA microarrays are used to determine gene expression profiles.

© Woodhead Publishing Limited, 2013

4

Applying systems and synthetic biology approaches to the production of food ingredients, enzymes and nutraceuticals by bacteria

P. A. Hoskisson, University of Strathclyde, UK

DOI: 10.1533/9780857093547.1.81

Abstract: Humans have exploited microorganisms for the production of foodstuffs and food additives for thousands of years. Empirical selection of strains for better or improved production has been a continuous process. Currently molecular biology, genetic engineering, genomics and systems biology are being exploited to speed up the process of selection in a rational fashion. In this chapter I discuss examples of how systems biology and related approaches have been used to inform the production of food-relevant ingredients. The chapter ends with a perspective on how systems biology will continue to develop and become more influential in this area of microbiology.

Key words: *Corynebacterium*, genome scale metabolic maps, metabolic engineering, next-generation sequencing, systems biology.

4.1   Introduction

Microorganisms have been used by humans for the production of foodstuffs and for food additives for many thousands of years. Through the selection of specific strains for enhanced production by bakers, brewers and food producers, humans have been intentionally engineering strains for their own benefit. In the current era of molecular biology, genetic engineering, directed evolution, genomics and systems biology, we are entering an exciting arena, being able to use a range of modern cutting edge technologies to speed up the process of selection. This will enable us to move from what is currently a largely empirical process of strain improvement, to a rationally designed system for production of food products, enzymes and nutraceuticals.

In this chapter, I aim to define systems biology, its utility in understanding microbial production processes and some of the major methodological

© Woodhead Publishing Limited, 2013

advances that have been made. I will also discuss specific examples of how systems biology and related approaches have been used to inform the production of food-relevant ingredients and how these can be exploited in the production of food ingredients, enzymes and nutraceuticals production, which are rapidly developing areas of interest (Basu *et al.*, 2007; Gupta *et al.*, 2010; Hugenholtz and Smid, 2002; Knorr, 1998). Finally the chapter will end with a perspective on how systems biology can continue to develop and become more influential in this area of food microbiology. It is clear that these approaches will expand their influence as the challenges of developing functional foods and food additives continue (Schwager *et al.*, 2008).

4.2   Definition and uses of systems biology in production

Systems biology has been defined in many different ways since it entered the psyche of the majority of scientists (Williamson, 2005). The most useful definition is that used by the Biotechnology and Biological Sciences Research Council (BBSRC) in the United Kingdom: 'Systems biology is an approach by which biological questions are addressed through integrating experiments with computational modelling and theory, in re-enforcing cycles'. Generally systems biology diverges from the traditional reductionist approach to understanding biological problems and is concerned with the incorporation and interpretation of diverse data sets to understand a biological 'system'. Noble (2006) refers to this as 'integration rather than reduction' and incorporates experimental, bioinformatics, computational and mathematical approaches ultimately to predict a system. These approaches generally tend to use 'omics'-based technologies, as the modelling aspect of whole systems generally requires large data sets to build reliable models, enhancing reliability and predictability. In addition to this, the global nature of 'omics' technologies coupled with modelling, facilitate a holistic approach to rational engineering of strains and the development of production processes. Several experimental techniques have advanced our ability to apply systems approaches to the production of food ingredients, enzymes and nutraceuticals and are applicable to wider systems biology.

The revolution in next-generation sequencing (Roche 454 and Illumina Solexa technologies for example) for the rapid sequencing and assembly of genomes of industrial organisms has allowed significant progress to be made (Parkhill and Wren, 2011). The exponential increase in the number of genome sequences available has enabled a greater understanding of industrial production strains of bacteria, yeast and fungi and to elucidate the genetic basis of their high yields compared to wild-type or low-producing strains. The speed and cost of these technologies are now so low that a standard bacterial genome of around 5 Mbps can be sequenced for as little as US $500. Leading on from this, the application of transcriptomics technology, through the use of DNA microarrays, facilitated rapid analysis

© Woodhead Publishing Limited, 2013

of important or highly expressed genes during production processes for industrial microorganisms. Recently the exploitation of next-generation sequencing technology (as used for genomics above) for sequencing cDNA libraries for the analysis of transcription, referred to as RNA-Seq (Wilhelm and Landry, 2009) has again revolutionized our ability to study changes in gene expression in industrially useful strains of microorganisms, allowing the identification of genes central to the production process. This is also particularly useful as *a priori* knowledge of the genome is not essential (although desirable to facilitate ease of mapping) to apply this technique, so it can be applied to a range of organisms rapidly.

Proteomics is a popular technique due to establishment of (two-dimensional) 2D-polyacrylamide gel electrophoresis (2D-PAGE) over the last 30 years or so. The availability of high-throughput mass spectrometry techniques such as matrix-assisted laser desorption/ionization-time of flight (MALDI-TOF) advanced this technique through spot identification and assignment to proteins via databases and search algorithms such as MASCOT percolator (Ma *et al.*, 2010). Recently advances in gel-free proteomics such as MudPIT have also increased the rate of exploration in this area (Fränzel and Wolters, 2011). The advantages of proteomic approaches over transcriptomics are that the proteome measures the protein components of the cell and can also provide information about the post-translational modification of proteins in the cells (through observed changes in protein mass or pI), rather than being a measure of translation potential (the presence of RNA molecules), adding value to whole organism analyses, especially when investigating pathway regulation. However there are some limitations to the ability to detect and identify low abundance proteins, multiple proteins in large 2D-PAGE spots and the under-representation of membrane proteins in analyses. Recently, metabolomics, the high-throughput identification and quantitative analysis of whole cell composition of small molecules 'metabolites', has become a new addition to the palette of techniques used in systems biology and more widely in biology. It is already attracting a great deal of interest in food applications for studying the quality and authenticity of food products (Wishart, 2008). However the ability to analyse and quantify the whole metabolome of an industrially relevant organism is currently challenging, although it allows knowledge of precursor supply, important pathway identification and product formation through metabolomics. These data, if collected through careful sampling, can feed directly into metabolic flux analysis techniques, which have been widely utilized in industry over the last 20 years or so. These data are also invaluable when trying to validate whole genome metabolic models (see below).

The key to exploiting these methods is the integration of these systems through mathematical modelling, simulation and bioinformatics. The ultimate aim of these models is to be predictive in the biological system, such that hypotheses can be generated and tested rapidly, without the

© Woodhead Publishing Limited, 2013

requirement to perform costly experiments. The description of individual methods of modelling is beyond the scope of this chapter, although I would refer the reader to several excellent reviews and texts, which cover the basics (Kanehisa *et al.*, 2011; Klipp *et al.*, 2009; Kugler *et al.*, 2010; Le Fèvre *et al.*, 2009; Sroka *et al.*, 2011; Santos *et al.*, 2011; Patil and Nielsen, 2005).

It is apparent from the diversity of methods described above and the ability to use computational and modelling approaches to draw the emerging data together, that they combine to provide a powerful approach to tackling production issues in bacterial processes. While these techniques are still relatively specialized, their influence will grow in the coming years and employing these approaches will become the standard way to tackle process development and strain improvement.

4.3   Advantages of systems biology in the production of food ingredients, enzymes and nutraceuticals by bacteria

Bacteria are widely used in the production of food ingredients, enzymes and nutraceuticals. Their rapid growth, small genomes, genetic tractability and diverse metabolic capabilities make them suitable hosts for industrial production processes either as native production strains or as heterologous producers of a wide range of useful products. Therefore industrial processes using microorganisms are ideal for application of systems biology-based approaches to gain understanding of the formation of a specific product.

The most obvious applications that emerge from using systems biology to understand and exploit process/production improvement are genome scale metabolic modelling and the defining of regulation and regulatory processes in industrially valuable strains. These processes rely on genomic data (or more recently RNASeq), which can be analysed using various computational methods such as RAST (rapid annotations using subsystems technology) (Aziz *et al.*, 2008). Using these algorithms, to gain a prediction of all the genes, the proteins and reactions present in a cell can be very powerful. These online resources link directly to the Kyoto Encyclopedia of Genes and Genomes, KEGG (Kanehisa *et al.*, 2011) enabling rapid evaluation of metabolism and how it compares across various strains and organisms. These computationally inferred reactions are also informed significantly by existing physiological data in the literature and from a laboratory's own unpublished data. In addition to this it is imperative that previously unknown reactions that maybe important are tested for functionality by traditional methods (such as enzymology), to ensure that the predictions are correct and prediction/modelling software are not infallible. Key to these analyses is the management and analysis of large amounts of data, which rely significantly on bioinformatic and computational biology

© Woodhead Publishing Limited, 2013

knowledge. The correct curation of these data requires significant consideration, computational power and storage.

Williamson (2005) expressed what a challenging task this data management and comprehension can be; if a typical microbial cell is considered, it has a complement of 5000 genes, of which during exponential growth 500 genes are expressed, encoding 400 metabolic enzymes and 100 regulatory proteins or transporters. Thus with more than 100 metabolites interacting in these pathways, it is conceivable that more than 1000 metabolic fluxes need to be considered (Takors *et al.*, 2007). Yet the power of understanding metabolism at this level can lead to great insight and advances in production processes.

The prediction of metabolic pathways from a new genome facilitates the incorporation of the metabolic reactions into a genome-scale metabolic model. These models can be constructed either at the pathway, the metabolic node or the whole genome level using a range of software which can build models specifically for your organism or strain (see Further sources of information); it is particularly useful if information is available for a range of 'improved' strains along a production lineage (comparing wild-type strains with a succession of 'improved' strains). It is also advantageous if gene deletions have been made or selected through improvement processes (classical random mutagenesis regimes, directed evolution) because it allows comparison of pathways, computation of maximum theoretical yields, potential by-product formation (Teusink *et al.*, 2011; Feist *et al.*, 2010).

Similarly these metabolic models can be exploited for mapping large-scale data sets such as that of Patil and Nielsen (2005) and Stevens *et al.* (2008) where genome-wide analysis of transcription in lactic acid bacteria from microarrays was mapped on to the genome-scale metabolic model. The outcome of this work from two separate groups was two-fold; directly facilitating improvement of aerobic growth (Stevens *et al.*, 2008) and identifying targets for engineering approaches through examination of the network topologies of the mapped strain. Interestingly, traditional mutagenesis approaches using ultraviolet light or *N*-methyl-*N'*-nitro-*N*-nitrosoguanidine (NTG), were previously considered as 'black-boxes' that were difficult to understand in detail owing to cumulative mutations. However the revolution of next generation sequencing has allowed comparison of these strains through genome sequencing and comparison of single nucleotide polymorphisms (SNPs) and insertions/deletions (indels) using a range of software such as the Variant Effect Predictor (http://bacteria.ensembl.org/tools.html) to compute the SNP variation between strains resulting in informed comparison of the metabolic potential of the strains.

The beauty of these approaches is that interesting genetic alterations to the strains can be investigated *in silico* and the results interrogated by comparing metabolic models of the modified and unmodified strains. If the hypothesis that a particular mutation event has lead to a beneficial mutation

© Woodhead Publishing Limited, 2013

appears to be correct *in silico*, mutagenesis can be undertaken to verify the genome modification and its effect on production. These iterations of computation/modelling and experiments can lead to predictability in terms of enhancing strains through rational design. These approaches have been referred to as 'reverse engineering' (Ikeda *et al.*, 2009) in *Corynebacterium* where up to 1000 mutations may have accumulated in empirically improved lysine-producing strains (Ikeda *et al.*, 2006).

Whilst the range of products that fall under the umbrella of food ingredients, enzymes and nutraceuticals is vast, often these products have been developed for industrial production from smaller scale traditional processes or established industrial processes which were developed empirically. Using systems biology approaches can significantly enhance this process in terms of cost effectiveness, speed and productivity. Below I have selected a few examples, which will serve to illustrate the power of systems biology approaches for the production of food ingredients, enzymes and nutraceuticals.

4.4   Production of food grade amino acids through the exploitation of systems biology and 'omics' approaches

Various bacterial strains have been used to produce amino acids on an industrial scale. Currently one of the 'work-horse' strains is *Corynebacterium glutamicum* (Dong *et al.*, 2011; Eggeling and Bott, 2005) although *Escherichia coli* has also been extensively exploited (Park *et al.*, 2010). This discussion will focus on *C. glutamicum*, a Gram-positive organism that can utilize a wide variety of sugars, organic acids and alcohols for the industrial production of l-glutamate, l-lysine and l-valine (Eggeling and Bott, 2005). There have been many years of investment in understanding the various pathways that lead to the production of these amino acids, which are used extensively in human and animal food production and in wider industrial processes (Blombach *et al.*, 2009).

There are several corynebacterial genomes available, although comparison of the natural glutamate producing strains *C. glutamicum* and *C. efficiens* has attracted much attention in understanding biotechnological output and to increase understanding of the central metabolism which can inform process and strain improvement (Eggeling and Bott, 2005). These strains have very similar sized genome and coding ability, however the main difference appears to be the differences in G+C content of the genome and substitution of lysine by arginine, serine by alanine and serine by threonine codons within protein coding regions of the genome of *C. efficiens* that contribute to industrial performance (Nishio *et al.*, 2003, 2004). These changes appear to have contributed to the greater stability of *C. efficiens* proteins at 40°C or above, making it more suited to certain industrial applications. Such genome analysis can be useful for selecting strains for a

© Woodhead Publishing Limited, 2013

particular application or for inter-strain comparison and can inform the molecular mechanisms of strain/pathway improvement.

Many approaches have been used to improve the production of amino acids and other metabolites by *Corynebacterium*. Traditional mutagenesis and screening have been successful in this organism along with molecular biology-based metabolic engineering, flux analysis, metabolism and metabolomics (Kjeldsen and Nielsen, 2009; Eggeling and Bott, 2005). However, as pointed out by Kjeldsen and Nielsen (2009), these approaches have largely focused upon isolated pathways rather than on more holistic approaches to understanding the organism and perhaps a whole genome *in silico* genome model would enable a greater understanding of the production process. Constructing a whole genome model, based on the annotated genome (Kalinowski *et al.*, 2003) and the extensive literature available for this organism was relatively straightforward.

Using a combination of KEGG (Kanehisa *et al.*, 2011), Biocyc (www.biocyc.com) and BioOptv4.9 software developed 'in-house' and available from the authors, a genome-scale, validated metabolic network model was developed. Analysis of this model led to some interesting insights into the biosynthesis of lysine, such as the drain on the tricasboxylic acid (TCA) cycle by using the succinylase route rather than the direct dehydrogenase route to lysine which reduced amino acid output. This indicates that considerations outside of a direct biosynthetic pathway are significant in contributing to production output. Similarly, the demand for NADPH is not limiting to the process until the yield of lysine exceeds 55% (mmol lysine(mmol glucose)$^{-1}$), and this finding goes some way to explain the lack of success in replacing the NADPH-dependent glutamate dehydrogenase (*gdh*) with the NADH-dependent *gdh* of *Peptostreptococcus asaccharolyticus* (Marx *et al.*, 1999), where it was hypothesized that the lower demand for NADPH would enhance amino acid output.

The validation of experimental data by this model is also important and interesting for developing production processes, but discrepancies were observed between the predictions from the model and experimental data. Specifically, differences in biomass yields, NADPH output from the pentose-phosphate pathway (PPP) and also the importance of the glyoxylate bypass during growth on glucose. Nevertheless, the greatest power of this model is the incorporation of data from whole genome experiments such as RNA-Seq, metabolomics and proteomics, which not only increased insight, but also allowed refinement of the model. This demonstrates that iterations of experiments and modelling facilitate greater accuracy and power to be achieved in systems biology.

Recently Ikeda *et al.* (2009) employed reverse engineering approaches, based on classically improved industrial strains of *C. glutamicum*. These authors examined the production strains genetically to identify the lesions in their genomes that contributed to increased lysine production. This approach then enabled the re-assembly of the important lysine production associated mutations in a robust wild-type strain background, which has been shown to be a useful strategy for improving production performance (Ohnishi *et al.*, 2002, 2003). The combination of knowledge accumulated over many years of strain improvement and analysis using modern high-throughput methods (sequencing, metabolomics and molecular biology) can enable rapid remaking of improved strains in genetic backgrounds that are amenable to specific fermentation processes.

In the example of Ikeda *et al.* (2009), five point mutations were identified in three arginine overproducing strains. All of the mutations interestingly fell within the arginine biosynthetic operon, three in structural genes, one in the arginine repressor and one in an intergenic region. Individually, when reproduced in a wild-type strain, to create a 'robust producer' these mutations did not contribute to overproduction. However, when combined an increase in arginine production was observed, albeit at lower levels than the best producing classically improved strain. To investigate this further, Ikeda *et al.* (2009) employed transcriptomics to look at gene expression profiles in the strains. In one arginine over-producing strain, a mutation associated with lysine over-production was observed that had previously been characterized (Hayashi *et al.*, 2006). The introduction of this mutation in to the wild-type strain resulted in genome-wide induction of amino acid biosynthetic pathways and concomitant over-production of arginine when combined in the robust producer with the arginine mutations. However, advancement of production beyond that of the classically improved strain was only achieved when *argB*, encoding acetylglutamate kinase, was replaced by *argB* from *E. coli* which is natively insensitive to arginine feedback, leading to a three-fold over-production of arginine. These kinds of approach require a firm understanding of physiology and biochemistry, such that enzyme targets can be identified for replacement by heterologous enzymes with better performance, or can be targeted for processes such as directed evolution (see Fig. 4.1).

The study of Ikeda *et al.* (2009) serves to illustrate that global investigation (genomics and transcriptomics), coupled with understanding of metabolic pathways can lead to rational strain improvement. Additional studies on *C. glutamicum*, with a view to enhancing amino acid biosynthesis, have approached rational strain improvement and metabolic engineering, through improving precursor supply by targeting central metabolic pathways to enhance production.

One such target is the TCA cycle, which has been underexploited in terms of its manipulation in *Corynebacterium* (Eggeling and Bott, 2005). In *C. glutamicum,* isocitrate dehydrogenase (ICD) is the highest expressed gene in the TCA cycle (Becker *et al.*, 2009). Downregulation of ICD through reduced translation, by altering the start codon, decreased ICD activity by 70%, with a concomitant 40% or greater increase in lysine production. Interestingly metabolic flux analysis indicates that reducing TCA cycle flux had the effect of increasing flux through anaplerotic carboxylation

© Woodhead Publishing Limited, 2013

Classically improved strains

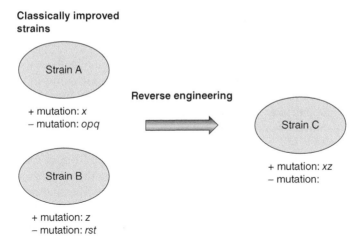

Fig. 4.1 Schematic representation of a reverse engineering approach to combine mutations, in two classically improved strains with multiple genetic lesions. In strains A and B there is a single positive mutation in each and multiple deleterious or inconsequential mutations. These are identified and combined in to a robust wild-type background to produce a strain expressing high production characteristics, without secondary deleterious mutations.

reactions and reduced loss of CO_2 through TCA cycle flux, which ultimately contributes to an increase in carbon yield for the desired product (Becker *et al.*, 2009).

A similar informed approach was used to engineer a *C. glutamicum* strain that produced additional NADPH during glycolysis. This was achieved through the replacement of the endogenous NAD-dependent glyceraldehyde 3-phosphate dehydrogenase (GAPDH) with a non-phosphorylating GAPDH from *Streptococcus*, which reduces $NADP^+$ to NADPH that will supply additional NADPH to the metabolism, which is also a critical cofactor in the biosynthesis of lysine (Takeno *et al.*, 2010). Initial replacement of the NAD-dependent GAPDH from *Sbreptococcus-mutans* was unsuccessful, although, isolation of a suppressor mutation that facilitated growth on glucose and other sugars as carbon sources enabled generation of additional NADPH in cells (Takeno *et al.*, 2010). This sort of approach is useful for the provision of cofactors for the biosynthesis of metabolically demanding substrates such as lysine, and the exploitation of genome scale models, coupled with extensive genetic tools (Nešvera and Pátek, 2011), 'omics' technologies and modelling will enhance our understanding of complex metabolic networks, enabling wider exploitation of the metabolic capabilities of *Corynebacterium* (Kind *et al.*, 2010; Schneider and Wendisch, 2010), through to heterologous expression of enzymes (Umakoshi *et al.*, 2011).

© Woodhead Publishing Limited, 2013

4.5   Using systems approaches to develop enzymes for use in food production

There is significant interest in the development of enzyme-based industrial processes owing to their high catabolic capability and low energy costs. The use of enzymes in the production of food ingredients and nutraceuticals has been established for millennia, whether this is through the direct addition of microorganisms (catalysis or probiotics) or purified enzymes as catalysts.

One approach to enhance catalysis is through the discovery of novel enzymes from 'sequence space', that is to exploit metagenomic screens for catalytic capability (Uchiyama and Miyazaki, 2009). The search for novel activity in uncultivable bacteria represents a potentially vast resource of undiscovered catabolic potential, which can be exploited through direct cloning of environmental DNA (metagenome) which can be screened based on sequence or activity and is independent of culturability. One limitation of these approaches is the availability of suitable hosts that can express the enzymes of interest. Thus the use of metabolically diverse microorganisms to screen metagenomic libraries is important (Troeschel *et al.*, 2010; Taupp *et al.*, 2011) and will facilitate the identification of novel activities. The expression of novel enzymes (and other products) may also require well-characterized and developed 'superhosts', which are well characterized genetically, physiologically and their fermentation characteristics are well understood (Medema *et al.*, 2010).

A good example of how novel activities can be discovered and exploited is the use of bacterial lipases in the food industry. Bacterial lipases (triacylglycerol hydrolases) have found wide application in biotechnology (Joseph *et al.*, 2008) owing to their chemo- and stereoselectivity, and offer endless possibilities for flavour enhancements, hydrolysis and preservation of diary products (Sun *et al.*, 2002). These applications and the diversity of these enzymes has led to searches for activity in diverse habitats. One study found high lipase diversity in glacial soils (Yuhong *et al.*, 2009), which would offer interesting proteins for cold applications in industrial settings. Another interesting approach is to use genome-wide interrogation of known strains to identify lipases and other enzymes that can be useful in isolation, should they be cloned and expressed outwith their native host. A good example of this is the recently sequenced bacterium *Propionibacterium freudenrichii*, which is known to contribute to the ripening of 'Swiss-type' cheeses. The extensive lipolysis of milk glycerides and release of free fatty acids during the growth of this organism throughout cheese maturation is essential for flavour characteristics to develop (Falentin *et al.*, 2010). The genome sequence revealed 12 genes putatively encoding lipases and esterases which may be responsible for the release of volatile esters, which contribute to aroma formation. The abundance of such enzymes in *P. freudenrichii* suggests that other cheese ripening strains and dairy associated strains, other

© Woodhead Publishing Limited, 2013

than lactic acid bacteria may offer potential for biotechnological exploitation outside of their normal roles. This is also certainly true for the emerging
transcriptomic and proteomic data derived from the study of bifidobacteria
(Gilad *et al.*, 2010, 2011), which has identified important proteins for the
prebiotic effects of these organisms enabling heterologous expression of
beneficial targets in superhosts, in addition to unravelling their roles within
their native host.

4.6   Future trends in the application of systems and synthetic biology to food microbiology

It is clear that global 'omics' technologies are at the forefront of the development of industrial microbiology processes. There are a limited range of
organisms in which these technologies are leading the way in terms of
process and strain development such as the corynebacteria (discussed
above) or the lactic acid bacteria (discussed elsewhere in this volume and
extensively in Teusink *et al.*, 2011). The ability to marry bacterial physiology,
'omics' technologies, modelling and production processes will be a powerful
force with which to address 21st century industrial microbiology.

From the early days of the Jacob and Monod operon model and the
development of the Monod equation (Yaniv, 2011) microbiologists have
always tried to embrace modelling processes and relate this to physiology.
We can therefore think of systems biology as quantitative physiology
(Teusink *et al.*, 2011), which will only grow in its influence as these methods
develop. It is only through the use of these approaches to global characterization of the cellular process that we will be able to understand fully the
subtle nuances of the global physiology of production strains. The incorporation of modelling into these areas can only serve to strengthen the outputs,
inform on experimental design and help the formulation of interesting
hypotheses.

One area that is still lacking is the integration of 'omics' data and, for
example, the integration of transcriptome data (the abundance of mRNA
transcripts) does not always correlate well with proteomic data. Many
different processes, such as the intrinsic individual half-life of an mRNA,
mRNA stability, post-transcriptional regulation, translational regulation
and efficiency, protein stability and turnover and post-translational regulation, can influence this correlation. The integration of these diverse processes is a challenge to biologists and modellers, but is key to comprehending
the complexity of living cells. Some progress has been made in this area
using lactic acid bacteria (Dressaire *et al.*, 2009; Picard *et al.*, 2009), and these
methods will be widely applicable throughout biology, but it serves to
highlight our future reliance on modelling science to help understand
complex biological data.

© Woodhead Publishing Limited, 2013

Finally, the influence of synthetic biology on the production of enzymes, food ingredients and nutraceuticals is only now beginning to emerge. The use of 'superhost' organisms with extensive metabolic capabilities for expressing various heterologous biosynthetic pathways has been recognized in the antibiotic field for many years (Alduina *et al.*, 2005; Donadio *et al.*, 2002). These kinds of approach can be adopted much more widely across industrial microbiology with the identification of minimal bacterial genome complements (Glass *et al.*, 2006), from which 'superhosts' can be developed. The technologies are advancing at a rapid rate, such that 'superhost' organisms can be rapidly assembled to meet the needs of a particular process using transformation-associated recombination cloning in yeast (Gibson *et al.*, 2008). However, in the near future this synthetic assembly of biosynthetic routes for expression in 'superhost' organisms will negate the need for multiple strains. These approaches can be carried out using standard molecular biology routes to create synthetic gene networks (Marchisio, 2012; Marchisio and Stelling, 2011) that are robust, reproducible and reliable (Lebiedz *et al.*, 2012) and can be predicted in a production scenario. The standardization of these synthetic and process optimized parts will also enable rapid assembly of synthetic constructs for expression of a desired biological product (Müller and Arndt, 2012), which can even be 'watermarked' to protect intellectual property (Gibson *et al.*, 2008). It is with advancements such as these that we really are at the point of design-based engineering of industrial microbes (Tyo *et al.*, 2010).

4.7   Sources of further information

There are many texts that cover the practical aspects of systems and synthetic biology. An excellent introductory text is *Systems Biology: An Introduction* by Klipp *et al.* (2009). Also of particular use are the excellent volumes of the Methods in Enzymology series, specifically the two recent volumes on *Synthetic biology*, Part A (ed. Christopher Voigt, Vol. 497, 2011) and Part B, *Computer Aided Design and DNA Assembly* (ed. Christopher Voigt, Vol. 498, 2011) and *Methods in Systems Biology* (Ed. Daniel Jameson, Malkhey Verma and Hans V. Westerhoff, Vol. 500, 2011).

There are also some excellent web resources for genome scale modelling and genome analysis which are listed below,

- RAST – rapid annotation software: rast.nmpdr.org/
  This takes next generation sequencing data and automatically annotates it.
- SEED – for analysis of RAST data: www.theseed.org
  This takes RAST data and feeds to KEGG database for visualization of pathways.

© Woodhead Publishing Limited, 2013

- KEGG – Kyoto Encyclopedia of Genes and Genomes: www.genome.jp/kegg/
  This is a vast resource of metabolic pathway data.
- Copasi – Biochemical Network Simulator: www.copasi.org/tiki-view_articles.php
  This is software for building metabolic models
- Biocyc–Pathway/Genome database:biocyc.org/
  This is a genome-based metabolic pathway database.
- BioMet Toolbox: www.sysbio.se/BioMet/
  This is genome scale metabolic modelling software.

4.8   References

ALDUINA, R., GIARDINA, A., GALLO, G., RENZONE, G., FERRARO, C., CONTINO, A., SCALONI, A., DONADIO, S. and PUGLIA, A. M. (2005). 'Expression in *Streptomyces lividans* of Nonomuraea genes cloned in an artificial chromosome'. *Appl Microbiol Biotechnol*, **68**, 656–62.

AZIZ, R. K., BARTELS, D., BEST, A. A., DEJONGH, M., DISZ, T., EDWARDS, R. A., FORMSMA, K., GERDES, S., GLASS, E. M., KUBAL, M., MEYER, F., OLSEN, G. J., OLSON, R., OSTERMAN, A. L., OVERBEEK, R. A., MCNEIL, L. K., PAARMANN, D., PACZIAN, T., PARRELLO, B., PUSCH, G. D., REICH, C., STEVENS, R., VASSIEVA, O., VONSTEIN, V., WILKE, A. and ZAGNITKO, O. (2008). 'The RAST Server: rapid annotations using subsystems technology', *BMC Genomics*, **9**, 75.

BASU, S., THOMAS, J. and ACHARYA, S. N. (2007). 'Prospects for growth in global nutraceutical and functional food markets: a Canadian perspective'. *Aust J Basic Appl Sci*, **1**, 637–49.

BECKER, J., KLOPPROGGE, C., SCHRÖDER, H. and WITTMANN, C. (2009). 'Metabolic engineering of the tricarboxylic acid cycle for improved lysine production by *Corynebacterium glutamicum*', *Appl Environment Microbiol*, **75**, 7866–9.

BLOMBACH, B., ARNDT, A., AUCHTER, M. and EIKMANNS, B. J. (2009). 'L-valine production during growth of pyruvate dehydrogenase complex-deficient *Corynebacterium glutamicum* in the presence of ethanol or by inactivation of the transcriptional regulator SugR', *Appl Environment Microbiol*, **75**, 1197–1200.

DONADIO, S., SOSIO M. S. and LANCINI, G. (2002). 'Impact of the first Streptomyces genome sequence on the discovery and production of bioactive substance'. *Appl Microbiol Biotechnol*, **60**, 377–80.

DONG, X., QUINN, P. J. and WANG, X. (2011). 'Metabolic engineering of *Escherichia coli* and *Corynebacterium glutamicum* for the production of L-threonine', *Biotechnol Adv*, **29**, 11–23.

DRESSAIRE, C., GITTON, C., LOUBIÈRE, P., MONNET, V., QUEINNEC, I. and COCAIGN-BOUSQUET, M. (2009). 'Transcriptome and proteome exploration to model translation efficiency and protein stability in *Lactococcus lactis*', *PLoS Comput Biol*, **5**, e1000606.

EGGELING, L. and BOTT, M. (2005). *Handbook of Corynebacterium Glutamicum*. CRC Press, Boca Raton, Florida.

FALENTIN, H., DEUTSCH, S.-M., JAN, G., LOUX, V., THIERRY, A., PARAYRE, S., MAILLARD, M.-B., DHERBÉCOURT, J., COUSIN, F. J., JARDIN, J., SIGUIER, P., COULOUX, A., BARBE, V., VACHERIE, B., WINCKER, P., GIBRAT, J.-F., GAILLARDIN, C. and LORTAL, S. (2010). 'The complete genome of *Propionibacterium freudenreichii* CIRM-BIA1, a hardy actinobacterium with food and probiotic applications', *PLoS ONE*, **5**, e11748.

© Woodhead Publishing Limited, 2013

FEIST, A. M., ZIELINSKI, D. C., ORTH, J. D., SCHELLENBERGER, J., HERRGARD, M. J. and PALSSON, B. Ø. (2010). 'Model-driven evaluation of the production potential for growth-coupled products of *Escherichia coli*', *Metab. Eng*, **12**, 173–186.

FRÄNZEL, B. and WOLTERS, D. A. (2011). 'Advanced MudPIT as a next step toward high proteome coverage', *Proteomics*, **11**, 3651–6.

GIBSON, D. G., BENDERS, G. A., ANDREWS-PFANNKOCH, C., DENISOVA, E. A., BADEN-TILLSON, H., ZAVERI, J., STOCKWELL, T. B., BROWNLEY, A., THOMAS, D. W., ALGIRE, M. A., MERRYMAN, C., YOUNG, L., NOSKOV, V. N., GLASS, J. I., VENTER, J. C., HUTCHISON, C. A. and SMITH, H. O. (2008). 'Complete chemical synthesis, assembly, and cloning of a *Mycoplasma genitalium* genome', *Science*, **319**, 1215–20.

GILAD, O., JACOBSEN, S., STUER-LAURIDSEN, B., PEDERSEN, M. B., GARRIGUES, C. and SVENSSON, B. (2010). 'Combined transcriptome and proteome analysis of *Bifidobacterium animalis* subsp. lactis BB-12 grown on xylo-oligosaccharides and a model of their utilization', *Appl Environment Microbiol*, **76**, 7285–91.

GILAD, O., SVENSSON, B., VIBORG, A. H., STUER-LAURIDSEN, B. and JACOBSEN, S. (2011). 'The extracellular proteome of *Bifidobacterium animalis* subsp. lactis BB-12 reveals proteins with putative roles in probiotic effects', *Proteomics*, **11**, 2503–2514.

GLASS, J. I., ASSAD-GARCIA, N., ALPEROVICH, N., YOOSEPH, S., LEWIS, M. R., MARUF, M., HUTCHISON, C. A., SMITH, H. O. and VENTER, J. C. (2006). 'Essential genes of a minimal bacterium', *Proc Natl Acad Sci USA*, **103**, 425–30.

GUPTA, S., CHAUHAN, D., MEHLA, K. and SOOD, P. (2010). 'An overview of nutraceuticals: current scenario', *J Basic Clin Pharm*, **1**, 55–62.

HAYASHI, M., OHNISHI, J., MITSUHASHI, S., YONETANI, Y., HASHIMOTO, S.-I. and IKEDA, M. (2006). 'Transcriptome analysis reveals global expression changes in an industrial L-lysine producer of *Corynebacterium glutamicum*', *Biosci Biotechnol Biochem*, **70**, 546–50.

HUGENHOLTZ, J. and SMID, E. J. (2002). 'Nutraceutical production with food-grade microorganisms', *Curr Opin Biotechnol*, **13**, 497–507.

IKEDA, M., OHNISHI, J., HAYASHI, M. and MITSUHASHI, S. (2006). 'A genome-based approach to create a minimally mutated *Corynebacterium glutamicum* strain for efficient L-lysine production', *J Ind Microbiol Biotechnol*, **33**, 610–15.

IKEDA, M., MITSUHASHI, S., TANAKA, K. and HAYASHI, M. (2009). 'Reengineering of a *Corynebacterium glutamicum* L-arginine and L-citrulline producer', *Appl Environment Microbiol*, **75**, 1635–41.

JOSEPH, B., RAMTEKE, P. W. and THOMAS, G. (2008). 'Cold active microbial lipases: some hot issues and recent developments', *Biotechnol Adv*, **26**, 457–470.

KALINOWSKI, J., BATHE, B., BARTELS, D., BISCHOFF, N., BOTT, M., BURKOVSKI, A., DUSCH, N., EGGELING, L., EIKMANNS, B. J., GAIGALAT, L., GOESMANN, A., HARTMANN, M., HUTHMACHER, K., KRÄMER, R., LINKE, B., MCHARDY, A. C., MEYER, F. MÖCKEL, B., PFEFFERLE, W., PÜHLER, A., REY, D. A., RÜCKERT, C., RUPP, O., SAHM, H., WENDISCH, V. W., WIEGRÄBE, I. and TAUCH., A. (2003). 'The complete *Corynebacterium glutamicum* ATCC 13032 genome sequence and its impact on the production of L-aspartate-derived amino acids and vitamins', *J Biotechnol*, **104**, 5–25.

KANEHISA, M., GOTO, S., SATO, Y., FURUMICHI, M. and TANABE, M. (2011). 'KEGG for integration and interpretation of large-scale molecular data sets', *Nucleic Acids Res*, **40**, D109–D114.

KIND, S., JEONG, W. K., SCHRÖDER, H. and WITTMANN, C. (2010). 'Systems-wide metabolic pathway engineering in *Corynebacterium glutamicum* for bio-based production of diaminopentane', *Metab Eng*, **12**, 341–51.

KJELDSEN, K. R. and NIELSEN, J. (2009). '*In silico* genome-scale reconstruction and validation of the *Corynebacterium glutamicum* metabolic network', *Biotechnol Bioeng*, **102**, 583–97.

© Woodhead Publishing Limited, 2013

KLIPP, E., LIEBERMEISTER, W. and WIERLING, C. (2009). *Systems Biology, A Textbook.* VCH Publishers, Weinheim, Germany.

KNORR, D. (1998). 'Technology aspects related to microorganisms in functional foods', *Trends Food Sci Technol*, **9**, 295–306.

KUGLER, H., LARJO, A. and HAREL, D. (2010). 'Biocharts: a visual formalism for complex biological systems', *J R Soc Interface*, **7**, 1015–24.

LE FÈVRE, F., SMIDTAS, S., COMBE, C., DUROT, M., D'ALCHÉ-BUC, F. and SCHACHTER, V. (2009). 'CycSim–an online tool for exploring and experimenting with genome-scale metabolic models', *Bioinformatics*, **25**, 1987–88.

LEBIEDZ, D., REHBERG, M. and SKANDA, D. (2012). 'Robust optimal design of synthetic biological networks', *Methods Mol Biol*, **813**, 45–55.

MA, J., ZHANG, J., WU, S., LI, D., ZHU, Y. and HE, F. (2010). 'Improving the sensitivity of MASCOT search results validation by combining new features with Bayesian nonparametric model', *Proteomics*, **10**, 4293–300.

MARCHISIO, M. A. (2012). 'In silico implementation of synthetic gene networks', *Methods Mol Biol*, **813**, 3–21.

MARCHISIO, M. A. and STELLING, J. (2011). 'Automatic design of digital synthetic gene circuits', *PLoS Comput Biol*, **7**, e1001083.

MARX, A., EIKMANNS, B. J., SAHM, H., DE GRAAF, A. A. and EGGELING, L. (1999). 'Response of the central metabolism in *Corynebacterium glutamicum* to the use of an NADH-dependent glutamate dehydrogenase', *Metab Eng*, **1**, 35–48.

MEDEMA, M. H., BREITLING, R., BOVENBERG, R. and TAKANO, E. (2010). 'Exploiting plug-and-play synthetic biology for drug discovery and production in microorganisms', *Nat Rev Micro*, **9**, 131–7.

MÜLLER, K. M. and ARNDT, K. M. (2012). 'Standardization in synthetic biology', *Methods Mol Biol*, **813**, 23–43.

NEŠVERA, J. and PÁTEK, M. (2011). 'Tools for genetic manipulations in *Corynebacterium glutamicum* and their applications', *Appl Microbiol Biotechnol*, **90**, 1641–54.

NISHIO, Y., NAKAMURA, Y., KAWARABAYASI, Y., USUDA, Y., KIMURA, E., SUGIMOTO, S., MATSUI, K., YAMAGISHI, A., KIKUCHI, H., IKEO, K. and GOJOBORI, T. (2003). 'Comparative complete genome sequence analysis of the amino acid replacements responsible for the thermostability of *Corynebacterium efficiens*', *Genome Res*, **13**, 1572–9.

NISHIO, Y., NAKAMURA, Y., USUDA, Y., SUGIMOTO, S., MATSUI, K., KAWARABAYASI, Y., KIKUCHI, H., GOJOBORI, T. and IKEO, K. (2004). 'Evolutionary process of amino acid biosynthesis in Corynebacterium at the whole genome level', *Mol Biol Evolut*, **21**, 1683–91.

NOBLE, D. (2006). *The Music of Life: Biology Beyond Genes.* Oxford University Press, oxford.

OHNISHI, J., MITSUHASHI, S., HAYASHI, M., ANDO, S., YOKOI, H., OCHIAI, K. and IKEDA, M. (2002). 'A novel methodology employing *Corynebacterium glutamicum* genome information to generate a new L-lysine-producing mutant', *Appl Microbiol Biotechnol*, **58**, 217–23.

OHNISHI, J., HAYASHI, M., MITSUHASHI, S. and IKEDA, M. (2003). 'Efficient 40 degrees C fermentation of L-lysine by a new *Corynebacterium glutamicum* mutant developed by genome breeding', *Appl Microbiol Biotechnol*, **62**, 69–75.

PARK, J. H., KIM, T. Y., LEE, K. H. and LEE, S. Y. (2010). 'Fed-batch culture of *Escherichia coli* for L-valine production based on *in silico* flux response analysis', *Biotechnol Bioeng*, **108**, 934–46.

PARKHILL, J. and WREN, B. W. (2011). 'Bacterial epidemiology and biology – lessons from genome sequencing', *Genome Biol*, **12**, 230.

PATIL, K. R. and NIELSEN, J. (2005). 'Uncovering transcriptional regulation of metabolism by using metabolic network topology', *Proc Natl Acad Sci USA*, **102**, 2685–9.

© Woodhead Publishing Limited, 2013

PICARD, F., DRESSAIRE, C., GIRBAL, L. and COCAIGN-BOUSQUET, M. (2009). 'Examination of post-transcriptional regulations in prokaryotes by integrative biology', *Compt Rend Biol*, **332**, 958–73.

SANTOS, F., BOELE, J. and TEUSINK, B. (2011). 'A practical guide to genome-scale metabolic models and their analysis', *Methods Enzymol*, **500**, 509–32.

SCHNEIDER, J. and WENDISCH, V. F. (2010). 'Putrescine production by engineered *Corynebacterium glutamicum*', *Appl Microbiol Biotechnol*, **88**, 859–68.

SCHWAGER, J., MOHAJERI, M., FOWLER, A. and WEBER, P. (2008). 'Challenges in discovering bioactives for the food industry', *Curr Opin Biotechnol*, **19**, 66–72.

SROKA, J., BIENIASZ-KRZYWIEC, L., GWÓŹDŹ, S., LENIOWSKI, D., LĄCKI, J., MARKOWSKI, M., AVIGNONE-ROSSA, C., BUSHELL, M. E., MCFADDEN, J. and KIERZEK, A. M. (2011). 'Acorn: a grid computing system for constraint based modeling and visualization of the genome scale metabolic reaction networks via a web interface', *BMC Bioinformatics*, **12**, 196.

STEVENS, M. J. A., WIERSMA, A., DE VOS, W. M., KUIPERS, O. P., SMID, E. J., MOLENAAR, D. and KLEEREBEZEM, M. (2008). 'Improvement of *Lactobacillus plantarum* aerobic growth as directed by comprehensive transcriptome analysis', *Appl Environ Microbiol*, **74**, 4776–8.

SUN, C. Q., O'CONNOR, C. J. and ROBERTON, A. M. (2002). 'The antimicrobial properties of milkfat after partial hydrolysis by calf pregastric lipase', *Chem Biol Interact*, **140**, 185–98.

TAKENO, S., MURATA, R., KOBAYASHI, R., MITSUHASHI, S. and IKEDA, M. (2010). 'Engineering of *Corynebacterium glutamicum* with an NADPH-generating glycolytic pathway for L-lysine production', *Appl Environ Microbiol*, **76**, 7154–60.

TAKORS, R., BATHE, B., RIEPING, M., HANS, S., KELLE, R. and HUTHMACHER, K. (2007). 'Systems biology for industrial strains and fermentation processes–example: amino acids', *J Biotechnol*, **129**, 181–90.

TAUPP, M., MEWIS, K. and HALLAM, S. J. (2011). 'The art and design of functional metagenomic screens', *Curr Opin Biotechnol*, **22**, 465–72.

TEUSINK, B., BACHMANN, H. and MOLENAAR, D. (2011). 'Systems biology of lactic acid bacteria: a critical review', *Microb Cell Fact*, **10** Suppl 1, S11.

TROESCHEL, S. C., DREPPER, T., LEGGEWIE, C., STREIT, W. R. and JAEGER, K.-E. (2010). 'Novel tools for the functional expression of metagenomic DNA', *Methods Mol Biol*, **668**, 117–39.

TYO, K. E. T., KOCHARIN, K. and NIELSEN, J. (2010). 'Toward design-based engineering of industrial microbes', *Curr Opin Microbiol*, **13**, 255–62.

UCHIYAMA, T. and MIYAZAKI, K. (2009). 'Functional metagenomics for enzyme discovery: challenges to efficient screening', *Curr Opin Biotechnol*, **20**, 616–22.

UMAKOSHI, M., HIRASAWA, T., FURUSAWA, C., TAKENAKA, Y., KIKUCHI, Y. and SHIMIZU, H. (2011). 'Improving protein secretion of a transglutaminase-secreting *Corynebacterium glutamicum* recombinant strain on the basis of (13)C metabolic flux analysis', *J Biosci Bioeng*, **112**, 595–601.

WILHELM, B. and LANDRY, J.-R. (2009). 'RNA-Seq-quantitative measurement of expression through massively parallel RNA-sequencing', *Methods*, **48**, 249–57.

WILLIAMSON, M. P. (2005). 'Systems biology: will it work?' *Biochem Soc Trans*, **33**, 503–6.

WISHART, D. (2008). 'ScienceDirect – Trends in Food Science and Technology : Metabolomics: applications to food science and nutrition research', *Trends Food Sci Technol*.

YANIV, M. (2011). 'The 50th anniversary of the publication of the operon theory in the *Journal of Molecular Biology*: past, present and future', *J Mole Biol*, **409**, 1–6.

YUHONG, Z., SHI, P., LIU, W., MENG, K., BAI, Y., WANG, G., ZHAN, Z. and YAO, B. (2009). 'Lipase diversity in glacier soil based on analysis of metagenomic DNA fragments and cell culture', *J Microbiol Biotechnol*, **19**, 888–97.

© Woodhead Publishing Limited, 2013

5

Production of foods and food components by microbial fermentation: an introduction

R. J. Seviour, La Trobe University, Australia and L. M. Harvey, M. Fazenda and B. McNeil, Strathclyde University, UK

DOI: 10.1533/9780857093547.1.97

Abstract: Many of the foods we eat are of microbial origin or contain constituents produced by microbial fermentation. This chapter considers the biotechnological problems associated with growing these producing strains, most of which are aerobic organisms, in large scale fermenters. It discusses how rational fermenter design is a crucial component in providing them with the physical environment for optimal desirable metabolite production, which with the filamentous fungi is often related to their morphological form. It also discusses how some of the problems in achieving this can be ameliorated by innovative fermenter configuration. Carefully selected examples are used to illustrate the practical application on an industrial scale of these philosophies and suggestions are made about how fermentation technology is likely to change in the future to satisfy the increasing demands for highly nutritious and safe foods for human consumption.

Key words: citric acid production, fermentation technology, food biotechnology, microbial exopolysaccharides, single cell protein.

5.1   Introduction

Humans have unwittingly and largely empirically been exploiting microbes for the production of foods for thousands of years and thus the food bio-technology industry is not a recent invention of western science. Evidence points to the exploitation of *Saccharomyces cerevisiae* in the production of beer, probably first by the Sumerians and the Chinese in 7000 BC (Legras *et al.*, 2007; Lodolo *et al.*, 2008). It is also clear that wine production existed in Iran in 6000 BC and in Egypt in 3000 BC, gradually spreading later to southern Europe. Production of bread in Egypt in 4000 BC has been docu-mented and records also show cheese manufacture was widespread several thousand years ago. Of course these processes, mostly anaerobic, were

© Woodhead Publishing Limited, 2013

carried out on a small scale under poorly controlled conditions with no understanding of which factors were responsible for the chemical changes taking place (Bentley and Bennett, 2008).

The availability of pure cultures and our current understanding of microbial physiology and molecular biology have led to the development of highly controlled fermentation processes. In the early days of the fermentation industry, many metabolites were produced industrially by solid state fermentations (Krishna, 2005) or static surface liquid culturing techniques (Bentley and Bennett, 2008), with all the accompanying disadvantages associated with an inability to standardize, monitor and control these systems precisely. Product recovery can also be problematic. Far more sophisticated now, solid state fermentations are still used profitably for production of some foodstuffs including citric acid (Krishna, 2005). However, we exploit microbes for their large-scale commercial production, mainly using aerobic submerged axenic culture fermentation processes carried out in highly sophisticated bioreactors, where exocellular products are more readily recoverable (Bentley and Bennett, 2008). Culture conditions are controlled meticulously and often manipulated, steps necessary to ensure an economically attractive high product yield. Furthermore, the manufacturing process can better guarantee a consistent product of a suitable quality to satisfy the increasingly stringent food safety standards needed for products for eventual human consumption (Seviour et al., 2011b).

Bull et al. (1979) suggested what the main stimuli for innovation in the fermentation industry have been and each of these has clearly played an important direct or indirect role in food biotechnology advances, as this chapter will hope to demonstrate. They include:

- Availability of large amounts of a cheap fermentable substrate, often a waste product of an existing industry whose disposal otherwise is costly. For example, the exploitation of n-alkanes and methane and methanol, waste products of the oil refinery, as substrates for the production of protein rich microbial biomass (single cell protein, SCP) for animal and human consumption led to an outburst of research activity into the basic microbiology of organisms capable of utilizing these substrates, but also the design and construction of highly novel chemostat fermentation systems able to grow these organisms at very high levels of productivity (Mateles, 1979; Olsen and Allermann, 1987).
- Scarcity of a product of high economic value. Citric acid, widely used as an additive in many foods (see other chapters), is a clear example of this. The only source of citric acid for nearly a century was from lemons imported solely from Italy, who thus monopolized global production. Hence price and supply could be manipulated to ensure demand was never met (Papagianni, 2007; Sauer et al., 2008). Then the discovery in Belgium that pure cultures of Aspergillus niger could be encouraged to synthesize this organic acid in large amounts changed the politics of its

© Woodhead Publishing Limited, 2013

production totally. However, nearly a century later, its supply still fails to meet demand and much secrecy surrounds its industrial production methods by fermentation. More details of the microbial production of organic acids, including citric acid, are given in Chapter 12 by Sauer and colleagues.

- Exploitation of novel products. Without doubt the requirement to grow in the aerobic fungus *Penicillium chrysogenum* in submerged culture for penicillin production in the 1940s gave a then moribund fermentation industry a massive injection of new creative energy and the incentive to understand better the role of bioreactor design in microbial metabolite production (Kardos and Demain, 2011). The increasing importance of microbial enzymes and heterologous proteins in foods and food production has had an equally profound influence on downstream methods of product recovery.

5.2   Food and food ingredients produced by microbial fermentation

There are clear advantages in producing foods and food ingredients from microbes grown in bioreactors, not the least being that an uninterruped supply of a safe and consistent product is possible, which is not always guaranteed with those obtainable from natural sources, where climate and seasonal changes can affect yield and quality. Process development is an interdisciplinary activity, with contributions from microbiologists and engineers to convert an initial small-scale laboratory system into one operating effectively on an industrial scale. Their joint aims are to optimize process yield and productivity, but in the past this exercise was often undertaken with little discussion or interaction between them (Mateles, 1979) and the physiology of the organism was largely ignored in bioreactor design.

The role of the microbiologist is to discover the strain of interest by a suitable screening protocol and then to optimize its performance in terms of how its desired attribute is affected by culture conditions, medium composition, pH, temperature, pO_2 and so on (Seviour *et al.*, 2011b). Generally this optimization is performed empirically since rarely do we understand the physiology of metabolite overproduction or its regulation of synthesis at the molecular level. Citric acid production by *A. niger* is a notable exception, although considerable uncertainties still exist about the biochemistry of the system (Karaffa and Kubicek, 2003; Goldberg *et al.*, 2006; Papagianni, 2007). This situation is likely to improve as whole genome sequencing of organisms of industrial importance is becoming more common and the information forthcoming will be invaluable. Although several *Aspergillus* genomes have now been sequenced (Pel *et al.*, 2007; Meyer *et al.*, 2011), allowing genomic information to relate to strain physiology, this information illustrates not only how metabolically versatile this fungus is, but it also

© Woodhead Publishing Limited, 2013

points the way towards reverse engineering approaches for developing novel high citric acid producers. This would require a molecular comparison of a high citric acid-producing strain with a low producer. Differences between them should reveal any target genetic modifications needed to create new strains. A word of caution is needed here, as this structured metabolic engineering approach has often been attempted with *A. niger* with the aim of generating high citric acid accumulating strains. Generally the results have been disappointing. Some attempts commence with modest citric acid accumulating strains, perform suitable genetic alterations and then announce high percentage increases in citrate to levels which, in industrial production terms, are still unconvincing. Citric acid production in *A. niger* submerged culture is perhaps an ideal model system to demonstrate that genetic potential is not the end of the citrate production story. Very precise control of culture physiology achieved by manipulating the feed, intelligent and informed fermenter design and operational conditions, and culture morphology are all absolutely essential to achieve high yields of citric acid. These aspects are covered in greater detail elsewhere in this book. Thus, while in Chapter 2 Vongsangnak and Nielsen deal with molecular and metabolic engineering of fungi in general, Sauer *et al.* in Chapter 12 discuss the impact of such changes on organic acid synthesis from fungi.

Whole genome sequence data are also available for the curdlan producing *Agrobacterium* sp. strain ATCC 31749 (Ruffing *et al.*, 2011) where nitrogen limitation is essential for production of curdlan (Zhan *et al.*, 2012) and the scleroglucan producing *Schizophyllum commune* (Ohm *et al.*, 2010) (see later).

Whether metabolite formation is growth or non-growth-related will have an impact on deciding what fermentation system is appropriate to use. Thus, a choice between batch, semi-continuous fed batch and continuous culture systems will need to be made, but with the exception of SCP production (Mateles, 1979; Trinci, 1992, 1994; Wiebe, 2002) where biomass productivity is much higher and downtimes shorter (Bull 2010), batch and fed batch culture systems are generally preferred even for growth-related processes. Reasons include an innate conservatism in the fermentation industry and a higher risk of contamination in continuous open culture systems. However, chemostat conditions may select for faster growing and, in some cases, morphological mutants (Bull, 2010; Wiebe, 2002). Their eventual dominance may lead to process shutdown where a stable organism morphotype is critical, as detailed later for the mycoprotein Quorn® (Wiebe, 2002).

Safety is also a key issue, especially for foods for human consumption produced involving fungi. Many fungi produce mycotoxins some of which are highly toxic to humans, so part of any initial screening would be directed at eliminating all strains where these are detected and all fermentations where fungi are used are sampled over their course for mycotoxin assays.

Somewhere in the evolution of any such process, the skills of the geneticist/molecular biologist will be exploited, to increase yields to levels where

© Woodhead Publishing Limited, 2013

the fermentation will become more economically attractive. Hence the ultimate adoption of the process may depend ultimately on how successful this is. In the early days of fermentation, strain improvement involved empirically based programs of mutation and selection of surviving mutants showing increased yields after screening, usually without understanding the physiological/biochemical bases for these improvements. *In vitro* genetic recombination methods are much more targeted in their application and can be decisive, as a few examples will illustrate. After replacing the genes encoding the ATP-requiring glutamine synthase/glutamine:oxoglutarate aminotransferase (GS/GOGAT) in the bacterium *Methylophilus methylotrophus*, which was the organism chosen by ICI for the production of SCP Pruteen, with that encoding the enzyme glutamate dehydrogenase for ammonia assimilation, cell yields improved by up to 5% (Gow *et al.*, 1975). However, regulatory authorities in general are nervous of approving genetically modified organisms for food manufacture and, in addition, many attempts at metabolic engineering of industrial strains (Sauer *et al.*, 2008) reportedly fail during the scale-up process. In addition, in many countries, public mistrust and unease about the involvement of any genetically modified microbes for food, feed or food components is deeply ingrained, adding to any regulatory body's caution.

Yet innovative cutting edge science does not by itself guarantee commercial success and sometimes fermentation-based production costs cannot compete with those associated with obtaining the same product from natural sources, as the story of SCP production, told later, demonstrates. Often the need to rely on expensive substrates for metabolite production can make the process economically unviable (Sauer *et al.*, 2008), as seen with citric acid production in Australia, where the need to use expensive sugar cane sucrose molasses led to its eventual demise. A similar pattern was noted for citric acid production in the EU resulting from elevated sugar prices.

5.3   Principles of bioreactor design and operation

The problems encountered in satisfying the needs of any axenic aerobic fermentation were discussed briefly above. These are difficult enough to overcome in small laboratory scale bioreactor systems, but become major challenges in industrial scale reactors.

Many challenges need to be met by the engineer, preferably working closely with the microbiologist who alone understands the physiological and biochemical details of the organism to be cultured, before a promising small-scale laboratory process can be brought to fruition on an industrial scale by scale up (Hewitt and Nienow, 2007), and the conditions ideally suited to the chosen culture reproduced. Maintaining the chemical environment in a stable state is not difficult, but replicating the same physical

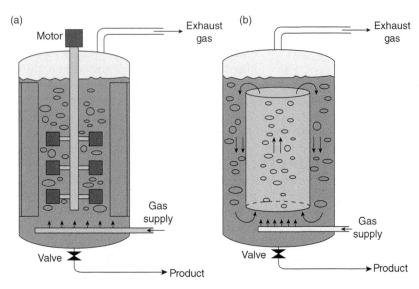

Fig. 5.1 Diagrammatic representation of different bioreactor designs. (a) Continuous stirred tank reactor (CSTR), (b) airlift reactor, adapted from Williams (2002).

conditions established in a small laboratory reactor in a vessel many thousand times larger is a problem which taxes the biochemical engineer. The highly complex problems in large scale industrial bioreactors of maintaining sterility, satisfying adequately the oxygen demands of aerobic cells, improving bulk liquid mixing to ensure culture homogeneity and controlling the physical environment (Chisti, 2006; Garcia-Ochoa and Gomez, 2009; Seviour *et al.*, 2011b) have encouraged the design of a wide range of bioreactors, where these aims are achieved more efficiently than in conventional baffled continuously stirred tank reactors (CSTR) (Fig. 5.1).

All industrial bioreactors are manufactured from expensive high quality stainless steel, constructed using polished welds in such a way that they can be cleaned efficiently and no reservoirs of hidden contamination can develop (Chisti, 2006; Matthews, 2008). They must allow for the aseptic additions of inocula and other solutions, including those required for pH and foam control and for aseptic sample removal. With continuous culture systems, opportunities for medium addition and removal need to be incorporated. The vessels must be mixed efficiently to achieve chemical and physicochemical homogeneity of reactor contents and to eliminate any stagnant regions where cells may suffer irreversible starvation of substrates, especially oxygen for aerobic organisms (Chisti, 2006; Harvey and McNeil, 2008). CSTRs use mechanical agitation, where impellers attached to a rotating drive shaft, its entry point a serious contamination risk, pump the bioreactor contents around. Rushton impellers, commonly used in CSTRs represent high shear mixing systems, capable of damaging shear sensitive

© Woodhead Publishing Limited, 2013

cells, although a wide range of impeller designs, some low shear, are avail-able (Chisti, 2006). Bulk mixing efficiency is highest with a slow moving large diameter single bladed impeller, especially for shear thinning and viscous systems like xanthan, gellan or high cell density fungal fermentation broth (Meyer *et al.*, 2011).

The agitation system has a second important function, in dispersing the continuously injected air as a fine stream of small bubbles to satisfy the oxygen demands of the growing cells, with baffles to generate turbulence. This aim is best achieved with fast moving multiblade small diameter impel-lers, emphasizing that all such configurations are a compromise between optimizing mixing and air supply efficiencies (Gibbs *et al.*, 2000). In fact some fermenters have been designed with two agitation systems, each designed for the separate functions, but of course with double the contami-nation risk (Gibbs *et al.*, 2000; Solomons, 1980).

A wide variety of impeller designs have been used in attempts to improve mixing efficiency (Matthews, 2008; Seviour *et al.*, 2011b). Most large indus-trial scale bioreactors are mixed by multiple impellers, whose individual spacings on the driveshaft are crucial. If too close together they act as a single mixing device, while if too far apart, unmixed stagnant regions may form in the vessel. The empirical recommendation is for these to be one impeller diameter apart and about one-third the vessel diameter above the vessel base for the bottom impeller (Chisti, 2006). Their relative dimensions are also important in attempts to optimize mixing efficiency and a diameter about one-third that of the vessel diameter is often chosen for Rushton impellers (Chisti, 2006). For viscous fungal and filamentous bacterial cul-tures (e.g *Streptomyces*) an impeller with a diameter of 0.5 vessel width has some reported advantages in terms of bulk mixing efficiencies. This point again illustrates that an ideal bioreactor design and operational strategy does not exist; most represent a compromise between competing requirements.

Some bioreactors rely solely on gas velocity to mix their contents. These are low shear systems and include airlift bioreactors, where mixing is achieved by establishing a difference in hydrostatic pressure between the aerated and non-aerated physical compartment, with only the former (draught tube) being aerated, reducing broth density there (Fig. 5.1). A large headspace allows efficient gas disentrainment, so there is minimal gas carry over to the non-aerated downcomer. Mixing is less efficient than that achieved in mechanically stirred bioreactors, but with no drive shaft entry points, airlift reactors are less susceptible to contamination and cheaper to operate and construct (Merchuk, 1990; Merchuk *et al.*, 1994). This is an important consideration especially for any product intended for food use where profit margins are low by comparison with therapeutic or diagnostic proteins. Mixing can be improved by increasing gas velocity, but eventually individual air bubbles coalesce into 'slugs', with concomitant decreases in volumetric oxygen gas transfer rates (see below). This problem can be

© Woodhead Publishing Limited, 2013

rectified to some extent by incorporating horizontal baffles or sieve plates into the vessel to redisperse these slugs, but this runs the eventual risk of further increases in gas velocity of emptying vessel contents.

Probably the main challenge in industrial scale bioreactors is to satisfy the oxygen (O_2) demands of aerobic organisms, otherwise growth rates and metabolite yields may be reduced (Garcia-Ochoa and Gomez, 2009; McNeil and Harvey, 1993; Seviour *et al.*, 2011b). O_2 is only sparingly soluble in aqueous solution (8.69 mg L^{-1} at 25°C) and the bioreactor needs to be designed and operated to maximize its mass transfer rate, N_a (kg O_2 m^{-3} h^{-1}):

$$N_a + k_L a\ (C^* - C_L)$$

where $k_L a$ = O_2 mass transfer coefficient (h^{-1}), a = air/liquid bubble interfacial surface area (cm^2 cm^{-3}), C^* = O_2 equilibrium concentration at liquid / air bubble interface (kg O_2 m^{-3}) and C_L = O_2 concentration measured in the bulk liquid (kg O_2 m^{-3}).

To achieve this, the following options are available (Garcia-Ochoa and Gomez, 2009; Matthews, 2008);

a. Increase $k_L a$ whose determination is problematic (Patel and Thibault, 2009). Many factors, including medium composition, presence of oils, surfactants or antifoams, operating temperature, culture rheology and nature of the metabolite produced affect $k_L a$, often negatively, and since the reasons for the decrease are not often known, most are not readily manipulated or controlled. Increasing agitation rates or gas velocities (both contributing to improved mixing) are probably the favoured approaches, but these too are not always appropriate (Patel and Thibault, 2009) especially for shear sensitive cells.
b. Increase a by decreasing the bubble volume, and avoid formation of slugs, an outcome increased at high bulk liquid turbulence.
c. Maximize ($C^* - C_L$) by increasing C^*. This may be achieved by replacing air with O_2 or O_2 enriched air, which is expensive, or by pressurizing the bioreactor to >1 atmosphere, again a strategy requiring expensive modifications. Both will increase C^* by increasing the bubble pO_2, an outcome achievable by appropriate bioreactor design. Using a tall, thin vessel with a high height to diameter aspect ratio (for CSTRs this is usually 4–5 : 1) ensures a high hydrostatic pressure exists at its base, the location for injection of air. Hence C^* will be substantially higher than in a short fat vessel. This configuration may also prolong the escape time of individual gas bubbles as they rise, allowing longer for O_2 transfer across the air–liquid interface. However, bubble pCO_2 will also increase concomitantly, reducing C^*.
d. Maximize ($C^* - C_L$) by reducing C_L, a strategy not to be recommended since the risk of establishing O_2 starvation conditions exists.

Much still needs to be learned about the influence of any physical parameters of the bioreactor on organism behaviour and especially the link

© Woodhead Publishing Limited, 2013

between growth conditions, culture morphology and the consequences for metabolite production (Papagianni, 2004; Junker *et al.*, 2004). The systematic approach to understanding the morphogenetic machinery of *Aspergillus* by applying tools of modern systems biology by Meyer and colleagues (Meyer, 2008; Meyer *et al.*, 2011) may help resolve these complexities.

These are not trivial considerations, especially when filamentous fungi are the organisms used in the fermentation. Thus, submerged fungal cultures growing as a pulpy diffuse mycelium exhibit a highly viscous non-Newtonian broth rheology (Wucherpfennig *et al.*, 2010), quite different to the non-viscous Newtonian rheology of a pelleted fungal culture medium, a form thought to be associated with increased yields of many fungal metabolites (Gibbs *et al.*, 2000; Grimm *et al.*, 2005; McNeil and Harvey, 1993; Seviour *et al.*, 2011a,b). As well as important physiological differences between the pulpy mycelium and pelleted growth, efficient mixing of the culture becomes difficult to achieve in the pulpy mycelium (Wucherpfennig *et al.*, 2010). If the fungus also synthesizes viscous metabolites like the exopolysaccharides used in the food industry (see later), these problems are exacerbated substantially (Seviour *et al.*, 2011a,b). Hence an ability to run bioreactors where culture morphology can be manipulated and controlled, in cases where it is known to increase metabolite yield, and avoid the operational problems associated with non-Newtonian broth rheology is highly desirable but as yet largely unachievable (Gibbs *et al.*, 2000; Junker *et al.*, 2004; Papagianni, 2004; Wucherpfenning *et al.*, 2010).

Attempts to deliberately engineer fungal morphology by microparticle addition have been successful in increasing enzyme production in *A. niger* (Driouch *et al.*, 2010), but this novel approach has not been applied widely to fungal fermentations. Driouch *et al.* (2012) also showed that addition of titanate microparticles of about 8 μm diameter to suspended cells of *A. niger* enhanced substantially fructofuranosidase and glucoamylase production. This work certainly demonstrates the potential of using such supports to tailor fungal morphology and hence maximize mass transfer into fungal pellets. It seems to have important implications for many industrial fungal processes based on a desirable pelleted morphology, including citric acid synthesis.

If the filamentous organism itself is the product of the fermentation, and its morphology is critical to its eventual application, as with *Fusarium venenatum* and mycoprotein SCP (Wiebe, 2002), then strict morphological control is imperative.

5.4   Examples of fermentation processes used for the production of foods and foodstuffs

The fermentation industry guards the details of its process secrets very closely, which makes this section difficult to write. Consequently, most of

© Woodhead Publishing Limited, 2013

what we know about how innovative fermentation technology might improve biomass or metabolite yields and productivity is derived from scientific publications emerging from academic research laboratories. The industry is inherently highly conservative and, with some exceptions discussed later, is generally unlikely to invest in expensive new major unproven fermentation equipment as a means to increase profit margins. This is especially the case in highly competitive markets with bulk chemicals like citric acid. Instead, cost savings are directed mainly at improving or changing the producing strain, which may also reduce the synthesis of other undesirable metabolites, or by using cheaper production substrates and product recovery. However, these can increase the subsequent purification costs, increasing the already substantial financial impact of downstream processing on profits (Sauer *et al.*, 2008). The scientific literature is full of reports claiming improved performances from bioreactors of a wide range of configurations (e.g. Cheng *et al.*, 2010) but how widely these have been adopted by industry is largely unknown.

5.4.1 Single cell protein production

People have consumed microbes as a source of food probably for thousands of years, harvesting fruiting bodies of *Basidiomycota* and *Ascomycota*. As well as having high nutritional value, many of these are now known to be highly beneficial to human health since they contain non-cellulosic β-glucans, which possess powerful anti-tumour and antimicrobial properties through their immunomodulating abilities (Chen and Seviour, 2007; Fazenda *et al.*, 2008; Seviour *et al.*, 2011a,b). Producing microbial biomass by fermentation in order to supply cheap protein rich food stocks to satisfy the increasing demands of animals and humans has long held attractions for industrial microbiologists (Olsen and Allermann, 1987; Anupama and Ravindra, 2000). Euphemistically referred to as single cell protein (SCP) to overcome the uninformed prejudices of consumers, these were intended mainly to tackle the increasing problems of relying on conventional agricultural techniques to supply expensive protein sources like soya meal or fishmeal for animal production to generate meat to feed an ever-expanding global population.

High growth rates mean high levels of proteins containing most of the essential amino acids like lysine, although sulphur-containing amino acids like methionine are often present only in low amounts. However, high RNA levels, also a consequence of high cell growth rates, can lead to increases in uric acid levels in consumers and are a characteristic of SCP. These levels need to be reduced during production (Mateles, 1979). Vitamin contents are generally adequate, but extensive and costly long-term nutritional and toxicological trials are demanded by all government bodies before any SCP product can be released onto the market.

© Woodhead Publishing Limited, 2013

As mentioned above, economics determines that fermentations are carried out on a large scale under continuous conditions where higher cell productivity than is achievable under batch conditions is possible. To ensure cells are grown close to their μ_{max} values, these aerobic industrial scale fermentations require bioreactors designed to achieve very efficient mixing of the medium to provide extremely high oxygen mass transfer rates, to dissipate the high levels of heat generated under these culture conditions and disperse rapidly and homogenously continuously added substrates allowing their rapid assimilation by the cells. These design features are discussed in more detail below.

Although yeast cells had been grown as a human food in wartime countries, the level of interest increased substantially in the 1960s. Despite successful industrial process development and product safety outcomes, the initial efforts using waste n-alkanes from the petroleum industry as cheap substrates for yeast SCP production for animal consumption by BP in Europe under the brand name Toprina (Olsen and Allermann, 1987) failed to materialize. Planned industrial production of 100,000 tonnes in Italy using *Candida lipolytica* and in Japan was met with considerable government and public resistance. Perceived risks associated with consuming products of petroleum fractionation were fatal to any planned industrial exploitation and despite its proven potential, the process had been abandoned by 1980 (Olsen and Allermann, 1987).

However, other processes based on using methane, another waste product of petrol refineries, as substrate were also under development. Although it has its attractions, there are safety and operational risks associated with using methane as a substrate, and methanol, readily converted from it, was soon preferred. A wide range of bacteria, the methylotrophic bacteria, can use methanol as their sole source of carbon and energy and their possible adoption as a source of SCP stimulated considerable research interest in understanding their basic biochemistry and physiology. Much of the data generated was applied by ICI to the development of commercial production of 'Pruteen' using *Methylophilus methylotrophus* chosen from about 10,000 candidate strains, for a planned output of about 50,000 tonnes per annum.

ICI designed a revolutionary 'pressure cycle' bioreactor based on an airlift configuration, since the power demands required for effective mechanical agitation for a vessel of the size required (1000 m^3 working volume) were not achievable. Considerable input from microbiologists into its design was an equally novel strategy, but considered essential in optimizing the conditions needed to support a high growth rate of the selected SCP bacterium. The main features, shown in Fig. 5.2, illustrate how many of the concepts mentioned above were incorporated into its design. Thus the vessel was configured deliberately to be tall (about 45 m) and slim, ensuring a very high hydrostatic pressure and hence C^* at its base at the point of aeration. The reactor was operated aseptically as a chemostat, and to reduce

© Woodhead Publishing Limited, 2013

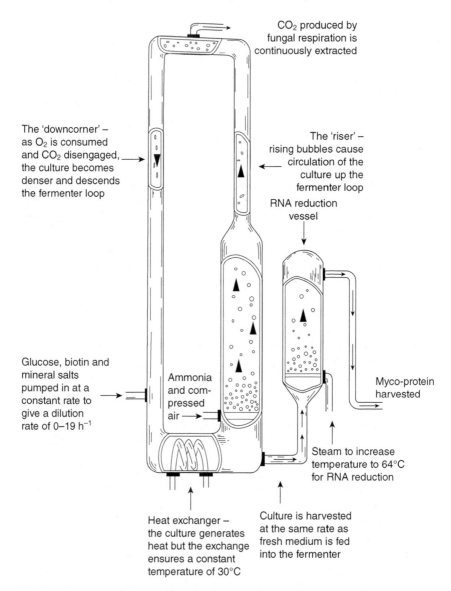

Fig. 5.2   Pressure cycle reactor used in the production of Quorn™ mycoprotein (adapted from Kavanagh, 2005).

cooling costs in a highly exothermic fermentation, the reactor was run at 35–40°C, at which the Pruteen bacterium grew best. The considerable problems of achieving efficient mixing in such large airlift vessels were ameliorated by employing several thousand inlets for the fresh medium. This also avoided any problems of methanol toxicity to the culture encountered if

© Woodhead Publishing Limited, 2013

added from a single addition point (Olsen and Allermann, 1987). Horizontal baffles were also incorporated to assist in mixing and air bubble dispersal (see above). The bioreactor was referred to as a pressure loop, since cells cycle through regions of high and low hydrostatic pressure (in this case about 5 atmospheres) and so these changes must have no effect on cell yield. Because of the small size of the bacterial cells and large process volumes, filtration was not suitable for biomass recovery and so a flotation process involving a preconcentration of cell mass by a factor of about ten, followed by centrifugation was used to harvest the biomass and the cooling water was collected and recycled (Olsen and Allermann, 1987).

Yet despite representing 'state of the art' bioreactor design and fermentation technology, the Pruteen process no longer operates, a victim of harsh commercial reality. It could not compete economically as an animal feedstock with natural proteins and so in 1994 ICI closed it down. However, the 'pressure cycle' story and ICI's involvement in SCP does not end there. Essentially the same technology is now used to produce Quorn®, a fungal SCP (mycoprotein) using *Fusarium venenatum* (previously named *Fusarium graminearum*). The strain (A 3/5) was selected from 3000 isolates mainly from soil on the basis of its high protein content, its safety (no mycotoxins could be detected, although many *Fusarium* species produce them), high growth rate and yield coefficient of glucose, and desirable organoleptic properties suiting its use as a meat substitute (Trinci, 1992). The evolution of this mycoprotein process has been reviewed in detail by Trinci (1992; 1994) and Wiebe (2002), and so only a brief account is given here.

Mycoprotein was first considered in the 1960s by Rank Hovis McDougall (RHM), producers of flour, as a potential source of protein. Initially wheat starch, a waste product, was used as the substrate, but was subsequently replaced with food grade glucose generated from the wheat starch. Choosing a fungus for this purpose instead of a bacterium was deliberate, since RHM thought consumer prejudice against it would be much less. Biomass recovery is also much easier for fungi than bacteria. Mycoprotein was always intended for human consumption, but the high production costs ensured it would never be suitable to feed people in poor parts of the world, as originally intended. Instead Quorn® is now marketed as a high value product targeting health conscious people and represents the last survivor of the many SCP production programs first undertaken in the 1960s (Trinci, 1994; Wiebe, 2002). Its success and popularity can be gauged by the fact that it is now available as a high protein low cholesterol meat substitute in many countries around the world (www.quorn.com) with many other desirable nutritional attributes (Wiebe, 2002). Its safety for human consumption is unquestioned after intensive stringent and prolonged toxicological and dietary evaluations (Trinci, 1992).

The same fermentation philosophy discussed previously for SCP processes generally was adopted for Quorn® production, namely, continuous

© Woodhead Publishing Limited, 2013

culture systems were preferred to batch culture since biomass productivity is increased by maintaining an exponential growth rate (five times that achievable in batch culture; Trinci, 1992) and more reproducible culture conditions are provided. One unique feature of the Quorn® process is the need to control the fungal morphology, especially the branching frequency of the growing fungal fragments.

Quorn® was produced by RHM in its early days in continuous culture under glucose limitation with ammonium as nitrogen sources using CSTRs (1.3 m³ capacity) fitted with modified impellers to improve mixing and to satisfy the high oxygen demands of the fungus (0.78 g O_2 for every gram biomass) needed to maintain a high growth rate and avoid production of ethanol as a fermentation end product (Trinci, 1992). The reactor operated at pH 6 and 28–30°C. Then, after product safety approval in the UK in 1980, the formation in 1984 of Marlow Foods, a joint venture between RHM and ICI, led to further scale up. This involved carrying out production in a pressure loop fermenter (40 m³), constructed originally by ICI as the pilot plant for its Pruteen process (Trinci, 1992, 1994; Wiebe, 2002). Further product demand in a rapidly expanding international market over the past 25 years saw a further substantial production increase in newly constructed highly sophisticated 140,000 m³ working volume pressure loop bioreactors by 1996, where rates of CO_2 evolution are used to control the feed rate of glucose. Wiebe (2002) states under these conditions, that *F. venenatum* has a growth rate of 0.17–0.20 h⁻¹ producing 300–350 kg biomass h⁻¹.

When recovered from the fermenter, the level of RNA in *F. venenatum* is about 8–9% (Trinci, 1992), which needs to be reduced to about 1% before it is considered suitable for human consumption in the amounts envisaged (see above). A heat shock process on the effluent culture from the pressure loop reactors, originally involving rapidly increasing temperature to 64°C for about 30 min allows RNAses to act and leads to rapid decreases in RNA levels. As Trinci (1994) and Wiebe (2002) emphasize, this treatment step is costly, since substantial amounts of biomass are lost (35–38%) and protein levels fall. However, these losses can be reduced by increasing the incubation temperature to 72–74°C (Wiebe, 2002). After heating to 90°C, the biomass is harvested by centrifugation yielding a paste with a solids concentration of >20% (w/w), mixed with egg albumen as a binding agent and then texturized to align the individual hyphae into a meat-like product (Wiebe, 2002).

In theory, continuous cultures can be operated indefinitely (Bull, 2010), but production of Quorn® is stopped after 1000–1200 hours because of the consistent eventual dominance of 'colonial' morphological mutants, where the hyphal fragments are more highly branched than those of the wild type (reviewed by Trinci, 1994 and Wiebe 2002). This phenotypic change adversely affects the required organoleptic properties of Quorn® having a 'crumbly' not the desirable 'chewy' texture (Trinci, 1994). Colonial mutants isolated from laboratory chemostats do not always behave kinetically in the same

© Woodhead Publishing Limited, 2013

way as those obtained from the full scale bioreactor, probably reflecting differences in their mixing efficiencies although more than one mutant genotype appears to exist. Wiebe (2002) discusses several strategies for their control in the bioreactor, involving reducing the dilution rate, changing the nitrogen source or reducing the pH, not all of which are appropriate for high biomass production rates. Alternatively periodically changing the selective pressures thought to favour their competitive advantages has been suggested as an approach to their control. Success has been reported in evolving the A3/5 wild type into morphologically more stable strains, but whether these are used currently for Quorn® production is not known. The interested reader is encouraged to read Wiebe (2002). Since fungi are apically growing branching organisms, any culture system operated continuously at high μ is likely to favour mutants with higher branching frequencies and may well outgrow their rivals. Marlow Foods simply take the pragmatic view and stop the process when the proportion of mutants becomes problematic.

5.4.2   Citric acid production

As mentioned earlier, citric acid was once recovered from lemons, but is now produced by microbial fermentation, using surface, solid state and largely submerged culture methods (Goldberg *et al.*, 2006; Papagianni, 2007; Soccol *et al.*, 2006). Although many microbes synthesize citric acid in large amounts (Papagianni, 2007), *A. niger* is by far the major industrial producer and this section will concentrate on it. As much as 95 g citric acid can be produced from 100 g of sugar substrate (Karaffa and Kubicek, 2003). Other industrial producers include *Yarrowia lipolytica* and *Candida* spp., but despite their advantages, which include attractively high productivities, (Goldberg *et al.*, 2006), as discussed in more detail later, these are probably not used widely on an industrial scale.

Production of citric acid, most of which is used as an additive in the food industry (Soccol *et al.*, 2006) is a highly competitive and expanding fermentation industry, with current (2010) production exceeding 1.7 million tonnes (Rywinska and Rymowicz, 2010) and a still unsatisfied demand of about 3.5–4% per annum. Most is now produced in China and many of the smaller producers have dropped out because of high production costs and falling prices (Soccol *et al.*, 2006). It is an excellent example of an industry where process details are not made available by companies to academic research workers interested in understanding this process better. Thus, fermentation conditions and strain improvement strategies are not in the public domain (Goldberg *et al.*, 2006). Nor is it clear how much of the publicly available bioreactor based research data have been adopted by the industry. It appears that many of these data have not been generated by the hyper citric acid producing *A. niger* strains used for its industrial production (Karaffa and Kubicek, 2003; Papagianni, 2007). Consequently, none of the excellent

© Woodhead Publishing Limited, 2013

reviews available on citric acid can provide much detailed information on industrial production details.

However, it is now generally agreed that citric acid overproduction in *A. niger* only takes place under a set of highly specific culture conditions. These were determined empirically, but most can be explained in basic biochemical terms from the more recent but still incomplete unravelling of the complex primary metabolism of high citrate yielding *A. niger* strains. In this organism, compartmentalization of metabolism adds an interesting dimension to its production, since the citrate synthase producing citrate is located in the mitochondrion. The oxaloacetate is formed in the cytosol and then transported into the mitochondrion before being involved in a condensation reaction with mitochondrial acetyl CoA (derived from decarboxylation of pyruvate also produced in the cytosol). Thus, the citrate has to be transported out into the cytosol before excretion into the medium (Karaffa and Kubicek, 2003). This step, rather than inhibition of an individual enzyme creating a 'bottleneck' in the tricarboxylic acid (TCA) cycle, probably controls citric acid production rates in *A. niger,* as does its transport out of the cell.

What is known about the biochemistry of its overproduction has been reviewed critically by Karaffa and Kubicek (2003), Legisa and Mattey (2007) and Papagianni (2007). By general agreement, for high yields, the following all appear to be required:

* High concentrations of rapidly utilized and preferably cheap carbon (molasses sucrose?) source (>50g l^{-1}) and its very rapid assimilation by *A. niger* by passive diffusion;
* High dissolved oxygen concentrations, which have to be maintained throughout the fermentation to achieve high product yields;
* Low pH (<2.5), which inhibits synthesis of oxalic acid by this fungus;
* Limiting availability of trace metal ions, especially Mn^{2+};
* A sub-optimal concentration of phosphate;
* Unrestricted metabolic flux through the Embden-Meyeroff-Parnas pathway ensuring unhindered provision of TCA cycle intermediates;
* Uncoupled reoxidation of NADH generating lower levels of ATP and hence lowered shunt of TCA intermediates into anabolic reactions;
* Anaplerotic cytosolic pyruvate carboxylase activity capturing generated CO_2 into oxaloacetate to replenish the TCA cycle has an important role, but other evidence suggests that citrate is also formed from previously accumulated cellular glycerol and erythritol.

How extensively this fundamental information has been exploited by industry is unknown. Furthermore, our understanding of the biochemistry of citrate overproduction by other potentially industrial organisms like *Candida* spp. and *Y. lipolytica* is not as well understood, except that again conditions of carbon excess and nitrogen limitation are required (Anastassiadis *et al.*, 2002, 2005; Rymowicz *et al.*, 2010 ; Rywiñska *et al.*, 2011).

© Woodhead Publishing Limited, 2013

The *A. niger* citric acid fermentation is an example of one where a clear link exists between cell morphology in metabolite production (Papagianni, 2007), although often the evidence from many studies is contradictory. This is not too surprising in such a complex process, with so many interacting variables where 'cause' and 'effect' relationships are difficult to resolve. Culture morphology determines both rheology, which in turn affects bioreactor mixing efficiencies and crucially oxygen mass transfer rates (Seviour *et al.*, 2011a; Garcia-Ochoa and Gomez, 2009), which also affects metabolite production. Which fungal morphotype develops depends on several factors including the strain used, its growth rate, spore inoculum concentration, culture medium, pH and in some, but not all studies (Papagianni *et al.*, 1998), the agitation speed used in the fermentation (Papagianni, 2007).

Many early reports used subjective qualitative descriptions of morphotypes, but applying image analytical methods (Papagianni *et al.*, 1998, 1999; Paul *et al.* 1999) has allowed quantitative assessments of culture morphology. In fact Paul *et al.* (1999) suggested that the industry might benefit by using such methods to monitor their large scale citric acid fermentations. Whether this idea has been adopted is unknown, but the general view now is high citric acid yields are associated with *A. niger* growing as vacuolated, highly branched hyphae organized into small clumps with 'hairy' edges. The sophisticated methodology described by Paul *et al.* (1999) is unlikely to appeal to industrial producers of citric acid when a more rapid microscopic description of fungal macromorphology (pellet size, shape, roundness, roughness) is probably adequate to allow process control decisions to be taken.

Details of the bioreactor designs used for commercial citrate production are not made available to academic researchers, but the requirements for sufficiently high productivities to allow companies to compete in the market might suggest that CSTR and airlift configurations with modified agitation/aeration systems are widely used, since the costs of investing in new and largely untested bioreactor configurations are prohibitive. The present authors' experience is that most industrial bioreactors have some form of mechanical agitation, but airlift systems are also in use at some sites. As expected, operators of both types insist their system is the best, and that only the morphology associated with their particular strain and the fermentation conditions they use should be considered optimal for citric acid production. This is despite much published evidence to the contrary. Continuous culture systems have been described for citrate production operating under nitrogen limitation and at very low pH, 1.8 to 2.0 (Kristiansen and Sinclair, 1979), but these are unlikely to contribute substantially at an industrial level. Chapter 12 on organic acid production by Sauer and Marx in this volume should be consulted for a more detailed discussion.

Equally likely is that this industry with its continued need to reduce production costs has probably considered using substrates other than the

© Woodhead Publishing Limited, 2013

usual sugar cane or beet molasses. Depending on their source, both of these might need some purification to reduce trace metal levels below those that inhibit citric acid production (Soccol *et al.*, 2006). *A. niger* can utilize a wide range of substrates (Dhillon *et al.*, 2011) under both submerged culture and solid state fermentation conditions (see Soccol *et al.*, 2006 for summary), but again details on which, if any, of these is used industrially are unavailable.

The search for cheaper substrates has led to interest in other citrate overproducing organisms, especially *Y. lipolytica* and *Candida* species. Some *Y. lipolytica* strains can use a diverse range of carbon sources, among which is glycerol, a product of the biodiesel industry, which competitively and consistently supports at laboratory scale high citrate yields under a range of fermentation conditions including repeat fed batch systems (Moeller *et al.*, 2011; Rywinska and Rymowicz, 2010; Rywinska *et al.*, 2011). The main disadvantage of using these organisms is that isocitrate is also synthesized, often in large amounts. As well as being a less attractive food additive, its removal increases purification costs.

5.4.3   Exopolysaccharide production

Many microbes, both bacteria and fungi synthesize large amounts of chemically diverse exopolysaccharides under appropriate culture conditions, which often differ for each producing strain. Some are homopolymers, which consist of one monomeric 'building block' which in most cases is glucose, while others are highly complex heteropolymers, made up of several monomeric sugar residues (Seviour *et al.*, 2011a,b). The former include the α-glucan pullulan, synthesized by the fungus *Aureobasidium pullulans* (Cheng *et al.*, 2011; Leathers, 2003; Seviour *et al.*, 2011a,b; Shingel 2004; Singh *et al.*, 2008). It consists of maltotriose units linked by (1-4)-α-glycosidic bonds, with differing extents of (1-6)-α-linked maltotriose substitutions (Fig. 5.3). Many are (1-3)-β-linked glucans produced by fungi, which in a few cases are unbranched linear polymers like the water-insoluble bacterial glucan curdlan synthesized by *Agrobacterium tumefacians*

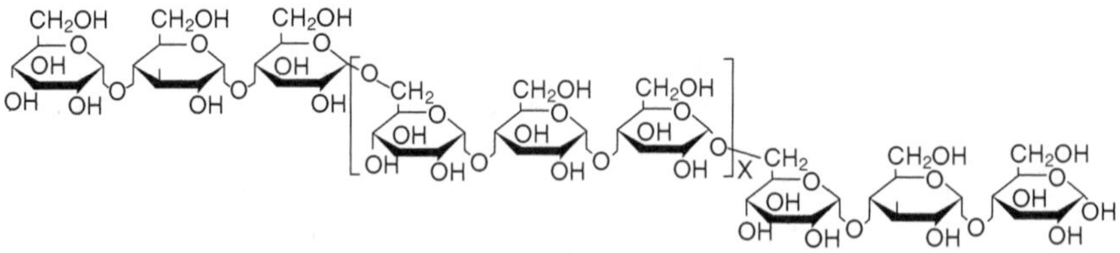

Fig. 5.3   Chemical structure of a representative portion of pullulan, illustrating the maltotriose units (α-glucan) repeating linkages. The maltotriose units are connected to each other by an α-1.6 glycosidic bond, while the three glucose units in the maltotriose are connected by an α-1,4-glycosidic bond.

© Woodhead Publishing Limited, 2013

(McIntosh *et al.*, 2005; Zhan *et al.*, 2012). Most (1-3)-β-linked glucans are branched to differing extents. Many have (1-6)-β-linked glycosidic residues attached to one in three of the (1-3)-β-linked glucose backbone residues, as with the scleroglucan family of fungal β-glucans (Schmid *et al.*, 2011; Seviour *et al.*, 2011a; Wang and McNeil, 1996). In others, like epiglucan from *Epicoccum* spp., the branching frequencies are one (1-6)-β-linked glycosidic residue to every two of three backbone glucose residues (Seviour *et al.*, 2011a). Such structural differences determine many of their exploitable important physicochemical and biological properties, including their water solubility (Seviour *et al.*, 2011a,b).

Fungi synthesizing these (1-3), (1,6)-β-linked glucans often produce enzymes capable of degrading them (Martin *et al.*, 2007). Their activities may reduce final exopolysaccharide yields substantially during fermentation and should be assayed for in any initial strain screening program. While the synthesis of most appears to be regulated by catabolite repression, some are constitutive (Martin *et al.*, 2007). Equally, high levels of formation of unwanted metabolites, increasing fermentation and downstream costs, needs to be considered. Thus, many *A. pullulans* isolates produce melanin pigments under the nitrogen-limiting conditions that favour pullulan production and the same conditions that encourage high scleroglucan yields by *Sclerotium glucanicum* also favour the synthesis of oxalate (Wang and McNeil, 1996). Of the heteropolymeric exopolysaccharides, the most important industrially is xanthan, a chemically complex polymer produced by *Xanthomonas campestris* strains (Garcia-Ochoa *et al.*, 2000; Seviour *et al.*, 2011b).

All exopolysaccharides share one attractive feature in common. They impart high viscosity to solutions, making them attractive thickening and gelling agents in foods. These properties unfortunately also ensure that optimizing their production by microbial fermentation is highly complex and challenging. Although alternative traditional sources of natural food thickening agents are available and are still used, microbial production is now more common. Reliability of supply, manufacture under carefully controlled and monitored fermentation conditions allowing yield optimization and product consistency and their often unique physicochemical properties are attractive features (Seviour *et al.*, 2011a,b). Yet, in spite of these attributes, each glucan must satisfy stringent government safety regulations before it can be used in foods. Thus, scleroglucan, with a higher thermostability than xanthan has not yet been approved as a food additive in USA or Europe, despite its acceptance in Japan (Schmid *et al.*, 2011).

The problems and challenges of producing microbial exopolysaccharides by fermentation have been the subject of several reviews (Richard and Margaritis, 2002; Garcia-Ochoa and Gomez, 2009; Gibbs *et al.*, 2000; McNeil and Harvey, 1993; Seviour *et al.*, 2011a,b; Wang and McNeil, 1996) and the interested reader is encouraged to read these and Chapter 16 on biopolymer production by Giavasis in the present volume. Briefly, their

© Woodhead Publishing Limited, 2013

formation in submerged culture changes broth rheology from uninoculated non-viscous Newtonian behaviour to a highly viscous non-Newtonian rheology, a change exacerbated if the producing fungus grows as dispersed hyphal elements (hyphal 'trees') (Seviour et al., 2011b).

With Newtonian fluids, the situation with unicellular organisms and pelleted fungal growth (Papagianni, 2004), the relationship between the applied shear stress and consequential shear rate is linear. With a non-Newtonian culture medium, this relationship is not linear, but instead is shear stress (agitation rate) dependent. Thus, as the agitation rate is increased, the viscosity of the medium decreases, a phenomenon known as shear thinning (Richard and Margaritis, 2002; Garcia-Ochoa and Gomez, 2009; McNeil and Harvey, 1993; Seviour et al., 2011b). Therefore, medium viscosity will be much lower near the impeller than in a region of the bioreactor distant from it, where no mixing may occur. Hence, heterogeneous, sub-optimal conditions will become established with serious consequences for transport processes including oxygen mass transfer. This shear thinning results from the individual exopolysaccharide polymer chains and fungal hyphae becoming aligned along the direction of the applied shear force (McNeil and Harvey, 1993; Seviour et al., 2011b; Wucherpfenning et al., 2010).

The story becomes more complex as agitation rates may affect organism morphology in submerged culture, so morphology is affected by and in turn affects directly and indirectly several other bioreactor operational parameters (Fig. 5.4). This can be illustrated, for example, by A. pullulans where interpretation of experimental data becomes problematic.

This organism has a highly complex life cycle, but it now seems likely that swollen cells are those which synthesize the α-glucan pullulan (Seviour et al., 2011a). The dominant morphological form produced in submerged culture by this fungus is affected by a range of factors, including the nitrogen source and its concentration. Thus, unicell production is favoured by nitrate, while ammonium ions favour a mycelial growth. Utilization of both leads to marked changes in the culture pH, increasing with the former and falling with the latter, depending on their initial concentrations. Increasing agitation speeds in CSTRs, which increases the pO_2, also favours unicell production. Its role in influencing pullulan production is unclear (Seviour et al., 2011a,b). Changing the reactor configuration from a CSTR to an airlift configuration affects the chemical composition of the exopolysaccharide synthesized by A. pullulans. At low levels of both nitrate and ammonium ions, pullulan is the major exopolysaccharide produced with both configurations. Increasing ammonium levels in the CSTR leads to mycelial growth, even at high agitation speeds and furthermore no pullulan is formed, a marked behavioural shift not seen when high nitrate concentrations are used instead. In an airlift bioreactor, high nitrate and ammonium levels in the medium support a unicellular population in both, and pullulan is again the major product (Orr et al., 2009).

© Woodhead Publishing Limited, 2013

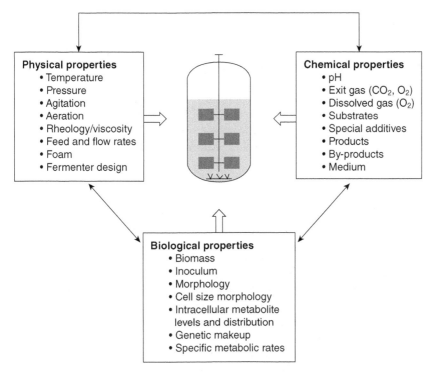

Fig. 5.4 System properties (physical, chemical and biological) that affect the performance of a bioprocess performance (adapted from Vaidyanathan *et al.*, 1999).

Therefore the normally difficult problems of process scale up and bioreactor design can become even more demanding. To design and operate bioreactors where mixing efficiencies achievable in non-Newtonian lab scale systems can be replicated may not simply involve increasing agitation rates if organism morphology changes to a form that does not support high metabolite productivity.

Nevertheless, much effort has been directed at designing and assessing bioreactors and especially aeration/mixing systems able to cope better with the problems of viscous non-Newtonian cultures. This extensive work, published in the scientific literature, is reviewed by Seviour *et al.* (2011a,b), but as with citric acid production, whether these have been taken up by industry is unclear. They include novel designs like the oscillatory baffled bioreactor (OBR), first used for pullulan production (Gaidhani *et al.*, 2003), but under investigation for more general use. Such pulsed baffled reactors are used widely in chemical polymerization studies where they have advantages over STR reactors. These include substantially better bulk mixing at lower power inputs, but the reality in fermentations for biopolymer (exopolysaccharide,

© Woodhead Publishing Limited, 2013

EPS) production from microbial cultures is that the complex three phase viscous system (liquid, solid and gas interactions) is highly challenging for a reactor type whose major advantage is a relatively gentle low shear agitation. Equally, publications on OBR applications tend to emphasize the modular nature of scale up, but again the fermentation industry tends to prefer fewer larger vessels, thus reducing labour demands. Bearing in mind its innate conservatism, the reported advantages of small scale OBR studies are considered unlikely to lead to their industrial adoption for microbial EPS production in the near future. Xanthan production in an OBR was about equal to that achievable in a CSTR in terms of its final yield and productivity and both systems had similar power consumptions, indicating no clear advantages over CSTR technology (Jambi, 2012) For industry to have sufficient confidence in its capabilities at an industrial scale, feasibility studies are needed at a scale larger than the 2 litres used in that study (Jambi, 2012). As Matthews (2008) has pointed out, standard bioreactors are usually less costly to manufacture than specialist one-off configurations.

What little detail is available on large-scale exopolysaccharide production is restricted to scleroglucan, but similar processes are probably used for pullulan and xanthan, both of which are produced on an industrial scale. Scleroglucan is manufactured using aerobic pH controlled batch or fed batch CSTR systems (Schmid et al., 2011). Product recovery involves an ethanol precipitation step followed by filtration. Because of the highly viscous nature of the medium, cell removal is difficult and technical grade scleroglucan would contain fungal biomass. Further filtration is required to produce a finer grade polymer, which requires medium dilution and then probably cross flow filtration (Schmid et al., 2011).

As with the other examples discussed in this chapter, production costs must be minimized in a highly competitive industry, meaning cheaper substrates and downstream processing. The influence of medium composition and other culture conditions on microbial exopolysaccharide production has been reviewed critically by Seviour et al. (2011a,b). There is no single set of parameters suitable for optimizing yields in all producing strains, so those for each have to be determined.

5.5   Dealing with fermentation waste

Eventually the spent culture medium and cultured biomass need to be disposed of. With small volumes, these are often discharged, after permission, into the domestic sewerage system, with little or no pretreatment. For larger plants, treatment is usually 'in-house' and involves a dedicated and often costly waste treatment process. To describe these as disposal systems is probably now inappropriate. Instead, most should be seen as environmentally friendly recycling systems, exploitable for regenerating water and other materials for reuse.

© Woodhead Publishing Limited, 2013

The properties of these bioreactor wastes are clearly very different from those of domestic sewage and so the aims of any treatment are different. Thus, there is no need to reduce the high levels of faecal pathogenic bacteria found in sewage (Seviour and Nielsen, 2010) to levels which no longer pose any serious environmental and health risks. Consequently a final disinfection step is not required before discharge. Nor are high levels of toxic heavy metals likely to be a concern. Unlike sewage, where the levels of organic carbon, nitrogen (N) and phosphorus (P) are sufficient to facilitate the more environmentally friendly biological treatment processes (Seviour and Nielsen, 2010) these bioreactor wastes are likely to be limited in either or all of these. Therefore, supplementation with cheap N and P sources may be necessary before any biological process can be employed successfully. Such nutrient imbalances may lead to proliferation of organisms like *Zoogloea* spp. which can cause non-filamentous bulking problems, which disrupt biomass settling in the clarifiers in processes like activated sludge (see later) (McIlroy *et al.*, 2011). With high waste N and P levels, the risk of eutrophication and development of toxigenic cyanobacteria and algal blooms may require biological treatment designed to reduce them to environmentally harmless levels (Seviour and Nielsen, 2010). Even so, the liquid waste fraction would generally have a low chemical/biochemical oxygen demand (BOD/COD) and so pose little threat to the self-purification properties of any receiving body of water (Seviour and Nielsen, 2010). Any downstream processing used for product recovery may change the chemical nature of the waste and this will need consideration in any treatment process choice. So will capital and operating costs and, in many countries, land availability.

Several options exist for biological waste treatment of the culture medium. Anaerobic systems are cheaper, but much slower and so may not be appropriate for large volumes. Aerobic treatment involves the additional cost of generating and supplying air, but requires a smaller space and is much more rapid.

Because of their proven flexibility and durability, activated sludge-based systems, developed initially for treating domestic sewage, are the methods of choice for handling bioreactor culture medium. These aerobic systems recycle the harvested microbial biomass from the reactors and reinoculate it into the incoming untreated waste, thus ensuring rapid and efficient treatment from an already adapted community (Seviour and Nielsen, 2010). Plant configurations can be modified readily to achieve not only removal of organic carbonaceous material, but also N and P, as detailed by Seviour and Nielsen (2010).

For small volumes of waste like those generated by isolated wineries, aerobic or anaerobic sequence batch reactor (SBR) systems are commonly used in Australia. These are very versatile and occupy a small space. Again the reader is referred to Seviour and Nielsen (2010) for details of their principles of operation.

© Woodhead Publishing Limited, 2013

Little information is available on the composition of the microbial communities developing in these systems. The attributes of the feed and the plant configuration are the two major selective pressures brought to bear on the community (Seviour and Nielsen, 2010), so any generalizations are dangerous. Furthermore, very little information is available on the microbial ecology of treatment processes for such wastes. McIlroy *et al.* (2011) showed that the community in an SBR treating winery wastes, characterized by high levels of complex carbohydrates and low pH, but low N and P levels, was quite different from that seen in similar processes treating domestic sewage. It was dominated by members of the genus *Amaricoccus* and *Defluviicoccus*. Both are Gram negative tetrad forming organisms, seen elsewhere in plants treating wastes with high carbon/N or P ratio feeds. *Zoogloea* was also dominant. Clearly similar studies need to be performed on systems treating different fermentation wastes before we are in a position where they can be better designed for each individual waste type.

Little information is available on what strategies are used to dispose of the microbial cells from spent fermentations. The biomass, rich in proteins, vitamins and other essential nutrients, is suitable as a source of animal feedstock or after suitable treatment, for human consumption, as with yeast extract products like Marmite® and Vegemite®.

5.6   Conclusions

The design and operation of bioreactors represents a series of compromises. This is especially the case with applications in the food industry where profit margins are generally low. Hence pressure is greater to run bioreactors at maximal efficiency to produce organic acids and polysaccharides at a profit. In most cases this requires using large conventional and well-tried bioreactor designs like CSTRs. There are a few exceptions, most notably the production of mycoprotein using large scale pressure cycle reactors. As the human population grows, intense interest will be shown in increasing the supply of conventional sources of proteins and microbial protein sources from non-protein materials will be exploited increasingly. The technology and fermenter design are already available and this resource-efficient model for protein production will become far more widely used. Combined with the rapid advances in systems biology discussed in earlier chapters, such expertise will provide us with the skills to produce microbial biomass with greatly enhanced nutritional characteristics compared to those produced and consumed today.

Modern molecular biology approaches have the potential to generate rationally new super producing microbial strains, but in order to realize fully their industrial potential, it is fundamentally important to improve our understanding of the highly complex interactions between fermenter design and operation and organism physiology.

© Woodhead Publishing Limited, 2013

5.7 References

ANASTASSIADIS, S., AIVASIDIS, A. and WANDREY, C. (2002). 'Citric acid production by *Candida* strains under intracellular nitrogen limitation'. *Appl Microbiol Biotechnol*, **60**, 81–7.

ANASTASSIADIS, S., WANDREY, C. and REHM, H.-J. (2005). 'Continuous citric acid fermentation by *Candida oleophila* under nitrogen limitation at constant C/N ratio'. *World J Microbiol Biotechnol*, **21**, 695–705.

ANUPAMA, P. and RAVINDRA, P. (2000) 'Value added food: single cell protein'. *Biotechnol Adv*, **18**, 459–79.

BENTLEY, R. and BENNETT, J.W. (2008). 'A ferment of fermentations: reflections on the production of commodity chemical using microorganisms'. *Adv Appl Microbiol*, **63**, 1–32.

BULL, A. (2010). 'The renaissance of continuous culture in the post-genomics age'. *J Ind Microbiol Biotechnol*, **37**, 993–1021.

BULL, A.T., ELLWOOD, D.C. and RATLEDGE, C. (1979). 'The changing scene in microbial technology'. In *Microbial Technology: Current State, Future Prospects*, A.T. Bull, D.C. Ellwood & C. Ratledge (eds). Cambridge University Press, Cambridge, UK, pp 1–28.

CHEN, J.Z. and SEVIOUR, R. (2007). 'Medicinal importance of fungal beta-(1->3), (1->6)-glucans'. *Mycolog Res*, **111**, 635–52.

CHENG, K.-C., DEMIRCI, A. and CATCHMARK, J.M. (2010). 'Advances in biofilm reactors for production of value-added products'. *Appl Microbiol Biotechnol*, **87**, 445–56.

CHENG, K.-C., DEMIRCI, A. and CATCHMARK, J.M. (2011). 'Pullulan: biosynthesis, production and applications'. *Appl Microbiol Biotechnol*, **92**, 29–44.

CHISTI, Y. (2006). 'Bioreactor design'. In *Basic Biotechnology*, 3rd edition, C. Ratledge & B. Kristiansen (eds). Academic Press, London, pp 181–200.

DHILLON, G.S., BRAR, S.K., VERMA, M. and TYAGI, R.D. (2011). 'Untilization of different agro-industrial wastes for sustainable bioproduction of citric acid by *Aspergillus niger*'. *Bioresource Technol*, **54**, 83–92.

DRIOUCH, H., SOMMER, B. and WITTMANN, C. (2010). 'Morphology engineering of *Aspergillus niger* for improved enzyme production'. *Biotechnol Bioeng*, **105**, 1058–68.

DRIOUCH, H., HÄNSCH, R., WUCHERPFENNIG, T., KRULL, R. and WITTMANN, C. (2012). 'Improved enzyme production by bio-pellets of *Aspergillus niger*: Targeted morphology engineering using titanate microparticles'. *Biotechnol Bioeng*, **109**, 462–71.

FAZENDA, M.L., SEVIOUR, R., MCNEIL, B. and HARVEY, L.M. (2008). 'Submerged culture fermentation of 'higher fungi': The macrofungi'. *Adv Appl Microbiol*, **63**, 33–103.

GAIDHANI, H.K., MCNEIL, B, and NI, X.W. (2003). 'Production of pullulan using an oscillatory baffled bioreactor'. *J Chem Technol Biotechnol*, **78**, 260–64.

GARCIA-OCHOA, F. and GOMEZ, E. (2009). 'Bioreactor scale-up and oxygen transfer rate in microbial processes: an overview'. *Biotechnol Adv*, **27**, 153–76.

GARCIA-OCHOA, F., SANTOS, V.E., CASAS, J.A. and GOMEZ, E. (2000). 'Xanthan gum: production, recovery, and properties'. *Biotechnol Adv*, **18**, 549–79.

GIBBS, P.A., SEVIOUR, R.J. and SCHMID, F. (2000). 'Growth of filamentous fungi in submerged culture: problems and possible solutions'. *Crit Rev Biotechnol*, **20**, 17–48.

GOLDBERG, I., ROKEM, J.S. and PINES, O. (2006). 'Organic acids: old metabolites, new themes.' *J Chem Technol Biotechnol*, **81**, 1601–11.

GOW, J.S., LITTLEHAILES, J.D., SMITH, S.R.L. and WALTER, R.B. (1975). 'SCP production from methanol:bacteria'. In *Single Cell Protein II*, S.R. Tannenbaum & D.I.C Wang (eds). MIT Press, Cambridge Massachusetts, pp 370–84.

© Woodhead Publishing Limited, 2013

GRIMM, L.H., KELLY, S., KRULL, R. and HEMPEL, D.C. (2005). 'Morphology and productivity of filamentous fungi'. *Appl Microbiol Biotechnol*, **69**, 375–84.

HARVEY, L.M. and MCNEIL, B. (2008). 'The design and preparation of media for bioprocesses. In *Practical Fermentation Technology*, B. McNeil & L.M. Harvey (eds). Wiley, Chichester, UK, pp 97–123.

HEWITT, C.J. and NIENOW, A.W. (2007). 'The scale up of microbial batch and fed batch fermentation processes.' *Adv Appl Microbiol*, **62**, 105–36.

JAMBI, E. (2012). *Production of Microbial Biopolymers in the Oscillatory Baffled Bioreactor, in SIPBS*. PhD Thesis, University of Strathclyde, Glasgow.

JUNKER, B.H., HESSE, M., BURGESS, B., MASUREKAR, P., CONNORS, B.H. and SEELEY, A. (2004). 'Early phase process scale-up challenges for fungal and filamentous bacterial cultures'. *Appl Biochem Biotechnol*, **119**, 241–77.

KARAFFA, L. and KUBICEK, C.P. (2003). '*Aspergillus niger* citric acid accumulation: do we understand this well working black box?' *Appl Microbiol Biotechnol*, **61**, 189–96.

KARDOS, N. and DEMAIN, A.L. (2011). 'Penicillin: the medicine with the greatest impact on therapeutic outcomes'. *Appl Microbiol Biotechnol*, **92**, 677–87.

KAVANAGH, K. (2005). 'Fungal fermentation systems and products'. In *Fungi: Biology and Applications*, K. Kavanagh (ed). Wiley, West Sussex, 89–111.

KRISHNA, C. (2005). 'Solid-state fermentation systems – an overview'. *Crit Rev Biotechnol*, **25**, 1–30.

KRISTIANSEN, B. and SINCLAIR, C.G. (1979). 'Production of citric acid in continuous culture'. *Biotech Bioeng*, **21**, 297–315.

LEATHERS, T.D. (2003). 'Biotechnological production and applications of pullulan'. *Appl Microbiol Biotechnol*, **62**, 468–73.

LEGISA, M. and MATTEY, M. (2007). 'Changes in primary metabolism leading to citric acid overflow in *Aspergillus niger*'. *Biotechnol Lett*, **29**, 181–90.

LEGRAS, J.-L., MERDINGLU, D., CORNUET, J.-M. and KARST, F. (2007). 'Beer bread and wine: *Saccharomyces cerevisiae* diversity reflects human history'. *Mole Ecol*, **16**, 2091–102.

LODOLO, E.J., KOCK, J.L., AXCELL, B.C. and BROOKS, M. (2008). 'The yeast *Saccharomyces cerevisiae* – the main character in beer brewing'. *FEMS Yeast Res*, **8**, 1018–36.

MARTIN, K., MCDOUGALL, B.M., MCILROY, S., JAYUS, CHEN, J.Z, and SEVIOUR, R.J. (2007). 'Biochemistry and molecular biology of exocellular fungal beta-(1,3)- and beta-(1,6)-glucanases'. *FEMS Microbiol Rev*, **31**, 168–92.

MATELES, R.I. (1979). 'The physiology of single-cell protein (SCP) production'. In *Microbial Technology: Current State, Future Prospects*, A.T. Bull, D.C. Ellwood & C. Ratledge (eds). Cambridge University Press, Cambridge, UK, pp 29–52.

MATTHEWS, G. (2008). 'Fermentation equipment selection: Lab scale bioreactor design considerations'. In *Practical Fermentation Technology*, B. McNeil & L.M. Harvey (eds). Wiley, Chichester, pp 3–36.

MCILROY, S., SPEIRS, L.B.M., TUCCI, J. and SEVIOUR, R.J. (2011). '*In situ* profiling of microbial communities in full scale aerobic sequencing batch reactors treating winery waste in Australia'. *Environ Sci Technol*, **45**, 8794–803.

MCINTOSH, M., STONE, B.A. and STANISICH, V.A. (2005). 'Curdlan and other bacterial (1->3)-beta-D-glucans'. *Appl Microbiol Biotechnol*, **68**, 163–73.

MCNEIL, B. and HARVEY, L.M. (1993). 'Viscous fermentation products'. *Crit Rev Biotechn*, **13**, 275–304.

MERCHUK, J.C. (1990). 'Why use air-lift bioreactors?' *Trends Biotechnol*, **8**, 66–71.

MERCHUK, J.C., BEN-ZVI, S. and NIRANJAN, K (1994). 'Why use bubble-column bioreactors?' *Trends Biotechnol*, **12**, 501–11.

MEYER, V. (2008). 'Genetic engineering of filamentous fungi – progress, obstacles and future trends'. *Biotechnol Adv*, **26**, 177–85.

© Woodhead Publishing Limited, 2013

MEYER, V., WU, B. and RAM, A.F.J. (2011). 'Aspergillus as a multi-purpose cell factory: current status and perspectives'. Biotechnol Lett, 33, 469–76.

MOELLER, L., GRÜNBERG, M., ZEHNSDORF, A., AURICH, A., BLEY, T. and STREHLITZ, B. (2011). 'Repeated fed-batch fermentation using biosensor online control for citric acid production by Yarrowia lipolytica'. J Biotechnol, 153, 133–7.

OHM, R.A., DE JONG, J.F., LUGONES, L.G., AERTS, A., KOTHE, E., STAJICH, J.E., DE VRIES, R.P., RECORD, E., LEVASSEUR, A., BAKER, S.E., BARTHOLOMEW, K.A., COUTINHO, P.M., ERDMANN, S., FOWLER, T.J., GATHMAN, A.C., LOMBARD, V., HENRISSAT, B., KNABE, N., KÜES, U., LILLY, W.W., LINDQUIST, E., LUCAS, S., MAGNUSON, J.K., PIUMI, F., RAUDASKOSKI, M., SALAMOV, A., SCHMUTZ, J., SCHWARZE, F.W.M.R., VANKUYK, P.A., HORTON, J.S., GRIGORIEV, I.V. and WÖSTEN, H.A.B. (2010). 'Genome sequence of the model mushroom Schizophyllum commune'. Nature Biotechnol, 28, 957–63.

OLSEN, J. and ALLERMANN, K. (1987). 'Microbial biomass as a protein source'. In Basic Biotechnology, C. Ratledge & B. Kristiansen (eds). Academic Press, London, pp 285–308.

ORR, D., ZHENG, W., CAMPBELL, B.S., MCDOUGALL, B.M. and SEVIOUR, R.J. (2009). 'Culture conditions affect the chemical composition of the exopolysaccharide synthesized by the fungus Aureobasidium pullulans'. J Appl Microbiol, 107, 691–8.

PAPAGIANNI, M. (2004). 'Fungal morphology and metabolite production in submerged mycelial processes'. Biotechnol Adv, 22, 189–259.

PAPAGIANNI, M. (2007). 'Advances in citric acid fermentation by Aspergillus niger: biochemical aspects, membrane transport and modeling'. Biotechnol Adv, 25, 244–63.

PAPAGIANNI, M., MATTEY, M. and KRISTIANSEN, B. (1998). 'Citric acid production and morphology of Aspergillus niger as functions of the mixing intensity in a stirred tank and a tubular loop bioreactor'. Biochem Eng J, 2, 197–205.

PAPAGIANNI, M., MATTEY, M. and KRISTIANSEN, B. (1999). 'The influence of glucose concentration on citric acid production and morphology of Aspergillus niger in batch culture'. Enzyme Microb Technol, 25, 710–17.

PATEL, N. and THIBAULT, J. (2009). 'Enhanced in situ dynamic method for measuring k_La in fermentation media'. Biochem Eng J, 47, 48–57.

PAUL, G.C., PRIEDE, M.A. and THOMAS, C.R. (1999). 'Relationship between morphology and citric acid production in submerged Aspergillus niger fermentations'. Biochem Eng J, 3, 121–9.

PEL, H.J., DE WINDE, J.H., ARCHER, D.B., DYER, P.S., HOFMANN, G., SCHAAP, P.J., TURNER, G., DE VRIES, R.P., ALBANG, R., ALBERMANN, K., ANDERSEN, M.R., BENDTSEN, J.D., BENEN, J.A., VAN DEN BERG, M., BREESTRAAT, S., CADDICK, M.X., CONTRERAS, R., CORNELL, M., COUTINHO, P.M., DANCHIN, E.G., DEBETS, A.J., DEKKER, P., VAN DIJCK, P.W., VAN DIJK, A., DIJKHUIZEN, L., DRIESSEN, A.J., D'ENFERT, C., GEYSENS, S., GOOSEN, C., GROOT, G.S., DE GROOT, P.W., GUILLEMETTE, T., HENRISSAT, B., HERWEIJER, M., VAN DEN HOMBERGH, J.P., VAN DEN HONDEL, C.A., VAN DER HEIJDEN, R.T., VAN DER KAAIJ, R.M., KLIS, F.M., KOOLS, H.J., KUBICEK, C.P., VAN KUYK, P.A., LAUBER, J., LU, X., VAN DER MAAREL, M.J., MEULENBERG, R., MENKE, H., MORTIMER, M.A., NIELSEN, J., OLIVER, S.G., OLSTHOORN, M., PAL, K., VAN PEIJ, N.N., RAM, A.F., RINAS, U., ROUBOS, J.A., SAGT, C.M., SCHMOLL, M., SUN, J., USSERY, D., VARGA, J., VERVECKEN, W., VAN DE VONDERVOORT, P.J., WEDLER, H., WÖSTEN, H.A., ZENG, A.P., VAN OOYEN, A.J., VISSER, J. and STAM, H. (2007). 'Genome sequencing and analysis of the versatile cell factory Aspergillus niger CBS 513.88'. Nature Biotechnol, 25, 221–30.

RICHARD, A. and MARGARITIS, A. (2002). 'Production and mass transfer characteristics of non-Newtonian biopolymers for biomedical applications'. Crit Rev Biotechnol, 22, 355–74.

RUFFING, A.M., CASTRO-MELCHOR, M., HU, W.-S. and CHN, R.R. (2011). 'Genome sequencing of the curdlan-producing Agrobacterium sp. ATCC 31749'. J Bacteriol, 193, 4294–5.

© Woodhead Publishing Limited, 2013

RYMOWICZ, W., FATYKHOVA, A.R., KAMZALOVA, S., RYWIŃSKA, A. and MORGUNOV, I.G. (2010). 'Citric acid production from glycerol-containing waste of biodiesel industry by *Yarrowia lipolytica* in batch, repeated batch and cell cycle regimes'. *Appl Microbiol Biotechnol*, **87**, 971–9.

RYWIŃSKA, A. and RYMOWICZ, W. (2010). 'High-yield production of citric acid by *Yarrowia lipolytica* on glycerol in repeated-batch bioreactors'. *J Indu Microbiol Biotechnol*, **37**, 431–5.

RYWIŃSKA, A., JUSZCZYK, P., WOJTATOWICZ, M. and RYMOWICZ, W. (2011). 'Chemostat study of citric acid production from glycerol by *Yarrowia lipolytica*'. *J Biotechnol*, **152**, 54–7.

SAUER, M., PORRO, D., MATTANOVICH, D. and BRANDUARDI, P. (2008). 'Microbial production of organic acids: expanding the markets'. *Trends Biotechnol*, **26**, 100–8.

SCHMID, J., MEYER, V. and SIEBER, V. (2011). 'Scleroglucan: biosynthesis, production and application of a versatile hydrocolloid'. *App Microbiol Biotechnol*, **91**, 937–47.

SEVIOUR, R.J. and NIELSEN, P.-H. (2010). *Microbial Ecology of Activated Sludge*. IWA Publishing, London.

SEVIOUR, R.J., SCHMID, F. and CAMPBELL, B.S. (2011a). 'Fungal exopolysaccharides'. In *Polysaccharides in Medicinal and Pharmaceutical Applications*, V. Popa (ed). iSmithers, Shrewsbury, pp 89–144.

SEVIOUR, R.J., MCNEIL, B., FAZENDA, M.L. and HARVEY, L.M. (2011b). 'Operating bioreactors for microbial exopolysaccharide production'. *Crit Rev Biotechnol*, **31**, 170–85.

SHINGEL, K.I. (2004). 'Current knowledge on biosynthesis, biological activity, and chemical modification of the exopolysaccharide, pullulan'. *Carbohydr Res*, **339**, 447–60.

SINGH, R.S., SAINI, G.K. and KENNEDY, J.F.. (2008). 'Pullulan: microbial sources, production and applications'. *Carbohydr Polym*, **73**, 515–31.

SOCCOL, C.R., VANDENBERGHE, L.P.S., RODRIGUES, C. and PANDEY, A. (2006). 'New perspectives for citric acid production and application'. *Food Technol Biotechnol*, **44**, 141–9.

SOLOMONS, G.L. (1980). 'Fermenter design and fungal growth'. In *Fungal Biotechnology*, J.E. Smith, D.R. Berry & B. Kristiansen (eds). Academic Press, London.

TRINCI, A.P.J. (1992). 'Mycoprotein: a twenty year overnight success story'. *Mycolog Res*, **96**, 1–13.

TRINCI, A.P.J. (1994). 'Evolution of Quorn® myco-protein fungus *Fusarium graminearum* A3/5'. *Microbiol*, **140**, 2181–8.

VAIDYANATHAN, S., MACALONEY, G., VAUGHAN, J., MCNEIL, B. and HARVEY, L.M. (1999). 'Monitoring of submerged bioprocesses'. *Crit Rev Biotech*, **19**, 277–316.

WANG, Y.C. and MCNEIL, B. (1996). 'Scleroglucan'. *Crit Rev Biotechnol*, **16**, 185–215.

WIEBE, M.G. (2002). 'Mycoprotein from *Fusarium venenatum*: a well established product for human consumption'. *Appl Microbiol Biotechnol*, **58**, 421–7.

WILLIAMS, J.A. (2002). 'Keys to bioreactor selections'. *Chem Eng Prog*, **98**, 34–41.

WUCHERPFENNIG, T., KLEP, K.A., DRIOUCH, H., WITTMANN, C. and KRULL. R. (2010). 'Morphology and rheology in filamentous cultivations'. *Adv Appl Microbiol*, **72**, 89–136.

ZHAN, X.-B., LIN, C.-C. and ZHANG, H.-T (2012). 'Recent advances in curdlan biosynthesis, biotechnological production and applications'. *Appl Microbiol Biotechnol*, **93**, 525–31.

© Woodhead Publishing Limited, 2013

6

Fermentation monitoring and control of microbial cultures for food ingredient manufacture

B. McNeil and L. M. Harvey, Strathclyde University, UK, N. J. Rowan, Athlone Institute of Technology, Ireland and I. Giavasis, Technological Educational Institute of Larissa, Greece

DOI: 10.1533/9780857093547.1.125

Abstract: This chapter focuses upon how and why we monitor fermentation processes for production of food ingredients. It also emphasises the critical importance of using monitoring technologies which are appropriate and which aid us in understanding the physiology of the microbial cultures we use. A range of monitoring strategies and techniques in current use are critically discussed in the context of the specific needs of the fermentation industry. Finally, some future trends and developments are introduced and their potential utility in this context is evaluated.

Key words: future developments, monitoring strategies, process analytical technology, process understanding, off-gas analysis, spectrometry.

6.1 Introduction

It is probably quite important to distinguish between fermentation processes for production of food ingredients and fermented foods. Although there is clear overlap between the two concepts, perhaps the easiest way to separate them is to look back upon the original aims. Fermented food processes, for example, yoghurt and beer, start with materials which are themselves foods (milk or cereals), but in the final product the original characteristics have been substantially changed by the action of microorganisms and the keeping qualities or microbiological safety has generally been improved relative to the starting material (see Sources of further information and advice, texts 1 and 2). By contrast, fermentation processes involve closely controlled fermentation (cultivation) of selected strains of microorganisms, plant cells or animal cells upon a nutrient medium (Sources of further reading and advice text 3), which often involves complete

© Woodhead Publishing Limited, 2013

consumption of the nutrients in the feed (medium) and is carried out to produce a wide range of products from antibodies to foodstuffs, such as the mycoprotein, Quorn. The key word phrase here is 'controlled cultivation', and clearly in order to effect control, there has to be effective and appropriate process monitoring. Here we will focus upon monitoring fermentation processes, further background reading on fermented foods is included in the Sources of futher information and advice section.

6.1.1   Types of fermentation processes

Despite the implications of the word 'fermentation' it is important to realise that most so-called fermentation processes are highly aerobic in their nature and, as discussed in Chapter 5, the need to provide air to the culture dominates the design and operation of fermenter vessels.

Broadly speaking, two distinct types of fermentation processes are operated for manufacture of food ingredients. The oldest is solid substrate fermentation (SSF) or low water fermentation; by contrast, submerged liquid fermentation (SLF) is more commonly used in the modern bioprocessing industry. SSF has long been successfully employed for production of food ingredients such as citric acid. The major advantages usually cited are the low energy inputs relative to SLF, the simplicity of the reactor systems, the suitability for small local scale production, the ease of product recovery from the concentrated low volume liquid phase and the ability to utilise waste materials as substrates for microbial growth.

However, SSF systems are heterogeneous in nature and this presents real difficulties in taking representative samples, and in turn feeds through into real challenges in achieving effective process control throughout the reactor in SSF systems. Productivity (amount of product made per unit reactor volume per hour) is an especially critical factor in the economics of fermentation systems in high wage economies. For these reasons, most microbially produced food ingredients and foods are now made using SLF. Although there is great overlap between the technologies employed to monitor both types of fermentation processes, for the rest of this chapter the focus will be upon SLF processes. This should not be taken as indicating that SSF is in any way inferior to SLF, in fact, as the suggested sources in the Sources of further information and advice section show, recent years have been a very dynamic period in the development and deployment of novel SSF processes for a wide range of products. Those wishing to read in more detail are directed to the Sources of further information and advice section.

6.1.2   Sterility or monosepsis

Before discussing how monitoring is achieved in fermenter systems, it is as well to understand that such systems usually have only one specific

© Woodhead Publishing Limited, 2013

microbial species present throughout the reactor volume for the duration of the process. This state is variously described as monosepsis or sterility. The need to generate a sterile medium and to maintain monosepsis in the fermenter dominates design and operation of nearly all SLF processes. It also profoundly influences the development and use of monitoring and sampling technologies in SLF. Any monitoring or sampling technique which, in the view of production staff (rightly or wrongly), increases the risk of contamination even slightly will not gain acceptance at production scale. Thus, many promising technologies which have led to numerous scientific papers from laboratory-based studies, have little or no prospect of ever being deployed at large scale, as their use makes production staff nervous! This is perhaps understandable given that monosepsis is a highly unnatural state, and some fermenters for production of, for example, citric acid, can have up to 450 m^3 capacity, thus, contamination by another microorganism can represent a significant economic loss, with associated difficulties in terms of eliminating the contaminating culture.

6.2   Monitoring bioprocesses for food fermentation: an overview

6.2.1   Rationale for monitoring

Monitoring of SLF processes is essential to ensure product consistency and safety. It is also a key tool in optimising the process economics, as the efficiency of nutrient utilisation and conversion efficiency or yield can be measured, as can productivity, specific productivity (productivity per gram of cell mass) and a structured programme to improve process economics can be centred around such key parameters. Usually such an optimisation programme would be carried out at laboratory scale (which is difficult to define, but usually from 5 to 50 litres total volume fermenters) and involves the use of suitable design of experiment (DoE) software. Owing to the inherent complexity of fermentation processes with many interacting variables, the use of DoE software allows the optimisation process to be carried out both more efficiently (fewer fermentations) and effectively (greater insight into interactions between process variables) than the traditional experimental plan of changing one process variable at a time.

However, the traditional approach usually involves making use of all the experience the process operators have developed in fermentations, such as that being optimised, including significant insights into the culture physiology within fermenters, whereas DoE assumes and requires no such insights. Since one major challenge in all SLF fermentation processes is scale transfer and that process is still rather empirical in nature and heavily dependent upon skill and experience, there is still a pressing need for those involved in fermentation process monitoring to seek to understand the process they are working with. This need to understand the culture behaviour and the

© Woodhead Publishing Limited, 2013

overall fermentation process has profound implications for the types of monitoring to be used. In principle, it is increasingly important, especially in an era when ever more powerful multivariate monitoring technologies are being introduced, to keep asking the question 'What do I really need to measure, and why?' The mere acquisition of data is not always a route to better understanding of a fermentation process.

6.2.2 Modes of monitoring
Fermentation systems can be monitored in different modes as shown below in Fig. 6.1.

Traditionally, most fermentation processes have been monitored off line. In this mode a representative sample is taken aseptically from the fermenter and analysed in the laboratory at some point later. In large scale fermenters and in those involving cultivation of filamentous microorganisms, for example, fungi, there can be considerable inhomogeneity (Fig. 6.2), so obtaining a representative sample can be challenging. The potential draw-back to off line analysis is that this method tells us what was happening in the fermenter, not what is happening. One major advantage of off line moni-toring is that, when combined with suitable sample storage (chilling or freez-ing, or addition of preservative agents) off line analysis such as gas liquid chromatography (GLC), high performance liquid chromatography (HPLC), gas chromatography-mass spectrometry (GC-MS) and nuclear magnetic resonance (NMR) can be used very efficiently to analyse large numbers of fermentation process samples (usually filtrates without cells present) for multiple analytes. Most fermentation laboratories operating multiple

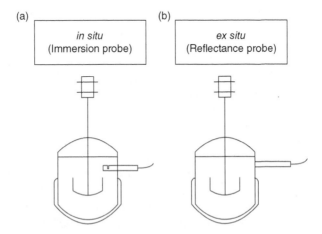

Fig. 6.1 On-line sampling configuration. From left to right: (a) *in situ* measure-ments using an immersion probe; (b) *ex situ* measurements using a reflectance probe on the glass wall of the reactor (adapted from Cervera *et al.*, 2009).

© Woodhead Publishing Limited, 2013

(a)   (b)

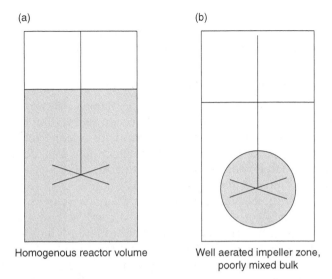

Homogenous reactor volume   Well aerated impeller zone,
poorly mixed bulk

Fig. 6.2   Mixing in STR fermenters with mechanical agitation by Rushton turbine. (a) Yeast culture, Newtonian behaviour; (b) EPS producing culture, non-Newtonian behaviour (Seviour *et al.*, 2011).

fermenters use multi-analysers such as those sold by YSI and Groton Systems, which are based upon clinical blood analysers and can simultaneously measure concentrations of sugars, waste products and product.

A variant on this is rapid off line or in line, which again involves aseptic sampling, then quick analysis of the fermentation fluid, often in the close vicinity of the fermenter. This gives information which approaches real time monitoring in terms of quality and availability. Ideally, monitoring the process in real time gives us the best understanding of what is happening in the fermenter and thus the widest range of options in achieving the desired process control. Real time information on the process state can be achieved by either *in situ* or *ex situ* sensors or analysers, that is, sensors actually within the fermenter system or analysers located outside the fermenter.

6.2.3   The 'Ideal' fermentation sensor
If we could design an idealised sensor for monitoring a fermentation process, what features would it possess? Most of the features of such an idealised fementation sensor are listed below:

- reliable
- cheap
- simple to implement and simple to operate
- physically tough

© Woodhead Publishing Limited, 2013

- capable of generating real time information
- capable of generating multi-analyte information
- capable of monitoring multiple fermenters simultaneously (multiplexing)
- data generated should be simple to interpret.

The need for sensor reliability needs no comment. However, although probably obvious, the need for relatively cheap sensors for fermentation monitoring is essential in the processes used for making food and food ingredients. Such processes suffer the same limitations as all fermentations in SLF, namely, large volumes of water with relatively low concentrations of product. This implies that fermentation is a costly business owing to the need to process these dilute streams (and for other reasons discussed in Chapter 5). Thus, where the product is relatively low cost, as most food products arising from microbial fermentation are, all costs have to be looked at very carefully and this includes the costs related to process monitoring. The principle here is simple: monitor only those process factors which are critically important to product safety or process economics and do so as inexpensively as possible.

Owing to the repeated exposure to sterilisation temperatures (usually achieved using pressurised steam), repeated cycles of cooling and heating and in many cases (e.g. citric acid production) intense agitation and aeration, any such sensor within the reactor must be physically very robust. Ideally, the sensor should be at least as robust as the stainless steel polarographic dissolved oxygen electrodes widely employed in food fermentation processes (Fig. 6.3). This type of sensor is in many respects the 'ideal' fermentation sensor and has been used successfully for decades. Not only is such a sensor cheap, robust and simple to use *in situ*, but the output from such a sensor needs no preprocessing and most process operators understand the information being presented to them.

6.3   On line bioprocess monitoring for food fermentation

Despite the arguments presented above, routine analysis of most fermentations for food or food ingredient manufacture is still carried out off line. In part this is due to the special concerns of the industry (sterility) and the perceived lack of physical toughness of many technologies which show promise at the laboratory scale. However, a great deal is due to the innate conservatism of production staff.

6.3.1   Exit or off gas analysis
However, most fermentations can be readily monitored *ex situ* in real time using off gas analysis. Most fermentation processes are aerobic and some, such as production of baker's yeast or citric acid, are highly aerobic,

© Woodhead Publishing Limited, 2013

Fig. 6.3 A stainless steel dissolved oxygen electrode (courtesy of Broadley James). Most other *in situ* sensors use this type of construction as a model.

typically such processes are vigorously gassed with large volumes of filtered sterile air. In such processes cell growth and product formation are stoichiometrically linked to oxygen consumption and also to carbon dioxide evolution via the process of cellular respiration. Simple combination sensors analysing the oxygen and CO_2 content (concentration) in the fermenter exit gas are readily available and inexpensive (€2–3000). These sensors do not have to be physically robust as they sit outside the fermenter's sterile envelope, but they do generate real time information about critical process parameters, such as oxygen uptake rate (OUR), carbon dioxide evolution rate (CER) and respiratory quotient (RQ). The latter coefficient is very widely used in baker's yeast fermentations to control sugar feed rates in order to maintain the culture in the desired fermentation state (Finn *et al.*, 2006, 2010). Since OUR and CER are mathematically linked to culture growth and substrate consumption, it is relatively straightforward to formulate models for specific growth rate of cultures and substrate consumption indirectly based upon these off gas measurements. Thus, in principle, most of the information needed to control fed-batch or batch cultures can be obtained from such sensors. Great care has to be taken with two aspects of supplying the off gas to such analysers. Firstly, the exit gas from a fermenter

© Woodhead Publishing Limited, 2013

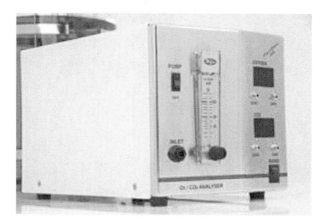

Fig. 6.4   A combination oxygen/carbon dioxide off gas analyser integral pump (courtesy of Electrolab Ltd).

usually has entrained water droplets in it which would lead to inaccurate measurements and possibly to analyser damage. So, the gas must be first passed through an effective drying column situated upstream of the analysers. The condition and function of the column must be checked daily. Fermenters usually run under positive pressure (usually +2–3 psi; 13.8–20.7 kPa) but foaming and wetting of the exit filter and partially blocked drying columns, may increase the pressure in the exit gas line. Pressure regulators have to be used to make sure the analysers are supplied with a reasonably stable supply gas pressure (see Fig. 6.4).

However, although these simple cheap off gas analysers, based on the paramagnetic analyser for oxygen content and infrared analysis for CO_2, are suitable for most routine fermentation monitoring purposes, these analytical techniques are not sufficiently accurate and precise to close a mass balance. In other words, they cannot be used to account fully for, for example, what all of the carbon and energy source (typically a sugar) fed to a culture has been used for. It has been estimated that the error associated with such combi gas analysers may be up to 10% (Van der Aar *et al.*, 1989), but this figure should be looked at with caution as the technology involved has matured. To gain the required level of precision in order to account accurately and reliably for the fate of all the carbon source requires the use of a mass spectrometer (MS) to measure off gas oxygen and CO_2 levels. Although relatively costly (€70,000) these systems are designed for multiplexing (up to 32 fermenters) and have sophisticated gas handling manifolds to ensure gas streams from multiple fermenters can be reliably analysed. Thus, the cost per fermenter becomes relatively modest and may well be less than the cost of supplying a combination off gas analyser for each fermenter if many fermenters are in use. Such MS systems are now both far more reliable and need far less technical support than previously,

© Woodhead Publishing Limited, 2013

Fig. 6.5   A typical magnetic sector mass spectrometer and gas manifold used for off gas analysis (courtesy of Thermo VG).

and can operate for many months maintenance free. Being able to account accurately for the fate of all the carbon source means that by-product or waste product formation can be minimised making the process efficiency higher and improving process economics. This is particularly important in large volume low cost fermentations typically used for food ingredient manufacture. In fact, efficient use of sugar is critical to the process economics of such bulk food products as baker's yeast and citric acid. In addition, these MS systems can be used to measure other species in the off gas flow, including simple alcohols such as ethanol and other volatile species. Such systems have gained widespread acceptance in the biofuels area (see Fig. 6.5).

6.4   Spectrometric monitoring of fermentation

Almost any chemically, enzymically or electrochemically based technology used for measurement purposes can be adapted to *in situ* operation in fermenters (Vaidyanathan *et al.*, 1999). Over the years, enzyme-based sensors have been widely developed for monitoring a wide range of fermentation

© Woodhead Publishing Limited, 2013

analytes including glucose (the commonest measurement target). These enzyme-linked sensors usually involve the immobilisation or entrapment of a suitable enzyme (plus co-factors if needed) and the use of a selectively permeable membrane between the enzyme and the fermentation fluid. Thus, the analyte of interest typically diffuses through the membrane, reacts with the enzyme and this reaction is linked electrochemically to a signal output from the probe proportional to the extent of the reaction. Although very widely investigated in research laboratories, such membrane-based sensors find little favour at production scale in food ingredient fermentations. This is mostly because their physical robustness is suspect, which concerns most production scale fermentation staff because of their perception of the increased risk to process sterility. Similar concerns surround devices which auto-sample from the fermenter, filter the fermentation fluid and pass the filtrate to systems like HPLC for detailed analysis of the levels of multiple analytes. This level of sophistication is also not always necessary or appropriate in many food ingredient fermentation processes.

Since the early 1990s there have been numerous investigations into the use of spectrophotometric methods of monitoring fermentation processes. Most of these have centred around near (Scarff et al., 2006; Cervera et al., 2009; Landgrebe et al., 2010) and mid infrared spectroscopy (Roychoudhury et al., 2007). Other techniques which have shown some utility include two dimensional (2D) fluorescence spectroscopy and recently there have been some studies on Raman spectroscopy which will be discussed later in the chapter.

6.4.1   Near infrared spectroscopy (NIRS)

The near infrared region of the electromagnetic spectrum occupies the wavelength range from 700–2500 nm. Absorbances in this region arise from combinations or overtones of absorbances arising in the fundamental IR region and are typically much weaker than absorbances in the IR. NIR absorbance peaks also tend to overlap and to have contributors from multiple analytes, making their interpretation more challenging. Most absorbances arise from X–H bonds (e.g C–H and OH), so since such bonds are abundant in biological systems, in principle, NIR spectra should contain information about most, if not all, analytes in a fermentation system (Scarff, et al., 2006).

Advantages and limitations
NIR spectroscopy (NIRS) is rapid (less than 2 min per analysis), multi-analyte, non-invasive, non-destructive and, via the use of robust stainless steel fibre optic probe systems, can be implemented readily *in situ*. It is also readily multiplexed to multiple fermenters. In many respects, it therefore possesses many of the attributes of the 'ideal' biosensor technology. Further, the relatively weak absorbances in the NIR region make it suitable for the

© Woodhead Publishing Limited, 2013

analysis of samples which are both highly light scattering and strongly light absorbing, such as the typical fermentation fluid, without in most instances any pretreatment. NIRS can replace conventional assays (e.g wet chemistry) and so save technical support resource, reduce reagent use and lower process environmental impact (Cervera *et al.*, 2009). Conversely, it has a number of inherent limitations, some of which can be circumvented.

One major limitation relates to the weak absorbances in the NIR region, which in turn translates into a relatively low detection threshold for analytes. This implies that very low (100s of mg/L) analyte concentrations may not be readily measurable by NIRS, though Mattes *et al.* (2009) has shown that with skilled use NIRS can be used to measure changes in fermentation process analyte levels which do not themselves have an NIRS signature.

Another limitation centres on the way NIRS is generally employed: commonly it is used as a secondary analytical technique, correlated in some way to a conventional reference assay. This is generally assumed to imply that the measurement error using NIRS can at best be only as good as that of the reference assay itself (Arnold *et al.*, 2002). However, Ralf Marbach (2007) at VTT in Finland has argued cogently against this simplistic view. He divides NIRS error contributions into system (spectrometer) and reference, and has developed effective methods to reduce significantly overall error when applied to NIRS in chemical reactions. Such an approach has direct applicability to the more complex biological systems being discussed here. One major drawback relates to multiple analytes contributing to absorbances at given wavelengths and overlapping peaks which are typical of NIRS spectra. Although appropriate chemometrics software can assist in deconvoluting spectra, sometimes strongly absorbing species absorbing in the same regions as another more weakly absorbing species can completely mask the second analyte, making it difficult to monitor it in NIR (Arnold *et al.*, 2004).

One common issue when using NIRS (and mid-infrared spectroscopy, MIRS) to monitor fermentation processes is collinearity in the data. All substrates and products in a fermentation system are stoichiometrically linked to each other. So, when using NIRS as a secondary technique, it is possible to build a model for one analyte based on the spectral absorbance of another. This can be avoided by careful attention to the analytical basis of the model via interrogation of the spectral contributors (scores and loadings plots), spiking prescanning to indentify which regions the particular analyte absorbs in. Another useful approach is to carry out an adaptive calibration approach such as that described by Arnold *et al.* (2004). Here a small set of fermentation samples are made by adding to aliquots of fermentation fluid known concentrations of analytes of interest. This not only increases the number of samples, but may also expand the analyte concentration range and, finally, since the concentration of the analyte being measured is now not linked stoichiometrically to other process analytes, the danger of collinearity is avoided.

© Woodhead Publishing Limited, 2013

In situ NIRS

Stainless steel probes containing optical fibres to illuminate the sample and to collect light transmitted through or reflected back from it are now available, which permit NIRS to be readily used *in situ* (in the fermenter). These probes are steam sterilisable in place and in principle as robust as a dissolved sxygen (DO) probe (see above). NIR spectrometers essentially can involve two types of instrumentation, either dispersive or Fourier transform (FT). FT instruments are commonly held to have a number of theoretical advantages over dispersive, as discussed by Scarff *et al.* (2006). However, in the authors' experience most of these advantages are of less importance than the operator's knowledge of the instrument they are using. That said, probes linked to an FT spectrometer, can use single fibres (one illuminating, one collecting) so that their dimensions can be much smaller than dispersive probes using a fibre bundle for illumination and collection of light. Figure 6.6 shows stainless steel NIRS transflectance probes and their construction. Using a normal (off line) mode spectrometer, there are two methods of employment: reflectance, where the light is reflected back to the detector off the sample, or transmittance, where the light passes through the sample and then is received by the detector. The former is best suited to solid and high density fluids, whereas the latter is best for thin liquid samples or liquid samples with suspended or dissolved solids. In transflectance (*in situ*) elements of reflectance and transmission are present offering flexibility of use.

Employing the type of probe shown in Fig 6.6 allowed simultaneous measurement of up to six key analytes every two minutes. This is a tremendous flow of real time process related data, which can greatly speed up process development.

Finn *et al.* (2006) discuss the use of NIRS in the production of baker's yeast. The authors point out that the challenges of monitoring with NIRS in a very highly aerated and vigorously agitated *Saccharomyces cerevisiae* process for manufacturing baker's yeast are much greater than those in ethanol or biofuel (bioethanol) production processes which have minimal gassing and modest or no mechanical agitation (Ge *et al.*, 1994). Baker's yeast is part of one of the most important industrial fermentation processes (Pretorius *et al.*, 2003) which is a classic high cell density fed-batch process where very precise process control is essential. The aim in this high cell density production process is to balance the feed rate of the carbon source (sugar in some form, often molasses) with the ability of the fermenter system to supply O_2 to the rapidly multiplying cells such that the maximum cell mass is achieved as rapidly as possible, with minimal by-product formation (ethanol), as this would reduce the efficiency of substrate use.

One key quality indicator in baker's yeast is protein content, which is taken to be a fair indicator of the diastatic potential of the cells and hence their ability to evolve CO_2 in baked goods. Finn *et al.* (2006) therefore used NIRS in line to monitor cell mass (biomass), from which specific growth

© Woodhead Publishing Limited, 2013

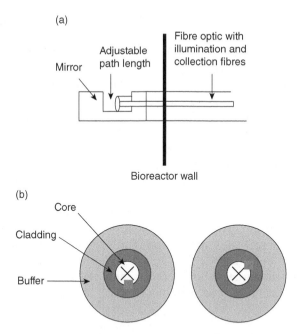

Fig. 6.6 Stainless steel transflectance NIRS probes for use in fermenters and details of construction (a) Diagram of a transflectance NIR probe. (b) Top view of NIR probe highlighting how to mark the probe casing to maintain probe orientation from run to run. (Fazenda *et al.*, 2011).

rates of the cultures could be inferred, protein content, glucose concentration and ethanol formation. An excellent model for biomass formation was formulated which was superior to many previously reported NIR models for this analyte. The authors argued that the relatively large dataset used in the modelling exercise assisted in the formulation of robust models capable of actual industrial utility in production systems. This point is often ignored in papers on NIRS monitoring of food fermentations, which are frequently characterised by quite small sample sets, so that formulated models for analytes may well be over-fitted and fail to predict analyte concentrations well in practice when applied to data not used in the original modelling exercise (Arnold *et al.*, 2003). A useful model of protein content was also formulated in this study. Modelling glucose and ethanol on whole broth (with cells present) was noted to be problematic in this study (Arnold *et al.*, 2003) and the authors modelled these two analytes in filtered (cell free) samples. Obviously, this necessity might well limit the potential of NIRS *in situ* in such high cell density systems and the risk of detector saturation in transmission mode was observed in high cell density yeast cultures by Crowley *et al.* (2005) who overcame this by switching to reflectance. In the study by Finn *et al.* (2006), adaptive calibration was used to extend the concentration range of the

© Woodhead Publishing Limited, 2013

ethanol (a low concentration range can lead to a poor NIRS model) and this coincidentally dealt with the concern over collinearity in the dataset (as discussed above). The results of this study showed that NIRS could be effectively used in this highly challenging matrix, to control the glucose feed rate and hence culture specific growth rate, to minimise ethanol formation and to monitor protein content in near real time.

Giavasis *et al.* (2003) investigated the application of NIRS to monitoring gellan concentration in fermentations of *Sphingomonas paucimobilis*. Gellan is a widely used food thickener, gelling agent and stabiliser. Despite the highly viscous and pigmented (dark yellow) nature of the process fluid, it was possible to monitor gellan concentration accurately in on-line near real time using FT NIR. Gellan models showed a very good correlation coefficient (0.98), based on spectra taken in transmittance mode, after a second derivative pre-process was applied to the sample spectra in order to make key absorbance regions more readily distinct. This and other related studies applying NIRS in industrial fermentation and cell culture point towards the future utility of this technology in monitoring fermentations for production of food ingredients.

6.4.2   Mid-infrared spectroscopy

In principle, as Roychoudhury *et al.* (2006a) argue, owing to the much stronger absorbances in the fundamental IR much lower analyte levels are measurable using MIRS, than NIRS. However, it is essential to keep in mind that in fermentation systems, especially in a food context, the matrix is usually very complex, with multiple nutrients, cells, product(s) and by-products present. In this context, the spectra in the MIR may well be dominated by very strong peaks arising from the dominant analytes masking analytes present at lower concentrations or with much lower molar absorptivities.

Macauley-Patrick *et al.* (2003) showed just what MIRS is capable of in a microbially catalysed process for the manufacture of vitamin C. They monitored on line, using an attenuated total reflectance (ATR) sample presentation accessory, a key step in the industrial production of vitamin C, namely the conversion by *Gluconobacter* of sorbitol to sorbose. These two chemical species are closely structurally related but models for both analytes were formulated allowing monitoring of the process status every two minutes. In all probability this would not have been possible with NIRS owing to the inability of NIRS to distinguish effectively between two or more species with a high degree of structural similarity.

Although this and similar studies (Roychoudhury *et al.*, 2006b, 2007) indicate the potential of MIRS for monitoring food ingredient fermentation processes, MIRS has generally been far less widely employed in this role than NIRS. NIRS is simpler to deploy in fermentation systems using cheap fibre optic systems, whereas while MIRS *in situ* probes can be used, they

© Woodhead Publishing Limited, 2013

need expensive wave guides and multiplexing would presumably be challenging and perhaps costly. Issues with sample presentation and temperature equilibration are also possible. Despite the use of ATR to overcome many problems surrounding sample presentation, there may still be concerns over the transferability of models built using one ATR crystal to another instrument with another crystal. Macauley-Patrick (2003) also noted that clouding of the ATR crystal in the very high sugar-containing media became a serious concern after analysing only a modest number of samples. This led quickly to the ATR crystal being non-functional. High sugar concentrations (100s of g/L) and the presence of salts and weak acids are all characteristic of many bacterial fermentations in the food industry, so the ability of such fluids to damage, cloud or deplete ATR accessories may be a further concern in the context of deploying MIRS in industrial food ingredient fermentation processes.

6.4.3   Raman spectroscopy

Raman spectroscopy has been even less frequently used than MIRS in fermentation monitoring. In part this is due to the relatively high costs of Raman systems compared to NIRS/MIRS. The other issue is the fluorescence of many biological molecules, although this is less of an issue with more modern Raman instruments. A recent study (Abu-Absi *et al.* 2011) in a cell culture system shows the ability of Raman to deliver multi-analyte information on the state of the bioprocess, with information about the levels of a number of key analytes arising from an in line probe.

Another recent study used Raman to monitor the intracellular concentrations of carotenoids in fermentations of *Blakeslea trispora* (Papaioannou *et al.*, 2009). Although promising, the Raman prediction of total carotenoids content was not very accurate relative to the reference assay (HPLC). This points to the use of Raman as a tool for fermentation monitoring and control, but balanced against this is the added cost and complexity of Raman systems. It will be a long time before Raman is a routine tool in food ingredient manufacturing by fermentation given the low profit margins in the industry.

6.5   Future trends

Within the food production industry, fermentation processes occupy a key role, delivering pure ingredients with consistent character and functionality. This inherent advantage of fermentation systems is often overlooked. By comparison with many other bio(logical) production processes, fermentation, especially SLF systems, is relatively closely controllable.

In future relatively novel promising techniques, such as measurement of process heat evolution (Marison and Senthilkumar, 2009) may find

© Woodhead Publishing Limited, 2013

application in food fermentations. These measurements are relatively simple to make, give a direct indication of culture activity and status and also, since at large scale, removal of biologically generated heat can be the limiting factor, heat evolution measurements offer the ability to match bio heat generation to the heat removal capacity of the fermenter. However, despite being examined over many years, so far heat evolution measurements have not yet been widely applied in the food fermentation industry. Much the same can be said of techniques such as 2D fluorescence (Boehle *et al.*, 2003) and *in situ* imaging technologies (Bluma *et al.*, 2010). Both techniques have distinct advantages, including in the latter case information about shape, size and distribution of particles (cells and product aggregates), but in the immediate future their deployment in industrial food fermentations will be limited.

Instead the food fermentation industry will continue to deploy and develop monitoring techniques appropriate to its particular needs and drivers, using robust inexpensive sensor types to give it the information required to control its processes for manufacturing pure consistent food ingredients. As argued above, the food fermentation industry is something of a specialised sub-set of the general fermentation and cell culture industry and many of the monitoring solutions applicable in the wider industry will not readily be applied soon within the smaller context.

6.6   Sources of further information and advice

6.6.1   Food fermentation
1. *Food, fermentation and microorganisms.* C.W. Bamforth, Blackwell, London, 2005.
   An excellent starter read in the area of food fermentations. This book covers the origins of fermentation processes, most major food fermentations and even has a chapter on mycoprotein (Quorn) production.
2. *Handbook of Food and Beverage Fermentation Technology* (Food Science and Technology, Vol. 134). Y.H. Hui, L. Meunier-Goddik, J. Josephsen, W-K. Nip, P.S. Stanfield, Fidel Toldra (eds). Marcel Dekker, New York, 2006.
   Good wide ranging coverage of the subject.

6.6.2   Submerged liquid fermentations (SLF) and process monitoring
3. *Practical Fermentation Technology.* B. McNeil and L.M. Harvey (eds). Wiley, London, 2008.
   This multi-authored book has chapters on fermentation equipment, reactor design, modes of culture, statistically based process control, continuous cultures and real time process monitoring and is largely aimed at introducing the subject of fermentation processes to relative

© Woodhead Publishing Limited, 2013

beginners. It is largely experience based, giving guidance on many practical aspects of SLF.

6.6.3 Solid substrate fermentation processes

4. *Current Developments in Solid Substrate Fermentation*. A. Pandey, C.R. Soccol and C. Larroche (eds). Springer, Heidelberg, 2010.

    This is a multi-authored, very good introduction to the subject, covering general principles first and characteristics of SSF systems (kinetics, design, informatics), including monitoring, before moving on to various applications including discussion of recent and developing trends in SSF – bioremediation, biodegradation of pollutants, integration of SSF into the biorefinery concept, high value low volume products from SSF and biofuel applications. Written by experts and very readable, this book gives real insight into the recent reinvigoration of SSF in its own right and as a partner to SLF.

5. 'Potential of biofilm based biofuel production'. Wang, Z.W. and Chen, S.L. *Appl Microbiol Biotechnol*, **83**, 1–18, 2009.

    This review argues strongly that the traditional highly effective deployment of SSF in waste water treatment or environmental applications could readily be extended to the use of biofilms (immobilised cells or consortia) for production of a variety of products including biofuels. The advantages of biofilms are discussed in this application, including their ability to implement effectively complex multistage processes (saccharification, fermentation, recovery) using biofilms in conjunction with membranes. Most of the techniques described would be applicable to food production processes for enzymes or bioconversions.

6. 'Recent advances in solid state fermentation'. Singhania, R.R., Patel, A.K., Soccol, C.R. and Pandey, A., *Biochem Eng J* **44**, 13–18, 2009.

    This is a concise position statement type review which points out that SSF has traditionally been a method of effectively dealing with waste materials (agro-industrial residues) and that since we must now seek both new sources of energy/feedstocks based upon such wastes, and also new low energy production processes, the time has come to re-evaluate SSF. It is proposed that new developments in mathematical modelling, monitoring and reactor design could fuel the use of SSF in a wide range of new fermentation processes.

6.7 References

ABU-ABSI N.R., KENTY, B.M., CUELLAR, M.E., SAKHAMURI, S., STRACHAN, D.J., HAUSLADEN, M.C., JIAN, Z. and BORYS, M.C. (2011). 'Real time monitoring of multiple parameters in mammalian cell culture bioreactors using an in-line Raman spectroscopy probe'. *Biotechnol Bioeng*, **108**, 1215–21.

© Woodhead Publishing Limited, 2013

ARNOLD, S.A., HARVEY, L.M., MCNEIL, B. and HALL, J.W. (2002). 'Employing near-infrared spectroscopic methods of analysis for fermentation monitoring and control. Part 1: Method Development'. *Biopharm Int*, **15**, 26–32.

ARNOLD, S.A., HARVEY, L.M., MCNEIL, B. and HALL, J.W. (2003). 'Employing near-infrared spectroscopic methods of anlaysis for fermentation monitoring and control. Part 2: Implementation strategies'. *Biopharm Int*, **16**, 47–9.

ARNOLD, S.A., CROWLEY, J., WOOD, N., HARVEY, L.M. and MCNEIL, B. (2004). '*In situ* near infrared spectroscopy to monitor key analytes in mammalian cell cultivation'. *Biotechnol Bioeng*, **84**, 13–9.

BLUMA, A., HOPFNER, T., LINDNER, P., REHBOCK, C., BEUTEL, S., RIECHERS, D., HITZMANN, B. and SCHEPER, T. (2010). '*In situ* microscopy sensors for bioprocess monitoring: the state of the art'. *Anal Bioanal Chem*, **398**, 2429–38.

BOEHLE, B., SOLLE, D., HITZMANN, B. and SCHEPER, T. (2003). 'Chemometric modelling with 2 dimensional fluorescence data for *Claviceps purpurea* bioprocess characterisation'. *J Biotechnol*, **105**, 179–88.

CERVERA, A.E., PETERSEN, N., LANTZ, A.E., LARSEN, A., GERNAEY, K.V. and KRIST, V. (2009). 'Application of near-infrared spectroscopy for monitoring and control of cell culture and fermentation'. *Biotechnol Prog*, **25**, 1561–81.

CROWLEY, J.C., ARNOLD, S.A., WOODS, N., HARVEY, L.M. and MCNEIL, B. (2005). 'Monitoring a high cell density recombinant *Pichia pastoris* fed-batch fermentation using transmission and reflectance NIRS'. *Enzyme Microb Technol*, **36**, 621–8.

FAZENDA, M., HARVEY, L.M. and MCNEIL, B. (2011). 'Cell culture PAT multiplexing NIR'. In *Process Analytical Technology Applied in Biopharmaceutical Process Development and Manufacturing*, Undey, D.L.C., Menezes, J.C. and Koch M. (eds), Taylor and Francis, London.

FINN, B. L. M HARVEY, L.M. and MCNEIL, B. (2006). 'Near infrared spectroscopic monitoring biomass, glucose, ethanol and protein content in a high cell density baker's yeast fed-batch culture'. *Yeast*, **23**, 507–17.

FINN, B., HARVEY, L.M. and MCNEIL, B. (2010). 'The effect of dilution rate upon protein content and cellular amino acid profiles in chemostat cultures of *Saccharomyces cerevisiae* CABI 039916'. *Int J Food Eng*, **6**, 1-21

GE, Z., CAVINATO, A.G. and CALLIS, J.B. (1994). 'Noninvasive spectroscopy for monitoring cell density in a fermentation process'. *Anal Chem*, **66**, 1354–62.

GIAVASIS, I., MCNEIL, B. and HARVEY, L.M. (2003). 'Simultaneous and rapid monitoring of gellan and biomass using Fourier transform near infrared spectroscopy'. *Biotechnol Lett*, **25**, 975–9.

LANDGREBE, D., HAAKE, C., HOPFNER, T., BEUTEL, S., HITZMANN, B., SCHEPE, T., RHIEL, M. and REARDON, K.F. (2010). 'On line near infrared spectroscopy for bioprocess monitoring'. *Appl Microbiol Biotechnol*, **8**, 11–22.

MACAULEY-PATRICK, S.E. (2003). *Physiological Studies on the Biotransformation of D-sorbitol to L-sorbose by* Gluoconobacter suboxydans. PhD Thesis, University of Strathclyde.

MACAULEY-PATRICK, S.E., ARNOLD, S.A., MCCARTHY, B., HARVEY, L.M. and MCNEIL, B. (2003). 'Attenuated total reflectance Fourier transform mid-infrared spectroscopic quantification of sorbitol and sorbose during a Gluconobacter biotransformation process'. *Biotechnol Lett*, **25**, 257–60.

MARBACH, R. (2007). 'Multivariate calibration: a science based method'. *Pharm Manufact*, **Jan**, 42-46.

MARISON, I. and SENTHILKUMAR, S. (2009). 'Progress and challenges on design of high sensitivity biocalorimeter for inline bioprocess monitoring'. *New Biotechnol*, **25**, S325–6.

MATTES, R., ROOT, D., MISA, A., SUGUI, M.A., CHEN, F., SHI, X., LIU, J. and GILBERT, PA. (2009). 'Real-time bioreactor monitoring of osmolality and pH using near-infrared spectroscopy'. *Bioprocess Inter*, **April**, 44–50.

© Woodhead Publishing Limited, 2013

PAPAIOANNOU, E.H., LIAKOPOULOU-KYRIAKIDES, M., CHRISTOFILOS, D., ARVANITIDIS, I and KOUROUKLIS, G. (2009). 'Raman spectroscopy for intracellular monitoring of carotenoid in *Blakeslea trispora*'. *Appl Biochem Biotechnol*, **159**, 478–87.

PRETORIUS, I.S., DU TOIT, M. and VAN RENSBURG, P. (2003). 'Designer yeasts for the fermentation industry of the 21st century'. *Food Technol Biotech*, **41**, 3–10.

ROYCHOUDHURY, P., HARVEY, L.M. and MCNEIL, B. (2006a). 'The potential of mid infrared spectroscopy for real time bioprocess monitoring'. *Anal Chim Acta*, **571**,159–66.

ROYCHOUDHURY, P, HARVEY L.M. and MCNEIL, B. (2006b). 'At-line monitoring of ammonium, glucose, methyl oleate and biomass in a complex antibiotic fermentation process using attenuated total reflectance mid-infrared spectroscopy.' *Analy Chim Acta*, **561**, 218–24.

ROYCHOUDHURY, P., MCNEIL, B. and HARVEY, L.M. (2007). 'Simultaneous determination of glycerol and clavulanic acid in an antibiotic bioprocess using attenuated total reflectance mid-infrared spectroscopy.' *Analy Chim Acta*, **585**, 246–52.

SCARFF, M., HARVEY, L.M. and MCNEIL, B. (2006). 'Near infrared spectroscopy: current status and future developments'. *Crit Rev Biotechnol*, **26**,17–39.

SEVIOUR, R.J., MCNEIL, B., FAZENDA, L.M. and HARVEY, L. (2011). 'Operating bioreactors for microbial exopolysaccharide production'. *Criti Rev Biotechnol*, **31**(2), 170–85.

VAIDYANATHAN, S., MACALONEY, G., VAUGHAN, J., MCNEIL, B. and HARVEY, L.M. (1999). 'Monitoring of submerged bioprocesses'. *Crit Rev Biotechnol*, **19**, 277–316.

VAN DER AAR, P.C., STOUTHAMER, A.H. and VAN VERSEVELD, H.W. (1989). 'Possible misconceptions about O_2 consumption and CO_2 production measurements in stirred microbial cultures'. *J Microbiol Methods*, **9**, 281–6.

© Woodhead Publishing Limited, 2013

7

Industrial enzyme production for the food and beverage industries: process scale up and scale down

S. M. Stocks, Novozymes A/S, Denmark

DOI: 10.1533/9780857093547.1.144

Abstract: Enzymes have many applications in the food and beverage industries and are produced in large quantities by large scale aerobic fermentation processes. Scale up of these fermentation processes is discussed in the context of process development for large scale enzyme production for the food and feed industry. Gassed agitated aerobic fermentations of differing rheological complexity are considered. Equations related to oxygen transfer and 'mixing' are presented and used to illustrate some non-intuitive phenomena that are often overlooked in consideration of 'scale up'. A 'scale down' mentality is presented where the final production scale is defined and used to find relevant conditions for experiments in smaller scale equipment. These results can then be used for development of empirical or mechanistic models to facilitate reduced risk on subsequent scale up. It is demonstrated that a perfect scale down cannot be accomplished and that scale up therefore remains an exercise in risk mitigation that is a mixture of both art and science.

Key words: enzyme production, fermentation, mass transfer, mixing, scale down, scale up, stirred tank reactor.

7.1 Introduction

There are many articles and book chapters on the subject of 'scale up'; why yet another? Readers of this chapter, in this particular book, almost certainly have no need for an introduction to biotechnology, or what is now established as the standard way to produce almost any of the products established in the biotech market, that is submerged aerobic fermentation (almost universally adopted). This subject has been covered comprehensively in other texts such as Stanbury *et al.* (1995). There are also many well-regarded courses on the subject of fermentation and biotechnology, which an interested reader may well have attended (for example, Berovic

© Woodhead Publishing Limited, 2013

and Nienow, 2005). Design of bioreactors has been extensively covered too (examples include Schugerl, 1991; Van't Riet and Tramper, 1991; or Atkinson and Mavituna, 1991). As such, scale up is something about which a great deal has been written, but there appears to be no one universally agreed approach. This lack of agreement leads to some confusion for those entering the discipline; what to believe?

The lack of agreement probably reflects the differences in physiology and genetic or metabolic regulation around production of the desired product, that is to say, there can be no agreement when the products and methods of production are necessarily so diverse. For this reason the scope of this chapter will be limited to the subject of submerged aerobic fermentation for the production of industrial enzymes, those enzymes sold from business to business in large quantities, amounting to several tens of thousands of tons each year (more precise figures can possibly be extracted from the financial reports of Novozymes A/S and its competitors). The bulk of these enzyme preparations end up in laundry detergents and an increasing amount is used for the production of fuel from corn or agricultural residues like corn stover or straw. Somewhere in between and representing a significant and strong business for enzyme production companies, are a diverse number of enzyme preparations sold to the food sector for the degradation of starch into sweeter substances, their isomerisation and modification, for juice or oil extraction, wine production, brewing and so on. Proteases are also an obvious class of use to the food industry, for de-bittering or flavour modification.

Juice and wine production is probably the most traditional area where enzymes have improved juice yields, or reduced the residual starch in the production of 'lite' brews, or improved clarity or productivity. There is now a range of enzymes which allows beer production without malt, which is good news for brewers facing a shortage of quality malts. There are enzymes for baking too, which can increase shelf life of bread or cake by altering starch crystallisation, slowing the rate of staling, or which can remove asparagine before the cooking process, to reduce acrylamide formation. Lipases are also very interesting for bakers, they can also be used in the dairy for better control of the cheese making process. And, since meat is also a big part of the food industry, we might include enzymes added to animal feeds in this book chapter; these enzymes improve the nutritional value of the feed by, for example, freeing phosphate from phytate, allowing reduced inorganic phosphate consumption at the farm, and therefore reduced amounts in waste water, yielding financial benefits for the farmer and also the consumer.

Generally speaking, enzymes are used when some customer benefit is obtained that cannot be obtained more economically by other means. Enzymes are so successful because they are relatively inexpensive natural products with great potency that eliminate or reduce the need for chemical additives or high temperatures and pressures seen in other process

© Woodhead Publishing Limited, 2013

industries. Enzymes are a natural choice and they are big business. If further information on enzymes themselves is required, then the reader is encouraged to consult a well-known Novozymes information book (Olsen, 2004).

There is nothing new about the enzyme business (Aunstrup, 1979). In fact, the only changes since the cited 1979 review article, seem to be the ones predicted within it: proliferation and commoditisation of genetic information and computational power has made it possible to produce almost any desired enzyme in one host organism or another, and through similarly enabled automation of enzyme and strain screening technology, production in larger quantities with higher yields is now possible. The enzyme producer is still faced with the challenge of producing more of everything as each year passes: thousands of large scale batches each year, from a diverse portfolio. This means that each piece of production equipment should be either in use or under brief planned maintenance 100% of the time and every batch should result at least in the planned amount of product, so the chosen technology for production needs to be robust and easily serviceable. Also a large optimisation effort is applied to avoid the need for constant investment in new production capacity. This constant drive for capacity increase at low cost probably explains the prevalence of the stirred tank reactor for enzyme production. It is a well established and reliable technology with proven commercial success that is very versatile, easily understood and operated and large versions are well supported by a diverse number of engineering companies. More exotic forms of production equipment are perhaps best left to more exotic types of business.

Also, it is not surprising that production of enzymes for food, feed, or even technical applications, is still done with GRAS (generally regarded as safe) microorganisms (not *E.coli* which features heavily in the literature on fermentation technology). Choosing a GRAS organism may increase the workload in R&D, and development as off the shelf tools or texts are harder to find or create than those for the molecular biologists' most favourite work horse (*E.coli*). GRAS organisms include *Aspergillus oryzae* (famous for its role in soy sauce production), *Aspegillus niger* (citric acid production) and harmless *Bacillus* species that can be found in a variety of fermented foods.

Where does scale up come into this? In this context, the above-mentioned production equipment is considered 'large scale' and, as already mentioned, nothing stands still in a competitive industry. In order to make the most of the capital investment and to stay competitive on price, constant improvements in yields and productivity are needed, as well as the introduction of new products. Thus, scale up, or perhaps more appropriately, scale down, is required for optimisation of existing products and for development of new ones. Of course, scale up as a discipline is not confined to an existing enzyme production business. Many new start up companies or young researchers with good candidate products may struggle with scale up – and

© Woodhead Publishing Limited, 2013

this is an important point in dealing with the contradictions and confusion surrounding scale up that exists in the literature – it is one thing to scale up or down surrounded by professionals in a successful company that has existing and well-defined production equipment with a high number of product introductions each year (i.e. a large experienced peer group can be called upon) and quite another situation to face the problem almost alone and for the first time, without existing production equipment in mind and with the barrier of secrecy agreements preventing an open discussion with potential help.

In the following sections I hope to bring some of the perspective gained in my few years working in the pilot plant at Novozymes, a large pilot plant based at head office in Denmark, serving the existing factories and business of Novozymes A/S. The overall aim will be to help clarify and explain some of the apparent contradictions in the scale up literature to a general reader based upon my direct experience. By way of a disclaimer, the content of these sections represents the personal views of the author and not necessarily the consensus of my peer group of experienced and talented colleagues.

7.2   Difficulties of the scale up approach

As mentioned above, there is no real consensus on a single satisfactory approach to scale up and there is some lack of clarity and apparent contradiction in the literature that can be confusing. The best approach to this is to identify first some common ground about which there is consensus, these being the basic concepts, and then use them to illustrate the challenges faced on scale up.

7.2.1   Basic concepts

All of the following concepts are covered much more extensively in the sources already cited, but the following texts are particularly useful, *Mixing in the Process Industries* (Harnby *et al.*, 1992), *Handbook of Industrial Mixing* (Paul *et al.*, 2004), *Biochemical Engineering and Biotechnology Handbook* (Atkinson and Mavituna, 1991), *Fluid Mixing and Gas Dispersion in Agitated Vessels* (Tatterson, 1991) and *Bioreaction Engineering Principles* (Nielsen *et al.*, 2003).

7.2.2   Geometric similarity

Geometric similarity means that the ratios of the dimensions of the tanks at various scales are similar, for example the height divided by the diameter (H/T = aspect ratio) is the same, or the agitator diameter is always half the tank diameter ($D/T = 0.5$). In real life this is rarely true, but needs to be

© Woodhead Publishing Limited, 2013

approximately correct or there can be difficulties with some of the calculations that follow. In practice, the actual aspect ratios of steam sterilisable (pressurised) large scale equipment vary somewhat. It is possible to get an idea of what is tried and tested by surveying the literature, but it is perhaps something of a side track to go hunting for these details as the diameter of a vessel is often specified by the footprint available in the factory or the height of a nearby bridge, under which the tanks or its sections must be transported. The final height of the tank then depends on the required volume and the final aspect ratio is something that just has to be lived with. Values from 2–3 are common and even higher for airlift reactors. So, not too much attention will be paid to aspect ratios.

7.2.3 Power draw

If there is one thing that gets fermentation scientists agitated it is the question of how much stirring power to add to a tank. If you have some reliable measurements to work from this is relatively easy to calculate at different scales. The power draw is given by equation [7.1]. This is derived from a dimensional analysis which is better described by others (e.g. Tatterson, 1991). Similar equations appear for boat and aircraft propellers as well as wind turbines:

$$P = P_0 \rho N^3 D^5 \tag{7.1}$$

where P is the power (W), P_0 the 'power number' (dimensionless), N is the impeller speed (revolutions per second) and D is the agitator diameter (m). One difficulty with this equation is which value to use for the power number P_0 (or under gassed conditions, P_0 is substituted by P_g). P_0 can vary as a function of the Reynolds number (Re), which is an important quantity that allows us to determine if the flow caused by the agitator is expected to be turbulent or not (equation [7.2] where ρ is the density (kg m^{-3}), and μ the viscosity (Pa s).). In selecting a value for P_0, it is advisable to choose sources in the literature carefully, since close geometric similarity can be important, or preferably make real measurements in the system being studied. Be careful to check that N and D are accurate.

$$Re = \frac{\rho N D^2}{\mu} \tag{7.2}$$

Generally, for industrial enzyme fermentations, the aim is to have the impellers achieve turbulent flow, with a Re of 10^3 or 10^4 or more. Most usefully, for the most relevant impeller types, once turbulent, the P_0 is constant as a function of the Reynolds number. This results in the first counterintuitive phenomenon that confuses many researchers. For turbulent flow, at constant impeller speed, increased viscosity does not increase power draw. This is confusing, but is obviously very useful to know when considering doing something expensive at large scale. Now comes the second most important

© Woodhead Publishing Limited, 2013

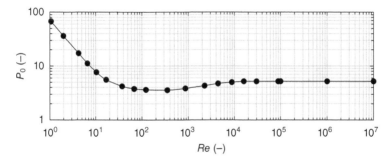

Fig. 7.1   Un-gassed power number, P_0 as a function of Re for a traditional Rushton disc turbine. Modified from Bates *et al.* (1963).

fact: at Re valves less than 10^3, P_0 varies as a function of Re, and again counterintuitively, for this type of impeller (Rushton disc turbine), as viscosity increases, the Re decreases *and so does the power number and therefore the power draw*. At least until Re values of less than 100 are reached (see Fig. 7.1).

7.2.4   Importance of impeller placement

If the impeller is located too close to the bottom of the vessel, there could be some unexpectedly low power numbers (Albaek *et al.*, 2008). If adding more than one impeller, they must be sufficiently separated before the power draws can be simply added as if they acted independently, although with a clearance of one tank diameter this can be done (Hudcova, 1989). However, it seems unlikely that tanks for liquid height aspect ratios of 2, that only two impellers would be mounted. Some consideration of these factors is needed and for practical reasons, the spacing will most likely be less than one tank diameter.

7.2.5   Gassed power

Power under gassed conditions, at equal impeller speed to ungassed conditions, has a tendency to be lower (Fig. 7.2). How much lower depends again on the agitator type and the source of the data. Most useful literature reports the ratio of P_g/P_0 as a function of Fl_g, the gas flow number (equation [7.3]):

$$Fl_g = \frac{Q}{ND^3} \qquad\qquad [7.3]$$

where Q is the gas flow rate ($m^3\ s^{-1}$). Again there is much data in the literature to choose from, but for fine tuning, caution on the source and geometry is advised. Generally though, in turbulent conditions where the

© Woodhead Publishing Limited, 2013

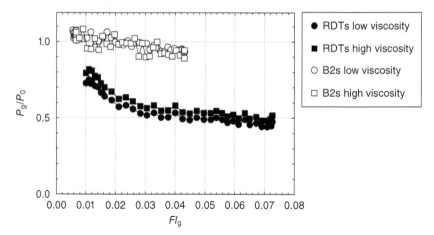

Fig. 7.2 P_g/P_0 as a function of Fl_g at two different viscosities (Stocks, 2005). Under relevant gassed conditions the power draw for Rushton disc turbines (RDTs) tends to be approx half that for ungassed conditions. Data is from 550Litre Pilot plant tanks. More modern impeller types like B2s do not suffer from this phenomenon.

P_0 is approximately constant, for Rushton disc turbines (RDT's) with fermentation relevant gassing, P_g is about 0.5 that of P_0, with the least gas sensitive most modern impeller types at about 0.8 to 0.9 (Albaek *et al.*, 2008). It should be noted that most manufacturers of very expensive large scale equipment know this and so fermenters are designed to deliver their power under the relevant gassed conditions, but the plethora of small scale equipment manufacturers are not as aware. Note, that unless very modern impeller types are installed in a fermenter, this effect also means that large scale equipment probably cannot be operated at full speed without gassing. More modern impellers (hydrofoils), pumping upwards do not suffer such a large fall in power, offering certain advantages for flexibility in operation, as well as other advantages such as reduced mixing times and better foam tolerance (Nienow and Bujalski, 2004). In industry, examples of all kinds of impeller types can be found, making agitation system design a subject worthy of a separate chapter, for clarity here, only the well-known and historically widely adopted Rushton disc turbines will be considered.

$k_L a$

Much has been said on the subject of $k_L a$ and various reviews exist (a recent one was Garcia-Ochoa, 2009). Essentially $k_L a$ measures the ease (or difficulty) with which oxygen makes its way through the interface between the bubbles and the liquid in the fermenter–'k_L' is the Liquid side gas/liquid interface mass transfer coefficient and 'a' is the area for transfer available in a unit volume of broth–so high $k_L a$ values are most desirable. Chemical

© Woodhead Publishing Limited, 2013

engineers seem quite fond of calculating k_L and 'a' separately, but then they often have the luxury of working with a few well-defined reactants, while fermentation broth (the whole process fluid), is very complex. k_La in fermentation is therefore best treated as the single entity most biotechnologists are familiar with. Various arguments about the other resistances to transfer (e.g. gas–liquid & liquid-cell side) are sometimes heard, but most biochemical engineers, rightly or wrongly, have been taught to ignore them (Nielsen *et al.*, 2003). k_La is then used to explain quantitatively how oxygen transfer can increase as a function of power or gas added in empirical relationships such as equation [7.4].

$$k_L a = K \left(\frac{P}{V}\right)^{\alpha} (v_s)^{\beta} \qquad [7.4]$$

In this equation, K, α and β are constants, V is the broth volume (m^3). Here, P/V is the power per unit volume of broth (W kg^{-1} or kW ton^{-1} if the density is water like). Some authors like to include power from the gas here, but it is not a universally adopted practice, probably because it is only significant at larger scales. v_s is the 'superficial gas velocity' (m s^{-1}), that is the velocity of the gas rising through the reactor as if it was a single plug flowing upwards through the tank. There is not much agreement on the values for the exponents K, α or β although it is generally agreed they need adjusting whether the medium is coalescing or non-coalescing, most fermentation broth is not coalescing, as the bubbles do not coalesce easily. Sometimes there is criticism of the measurement methods applied and speculation over what the real values might be. However, plotting the correlations usually reveals that despite this, the predicted k_La values are quite similar. This is explained because small reductions in α can be compensated for by an increase in β, without affecting the predicted k_La too much as α and β are correlated to each other.

Despite what some vendors or academics may state, within the accuracy of this type of correlation, there is no dependency on, or at best only a weak dependency on agitator type (Cooke *et al.*, 1988; Zhu *et al.*, 2001). It is important to note here that much of the literature on k_La completely ignores the effects of viscosity, which in the author's experience for most of the industrially important enzyme production processes, is a mistake that simply cannot be ignored. The viscosity can mostly be seen to appear in the form of equation [7.5].

$$k_L a = K \left(\frac{P}{V}\right)^{\alpha} (v_s)^{\beta} (\mu_{app})^{\delta} \qquad [7.5]$$

Here μ_{app} is the apparent viscosity (Pa s) and δ is another empirically determined constant. So, before this equation can be used, some understanding of broth rheology and viscosity is needed. For Newtonian systems, like oils or syrups, it is relatively trivial to determine a relevant viscosity.

© Woodhead Publishing Limited, 2013

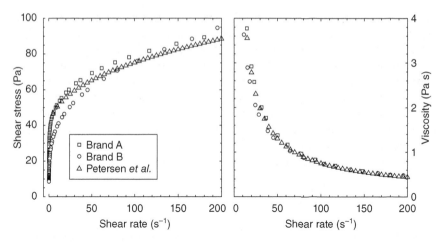

Fig. 7.3 Stress vs. rate and viscosity for two brands of ketchup, popular in Denmark; a shear stress sweep from 80 to 0 Pa over 2 min was used with a 14-mm radius vane with 42-mm immersion in a 15-mm radius cup at 28°C (AR-G2 Rheometer from TA instruments). It was then compared to some most viscous rheological data from *Aspergillus oryzae* (Petersen *et al.*, 2008: Ty 35 Pa, K 6 Pa. sn, *n* 0.43). It can be seen that the properties of the most viscous broth and ketchup are approximately similar.

Typically, and unfortunately, many fermentation types have broth of a non-Newtonian type, although thankfully, of the shear thinning variety. This is most often described by a simple power law (equation [7.6]).

$$\mu_{app} = \frac{\tau}{\gamma} = \frac{K\gamma^n}{\gamma} \qquad [7.6]$$

Here τ is the shear stress (Pa) and γ the shear rate (s^{-1}). K and n are some new constants. A value for n of 1 would indicate Newtonian behaviour, less than 1 and the broth becomes thinner (μ_{app} reduces) as the shear rate increases. In the author's direct experience, 'viscosity' or rheological properties worse than those of popular brands of ketchup are not especially favourable for trouble free scale up/down (see Fig. 7.3).

The non-Newtonian shear thinning nature of most fermentation broth types means that the viscosity experienced by the stirrer in the tank reduces the faster it is stirred (Fig. 7.3), some find this a difficult concept to grasp. The simple shear thinning equation (equation [7.6]) can also be modified to include a shear stress 'τ_y', which indicates that a certain stress must be applied to the fluid before it starts to flow (equation [7.7], the Herschel–Bulkley equation):

$$\tau = \tau_y + K\gamma^n \qquad [7.7]$$

This is a classic property of ketchup, which does not flow out of the bottle until you help it by shaking or, more recently, squeezing it. Fermentation

© Woodhead Publishing Limited, 2013

scientists can experience the same problem with fermentation samples. As can be seen, this yield stress value is approximately 20–40 Pa (Fig. 7.3), but it is notoriously difficult to get very good measurements of yield stress as measurements at low shear rates are tricky for various reasons, especially in suspensions of particles like ketchup or fermentation broth (Barnes *et al.*, 1989; Mezger, 2006). As long as rheological data is available, either model (equation [7.6] or [7.7]) usually provides a good fit to the data in the relevant shear rate range. The choice of model depends mostly upon whether or not an estimate of the yield stress is required. Other models exist, in which case texts on rheology should be consulted (Barnes *et al.*, 1989; Mezger, 2006).

The final complication here is to decide at what shear rate to evaluate the viscosity in the fermenter. Again, there is some controversy surrounding this calculation, but ignoring criticisms and proceeding to make this estimation with the Metzner and Otto correlation is what is typically done (equation [7.8]). Here the relevant shear rate is simply proportional to the impeller speed, irrespective of scale. Depending on agitator type, the constant is between 10 and 13 (reviewed by Tatterson, 1991). Looking at Fig. 7.3, it can be seen that the value of the constant is not extremely critical as it confines the result to a fairly narrow range, the position of which is more sensitive to the impeller speed (N, rps). So again, a complicated scale up or down problem arises: as N tends to decrease as scale increases, so does the calculated shear rate and so the viscosity in the same broth will tend to increase (more on this later).

$$\gamma = K \cdot N \qquad\qquad\qquad [7.8]$$

7.2.6 The fed batch

There are a lot of articles that give various complicated descriptions of the fed batch process, but the basic concept really is very simple. Simply put, in a fed batch process, a carbohydrate solution or syrup containing the primary carbon source is fed to the reactor at a rate that ensures that the oxygen demand in the tank is no more than the rate at which oxygen can be transferred from the gas to the liquid (equation [7.9]). In this way, the oxygen level, p_{O_2}, in the reactor remains above zero, preferably above a specified value which is known to be important for productivity. This can then become the 'scale up rule'; assuming that the carbon source is used at the same rate as it is supplied, it can be scaled up or down in such a way that the p_{O_2} does not fall below a desired value by manipulating the feed rate or the other parameters affecting oxygen transfer. Note, that not all fermentation scientists agree with this basic description, nor do all fermentations follow this empirical rule, but the present author has had considerable success applying it generally, whether it is completely accurate or not (e.g. Albaek *et al.*, 2008, 2012).

© Woodhead Publishing Limited, 2013

$$F \propto OUR \leq K \times \left(\frac{P}{V}\right)^{\alpha} (v)^{\beta} (\mu_{app})^{\gamma} (c^* - c)$$ [7.9]

The feed rate (F), is proportional to the oxygen uptake rate (OUR), which is less than the rate that can be supported by the fermentation system. c^* is the saturation concentration of oxygen and c is the actual concentration (various units are applied in the literature); $c^* - c$ provides a driving force for transfer from the assumed to be saturated gas side of the gas bubbles' liquid interface to the liquid side where the fermentation takes place.

The fed batch is an elegant trick allowing considerably more carbohydrate to be converted to product than in a simple batch process, where all the carbohydrate is included in the initial substrate mixture and the oxygen demand can quickly rise above the rate at which it can be supplied, resulting in short processes of low product concentration. In the fed batch, the process runs at different average rate, but for longer and achieving higher product concentrations. Also, by eliminating the downtime from restarting short batches, more product per unit time can be made and with less labour. High product concentrations require volumetrically less downstream processing, which can be done with higher yields, and less waste water is produced. If not already doing it, the fed batch is probably the single most simple and elegant trick you can play to increase productivity, boost product concentration and harvest savings on recovery and water recycle/disposal. This depends a little on the microorganism and the product, but for the aerobic processes exploited in enzyme production, it cannot be beaten. There are some who eloquently advocate the use of continuous processes, but there are some practical, capital and operating cost constraints that need to be considered and these are specific to the business case, strain stability and product. Therefore, the final choice between batch, continuous, fed batch or different variations of these, is a business consideration without one universal answer, but fed batch tends to be the one that is chosen most frequently (Stanbury et al., 1995; Nielsen, 2006).

7.3   Consequences of changing scale

The reader is now armed with a few equations and some knowledge that can be applied quantitatively to the problems of scale up.

Many of the pitfalls experienced by some during scale up can be avoided by ignoring it completely and adopting instead a scale down mentality. This is a highly effective technique reflected not only in biotechnology, but across the whole field of commerce and personal development or achievement, where it is sometimes described as 'begin with the end in mind' (Covey, 1989). To illustrate this some reactor specifications from *Mixing in the Process Industries*, Chapter 15 in Harnby et al. (1992) can be used. The 'end'

© Woodhead Publishing Limited, 2013

will be some 100 m^3 working volume production fermenters that are equipped with large 700 kW motors and are connected to an air supply of sufficient over pressure to deliver 0.5 normal volumes of gas per liquid volume per minute (NVVM). When doing this sort of thing, keep in mind that the largest steam sterilisable fermenter ever constructed had a working volume of 2300 m^3 (Smith, 1980) and although it was an airlift fermenter, it is a good indication that we will be able to work comfortably in the much smaller stirred 100 m^3 tanks (see Fig. 7.4). Some further assumptions can be made about these fermenters, such as, the ungassed liquid height to diameter ratio is approximately 2, the power from the motor is delivered by some old fashioned, but tried and trusted Rushton disc turbines. What happens to the described equations as scale changes can now be calculated (see following sections).

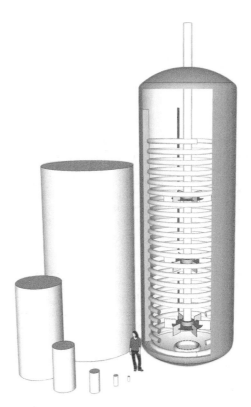

Fig. 7.4 To give a better feeing of scale, right circular cylinders of $H/T = 2$ are shown representing ungassed broth of volumes of 1 litre, 10 litres, 100 litres, 1 m^3, 10 m^3 and 100 m^3, with the final 100 m^3 fitting into the depicted production scale fermentation tank, which has space for gas hold up.

© Woodhead Publishing Limited, 2013

7.3.1   Reynolds number as a function of scale

It is well known that the agitation speeds in small tanks can get quite extreme (more than 1000 rpm, for example). Many running these small tanks are unaware of why this happens, although hopefully it will become clearer in time. Table 7.1 shows some typically observed speeds for various scales of equipment and the expected Reynolds number ranges based on water or broth. These are then plotted in Fig. 7.5 to show how they compare

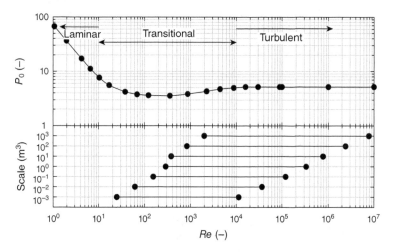

Fig. 7.5   The power number (P_0) is reproduced once more, with the ranges for expected Reynolds numbers at various scales indicated beneath it. The upper limit for Re is calculated as if water is in the tank and the lower limit as if *Aspergillus oryzae* broth is in the tank (parameters from the legend of Fig. 7.3). It is not possible to scale up or down with the same Reynolds numbers, but there is some overlap between scales and, counterintuitively, Reynolds numbers tend to be confined to the transitional region as scale decreases.

Table 7.1   Assuming an ungassed liquid filling with an aspect ratio of 2, we can calculate tank diameter and height for various fillings or scales. Typical agitation speeds are shown and Re is calculated for water (Re_{max}) or for a fill of *A.oryzae* broth (parameters from Fig. 7.3) at the shear rate calculated in equation [7.8]

Scale (m³)	T (m)	D (m)	N (rpm)	Re (min)	Re (max)
0.001	0.09	0.026	1000	2.5×10^1	1.1×10^4
0.01	0.19	0.056	700	6.1×10^1	3.7×10^4
0.1	0.40	0.12	500	1.5×10^2	1.2×10^5
1	0.86	0.26	300	2.9×10^2	3.3×10^5
10	1.85	0.56	150	3.8×10^2	7.7×10^5
100	3.99	1.2	100	8.3×10^2	2.4×10^6
1000	8.60	2.6	70	2.0×10^3	7.8×10^6

© Woodhead Publishing Limited, 2013

to the power number curve and the regions of laminar or turbulent flow. Note that it is not possible to scale down at the same Reynolds number, as Re tends to decrease with scale and enter deep into the transitional region.

This is an important point as it is intuitive to assume that 'mixing is better at small scales'. This is probably true in terms of concentration gradients, but only up to a certain viscosity, when flow in the small tank will be moving towards the laminar region while a large tank is still turbulent. In a small glass tank where visual inspection is possible, a lack of motion in the broth may be seen and the conclusion drawn that the broth is too thick which, in turn, leads the operator to raise the speed. This partly explains why extreme rpm's are often encountered in small tanks, at least for anything which is industrially relevant. Having lots of probes sticking into a small broth volume also helps to reduce motion, especially in small tanks. So, at small scales the need to raise stirrer speeds can arise directly as a result of observed poor motion in the broth, other reasons will follow.

7.3.2 v_s as a function of scale

In order to make things simple here, the assumption will be made that as scale changes similar 'VVM' of gas will be used, where VVM is 'volume of gas per volume of liquid per minute'. So, for 0.5 VVM, which is not an uncommon value (Harnby *et al.*, 1992), at 10 L scale, 5 L min^{-1} will be added, and to keep things simple this will be equated to 5 N L min^{-1}, where the N tells us that this is 5 L of air at normal pressure and temperature. This is a useful way to scale up or down, as not only must we add oxygen, we must also remove CO_2. High CO_2 partial pressures are not a good idea as they can have some inhibitory effects on metabolism, so it is a good idea to keep the concentrations approximately the same across scales by using approximately the same amounts of gas (some consideration of the partial pressure, $p_{CO2,}$ may be needed for fine tuning (McIntyre and McNeil, 1997; Albaek *et al.*, 2012)).

v_s is an intrinsically simple concept, but it is not a completely trivial one: the volume of a gas changes as a function of pressure. In small non-pressurised tanks, the volume per unit time at the inlet can be calculated and then divided by the cross sectional area of the reactive volume, using this value of velocity without too much criticism. But on scale up the pressure in the bottom of the tank is not the same as the pressure at the top. For a 10 m liquid column, one extra atmosphere of pressure at the base would be expected, or twice the difference and so only half the volume of gas. The question then is, should an average v_s be used or v_s at the inlet or outlet? It is intuitive to think that there is a simple answer to this question, but the author has not yet been convinced of a truly correct method. It is more important to be consistent, so when considering scale up or scale down, define the rules and apply them consistently to all scales. The author routinely uses the inlet v_s as seen in Fig 7.6. Note, that for the k_La

© Woodhead Publishing Limited, 2013

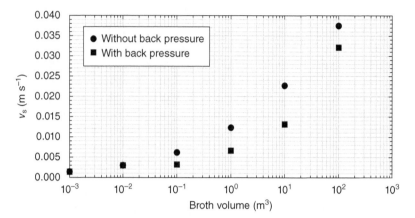

Fig. 7.6   Inlet superficial gas velocity as a function of scale with fixed 0.5 N VVM and ungassed aspect ratio $H/T = 2$, without back pressure and with some back pressure. Superficial gas velocity tends to increase, providing some extra k_La compared to smaller scales.

calculation, the value of v_s to the selected exponent is usefully flattened, so while v_s might increase by more than 10 times on scale up, v_s^β will not more than double.

7.3.3   Pressure as a function of scale

Pressure is important as it has a large influence on the value of c^*, which is a major feature of the oxygen transfer equation (equation [7.9]). For 21% O_2 (air) we can say that the partial pressure of oxygen is 21% of the actual pressure (back pressure + atmospheric pressure + pressure from depth) and that c^* is proportional to this partial pressure. As discussed in the previous section, the inlet pressure in the tank tends to increase as the liquid height increases (this is something often forgotten), if the pressure in a large tank is twice that at small scale, about twice the oxygen transfer will be available if everything else is equal. Often some back pressure is applied to the reactors. There are a number of reasons for this, first, the volumes of gas involved and sensible exhaust diameters means that some back pressure in the larger tanks will be inevitable, but another reason is that to maintain sterility, having some back pressure is a good idea as the increased positive pressure will tend to exclude potential contaminants (Stanbury *et al.*, 1995). On the other hand, adding significant back pressure to glass (or more recently, plastic) reactors in small scale, is not usually a good idea for safety and containment reasons. Adding too much back pressure at larger scales will increase energy consumption for gas compression and is especially wasteful if throttling with a back pressure control valve. This 'waste' might be tolerated by a good business case, where the extra productivity pays for the

© Woodhead Publishing Limited, 2013

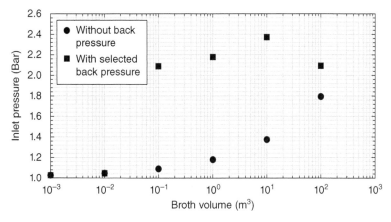

Fig. 7.7 Inlet pressure without back pressure as a function of scale and what it might be if some back pressure is added by a control valve on the exhaust. Large scale pressures can be simulated at pilot scales, but can be more difficult in laboratory scale glass or plastic reactors where adding pressure can be more difficult.

energy consumed, but there is increasing concern over fossil fuel and its associated emissions which is now a serious consideration for many companies. The pressure in the bottom of the tank can easily be calculated as a function of scale, as can what it might be if some back pressure is added (Fig 7.7). From Fig. 7.7 it is clear that adding back pressure to stainless steel pilot tanks, from say 10 L and upwards, is one way to reproduce large scale conditions in smaller scales. This is one parameter that can be scaled down to pilot without too much complication, but it is more tricky for laboratory scale glass or plastic reactors, so some consideration is needed.

7.3.4   Oxygen transfer and the fed batch as a function of scale

We already mentioned that for a fed batch process, ideally the feed would be added at a rate no greater than that at which it can be consumed and this rate is limited by the rate of oxygen transfer. The right hand side (rhs) of equation [7.9] describes this condition; it can be used for scale down as follows:

- Keep NVVM constant;
- Match the average pressure in the tank or choose a more relevant one;
- Calculate v_s at the inlet;
- Use a solver or manually iterate rpm of intermediate scales until the rhs of equation [7.9] is equal at all scales.
- Do this for the thinnest and the thickest broth anticipated.

The value of constants can be taken from the literature, here values of 0.41, 0.16 and −0.39 for α, β and δ have been taken from a recent

© Woodhead Publishing Limited, 2013

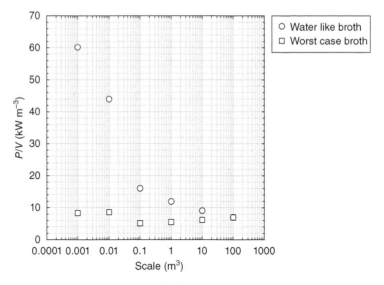

Fig. 7.8   Scale down from 100 m³, 7 kW m⁻³ for two broth types while keeping the expected mass transfer per unit volume for 'thin' and 'thick' broth constant. For viscous but highly shear thinning broths, the differences between scales are minimised, while for thin broth more power per unit volume is needed as the scale decreases. This is especially true at 10 litres or less, where back pressure is not typically added to the tanks. Broth rheology is not constant during fermentation, making scale down models quite difficult to set up as conditions in the small tanks must be tracked to equal those that develop in the larger tank.

publication (Albaek *et al.*, 2012). The results of this exercise are now plotted in Fig 7.8.

Now it can be seen that for thin broth, to keep the *OUR* per unit mass of broth constant, the power input per unit volume required at small scales rises considerably compared to the target at larger scale. This is something that many fail to recognise and is somewhat counterintuitive. There is an intuitive and almost unshakeable belief that 'oxygen transfer is less at scale', but in reality the exact opposite tends to be true. Counterintuitively it is in small scale that more power per unit mass of broth needs to be added to match the oxygen transfer available at scale. Further, this is not worse for rheologically complex processes; if anything, having complex rheology makes scale differences less challenging (but there is always less oxygen transfer available than for thinner broths). While this is broadly speaking the truth, simply solving equation [7.9] is not the complete story for scale up/down of fed batch processes considering oxygen transfer. The broth rheology is not constant as a function of time, as even the most rheologically complex fermentations tend to start relatively Newtonian with low viscosity, not becoming like ketchup until the very end of the process. This means that a true scale down model would have to track the OUR capability of the large scale system in a way that it is not trivially defined, especially if trying to consider the influence of stirrer power on morphology and its complicating feedback relationship with rheology and, simultaneously trying to do process development or optimisation.

By far the best approach to scale up or down, is not to jump from the laboratory to full production scale or *vice versa*, but to pass through a relevant scale of pilot plant and do some trials there (a simple way to define pilot plant now is to say that it is anything that cannot be placed in an autoclave, say from 20 litres upwards). By doing this the risk of making a fatal business decision is reduced, but as hinted at in the introduction, not all practising the art of scale up/down have the luxury of a pilot plant and some are therefore required to take risks with business models. Still, this risk can be mitigated in two ways: in the first, for a company that has done a lot of scale up or down of this type previously, experience can guide expectations and this reduces concerns. In the second case, where no such experience is available, the influences of the various factors and their expected ranges can be investigated systematically, for example a design of experiments (DoE) where aeration, power and preferably back pressure are varied is deployed and the process responses observed, finally assuming that the ranking of process A > process B > process C and so on remains constant as scale changes, or even attempting construction of a mathematical model.

7.4   Further complexities when changing scale

7.4.1   Computer modelling to cope with inevitable changes

Relevant equations have been presented for most items of interest in scale translation, and these can all be solved at 'time t', but fermentations move on through time and become more or sometimes less viscous as they proceed. We could choose to construct all kinds of laboratory-based scale down models, trying to make the equipment deliver constant oxygen transfer per unit volume by iterating the agitator speeds or adding pure O_2 to improve the driving forces. Or other techniques might be applied. But will this help? What typically happens during the fed batch process is that the viscosity changes, starting low and then rising, but rising by different amounts for different reactors operating at different speeds or oxygen transfer rates (OTRs) and with different organisms (different morphology). This is too complex to match from one scale to the next if the rheology is changing as a function of speed, biomass concentration and morphology and there is a need simultaneously to change some other process parameter to optimise the process. But, if there is a numerical approximation for each of these considerations, it is possible to manage the scale up/down complexity by construction of a computer model. This has been done with some success and is something that will continue to be worked upon by various

© Woodhead Publishing Limited, 2013

persons around the globe (examples include Yang *et al.,* 1992; Yang and Allen, 1999; Albaek *et al.,* 2008, 2012).

Computer modelling can be an enormously difficult area and has many pitfalls. One of the worst pitfalls is thinking that it can not be done except by very clever or expert individuals. It is true that to do it at a world class level or publish in the area requires well above average intelligence and skills, but modern programming environments are not prohibitively complicated and it is not compulsory to publish anything. Numerical integration, or 'ODE (ordinary differential equations) solvers', are available with free software like GNU Octave, or old favourites like MatLab, Mathematica and probably others. Anyone who can solve the equations presented so far, using some paper and a calculator, or in Excel, for one time point in the fermentation process, can, with some investment of time, also learn how to use these tools to solve the equations throughout the entire duration of the process. By doing this it is possible to predict with more confidence what happens on scale up or scale down.

Interestingly, the most uncertainty in this type of model seems to arise from the biological parts of the numerical system–yield coefficients, or the type of maintenance model selected are more important than the exponents in the $k_L a$ correlation or the viscosity models (Albaek *et al.,* 2012). It is the complexity of the biomass itself that presents the challenge, how should it be represented numerically? How does it react to overfeeding? How do the yield coefficients change as a function of the growth rate? How much of the biomass is active? There is still plenty of scope for work in this area.

7.4.2   Cavern formation

When considering the yield stress (equation [7.7]), a possible problem arises and this problem can also be observed in reality. If a fluid has a yield stress, it will not flow unless that stress is exceeded. Now, imagining a large fermenter, it seems likely that at larger distances from the impellers, the yield stress may not be exceeded and that the fluid will cease to flow, giving rise to a so called 'cavern' around the agitators, which is mixed, while the rest of the vessel is not. This is a well-known phenomenon in xanthan gum fermentations, which is a food ingredient (Amanullah *et al.,* 1998a). But this can also occur in filamentous fermentations where the mycelial nature of the broth imparts inherent yield stress behaviour, or it can happen in very shear thinning broth, where at low shear stresses, the broth is moving almost imperceptibly at a very low shear rate.

Various correlations for the prediction of cavern formation have been reported. They all allow prediction of a cavern dimension, which can then be compared to the dimensions of the tank. If the predicted size of the cavern is more than the size of the tank, there will essentially be no cavern. If a yield stress has been measured, the problem can be treated in a number

© Woodhead Publishing Limited, 2013

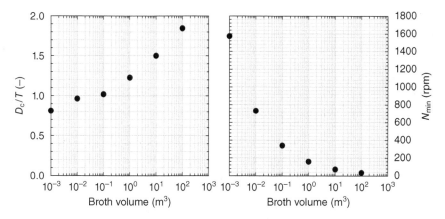

Fig. 7.9 Evaluation of cavern diameter compared to tank diameter with yield stress set to 40 Pa s. Left, for an equal OTR scenario made for viscous broth (data from Fig. 7.8). Right, the minimum agitator speed to obtain a cavern diameter equal to vessel diameter. Counterintuitively, it is more likely that cavern formation will occur as tanks are reduced in size.

of ways, for example as described by equation [7.10] (Elson *et al.*, 1986), which is for a right circular cylindrical cavern centred on an agitator:

$$\left(\frac{D_c}{D}\right)^3 = \frac{1.36 P_o}{\pi^2} \cdot \frac{\rho_L N^2 D^2}{\tau_y} \qquad [7.10]$$

Here the ratio D_c/D is the cavern diameter compared to the impeller diameter. From Fig. 7.3 it can be seen that the most viscous broth has a yield stress of approximately 40 Pa s. This can be applied to the scenarios from Fig. 7.8, and from this it is apparent that cavern formation will occur at scales less than 10 litres (Fig. 7.9). This adds to the complications of scale down/up. Applying at all scales, it is also possible to re-evaluate the minimum agitator speed needed to prevent cavern formation across the chosen scale range (Fig. 7.9). It can be seen that contrary to intuition, it actually seems more likely that cavern formation will occur in smaller vessels, unless very high mixing speeds are adopted or the D/T ratio is changed (see Fig 7.9).

For a highly shear thinning broth another approach is needed as there is no real measurable yield stress. One successful approach has been to specify the cavern boundary as corresponding to a low velocity (e.g. 1 mm s^{-1}) and evaluate the stress at which this occurs on the flow curve (Amanullah *et al.*, 1998b). This is a technique that can be adapted, for example, this velocity can be altered by factors of 10 to re-evaluate the likelihood of cavern formation for a range of conditions at any scale, getting a better feeling for whether or not cavern formation is likely to be a problem. The same results are obtained: it seems more likely for cavern

© Woodhead Publishing Limited, 2013

formation to occur at smaller scales. Cavern formation is another example of how the real world refuses to behave the way we expect it to.

7.4.3   Shear stress and scale

So far two things which seem to concern many who practice or consider the art of fermentation and its scale up or down have not been considered. These are shear stress and mixing time. The sensitivity of fermentation organisms to shear stress is still slightly controversial. The proteins themselves are unlikely to be sensitive to shear stress (Thomas and Geer, 2011). Mammalian cells in submerged culture were thought to be sensitive to shear stresses, but then it was found that the gas–liquid interface and bubble bursting was a cause of damage, not agitation itself (Oh *et al.*, 1989, 1992, Mollet *et al.*, 2004). Single cell organisms, like bacteria or yeast, might respond in some way to the shear stresses in the tank, but it seems highly unlikely that anything harmful to productivity will occur (Hewitt *et al.*, 1998). The morphology of filamentous organisms is almost certainly under the influence of shear stresses, although productivity is not always affected negatively (Amanullah *et al.*, 1999, 2000, 2002). *Streptomyces* productivity was sensitive only at very high power dissipation rates (Heydarian *et al.*, 1999); *Penecillium* may be sensitive (Smith *et al.*, 1990), while Humphrey (1998) states 'The time has come to give up on shear as an excuse for scale up failure'.

If proteins and bacterial or yeast fermentations are removed from the shear sensitive category and only *some* filamentous organisms are included, what is the explanation for this apparent indestructible tendency? One explanation that has been proposed is that older compartments of the hyphae or mycelium are not productive and are 'shut down' or subject to 'apoptosis'. These represent weak links and therefore are the sites of mechanical breakage from stirring, that is only non-productive regions are subject to shear sensitivity (Stocks and Thomas, 2001). Since a reduction in dimensions will reduce viscosity, this may even be beneficial for productivity. These few citations by no means represent a complete analysis of the literature, but are illustrative of a general lack of agreement. There is a steady trickle each year of publications on all types of fermentation process claiming that a cell type or process is sensitive to shear stress, while there are a few showing that they are not and those that show no influence of shear stress rarely point this out as it was not the original objective of the work. So again, the reader is warned to pick their sources carefully and exercise caution in what their intuition might tell them; 'This world is quite different from the one that we have developed our intuitions in' (Purcell, 1977). Instead, read about the organism type (perhaps start with Thomas and Zhang, 1998) and remember that simply typing 'shear sensitivity' into a search engine will only turn up examples where the work is presented as having a shear sensitivity, it will turn up no claims for 'no sensitivity to shear stress'.

© Woodhead Publishing Limited, 2013

The overall lack of production sensitivity to shear stress for the enzyme products makes a discussion of shear stresses on scale up partially superfluous, but it is important for morphology and again what happens to stress is not what is intuitively expected. Intuitively, the size of the agitators and the torque on the shaft increases, so surely the stresses must increase too? This tends not to be the case. As discussed above, in order to imitate the oxygen transfer available in larger tanks, more specific power per unit volume is required at the smaller scale, that is shear stresses ought to be largest in smaller tanks. Some talk about 'tip speed' can often be heard and while this is a successful correlating parameter for particle size at a single scale with different shaft speeds, it is not a good correlating parameter between scales (Justen et al., 1996, 1998; Amanullah et al., 2000). This is partly because the tip speed does not actually change that much as a function of scale, but it is also because, at the length scales in which the organisms live, it does not have much meaning. The best correlating parameter to date has been the 'energy dissipation circulation function' or EDCF, which does correlate between scales, and also predicts a tendency for shear stress to fall as scale increases (Justen et al., 1996, 1998; Amanullah et al., 2000). Why this parameter is so successful is not completely clear; it includes a circulation time, which implicates some kind of fatigue mechanism which seems very unlikely (equation [7.11]).

$$EDCF = \left(\frac{P}{kD^3}\right)\left(\frac{1}{t_c}\right)$$ [7.11]

Here, the engery dissipation circulation function is calcualted from the power (W), k, a constant related to impeller volume (e.g. $k = (\pi/4) \cdot (W/D)$, W being the blade width) and t_c the circulation time. In short, physiologically speaking, the reader should be sceptical about the effects of 'shear damage' and not expect to see more 'shear damage' as scale increases, but less. While this is at first reassuring, it also means that there will be longer hyphae, or larger mycelia and so a higher viscosity as scale increases. Any attempt to model between scales would ideally consider this, numerically.

7.4.4   Mixing time and scale

It is obvious that truly perfect mixing cannot be achieved at any scale where a reaction is taking place. Intuitively this can be ignored in small scales while is especially untrue at large scales. A regime of analysis, or actually checking by measurement usually confirms that concentration gradients in p_{O_2}, pH and nutrient will exist at the large scale (see for example Larsson et al., 1996; Marten et al., 1997; Enfors et al., 2001). This knowledge is fuel for the natural worry that the longer mixing times at large scales will have a negative influence on cell physiology and therefore productivity and potentially product quality. How much of a concern should this be? One way to think about this is to consider that broth pH, temperature and p_{O_2} measurements,

© Woodhead Publishing Limited, 2013

as well as samples, are typically taken at only one point in the vessel, even in very large vessels. Yet various fermentation businesses remain successful and the product remains within quality specifications. Another way to think about this is to consider that all biological processes tend to contain some variance, not as much as most people think, but nonetheless, there is a mean and standard deviation in yield or product concentration from 'n' perfectly executed batches. Now, if we make 'mixing time' progressively 'worse' somehow, but independent of oxygen transfer, how much worse does it have to be before we can detect reduced quality or yields?

The answer is usually 'quite a lot worse'. Whether or not a large scale system will react negatively to the inevitable reduction in mixing quality or increased mixing time on scale up can be investigated in the laboratory by application of various methods. One is to set up a plug flow loop outside the fermenter to represent a poorly mixed zone and is probably the best established method (Amanullah et al., 2003), although, some surprising results can be observed, such as increased viability at large scale and in simulated large scale (see for example Hewitt et al., 2000). The use of pulsed then paused feed is another way to imitate the circulation loop of a large fermenter in a much smaller one and in this case some surprising results can also be observed, for example, cultures of A.oryzae and other filamentous organisms can actually respond positively to these feeding techniques (Bhargava et al., 2003a, 2003b, 2005), techniques that are patent protected (Marten et al., 2002). It has also been proposed that 'poor' mixing in large tanks can be imitated in smaller ones by stopping and starting the agitator after a scheme based on circulation times, or the speed can be varied in a more sophisticated way to simulate the actual circulation time distribution of a larger fermenter.

Testing at the small scale can be done approximately as follows: first estimate the circulation time of the large tank either by measuring it or estimating it. To estimate it, find a suitable flow number (Fl) for something approximating the impeller geometry and calculate Q the amount of fluid pumped per second by the agitator (equation [7.12]). This can then be compared to the liquid volume in the vessel, where one circulation time (t_c) is approximately one divided by the other (equation [7.12]). It is now possible to design a down scaled model with the time for flow through the loop specified. Determining the volume ratio of the loop and the stirred tank is not so trivial, but experiments can be done with ratios like ½, 1/3 and so on; or the reader is referred to Amanullah et al. (2003), for a more comprehensive review of the methodology:

$$t_c \approx \frac{V}{Q} = \frac{V}{Fl.ND^3} \qquad\qquad [7.12]$$

In summary, 'mixing' or 'mixing time' needs to be good enough, or acceptable for the quality demanded by the business case; it can never be

© Woodhead Publishing Limited, 2013

perfect. To find out if a system is very sensitive to 'mixing times', a laboratory investigation should focus as a bare minimum on one set of scaled conditions, but the conditions should preferably be varied to discover at what point the system fails and thereby whether it is likely to be a concern on scale up. Luckily, the conditions for a total failure are usually far from that which can be achieved in a large gassed fermenter.

7.4.5   Customer requirements and scale

Most customers are interested in common sense quality, namely low product variation where the product performs repeatedly and predictably and is more or less guaranteed to be risk free from a business and health point of view. But there are further expectations. Globally speaking, many customers are concerned about concepts such as GM (genetically modified), halal, kosher and allergies or intolerances. Since the customers' concerns are a number one priority for any successful business, these become major concerns for scale up and scale down. As a general rule of thumb most problems can be circumvented by selecting only fermentation ingredients and processing aids that are non-animal in origin and 'GM' free. In fact, it is a generally good idea not to even warehouse anything of animal origin. In any case, large scale fermentation with such materials is not very feasible because there are not sufficient volumes available and the costs are too high. Even the microbiologist's favourite, yeast extract, starts to look expensive when it is required in ton quantities. Therefore, most of the very large scale fermentations are done with media that are largely defined, containing some simple salts and carbohydrates of agricultural origin. Inclusion of some in-house tryptone from agricultural protein sources such as soy or potato can be made with some of the in-house enzyme products, if any is found to be truly needed. In any case, in order to be in compliance with the various demands of the customers, an on site audit can be expected from the relevant authoritative body or customer, so there will need to be very tight control of inventory, certificates and warehousing.

With respect to scale up, the scale down rule once more applies. Fermentation development and optimisation work in the laboratory or pilot plant must be done in realistically plausible media that can be produced at the larger scale and with ingredients that are acceptable to customer demands and that can be handled in a logistically scaleable way.

7.5   Future trends and scale

Sustainability is a word that features increasingly in many business ventures, existing or new. Biotechnology is by nature a sustainable proposition, especially if the carbon or carbohydrates for the processes are agriculturally sourced. Biotechnological businesses are therefore in a good position for

© Woodhead Publishing Limited, 2013

future business. The use of genetic improvements to enhance yields will help to resist the need for new capital investments and mitigate against increased energy and water consumption (as it has for the last few decades), but eventually this expansion will see the need for new capacity installation in the form of new factories or larger scales of operation. The parameters of equation [7.9] are interesting in this respect. They indicate that doubling the amount of gas or agitation power added to a fermenter will not double the gas transfer rates, or the productivity. This is far from the case, especially if the resulting increase in viscosity minimises further the increase in gas liquid mass transfer. Therefore, product concentrations and productivity will not double; as we add more power to a production reactor, the energy efficiency of the system (kg product / kWh) reduces. As we become more concerned over energy consumption we may see a shift in the production technology to systems of greater efficiency, but almost inevitably, lower volumetric productivity. There is a great challenge here. Perhaps there is a special technology, airlift technology or static mixer technology or some-thing new, that can offer a balanced business case with acceptable product concentrations, productivity and improved efficiency. But, in identifying these technologies, the complete life cycle must be considered. It is rela-tively easy to optimise the efficiency of the fermentation unit operation by simply running at lower power and accepting lower product concentrations, but this results in the need for increased downstream processing. Even acceptance of more crude lower strength products will require increased road, rail or sea transportation volumes. Production at customer or even consumer sites might be considered, but managing a diverse portfolio of products across numerous sites is not trivial, especially when considering the inevitable introduction of optimised products in new strains that will result in idle customer side capacity. How the industry will look in 10 or 20 years is not carved in stone, but it will involve improvements in overall process energy efficiency and it will involve larger volumes of product and larger accumulated worldwide fermentation volumes.

7.6   Conclusion: scale up is scale down

Some basic and well known, but often ignored, concepts have been intro-duced. The wider literature can elaborate and refine these ideas but the overall message remains true: scale up is best approached with an 'end in mind' scale down mentality. My own view on this, which I hope I have justi-fied with examples, is that scaled down systems should not be too many orders of magnitude different in volume from the final scale, or the engi-neering parameters that are sensitive to viscosity (especially Re, k_La and consequent OTR) will be too far away for a reliable process transfer. We have seen that we cannot simply match the OTR of a large reactor to a much smaller one because the amounts of power or other tricks needed to do this change in a complex way as a process executes and the rheology changes. DoEs are preferably deployed at a close relevant scale with subsequent preparation of some kind of model, statistical or more mechanistic, to predict what happens as scale changes. In some cases, perhaps with a well-known host/product system, or perhaps at a company with a lot of scale up/down experience and a strong peer group, some of the confidence gained from previous DoEs or experiences might be used to move rapidly to a larger scale, but only because there is historical evidence available to provide such confidence. New beginners in this area should be aware that despite claims to the contrary in the literature, there are few shortcuts here: doing relevant experiments at appropriate scales is still the only consistently proven way forwards. Deciding what is relevant is more art or experience based, than science based. As stated previously by others, 'scale up is an art, not a science' (Humphrey, 1998) and it remains so today.

7.7   Acknowledgements

Many thanks go to Novozymes A/S for granting permission to write this chapter and special thanks to Mads O Albæk, Alvin Nienow, Colin Thomas and Mike Cooke for greatly influencing the opinions presented.

7.8   References

ALBAEK MO, STOCKS S and GERNAEY KV (2008). 'Gassed and ungassed power draw in a pilot scale 550 litre fermentor retrofitted with up-pumping hydrofoil B2 impellers in media of different viscosity and with very high power draw'. *Chem Eng Sci*, **63**, 5813–20.

ALBAEK MO, GERNAEY KV, HANSEN MS and STOCKS SM (2012). 'Evaluation of the energy efficiency of enzyme fermentation by mechanistic modeling'. *Biotechnol Bioeng*, **109**, 950–61.

AMANULLAH A, SERRANOCARREON L, CASTRO B, GALINDO E and NIENOW AW (1998a). 'The influence of impeller type in pilot-scale xanthan fermentations'. *Biotechnol Bioeng*, **57**, 95–108.

AMANULLAH A, HJORTH SA and NIENOW AW (1998b). 'A new mathematical-model to predict cavern diameters in highly shear thinning, power-law liquids using axial-flow impellers'. *Chem Eng Sci*, **53**, 455–69.

AMANULLAH A, BLAIR R, NIENOW AW and THOMAS CR (1999). 'Effects of agitation intensity on mycelial morphology and protein production in chemostat cultures of recombinant *Aspergillus oryzae*'. *Biotechnol Bioeng*, **62**, 434–46.

AMANULLAH A, JUSTEN P, DAVIES A, PAUL GC, NIENOW AW and THOMAS CR (2000). 'Agitation induced mycelial fragmentation of *Aspergillus oryzae* and *Penicillium chrysogenum*'. *Biochem Eng J*, **5**, 109–14.

AMANULLAH A, CHRISTENSEN LH, HANSEN K, NIENOW AW and THOMAS CR (2002). 'Dependence of morphology on agitation intensity in fed-batch cultures of *Aspergillus oryzae* and its implications for recombinant protein production'. *Biotechnol Bioeng*, **77**, 815–26.

© Woodhead Publishing Limited, 2013

AMANULLAH A, BUCKLAND BC and NIENOW A (2003). 'Mixing in the fermentation and cell culture industries'. In: *Handbook of Industrial Mixing: Science and Practice*, Paul EL, Kresta SM, Atiemo-Obeng AA, (eds). Wiley-Interscience; New Jersey, 1071–170.

ATKINSON B and MAVITUNA F (1991). *Biochemical Engineering and Biotechnology Handbook*, Macmillan Publishers, New York.

AUNSTRUP K (1979). 'Production, isolation, and economics of extracellular enzymes'. *Appl Biochem Bioeng*, **2**, 27–69.

BARNES HA, HUTTON JF and WALTERS K (1989). *An Introduction to Rheology*. Elsevier Science, Amsterdam.

BATES RL, FONDY PL and CORPSTEIN RR (1963). 'Examination of some geometric parameters of impeller power'. *Ind Eng Chem Process Des Dev*, **2**, 310–14.

BEROVIC M and NIENOW AW (2005). *Biochemical Engineering Principles*. European Federation of Biotechnology, Kemijsko Inzenirstvo.

BHARGAVA S, WENGER KS and MARTEN MR (2003a). 'Pulsed feeding during fed-batch *Aspergillus oryzae* fermentation leads to improved oxygen mass transfer'. *Biotech Progr*, **19**, 1091–94.

BHARGAVA S, WENGER KS and MARTEN MR (2003b). 'Pulsed addition of limiting-carbon during *Aspergillus oryzae* fermentation leads to improved productivity of a recombinant enzyme'. *Biotechnol Bioeng*, **82**, 111–17.

BHARGAVA S, WENGER KS, RANE K, RISING V and MARTEN MR (2005). 'Effect of cycle time on fungal morphology, broth rheology, and recombinant enzyme productivity during pulsed addition of limiting carbon source'. *Biotechnol Bioeng*, **89**, 524–9.

COOKE M, MIDDLETON JC and BUSH JR (1988). 'Mixing and mass transfer in filamentous fermentations'. *Proceedings 2nd International Conference Bioreactors*, BHRA, Cambridge, UK, 37–64.

COVEY SR (1989). *The 7 Habits of Highly Effective People*. Free Press, New York.

ELSON TP, CHEESEMAN D and NIENOW AW (1986). 'X-ray studies of cavern sizes and mixing performance with fluids possessing a yield stress'. *Chem Eng Sci*, **41**, 2555.

ENFORS SO, JAHIC M, ROZKOV A, XU B, HECKER M, JURGEN B, KRUGER E, SCHWEDER T, HAMER G, O'BEIRNE D, NOISOMMIT-RIZZI N, REUSS M, BOONE L, HEWITT C, MCFARLANE C, NIENOW A, KOVACS T, TRAGARDH C, FUCHS L, REVSTEDT J, FRIBERG PC, HJERTAGER B, BLOMSTEN G, SKOGMAN H, HJORT S, HOEKS F, LIN HY, NEUBAUER P, VAN DER L, LUYBEN K, VRABEL P and MANELIUS A (2001). 'Physiological responses to mixing in large scale bioreactors'. *J Biotechnol*, **85**, 175–85.

GARCIA-OCHOA F (2009). 'Bioreactor scale-up and oxygen transfer rate in microbial processes: An overview'. *Biotechnol Adv*, **27**, 153–76.

HARNBY N, EDWARDS MF and NIENOW AN (1992). *Mixing in the Process Industries*. Butterworth-Heinemann, Oxford UK.

HEWITT CJ, BOON LA, MCFARLANE CM and NIENOW AW (1998). 'The use of flow cytometry to study the impact of fluid mechanical stress on *Escherichia coli* W3110 during continuous cultivation in an agitated bioreactor'. *Biotechnol Bioeng*, **59**, 612–20.

HEWITT CJ, NEBE-VON-CARON G, AXELSSON B, MCFARLANE CM and NIENOW AW (2000). 'Studies related to the scale-up of high-cell-density *E. coli* fed-batch fermentations using multiparameter flow cytometry: Effect of a changing microenvironment with respect to glucose and dissolved oxygen concentration'. *Biotechnol Bioeng*, **70**, 381–90.

HEYDARIAN SM, MIRJALILI N and ISON AP (1999). 'Effect of shear on morphology and erythromycin production in *Saccharopolyspora erythraea* fermentations'. *Bioprocess Eng*, **21**, 31–9.

HUDCOVA V (1989). 'Gas-liquid dispersion with dual Rushton turbine impellers'. *Biotechnol Bioeng*, **34**, 617–28.

HUMPHREY A (1998). 'Shake flask to fermenter – what have we learned'. *Biotech Progr*, **14**, 3–7.

© Woodhead Publishing Limited, 2013

JUSTEN P, PAUL GC, NIENOW AW and THOMAS CR (1996). 'Dependence of mycelial morphology on impeller type and agitation intensity'. *Biotechnol Bioeng*, **52**, 672–84.

JUSTEN P, PAUL GC, NIENOW AW and THOMAS CR (1998). 'Dependence of *Penicillium chrysogenum* growth, morphology, vacuolation, and productivity in fed-batch fermentations on impeller type and agitation intensity'. *Biotechnol Bioeng*, **59**, 762–75.

LARSSON G, TORNKVIST M, STAHL-WERNERSSON E, TRAGARDH C, NOORMAN H and ENFORS SO (1996). 'Substrate gradients in bioreactors: Origin and consequences'. *Bioprocess Eng*, **14**, 281–9.

MARTEN MR, WENGER KS and KHAN SA (1997). 'Rheology, mixing time, and regime analysis for a production-scale *Aspergillus oryzae* fermentation'. *4th International Conference on Bioreactor and Bioprocess Fluid Dynamics*. Bury St. Edmonds. Nienow, A. W. (ed.). Mechanical Engineering Publications, 295–313.

MARTEN MR, WENGER KS and STOCKS SM (2002). *Cyclic Pulse-pause Feeding*. Patent WO1028.

MCINTYRE M and MCNEIL B (1997). 'Dissolved carbon dioxide effects on morphology, growth, and citrate production in *Aspergillus niger* A60'. *Enzyme Microb Technol*, **20**, 135–42.

MEZGER TG (2006). *The Rheology Handbook: For users of rotational and oscillatory rheometers*. Vincentz Network GmbH, Hanover.

MOLLET M, MA NN, ZHAO Y, BRODKEY R, TATICEK R and CHALMERS JJ (2004). 'Bioprocess equipment: Characterization of energy dissipation rate and its potential to damage cells'. *Biotech Progr*, **20**, 1437–48.

NIELSEN J (2006). 'Microbial process kinetics'. In: *Basic Biotechnology*, Ratledge C and Kristiansen B (eds). Cambridge University Press, Cambridge, UK, pp 155–80.

NIELSEN J, VILLADSEN J and LIDEN G (2003). *Bioreaction Engineering Principles*. Kluwer Academic/Plenum, London.

NIENOW AW and BUJALSKI W. (2004). 'The versatility of up-pumping hydrofoil agitators'. *Chem Eng R*, **82**, 1073–81.

OLSEN HS (2004). *Enzymes at Work*. Novozymes A/S, Denmark (available from www. novozymes.com).

OH SKW, NIENOW AW, AL RUBEAI M and EMERY AN (1989). 'The effects of agitation intensity with and without continuous sparging on the growth and antibody production of hybridoma cells'. *J Biotechnol*, **12**, 45–61.

OH SKW, NIENOW AW, AL RUBEAI M and EMERY AN (1992). 'Further studies of the culture of mouse hybridomas in an agitated bioreactor with and without continuous sparging'. *J Biotechnol*, **22**, 245–70.

PAUL EL, KRESTA SM and ATIEMO-OBENG AA (eds). (2004). *Handbook of Industrial Mixing: Science and Practice*. Wiley-Interscience, New Jersey, pp 1071–170.

PETERSEN N, STOCKS S and GERNAEY KV (2008). 'Multivariate models for prediction of rheological characteristics of filamentous fermentation broth from the size distribution'. *Biotechnol Bioeng*, **100**, 61–71.

PURCELL EM (1977). 'Life at low Reynolds number'. *Am J Phys*, **45**, 1–11.

SCHUGERL K (1990). *Bioreaction Engineering, Volume 2: Characteristic Features of Bioreactors*. John Wiley & Sons.

SMITH S (1980). Single cell protein. *Philo Transa Ro Soc: Biol Sci*, **290**, 341–54.

SMITH JJ, LILLY MD and FOX RI (1990). 'The effect of agitation on the morphology and penicillin production of *Penicillium chrysogenum*'. *Biotechnol Bioeng*, **35**, 1011–23.

STANBURY PF, WHITTAKER A and HALL SJ (1995). *Principles of Fermentation Technology*. Butterworth-Heinemann, Oxford, UK.

© Woodhead Publishing Limited, 2013

STOCKS SM (2005). 'Traditional Rushton disc turbines vs. up-pumping axial flow impellers in 550 litre pilot scale aerobic submerged fermentations'. *7th World Congress of Chemical Engineering*, Glasgow UK, IChemE, London, UK.

STOCKS SM and THOMAS CR (2001). 'Strength of mid-logarithmic and stationary phase *Saccharopolyspora erythraea* hyphae during a batch fermentation in defined nitrate-limited medium'. *Biotechnol Bioeng*, **73**, 370–8.

TATTERSON GB (1991). *Fluid Mixing and Gas Dispersion in Agitated Tanks*. Library of Congress, USA.

THOMAS CR and GEER D (2011). 'Effects of shear on proteins in solution'. *Biotechnol Lett*, **33**, 443–56.

THOMAS CR and ZHANG Z (1998). 'The effect of hydrodynamics on biological materials'. In: *Advances in Bioprocess Engineering II*. Galindo E and Ramirez OT (eds). Kluwer Academic Publishers, Boston/London, pp 137–70.

VAN'T RIET K and TRAMPER H (1991). *Basic Bioreactor Design*. Marcel Dekker, New York.

YANG H and ALLEN DG (1999). 'Model-based scale-up strategy for mycelial fermentation processes'. *Can J Chem Eng*, **77**, 844–54.

YANG H, KING R, REICHL U and GILLES ED (1992). 'Mathematical model for apical growth, septation, and branching of mycelial microorganisms'. *Biotechnol Bioeng*, **39**, 49–58.

ZHU YG, BANDOPADHYAY PC and WU J (2001). 'Measurement of gas-liquid mass transfer in an agitated vessel–A comparison between different impellers'. *J Chem En J*, **34**, 579–84.

© Woodhead Publishing Limited, 2013

Part II

Use of microorganisms for the production of natural molecules for use in foods

© Woodhead Publishing Limited, 2013

8

Microbial production of food flavours

Y. Waché, AgroSup Dijon, France

DOI: 10.1533/9780857093547.2.175

Abstract: This chapter presents the field of microbial flavour compounds through the main principles governing strategies to obtain them. The production of microbial flavours in fermented products is presented, as this is generally the basis of aroma compounds production. The traditional ways that enabled the production of 'easy-to-produce' compounds, which can be considered the first generation of biotechnological aroma compounds is described. The utilisation of progress in biology technologies to give rise to a second generation of aroma compounds is then discussed. At the end of the chapter, the idea of naturalness is discussed with regard to regulation, ethics and analysis followed by perspectives on the field.

Key words: biotechnology, biotransformation, fermented food, flavour, naturalness.

8.1 Introduction

The aroma of food depends on a number of compounds which can be many for certain products (up to 650 different compounds impact on the aroma of coffee) or relatively few for others. The particular nature of these compounds is that some can be volatile at the temperature of food ingestion so they can be perceived by the consumer through the nasal or retronasal route. They are thus small molecules belonging to various chemical families with different physical and chemical properties. Depending on these properties, they can be retained well in the food until ingestion or they can be released during processing, especially in drastic processing such as extrusion processes. They can also react chemically with other food molecules changing their chemical impact on the food. For all these reasons, it is often necessary to correct, reinforce or simply create a new aroma during food production. This requires that aroma compounds be available to take part in the formulation of aromas. These compounds can be extracted from plant or animal tissues as a flavour preparation or as an isolated molecule they can be chemically synthesised to obtain nature-identical compounds if the

© Woodhead Publishing Limited, 2013

compound has already been detected in nature (although in Europe, following regulation CE 1334/2008, the nature-identical label is no longer used) or from artificial sources otherwise.

Data about the market in flavours and fragrances can be found on the site of Leffingwell (2012). The global market in flavours was estimated in value for 2006 to be US\$ 6375 million in four broadly equivalent groups: North America (30.6%), Asia-Pacific (27%), Western Europe (23.2%) and the rest of the world (19.1%). Relatively similar data described the fragrance market. In 1999, the market for flavours was divided into beverages (31%), savory (23%), dairy (14%) and other (32%). As the most important property of biotechnological flavours is to be natural, it is of interest to get some data on the market for natural flavours. This can be found on the web site of RTS especially in a document called *The Market for Flavours and Flavour Trends* (RTS, 2010). In this study, the European market for synthetic compared with natural flavours is estimated to be €726 million compared with €570 million in 2009 with an expected value in 2014 of €731 million compared with €671 million, respectively with the best opportunities expected in the beverage, bakery, dairy and snack markets.

One way to produce flavour compounds is through microbial metabolism or biocatalysis. Molecules are then classified as natural in a way similar to compounds extracted from plant or animal tissues. Several reviews have already presented the global production of aroma compounds and specific production of some families of compounds (Krings and Berger, 1998; Aguedo et al.; 2004; Feron and Waché, 2005; Gatfield, 1988, 1999; Rabenhorst, 2008; Belin et al., 1992). Therefore, the goal of this chapter is not to give an exhaustive view of microbial flavour compounds but to present them through the main principles governing strategies to obtain microbial aroma compounds. To attain this goal, the production of microbial flavours in fermented products will be presented, as this is generally the basis of aroma compounds production. The traditional methods that enabled the production of 'easy-to-produce' compounds, which can be considered the first generation of biotechnological aroma compounds will be described. We will then discuss the utilisation of progress in biology technologies to give rise to a second generation of aroma compounds. The chapter ends by discussing, 'naturalness' with regard to regulation, ethics and analysis followed by perspectives on the field.

8.2   Production of flavours by microorganisms in their classical environment

8.2.1   Origin of flavour compounds in fermented food

Fermented foods are usually products in which many aroma compounds have been detected. Among them, some compounds were already present in the raw unfermented material or result from the chemical degradation

© Woodhead Publishing Limited, 2013

of precursors initially present, but for many others, synthesis proceeds through microbial catalysis. The ability of microorganisms to produce food flavours has been observed in fermented food where many compounds are produced in universal or strain specific metabolic pathways. The production can be 'substrate independent', that is *de novo* or by biotransformation of a specific precursor. The main pathways involved in the accumulation of aroma compounds in cheese are reviewed by McSweeney and Sousa (2000).

8.2.2   *De novo* production and biotransformation

Examples of compounds from the carbohydrate metabolism are shown in Fig. 8.1. Compounds can result from metabolism directly or through an additional chemical step, like diacetyl which can be synthesised from the bacterial or chemical oxidative decarboxylation of α-acetolactate. Besides this family of compounds which are end products of the metabolism of most microbial cells, some other more specific *de novo* aroma compounds can be produced by strains possessing an enzymatic particularity. For instance, strains able to hydroxylate fatty acids and then β-oxidise them can produce lactones *de novo* (Fig. 8.2) (Alchihab *et al.*, 2009). Strains possessing a specific enzymatic potential for aminoacids metabolism can also produce original *de novo* aroma compounds (Ayad *et al.*, 1999).

Some examples of aroma compounds from biotransformation of lipids are shown in Fig. 8.2. In this case, lipids can be present in the unfermented material or come from the presence of microorganisms. It is noteworthy that, as the sensorial detection threshold of flavour compounds is usually in the order of ppb to ppm, a low concentration of lipids in the raw material and a low ratio of bioconversion may be sufficient to provide a product with characteristic lipid degradation aromas.

An example of a fermented product for which main flavour compounds are the result of lipid degradation, whereas the amount of lipids is low, is

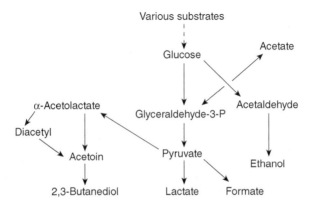

Fig. 8.1   Metabolic pathways that give rise to compounds with a sensorial impact.

© Woodhead Publishing Limited, 2013

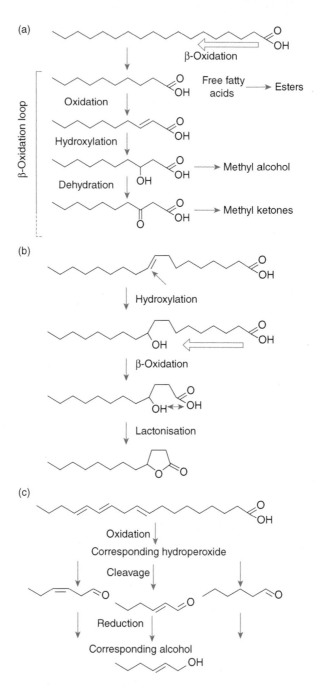

Fig. 8.2 (a) Generation of medium chain-length volatile compounds in fermented products. Fatty acids in the medium are (1) β-oxidised by yeasts or fungi. They exit the β-oxidation pathway between two cycles or during the cycle giving rise to free fatty acids, methyl esters, methyl alcohol or methyl ketones. (b) Generation of lactones in fermented products. Fatty acids in the medium are (1) hydroxylated by bacterial hydroxylases or fungi ROS-generating enzymes, (2) β-oxidated by yeasts and (3) when the distance between the two hydroxyl is short enough, lactonisation occurs. (c) Generation of green notes (aldehydes and alcohol) in fermented products. Polyunsaturated fatty acids in the medium are (1) oxidised into hydroperoxides by the action of fungi ROS-generating enzymes, (2) cleaved by fungal hydroperoxide lyase giving rise to various aldehydes depending on the precursor and (3) reduced by microbial alcohol dehydrogenase hydroperoxides into alcohols.

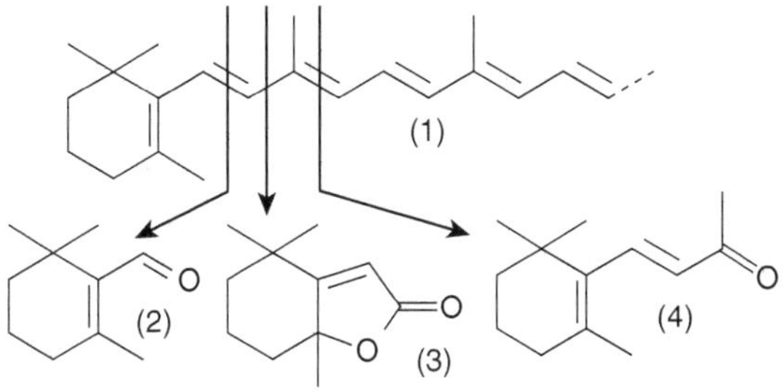

Fig. 8.3 Examples of flavour compounds resulting from the cleavage of β-carotene
(1). (2) β-cyclocitral, (3) dihydroactinidiolide and (4) β-ionone.

tea. Typical tea aroma compounds are derived from the oxidation of polyunsaturated fatty acids (PUFA) and carotenoids following the green note
and carotenoid-derived product pathways (Fig. 8.2 and 8.3). The origin of
oxidation flavour compounds is often multiple. For instance, some of these
compounds are already present in green tea which is not fermented but
undergoes a plant systemic response to harvesting stress inducing an oxidative burst with activities of reactive oxygen species (ROS) generating
enzymes. Typical compounds of these pathways are ionone (Fig. 8.3), green
note aldehydes and theaspirone. For black and oolong teas, the production
may be completed by the presence of microbial ROS generating enzymes,
yielding more oxidised compounds such as dihydroactinidiolide. It can be
noted that the degradation of polyunsaturated lipidic compounds such as
PUFA and carotenoids may result from several treatments in the biological
production of ROS owing to the exposure of enzymes of *Camilia sinensis*
or fungal cells to chemical heat- or light-induced oxidation during firing of
tea. To separate the microbial impact from the other impacts is thus quite
difficult.

Besides this general degradation pathway which leads to many aroma
compounds, some pathways are specific and depend on specific enzymatic
activities. The presence of lactones in many fermented products such as
beer, wine, distilled alcohol, bread and so on would thus be the result of a
multispecies metabolism with a first group hydroxylating fatty acids and a
second degrading them thereafter (Fig. 8.2). Although it is difficult to establish a relationship between one strain and activity, hydroxylation could be
catalysed by *Lactobacillus* sp. while the degradation of the hydroxylated
fatty acid until lactonisation by a yeast is likely (Ogawa *et al.*, 2001;
Wanikawa *et al.*, 2000a, 2000b).

8.2.3   Controlling aroma genesis in fermented food
Fermented foods are very ancient. It is likely that they initially appeared
from a spontaneous fermentation which gave them an advantage for

© Woodhead Publishing Limited, 2013

conservation and safety and thus stimulated their development. There is a great diversity among fermented products and, in some cases, the microbial ecosystem of fermented food is very complex (Demarigny, 2012). In the era of industrialisation, technologists tried to increase safety and control their organoleptic properties. The first step in increasing microbial safety was negative for the taste of those products as it consisted of pasteurising the raw material to prevent the development of pathogen microorganisms and to inoculate a controlled ecosystem. With the contraction of the starter industry into some huge groups investing mainly in the probiotic and bioprotection fields (Hansen, 2002), attempts to find a commercially available starter ecosystem corresponding to a low-production traditional product have become difficult and typically have tended to decline for certain classes of products. However, consumers' demand for tasty food has stimulated research to control the organoleptic properties of products.

Impacting on the flavour of fermented products is not as simple as it seems. Three main strategies are possible, the addition of aroma precursors, or microbial biocatalysts, and modification of fermentation conditions. However, these strategies are only rarely applicable. First, many fermented products are precisely defined in regulatory terms and the addition of strains or precursors may be prohibited. Then, fermented products are often complex in terms of microscopic structure as well as in terms of their ecosystem. This means that the addition of a precursor or a biocatalyst may not work as expected. Indeed, taking the example presented above about the genesis of lactone in fermented products, lactobacilli have to be in contact with lipids, which may be difficult if their surface properties do not correspond to those of lipid droplets or if other strains have a higher affinity for these droplets (as shown by Ly-Chatain et al., 2010). Moreover, in complex microbial ecosystems, strains may collaborate or exhibit antagonistic properties (production of antimicrobial peptides or organic acids etc.) thus inhibiting the development of the desired strain. Such an example has been shown by Goerges et al. (2008) who observed no colonisation of the surface of a German cheese by starter strains owing to the presence of the resident microbial ripening consortia. Despite these restrictions, it is possible to show a relationship between the presence of microorganisms and flavours (Imhof and Bosset, 1994) and to select strains exhibiting a particular metabolism.

In order to enhance the production of diacetyl, a flavour exhibiting cream and nut notes in milk products, several strategies were followed to increase the concentration of diacetyl, with genetic engineering of the NAD/NADH ratio being the most efficient (genetic engineering strategies are described by Feron and Waché, 2005). Another way to increase this production is by deleting the gene coding for the α-acetolactate decarboxylase. Interestingly, by characterising physiological properties of mutants obtained through directed mutagenesis (genetically modified

© Woodhead Publishing Limited, 2013

organisms), it has been possible to select spontaneous mutants able to produce three times more diacetyl than the wild type (Monnet and Corrieu, 2007).

Another example is the production of a chocolate-like flavour in milk from the presence of lactococcal strains with an enhanced flavour corresponding to the combination of strains with proteolytic and amino acid decarboxylating activities (Ayad *et al.*, 2001). Moreover, some treatments can be used to modify metabolic fluxes in fermented food and increase aroma production. After a model study using redox active compounds to modify metabolic fluxes in *Escherichia coli* (Riondet *et al.*, 2000), Martin *et al.* (2011) modified the aroma compound production in yoghurt by applying different redox active-gas combinations.

8.3 Microorganisms for biotechnological flavour production: first generation of biotechnological flavour compounds

In food or plants, flavours are present at low concentration which makes their cost after extraction very high. They can usually be produced by chemical synthesis but these compounds cannot be called 'natural'. Biotechnological pathways have thus been developed mainly using microorganisms. With an adequate couple composed of an available substrate and an active biocatalyst, some pathways are relatively easy to proceed and some compounds can be produced in economically viable processes. Some compounds were successfully produced in this way at the end of the 1980s and beginning of the 1990s in the first period of demand for natural flavour compounds. Other highly demanded compounds have been the subject of many attempts but without enabling technologists to propose an economically viable process. The main examples will be presented here.

The typical strategy for producing the first generation of aroma compounds by biotransformation was by mimicking reactions occurring in nature. For instance, γ-decalactone, which is encountered in many fruits, probably comes from the same pathway described in Fig. 8.2(a), that is by β-oxidation of hydroxylated fatty acids. In fruits, the amount of γ-decalactone is in the order of ppb to ppm thereby preventing all extraction strategies. However, by selecting in nature a concentrated source of hydroxylated fatty acids and an active biocatalyst for β-oxidation, it should be possible to obtain high amounts of lactones. In the history of lactone production, an 'ideal' precursor has been identified from the beginning as it is a particularity of evolution. Indeed, castor oil contains between 70 and 90% ricinoleyl (9-*cis*, 12-hydroxyoctadecenoyl) moieties. This makes ricin an interesting model for lipid synthesis and a cheap source of hydroxylated fatty acid. For this second reason ricinoleic acid was used in the 1960s in studies on the catabolism of hydroxylated fatty acids by the group of Okui,

© Woodhead Publishing Limited, 2013

who observed for a yeast strain that an interesting compound accumulated during metabolism and was later degraded, γ-decalactone (Okui *et al.*, 1963a, 1963b, 1963c; Uchimaya *et al.*, 1963). From this pioneer work, many biotechnologists screened better biocatalysts and strains belonging to several genera were selected such as *Yarrowia, Candida, Pichia, Rhodotorula, Sporidiobolus* (Endrizzi *et al.*, 1996; Romero-Guido *et al.*, 2011; Waché *et al.*, 2003a; Gatfield, 1999).

Although the general pathway is the same for the various yeast species, some differences occur as certain strains are able to degrade the newly accumulated γ-decalactone while others cannot (Endrizzi *et al.*, 1996) and, for many strains, the accumulation of a C10 intermediate compound from the β-oxidation pathway which should normally convert long chain fatty acids to acetyl moieties reflects an alteration in the pathway (Escamilla García *et al.*, 2009, 2007a, 2007b) while, for other strains, accumulation increases with optimisation of β-oxidation conditions (Alchihab *et al.*, 2009).

Depending on the type of metabolism, the strategy for optimising the pathway is thus different. It should be noted that a genetic engineering strategy is difficult to apply as the metabolic pathway concerned is β-oxidation with a group of four enzymes catalysing several cycles of oxidation for the same precursor. Deleting a gene coding for an oxidation enzyme would thus block the entire metabolism. However, a strategy was carried out with *Yarrowia lipolytica*, a species which has multiple gene families coding for most lipid metabolism enzymes. The identification of long- and short-chain selective enzymes (Waché *et al.*, 2001, 2000) enabled us to engineer the production of lactone increasing production significantly (Groguenin *et al.*, 2004; Waché *et al.*, 2002, 2003a). However, as production of this lactone can be considered to be relatively easy, it is carried out with non-genetically engineered strains with limitations that result at least partly from toxicity of the end-product toward the catalyst and the price of natural γ-decalactone is now relatively low (Gatfield, 1999).

Other examples of compounds belonging to this group are vanillin (see Priefert *et al.*, 2001 for a review) from the most popular aroma compounds and esters from the most numerous family. Indeed, numerous flavour compounds from biotechnology are produced from essential oils. These extracts can then be distilled to separate valuable compounds which, to increase their value further, can be esterified differently.

Besides these successful examples, microbial production of some highly demanded compounds is more difficult. The rose-note-exhibiting 2-phenylethanol is, for instance, a very popular compound for fragrances and flavour. Thousands of tons are produced yearly in chemical synthesis and some yeast-catalysed processes from L-phenylalanine are proposed (reviewed by Etschmann *et al.*, 2002) but this method of production still seems to be marginal in particular because of the toxicity of the final product towards the producing yeast.

© Woodhead Publishing Limited, 2013

8.4   New attempts to produce flavour compounds when precursors are unavailable

The first period of natural flavour production responded to the demand from consumers from some European countries. The present quest of consumers for naturalness has reached northern America and, from there, the whole world. With this growing demand, flavour compounds for which the process in the first period was not economically satisfactory are being revisited in the present context. Between both waves for Nature, the world has changed with the opening of new economies and the appearance of new actors making huge efforts to harvest and extract at low price compounds that were not available up to now. In the meantime, genetic techniques have developed enabling non-specialists to construct genetically engineered strains but also, through the development of screening and directed-evolution technologies, the first examples of applications of chimera biocatalysts have arisen (reviewed by Turner, 2003). These changes have resulted in the evolution of the fine chemistry and pharmaceutical industries (see Pollard and Woodley, 2007; Woodley, 2008 for reviews), however, owing to some specificities of the food domain such as consumers' demand for naturalness and products free from genetically modified organisms (GMOs) and the limitation of prices, microbial production of aroma compounds have not experienced a revolution in this period.

8.4.1   Hydroxylation

An important property of biocatalysts that made them popular among chemists is the ability to functionalise a carbon chain at a specific site. Particularly, the oxidation of alkanes is of interest. The question of enzymes able to hydroxylate fatty acids is an important concern in modern biotechnology and many articles deal with the screening of new activities. However, for most microorganisms, hydroxylation is still rather unspecific and yields are low (Romero-Guido *et al.*, 2011). Actually, the only hydroxylating enzymes that are site specific and efficient are ω-oxidising monooxygenases and they are used for the production of α,ω-diacids which are precursors of polymers and, in the fragrance field, of the musk macrocyclic lactones (reviewed by Waché, 2010) (Fig. 8.4). Besides this important screening of hydroxylating enzymes, other strategies are used in the field of directed evolution. Indeed, an interesting example exists in nature when an oleate desaturase becomes an oleate hydroxylase after a minor evolution step (Arnold, 2001). The change of activity can be obtained with a change of four amino acids in the protein. This change occurred in the ricin plant giving rise to high ratio of ricinoleic acid in castor oil which thus became the cheap substrate used to produce γ-decalactone (see above). If a similar evolution step could be achieved for a stearate or a linoleate desaturase, it would give rise to δ-tetradecalactone and δ-octalactone, respectively.

© Woodhead Publishing Limited, 2013

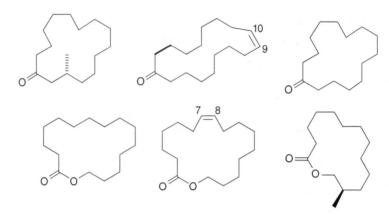

Fig. 8.4   Structure of macrocyclic musk compounds resulting from the cyclisation of diacids of hydroxy acids.

Then, if we understand which parameters control the site of hydroxylation, this could lead to the production of precursors of other lactones using cheaper methods of production of the already produced lactones γ-octalactone, γ-nonalactone, δ-decalactone, 6-pentyl-α-pyrone, γ- and δ-dodecalactone and possibilities for production of new lactones.

8.4.2   Cleavage of carotenoids

Carotenoids are precursors of many valuable compounds (Wintherhalter and Rouseff, 2001) among which some are difficult to produce in a natural way (β- or α-ionone) and others which have been up to now impossible to produce (β-damascenone) (Aguedo et al., 2004). Up to the beginning of the 21st century, biotechnological ways of production of ionone were based on the cleavage of carotene through oxidation by enzymatically generated unspecific reactive oxygen species (Belin et al., 1994; Bosser and Belin, 1994; Waché et al., 2003b; Zorn et al., 2003a, 2003b). The discovery of site-specific cleavage enzymes was expected to revolutionise production of carotenoid-derived aroma compounds but, owing to problems of dispersion of carotenoids and low yields (Schmidt et al., 2006), the activity of these enzymes has not yet reached expected results and better yields are still available with unspecific enzymes (Ly et al., 2008). However, although cleavage enzymes are still promising for increasing production yields of some compounds, for others like damascenone, the solution would require terpene-specific enzymes and these have still to be developed. Considering now an aromatic phenolic cycle, such enzyme development to transform some phenolic precursors into valuable aroma compounds is also a source of future development as the biocatalytic chemistry of phenolic compounds is still poor compared with what exists in nature.

© Woodhead Publishing Limited, 2013

8.5   Analysing natural flavours in food

As presented in the introduction, the main economic interest in producing aroma compounds through microbial catalysis is the opportunity to obtain a natural label for the resulting flavours. The word 'natural' is defined legally by the regulations of most countries. For aromas, the regulations concerned are CFR 1990 and the European regulation CE 1334/2008. With small differences, a substance can be considered to be natural when it comes from a plant, animal or microbial origin with a physical, microbial or enzymatic process. Therefore, biotechnological routes may be, if they exclude any chemical step, a way to obtain natural products.

It must be kept in mind that, although the concept of naturalness seems rather clear to most consumers, definitions can vary between people and particularly between disciplines. Several definitions can be found for 'natural'. One definition is 'present in or produced by nature'. However, other definitions are proposed and among them, some introduce the concept of absence of 'non-natural action', 'conforming to the usual or ordinary course of nature' or 'not altered, treated, or disguised' and particularly the meaning given in biology: 'not produced or changed artificially; not conditioned'. In these latter definitions, 'natural' is the opposite of artificial or artefactual. As discussed by Jacques Monod in the first chapter of his book *Chance and Necessity* (Monod, 1971), the definition and hallmarks of naturalness are not so easy to choose, particularly when biological techniques may be used. Can we limit the definition to what is not literally artefactual or do we have to consider natural techniques as belonging to a project or purpose. As an example, yeasts transforming a natural precursor to a flavouring compound are part of a natural process, whereas flavouring compounds produced by cells for a human purpose in a reactor is not literally artefactual but would have to be considered as belonging to a project.

This case may seem extreme as, of course, according to this definition agricultural products would not be natural and this term would be reserved for products of hunting and gathering in non-cultivated areas. However, when discussing the subject with consumers, at least in several European countries, we often observe an aversion to flavouring compounds from biotechnological reactors, showing that this definition taking into account the belonging to a project is considered valid by some consumers. But as discussed by Monod (1971), there are also several hallmarks suggesting that Nature itself is part of a project, a 'pure postulate, for evermore indemonstrable' but which may impact on the definition of natural.

A point that has to be addressed in this discussion is the possible genetic engineering of microorganisms to obtain flavours. Two cases should be considered. Starters that are used for fermentation are usually present in the consumed product and if they are genetically modified, consumers will eat them. Contrary to open field genetically modified organisms (GMOs), they will not be disseminated but they will be ingested. For their part,

© Woodhead Publishing Limited, 2013

186 Microbial production of food ingredients, enzymes and nutraceuticals

biocatalysts used in reactors to produce aroma compounds will neither be disseminated nor be ingested. The risk of the utilisation of these two types of GMOs is thus different. However, there is once again an ethical dimension, considering again the project, can GMOs produce 'natural compounds'? In certain cases, a social dimension is also present generated by the comfort of producing in the reactors of rich countries what was formerly produced by agriculture in poor countries. However, aroma compounds produced in reactors are up to now not often competitors of agriculture and if we take the example of vanillin, biotechnological production is usually used to complete production when economic (speculation) and environmental conditions are not favourable (Loeillet, 2003).

Whatever the definition selected, it must be kept in mind that regulations are the product of several influences among which human lobbying and thinking streams are not to be neglected. Regulations are therefore not fixed and they may change, taking another definition as a basis. In this context, aroma producers are attentive to the evolution of regulation and they try to have extra processes that they could use in case of modification of the rules.

For chemists, a flavouring compound extracted from plants or synthesised in a laboratory is the same if the structure is the same. And this introduces a second point in the regulation process. To be applicable, a regulation must be based on clearly identifiable differences. When compounds have never been detected in Nature, they are clearly artificial (as long as they are not detected in plants or animals). However, how can we differentiate between the same aroma compound from natural or natural-like origin? There is sometimes a way to differentiate between compounds resulting from enzymes catalysis and chemical synthesis. Indeed, enzymes usually have different affinities for enantiomers of the same compounds whereas classical chemistry is in favour of parity. As a result, chemical synthesis will give rise to racemics (an equal amount of each enantiomer) whereas nature will give rise to unequal amounts. This latter solution can also be found for asymmetric chemical synthesis but owing to its high cost, there is unlikely to be such a synthesis for aroma compounds. From the development of chiral columns for chromatography, it is possible to separate enantiomers and appreciate whether chiral compounds are present in a racemic form or not. Unfortunately, this information is not absolute evidence of naturalness as for instance γ-dodecalactone from raspberry has been reported to be present in a racemic form (Dufossé et al., 1994). Reports are lacking for other compounds and it is difficult to come to any conclusions. When compounds are asymmetric, analysis is based on isotopic spectrometry (see Mosandl, 1999 for a review of analytical methods). Frédéric Saltron in a presentation in December 2006 (Saltron, 2006) on the control of aromas gives a fictional example of aroma analysis. Although all the major components are shown to be natural, during analysis, the existence of a minor compound, α-ionone, in a racemic form engendered a prejudice against this

© Woodhead Publishing Limited, 2013

product. Interestingly, a similar case showing the presence of a racemic mixture of α-ionone in a fruit natural aroma was brought to me by a natural aroma producer who had found it on the market. As mentioned by Saltron, control by European administrations is difficult in European countries and almost impossible outside Europe. What would be required would be a consensual database but this only existed for vanillin.

A second advantage of producing flavouring compounds through microbial catalysis is that the impact of biocatalysis on the environment is generally far lower than the impact of a similar chemical process and even, in the case of a flavouring compound, than the impact of extraction.

8.6   Conclusion and future trends

The world of flavours and fragrances is in a quest for naturalness. In this process, microbial production is an interesting response that is developed when extraction is not possible. The general flowchart explaining how to proceed to produce a natural flavouring compound is given in Fig. 8.5. As for all biocatalytic pathways, the production of flavour compounds is subject to the following constraints: existence of an available precursor, a corresponding biocatalytic pathway, possibility to increase concentrations (concentrations of metabolites in living organisms are usually maintained low), and so on. As the success of microbial flavours is related to consumers'

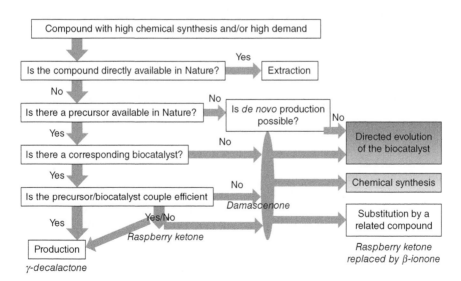

Fig. 8.5   Flow chart of natural flavour compounds production. Grey boxes indicate that the solution is not adequate (not natural or GMO). Some examples are given in italics.

© Woodhead Publishing Limited, 2013

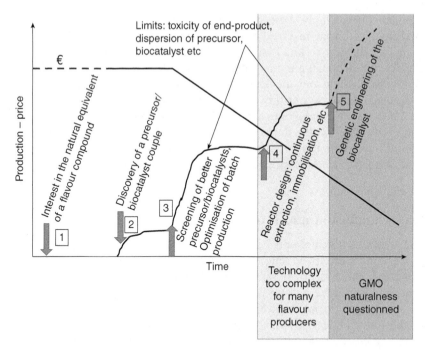

Fig. 8.6 Different stages of production of a microbial aroma compound. The black curve indicates the volume produced, the dotted and straight line with a € symbol indicates the price, the shaded arrows explain the different phases (numbered from 1 to 5), the grey shading indicates that these steps are not well accepted and the commentaries at the bottom explain why.

demand, additional constraints are present: the process must avoid any chemical step and genetically modified organisms are usually undesirable. Finally, specific market constraints also exist as prices usually decrease when production is possible so complex technological processes comparable with what exists in the pharmaceutical industry are usually not adequate.

The strategy followed to produce an aroma compound when a precursor/biocatalyst couple is identified is shown in Fig. 8.6. Several compounds were relatively easy to produce microbially and passed the three first stages. But as a result of naturalness, GMO and economic constraints, compounds blocked at stage 2 are unlikely to pass this phase in the future. However, despite the fact that all 'easy-to-produce' aroma compounds have already been produced, microbial natural aroma compounds still have a future, given the high demand for these products. But to develop new production, researchers have to follow the very specific constraints given above which lead to a method with specific rules. I therefore believe that reactor approaches modifying metabolic fluxes as well as developments in 'directed

© Woodhead Publishing Limited, 2013

spontaneous mutations' could lead to the appearance of several new flavouring compounds in the market. Other strategies based on genetic engineering could induce a bad public perception of all microbial flavour compounds, at least in several European countries, and it could eventually result in a change in the philosophy of regulation.

From the point of view of the market, trends are still notable towards naturalness and the objective is to substitute all synthetic compounds by their natural equivalents. Of course, during formulation, several methods are available and sometimes, when possible, a synthetic compound can be changed by one or two different compounds giving the same sensorial result. However, it is likely that the quest for biocatalytic ways to produce damascenone or to better produce raspberry ketone will still occupy flavour technologists.

8.7   Sources of further information and advice

Flavours and fragrances are the subject of several internet sites. One site is a must on the subject: http://www.leffingwell.com/ owing to the great deal of free documentation and links on flavour, fragrances, aromatherapy and so on to different scientist, technology, toxicology, economic, medical points of views. A lot of documentation can also be found on the Perfume & Flavorist website (Perfume & Flavorist, 2012). This field has also been the subject of many journal reviews and books. We mention some examples of interest. A comprehensive review of fragrance chemistry has been published by Frater (Frater *et al.*, 1998). A journal named the *Flavour and Fragrance Journal* also deals with the subject. Several reviews deal more precisely with microbial production of flavours and some interesting examples are given here (Hagedorn and Kaphammer, 1994; Berger, 2009; Feron and Waché, 2005; Gatfield, 1988, 1999; Krings and Berger, 1998; Berger, 1995). As the field is relatively wide, there are several reviews on more precise topics like vanillin, raspberry ketone, carotenoid-derived compounds and lactones (Etschmann *et al.*, 2002; Romero-Guido *et al.*, 2011; Waché *et al.*, 2003a; Priefert *et al.*, 2001; Beekwilder *et al.*, 2007; Rodriguez-Bustamante and Sanchez, 2007).

The main actors in the market of microbial aroma compounds are the starter and the aroma industries. Some reviews written by scientists belonging to these groups give information on these industries. For instance, Hansen (2002) and Pedersen (Pedersen *et al.*, 2005) for the starter industry and Rabenhorst (2008) or Schrader *et al.* (2004) for the aroma producers show the trends in these industries and illustrate the innovation process in this field.

Finally, people interested in a novel to be introduced into the field of fragrance can read (or watch the movie) 'Perfume: The Story of a Murderer' by Patrick Süskind and people interested in a philosophical approach to

© Woodhead Publishing Limited, 2013

Nature and biology would be interested in reading *Chance and Necessity: An Essay on the Natural Philosophy of Modern Biology* by Jacques Monod.

8.8   References

AGUEDO, M., LY, M. H., BELO, I., TEIXEIRA, J., BELIN, J.-M. and WACHÉ, Y. (2004). 'The use of enzymes and microorganisms for the production of aroma compounds from lipids'. *Food Technol Biotechnol*, **42**, 327–36.

ALCHIHAB, M., DESTAIN, J., AGUEDO, M., MAJAD, L., GHALFI, H., WATHELET, J. P. and THONART, P. (2009). 'Production of gamma-decalactone by a psychrophilic and a mesophilic strain of the yeast *Rhodotorula aurantiaca*'. *Appl Biochem Biotechnol*, **158**, 41–50.

ARNOLD, F. H. (2001). 'Combinatorial and computational challenges for biocatalyst design'. *Nature*, **409**, 253–7.

AYAD, E. H. E., VERHEUL, A., DE JONG, C., WOUTERS, J. T. M. and SMIT, G. (1999). 'Flavour forming abilities and amino acid requirements of *Lactococcus lactis* strains isolated from artisanal and non-dairy origin'. *Int Dairy J*, **9**, 725–35.

AYAD, E. H., VERHEUL, A., ENGELS, W. J., WOUTERS, J. T. and SMIT, G. (2001). 'Enhanced flavour formation by combination of selected lactococci from industrial and artisanal origin with focus on completion of a metabolic pathway'. *J Appl Microbiol*, **90**, 59–67.

BEEKWILDER, J., VAN DER MEER, I. M., SIBBESEN, O., BROEKGAARDEN, M., QVIST, I., MIKKELSEN, J. D. and HALL, R. D. (2007). 'Microbial production of natural raspberry ketone'. *Biotechnol J*, **2**, 1270–9.

BELIN, J.-M., BENSOUSSAN, M. and SERRANO-CARREON, L. (1992). 'Microbial biosynthesis for the production of food flavours'. *Trends Food Sci Technol*, **3**, 11–14.

BELIN, J. M., DUMONT, B. and ROPERT, F. (1994). *Procédé de fabrication, par voie enzymatique, d'arômes, notamment des ionones et des aldéhydes en C6 à C10*. France patent application: WO 94/08028.

BERGER, R. (1995). *Aroma Biotechnology*, Springer-Verlag, Berlin, Heidelberg.

BERGER, R. G. (2009). 'Biotechnology of flavours–the next generation'. *Biotechnol Lett*, **31**, 1651–9.

BOSSER, A. and BELIN, J.-M. (1994). 'Synthesis of beta-ionone in an aldehyde/xanthine oxidase/beta-carotene system involving free radical formation'. *Biotechnol Prog*, **10**, 129–33.

DEMARIGNY, Y. (2012). 'Fermented food products made with vegetable materials from tropical and warm countries: microbial and technological considerations'. *International J Food Sci Technol*, **47**, 2469–76.

DUFOSSÉ, L., LATASSE, A. and SPINNLER, H. E. (1994). 'Importance des lactones dans les arômes alimentaires: Structure, distribution, propriétés sensorielles'. *Sci Aliments*, **14**, 19–50.

ENDRIZZI, A., PAGOT, Y., LE CLAINCHE, A., NICAUD, J.-M. and BELIN, J.-M. (1996). 'Production of lactones and peroxisomal beta-oxidation in yeasts'. *Crit Rev Biotechnol*, **16**, 301–29.

ESCAMILLA GARCÍA, E., BELIN, J.-M. and WACHÉ, Y. (2007a). 'Use of a Doehlert factorial design to investigate the effects of pH and aeration on the accumulation of lactones by *Yarrowia lipolytica*.' *J Appl Microbiol*, **103**, 1508–15.

ESCAMILLA GARCÍA, E., NICAUD, J.-M., BELIN, J.-M. and WACHÉ, Y. (2007b). 'Effect of acyl-CoA oxidase activity on the accumulation of gamma-decalactone by the yeast *Yarrowia lipolytica*: A factorial approach'. *Biotechnol J*, **2**, 1280–85.

ESCAMILLA GARCÍA, E., AGUEDO, M., GOMES, N., CHOQUET, A., BELO, I., TEIXEIRA, J., BELIN, J. and WACHÉ, Y. (2009). 'Production of 3-hydroxy-gamma-decalactone, the

© Woodhead Publishing Limited, 2013

precursor of two decenolides with flavouring properties, by the yeast *Yarrowia lipolytica*'. *J Mol Catal B*, **57**, 22–6.

ETSCHMANN, M. E., BLUEMKE, W. B., SELL, D. S. and SCHRADER, J. S. (2002). 'Biotechnological production of 2-phenylethanol'. *Appl Microbiol Biotechnol*, **59**, 1–8.

FERON, G. and WACHÉ, Y. (2005). 'Microbial biotechnology of food flavor production'. In: *Food Biotechnology*, Dominick, T. (ed.). 2nd edition Dekker, New York.

FRATER, G., BAJGROWICZ, J. and KRAFT, P. (1998). 'Fragrance chemistry'. *Tetrahedron*, **54**, 7633–703.

GATFIELD, I. L. (1988). 'Production of flavor and aroma compounds by biotechnology'. *Food Technology*, **10**, 110–22.

GATFIELD, I. L. (1999). 'Biotechnological production of natural flavor materials'. In: *Flavor Chemistry, Thirty Years of Progress*. Teranishi, R., Wick, E. L. & Hornstein, I. (eds). Kluwer Academic, Plenum Publishers, New York.

GOERGES, S., MOUNIER, J., REA, M. C., GELSOMINO, R., HEISE, V., BEDUHN, R., COGAN, T. M., VANCANNEYT, M. and SCHERER, S. (2008). 'Commercial ripening starter microorganisms inoculated into cheese milk do not successfully establish themselves in the resident microbial ripening consortia of a South german red smear cheese'. *Appl Environ Microbiol*, **74**, 2210–7.

GROGUENIN, A., WACHÉ, Y., ESCAMILLA GARCIA, E., AGUEDO, M., HUSSON, F., LEDALL, M., NICAUD, J. and BELIN, J. (2004). 'Genetic engineering of the beta-oxidation pathway in the yeast *Yarrowia lipolytica* to increase the production of aroma compounds'. *J Molec Catal B*, **28**, 75–9.

HAGEDORN, S. and KAPHAMMER, B. (1994). 'Microbial biocatalysis in the generation of flavor and fragrance chemicals'. *Annu Rev Microbiol*, **48**, 773–800.

HANSEN, E. B. (2002). 'Commercial bacterial starter cultures for fermented foods of the future'. *Int J Food Microbiol*, **78**, 119–31.

IMHOF, R. and BOSSET, J. O. (1994). 'Relationships between micro-organisms and formation of aroma compounds in fermented dairy products'. *Zeit Lebensmittel-Forsch A*, **198**, 267–76.

KRINGS, U. and BERGER, R. G. (1998). 'Biotechnological production of flavours and fragrances'. *Appl Microbiol Biotechnol*, **49**, 1–8.

LEFFINGWELL & ASSOCIATES (2012). http://www.leffingwell.com

LOEILLET, D. (2003). 'Le marché international de la vanille – Le prix comme handicap majeur'. *Fruitrop*, **98**, 4–7.

LY, M. H., CAO HOANG, L., BELIN, J.-M. and WACHÉ, Y. (2008). 'Improved co-oxidation of beta-carotene to beta-ionone using xanthine oxidase-generated reactive oxygen species in a multiphasic system'. *Biotechnol J*, **3**, 220–5.

LY-CHATAIN, M. H., LE, M. L., THANH, M. L., BELIN, J.-M. and WACHÉ, Y. (2010). 'Cell surface properties affect colonisation of raw milk by lactic acid bacteria at the microstructure level'. *Food Res Int*, **43**, 1594–602.

MARTIN, F., CACHON, R., PERNIN, K., DE CONINCK, J., GERVAIS, P., GUICHARD, E. and CAYOT, N. (2011). 'Effect of oxidoreduction potential on aroma biosynthesis by lactic acid bacteria in nonfat yogurt'. *J Dairy Sci*, **94**, 614–22.

MCSWEENEY, P. L. H. and SOUSA, M. J. (2000). 'Biochemical pathways for the production of flavour compounds in cheeses during ripening: A review'. *Lait*, **80**, 293–324.

MONNET, C. and CORRIEU, G. (2007). 'Selection and properties of α-acetolactate decarboxylase-deficient spontaneous mutants of *Streptococcus thermophilus*'. *Food Microbiol*, **24**, 601–6.

MONOD, J. (1971). *Chance and Necessity: An Essay on the Natural Philosophy of Modern Biology*, Alfred A. Knopf, New York.

MOSANDL, A. (1999). 'Analytical authentication of genuine flavor compounds'. In: *Flavor Chemistry. Thirty Years of Progress*. Teranishi, R., Wick, E. & Hornstein, I. (eds). Kluwer Academic/Plenum Publishers, New York.

© Woodhead Publishing Limited, 2013

OGAWA, J., MATSUMURA, K., KISHINO, S., OMURA, Y. and SHIMIZU, S. (2001). 'Conjugated linoleic acid accumulation via 10-hydroxy-12-octadecaenoic acid during microaerobic transformation of linoleic acid by *Lactobacillus acidophilus*'. *Appl Environ Microbiol*, **67**, 1246–52.

OKUI, S., UCHIYAMA, M. and MIZUGAKI, M. (1963a). 'Metabolism of hydroxy fatty acids: 1. Metabolic conversion of ricinoleic acid by a certain microorganism to 8-D-(+)-hydroxy tetradec-*cis*-5-enoic acid'. *J Biochem*, **53**, 265–70.

OKUI, S., UCHIYAMA, M. and MIZUGAKI, M. (1963b). 'Metabolism of hydroxy fatty acids: 2. Intermediates of the oxidative breakdown of ricinoleic acid by genus *Candida*'. *J Biochem*, **54**, 536–40.

OKUI, S., UCHIYAMA, M., MIZUGAKI, M. and SUGAWARA, A. (1963c). 'Characterization of hydroxy acids in depot fat after feeding of ricinoleic acid'. *Biochim Biophys Acta*, **70**, 344–6.

PEDERSEN, M. B., IVERSEN, S. L., SORENSEN, K. I. and JOHANSEN, E. (2005). 'The long and winding road from the research laboratory to industrial applications of lactic acid bacteria'. *FEMS Microbiol Rev*, **29**, 611–24.

PERFUME AND FLAVORIST MAGAZINE (2012). http://www.perfumerflavorist.com/

POLLARD, D. J. and WOODLEY, J. M. (2007). 'Biocatalysis for pharmaceutical intermediates: the future is now'. *Trends Biotechnol*, **25**, 66–73.

PRIEFERT, H., RABENHORST, J. and STEINBÜCHEL, A. (2001). 'Biotechnological production of vanillin'. *Appl Microbiol Biotechnol*, **56**, 296–314.

RABENHORST, J. (2008). 'Biotechnological production of natural aroma chemicals by fermentation processes'. *Biotechnology Set*. Rehm, H.-J. and Reed, G. (eds), Wiley-VCH, Germany.

RIONDET, C., CACHON, R., WACHÉ, Y., ALCARAZ, G. and DIVIÈS, C. (2000). 'Extracellular oxidoreduction potential modifies carbon and electron flow in *Escherichia coli*'. *J Bacteriol*, **182**, 620–6.

RODRIGUEZ-BUSTAMANTE, E. and SANCHEZ, S. (2007). 'Microbial production of C13-norisoprenoids and other aroma compounds via carotenoid cleavage'. *Crit Rev Microbiol*, **33**, 211–30.

ROMERO-GUIDO, C., BELO, I., TA, T. M., CAO-HOANG, L., ALCHIHAB, M., GOMES, N., THONART, P., TEIXEIRA, J. A., DESTAIN, J. and WACHE, Y. (2011). 'Biochemistry of lactone formation in yeast and fungi and its utilisation for the production of flavour and fragrance compounds'. *Appl Microbiol Biotechnol*, **89**, 535–47.

RTS (2010). *The Market for Flavours and Flavour Trends*. Research to Solutions Resource Ltd, Wolverhampton, UK. http://issuu.com/rtsresource/docs/flavour_trends_presentation_23_june_2010_

SALTRON, F. (2006). *Contrôle officiel des arômes et des denrées aromatisés*. http://www.economie.gouv.fr/files/directions_services/dgccrf/manifestations/colloques/aromes_alimentaires/11_saltron.pdf

SCHMIDT, H., KURTZER, R., EISENREICH, W. and SCHWAB, W. (2006). 'The carotenase AtCCD1 from *Arabidopsis thaliana* is a dioxygenase'. *J Biol Chem*, **281**, 9845–51.

SCHRADER, J., ETSCHMANN, M., SELL, D., HILMER, J.-M. and RABENHORST, J. (2004). 'Applied biocatalysis for the synthesis of natural flavour compounds-current industrial processes and future prospects'. *Biotechnol Lett*, **26**, 463–72.

TURNER, N. J. (2003). 'Directed evolution of enzymes for applied biocatalysis'. *Trends Biotechnol*, **21**, 474–8.

UCHIYAMA, M., SATO, R. and MIZUGAKI, M. (1963). 'Characterization of hydroxy acids in depot fat after feeding of ricinoleic acid'. *Biochim Biophys Acta*, **70**, 344–6.

WACHÉ, Y. (2010). 'Production of dicarboxylic acids and flavours by the yeast *Yarrowia lipolytica*.' In: *Yarrowia lipolytica, Molecular Biology and Biotechnology*. Barth, G. and Steinbüchel, A. (eds). Springer, Berlin.

© Woodhead Publishing Limited, 2013

WACHÉ, Y., LAROCHE, C., BERGMARK, K., MOLLER-ANDERSEN, C., AGUEDO, M., LE DALL, M.-T., WANG, H., NICAUD, J.-M. and BELIN, J.-M. (2000). 'Involvement of acyl-CoA oxidase isozymes in biotransformation of methyl ricinoleate into γ-decalactone by *Yarrowia lipolytica*'. *Appl Environ Microbiol*, **66**, 1233–6.

WACHÉ, Y., AGUEDO, M., CHOQUET, A., GATFIELD, I., NICAUD, J.-M. and BELIN, J.-M. (2001). 'Role of β-oxidation enzymes in the production of γ-decalactones from methyl ricinoleate'. *Appl Environ Microbiol*, **67**, 5700–4.

WACHÉ, Y., AGUEDO, M., LEDALL, M.-T., NICAUD, J.-M. and BELIN, J.-M. (2002). 'Optimization of *Yarrowia lipolytica*'s β-oxidation pathway for lactones production'. *J Molec Catal B Enzym*, **19–20**, 347–51.

WACHÉ, Y., AGUEDO, M., NICAUD, J.-M. and BELIN, J.-M. (2003a). 'Catabolism of hydroxyacids and production of lactones by the yeast *Yarrowia lipolytica*'. *Appl Microbiol Biotechnol*, **61**, 393–404.

WACHÉ, Y., BOSSER-DERATULD, A., LHUGUENOT, J. C. and BELIN, J. M. (2003b). 'Effect of *cis/trans* isomerism of β-carotene on the ratios of volatile compounds produced during oxidative degradation'. *J Agric Food Chem*, **51**, 1984–7.

WANIKAWA, A., HOSOI, K. and KATO, T. (2000a). 'Conversion of unsaturated fatty acids to precursors of gamma-lactones by lactic acid bacteria during the production of malt whisky'. *J Am Soc Brew Chem*, **58**, 51–6.

WANIKAWA, A., HOSOI, K., TAKISE, I. and KATO, T. (2000b). 'Detection of gamma-lactones in malt whisky'. *J Inst Brew*, **106**, 39–43.

WINTHERHALTER, P. and ROUSEFF, R. (2001). 'Carotenoid-derived aroma compounds: an introduction'. In: *Carotenoid-derived Aroma Compounds*. Wintherhalter, P. & Rouseff, R. (eds). ACS, Washington DC.

WOODLEY, J. M. (2008). 'New opportunities for biocatalysis: making pharmaceutical processes greener'. *Trends Biotechnol*, **26**, 321–7.

ZORN, H., LANGHOFF, S., SCHEIBNER, M. and BERGER, R. G. (2003a). 'Cleavage of beta, beta-carotene to flavor compounds by fungi'. *Appl Microbiol Biotechnol*, **62**, 331–6.

ZORN, H., LANGHOFF, S., SCHEIBNER, M., NIMTZ, M. and BERGER, R. G. (2003b). 'A peroxidase from *Lepista irina* cleaves beta, beta-carotene to flavor compounds'. *Biol Chem*, **384**, 1049–56.

© Woodhead Publishing Limited, 2013

9

Microbial production of carotenoids

S. Sanchez, B. Ruiz and R. Rodríguez-Sanoja, Universidad Nacional Autónoma de México, Mexico, DF and L. B. Flores-Cotera, Cinvestav-IPN, Mexico, DF

DOI: 10.1533/9780857093547.2.194

Abstract: This chapter focuses on carotenoids and covers their definition, main biological functions and general chemical characteristics. The chapter first reviews the microorganisms that are currently used to produce carotenoids, the main biosynthetic pathways involved in their production and the regulatory mechanisms used for modulating carotenoid production. The chapter then discusses the progress made in genetic improvements for carotenoid production and the advances in fermentation conditions for producing high concentrations of carotenoids. Finally, the chapter presents specific examples of commercially significant carotenoids.

Key words: astaxanthin, canthaxanthin, β-carotene, lutein, lycopene, zeaxanthin.

9.1 Introduction

Carotenoids are tetraterpenoid organic pigments produced by plants, algae, bacteria, fungi and yeasts. Among other roles in these organisms, carotenoids may function as photosynthetic or light-quenching pigments, antioxidants, colorants and precursors of vitamin A (Frengova and Beshkova, 2009). To date, more than 600 different naturally occurring carotenoids have been described. Animals require carotenoids but cannot synthesize them; therefore, these compounds must be obtained from plants and other natural sources where they are present. Several carotenoids, such as lycopene, zeaxanthin, astaxanthin, β-carotene, lutein and canthaxanthin (Fig. 9.1) are commercially important owing to their specific applications in the food (as nutrient supplements, food colorants and feed additives), pharmaceutical (antioxidants, anti-carcinogenics and immune-modulators) and cosmetic industries (skin care products, aftershave lotions, bath products, hair conditioners, shampoos and suntan products). The global market demand for carotenoids has been growing at 2.9% per annum and was estimated to be

© Woodhead Publishing Limited, 2013

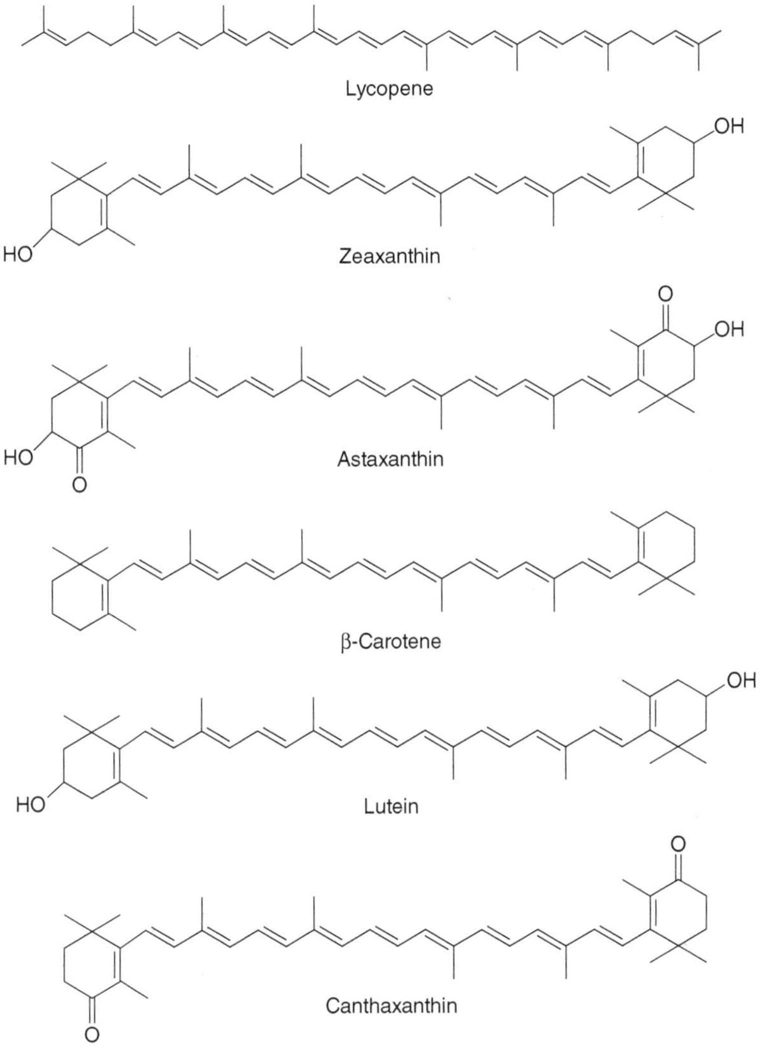

Fig. 9.1   Chemical structures of various carotenoids.

approximately US$1.2 billion in 2010, with a possible value of US$1.4 billion in 2018 (dependent on consumers continuing to look for natural ingredients) (Gu *et al.*, 2008; BCC Research, 2011). The United States and Europe collectively account for a major share of sales in the global carotenoids market (Global Industry Analysts, 2010).

The demand for carotenoids as nutraceutical compounds has triggered research into the exploration of commercially viable processes for economic production of carotenoids. Chemical synthesis of carotenoids is

© Woodhead Publishing Limited, 2013

challenging and costly. Extraction from plants is often limited by the source availability and the economics of the processes, that is, they are labour intensive and land expensive. Therefore, an increasing interest in microbial sources for carotenoids has been observed, owing to consumer preferences for natural additives and the potential cost effectiveness of creating carotenoids via microbial biotechnology (Dufosse *et al.*, 2005).

Chemically, carotenoids are a diverse class of C_{40} isoprenoids that are either linear or cyclized at one or both ends of the molecule. They consist of isoprenoid building blocks and are closely related to sterols, ubiquinones and terpenes and terpenoids (Namitha and Negi, 2010). Depending on the presence of an additional hydroxyl group in their structure, carotenoids can be classified as either carotenes or xanthophyls (Fig. 9.1). Carotenes contain only carbon and hydrogen (e.g. beta-carotene, alpha-carotene and lycopene), whereas xanthophylls contain an additional keto or hydroxyl group (e.g. lutein, zeaxanthin, astaxanthin, etc.). Most of the diversity in natural carotenoids arises from differences in types and levels of desaturation and other modifications of the C_{40} backbone.

9.2   Microbial sources of carotenoids

In recent years, there has been increasing interest in the production of natural carotenoids by microbial fermentation. Compared to chemical methods, microbial production of carotenoids is an environmentally friendly method and is expected to meet the increasing demand for natural carotenoids. Carotenogenic microbes comprise a wide variety of microbes including halotolerant fresh water algae, photosynthetic and phototropic bacteria, asporogenous yeasts and fungi. Among these microorganisms, *Dunalliela salina*, *Xanthophyllomyces dendrorhous* (formerly named *Phaffia rhodozyma*), *Haematococcus pluvialis* and *Blakeslea trispora* have been considered for production on a large scale (25,000 L) (Frengova and Beshkova, 2009; Mehta *et al.*, 2003; Olaizola, 2000; Raja *et al.*, 2007).

Because yeasts are unicellular organisms with a relatively high growth rate in low-cost fermentation media, they have an advantage over algae, fungi and bacteria (Malisorn and Suntornsuk, 2008). The characteristic red colour of yeast cells results from pigments that the yeast create that block certain wavelengths of light that would otherwise damage the cell. Typical concentrations reported in the literature for carotenoids produced from red yeasts range from 50–800 µg g dry weight[-1] (Davoli *et al.*, 2004; An *et al.*, 1999; Zeni *et al.*, 2011).

In addition to the above-mentioned microorganisms, other carotenogenic microbes that have been investigated for large-scale production include *Bradyrhizobium* sp. *Paracoccus zeaxanthinifaciens* (formerly *Flavobacterium aurantiacum*) and *Brevibacterium aurantiacum* (Chattopadhyay

© Woodhead Publishing Limited, 2013

et al., 2008). In various species of terrestrial *Streptomyces*, carotenogenesis occurs only following photo-induction in a mechanism that remains unclear (Takano *et al.*, 2006). To date, studies on carotenoid production in *Streptomyces* have been performed with *Streptomyces griseus*, *Streptomyces setonii* and *Streptomyces coelicolor* (Takano *et al.*, 2006). Indeed, production of carotenoids in marine sponge has been credited to *Streptomyces* (Dharmaraj *et al.*, 2009).

Microbial biosynthesis of carotenoids has been strongly influenced by other techniques such as genetic and metabolic engineering. The carotenoid biosynthetic pathway is especially amenable to manipulation by recombinant DNA techniques because all carotenoids share a common precursor. By combining biosynthetic genes, a much wider range of carotenoids can be produced (Dannert *et al.*, 2000). Non-carotenogenic microbes, for example, *Escherichia coli*, *Saccharomyces cerevisiae*, *Candida utilis* and *Zymomonas mobilis*, have been engineered to produce traditional and novel carotenoids following transformation with carotenoid genes from different carotenogenic microbes, for example, *X. dendrorhous*, *Rhodotorula glutinis*, *Erwinia herbicola* and so on, and by recombining these genes to produce new acyclic, cyclic and oxo-carotenoids for specific applications (Misawa and Shimada, 1998; Das *et al.*, 2007). In particular, *E. coli* has been widely used as a powerful genetic tool system for the production of various carotenoids, for example, zeaxanthin, astaxanthin and β-carotene have all been produced by this microorganism in amounts of grams per litre (Ruther *et al.*, 1997; Lemuth *et al.*, 2011; Yoon *et al.*, 2007a). Finally, to synthesize astaxanthin, *S. cerevisiae* was recently used as a host organism for the *X. dendrorhous*-derived astaxanthin-producing gene (Ukibe *et al.*, 2009).

9.3   Main biosynthetic pathways used for carotenoid production

All carotenoids are synthesized via two pathways that lead to the formation of isopentenyl diphosphate (IPP) and/or dimethylallyl diphosphate (DMAPP), the pathway building units. Eukaryotes (except *Euglenophyta*) and archeabacteria typically follow the mevalonate pathway (MVA) producing both IPP and DMAPP precursors. In contrast, bacteria use the 2C-methyl-D-erythritol 4-phosphate pathway (MEP) (Namitha and Negi, 2010). Plants and streptomycetes use both MVA and MEP (Das *et al.*, 2007).

The MEP pathway (Fig. 9.2) starts with the transketolase condensation of pyruvate and glyceraldehyde-3-phosphate (G3P) by 1-deoxy-D-xylulose 5-phosphate synthase (DXS), which results in 1-deoxy-D-xylulose 5-phosphate (DXP). Subsequently, DXP is reduced to methylerythritol 4-phosphate (MEP) by 1-deoxy-D-xylulose 5-phosphate reductoisomerase (IspC) (Namitha and Negi, 2010; Walter and Strack, 2011). MEP is then

© Woodhead Publishing Limited, 2013

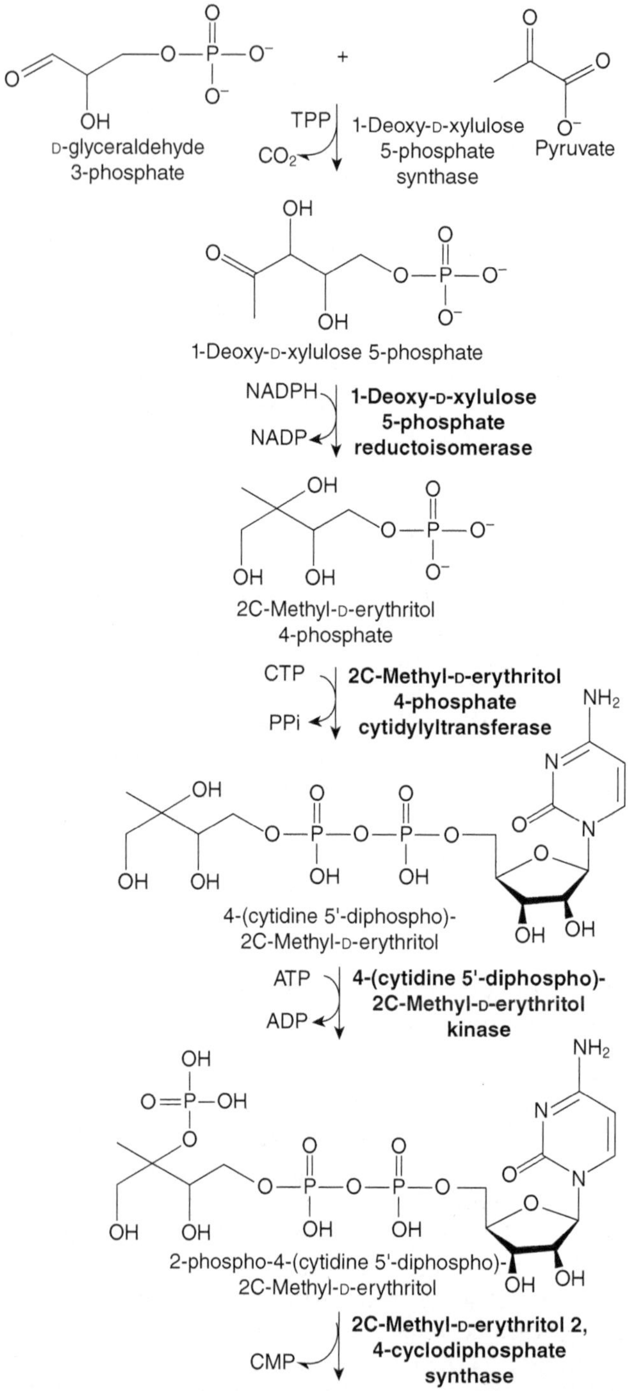

Fig. 9.2 Methyl erythritol pathway for the biosynthesis of IPP and DMAPP precursors. See the text for details. Abbreviations used are: TPP, thiamine pyrophosphate; CTP, cytidine triphosphate; CMP, cytidine monophosphate; HMBDP, 1-hydroxy-2-methyl-2-(*E*)-butenyl 4-diphosphate; IPP, isopentenyl diphosphate; DMAPP, dimethylallyl diphosphate.

© Woodhead Publishing Limited, 2013

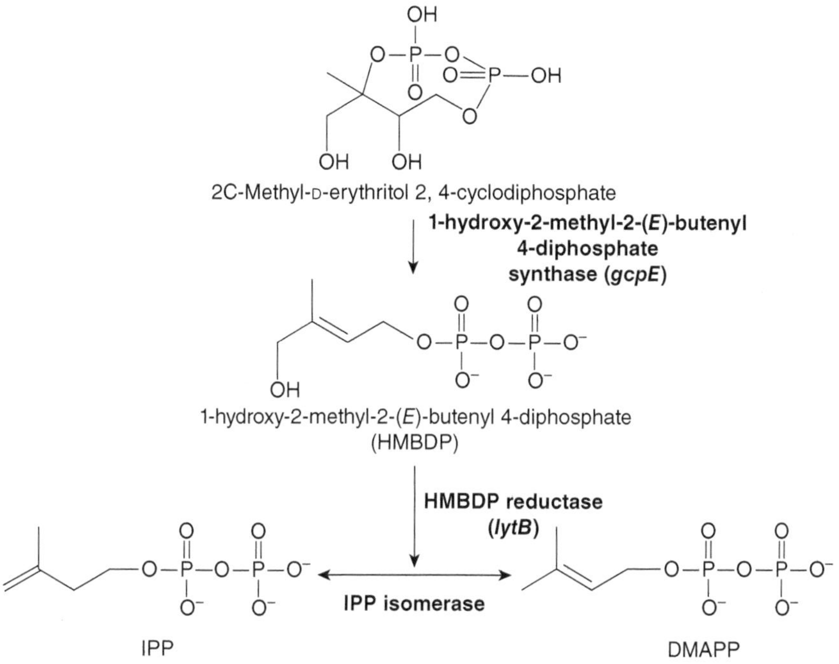

2C-Methyl-D-erythritol 2, 4-cyclodiphosphate

1-hydroxy-2-methyl-2-(E)-butenyl 4-diphosphate synthase (gcpE)

1-hydroxy-2-methyl-2-(E)-butenyl 4-diphosphate (HMBDP)

HMBDP reductase (lytB)

IPP isomerase

IPP

DMAPP

Fig. 9.2 *Continued*

cytidylylated by MEP cytidylyltransferase to generate 4-(cytidine 5'-diphospho)-2C-methyl-D-erythritol (CDP-ME), which is then phosphorylated by CDP-ME kinase (IspE) to produce 2-phospho-4-(cytidine 5'-diphospho)-2C-methyl-D-erythritol (CDP-ME2P). Next, CDP-ME2P is transformed to 2C-methyl-D-erythritol 2,4-cyclodiphosphate (MECDP) by MECDP synthase (Kuzuyama, 2002). 1-Hydroxy-2-methyl-2-(E)-butenyl 4-diphosphate (HMBDP) is then produced by reducing the MECDP ring in the presence of MECDP synthase, an enzyme encoded by the *gcpE* gene (Wang *et al.*, 2010). Finally, HMBDP is reduced to IPP by HMBDP reductase, which is encoded by the *lytB/ispH* gene (Adam *et al.*, 2002).

In the MVA pathway (Fig. 9.3), acetyl-CoA is condensed to form acetoacetyl-CoA by acetyl-CoA acetyltransferase. Then, an additional acetyl-CoA is condensed to acetoacetyl-CoA and together they produce 3-hydroxy-3-methyl-glutaryl-CoA (HMG), catalysed by 3-hydroxy-3-methyl-glutary-CoA synthase (HMG-CoA synthase). HMG is then reduced by HMG-CoA reductase finally to obtain mevalonate. Next, mevalonate is sequentially phosphorylated, first by mevalonate kinase and then by phosphomevalonate kinase. Once phosphorylated, it undergoes a decarboxylation that results in the formation of IPP. This IPP is then transformed to its isomer DMAPP by IPP isomerase (Kuzuyama, 2002).

© Woodhead Publishing Limited, 2013

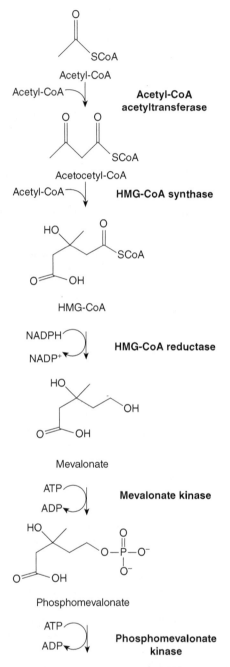

Fig. 9.3 The mevalonate pathway for the biosynthesis of IPP and DMAPP precursors. See the text for details. Abbreviations used are: HMG-CoA, hydroxymethylglutaryl CoA; IPP, isopentenyl diphosphate; DMAPP, dimethylallyl diphosphate.

© Woodhead Publishing Limited, 2013

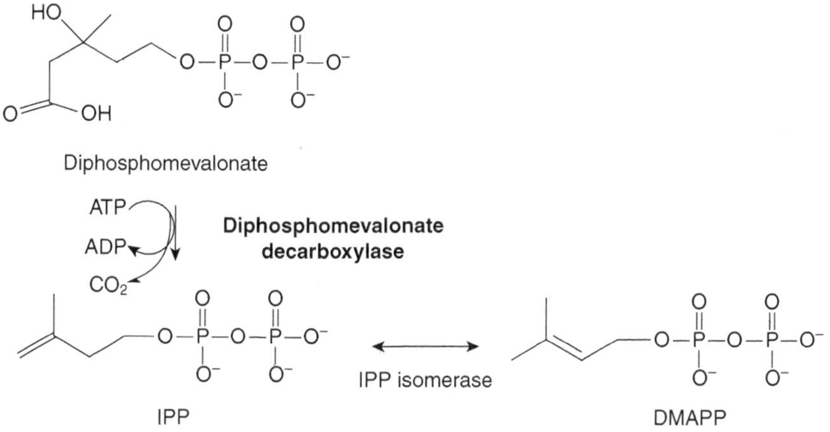

Diphosphomevalonate

Diphosphomevalonate decarboxylase

IPP isomerase

IPP   DMAPP

Fig. 9.3   *Continued*

Condensation of IPP with DMAPP produces geranyl diphosphate (GPP) through the action of GPP synthase (Fig. 9.4). The addition of IPP to GPP produces farnesyl diphosphate and the addition of one more IPP unit yields geranylgeranyl diphosphate (GGPP). The enzyme that catalyzes GGPP synthesis (GGPP synthase) is well conserved among proteobacteria and Gram-positive bacteria (Phadwal, 2005). In eukaryotes, GGPP is a frequent precursor for several groups of plastid isoprenoids (Namitha and Negi, 2010).

Once formed, two GGPP molecules are condensed head-to-head to produce the C-40 compound phytoene, a step catalysed by the enzyme phytoene synthase (PSY, CrtB). Phytoene is converted to phytofluene, ζ-carotene, neurosporene, and finally lycopene in four sequential desaturation reactions (Namitha and Negi, 2010, Walter and Strack, 2011).

There is more divergence in the enzymes that catalyse phytoene desaturation. In most bacteria this step is catalysed by the phytoene desaturase CrtI. However, in Cyanobacteria, Chlorobi and photosynthetic eukaryotes, phytoene desaturation is carried out by three different enzymes: CrtP (phytoene desaturase), which converts phytoene to ζ-carotene; CrtQ (ζ-carotene desaturase), which converts ζ-carotene into 7,9,7′,9′-*cis*-lycopene; and CrtH (7,9,7′,9′-*cis* lycopene isomerase), which converts 7,9,7′,9′-*cis*-lycopene into all-*trans* lycopene (Klassen, 2010). Desaturation reactions lead to an increase in conjugated double bonds, which transforms the colourless phytoene into the brilliant pink lycopene (Namitha and Negi, 2010; Walter and Strack, 2011).

The cyclization of lycopene to β-carotene, via a γ-carotene intermediate, is catalysed by lycopene cyclases (Fig. 9.5). The first β-bicyclase described was CrtY from *Erwinia uredovora* and CrtL was later found to be its

© Woodhead Publishing Limited, 2013

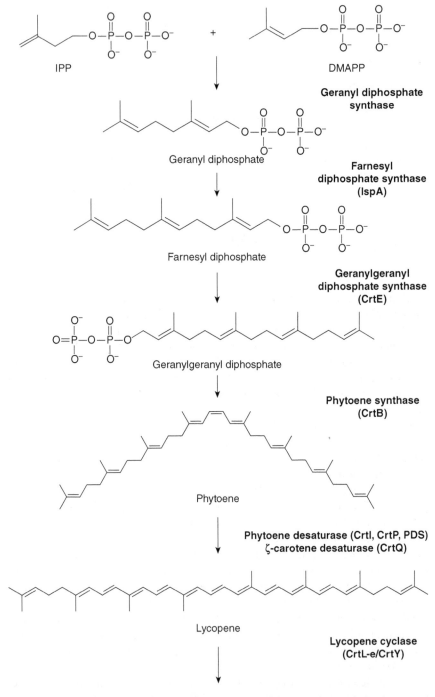

Fig. 9.4 Lycopene formation from IPP and DMAPP precursors and lutein biosynthesis. See text for details.

© Woodhead Publishing Limited, 2013

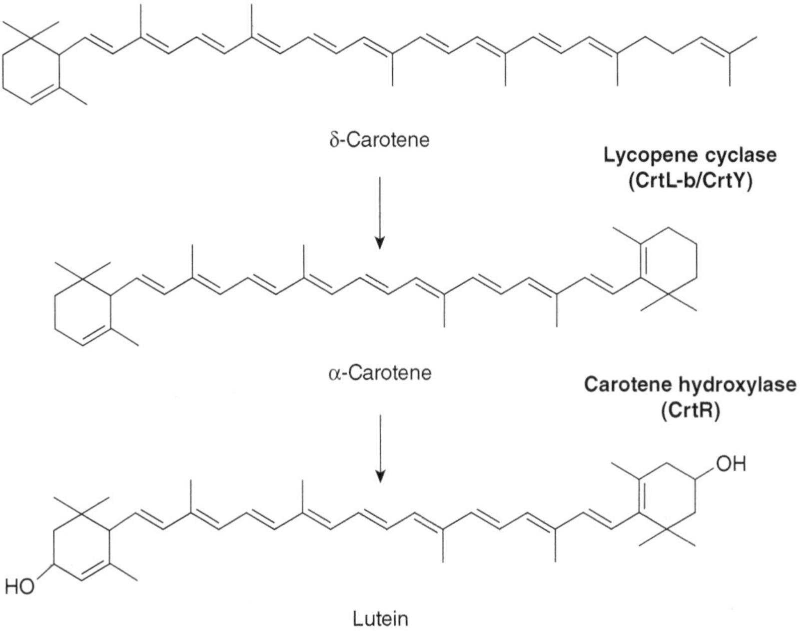

δ-Carotene

**Lycopene cyclase
(CrtL-b/CrtY)**

α-Carotene

**Carotene hydroxylase
(CrtR)**

Lutein

Fig. 9.4 *Continued*

homologue in Cyanobacteria. In some Cyanobacteria, two types of CrtL exist, called CrtL-b (β-cyclase) and CrtL-e (ε-cyclase), which have similar functions to those in eukaryotes (Klassen, 2010). Additionally, in Cyanobacteria and Chlorobi, there are three variants of cyclases named CruA (bi-cyclase), CruP (monocyclase) and CruB, which produce mainly γ-carotene (Walter and Strack, 2011).

δ-Carotene is produced by the addition of one ε-ring to lycopene in the presence of CrtL-e. Subsequently, a β-ring is added to the other end of δ-carotene, thus producing α-carotene (Fig. 9.4). In contrast, the addition of one β-ring to lycopene results in the formation of γ-carotene, and the further addition of one more β-ring to γ-carotene forms β-carotene (Chen *et al.*, 2007). Lutein is produced by hydroxylating the C3 site of each ring of α-carotene and is catalysed by a carotene hydroxylase (CrtR).

In the other branch (Fig. 9.5), the β-C-4-oxygenase/β-carotene ketolase (CrtO) adds one keto group at the C4 site, resulting in the formation of echinenone. Similar to CrtO, CrtW adds one more keto group to the C4′ site, leading to canthaxanthin production (Bhosale and Berstein, 2005; Phadwal, 2005; Chen *et al.*, 2007).

CrtR or CrtZ carotene hydroxylases carry out β-carotene hydroxylation at position C3, which results in the formation of β-cryptoxanthin. The addition of another hydroxyl group at the C3′ site of this compound produces

© Woodhead Publishing Limited, 2013

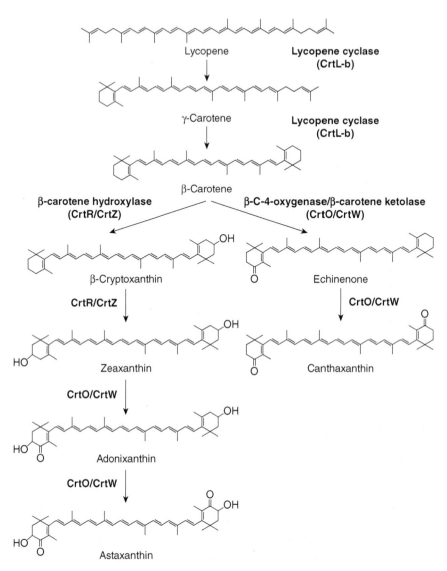

Fig. 9.5   Canthaxanthin and astaxanthin biosynthesis from lycopene. See text for details.

zeaxanthin. Both hydroxylations are catalysed by the carotenoid 3,3′-hydrolase (CrtR or CrtZ). The sequential introduction of keto groups by CrtO/CrtW at the C4 site and at the C4′ site to the β-ionone ring of zeaxanthin lead to the formation of astaxanthin (Ye *et al.*, 2006). The pathway of keto carotenoid formation from β-carotene through to echinenone and canthaxanthin is well characterized in algae, yeast and non-photosynthetic bacteria (Bhosale and Berstein, 2005).

© Woodhead Publishing Limited, 2013

There are some peculiarities in the pathways leading to the formation of fungal carotenoids, such as the presence of bifunctional enzymes, for example the lycopene cyclase/phytoene synthases that are encoded by the *crtYB* (*X. dendrorhous*), *carA* (*Phycomyces blakesleeanus*) and *al-z* genes (*Neurospora crassa*) (Sanz *et al.*, 2011). In addition, the *carB* gene product (phytoene desaturase) from *Fusarium fujikuroi* can carry out all desaturation steps between phytoene and lycopene (Prado-Cabrero *et al.*, 2009). Finally, the yeast *X. dendrorhous* contains a *crtS* gene which encodes cytochrome P450. This enzyme appears to have both hydroxylase and ketolase activities, which enable it to transform β-carotene into canthaxanthin and zeaxanthin (Martín *et al.*, 2008).

9.4   Regulation of carotenoid production

The presence of carotenoids is an absolute requirement for photosynthetic activity in photosynthetic organisms and they are synthesized by non-photosynthetic bacteria and fungi (Takaichi, 2011). Regardless of the cell type, carotenoid synthesis occurs under conditions that typically cause stress or inhibit cell growth, for example, high light intensity, high salinity, nutrient limitations (N and P) and inhibitory substances. In this section, we discuss the main types of carotenoid regulation observed in commercially important microbial producers.

9.4.1   Dissolved oxygen (DO) and oxidative stress

Carotenoids play important roles in light harvesting and photoprotection of photosynthetic organisms. In non-photosynthetic organisms, carotenoids are effective scavengers of the reactive oxygen species (ROS) that are generated by normal respiration, thus preventing oxidative damage to cell components. Oxygen plays a pivotal role in carotenoid biosynthesis in all known carotenogenic microorganisms. Many bacterial and fungal cultures accumulate carotenoids upon the transition from anaerobic to aerobic respiration (Frengova and Beshkova, 2009; Schmidt *et al.*, 2011). A critical dissolved oxygen tension (DOT), between 15–20% air saturation, is generally required for efficient carotenoid synthesis in fungi (Johnson and Lewis, 1979; Wang and Yu, 2009). In *X. dendrorhous*, a low DOT causes the accumulation of non-oxygenated carotenoids such as β-carotene, which is an astaxanthin precursor that cannot be effectively hydroxylated under such conditions (Frengova and Beshkova, 2009). The presence of photogenerated H_2O_2 and $^1O_2$ also artificially promotes astaxanthin synthesis in this yeast (Schroeder and Johnson, 1995a). However, cultures of *B. trispora* provided with intensive aeration generated higher levels of β-carotene (Kuzina and Cerda-Olmedo, 2007; Nanou *et al.*, 2011). ROS involvement in β-carotene synthesis has also been reported in the alga *D. salina* (Lamers

© Woodhead Publishing Limited, 2013

et al., 2008). In this alga, photosynthesis yields oxygen, which leads to formation of ROS and oxidative stress and correspondingly promotes the accumulation of β-carotene.

9.4.2   Nitrogen limitation

Carotenoid accumulation typically occurs during the latter stages of cultivation in many microorganisms, that is, after growth has ceased. Thus, nutrient
depletion appears to be associated with carotenogenesis. Thus, a high C/N
ratio in the culture medium has been used extensively to promote carotenoid accumulation. In *X. dendrorhous*, the concentration of total carotenoids and astaxanthin content increased when nitrogen was limited. Slow
protein synthesis and slow growth are generally associated with carotenogenesis (Flores-Cotera *et al.*, 2001; Chávez-Cabrera *et al.*, 2010; Yamane
et al., 1997a,b; Meyer and du Preez, 1994a). Nitrogen-limitation promotes
carotenoid synthesis in many other fungi (Bramley and Mackenzie, 1988;
Rodriguez-Ortiz *et al.*, 2009). In most of these organisms, carotenoid synthesis may be a response that makes use of the excess carbon and energy
that cannot be used for growth or protein synthesis, while nitrogen is limiting. This interpretation is also consistent with other published data. An *et
al.* (1989) have suggested that efficiency of nitrogen utilization is increasingly impaired in *X. dendrorhous* mutant strains with higher carotenoid
content. Furthermore, the study of different nitrogen sources for astaxanthin production has often led to the choice of amino acids that are metabolized slowly, possibly because they induce nitrogen limitation (Meyer *et al.*,
1993; Meyer and du Preez, 1994b).

Microalgae are sensitive to nutrient limitation as well. *H. pluvialis* shows
a vegetative growth phase when sufficient nitrogen is available, but no
significant astaxanthin accumulation occurs during this phase. In contrast,
moderate nitrogen limitation results in lower biomass, but favours early
encystment and astaxanthin formation (del Rio *et al.*, 2005; Tripathi *et al.*,
2002; Jin *et al.*, 2006; Dominguez-Bocanegra *et al.*, 2004). Nitrogen
limitation has been used to stimulate β-carotene accumulation in
Dunaliella (Mogedas *et al.*, 2009) and canthaxanthin in *Chlorella emersonii*
(Arad *et al.*, 1993).

9.4.3   Photoregulation

Photoregulation of carotenoid synthesis has been reported in all photosynthetic organisms, and in fungi. It is thought that translation of the light signal
is mediated by ROS that are formed by the interaction of light with oxygen
and light-sensitive molecules such as protoporphyrin IX, heme and other
compounds (Bramley and MacKenzie, 1988; Johnson and An, 1991). *B. trispora* grown in dark conditions accumulates β-carotene following illumination with white light (Quiles-Rosillo *et al.*, 2005). Carotenoid accumulation

© Woodhead Publishing Limited, 2013

has also been reported to be stimulated by light in many other bacteria and fungi (Bhosale, 2004; Johnson and An, 1991; Khodaiyan *et al.*, 2007).

H. pluvialis can accumulate astaxanthin (4–5% dry weight) in strong light irradiation conditions, a condition regularly associated with cell encystment (Jin *et al.*, 2006; Yuan *et al.*, 2011; Park and Lee, 2001; Tripathi *et al.*, 2002). The expression of genes involved in astaxanthin synthesis, that is, lycopene cyclase, phytoene synthase, phytoene desaturase and carotenoid hydroxylase, was up-regulated by blue and red light (Steinbrenner and Linden, 2001, 2003; Vidhyavathi, *et al.*, 2008). Many other algae such as *Dunaliella* and *Muriellopsis* accumulate carotenoids in response to light (Mogedas *et al.*, 2009; Del Campo *et al.*, 2001).

Optimization of fermentation parameters provides maximal astaxanthin titers of 4.7 mg g^{-1} dry cell weight (DCW) when *X. dendrorhous* is fermented under continuous white light (Rodriguez-Saiz *et al.*, 2010; Schmidt *et al.*, 2011).

9.4.4   Temperature

Cultivation temperature has a strong influence on intracellular astaxanthin content, carotenoid profile and growth rate of *X. dendrorhous* (Johnson and Lewis 1979; Ramírez *et al.*, 2001; Schmidt *et al.*, 2011). *Dunaliella* sp. and *H. pluvialis* are also both affected by temperature. *Dunaliella* sp. showed a 7.5-fold increase in cellular levels of β-carotene when the temperature was reduced from 34 to 17°C, but the quantity was maximized at between 24 and 29°C. High temperature-mediated enhanced ROS formation was suggested as the effector behind a 15- to 20-fold increase in the cellular accumulation of carotenoids in *Haematococcus* (Bhosale, 2004).

9.4.5   Tricarboxylic acid (TCA) intermediates

Intermediates of the tricarboxylic acid (TCA) cycle are usually carbon skeletons that can be directed to carotenoid biosynthesis in carotenogenic microorganisms. Thus, *Flavobacterium* sp. displayed increased zeaxanthin production following supplementation with oxaloacetate or a mixture of malic acid, isocitric acid and α-ketoglutarate (Alcantara and Sanchez 1999; Bhosale, 2004). In *Dietzia natronolimnaea*, α-ketoglutarate, oxaloacetate and succinate were selected for canthaxanthin production (Nasrabadi and Razavi, 2010b). Citrate and succinate have been used for similar purposes in *X. dendrorhous* (Flores-Cotera *et al.* 2001; Wosniak *et al.*, 2011), whereas malate was used in *B. trispora* (Choudhari *et al.*, 2008).

9.4.6   Salt stress

Many microorganisms accumulate carotenoids when exposed to salt stress. *Mycobacterium aurum* accumulates more lycopene when cultured in a

© Woodhead Publishing Limited, 2013

medium with 85 mM NaCl. However, this was at expense of a 40% reduction in biomass (Kerr *et al.*, 2004). Similarly, *R. glutinis* and *Sporidiobolus salmonicolor* accumulated higher levels of carotenoids when grown in 5–10% NaCl (Marova *et al.*, 2004).

The influence of salt stress has been widely studied in algae. At concentrations of above 1.0% NaCl, *H. pluvialis* dies, but vegetative cells stress-treated with 0.1% NaCl generally encyst while astaxanthin accumulation is promoted (Kobayashi *et al.*, 1997; Sarada *et al.*, 2002 a,b). Salt stress promotes the expression of carotenogenic genes encoding for phytoene synthase, phytoene desaturase, lycopene cyclase, β-carotene ketolase and β-carotene hydroxylase (Vidhyavathi, *et al.*, 2008; Steinbrenner and Linden, 2001).

9.4.7   Effect of other chemical compounds

Several chemical agents, particularly those that inhibit growth, have a generally positive effect on carotenogenesis. Ethanol supplementation (0.2–2%, v/v) was reported to stimulate β-carotene and torulene formation in *R. glutinis* (Margalith and Meydav, 1968) and astaxanthin production in *X. dendrorhous* cultures (Gu *et al.*, 1997; Kim and Chang, 2006; Yamane *et al.*, 1997b). Ethanol activates oxidative metabolism with parallel induction of HMG-CoA reductase. Other genes whose expression is associated with ethanol metabolism include those coding for phytoene synthase, phytoene dehydrogenase and astaxanthin synthase (Lodato *et al.*, 2007). Other chemical compounds known for their positive effects on carotenogenesis include terpenes, ionones, amines, penicillin and alkaloids (Schmidt *et al.*, 2011; Bhosale, 2004; de la Fuente *et al.*, 2005, 2010). *P. blakesleeanus* and *B. trispora* both accumulate carotenoids upon mating (Kuzina and Cerda-Olmedo, 2007). This effect has been attributed to the production of the sex hormone trisporic acid, or to compounds with structural similarity to trisporic acid (Bhosale, 2004).

9.4.8   Inhibitors

Compounds known to inhibit the formation of β-carotene, such as pyridine (500 mg l^{-1}) and imidazole (50 mg l^{-1}), have been used to improve lycopene production by *B. trispora* (by seven-fold) (Choudhari *et al.*, 2008; Pegklidou *et al.*, 2008; Gavrilov *et al.*, 1996). Limitation of ergosterol synthesis is also effective in increasing lycopene formation in this fungus, for example, when ketoconazole (30 mg l^{-1}) is added to the culture medium, the lycopene content is increased by 277% (Sun *et al.*, 2007). In *X. dendrorhous*, astaxanthin synthesis is stimulated in the presence of cellular respiration inhibitors including antimycin (An *et al.*, 1989; Schroeder and Johnson, 1995a,b). Antimycin blocks the electron transfer from complex III to complex IV and

© Woodhead Publishing Limited, 2013

as a consequence, astaxanthin synthesis occurs in parallel with the induction of cyanide-insensitive respiration. Antimycin also stimulates ROS production, which could act as mediator for carotenogenesis (An and Johnson, 1990). Finally, application of different inhibitors of photosynthetic electron flow in *Haematococcus* indicates that the photosynthetic plastoquinone pool functions as the redox sensor for the up-regulation of the carotenoid biosynthetic genes (Steinbrenner and Linden, 2003).

9.4.9   Catabolite repression

In yeast and other organisms, the overall control of the mevalonate pathway occurs in the initial stages of isoprenoid synthesis, in particular at the level of the enzyme responsible for mevalonate formation, hydroxymethylglutaryl-CoA reductase (Goldstein and Brown, 1990; Gu *et al.*, 1997). This enzyme is regulated by catabolite repression and DOT (Hampton *et al.*, 1996; Brown and Goldstein, 1980). Glucose catabolic repression has been observed in native strains of *X. dendrorhous* (Johnson and Schroeder, 1996) and *Myxococcus xanthus* (Armstrong, 1997) and in some strains of *E. herbicola* (Armstrong, 1997), and in algae.

9.4.10   Feedback regulation by end products

End product regulation is a common regulatory mechanism in carotenogenesis. However, in the case of the enzyme HMGR, it is also regulated by late intermediates in the carotenoid pathway (Hampton *et al.*, 1996; Brown and Goldstein, 1980). In *P. blakesleeanus*, the regulatory mechanism functions at the level of protein synthesis (Bejarano *et al.*, 1988). The phytoene desaturases of *P. blakesleeanus* and *Synechococcus* are modulated by the subsequent carotenoids in their pathways (Sandmann, 1994). Lastly, in *X. dendrorhous*, astaxanthin production is regulated by its end product (Johnson and Schroeder, 1996; Schroeder and Johnson, 1995b).

9.5   Genetic improvement of carotenoid production

In recent years, many efforts have been made to develop biological environmentally friendly sources for the supply of commercial carotenoids. These efforts can be divided into optimizing culture conditions and in the generation of mutant microalgae and yeast to produce carotenoids naturally (Rodriguez-Saiz *et al.*, 2010; Lamers *et al.*, 2008), introducing carotenoid genes into non-carotenogenic microbes, especially *E. coli*, but also *S. cerevisiae*, *C. utilis* and *Z. mobilis* and creating carotenoid biosynthetic pathways in engineered bacteria (Kim *et al.*, 2011; Schmidt-Dannert *et al.*, 2000).

© Woodhead Publishing Limited, 2013

9.5.1   Classical genetics

Some hyperproducing mutants have been selected using classical mutagenesis and screening. *X. dendrorhous* astaxanthin-overproducing mutants have been isolated by growing cultures in the presence of antimycin A, *N*-methyl-*N*′-nitro-*N*-nitrosoguanidine, or by treating with low-dose (below 10 kGy) gamma irradiation. Some of the mutants obtained showed a two-to five-fold increase in astaxanthin content when compared with their parental strain (Schmidt *et al.*, 2011).

Improved astaxanthin yield in *X. dendrorhous* was obtained by treating with precursors of carotenogenesis such as mevalonate (Calo *et al.*, 1995), glutamate (de la Fuente *et al.*, 2005) or citrate (Flores-Cotera *et al.*, 2001). The increased yield resulted from increased metabolic fluxes to carotenoid-producing pathways. This result confirms that the precursor supply for carotenogenesis is limiting. On a genetic level, the expression of required enzymes can be modified by directed pathway engineering to overcome crucial bottlenecks in astaxanthin synthesis. Overexpression of the *crtI* gene encoding phytoene desaturase in *X. dendrorhous* caused an increase in monocyclic carotenoids such as torulene and a decrease in bicyclic carotenoids such as echinenone, β-carotene and astaxanthin (Visser *et al.*, 2003). Alternatively, transformants containing multiple copies of *crtYB*, which encode the bifunctional protein involved in the biosynthesis of phytoene and the cyclization of lycopene to β-carotene, showed an increased total carotenoid content.

9.5.2   Genetic engineering approaches

Some new approaches for the efficient biosynthesis of carotenoids and the production of new carotenoid molecules include the recruitment of foreign catalytic machinery that add one or several new branches to generate a 'mini-pathway' of engineered steps. A pathway can also diversify when the function of an existing enzyme is changed by mutation or other genetic modification. In directed evolution, libraries of mutant alleles are expressed in host cells along with other genes of a particular pathway and are then screened or selected for the desired properties (Walter and Strack, 2011; Umeno *et al.*, 2005).

Isoprenoids serve numerous biochemical functions: as quinones in electron transport chains, as membrane components (prenyl-lipids in archaebacteria and sterols in eubacteria and eukaryotes), in subcellular targeting and regulation (prenylation of proteins) (Lange *et al.*, 2000), among others. All isoprenoids are derived from the common five-carbon (C_5) building units isopentenyl diphosphate (IPP) and its isomer dimethylallyl diphosphate (DMAPP). It is thus feasible partially to direct the carbon flux for the biosynthesis of these isoprenoid compounds to the pathway for carotenoid production by introducing carotenogenic genes. In virtually all photosynthetic organisms, archaea, some bacteria and fungi, carotenoid

© Woodhead Publishing Limited, 2013

biosynthesis depends on the supply of these building blocks via the MEPP pathway (Walter and Strack, 2011). The pathway initiates with the reaction between pyruvate and glyceraldehyde 3-phosphate, as shown in Fig. 9.3. DMAPP and IPP are then condensed (Fig. 9.4) to generate geranyl diphosphate (GPP, C10), which is further converted, by FPP synthase (*ispA* in *E. coli*) and with IPP, into farnesyl diphosphate (FPP, C15) (Harada and Misawa, 2009).

However, because *E. coli* lacks a C20PP synthase function, a geranylgeranyl diphosphate (C20PP) synthase gene must be heterologously expressed along with the other carotenoid biosynthetic genes to generate C40 carotenoids in this organism (Umeno *et al.*, 2005). From this point in the pathway, new routes may be assembled to produce natural carotenoids or perhaps new carotenoid molecules.

The *crt* genes derived from *Pantoea ananatis* (formerly *E. uredovora*) or *Pantoea. agglomerans* (formerly *E. herbicola*) were successfully used for the *de novo* biosynthesis of lycopene, β-carotene astaxanthin and zeaxanthin in *E. coli* (Yoon *et al.*, 2007a; Lemuth *et al.*, 2011), zeaxanthin in *Pseudomonas putida* (Beuttler *et al.*, 2011), and other bacteria and some yeasts (Miura *et al.*, 1998a, b; Hunter *et al.*, 1994).

Yoon and co-workers (2007a,b) investigated which of the two isoprenoid production pathways, the MEPP pathway (naturally present in *E. coli*) or the mevalonate pathway from *Streptococcus pneumoniae* (*mvaK1*, *mvaK2* and *mvaD*), were better for lycopene production in *E. coli* (Fig. 9.4 and 9.5). They reported that lycopene production was significantly increased when the mevalonate pathway was used. This must surely have been achieved by increasing carbon flux and therefore, the building blocks (see above).

However, the effect of non-carotenogenic genes on lycopene production was also investigated. A genomic *E. coli* DNA library was transformed into cells previously transformed with a plasmid conferring lycopene accumulation. Thirteen genes were found to be involved in increasing lycopene production. DXP (1-deoxy-D-xylulose-5-phosphate) synthase is encoded by the *dxs* gene and participates in the synthesis of isopentenyl diphosphate (IPP). Sigma S factor, encoded by *rpoS*, regulates transcription of genes induced during the stationary growth phase. The *appY* gene encodes a transcriptional regulator associated with anaerobic energy metabolism. *E. coli*-harboring *appY* plasmids produced 2.8 mg lycopene g DCW^{-1}, which is the same amount obtained with *dxs* despite the fact that *appY* is not directly involved in the lycopene synthesis pathway. The co-expression of *appY* in addition to *dxs* produced eight times the amount of lycopene (4.7 mg g DCW^{-1}) that was produced without expression of both genes (0.6 mg g DCW^{-1}) (Kang *et al.*, 2005a). Overexpression of the stress-responsive sigmaB-like protein CrtS from *S. setonii* also activated the cryptic *crt* genes in *S. griseus* and conferred pigmentation (Lee *et al.*, 2001).

In recent years, researchers have constructed several biosynthesis pathways of carotenoids in *E. coli* that do not naturally exist. Combinatorial

© Woodhead Publishing Limited, 2013

chemistry or synthetic biology can provide new unnatural molecular scaffolds, from which whole new families of 'natural' products can be generated. Directed evolution of enzymes can be used to reconstruct natural evolutionary steps and can help to comprehend them better. It can also provide novel biochemical properties that have never occurred naturally or have been discontinued by natural selection. The backbones are desaturated, cyclized, oxidized, cleaved, glycosylated or otherwise modified in a specific order (Walter and Strack, 2011; Albrecht *et al.*, 2000).

Some bacteria, such as *Haloferax mediterranei, Dietzia* sp, *Corynebacterium glutamicum, Agromyces mediolanus* (formerly *Flavobacterium dehydrogenans*), *Micrococcus luteus* (formerly *Sarcina lutea*) and *Curtobacterium flaccumfaciens* (formerly *Corynebacterium poinsettiae*) are known to synthesize C45 and C50 carotenoids (Fang *et al.*, 2010; Tao *et al.*, 2007). These longer chain structures are biosynthesized by adding one or two C5 (isoprene) units to the C40 structure (Umeno and Arnold, 2004). One unique feature of the C50 carotenoid synthesis genes in *Dietzia* sp. CQ4 is that a C50 carotenoid β-cyclase subunit has fused with a lycopene elongase. The diversity of genes involved in carotenoid synthesis has been enriched by new genes derived from gene fusions. The fusion of the heterodimeric lycopene β-cyclases (CrtYcd) was also reported in bacteria and archaea. In fungi, fusion of lycopene β-cyclase with phytoene synthase (CrtYB) has been reported (Tao *et al.*, 2007).

Some unusual acyclic carotenoids have been obtained by fusion of carotenogenic genes from *P. ananatis* and *Rhodobacter capsulatus*. Co-expression of the C20PP synthase, CrtE, and the phytoene synthase, CrtB, from *P. ananatis* along with various combinations of three different carotenoid desaturases (a carotenoid hydratase, a β-cyclase and a hydroxylase) resulted in the production of four previously unidentified carotenoids. In another example, the addition of CrtW, a β-end ketolase from *Paracoccus* sp., extended the zeaxanthin β-D-diglucoside pathway from *Pantoea* and led to the synthesis of the novel compounds astaxanthin β-D-diglucoside and adonixanthin 3'-β-D-glucoside (Umeno *et al.*, 2005). Unnatural backbones have been generated in *E. coli* by the coexpression of an FPP (C15) synthase mutant gene (Y81A) from *Bacillus stearothermophilus* and a double mutant (F26A and W38A) of the C30 diapophytoene synthase (CrtM) gene from *Staphylococcus aureus*. These *E. coli* synthesized a mixture of C15, C20 and C25 precursors, leading to the production of novel C35, C40, C45 and C50 backbones (Umeno and Arnold, 2004).

In another approach, a library of phytoene desaturases (*crtI*), generated by DNA shuffling of genes from *P. agglomerans* and *P. ananatis*, was transformed into phytoene-synthesizing *E. coli*. One of the desaturase chimeras obtained introduced two double bonds to lycopene that led to the accumulation of the terminal desaturation product 3,4,3',4'-tetradehydrolycopene, in addition to lycopene. A second chimera synthesized neurosporene and lycopene in addition to the main product, ζ-carotene. To access the

© Woodhead Publishing Limited, 2013

cyclization products of the extended desaturase pathway, a library of lyco-pene cyclases was created by shuffling the *crtYEU* (*crtY* from *P. ananatis*) and *crtYEH* (*crtY* from *P. agglomerans*) genes. Using the shuffled cyclase to extending the pathway to generate 3,4-didehydrolycopene led to the first reported biosynthesis of torulene in *E. coli* (Schmidt-Dannert *et al.*, 2000).

9.6   Fermentation conditions

Biotechnological production of carotenoids has been widely analysed, espe-cially for the commercial production of β-carotene by either the fungus *B. trispora* or the microalga *Dunaliella*. Production of astaxanthin by *Haema-tococcus* sp. and *X. dendrorhous* has also been seriously considered. Since the 1970s, several processes have been commercially exploited in the pro-duction of β-carotene, whereas the production of astaxanthin has been more recent. Because all microorganisms accumulate carotenoids in response to stress conditions that may slow cell growth, the following section will review some of the most relevant features of carotenoid produc-tion with regard to cell growth (Valduga *et al.*, 2009). The main focus will be on β-carotene production from *B. trispora* and *Dunaliella*, and on astax-anthin production in *X. dendrorhous* and *Haematococcus* (Table 9.1).

9.6.1   Growth conditions

Biotechnological production of β-carotene was achieved more than 40 years ago using the fungus *B. trispora*. It was later discontinued because synthetic manufacture provided a more economical product (Bhosale, 2004; Ribeiro *et al.*, 2011). A semi-industrial process (800 L fermenter) for lyco-pene production by fermentation of mated *B. trispora* plus (+) and minus (−) strains was also developed. Fermentation involves separate vegetative growth phases for (+) and (−) strains and the subsequent inoculation of the production medium with a mix of both *B. trispora* strains without lycopene cyclase (Mehta and Cerdá-Olmedo, 1995) and specific inhibition of this enzymatic activity with imidazole or pyridine (0.8 g l^{-1}), improved lycopene accumulation (>5% dry matter). Different raw materials and physical parameters were analysed to obtain maximal lycopene production (Bhosale 2004; López-Nieto *et al.*, 2004).

Profitable industrial production of astaxanthin by *X. dendrorhous* was possible following the development of mutant strains of at least 3000 µg g^{-1} astaxanthin content (Dufossé, 2006). Astaxanthin production can be signifi-cantly improved following illumination with white and UV light. Further, a scale-up, to 800 litres, of a fed-batch process was developed using glucose feeding (de la Fuente *et al.*, 2010).

Dunaliella has been exploited commercially since the 1980s (Jin and Melis, 2003; Ye *et al.*, 2008). Autotrophically grown microalgae require

© Woodhead Publishing Limited, 2013

Table 9.1 Some culture conditions and compounds used to promote carotenoid synthesis in different microorganisms

Condition/ substance	Microorganism	Product	Possible mechanism	References
Leucine, valine	Algae and P. blakesleeanus X. dendrorhous	Carotenoids astaxanthin	HMG-CoA precursors	Bramley and Mackenzie (1988); Johnson and An (1991); Meyer et al. (1993); Meyer and du Preez (1994b)
Oxygen	X. dendrorhous R. gracilis	Astaxanthin β-carotene	Increases ROS generation	An et al. (1989); Bramley and Mackenzie (1988); Frengova and Beshkova (2009); Schmidt et al. (2011)
Intense aeration	X. dendrorhous R.glutinis B. trispora	Astaxanthin β-carotene	Increases ROS generation	Johnson and An (1991); Kuzina and Cerda-Olmedo (2007); Nanou et al. (2011)
H_2O_2, $^-O_2$, OH, $^1O_2$	X. dendrorhous D. salina H. pluvialis Fusarium	Astaxanthin carotenoids	Oxidative stress	Schroeder and Johnson (1995a, b); Lamers et al. (2008); Bramley and Mackenzie (1988)
Duroquinone	X. dendrorhous Rhodotorula mucilaginosa	Astaxanthin, carotenoids	Induces superoxide generation	Johnson and An (1991)
Nitrogen limitation	H. pluvialis X. dendrorhous Dunaliella	Astaxanthin β-carotene	Limited protein synthesis. Unfavourable conditions for growth.	Chavez-Cabrera et al. (2010); Yamane et al. (1997a); Bramley and Mackenzie (1988); Rodriguez-Ortiz et al. (2009); del Rio et al. (2005); Jin et al. (2006); Mogedas et al. (2009)
Phosphate limitation	H. pluvialis Other	Astaxanthin	Limited protein synthesis. Unfavourable conditions for growth. Excess of phosphate suppresses the synthesis of enzymes of secondary metabolism.	Flores-Cotera et al. (2001)
Nutrient limitations	Bacteria, algae and fungi	Carotenoids	Carotenoid biosynthesis occurs with slow growth or after cessation of growth. Could be a response to nutrient limitation (non specific).	Johnson and An (1991); Bramley and Mackenzie (1988); Bubrick, (1991); Garcia-Malea et al. (2009); Harker et al. (1996); Kang et al. (2005b)

© Woodhead Publishing Limited, 2013

Factor	Organism	Carotenoid(s)	Effect/Mechanism	References
Light	Bacteria, algae and fungi; H. pluvialis; X. dendrorhous	Carotenoids, β-carotene, astaxanthin	Induction of enzymes involved in carotenoid synthesis (e.g. GGPP sintetase, HMG-CoA reductase).	Bramley and Mackenzie (1988); Takaichi (2011); Quiles-Rosillo et al. (2005); Bhosale (2004); Yuan et al. (2011); Steinbrenner and Linden (2003); Vidhyavathi, et al. (2008); Mogedas et al. (2009); Kim et al. (2006)
Light + oxygen	Bacteria, fungi and yeast	Carotenoids	Induction of enzymes involved in carotenoid synthesis	Johnson and An (1991)
Trisporic acid β-Ionona	Mucorales; Blakeslea trispora	Carotenoids	Induction of enzymes involved in carotenoid synthesis. β-Ionona has a similar structure to trisporic acid	Bramley and Mackenzie (1988)
Temperature	Bacteria, algae and fungi; H. pluvialis	Astaxanthin	Stimulates the formation or increased reactivity of ROS. Inhibits growth.	Bramley and Mackenzie (1988); Ramírez et al. (2001); Schmidt et al. (2011); Bhosale (2004)
Ethanol	X. dendrorhous; R. glutinis	Astaxanthin, carotenoids	Inhibits growth and TCA function	Gu et al. (1997); Kim and Chang (2006); Yamane et al. (1997b); Lodato et al. (2007)
Antimycin A	Fungi; X. dendrorhous	Astaxanthin	Blocks electron transport through the respiratory chain. Inhibits growth	An et. al. (1989)
Salt (NaCl)	Algaes; R. glutinis	Astaxanthin, carotenoids	Unfavourable conditions for growth	Marova et al. (2004); Sarada et al. (2002a,b); Kobayashi et al. (1997); Ye et al. (2008); Jin and Melis (2003)
Yeast extract	X. dendrorhous	Astaxanthin	Amino acids availability may imply an excess of carbon and energy that can be used for other biosynthetic processes	Johnson and Lewis (1979)
Acetate Mevalonate	Fungi; Algae	Carotenoids	Carotenoid precursors	Johnson and An (1991); Hata et al. (2001)
TCA cycle intermediaries	B. trispora; Flavobacterium; X. dendrorhous	β-carotene, zeaxanthin, astaxanthin, canthaxanthin	Carotenoid precursors	Alcantara and Sanchez (1999); Flores-Cotera et al. (2001); Nasrabadi and Razavi (2010b); Wozniak et al. (2011)
Ketoconazole	Fungus	Lycopene	Limits ergosterol synthesis	Sun et al. (2007)

© Woodhead Publishing Limited, 2013

water, mineral salts, CO_2 and light. The production of such microalgal genera is commonly carried out in outdoor ponds near lakes with high salinity. Major farming systems have capacities of 10 million litres. When *D. salina* is grown in media with limited nitrogen, high salinity and intense light, the cell β-carotene content can reach 14% (of dry weight).

Commercial astaxanthin production by *H. pluvialis* has been described previously in detail (Bubrick, 1991; Lorenz and Cysewski, 2000; Olaizola, 2000). Enclosed photobioreactors and open culture ponds are both used, resulting in cells with 2.8%–3.0% astaxanthin content. The effect of light is undoubtedly the most important factor for astaxanthin accumulation in *H. pluvialis* (Kim *et al.*, 2006). However, production is also commonly performed under conditions of stress such as nutrient (N or P) deficiency or high NaCl concentrations (Bubrick, 1991; García-Malea *et al.*, 2009; Harker *et al.*, 1996).

9.6.2 Open versus closed cultures

Most microalgal production systems built before the 1990s were essentially open pond/raceway systems that allow cell densities of up to 0.7 g cells l^{-1} (dry basis). Open pond/raceway systems have been particularly successful in culturing *Dunaliella*, *Chlorella* and *Spirulina*, species that are able to withstand adverse growing conditions (high salinity, high light intensity) and consequently able to outcompete other microorganisms (Ye *et al.*, 2008; Jin and Melis, 2003). However, a low cell density results in low productivity, costly product recovery, large water requirements and temperature control difficulties. These drawbacks promoted the development of photobioreactors constructed from transparent materials such as glass and polycarbonate. Photobioreactors have advantages in the intensive cultivation of microalgae and provide lower production costs in some geographic areas. In particular, tubular photobioreactors have received a lot of attention because they allow high cell density cultures, with three-fold or more biomass when compared with conventional open ponds. The main advantages are as follows: (1) easy biomass harvesting; (2) continued cultivation without risk of contamination; (3) better control of growing conditions (less dependent on climate); and (4) lower capital investment in the photobioreactor. In addition, because light penetrates poorly in dense cultures and cells are only briefly exposed to light (in cycles) in a medium in constant motion, these systems can be used at high incident light intensities without risk of photoinhibition.

9.6.3 Heterotrophic and mixotrophic cultivation of algae

Some algae can grow well as a mixotroph (they use an organic carbon source and light energy as a source of electrons) and a heterotroph (using a source of organic carbon in the dark). Organic carbon sources may include

© Woodhead Publishing Limited, 2013

ethanol, acetate and glucose. *H. pluvialis* can synthesize astaxanthin under mixotrophic (capable of deriving energy from multiple sources) or heterotrophic conditions, using sodium acetate as a carbon source. Astaxanthin production by *H. pluvialis* grown in sequential heterotrophic/photoautotrophic culture has been reported (Hata *et al.*, 2001). Initially, this alga was grown heterotrophically to a high cell concentration. Next, astaxanthin accumulation was achieved by illuminating the culture. When encysted cells were induced, the cell number decreased but astaxanthin accumulation was nevertheless very high. However, more recently Kang *et al.* (2005b), reported that photoautotrophic induction under nitrogen-deprivation, followed by bicarbonate (HCO_3^-) supplementation was more effective (3.4-fold higher) than heterotrophic induction using acetate.

These culture methods have been effectively used with other potentially useful algae such as *Scenesdesmus* sp. (Yen *et al.*, 2011) and *Chlorella prototheoides* for lutein production (Shi *et al.*, 1999, 2002) and for other carotenoids (Ip *et al.*, 2004; Ip and Chen, 2005).

9.6.4   One stage versus two stage cultures

For most microorganisms, optimal growth conditions are not usually optimal carotenoid synthesis conditions. Thus, a two-phase process is commonly used for maximal productivity in fungi and microalgal cultures. The traditional batch fungal cultures generally produce low pigment concentrations and low biomass yield. Therefore, the preferred industrial fermentation process for astaxanthin production by *X. dendrorhous* is performed in two phases: a cell growth phase and a maturation phase. In the first phase, the yeast grows rapidly in a culture medium with appropriate concentrations of nutrients (low C/N ratio). In the second phase, a carbon source, such as ethanol or glucose, is slowly and continuously supplied to induce accumulation of astaxanthin while the cell growth slows (de la Fuente *et al.*, 2005; Frengova and Beshkova, 2009; Schmidt *et al.*, 2011). Various feeding strategies for astaxanthin production have been reported (Hu *et al.*, 2005). β-Carotene production has also used batch and fed-batch cultures of *B. trispora*. A 5-fold higher concentration of β-carotene was obtained using the latter culture method (Kim *et al.*, 1999).

Because stress conditions cause *H. pluvialis* to overproduce astaxanthin, a production strategy has been developed that separates in time the growth phase from the pigment formation phase. In general, two-stage systems yield a richer astaxanthin product (4% of dry biomass) and higher astaxanthin productivity than one-stage systems (Aflalo *et al.*, 2007). Each stage can be independently optimized using different methods (Choi *et al.*, 2002; Harker *et al.*, 1996; Zhang *et al.*, 2009; Del Rio *et al.*, 2008). The feasibility of two step processes with separate stages for growth and carotenoid production for *D. salina*, *Spirulina* and *Chlorella* have also been reported (Benamotz, 1995; Prieto *et al.*, 2011; Zhang *et al.*, 2009).

© Woodhead Publishing Limited, 2013

9.6.5   Non-refined carbon sources

The use of non-refined carbon sources for carotenoid production by micro-organisms has been the subject of much research. Cost reduction of industrial scale fermentations, or issues related to the disposal of agro-industrial residues are the likely main driving forces. However, low-cost media may contain unknown inhibitors of carotenogenesis, which may make them unsuitable for a reproducible production process (Schmidt *et al.*, 2011). The use of non-refined carbon sources has been studied extensively for yeast-derived (*X. dendrorhous* and *Rhodotorula*) carotenoids including grape juice, peat extract and peat hydrolysate, juice of yucca, hemicellulose hydrolysates, sugar cane juice and corn syrup (Buzzini *et al.*, 2001; Frengova and Beshkova, 2009).

9.7   Commercially significant carotenoids

Owing to their outstanding antioxidant properties and health-related functions, xanthophylls such as lutein, zeaxanthin and astaxanthin constitute a multimillion-dollar market worldwide.

9.7.1   Astaxanthin

Astaxanthin (3,3'-dihydroxy-β, β-carotene-4,4'-dione) is a major ketocarotenoid that is produced by several microorganisms. This orange-red pigment is widely used as a colorant in animal diets including diets for salmonids and lobsters and to colour chicken egg yolks (Bjerkeng, 2008). Additionally, astaxanthin algae meal for poultry provides excellent pigmentation of egg yolk at concentrations as low as 2 ppm. Aquaculture output has grown at 11% annually over the past decade and is the fastest growing sector of the world food economy. Worldwide, fish farming is a huge business with some estimates placing its 2008 value at US$94 billion (Pauli, 2011). Astaxanthin has an estimated market size of US$252 million for fish food and over US$30 million for human uses (Global Industry Analysts, 2010). Its average price is US$2500.00 per kg for the synthetic form (Del Campo *et al.*, 2007) and over US$7000.00 per kg for the natural native form (Frucht and Kanon, 2005).

In addition to the applications of astaxanthin in aquaculture, because of its high antioxidant activity it has an extraordinary potential for protecting organisms against a wide range of ailments such as cardiovascular disease, various degenerative diseases including cancer (Hussein *et al.*, 2006), skin related illness, heart disease (Guerin *et al.*, 2003), *Helicobacter pylori* infections (Wang *et al.*, 2000) and immunological diseases (Park *et al.*, 2010). Its antioxidant properties have been reported to surpass those of β-carotene or even α-tocopherols (100–500 fold higher) (Naguib, 2000). Owing to its

© Woodhead Publishing Limited, 2013

petrochemical origin, synthetic astaxanthin is not accepted for human use and therefore is limited to a colour additive in aquaculture.

The estimated market size of astaxanthin for human uses surpasses US$30 million (Global Industry Analysts, 2010). The human consumption sector for this carotenoid includes a wide array of marketable products that are already being exploited. One group is the long-chain poly unsaturated fatty acids (vlcPUFAs), comprising eicosapentaenoic (EPA), docosahexaenoic acid (DHA) and arachidonic acid (AA). These compounds have been shown to be essential for brain development and aid the cardiovascular system (de Urquiza *et al.*, 2000; Colquhoun, 2001). For example, vlcPUFAs are found in many different products including infant formulas, adult dietary supplements, animal feed, food additives and pharmaceutical precursors. The world wholesale market for this compound for infant formula alone is estimated to be approximately US$10 billion per year (Ward and Singh, 2005).

Astaxanthin can be produced in open ponds (25,000 l photo bioreactor) using the freshwater microalga *H. pluvialis* (Chlorophyceae) (Guerin *et al.*, 2003) and by fermentation using the heterobasidiomycetous yeast *X. dendrorhous*, the perfect match for *P. rhodozyma* (Rodríguez-Sáiz *et al.*, 2010). Astaxanthin production levels of 4.7 mg g dry cell matter^{-1} (420 mg l^{-1}) have been reported for *X. dendrorhous* fermentations conducted under continuous white light (Rodríguez-Saiz *et al.*, 2010). Other microbial sources of astaxanthin include *Brevibacterium*, *Mycobacterium lacticola* (Nelis and DeLeenheer, 1991) and the marine bacterium *A. auratium* (Yokoyama *et al.*, 1994).

9.7.2   Canthaxanthin

Canthaxanthin (β, β-carotene-4,4′-dione) is a keto-carotenoid that is generated naturally in a wide variety of living organisms that have a strong antioxidant activity. It has been reported to be a more effective antioxidant than β-carotene (Naguib, 2001). From a commercial point of view, canthaxanthin is one of the most important xanthophylls because of its extensive applications (Perera and Yen, 2007). Canthaxanthin has been used as a food additive for egg yolk, in cosmetics and as a pigmenting agent for human skin applications and as a feed additive in fish and crustacean farms (Asker and Ohta, 2002). It has been reported to be effective in the treatment of polymorphous light eruptions and idiopathic photodermatosis and as a protective agent against skin cancers in experimental animals (Santamaría *et al.*, 1988). The global market for canthaxanthin is almost US$100 million (BCC Research, 2011).

Canthaxanthin is synthesized in relatively low concentrations by bacteria, green micro-algae and a halophilic archaeon (*Haloferax alexandrines*) and therefore these organisms cannot compete economically with synthetic

© Woodhead Publishing Limited, 2013

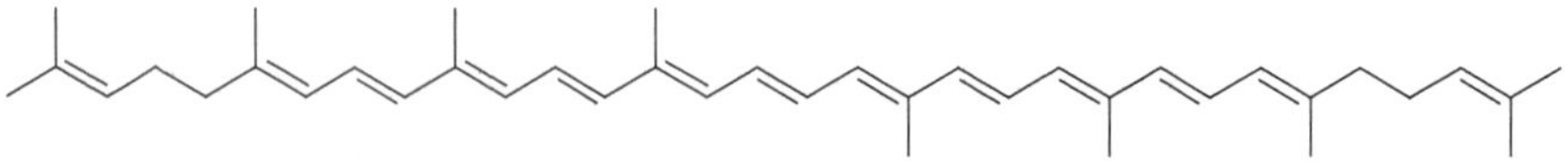

Fig. 9.6   Chemical structure of lycopene.

canthaxanthin production. The bacterium *Dietza natronolimnaea* HS-1 is reported to be one of the most promising sources for the microbial production of canthaxanthin (Khodaiyan *et al.*, 2008). The highest canthaxanthin production using this microorganism has been reported to be 9.6 mg l^{-1} using a fed-batch process (Nasrabadi and Razavi, 2010a, b).

9.7.3   Lycopene

Lycopene (Fig. 9.6), a strong natural antioxidant, is a symmetrical tetraterpene assembled from eight isoprene units ($C_{40}H_{56}$). It is a red-coloured intermediate of the β-carotene biosynthetic pathway. Lycopene has many uses including in beverages, dairy foods, surimi, confectionery, soups, nutritional bars, breakfast cereals, pastas, chips, sauces, snacks, dips and spreads (Kong *et al.*, 2010). It has a very high singlet oxygen-quenching activity and ability to suppress proliferation of MSF-7 tumor cells (Feofilova *et al.*, 2006). Commercial preparations of lycopene are formulated as oil suspensions or water-soluble powders. Of the carotenoids with nine or more conjugated double bonds, lycopene is the most effective singlet oxygen quencher (Di Mascio *et al.*, 1989). The global market for lycopene is US\$60 million (BCC Research, 2011).

Fermentation is, currently, the most effective known method for production of lycopene. Although several microorganisms have been reported to be lycopene producers, the carotenogenic fungus *B. trispora* (Tereshina *et al.*, 2010) and metabolically engineered *E. coli* or *Pichia pastoris* cells (Farmer and Liao, 2000; Bhataya *et al.*, 2009) are the only examples that are commercially tenable for lycopene production (Kim *et al.*, 2011). Lycopene has been manufactured commercially using *B. trispora* (Vitatene, Leon, Spain). The bacterial *crt* gene cluster (*crtE, crtI* and *crtB*), from different bacteria, has been introduced into *E. coli* and used in the heterologous biosynthesis of lycopene (Yoon *et al.*, 2007b). One of the metabolically engineered *E. coli* strains produced 1.35 g l^{-1} lycopene following 34 h in a glycerol fed-batch fermenter with 10 g l^{-1} glucose and 7.5 g l^{-1} L-arabinose. Under these conditions, the specific lycopene content and lycopene productivity were 32 mg g cells^{-1} and 40 mg l^{-1} h^{-1}, respectively (Kim *et al.*, 2011). Given that in *B. trispora* lycopene production reaches approximately 1 g l^{-1} after 120 h fermentation, with a productivity of 7.8 mg l^{-1} h^{-1} (Tereshina *et al.*, 2010), this engineered *E. coli* strain provides the highest reported concentration and productivity of lycopene (Kim *et al.*, 2011).

© Woodhead Publishing Limited, 2013

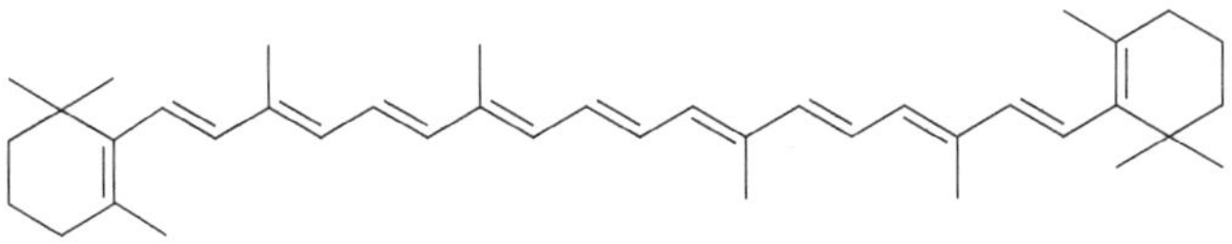

Fig. 9.7   Chemical structure of β-carotene.

9.7.4   β-Carotene

β-Carotene (Fig. 9.7) is a tetraterpene ($C_{40}H_{56}$) that is synthesized biochemically from eight isoprene units that are cyclized, thus forming beta-rings at both ends of the molecule. It is believed to be the most important carotenoid in human nutrition. Indeed, β-carotene is a safe source of vitamin A because it can be cleaved into two molecules of vitamin A by an intestinal monooxygenase (Biesalski *et al.*, 2007). β-Carotene has been applied in a wide range of food products (margarine, cheese, fruit juices, baked goods, dairy products, canned goods and confectionary and health products). It has also been used to enhance the colour of birds, fish and crustaceans, in addition to improving the appearance of pet food (Coma, 1991). New food applications such as the coloration of processed meats (sausage, ham), marine products (fish paste, surimi) and tomato ketchup have also been described (Cantrell *et al.*, 2003). In addition, β-carotene helps to mediate the harmful effects of free radicals that may cause dangerous diseases including various forms of cancer, coronary heart disease, premature ageing and arthritis (Tornwall *et al.*, 2004). β-Carotene has the largest share of the carotenoid market, estimated to be approximately US$250 million in 2007, increasing to almost US$261 million in 2010. This market is expected to grow to US$334 million by 2018, at an annual rate of 3.1% (BCC Research, 2011).

β-Carotene is mainly produced by a number of fungi (*Mucor, Phycomyces*) and commercial production exploits *B. trispora*, with a record β-carotene content of 9 g l^{-1} (Costa-Perez *et al.*, 1997). Microalgal sources include *Dunaliella*, which is the best carotenoid-producing alga and other organisms. Other commonly cultivated species are *D. salina* and *Dunaliella bardawil* (Duffose *et al.*, 2005).

9.7.5   Lutein and zeaxanthin

Lutein (3R,3'R,6'R)-βε-carotene-3,3'-diol) and zeaxanthin (3R,3'R)-β,β-carotene-3,3'diol) are known as xanthophylls or oxycarotenoids. These hydroxycarotenoids lack of provitamin A activity, but during the last decade they have attracted the attention of medicinal and biotechnological researchers owing to their possible role in the prevention of age-related macular degeneration (AMD) and cataracts (Beatty *et al.*, 2004). Indeed, lutein and zeaxanthin accumulate in the macula (Khachik *et al.*, 2002) and

© Woodhead Publishing Limited, 2013

act either as filters by absorbing blue light, which causes the generation of oxygen and reactive radicals that damage photoreceptor cells, or as antioxidants that protect the photoreceptor cells from potential free radical-induced damage (Roberts *et al.*, 2009). Both pigments may also serve to protect the skin from UV-induced damage and they may help reduce the risk of cardiovascular disease. Consumption of carotenoids dominated by lutein is believed to help maintain skin health by reducing UV-induced erythema (Roberts *et al.*, 2009; Stahl and Sies, 2002). Lutein has also received attention owing to its anticancer activity (Chew *et al.*, 1996) and its ability to enhance immune function (Hughes *et al.*, 2000). Zeaxanthin has applications in poultry pigmentation and as a nutritional supplement in a wide range of foods. In addition, it has a protective effect on cancer metastasis in mice (Kurihara *et al.*, 2002; Hussein *et al.*, 2006).

The lutein global market value was approximately US$233 million in 2010 and is expected to reach US$309 million by 2018, with an annual growth rate of 3.6%. The global market for zeaxanthin was close to US$10 million in 2010 (BCC Research, 2011).

Of the microbial sources of these pigments, the photosynthetic marine microalga *D. bardawil* produces lutein and zeaxanthin in considerable amounts. An outdoor 50-l tubular system using a *Murielopsis* sp. produced an annual maximum of 1800 g ha^{-1} day^{-1}, or 2900 g ha^{-1} day^{-1} for *Scenedesmus almeriensis* operating on a pilot scale. These productivities are 600–966 times higher than the estimated 3 g ha^{-1} day^{-1} for *Tagetes erecta* (Del Campo *et al.*, 2007). *Flavobacterium multivorum*, a non-fastidious, non-pathogenic and non-photosynthetic bacterium is considered to be an important potential microbial source of zeaxanthin (Alcantara and Sanchez, 1999; Masetto *et al.*, 2001). Lycopene production has also been reported in *B. trispora*. In this microorganism, β-carotene, but not lycopene, is synthesized as the main carotenoid under conventional fermentation conditions. To obtain lycopene, lycopene cyclase activity must be suppressed by chemical or genetic means to prevent the formation of β-carotene and promote the accumulation of lycopene. The optimized bioprocess in the presence of a cyclase inhibitor (2-methyl imidazole) yielded 360 mg l^{-1} lycopene (Pegklidou *et al.*, 2008).

9.8   Conclusion

Carotenoids play an ever-increasing role in human health in the developed world. They are valuable bioactive compounds of interest as pigments, nutraceuticals and medicines. Their antioxidant properties and their efficiency in the prevention of certain human diseases have been well established. The commercial requirement for many carotenoids has long been met by chemical synthesis. However, some of the by-products resulting from such chemical processes may have undesirable side effects upon

© Woodhead Publishing Limited, 2013

consumption. Plants represent another source of carotenoids. However, low extraction yields and the difficulties of seasonal and geographic variability in the production and marketing of pigments of plant origin are frequently encountered. For this reason, the production of carotenoids from microbial sources has been the focus of extensive research. Yeasts such as *Phaffia*, *Rhodotorula* and algae including *Dunaliella* have been identified as commercially significant sources of carotenoids. However, the types of carotenoids and their relative amounts may vary depending on the microbial culture conditions. Additionally, carotenoid extraction and recovery from microbial cells remains a costly process.

Various carotenoids of microbial origin, such as astaxanthin, lycopene, β-carotene and canthaxanthin are being commercialized to some extent and have found applications in beverages, dairy foods, cereal products, meats, cosmetics, pharmaceutical and aquaculture, among others. The growing demand for carotenoids has triggered research into the commercial production of carotenoids. The development of new strains that could withstand robust industrial conditions and utilize industrial waste as substrate would help to reduce the costs of the whole process. There is a need to improve fermentation strategies so that the accumulation of intracellular carotenoids in yeast is feasible at an industrial scale. Manipulation of the culture and external stimulants, as detailed in this review, will allow carotenoid production to be scaled-up for commercialization. A vast biodiversity of microorganisms from various habitats is yet to be explored for exploitation as potential factories of carotenoids. With the advancement in metabolic engineering protocols, xanthophyll pigment production by native and mutant microbes has become competitive with chemical synthesis for a few selected xanthophylls and, recently, xanthophyll production has been engineered in non-carotenogenic microbes using recombinant DNA technologies.

9.9   Acknowledgements

We are indebted to Marco A. Ortíz and Laura Escalante for their help in the elaboration of this manuscript.

9.10   References

ADAM P, HECHT S, EISENREICH W, KAISER J, GRÄWERT T, ARIGONI D, BACHER A and ROHDLICH F (2002), 'Biosynthesis of terpenes: studies on 1-hydroxy-2-methyl-2-(E)-butenyl 4-diphosphate reductase', *P Natl Acad Sci USA*, **99**, 12108–13.

AFLALO C, MESHULAM Y, ZARKA A and BOUSSIBA S (2007), 'On the relative efficiency of two- vs. one-stage production of astaxanthin by the green alga *Haematococcus pluvialis*', *Biotechnol Bioeng*, **98**, 300–5.

ALBRECHT M, TAKAICHI S, STEIGER S, WANG ZW and SANDMANN G (2000), 'Novel hydroxycarotenoids with improved antioxidative properties produced by gene combination in *Escherichia coli*', *Nat Biotechnol*, **18**, 843–6.

© Woodhead Publishing Limited, 2013

ALCANTARA S and SANCHEZ S (1999), 'Influence of carbon and nitrogen sources on *Flavobacterium* growth and zeaxanthin biosynthesis', *J Ind Microbiol Biotechnol*, **23**, 697–700.

AN G and JOHNSON EA (1990), 'Influence of light on growth and pigmentation of the yeast *Phaffia rhodozyma*', *Anton Leeuw Int J G*, **57**, 191–203.

AN GH, SCHUMAN DB and JOHNSON EA (1989), 'Isolation of *Phaffia rhodozyma* mutants with increased astaxanthin content', *Appl Environ Microbiol*, **55**, 116–24.

AN G-H, CHO M-H and JOHNSON EA (1999), 'Monocyclic carotenoids biosynthetic pathway in the yeast *Phaffia rohodozyma* (*Xanthophyllomyces dendrorhous*)', *J Biosci Bioeng*, **88**, 189–93.

ARAD S, COHEN E and BENAMOTZ A (1993), 'Accumulation of canthaxanthin in *Chlorella*-Emersonii', *Physiol Plantarum*, **87**, 232–6.

ARMSTRONG GA (1997), 'Genetics of eubacterial carotenoid biosynthesis: A colorful tale', *Annu Rev Microbiol*, **51**, 629–59.

ASKER D and OHTA Y (2002), 'Production of canthaxanthin by *Haloferax alexandrinus* under nonaseptic conditions and a simple, rapid method for its extraction', *Appl Microbiol Biot*, **58**, 743–50.

BCC RESEARCH (2011), *Food And Beverage. The global market for carotenoids.* September 2011: Report code: FOD025D Wellesley, MA. http://www.bccresearch. com/report/carotenoids-global-607market-fod025d.html

BEATTY S, NOLAN J, KAVANAGH H and O'DONOVAN O (2004), 'Macular pigment optical density and its relationship with serum and dietary levels of lutein and zeaxanthin', *Arch Biochem Biophys*, **430**, 70–6.

BEJARANO ER, PARRA F, MURILLO FJ and CERDA-OLMEDO (1988), 'End-product regulation of carotenogenesis in *Phycomyces*', *Arch Microbiol*, **150**, 209–14.

BENAMOTZ A (1995), 'New mode of *Dunaliella* biotechnology-2-phase growth for β-carotene production', *J Appl Phycol*, **7**, 65–8.

BEUTTLER H, HOFFMANN J, JESKE M, HAUER B, SCHMID RD, ALTENBUCHNER J and URLACHER VB (2011), 'Biosynthesis of zeaxanthin in recombinant *Pseudomonas putida*', *Appl Microbiol Biotechnol*, **89**, 1137–47.

BHATAYA A, SCHMIDT-DANNERRT C and LEE PC (2009), 'Metabolic engineering of *Pichia pastoris* X-33 for lycopene production', *Process Biochem*, **44**, 1095–102.

BHOSALE P (2004), 'Environmental and cultural stimulants in the production of carotenoids from microorganisms', *Appl Microbiol Biotechnol*, **63**, 351–61.

BHOSALE P and BERSTEIN PS (2005), 'Microbial xanthophylls', *Appl Microbiol Biotechnol*, **68**, 445–55.

BIESALSKI HK, CHICHILI GR, FRANK J, VON LINTIG J and NOHR D (2007), 'Conversion of β-carotene to retinal pigment', *Vitam Horm*, **75**, 117–30.

BJERKENG B (2008), 'Carotenoids in aquaculture: fish and crustaceans', in *Carotenoids. Natural Functions*, Britton G, Liaaen-Jensen S and Pfander H (eds). vol. 4, Birkhäuser, Basel, 237–54.

BRAMLEY PM and MACKENZIE A (1988), 'Regulation of carotenoid biosynthesis', *Curr Top Cell Regul*, **29**, 291–343.

BROWN MS and GOLDSTEIN KL (1980), 'Multivalent feedback regulation of HMG CoA reductase, a control mechanism coordinating isoprenoid synthesis and cell growth', *J Lipid Res*, **21**, 505–17.

BUBRICK P (1991), 'Production of astaxanthin from *Haematococcus*', *Biores Technol*, **38**, 237–9.

BUZZINI P, RUBINSTEIN L and MARTINI A (2001), 'Production of yeast carotenoids by using agro-industrial by-products', *Agro Food Industry HiTec*, **12**, 7–10.

CALO P, MIGUEL T, VELÁZQUEZ J B and VILLA TG (1995), 'Mevalonic acid increases *trans*-astaxanthin and carotenoid biosynthesis in *Phaffia rhodozyma*', *Biotechnol Lett*, **17**, 575–8.

© Woodhead Publishing Limited, 2013

CANTRELL A, MCGARVEY DJ, TRUSTCOTT G, RANCON F and BOHMM F (2003), 'Singlet oxygen quenching by dietary carotenoids in a model membrane environment', *Arch Biochem Biophys*, **412**, 47–54.

CHATTOPADHYAY P, CHATTERJEE S and SEN SK (2008), 'Biotechnological potential of natural food grade biocolorants', *Afr J Biotechnol*, **7**, 2972–85.

CHÁVEZ-CABRERA C, FLORES-BUSTAMANTE Z, MARSCH R, MONTES M, SÁNCHEZ S, CANCINO-DÍAZ J and FLORES-COTERA L (2010), 'ATP-citrate lyase activity and carotenoid production in batch cultures of *Phaffia rhodozyma* under nitrogen-limited and nonlimited conditions'. *Appl Microbiol Biotechnol*, **85**, 1953–60.

CHEN Q, JIANG J-G and WANG F (2007), 'Molecular phylogenies and evolution of *crt* genes in algae', *Crit Rev Biotechnol*, **27**, 77–91.

CHEW BP, WONG MW and WONG TS (1996), 'Effects of lutein from marigold extract on immunity and growth of mammary tumors in mice'. *Anticancer Res*, **16**, 3689–94.

CHOI YE, YUN YS and PARK JM (2002), 'Evaluation of factors promoting astaxanthin production by a unicellular green alga, *Haematococcus pluvialis*, with fractional factorial design', *Biotechnol Progr*, **18**, 1170–5.

CHOUDHARI SM, ANANTHANARAYAN L and SINGHAL RS (2008), 'Use of metabolic stimulators and inhibitors for enhanced production of β-carotene and lycopene by *Blakeslea trispora* NRRL 2895 and 2896', *Biores Technol*, **99**, 3166–73.

COLQUHOUN DM (2001), 'Nutraceuticals: Vitamins and other nutrients in coronary heart disease', *Curr Opin Lipidol*, **12**, 639–46.

COMA J (1991), *Dietary Reference Values for Food Energy and Nutrients for the United Kingdom. Report of the panel on dietary reference values*. Committee on Medical Aspects of Food and Nutrition Policy, HMSO, London.

COSTA-PEREZ J, RODRIGUEZ AT, DE LA FUENTE JL, RODRIGUEZ M, DIEZ B, PEIRO E, CABRI W and BARREDO J (2007), *Method of Production of beta-Carotene by Fermentation in Mixed Culture Using (+) and (–) Strains of* Blakeslea trispora, US patent 7,252,965.

DANNERT CS, UMENO D and ARNOLD FH (2000), 'Molecular breeding of carotenoid biosynthetic pathways', *Nature Biotechnol*, **18**, 750–3.

DAS A, YOON SH, LEE SH, KIM JY, OH DK and KIM SW (2007), 'An update on microbial carotenoid production: application of recent metabolic engineering tools', *Appl Microbiol Biotechnol*, **77**, 505–12.

DAVOLI P, MIERAU V and WEBER RWS (2004), 'Carotenoids and fatty acids in red yeasts *Sporobolomyces roseus* and *Rhodotorula glutinis*', *Appl Biochem Microbiol*, **40**, 392–7.

DE LA FUENTE JL, PEIRO E, DIEZ B, MARCOS AT, SCHLEISSNER C, RODRIGUEZ-SAIZ M, RODRIGUEZ OTERO C, CABRI W and BARREDO JL (2005), *Method of Production of Astaxanthin by Fermenting Selected Strains of* Xanthophyllomyces dendrorhous, United States Patent US20050124032A.

DE LA FUENTE JL, RODRIGUEZ-SAIZ M, SCHLEISSNER C, DIEZ B, PEIRO E and BARREDO, JL (2010), 'High-titer production of astaxanthin by the semi-industrial fermentation of *Xanthophyllomyces dendrorhous*', *J Biotechnol*, **148**, 144–6.

DEL CAMPO JA, RODRIGUEZ H, MORENO J, VARGAS MA, RIVAS J and GUERRERO MG (2001), 'Lutein production by *Muriellopsis* sp. in an outdoor tubular photobioreactor', *J Biotechnol*, **85**, 289–95.

DEL CAMPO JA, GARCÍA-GONZÁLEZ M and GUERRERO MG (2007), 'Outdoor cultivation of microalgae for carotenoid production: current state and perspectives', *Appl Microbiol Biotechnol*, **74**, 1163–74.

DEL RIO E, ACIEN G, GARCÍA-MALEA MC, RIVAS J, MOLINA-GRIMA E and GUERRERO MG (2005), 'Efficient one-step production of astaxanthin by the microalga *Haematococcus pluvialis* in continuous culture', *Biotechnol Bioeng*, **91**, 808–15.

© Woodhead Publishing Limited, 2013

DEL RIO E, ACIEN FG, GARCÍA-MALEA MC, RIVAS J, MOLINA-GRIMA E and GUERRERO, MG (2008), 'Efficiency assessment of the one-step production of astaxanthin by the microalga *Haematococcus pluvialis*', *Biotechnol Bioeng*, **100**, 397–402.

DE URQUIZA AM, LIU S, SJOBERG M, ZETTERSTROM RH, GRIFFITHS W, SJOVALL J and PERLMANN T (2000), 'Docosahexaenoic acid, a ligand for the retinoid X receptor in mouse brain', *Science*, **290**, 2140–4.

DHARMARAJ S, ASHOKKUMAR B and DHEVENDARAN K (2009), 'Fermentative production of carotenoids from marine actinomycetes', *Iran J Microbiol*, **1**, 36–41.

DI MASCIO P, KAISER S and SIES H (1989), 'Lycopene as the most efficient biological carotenoid singlet oxygen quencher', *Arch Biochem Biophys*, **274**, 532–8.

DOMINGUEZ-BOCANEGRA AR, LEGARRETA IG, JERONIMO FM and CAMPOCOSIO AT (2004), 'Influence of environmental and nutritional factors in the production of astaxanthin from *Haematococcus pluvialis*', *Biores Technol*, **92**, 209–14.

DUFOSSE L (2006), 'Microbial production of food grade pigments', *Food Technol Biotechnol*, **44**, 313–21.

DUFFOSE L, GALAUP P, YARON A, ARAD SM, BLANC P, MURTHY KNC and RAVISHANKAR GA (2005), 'Microorganisms and microalgae as sources of pigments for food use: a scientific oddity or an industrial reality?' *Trends Food Sci Technol*, **16**, 389–406.

FANG C-J, KU K-L, LEE M-H and SU N-W (2010), 'Influence of nutritive factors on C_{50} carotenoids production by *Haloferax mediterranei* ATCC 33500 with two-stage cultivation', *Biores Technol*, **101**, 6487–93.

FARMER WR and LIAO JC (2000), 'Improving lycopene production in *Escherichia coli* by engineering metabolic control', *Nature Biotechnol*, **18**, 533–7.

FEOFILOVA EP, TERESHINA VM, MEMORSKAYA AS, DUL'KIN LM and GONCHAROV NG (2006), 'Fungal lycopene, the biotechnology of its production and prospects for its application in medicine', *Microbiology*, **75**, 629–33.

FLORES-COTERA LB, MARTIN R and SANCHEZ S (2001), 'Citrate, a possible precursor of astaxanthin in *Phaffia rhodozyma*: influence of varying levels of ammonium, phosphate and citrate in a chemically defined medium', *Appl Microbiol Biotechnol*, **55**, 341–7.

FRENGOVA GI and BESHKOVA DM (2009), 'Carotenoids from *Rhodotorula* and PhaYa: yeasts of biotechnological importance', *J Ind Microbiol Biotechnol*, **36**, 163–80.

FRUCHT LE and KANON S (2005), 'Israel grows red algae in the desert to fight disease'. *Israel21c Newsletter*. Available from: www.israel21c.org/environment/israel-grows-red-algae-in-the-desert-to-fight-disease (Accessed 23 September 2011).

GARCÍA-MALEA MC, ACIEN FG, DEL RIO E, FERNANDEZ JM, CERON MC, GUERRERO MG and MOLINA-GRIMA, E (2009), 'Production of astaxanthin by *Haematococcus pluvialis*: taking the one-step system outdoors', *Biotechnol Bioeng*, **102**, 651–7.

GAVRILOV AS, KISELEVA AI, MATUSHKINA SA, KORDYUKOVA NP and FEOFILOVA EP (1996), 'Industrial production of lycopene by a microbiological method', *Appl Biochem Microbiol*, **32**, 492–4.

GLOBAL INDUSTRY ANALYSTS INC. (2010), *Carotenoids A Global Strategic Business Report San José, CA*. Available from: www.prweb.com/releases/carotenoids/β-carotene/prweb4688784.htm (Accessed 23 September 2011).

GOLDSTEIN JL and BROWN MS (1990), 'Regulation of the mevalonate pathway', *Nature*, **343**, 425–30.

GU WL, AN GH and JOHNSON EA (1997), 'Ethanol increases carotenoid production in *Phaffia rhodozyma*', *J Ind Microbiol Biotechnol*, **19**, 114–17.

GU Z, CHEN D, HAN Y, CHEN Z and GU F (2008), 'Optimization of carotenoids extraction from *Rhodobacter sphaeroides*', *LWT Food Sci Technol*, **41**, 1082–8.

GUERIN M, HUNTLEY ME and OLAIZOLA M (2003), '*Haematococcus* astaxanthin, health and nutritional applications', *Trends Biotechnol*, **21**, 210–16.

HAMPTON R, DIMSTER-DENK D and RINE J (1996), 'The biology of HMG-CoA reductase: the pros of contra-regulation', *Trends Biochem Sci*, **21**, 140–5.

© Woodhead Publishing Limited, 2013

HARADA H and MISAWA N (2009), 'Novel approaches and achievements in biosynthesis of functional isoprenoids in *Escherichia coli*', *Appl Microbiol Biotechnol*, **84**, 1021–31.

HARKER M, TSAVALOS AJ and YOUNG AJ (1996), 'Autotrophic growth and carotenoid production of *Haematococcus pluvialis* in a 30 liter air-lift photobioreactor', *J Ferment Bioeng*, **82**, 113–18.

HATA N, OGBONNA JC, HASEGAWA Y, TARODA H and TANAKA H (2001), 'Production of astaxanthin by *Haematococcus pluvialis* in a sequential heterotrophic-photoautotrophic culture', *J Appl Phycol*, **13**, 395–402.

HU ZC, ZHENG YG, WANG Z and SHEN YC (2005), 'Effect of sugar-feeding strategies on astaxanthin production by *Xanthophyllomyces dendrorhous*', *World J Microb Biotechnol*, **21**, 771–5.

HUGHES DA, WRIGHT AJA, FINGLAS PM, POLLEY ACJ, BAILEY AL, ASTLEY SB and SOUTHON S (2000), 'Effects of lycopene and lutein supplementation on the expression of functionally associated surface molecules on blood monocytes from healthy male nonsmokers', *J Infect Dis*, **182**, S11–S15.

HUNTER CN, HUNDLE BS, HEARST JE, LANG HP, GARDINER AT, TAKAICHI S and COGDELL RJ (1994), 'Introduction of new carotenoids into the bacterial photosynthetic apparatus by combining the carotenoid biosynthetic pathways of *Erwinia herbicola* and *Rhodobacter sphaeroides*', *J Bacteriol*, **176**, 3692–7.

HUSSEIN G, SANKAWA U, GOTO H, MATSUMOTO K and WATANABE H (2006), 'Astaxanthin, a carotenoid with potential in human health and nutrition', *J Nat Prod*, **69**, 443–9.

IP PF and CHEN F (2005), 'Production of astaxanthin by the green microalga *Chlorella zofingiensis* in the dark', *Process Biochem*, **40**, 733–8.

IP PF, WONG KH and CHEN F (2004), 'Enhanced production of astaxanthin by the green microalga *Chlorella zofingiensis* in mixotrophic culture', *Process Biochem*, **39**, 1761–6.

JIN ES and MELIS A (2003), 'Microalgal biotechnology: carotenoid production by the green algae *Dunaliella salina*', *Biotechnol Bioproc Eng*, **8**, 331–7.

JIN E, LEE CG and POLLE JEW (2006), 'Secondary carotenoid accumulation in *Haematococcus* (chlorophyceae): biosynthesis, regulation, and biotechnology', *J Microbiol Biotechnol*, **16**, 821–31.

JOHNSON, EA and AN, G (1991), 'Astaxanthin from microbial sources', *Crit Rev Biotechnol*, **11**, 297–326.

JOHNSON, EA and LEWIS, MJ (1979), 'Astaxanthin formation by the yeast *Phaffia rhodozyma*', *J Gen Microbiol*, **115**, 173–83.

JOHNSON, EA and SCHROEDER, WA (1996), 'Biotechnology of *astaxanthin* production in *Phaffia rhodozyma*', in *Biotechnology for Improved Foods and Flavors*, Takeoka GR, Teranishi R, Williams PJ, Kobayashi A (eds), American Chemical Society, Washington DC.

KANG MJ, LEE, YM, YOON SH, KIM JH, OCK SW, JUNG KH, SHIN JC, KEASLING JD, KIM SW (2005a) 'Identification of genes affecting lycopene accumulation in *Escherichia coli* using a shot-gun method', *Biotechnol Bioeng*, **91**, 636–42.

KANG CD, LEE JS, PARK TH and SIM SJ (2005b), 'Comparison of heterotrophic and photoautotrophic induction on astaxanthin production by *Haematococcus pluvialis*', *Appl Microbiol Biotechnol*, **68**, 237–41.

KERR S, CALE C, CABRAL JMS and VAN KEULEN F (2004), 'Factors enhancing lycopene production by a new *Mycobacterium aurum* mutant', *Biotechnol Lett*, **26**, 103–8.

KHACHIK F, DE MOURA FF, ZHAO DY, AEBISCHER CP and BERNSTEIN PS (2002), 'Transformations of selected carotenoids in plasma, liver, and ocular tissues of humans and in non-primate animal models', *Investig Ophthalmol Vis Sci*, **43**, 3383–92.

© Woodhead Publishing Limited, 2013

KHODAIYAN, F, RAZAVI, SH, EMAM-DJOMEH, Z, MOUSAVI, SMA, and HEJAZI, MA (2007), 'Effect of culture conditions on canthaxanthin production by *Dietzia natronolimnaea* HS-1', *J Microbiol Biotechn*, **17**, 195–201.

KHODAIYAN F, RAZAVI SH and MOUSAVI SH (2008), 'Optimization of canthaxanthin production by *Dietzia natronolimnaea* HS-1 from cheese whey using statistical experimental methods', *Biochem Eng J*, **40**, 415–22.

KIM JH and CHANG HI (2006), 'High-level production of astaxanthin by *Xanthophyllomyces dendrorhous* mutant JH1, using chemical and light induction', *J Microbiol Biotechnol*, **16**, 381–5.

KIM SW, LEE IY, JEONG JC, LEE JH and PARK YH (1999), 'Control of both foam and dissolved oxygen in the presence of a surfactant for production of β-carotene in *Blakeslea trispora*', *J Microbiol Biotechnol*, **9**, 548–53.

KIM ZH, KIM SH, LEE HS and CG (2006), 'Enhanced production of astaxanthin by flashing light using *Haematococcus pluvialis*', *Enzyme Microbiol Technol*, **39**, 414–19.

KIM YS, LEE JH, KIM NH, YEOM SJ, KIM SW and OH DK (2011), 'Increase of lycopene production by supplementing auxiliary carbon sources in metabolically engineered *Escherichia coli*', *Appl Microbiol Biot*, **90**, 489–97.

KLASSEN JL (2010), 'Phylogenetic and evolutionary patterns in microbial carotenoid biosynthesis are revealed by comparative genomics', *PLoSOne*, 5, e11257. doi:10.1371/journal.pone.0011257.

KOBAYASHI M, KURIMURA Y and TSUJI Y (1997), 'Light-independent, astaxanthin production by the green microalga *Haematococcus pluvialis* under salt stress', *Biotechnol Lett*, **19**, 507–9.

KONG KW, KHOO HE, PRASAD NK, ISMAIL A, TAN CP and RAJAB NF (2010), 'Revealing the power of the natural red pigment lycopene', *Molecules*, **15**, 959–87.

KURIHARA H, KODA H, ASAMI S, KISO Y and TANAKA T (2002), 'Contribution of the antioxidative property of astaxanthin to its protective effect on the promotion of cancer metastasis in mice treated with restraint stress', *Life Sci*, **70**, 2509–20.

KUZINA V and CERDA-OLMEDO E (2007), 'Ubiquinone and carotene production in the mucorales *Blakeslea* and *Phycomyces*', *Appl Microbiol Biotechnol*, **76**, 991–9.

KUZUYAMA T (2002), 'Mevalonate and nonmevalonate pathways for the biosynthesis of isoprene units', *Biosci Biotechnol Biochem*, **66**, 1619–27.

LAMERS PP, JANSSEN M, DE VOS RCH, BINO RJ and WIJFFELS RH (2008), 'Exploring and exploiting carotenoid accumulation in *Dunaliella salina* for cell-factory applications', *Trend Biotechnol*, **26**, 631–8.

LANGE BM, RUJAN T, MARTIN W and CROTEAU R (2000), 'Isoprenoid biosynthesis: The evolution of two ancient and distinct pathways across genomes', *Proc Natl Acad Sci USA*, **97**, 13172–7.

LEE HS, OHNISHI Y and HORINOUCHI S (2001), 'A sigma B-like factor responsible for carotenoid biosynthesis in *Streptomyces griseus*', *J Mol Microbiol Biotechnol*, **3**, 95–101.

LEMUTH K, STEUER K and ALBERMANN C (2011), 'Engineering of a plasmid-free *Escherichia coli* strain for improved *in vivo* biosynthesis of astaxanthin', *Microb Cell Fact*, **10**, 29–40.

LODATO P, ALCAÍNO J, BARAHONA S, NIKLITSCHEK M, CARMONA M, WOZNIAK A, BAEZA M, RETAMALES P, JIMENEZ A and CIFUENTES V (2007), 'Expression of the carotenoid biosynthesis genes in *Xanthophyllomyces dendrorhous*', *Biol Res*, **40**, 73–84.

LÓPEZ-NIETO MJ, COSTA J, PEIRO E, MÉNDEZ E, RODRÍGUEZ-SÁIZ M, DE LA FUENTE JL, CABRI W and BARREDO JL (2004), 'Biotechnological lycopene production by mated fermentation of *Blakeslea trispora*', *Appl Microbiol Biotechnol*, **66**, 153–9.

LORENZ RT and CYSEWSKI GR (2000), 'Commercial potential for *Haematococcus* microalgae as a natural source of astaxanthin', *Trends Biotechnol*, **18**, 160–7.

© Woodhead Publishing Limited, 2013

MALISORN C and SUNTORNSUK W (2008), 'Optimization of β-carotene production by *Rhodotorula glutinis* DM28 in fermented radish brine', *Biores Technol*, **99**, 2281–7.

MARGALITH P and MEYDAV S (1968), 'Some observation on carotenogenesis in yeast *Rhodotorula mucilaginosa*', *Phytochemistry*, **7**, 765–8.

MAROVA I, BREIEROVA E, KOCI R, FRIEDL Z, SLOVAK B and POKORNA J (2004), 'Influence of exogenous stress factors on production of carotenoids by some strains of carotenogenic yeasts', *Ann Microbiol*, **54**, 73–85.

MARTÍN JL, GUDIÑA E, and BARREDO JL (2008), 'Conversion of β–carotene into astaxanthin: two separate enzymes or a bifunctional hydroxylase-ketolase protein?' *Microb Cell Fact*, **7**, 3.

MASETTO A, FLORES-COTERA LB, DÍAZ C, LANGLEY E and SÁNCHEZ S (2001). 'Application of a complete factorial design for the production of zeaxanthin by *Flavobacterium* sp.', *J Bioscience Bioeng*, **92**, 55–8.

MEHTA BJ, and CERDA-OLMEDO E (1995), 'Mutants of carotene production in *Blakeslea trispora*', *Appl Microbiol Biotechnol*, **42**, 836.838.

MEHTA BJ, OBRAZTSOVA IN and CERDA-OLMEDO E (2003), 'Mutants and intersexual heterokaryons of *Blakeslea trispora* for production of β-carotene and lycopene', *Appl Environ Microbiol*, **69**, 4043–8.

MEYER PS and DU PREEZ JC (1994a), '*Astaxanthin* production by a *Phaffia rhodozyma* mutant on grape juice', *World J Microb Biotechnol*, **10**, 178–83.

MEYER PS and DU PREEZ JC (1994b), 'Effect of culture conditions on astaxanthin production by a mutant of *Phaffia rhodozyma* in batch and chemostat culture', *Appl Microbiol Biotechnol*, **40**, 780–5.

MEYER PS, DU PREEZ JC and KILIAN SG (1993), 'Selection and evaluation of *astaxanthin*-overproducing mutants of *Phaffia rhodozyma*', *World J Microbiol Biotechnol*, **9**, 514–20.

MIURA Y, KONDO K, SHIMADA H, SAITO T, NAKAMURA K and MISAWA N (1998a) 'Production of lycopene by the food yeast, *Candida utilis* that does not naturally synthesize carotenoid', *Biotechnol Bioeng*, **58**, 306–8.

MIURA Y, KONDO K, SAITO T, SHIMADA H, FRASER PD and MISAWA N (1998b) 'Production of the carotenoids lycopene, β-carotene, and astaxanthin in the food yeast *Candida utilis*', *Appl Environ Microbiol*, **64**, 1226–9.

MISAWA N and SHIMADA H (1998), 'Metabolic engineering for the production of carotenoids in non-carotenogenic bacteria and yeasts', *J Biotechnol*, **59**, 169–81.

MOGEDAS B, CASAL C, FORJAN E and VILCHEZ C (2009), 'β-carotene production enhancement by UV-A radiation in *Dunaliella bardawil* cultivated in laboratory reactors', *J Biosc Bioeng*, **108**, 47–51.

NAGUIB YMA (2000), 'Antioxidant activities of astaxanthin and related carotenoids', *J Agric Chem*, **48**, 1150–4.

NAGUIB YMA (2001), 'Pioneering astaxanthin'. *Nutrition Science News*. February, 2001. Boulder CO. Available from: http://www.chiro.org/nutrition/FULL/Pioneering_Astaxanthin.shtml (Accessed 28 October 2012).

NAMITHA KK and NEGI PS (2010), 'Chemistry and biotechnology of carotenoids', *Crit Rev Food Sci Nutr*, **50**, 728–60.

NANOU K, ROUKAS T and PAPADAKIS E (2011), 'Oxidative stress and morphological changes in *Blakeslea trispora* induced by enhanced aeration during carotene production in a bubble column reactor', *Biochem Eng J*, **54**, 172–7.

NASRABADI MRN and RAZAVI SH (2010a) 'Enhancement of canthaxanthin production from *Dietzia natronolimnaea* HS-1 in a fed-batch process using trace elements and statistical methods', *Braz J Chem Eng*, **27**, 517–29.

NASRABADI MRN and RAZAVI SN (2010b), 'Use of response surface methodology in a fed-batch process for optimization of tricarboxylic acid cycle intermediates to achieve high levels of canthaxanthin from dietzia natronolimnaea HS-1', *J Biosc Bioeng*, **109**, 361–8.

© Woodhead Publishing Limited, 2013

NELIS HJ and DELEENHEER AP (1991), 'Microbial sources of carotenoid pigments used in food and feeds', *J Appl Bacteriol*, **70**, 181–91.

OLAIZOLA M (2000), 'Commercial production of astaxanthin from *Haematococcus pluvialis* using 25,000-liter outdoor photobioreactors', *J Appl Phycol*, **12**, 499–506.

PARK EK and LEE CG (2001), 'Astaxanthin production by *Haematococcus pluvialis* under various light intensities and wavelengths', *J Microbiol Biotechnol*, **11**, 1024–30.

PARK JS, CHYUN JH, KIM YK, LINE LL and CHEW BP (2010), 'Astaxanthin decreased oxidative stress and inflammation and enhanced immune response in humans', *Nutr Metab*, **7**, 18.

PAULI G (2011), *Farming Fish Without Feed*. Vienna, The Club of Rome – European Support Centre. Available from: www.clubofrome.at/inspiration/article. php?id=162 (Accessed September 25 2011).

PEGKLIDOU K, MANTZOURIDOU F and TSIMIDOU MZ (2008),'Lycopene production using *Blakeslea trispora* in the presence of 2-methyl imidazole: yield, selectivity, and safety aspects', *J Agric Food Chem*, **56**, 4482–90.

PERERA CO and YEN GM (2007),'Functional properties of carotenoids in human health', *Int J Food Prop*, **10**, 201–30.

PHADWAL K (2005), 'Carotenoid biosynthetic pathway: molecular phylogenies and evolutionary behavior of *crt* genes in eubacteria', *Gene*, **345**, 35–43.

PRADO-CABRERO A, SCHAUB P, DÍAZ-SÁNCHEZ V, ESTRADA AF, AL-BABILI S and AVALOS J (2009), 'Deviation of the neurosporaxanthin pathway towards β–carotene biosynthesis in *Fusarium fujikuroi* by a point mutation in the phytoene desaturase gene', *FEBS J*, **276**, 4582–97.

PRIETO A, CANAVATE JP and GARCÍA-GONZALEZ M (2011), 'Assessment of carotenoid production by *Dunaliella salina* in different culture systems and operation regimes', *J Biotechnol*, **151**, 180–5.

QUILES-ROSILLO MD, RUIZ-VAZQUEZ RM, TORRES-MARTINEZ S and GARRE V (2005), 'Light induction of the carotenoid biosynthesis pathway in *Blakeslea trispora*', *Fungal Genet Biol*, **42**, 141–53.

RAJA R, HEMAISWARYA S and RENGASAMY R (2007), 'Exploitation of *Dunaliella* for β-carotene production', *Appl Microbiol Biot*, **74**, 517–23.

RAMÍREZ J, GUTIERREZ H and GSCHAEDLER A (2001), 'Optimization of astaxanthin production by *Phaffia rhodozyma* through factorial design and response surface methodology', *J Biotechnol*, **88**, 259–68.

RIBEIRO BD, BARRETO DW, BARRETO and COELHO MAZ (2011), 'Technological aspects of β-carotene production', *Food Bioprocess Technol*, **4**, 693–701.

ROBERTS RL, GREEN J and LEWIS B (2009),'Lutein and zeaxanthin in eye and skin health', *Clin Dermatol*, **27**, 195–201.

RODRIGUEZ-ORTIZ R, LIMON MC and AVALOS J (2009), 'Regulation of carotenogenesis and secondary metabolism by nitrogen in wild-type *Fusarium fujikuroi* and carotenoid-overproducing mutants', *Appl Environ Microbiol*, **75**, 405–13.

RODRÍGUEZ-SÁIZ M, DE LA FUENTE JL and BARREDO JL (2010), '*Xanthophyllomyces dendrorhous* for the industrial production of astaxanthin', *Appl Microbiol Biotechnol*, **88**, 645–58.

RUTHER A, MISAWA N, BOGER P and SANDMANN G (1997), 'Production of zeaxanthin in *Escherichia coli* transformed with different carotenogenic plasmids', *Appl Microbiol Biotechnol*, **48**, 162–7.

SANDMANN G (1994), 'Carotenoids biosynthesis in microorganisms and plants', *European J Biochem*, **223**, 7–24.

SANTAMARIA L, BIANCHI A, ARNABOLDI A, RAVETTO C, BIANCHI L, PIZZALA R, ANDREONI L, SANTAGATI G and BERMOND P (1988), 'Chemoprevention of indirect and direct

© Woodhead Publishing Limited, 2013

chemical carcinogenesis by carotenoids as oxygen radical quenchers', *Ann NY Acad Sci*, **534**, 584–96.

SANZ C, VELAYOS A, ÁLVAREZ MI, BENITO EP and ESLAVA AP (2011), 'Functional analysis of the *Phycomyces carRA* gene encoding the enzymes phytoene synthase and lycopene cyclase', *PlosOne*, **6**, e23102.

SARADA R, BHATTACHARYA S, BHATTACHARYA S and RAVISHANKAR GA (2002a), 'A response surface approach for the production of natural pigment astaxanthin from green alga, *Haematococcus pluvialis*: effect of sodium acetate, culture age, and sodium chloride', *Food Biotechnol*, **16**, 107–20.

SARADA R, TRIPATHI U and RAVISHANKAR GA (2002b), 'Influence of stress on astaxanthin production in *Haematococcus pluvialis* grown under different culture conditions', *Process Biochem*, **37**, 623–7.

SCHMIDT I, SCHEWE H, GASSEL S, JIN C, BUCKINGHAM J, HÜMBELIN M, SANDMAN G and SCHRADER J (2011), 'Biotechnological production of astaxanthin with *Phaffia rhodozyma/Xanthophyllomyces dendrorhous*', *Appl Microbiol Biotechnol*, **89**, 555–71.

SCHMIDT-DANNERT C, UMENO D and ARNOLD FH (2000), 'Molecular breeding of carotenoid biosynthetic pathways', *Nat Biotechnol*, **18**, 750–3.

SCHROEDER WA and JOHNSON EA (1995a), 'Singlet oxygen and peroxyl radicals regulate carotenoid biosynthesis in *Phaffia-rhodozyma*', *J Biol Chem*, **270**, 31, 18374–9.

SCHROEDER WA and JOHNSON EA (1995b), 'Carotenoids protect *Phaffia rhodozyma* against singlet oxygen damage', *J Ind Microbiol*, **14**, 6, 502–7.

SHI XM, LIU HJ, ZHANG XW and CHEN F (1999), 'Production of biomass and lutein by *Chlorella protothecoides* at various glucose concentrations in heterotrophic cultures', *Proc Biochem*, **34**, 341–7.

SHI XM, JIANG Y and CHEN F (2002), 'High-yield production of lutein by the green microalga *Chlorella protothecoides* in heterotrophic fed-batch culture', *Biotechnol Progr*, **18**, 723–7.

STAHL W and SIES H (2002), 'Carotenoids and protection against solar UV radiation', *Skin Pharmacol Appl Skin Physiol*, **15**, 291–6.

STEINBRENNER J and LINDEN H (2001), 'Regulation of two carotenoid biosynthesis genes coding for phytoene synthase and carotenoid hydroxylase during stress-induced astaxanthin formation in the green alga *Haematococcus pluvialis*', *Plant Physiol*, **125**, 810–17.

STEINBRENNER J and LINDEN H (2003), 'Light induction of carotenoid biosynthesis genes in the green alga *Haematococcus pluvialis*: regulation by photosynthetic redox control', *Plant Mol Biol*, **52**, 2, 343–56.

SUN Y, YUAN QP and VRIESEKOOP F (2007), 'Effect of two ergosterol biosynthesis inhibitors on lycopene production by *Blakeslea trispora*', *Proc Biochem*, **42**, 1460–4.

TAKAICHI S (2011), 'Carotenoids in algae: distributions, biosyntheses and function', *Marine Drugs*, **9**, 1101–18.

TAKANO H, ASKER D, BEPPU T and UEDA K (2006), 'Genetic control for light-induced carotenoid production in non-phototrophic bacteria', *J Ind Microbiol Biotechol*, **33**, 88–93.

TAO L, YAO H and CHENG Q (2007), 'Genes from a *Dietzia* sp. for synthesis of C40 and C50 β-cyclic carotenoids', *Gene*, **386**, 90–7.

TERESHINA VM, MEMORSKAYA AS and FEOFILOVA EP (2010), 'Lipid composition of the mucoraceous fungus *Blakeslea trispora* under lycopene formation-stimulating conditions', *Microbiology*, **79**, 34–9.

TORNWALL ME, VIRTAMO J, KORHONEN PA, VIRTANEN MJ, TAYLOR PR, ALBANES D and HUTTUNEN JK (2004), 'Effect of α-tocopherol and β-carotene supplementation on

© Woodhead Publishing Limited, 2013

coronary heart disease during the 6-year post-trial follow–up in the ATBC study', *Eur Heart J*, **25**, 1171–8.

TRIPATHI U, SARADA R and RAVISHANKAR GA (2002), 'Effect of culture conditions on growth of green alga – *Haematococcus pluvialis* and astaxanthin production', *Acta Physiol Plant*, **24**, 323–9.

UKIBE K, HASHIDA K, YOSHIDA N and TAKAGI H (2009), 'Metabolic engineering of *Saccharomyces cerevisiae* for astaxanthin production and oxidative stress tolerance', *Appl Environ Microb*, **75**, 7205–11.

UMENO D and ARNOLD FH. (2004), 'Evolution of a pathway to novel long-chain carotenoids', *J Bacteriol*, **186**, 1531–6.

UMENO D, TOBIAS AV and ARNOLD FH. (2005), 'Diversifying carotenoid biosynthetic pathways by directed evolution', *Microbiol Mol Biol Rev*, **69**, 51–78.

VALDUGA E, TATSCH PO, TIGGERMANN L, TREICHEL H, TONIAZZO G, ZENI J, DI LUCCIO M and FURIGO A (2009), 'Carotenoids production: microorganisms as source of natural dyes', *Quimica Nova*, **32**, 2429–36.

VIDHYAVATHI R, VENKATACHALAM L, SARADA R and RAVISHANKAR GA (2008), 'Regulation of carotenoid biosynthetic genes expression and carotenoid accumulation in the green alga *Haematococcus pluvialis* under nutrient stress conditions', *J Exp Bot*, **59**, 1409–18.

VISSER H, VAN OOYEN AJ and VERDOES JC (2003), 'Metabolic engineering of the astaxanthin-biosynthetic pathway of *Xanthophyllomyces dendrorhous*', *FEMS Yeast Res*, **4**, 221–31.

WALTER MH and STRACK D (2011), 'Carotenoids and their cleavage products: biosynthesis and functions', *Nat Prod Rep*, **28**, 663–92.

WANG W and YU L (2009), 'Effects of Oxygen Supply on Growth and Carotenoids Accumulation by *Xanthophyllomyces dendrorhous*', *Zeit Naturforsch Section C-A J Biosc*, **64**, 853–8.

WANG X, WILLEN R and WADSTROM T (2000), 'Astaxanthin-rich algal meal and vitamin C inhibit *Helicobacter pylori* infection in BALB/cA mice', *Antimicrob Agents Chem*, **44**, 2452–7.

WANG W, LI J, WANG K, HUANG C, ZHANG Y and OLFIELD E (2010), 'Organometallic mechanism of action and inhibition of the 4Fe-4S isoprenoid biosynthesis protein GcpE (IspG)', *Proc Natl Acad Sci USA*, **107**, 11189–93.

WARD OP and SINGH A (2005), 'Omega-3/6 fatty acids: Alternative sources of production', *Process Biochem*, **40**, 3627–52.

WOZNIAK A, LOZANO C, BARAHONA S, NIKLITSCHEK M, MARCOLETA A, ALCAÍNO J, SEPÚLVEDA D, BAEZA M and CIFUENTES V (2011), 'Differential carotenoid production and gene expression in *Xanthophyllomyces dendrorhous* grown in a nonfermentable carbon source', *FEMS Yeast Res*, **11**, 252–62.

YAMANE Y, HIGASHIDA K, NAKASHIMADA Y, KAKIZONO T and NISHIO N (1997a), 'Influence of oxygen and glucose on primary metabolism and astaxanthin production by *Phaffia rhodozyma* in batch and fed-batch cultures: kinetic and stoichiometric analysis', *Appl Environ Microbiol*, **63**, 4471–8.

YAMANE Y, HIGASHIDA K, NAKASHIMADA Y, KAKIZONO T and NISHIO N (1997b), '*Astaxanthin* production by *Phaffia rhodozyma* enhanced in fed-batch culture with glucose and ethanol feeding', *Biotechnol Lett*, **19**, 1109–11.

YE RW, STEAD KJ, YAO H and HE H (2006), 'Mutational and functional analysis of the β-carotene ketolase involved in the production of canthaxanthin and astaxanthin', *Appl Environ Microbiol*, **72**, 5829–37.

YE ZW, JIANG JG and WU GH (2008), 'Biosynthesis and regulation of carotenoids in *Dunaliella*: progresses and prospects', *Biotechnol Adv*, **26**, 352–60.

YEN HW, SUN CH and MA TW (2011), 'The comparison of lutein production by *Scenesdesmus* sp. in the autotrophic and the mixotrophic cultivation', *Appl Biochem Biotech*, **164**, 353–61.

© Woodhead Publishing Limited, 2013

YOKOYAMA A, IZUMIDA H and MIKI W (1994), 'Production of astaxanthin by the marine bacterium *Agrobacterium aurantiacum*', *Biosci Biotechnol Biochem*, **58**, 1842–4.

YOON SH, PARK HM, KIM JE, LEE SH, CHOI MS, KIM JY, OH DK, KEASLING JD and KIM SW (2007a) 'Increased β-carotene production in recombinant *Escherichia coli* harboring an engineered isoprenoid precursor pathway with mevalonate addition', *Biotechnol Prog*, **23**, 599–605.

YOON SH, KIM JE, LEE SH, PARK HM, CHOI MS, KIM JY, LEE SH, SHIN YC, KEASLING JD and KIM SW (2007b) 'Engineering the lycopene synthetic pathway in E. coli by comparison of the carotenoid genes of *Pantoea agglomerans* and *Pantoea ananatis*', *Appl Microbiol Biot*, **74**, 131–9.

YUAN JP, PENG JA, YIN K and WANG JH (2011), 'Potential health-promoting effects of astaxanthin: A high-value carotenoid mostly from microalgae', *Mol Nutr Food Res*, **55**, 150–65.

ZENI J, COLET R, CENCE K, TIGGEMANN L, TONIAZZO G, CANSIAN RL, DI LUCCIO M, OLIVEIRA D and VALDUGA E (2011), 'Screening of microorganisms for production of carotenoids', *CyTA – J Food*, **9**, 160–6.

ZHANG BY, GENG YH, LI ZK, HU HJ and LI YG (2009), 'Production of astaxanthin from *Haematococcus* in open pond by two-stage growth one-step process', *Aquaculture*, **295**, 275–81.

© Woodhead Publishing Limited, 2013

10

Microbial production of flavonoids and terpenoids

H. Dvora and M. A. G. Koffas, Rensselaer Polytechnic Institute, USA

DOI: 10.1533/9780857093547.2.234

Abstract: Flavonoids and terpenoids are two classes of natural products that are currently used as food ingredients, natural food colorants, nutraceuticals and pharmaceuticals. The chapter provides an overview of these two classes of compounds, their biosynthetic pathways, significance in human health and the importance of microbial production. It then discusses strategies for optimizing production in microorganisms, including host selection, redirection of carbon flux, protein engineering and production of non-natural analogs.

Key words: flavonoids, metabolic engineering microorganisms, novel and non-natural structures, protein engineering, terpenoids.

10.1 Introduction

Flavonoids and terpenoids are two classes of natural products that are currently used as food ingredients, natural food colorants and nutraceuticals. They are also under extensive investigation for prevention and treatment of many chronic diseases including obesity, diabetes and cancer, with several terpenoids and some polyphenols already marketed as drugs or in clinical trials. Since the biosynthetic pathways of flavonoids and terpenoids are well characterized, the cDNA of pathway proteins can be used to engineer recombinant strains of microorganisms that express the pathway enzymes. Such recombinant microorganisms can then be used in precursor-feeding experiments where chemically simple precursors are converted to derivatives of these natural products. In some cases, the intracellular metabolites themselves may serve as precursors. The products are typically excreted into the liquid culture, making purification a straightforward process.

In this chapter, we begin with an overview of flavonoids and terpenoids. In this section, we discuss the associated health benefits for which they are being used or studied, introducing the motivation for cheap, high-yield

© Woodhead Publishing Limited, 2013

production of these natural products. The relevant biosynthetic pathways for each of the classes are also reviewed. In the second section, we discuss various strategies that are used to improve natural product production in microorganisms especially in common platforms such as *Escherichia coli* and *Saccharomyces cerevisiae*. Once a platform is selected, the first objective is to direct carbon flux toward bioavailability of co-factors and precursors necessary for heterologous production in the microorganism. We review approaches that have been used to accomplish this. Next, we discuss approaches and applications for protein engineering applied in the area of polyphenol and terpenoid biosynthesis. Finally, we review techniques to diversify the structures and create non-natural analogs of flavonoids and terpenoids. The chapter concludes with a brief discussion of future trends in the field.

10.2   Overview of flavonoids and terpenoids

Flavonoids and terpenoids are two important classes of natural products. Flavonoids are found primarily in plants, while terpenoids can be found in all classes of organisms. With diverse and complex structures, these secondary metabolites possess interesting properties including pigmentation, flavor and/or fragrance that lend them great industrial value. This is especially true with growing public demand for natural ingredients in the food and cosmetic industry (Aburjai and Natsheh, 2003; Cabrita *et al.*, 2000; Chandi and Gill, 2011). Furthermore, these compounds have significant associated health benefits such that they have drawn extensive interest for application in the nutraceutical and pharmaceutical industries (Yao *et al.*, 2004; Iriti *et al.*, 2010; Harvey, 2008).

10.2.1   Health-promoting characteristics of flavonoids and terpenoids

Owing to their several health-promoting characteristics, natural products have long played a significant role in drug discovery and development. It comes as no surprise that at least half new drugs produced in the past 20 years are natural products or derivatives thereof (Newman and Cragg, 2007; Harvey, 2008). Furthermore, almost half of all drugs that are produced completely synthetically are natural product mimics and/or contain a pharmacophore derived from a natural product. Among natural products, flavonoids and terpenoids draw considerable interest for their potential as drug targets. Their abundance in fruit and vegetables is one of the reasons these food groups are so strongly touted as the key to a healthy diet.

The potent antioxidant activity of polyphenols such as flavonoids is arguably their most important property related to human health. Reactive oxygen species are important for the normal function of the human body. However, an abundance of these species can cause adverse effects that

© Woodhead Publishing Limited, 2013

could lead to cardiovascular disease and cancer (Allen and Tresini, 2000). The antioxidant effect of polyphenols has been studied extensively both *in vitro* and *in vivo* (Crozier *et al.*, 2009; Khan *et al.*, 2008; Ramirez-Tortosa *et al.*, 2001; Tabart *et al.*, 2009; Ohkatsu *et al.*, 2010). In one example, scientists studied the antioxidant potency of anthocyanins (a sub-class of the flavonoid family of molecules) *in vivo* using vitamin E-deficient rats. Vitamin E deficiency leads to oxidative damage susceptibility. The rats were fed with purified anthocyanins that were extracted from *Abies koreana*. This led to decreased concentrations of hydroperoxides and 8-oxodeoxyguanosine in the livers, indicating reduced lipid peroxidation and DNA damage that had been caused by the vitamin E deficiency (Ramirez-Tortosa *et al.*, 2001). Additionally, a recent report compared several tests of antioxidant capacity and recommended a standardized weighted average of these tests because of variability of results between the different tests. Using the weighted method, the flavonoids gallocatechin, flavan-3-ols, anthocyanins and flavonols displayed the highest antioxidant activity in the contents of different fruit juices (Tabart *et al.*, 2009).

Flavonoids have also been studied for activity against type II diabetes. An *in vitro* study determined the effect of the flavan-3-ols (+)-catechin and (+)-afzelechin on glucose-induced insulin secretion of pancreatic β-cells (Chemler *et al.*, 2007a). The results demonstrated that these flavonoids have the capacity to modulate insulin secretion, an important property for diabetes treatment. In a series of studies performed by Matsui and co-workers, the potential of anthocyanins to act as antidiabetic agents was again investigated. The studies focused on the capacity of anthocyanins to inhibit rat intestinal α-glucosidase. This is important because a slower release of dietary sugars to the bloodstream is more manageable for diabetic patients. In the first part of a two-part study reported in 2001, the authors showed that plant extracts of anthocyanins inhibited α-glucosidase activity against maltose (Matsui *et al.*, 2001a). Inhibition improved when the α-glucosidase was immobilized to mimic the natural membrane-bound state of the enzyme. The second part of the study confirmed that the α-glucosidase inhibition was due to the anthocyanins and not other compounds in the extracts. The most active compounds were acylated anthocyanins (Matsui *et al.*, 2001b). The following year, the group sought to demonstrate the *in vivo* effects of anthocyanins on blood glucose levels. They found that a single dose of anthocyanin extract reduced the rate of increase of the blood glucose level in rats following consumption of carbohydrates (Matsui *et al.*, 2002). These studies demonstrate the need to investigate these compounds further for treatment of diabetes.

Flavonoids are also involved in cancer prevention, probably owing to the additive effect of multiple cancer-fighting mechanisms. These mechanisms include antimutagenic activity, inhibition of oxidative DNA damage, inhibition of carcinogen activation, induction of apoptosis and antiangiogenesis (Duthie, 2007). In a recent high-throughput screening campaign for

© Woodhead Publishing Limited, 2013

inhibitors of a target enzyme in anticancer drug discovery, two β-hydroxychalcone derivatives and one flavanone were identified as strong inhibitors (Buchanan *et al.*, 2008). This study and the others mentioned here are only a small sampling of the vast research on flavonoid preventative effects against an array of chronic diseases and their potential as therapeutics (Cos *et al.*, 2004; He and Giusti, 2010; Khan and Mukhtar, 2007; Middleton *et al.*, 2000).

With even greater structural diversity than flavonoids, terpenoids have long been significant as pharmaceutical drug targets as well as nutraceuticals. These, too, have strong antioxidant, anti-inflammatory and anticancer activities, among others.

Inflammation is an immune response to tissue injury. However, it provides a favorable environment for tumor growth (Coussens and Werb, 2002). This emphasizes the importance of the anti-inflammatory activity of natural products such as terpenoids. One study looked specifically at linalool, a monoterpene found in essential oils and plants used in traditional (folklore) medicine. Researchers used the acetic acid-induced writhing model in mice and found that (-)-linalool effectively relieved the pain (Peana *et al.*, 2003). Since they also found that the motility of the mice improved with increased doses, they were certain that the pain relief was due to an anti-inflammatory effect and not a sedative effect, as had been previously suggested. In another study, the triterpene α-amyrin was studied for anti-inflammatory response to models of skin inflammation in mice. It was found to inhibit ear oedema (measured as ear thickness) and the migration of polymorphonuclear leukocytes in a dose-dependent manner, demonstrating its anti-inflammatory response (Otuki *et al.*, 2005).

Terpenoids are also important in cancer treatment and prevention. Paclitaxol is a diterpene that is currently used in cancer chemotherapy. It acts as a mitotic inhibitor by disrupting the disassembly of microtubules and thus inhibiting cell division in tumors (Rowinsky and Donehower, 1995). Several monoterpenes are also known to have anticancer activity. Among these is D-limonene, which was found to inhibit tumor progression of various carcinomas in animals, including mammary, kidney, skin and lung cancer (Wagner and Elmadfa, 2003). Squalene is yet another terpenoid demonstrating applicability in cancer treatment. Although it does not have strong anti-tumor effects, it improves the cancer-drug potency and has many promising qualities for drug delivery (Reddy and Couvreur, 2009).

Another important terpenoid that is currently used as a pharmaceutical is artemisinin. This sesquiterpene lactone is an antimalarial drug that is even effective against the drug-resistant *Plasmodium falciparum* strain. This terpenoid and its derivatives are fiercely sought after to control widespread malaria in Africa (Withers and Keasling, 2007; Enserink, 2005). The value of terpenoids for human health is immense and new studies and applications continue to emerge (Chandi and Gill, 2011; Wagner and Elmadfa, 2003; Oldfield, 2010). The tragedy is that in many cases, the amounts that

© Woodhead Publishing Limited, 2013

can be obtained from plant extracts are so low that the source plant cannot be grown fast enough for drug discovery or for treatment development, forcing patients to go without treatment (Enserink, 2005).

10.2.2   Significance of microbial production

Because of the diverse and potent health effects of natural products such as flavonoids and terpenoids, it is extremely important to find a time- and cost-efficient method to produce them in meaningful quantities. Currently, they are primarily obtained by plant extraction, but this method has several limitations. Since natural products are secondary metabolites, they are not essential for the growth of an organism. Therefore their concentrations are dependent on environmental conditions and can be very low. Large amounts of plant matter are then necessary in order to obtain usable amounts of product. Furthermore, plant extracts yield mixtures of compounds whose separation is often difficult and expensive. The mixture composition also varies from extraction to extraction. When used for research purposes, this may lead the researchers to overlook potentially potent compounds if they are only found in trace concentrations within the mixture. These inherent problems call for alternate strategies to obtain flavonoids and terpenoids.

One method that can be used to obtain pure samples of a compound of interest is chemical synthesis. The chemical synthesis of flavonoids and terpenoids has been extensively investigated (Banwell *et al.*, 2011; Hoeffler *et al.*, 2000; Kajiwara *et al.*, 1997; Khan *et al.*, 2010; Wang *et al.*, 2006). While the synthesis of some compounds is straightforward on a small scale, it relies on expensive precursors and toxic material; as a result, its scale-up remains an engineering conundrum. Furthermore, the organic synthesis of more elaborate structures, such as glycosylated compounds, requires several steps. As a result, production yields are low. Finally, the possible chemistry that can be generated is limited by the available methods. For example, specific hydroxylations and glycosylations are either not possible at all or only feasible after several steps, again resulting in low yields. Therefore, chemical synthesis does not alleviate the cost and time requirements of plant extraction.

Production in recombinant microorganisms provides an attractive alternative to plant extraction and chemical synthesis. Molecular biology techniques and fermentation of microorganisms are efficient and environmentally friendly. Microbes can be engineered to produce complex structures from simple, inexpensive precursors. Side products are usually minimal, especially when compared with plant extracts, such that the purification process is simple. With recent elucidation of primary pathway enzymes, the pathways can be grafted into microbial hosts for production of even complicated structures. This approach also yields biologically relevant racemic purity. As metabolic engineering strategies continue to emerge, production yields continue to improve.

© Woodhead Publishing Limited, 2013

10.2.3   Structures and biosynthetic pathways

Flavonoids

Flavonoids are a class of natural products found in plants, where they function as antimicrobial agents, colorants to attract pollinators, UV protectors and signaling molecules. As an essential part of the human diet, they have been found to be strong antioxidants with several health-promoting characteristics involved in prevention of chronic diseases including cancer, obesity and diabetes. These effects have led flavonoids to come under wide investigation as possible drug targets. Flavonoids are commercially important as nutraceuticals and natural food colorants and ingredients.

With over 9000 different structures discovered to date, flavonoids make up one of the most diverse groups of natural products. They are characterized by a 15-carbon skeleton (Figure 10.1, boxed structure), made up of two phenolic rings (A and B) connected by three carbons arranged either linearly or as a pyran or pyrone ring (C). Depending on the skeleton, flavonoids are classified into sub-classes including chalcones, flavones, flavonols, flavandiols and anthocyanins. The structural diversity stems from endless possible combinations of functional groups attached to the compound skeleton. These include hydroxylations, methylations and glycosylations, among others.

Flavonoids are made via the phenylpropanoid pathway, which has been elucidated to a large degree. The pathway begins with the amino acid phenylalanine, which is deaminated to cinnamic acid by phenylalanine ammonia lyase (PAL). The P450 monooxygenase cinnamate-4-hydroxylase (C4H) oxidizes cinnamic acid to 4-coumaric acid. This carboxylic acid is activated by the addition of a coenzyme A (CoA) unit, which is catalyzed by 4-coumarate:CoA ligase (4CL), yielding 4-coumaroyl-CoA. The sequential addition of three malonyl-CoA molecules by chalcone synthase (CHS) commits the resulting chalcone to the flavonoid biosynthetic pathway (Winkel-Shirley, 2001). Chalcone isomerase (CHI) isomerizes chalcones selectively to (2*S*)-flavanones, which are then hydroxylated by flavanone 3β-hydroxylase (FHT) at the 3-carbon position to give dihydroflavonols. These are reduced by dihydroflavonol 4-reductase (DFR) at the 4-carbon position, yielding the unstable leucoanthocyanidins. Leucoanthocyanidin reductase (LAR) catalyzes the subsequent reduction to flavan-3-ols (also called catechins). Both the leucoanthocyanidins and the flavan-3-ols are possible substrates for anthocyanidin synthase (ANS), which catalyzes the reaction to anthocyanidins. Finally, UDP-glucose:flavonoid 3-*O*-glucosyltransferase (3GT) catalyzes the glycosylation at the 3-carbon, yielding anthocyanins (Dixon and Steele, 1999; Winkel-Shirley, 2001; Yan *et al.*, 2008). Figure 10.1 depicts the main biosynthetic pathway. Additional enzymes can catalyze addition of functional groups or manipulation of the skeleton to lend structural diversity or related structures including isoflavonoids, condensed tannins, aurones and stilbenes. The reader is referred to excellent reviews for more information on the biosynthetic pathway and

© Woodhead Publishing Limited, 2013

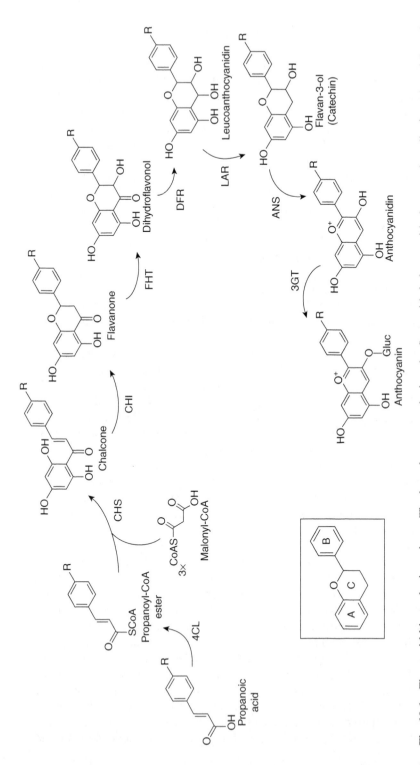

Fig. 10.1 Flavonoid biosynthetic pathway. The boxed structure depicts the flavonoid skeleton, with the two phenolic rings, A and B, and a connecting C ring. The simplified biosynthetic pathway from propanoic acid is also shown. 4CL, 4-coumarate:coenzyme A ligase; CHS, chalcone synthase; CHI, chalcone isomerase; FHT, flavanone 3β-hydroxylase; DFR, dihydroflavonol 4-reductase; LAR, leucoanthocyanidin reductase; ANS, anthocyanidin synthase; 3GT, UDP-glucose:flavonoid 3-O-glucosyltransferase.

© Woodhead Publishing Limited, 2013

structural diversity of flavonoids (Veitch and Grayer, 2011; Winkel-Shirley, 1999).

Terpenoids

Terpenoids, also called isoprenoids, make up the most diverse group of natural products, with over 30,000 different structures discovered in nature (Julsing *et al.*, 2006). They are found in all classes of living things, with a wide range of biological activities. In plants they function for growth, anti-microbial protection and to attract pollinators with their color and fragrance. Their color, fragrance and flavor also make terpenoids important for commercial use in food. Terpenoids have important health benefits and are sold as nutraceuticals, with several compounds already being exploited as drugs.

All terpenoids are made up of the 5-carbon building blocks isoprenes. These are produced through two different biosynthetic pathways: the mevalonate pathway (MVA) found in all eukaryotes and in some gram-positive prokaryotes and the non-mevalonate, or deoxyxylulose 5-phosphate (DXP) pathway, found in prokaryotes (Withers and Keasling, 2007). The biosynthetic pathways are depicted in Fig. 10.2.

The MVA pathway starts with two molecules of acetyl-CoA that are combined by acetoacetyl-CoA thiolase (AACT) to obtain acetoacetyl-CoA. The addition of another acetyl-CoA unit by 3-hydroxy-3-methylglutaryl-CoA synthase (HMGS) yields 3-hydroxy-3-methylglutaryl-CoA (HMG-CoA), which is reduced by HMG-CoA reductase (HMGR) to yield mevalonate. Two phosphate groups are added to the mevalonate sequentially, first by mevalonate kinase (MK), and then by phosphomevalonate kinase (PMK), leading to mevalonate 5-diphosphate. Finally, mevalonate 5-diphosphate decarboxylase (MPD) catalyzes the decarboxylation to isopentenyl diphosphate (IPP), which may also be converted to its isomer dimethylallyl diphosphate (DMPP) by IPP isomerase (IPI) (Withers and Keasling, 2007; Lange *et al.*, 2000; Eisenreich *et al.*, 2004).

The DXP pathway begins with the addition of pyruvate to glyceraldehyde 3-phosphate to yield 1-deoxy-D-xylulose-5-phosphate (DXP), catalyzed by DXP synthase (DXS). DXP is reduced to 2-*C*-methyl-D-erythritol 4-phosphate (MEP) by DXP reductase (DXR), then converted to 4-diphosphocytidyl-2-*C*-methyl-D-erythritol (CDP-ME) by CDP-ME synthase (CMS). A phosphate group addition by CDP-ME kinase (CMK) yields 2-diphosphocytidyl-2-*C*-methyl-D-erythritol 2-phosphate (CDP-ME2P). A cyclization gives 2-*C*-methyl-D-erythritol-2,4-cyclodiphosphate (MECDP), catalyzed by MECDP synthase (MCS). Next, 1-hydroxy-2-methyl-2-(*E*)-butenyl 4-diphosphate (HMBDP) is formed by HMBDP synthase (HDS). HMBDP is finally reduced to either IPP or DMPP by HMBDP reductase (HDR). IPP and DMPP are sequentially condensed by prenyl transferases to give the key intermediates geranyl diphosphate (GPP), farnesyl diphosphate (FPP) and geranylgeranyl diphosphate (GGPP), which are further

© Woodhead Publishing Limited, 2013

© Woodhead Publishing Limited, 2013

metabolized to give terpenoids. The reader is referred to more detailed reviews for additional information on the biosynthetic pathways of terpenoids (Eisenreich *et al.*, 2004; Lange *et al.*, 2000; Okada, 2011; Rohdich *et al.*, 2001).

With elucidation of biosynthetic pathways and continuing advances in genomics research and knowledge, microbial production of flavonoids and terpenoids has been increasingly studied. Once the feasibility of the heterologous expression of the biosynthetic pathways and exogenous production in microorganisms was proven, two clear aims emerged. The first was to engineer strategies to improve production in order to obtain large enough quantities for commercial production. The second was to discover ways to produce novel and non-natural flavonoid and terpenoid analogs, controlling the structure where possible. This would allow structure design for specific applications such as drug targets or a desired pigment. Furthermore, some engineering techniques may lead to production of libraries of natural product analogs. Such libraries can be studied for various applications, especially with the increasing development of high-throughput screening technologies.

10.3   Current and emerging techniques in microbial production of flavonoids and terpenoids

10.3.1   Selecting a production platform

The first consideration in production of natural products in microorganisms is the choice of the host organism. Several factors affect this decision and

---

Fig. 10.2   Terpenoid biosynthetic pathway. The terpenoid building blocks IPP and DMPP are made through the mevalonate pathway, left, and the deoxyxylulose 5-phosphate pathway, right. IPI catalyzes the reversible isomerization from IPP to DMPP. These are sequentially condensed by prenyl transferases to important terpenoid precursors GPP, FPP and GGPP. 1, Acetyl-CoA; 2, Acetoacetyl-CoA; 3, 3-hydroxy-3-methylglutaryl-CoA (HMG-CoA); 4, Mevalonate; 5, Mevalonate-5-phosphate; 6, Mevalonate-5-diphosphate; 7, Pyruvate; 8, Glyceraldehyde-3-phosphate; 9, 1-deoxy-D-xylulose-5-phosphate (DXP); 10, 2-*C*-Methyl-D-erythritol-4-phosphate; 11, 4-Diphosphocytidyl-2-*C*-methyl-D-erythritol (CDP-ME); 12, 4-Diphosphocytidyl-2*C*-methyl-D-erythritol 2-phosphate; 13, 2*C*-Methyl-D-erythritol 2,4-cyclodiphosphate (MECDP); 14, 1-Hydroxy-2-methyl-2-(*E*)-butenyl 4-diphosphate (HMBDP); IPP, isopentenyl diphosphate; DMPP, dimethylallyl diphosphate; GPP, geranyl diphosphate; FPP, farnesyl diphosphate; GGPP, geranylgeranyl diphosphate; AACT, acetoacetyl-CoA thiolase; HMGS, HMG-CoA synthase; HMGR, HMG-CoA reductase; MK, mevalonate kinase; PMK, phosphomevalonate kinase; DXS, DXP synthase; DXR, DXP reductase; CMS, CDP-ME synthase; CMK, CDP-ME kinase; MCS, MECDP synthase; HDS, HMBDP synthase; HDR, HMBDP reductase; IPI, IPP isomerase.

© Woodhead Publishing Limited, 2013

244   Microbial production of food ingredients, enzymes and nutraceuticals

each host offers its own advantages and disadvantages. Here we discuss the bacterial host *E. coli* and yeast host *S. cerevisiae* as these are most commonly chosen in microbial engineering for production of flavonoids and terpenoids. These organisms have been well characterized and molecular biology tools and techniques for them have been developed and continue to emerge, making these attractive choices for hosts.

Escherichia coli

E. coli is the most widely studied microbial platform for production of flavonoids and terpenoids. It has several advantages over yeast as a heterologous host. First, it has a particularly fast growth rate, with doubling times as short as 20 minutes. This allows for rapid experimentation, as well as shorter fermentation times for production. *E. coli* is also an extremely well characterized organism and this knowledge, along with easy manipulation of the genome, allows continual development of molecular biology tools and techniques. This facilitates the ability to express functionally many heterologous enzymes, episomally overexpress existing genes of the genome and knock out genes as needed (Datsenko and Wanner, 2000), making *E. coli* a favorable platform for engineering.

These advantages are demonstrated by the successful expression of the main flavonoid pathway from phenylpropanoic acids to anthocyanins in a single *E. coli* strain (Leonard *et al.*, 2008). Since *E. coli* will readily take up flavonoid intermediates fed to the fermentation culture, higher production levels can be achieved by only expressing part of the pathway and feeding with precursors, as in the production of anthocyanins from catechins (Yan *et al.*, 2008). Likewise, while *E. coli* harbors the DXP pathway for terpenoid precursor production, the exogenous mevalonate pathway was successfully expressed to produce these precursors as well (Martin *et al.*, 2003). Metabolic engineering of the homologous *E. coli* DPX pathway was effective in producing isoprenoid precursors for heterologous carotenoid production (Farmer and Liao, 2001; Kajiwara *et al.*, 1997; Kim and Keasling, 2001), but using the nonnative mevalonate pathway from *S. cerevisiae* allowed better control since the metabolic factors were known (Martin *et al.*, 2003; Ma *et al.*, 2011).

Recently, a new technique was reported which incorporates the gene of interest into the chromosome (Tyo *et al.*, 2009). Traditionally, protein expression in *E. coli* relies on the use of plasmids, which are easy to manipulate, and insertion into the host, but also have drawbacks leading to reduced productivity over time. However, one group used recombination events to insert the gene of interest with an antibiotic resistance gene into the chromosome itself. The *recA* gene acted as an on/off switch. Once a high enough copy number was achieved, *recA* was deleted to halt the recombination events. At this point, there was no longer a need for antibiotic selective pressure because the gene would automatically propagate to the next generation with the chromosome. According to the authors, this led to more

© Woodhead Publishing Limited, 2013

stable, longer term propagation of the inserted pathway. When applied to production of lycopene, a 60% yield increase was achieved (Tyo *et al.*, 2009).

One major limitation of the *E. coli* platform is that functional expression of P450 enzymes is particularly difficult. The cytochrome P450 superfamily is one of the most important groups of enzymes for production of flavonoids, terpenoids and other natural products. Heterologous expression of P450 enzymes in *E. coli* is a challenge for two primary reasons. First, many of these enzymes are membrane bound, making them difficult to express functionally in a prokaryotic organism that lacks an endoplasmic reticulum. Also, in most cases, P450 enzymes require P450-reductase enzymes, which are found natively in yeast, but not in *E. coli*. In some cases, by studying the differences in the pathway of different organisms, the need for a P450 can be bypassed by using another enzyme with similar activity, or which renders the P450 step unnecessary (Hwang *et al.*, 2003; Leonard *et al.*, 2006a). There have also been reports of successful P450 monooxygenase expression where the membrane domain was truncated and a fusion protein was created with a P450 reductase (Ajikumar *et al.*, 2010; Kim *et al.*, 2009; Leonard *et al.*, 2006b). Other strategies have also been used, but successful expression can only be achieved empirically, requiring many additional engineering strategies and trials than expression of other enzymes (Gillam, 2008).

Saccharomyces cerevisiae
While *S. cerevisiae* is slower growing than bacteria, it is still considered a relatively fast-growing platform, with several advantages over bacteria for production of plant natural products. These advantages are primarily due to its being a eukaryote, thus closer to the native plant cell environment, with capability for post-translational modification and expression of membrane-bound proteins.

The first report of heterologous expression of a metabolic pathway in a eukaryotic host was the production of the terpenoids lycopene and β-carotene in *S. cerevisiae* (Yamano *et al.*, 1994). This work was a successful proof of feasibility, but the yields were very low. More recently, a group took advantage of the fact that *S. cerevisiae* naturally makes farnesyl pyrophosphate (FPP), the precursor to sesquiterpene, for its own sterol biosynthesis (Jackson *et al.*, 2003). Since control of this pathway is well understood, the authors were able to improve production and achieve higher titers than had been reported in *E. coli*. The yeast platform was again chosen for the FPP precursor for the production of the sesquiterpenes valencene, cubebol, and patchoulol, which are under investigation for their potential as anticancer drugs (Asadollahi *et al.*, 2008).

The flavonoid pathway has been fully expressed, producing flavanones, isoflavones, and flavonols from phenylalanine, with all enzymes expressed in a single strain, although the yields were quite low (Trantas *et al.*, 2009). However, to our knowledge, production of anthocyanins in this platform has not been reported. Recently, it was found that *S. cerevisiae* harbors a

© Woodhead Publishing Limited, 2013

glucosidase that hydrolyzes flavonoid glucosides (Schmidt *et al.*, 2011), which may hinder heterologous production of anthocyanin since it contains a glycosylation step. This challenge may be overcome by inactivating the glucosidases by mutations and/or gene knockouts (Schmidt *et al.*, 2011).

Since *S. cerevisiae* is eukaryotic and has its own native P450 reductases, the expression of the P450 superfamily is much easier than in bacteria and has been employed for production of both flavonoids and terpenoids. For example, when the P450 monooxygenase cinnamate 4-hydroxylase (C4H) was expressed in *S. cerevisiae* with the three subsequent enzymes of the upper flavonoid pathway, flavanone production was observed at much higher levels than had been previously achieved using an *E. coli* strain that was engineered to bypass the P450 step (Hwang *et al.*, 2003; Yan *et al.*, 2005). Flavanone production increased when the yeast P450-reductase (CPR1) was overexpressed, pointing to the importance of pairing the two enzymes at this rate-limiting step (Leonard *et al.*, 2005). This capability in yeast also proved essential for heterologous production of terpenoid artemisinic acid, the immediate precursor to the antimalarial drug artemisinin (Ro *et al.*, 2006). In addition to expressing amorphadiene synthase, the recombinant expression of a P450 monooxygenase and its redox partner were required for the oxidation of amorphadiene to artemisinic acid.

Finally, there are strategies and tools that researchers should consider whenever a heterologous host is used, regardless of the host organism. One is the selection of the source species of the gene for the enzyme that is to be expressed recombinantly. Even with a high level of homology for a particular enzyme among several species, in many cases one will show much higher activity than the others in the heterologous host system. The codon usage is another important factor to consider (Welch *et al.*, 2009). These days, the codon usage for many host organisms, and certainly for *E. coli* and *S. cerevisiae*, has been determined, such that one may simply synthesize a particular gene with its codons optimized for the particular host organism. Such tools are important for the goal of soluble and functionally active expression of the recombinant proteins.

Since much more work has been done to produce flavonoids and terpenoids in *E. coli* than in yeast, it appears to be the current platform of choice despite the advantages of the eukaryotic yeast. This is likely to be due to the advantage of a better selection of molecular biology tools and a much faster growth rate. These factors lead to easier scale-up for commercial production.

10.3.2   Carbon flux manipulation for heterologous production pathways

Once the pathway is successfully expressed in the heterologous microbial host, the researcher should study the effects of the graft on the metabolism of the host and consider how the metabolism can be perturbed to give higher product yield while maintaining the host's growth rate.

© Woodhead Publishing Limited, 2013

Rational design: flavonoids

One of the most important co-factors for flavonoid biosynthesis is malonyl-CoA. Each molecule of chalcone (and subsequently, any flavonoid produced from the chalcone) requires three molecules of malonyl-CoA, which are added stepwise by chalcone synthase to commit the compound to the flavonoid pathway. To determine whether malonyl-CoA was, in fact, a limiting reactant, Miyahisa and co-workers overexpressed the enzyme acetyl-CoA carboxylase (ACC), which converts acetyl-CoA to malonyl-CoA in the fatty acid biosynthesis pathway (Miyahisa *et al.*, 2005). They added a plasmid containing the gene for ACC to an *E. coli* strain harboring a plasmid with the four genes for production of flavanones from amino acids. A three-fold increase in production of naringenin from tyrosine and a four-fold increase in pinocembrin production from phenylalanine were observed (Miyahisa *et al.*, 2005). This demonstrated the importance of malonyl-CoA bioavailability for flavonoid production.

Another group extended these efforts to increase malonyl-CoA bioavailability further. They found that the four-sub-unit ACC from *Photorhabdus luminescens* gave better enhancement of flavanone production than the two-sub-unit ACC from *Corynbacterium glutamicum* used previously (Leonard *et al.*, 2007). Neither group used the *E. coli* form of ACC for overexpression because it is known to be feedback inhibited by acyl–acyl carrier proteins (Davis and Cronan, 2001). Since ACC is a biotin-dependent enzyme, the gene for biotin ligase was co-overexpressed by ACC. The authors achieved a yet stronger drive of carbon flux for malonyl-CoA with the overexpression of acetate assimilation pathways by *ackA* and *pta* overexpression or *acs* overexpression in addition to ACC. The acetate assimilation pathways improved availability of acetyl-CoA for conversion to malonyl-CoA by ACC. This led to a marked improvement in flavanone production of up to 14 times higher than strains without the overexpressions (Leonard *et al.*, 2007).

Another report presented two alternate approaches to increase the pool of malonyl-CoA in the engineered *E. coli* (Leonard *et al.*, 2008). The first was to introduce the genes *matB* and *matC* from *R. trifolii* into the *E. coli* strain engineered for production of flavanones. These genes encode the malonate assimilation pathway, allowing conversion of malonate directly to malonyl-CoA, as opposed to the native conversion from glucose, which requires several steps. This approach led to an over 250% increase in flavanone production. Next, the authors attenuated the fatty acid biosynthesis pathway, which competes with the grafted flavonoid pathway for malonyl-CoA. In order to achieve this, they added cerulenin to inhibit fatty acid biosynthesis. This led to a more than 900% increase in flavanone levels. Flavanone production improvement was only observed with cerulenin concentrations up to 1 mM. Above this concentration, the growth continued to slow, but flavanone yield did not improve (Leonard *et al.*, 2008). Zha *et al.* (2009) combined several strategies into one strain to improve malonyl-CoA

© Woodhead Publishing Limited, 2013

levels in *E. coli*. The strategies included overexpressing ACC, increasing acetyl-CoA availability with overexpression of *acs* and deletion of *ackA-pta* and *adhE* and inactivation of fatty-acid synthesis by overexpression of β-ketoacyl-ACP synthase II (*fabF*) just before stationary growth is achieved. Combining the first two strategies led to a 15-fold increase in intracellular malonyl-CoA. However, while the FabF overexpression alone yielded a four-fold increase in malonyl-CoA, its overexpression in combination with the other strategies did not contribute to additional product yield (Zha *et al.*, 2009).

UDP-glucose has been identified as another important co-factor, although more relevant in the lower pathway for production of anthocyanins and glycosylated flavonoids. In order to improve its bioavailability, researchers overexpressed the genes *pgm* and *galU* for the enzymes that convert glucose-6-phosphate to UDP-glucose. They observed a 60% increase in anthocyanin yield in an *E. coli* strain expressing the pathway proteins from either flavanones or (+)-catechins. In a subsequent study, the nucleic acid biosynthesis pathway, starting from orotic acid, was targeted (Leonard *et al.*, 2008). The authors found that overexpression of *ndk* improved anthocyanin production when fermentation was supplemented with 0.1 mM orotic acid. To improve UDP-glucose bioavailability further, competing pathways were inhibited. The *E. coli* strain used was already lacking the genes *galE* and *galT* that convert UDP-glucose to UDP-galactose, but the gene *udg* for UDP-glucose 6-dehydrogenase, which converts UDP-glucose to UDP-gluconorate was still active. When the authors deleted it, they observed additional improvement to anthocyanin production, with overexpression of *ndk* and supplementation of orotic acid (Leonard *et al.*, 2008). In a separate study aimed at producing flavonoid glycosides in *S. cerevisiae*, researchers found that addition of orotic acid improved the yield of glycosides produced, probably due to increased production of UTP for UDP-glucose availability (Werner *et al.*, 2010).

Rational design: terpenoids
Since terpenoid biosynthesis always requires the same precursors, IPP and DMPP, which are produced either through the mevalonate (MVA) or non-mevalonate (DXP) pathway, most research has focused on improving these pathways, whether they are homologous or not. Other important common precursors are geranyl diphosphate (GPP), farnesyl pyrophosphate (FPP) and geranylgeranyl diphosphate (GGPP).

For production of terpenoids in *E. coli*, the research initially focused on the DXP pathway for isoprenoid production and the central metabolism. One study looked at the interface of the central metabolism with the isoprenoid pathway and determination of the precursor balance problem in isoprenoid biosynthesis (Farmer and Liao, 2001). The authors grafted the lycopene biosynthesis pathway into *E. coli* in order to use lycopene production levels as a measure of the effects of their perturbations. Precursor

© Woodhead Publishing Limited, 2013

availability was modulated by overexpressing or inactivating certain enzymes to control the carbon flux between pyruvate and glucose-3-phosphate (G3P). They found that directing flux to G3P significantly increased lycopene production, and vice versa, indicating that this is an important metabolite that is not selectively channeled towards isoprenoid biosynthesis (Farmer and Liao, 2001). In another study, the steps of the DXP biosynthetic pathway were studied to find the point of control of the flux (Kim and Keasling, 2001). More specifically, they studied the effects of overexpressing DXP synthase and DXP reductoisomerase, which catalyze the first two reactions that produce IPP. Using lycopene production as an indication of high IPP precursor bioavailability, they found that while DXP reductoisomerase affected production, DXP synthase controlled the flux to IPP. However, these findings also depended on choice of promoter and plasmid copy number, indicating that these factors significantly affect production as well (Kim and Keasling, 2001).

In order to avoid unknown regulation challenges posed when using the native *E. coli* isoprenoid biosynthetic pathway, Martin *et al.* (2003) grafted the mevalonate pathway into *E. coli*. This became a new platform for work to improve carbon flux toward heterologous terpenoid production in *E. coli*. First, it was discovered that mevalonate pathway intermediates limited the flux to IPP. Through metabolite analysis, the accumulation of HMG-CoA in particular was found to slow the culture growth (Pitera *et al.*, 2007). In order to balance the pathway, additional HMG-CoA reductase was expressed in a truncated form. This restored the growth rate and increased mevalonate levels (Pitera *et al.*, 2007). Further research determined that the mechanism of inhibition by HMG-CoA was membrane stress caused by inhibited fatty acid biosynthesis (Kizer *et al.*, 2008). This conclusion was reached using DNA microarray analysis and targeted metabolite profiling, using a strain with inactivated proteins as a control. It was determined that supplementation with palmitic acid and oleic acid ameliorated the cytotoxic effects (Kizer *et al.*, 2008). Recently, another approach was attempted to overcome the toxic accumulation of HMG-CoA (Ma *et al.*, 2011). The authors tested five variants of HMG-CoA reductase and found that using the variant from *Delftia acidovorans* yielded the best production of amorphadiene. Furthermore, as this enzyme is NADH-dependent, they increased NADH availability by expressing NAD+-dependent formate dehydrogenase and supplemented with formate. This yielded a 120% improvement in amorphadiene production over the original strain (Ma *et al.*, 2011).

There has also been work on *S. cerevisiae* to improve bioavailability. Yeast naturally produces FPP as the common precursor to all yeast isoprenoids and for sterol biosynthesis. One study overexpressed a truncated version of HMG-CoA reductase to yield a four-fold increase in sesquiterpene production, which relies on FPP availability (Jackson *et al.*, 2003). Another study sought to increase the amount acetyl-CoA available for use

© Woodhead Publishing Limited, 2013

in the mevalonate pathway for production of the sesquiterpene amorphadiene (Shiba *et al.*, 2007). The authors designed a bypass of pyruvate dehydrogenase by overproducing acetaldehyde dehydrogenase while expressing a heterologous acetyl-CoA synthase variant. They achieved a considerable increase in amorphadiene production and noted that this improved platform can be applied to the production of any other terpenoid requiring high levels of acetyl-CoA or FPP (Shiba *et al.*, 2007). Another study diverted carbon flux from sterol biosynthesis by repressing the gene encoding squalene synthase, which converts FPP to squalene for sterol biosynthesis in *S. cerevisiae*. This was accomplished by replacing the native promoter of squalene synthase with the repressible MET3 promoter. The authors observed that as a result of the repression, levels of endogenous squalene and ergosterol decreased while heterologous amorphadiene production increased five-fold (Paradise *et al.*, 2008).

Some terpenoid structures have been identified as particularly important and valuable in the pharmaceutical industry and have therefore drawn additional research into improving carbon flux beyond the common terpenoid precursor pathways. One important example is taxadiene production. Taxadiene is a precursor of the valuable cancer drug taxol. In order to optimize taxadiene production in *E. coli*, Ajikumar *et al.* (2010) attempted a multivariate approach. This approach divided taxadiene production into two modules: The 'upstream module' was the eight-gene native DXP pathway up to IPP production and the 'downstream module' included the two-step conversion of IPP to taxadiene through the heterologous proteins GGPP synthase and taxadiene synthase. Four of the upstream module genes were considered rate-limiting and their expression was modulated simultaneously with the two downstream genes in order to find the optimal balance for highest production. This global approach led to a significant increase in production, up to gram-quantities of taxadiene (Ajikumar *et al.*, 2010).

Computational design
Recently, total flux balance analysis has emerged as a method of predicting genetic modifications that could draw carbon from non-essential pathways towards the heterologous production pathway. Two such models were utilized to predict interventions for increased malonyl-CoA bioavailability. The cipher of evolutionary design (CiED) model couples the two goals of biomass accumulation and production. The resulting gene deletion predictions were tested experimentally, with the quadruple gene deletions implemented and added to overexpressions that were previously identified for improved malonyl-CoA bioavailability. The resulting strain was a significantly better flavanone producer (Fowler *et al.*, 2009). The OptForce model uses direct flux measurements of the wildtype strain to predict the minimal interventions required to obtain a target production level by comparing the maximum range of flux variability (Ranganathan *et al.*, 2010). When the

model was applied to malonyl-CoA levels in *E. coli* and the results were tested experimentally, a four-fold increase in intracellular malonyl-CoA was observed. The strain was used for flavanone production and yielded the highest production of naringenin ever reported (Xu *et al.*, 2011).

This systems-based approach was also applied to improve the bioavailability of NADPH, an important co-factor for reductases in many natural product pathways, including the flavonoid biosynthetic pathway (Chemler *et al.*, 2010). Based on results from the CiED model, the following three genes were deleted from *E. coli*: glucose-6-phosphate isomerase (*pgi*), phosphoenolpyruvate carboxylase (*ppc*) and phospholipase A (*pldA*). The first two genes had been individually identified as important knockouts for NADPH bioavailability in the past (Alper *et al.*, 2005; Chin *et al.*, 2009; Kabir and Shimizu, 2003), but the triple deletion strain yielded a four-fold increase in production of leucocyanidin and (+)-catechin from dihydroflavonol, demonstrating the significance of considering the global metabolic network.

Recently, a mathematical model predicting production levels in *S. cerevisiae* was reported (Varman *et al.*, 2011). The model inputs the number of steps in the heterologous biosynthetic pathway being introduced, the genetic modifications, the culturing conditions and the nutrient availability. Using this information, the program predicts the yield of the chemical production. The model was tested using data from recent publications of metabolic engineering efforts for heterologous production of terpenoids and other chemicals in *S. cerevisiae* and was found to be close to the experimental results. Therefore, this model could serve as a fast method to evaluate approaches, saving significant time and effort in constructing a strain with metabolic engineering strategies that may not work (Varman *et al.*, 2011).

10.3.3   Protein engineering

Protein engineering is another tool the researcher can use to affect production of the desired product. Proteins can be engineered to have higher specificity for improved activity, altered specificity to accept an altered substrate, or reduced specificity towards structure diversification. In some cases, the protein may be redesigned using rational engineering. For instance, if the active site is known, or if certain properties make the enzyme particularly difficult to express in high quantities (such as insolubility, or a membrane-binding segment when expressed in prokaryotes), the amino acid sequence can be modified rationally to improve the properties of the protein. In other cases, directed evolution of the enzyme can be performed for faster identification of improved enzymes. This strategy involves the creation of a library of randomly mutated forms of the target enzyme, followed by rapid screening, most commonly the production of a colored compound, for identification of the best form. While directed evolution is

© Woodhead Publishing Limited, 2013

a strong tool in protein engineering, it suffers from the limitation of requiring a high-throughput screening method to find effective mutants. Several studies have found combinatorial approaches that direct evolution without the need for a high-throughput selection process.

As discussed in earlier sections, the P450 superfamily of enzymes is a frequent target for rational engineering for functional expression in *E. coli* because they are particularly difficult to express in this platform. Many of these enzymes are membrane bound in their native organism and require a P450 reductase for activity. Active forms have been successfully expressed in *E. coli* for production of flavonoids and terpenoids by removing the membrane-binding segment and creating a protein fusion with a reductase (Ajikumar *et al.*, 2010; Kim *et al.*, 2009; Leonard *et al.*, 2006b). Sometimes it may be useful to change the specificity of a particular enzyme. In the case of *ispA*, the *E. coli* gene for FPP synthase, it was found that changing a particular tyrosine to a non-aromatic residue could change the specificity from FPP to GGPP. Since *E. coli* only harbors the DXP biosynthetic pathway, it does not yield high levels of GGPP, an important precursor for production of many different terpenoids. The expression of this altered enzyme in *E. coli* contributed to a six-fold increase in lycopene production (Reiling *et al.*, 2004). Wang *et al.* (2000) used directed evolution to improve the activity of GGPP synthase for terpenoid production in *E. coli*. By employing error-prone PCR and the staggered extension process, a library of randomly mutated genes was obtained. Since higher production of the carotenoid astaxanthin leads to deeper orange-colored colonies, it was used as a screening method for more active enzymes. Higher activity leads to higher production, which leads to a deeper color. Eight mutants were selected and the best strain improved lycopene production by 100% (Wang *et al.*, 2000).

Increasing the specificity of promiscuous enzymes for improved activity and product purity is another application of protein engineering. Yoshikuni *et al.* (2006) sought to find the residues with highest plasticity, or ability to change function when mutated, in order to achieve more specific activity in sesquiterpene synthase. Nineteen residues in the active site underwent saturation mutagenesis and the changes in the product distribution were profiled. Four residues were found to affect catalysis significantly. Mutations at these positions were recombined using an algorithm and the function was predicted based on the product distribution of each. By this method of divergent evolution, the authors developed several highly specific sesquiterpene synthases (Yoshikuni *et al.*, 2006).

Another study employed a combinatorial approach to protein engineering to alter levopimaradiene synthase (LPS) to a more specific, more productive enzyme for the production of levopimaradiene. Fifteen residues in the enzyme's binding pocket were mutated based on studies of paralogous LPS-type enzymes that had different functionalities. Once the mutations were made, both the product profile and the production levels were

© Woodhead Publishing Limited, 2013

analyzed. The two mutations that contributed to the greatest change in phenotype were then selected to undergo saturation mutagenesis and the production was analyzed again. The two most productive mutants were combined in one enzyme variant, which increased production 10-fold. The authors also targeted GGPP synthase and engineered an improved form, using lycopene production as a reporter for traditional directed evolution. Together, the two engineered proteins improved levopimaradiene production by almost 18-fold (Leonard *et al.*, 2010).

Yoshikuni *et al.* (2008) developed an adaptive evolution method for improving an enzyme's *in vivo* properties when expressed in a heterologous host. This method overcomes the dependence of directed evolution on a reporter because it can be applied even when there is no reporter protein or colored product. In order to improve sesquiterpene production in *E. coli*, the method was applied to improve the truncated HMG-CoA reductase (tHMGR) and the sesquiterpene synthase γ-humulene synthase (HUM). The approach was based on the knowledge that amino acids are more conserved, as enzymes are evolutionarily transferred from one organism to another if they are more essential to maintaining the *in vivo* properties of the enzyme. The authors studied the enzymes of the *E. coli* central metabolic network and found that glycine and proline were most conserved. They then ran a multiple sequence alignment of the target enzymes with the same enzymes in different species and calculated the probabilities of conservation for each glycine and proline at each residue position. Finally, the amino acid substitutions were made based on the results. These adjustments improved each of the two proteins, as evidenced by improved sesquiterpene production. Furthermore, the improvements were found to be additive when multiple mutations were made. When these adapted enzymes were expressed together in the same *E. coli* production strain, a 1000-fold increase in production was observed (Yoshikuni *et al.*, 2008).

Protein engineering approaches continue to emerge, with combinatorial methods developing to apply semi-randomized protocols while overcoming the requirement for a high-throughput screen of the traditional directed evolution approach. The methods depend on the motivation for an altered protein, that is, whether it is higher specificity, lower specificity, or altered specificity that is desired. These methods will become increasingly useful for development of non-natural analogs to natural products as well.

10.3.4   Producing non-natural derivatives of flavonoids and terpenoids

As the microbial platform carrying the flavonoid or terpenoid pathway has become established and continues to improve, there has been a push to develop tools for producing non-natural analogs of these structures. The vast interest in the potential of flavonoids and terpenoids for human health is driving researchers to tap into the full potential of the structures by either producing libraries of structures for high-throughput studies (Naesby *et al.*,

© Woodhead Publishing Limited, 2013

2009) or by developing tools for designing a particular structure as a target for study.

The enzymes of the flavonoid biosynthetic pathway have shown rather broad specificity in accepting non-natural substrates when expressed heterologously in microbials. For example, Chemler *et al.* (2007b) took advantage of the broad enzyme specificities to mutasynthesize non-natural flavonoid analogues in *S. cerevisiae*. The strain harbored a gene cluster of 4CL, CHS and CHI on a single plasmid. When cultures were fed with acrylic acid analogs, the corresponding flavanone analog was produced. Similarly, when a plasmid encoding the gene for FHT was added to the strain, the non-natural dihydroflavonol analogs were produced. However, some cinnamic acids were not accepted as substrates, suggesting some steric exclusions by 4CL (Chemler *et al.*, 2007b). Another report of a similar experiment in *S. cerevisiae* yielded a novel flavanone and three novel dihydrochalcones when fed with cinnamic acid analogs (Werner *et al.*, 2010). A similar approach was successfully attempted in *E. coli*. Using different combinations of several plasmids, each with a different stage of the biosynthetic pathway (substrate synthesis, polyketide synthesis and postpolyketide synthesis), the authors produced non-natural flavanones, flavones, flavonols, and the related compound stilbenes (Katsuyama *et al.*, 2007).

Schmidt-Dannert *et al.* (2000) introduced and demonstrated the concept of breeding to create new biosynthetic pathways. Since the host organism is not dependent on the heterologous pathway for survival, any and all options can be explored, including combining genes from different pathways to obtain novel structures and mixing genes to alter catalytic function. The authors demonstrated this concept with an example of production of novel carotenoids in *E. coli*. Key enzymes of the pathway were targeted for evolution by DNA shuffling, with the red color of lycopene production as a screening method, as was described for directed evolution. By directing enzyme activity at well-chosen steps in the pathway, novel carotenoids were produced in *E. coli*.

In contrast to the study of carefully selected branchpoints, another study reported a completely randomized approach for synthesizing novel structures using a yeast artificial chromosome (YAC) (Naesby *et al.*, 2009). Genes of the flavonol biosynthetic pathway were amplified and randomly ligated to form all different YACs with random variation of combinations of genes and different copy numbers of each gene on each YAC. These chromosomes were transformed into *S. cerevisiae* and flavonoid production was analyzed in randomly selected colonies. Some of the selected colonies produced the flavonols, while others accumulated intermediates or derivatives. When non-natural precursors were fed to a clone with a YAC containing the pathway only from 4CL (to prevent use of homologous amino acid precursors), the corresponding non-natural flavanones were produced. This concept can be applied to create randomized pathways and libraries of

© Woodhead Publishing Limited, 2013

non-natural analogs to natural products where the enzymes of the biosynthetic pathways are known.

There is still much room for diversification of structure using the microbial production platform. Diversification can be accomplished through a carefully directed process or by more randomized methods, as described above. In some ways, this may be dependent on the pathway. For instance, if researchers attempt to apply the YAC method to the terpenoid pathway, they still must start with the IPP building blocks. Therefore, while the YAC system may help optimize production of the upstream precursors, there is no diversification component in this portion of the pathway. An important application for non-natural analog production is the production of a particular structure of interest, which would be impossible or nearly impossible to produce by organic synthesis. For example, a particular non-natural structure may be deemed a likely drug target. Also, since flavonoids and terpenoids are used for natural flavors and food coloring, it may be interesting to create a particular compound with a desired characteristic. For example, it is known that anthocyanin pigment depends heavily on the functionalization of the B-ring (Tanaka *et al.*, 2008). Experimentation with non-natural functionality can lead to the ability to fine tune pigment and flavor for food production applications.

10.4   Future trends

There has been tremendous progress over the past two decades in microbial production of natural products, including flavonoids and terpenoids. With the infrastructure in place and feasibility proven, titers must now be improved for production on an industrial scale. The strategies described in this chapter provide the blueprints for continued improvement and diversification of production. Arguably the most important strategy is carbon flow for the pathway of interest. This is accomplished by increased precursor bioavailability and exploitation of regulatory mechanisms and removal of bottlenecks. Protein engineering has emerged as an indispensible tool and production of novel structures is a new and powerful application.

For terpenoids, there are two areas to focus on. The precursor pathways, both mevalonate and non-mevalonate, whether homologous or not, should be optimized for improved precursor bioavailability. Then, for each terpenoid, the downstream pathway should be optimized. The two parts should then be combined to balance the overall pathway through systematic approaches, as described above (Ajikumar *et al.*, 2010). For flavonoids there is still plenty of opportunity for production improvement of the lower pathway compounds. While there are many reports of metabolic engineering efforts for production of flavanones, there have not been many reports on improved catechin and anthocyanin production.

© Woodhead Publishing Limited, 2013

In developing additional methods using these tools, researchers should continue to use combinatorial approaches utilizing rational engineering, randomized evolution and computational approaches. Significant progress has been made by rational engineering of pathways and proteins, using the continued flow of available knowledge of genomics and proteomics and the relevant pathways. A randomized approach should then be used to identify additional productive perturbation that would not have been identified rationally. As additional high-throughput screens are developed, these randomized approaches will become increasingly useful. Finally, the power of a computational approach cannot be disputed. As the metabolic network of heterologous hosts improve with increased knowledge of the organisms, computer models will become more accurate. This should allow fast and cheap experimentation, with more predictions of productive metabolic engineering perturbations and product yields.

10.5   References

ABURJAI, T. and NATSHEH, F. M. (2003) 'Plants used in cosmetics'. *Phytotherapy Research*, **17**, 987–1000.

AJIKUMAR, P. K., XIAO, W. H., TYO, K. E. J., WANG, Y., SIMEON, F., LEONARD, E., MUCHA, O., PHON, T. H., PFEIFER, B. and STEPHANOPOULOS, G. (2010) 'Isoprenoid pathway optimization for taxol precursor overproduction in *Escherichia coli*'. *Science*, **330**, 70–4.

ALLEN, R. G. and TRESINI, M. (2000) 'Oxidative stress and gene regulation'. *Free Radical Biology and Medicine*, **28**, 463–99.

ALPER, H., MIYAOKU, K. and STEPHANOPOULOS, G. (2005) 'Construction of lycopene-overproducing *E coli* strains by combining systematic and combinatorial gene knockout targets'. *Nature Biotechnology*, **23**, 612–16.

ASADOLLAHI, M. A., MAURY, J., MOLLER, K., NIELSEN, K. F., SCHALK, M., CLARK, A. and NIELSEN, J. (2008) 'Production of plant sesquiterpenes in *Saccharomyces cerevisiae*: Effect of ERG9 repression on sesquiterpene biosynthesis'. *Biotechnology and Bioengineering*, **99**, 666–77.

BANWELL, M. G., LEHMANN, A. L., MENON, R. S. and WILLIS, A. C. (2011) 'New methods for the synthesis of certain alkaloids and terpenoids'. *Pure and Applied Chemistry*, **83**, 411–23.

BUCHANAN, M. S., CARROLL, A. R., FECHNER, G. A., BOYLE, A., SIMPSON, M., ADDEPALLI, R., AVERY, V. M., FORSTER, P. I., GUYMER, G. P., CHEUNG, T., CHEN, H. and QUINN, R. J. (2008) 'Small-molecule inhibitors of the cancer target, isoprenylcysteine carboxyl methyltransferase, from *Hovea parvicalyx*'. *Phytochemistry*, **69**, 1886–9.

CABRITA, L., FOSSEN, T. and ANDERSEN, O. M. (2000) 'Colour and stability of the six common anthocyanidin 3-glucosides in aqueous solutions'. *Food Chemistry*, **68**, 101–7.

CHANDI, G. K. and GILL, B. S. (2011) 'Production and characterization of microbial carotenoids as an alternative to synthetic colors: a review'. *International Journal of Food Properties*, **14**, 503–13.

CHEMLER, J. A., LOCK, L. T., KOFFAS, M. A. G. and TZANAKAKIS, E. S. (2007a) 'Standardized biosynthesis of flavan-3-ols with effects on pancreatic beta-cell insulin secretion'. *Applied Microbiology and Biotechnology*, **77**, 797–807.

© Woodhead Publishing Limited, 2013

CHEMLER, J. A., YAN, Y. J., LEONARD, E. and KOFFAS, M. A. G. (2007b) 'Combinatorial mutasynthesis of flavonoid analogues from acrylic acids in microorganisms'. *Organic Letters*, **9**, 1855–8.

CHEMLER, J. A., FOWLER, Z. L., MCHUGH, K. P. and KOFFAS, M. A. G. (2010) 'Improving NADPH availability for natural product biosynthesis in *Escherichia coli* by metabolic engineering'. *Metabolic Engineering*, **12**, 96–104.

CHIN, J. W., KHANKAL, R., MONROE, C. A., MARANAS, C. D. and CIRINO, P. C. (2009) 'Analysis of NADPH supply during xylitol production by engineered *Escherichia coli*'. *Biotechnology and Bioengineering*, **102**, 209–20.

COS, P., DE BRUYNE, T., HERMANS, N., APERS, S., VANDEN BERGHE, D. and VLIETINCK, A. J. (2004) 'Proanthocyanidins in health care: Current and new trends'. *Current Medicinal Chemistry*, **11**, 1345–59.

COUSSENS, L. M. and WERB, Z. (2002) 'Inflammation and cancer'. *Nature*, **420**, 860–7.

CROZIER, A., JAGANATH, I. B. and CLIFFORD, M. N. (2009) 'Dietary phenolics: chemistry, bioavailability and effects on health'. *Natural Product Reports*, **26**, 1001–43.

DATSENKO, K. A. and WANNER, B. L. (2000) 'One-step inactivation of chromosomal genes in *Escherichia coli* K-12 using PCR products'. *Proc Natl Acad Sci U S A*, **97**, 6640–5.

DAVIS, M. S. and CRONAN, J. E. (2001) 'Inhibition of *Escherichia coli* acetyl coenzyme A carboxylase by acyl-acyl carrier protein'. *Journal of Bacteriology*, **183**, 1499–503.

DIXON, R. A. and STEELE, C. L. (1999) 'Flavonoids and isoflavonoids – a gold mine for metabolic engineering'. *Trends in Plant Science*, **4**, 394–400.

DUTHIE, S. J. (2007) 'Berry phytochemicals, genomic stability and cancer: Evidence for chemoprotection at several stages in the carcinogenic process'. *Molecular Nutrition & Food Research*, **51**, 665–74.

EISENREICH, W., BACHER, A., ARIGONI, D. and ROHDICH, F. (2004) 'Biosynthesis of isoprenoids via the non-mevalonate pathway'. *Cellular and Molecular Life Sciences*, **61**, 1401–26.

ENSERINK, M. (2005) 'Infectious diseases – Source of new hope against malaria is in short supply'. *Science*, **307**, 33.

FARMER, W. R. and LIAO, J. C. (2001) 'Precursor balancing for metabolic engineering of lycopene production in *Escherichia coli*'. *Biotechnology Progress*, **17**, 57–61.

FOWLER, Z. L., GIKANDI, W. W. and KOFFAS, M. A. G. (2009) 'Increased malonyl coenzyme A biosynthesis by tuning the *Escherichia coli* metabolic network and its application to flavanone production'. *Applied and Environmental Microbiology*, **75**, 5831–9.

GILLAM, E. M. J. (2008) 'Engineering cytochrome P450 enzymes'. *Chemical Research in Toxicology*, **21**, 220–31.

HARVEY, A. L. (2008) 'Natural products in drug discovery'. *Drug Discovery Today*, **13**, 894–901.

HE, J. A. and GIUSTI, M. M. (2010) 'Anthocyanins: natural colorants with health-promoting properties'. In *Annual Review of Food Science and Technology, Vol 1*. Doyle, M. P. and Klaenhammer, T. R. (eds) Palo Alto, Annual Reviews, Springer.

HOEFFLER, J. F., PALE-GROSDEMANGE, C. and ROHMER, M. (2000) 'Chemical synthesis of enantiopure 2-*C*-methyl-*D*-erythritol 4-phosphate, the key intermediate in the mevalonate-independent pathway for isoprenoid biosynthesis'. *Tetrahedron*, **56**, 1485–9.

HWANG, E. I., KANEKO, M., OHNISHI, Y. and HORINOUCHI, S. (2003) 'Production of plant-specific flavanones by *Escherichia coli* containing an artificial gene cluster'. *Applied and Environmental Microbiology*, **69**, 2699–706.

IRITI, M., VITALINI, S., FICO, G. and FAORO, F. (2010) 'Neuroprotective herbs and foods from different traditional medicines and diets'. *Molecules*, **15**, 3517–55.

© Woodhead Publishing Limited, 2013

JACKSON, B. E., HART-WELLS, E. A. and MATSUDA, S. P. T. (2003) 'Metabolic engineering to produce sesquiterpenes in yeast'. *Organic Letters*, **5**, 1629–32.

JULSING, M. K., KOULMAN, A., WOERDENBAG, H. J., QUAX, W. J. and KAYSER, O. (2006) 'Combinatorial biosynthesis of medicinal plant secondary metabolites'. *Biomolecular Engineering*, **23**, 265–79.

KABIR, M. M. and SHIMIZU, K. (2003) 'Fermentation characteristics and protein expression patterns in a recombinant *Escherichia coli* mutant lacking phosphoglucose isomerase for poly(3-hydroxybutyrate) production'. *Applied Microbiology and Biotechnology*, **62**, 244–55.

KAJIWARA, S., FRASER, P. D., KONDO, K. and MISAWA, N. (1997) 'Expression of an exogenous isopentenyl diphosphate isomerase gene enhances isoprenoid biosynthesis in *Escherichia coli*'. *Biochemical Journal*, **324**, 421–6.

KATSUYAMA, Y., FUNA, N., MIYAHISA, I. and HORINOUCHI, S. (2007) 'Synthesis of unnatural flavonoids and stilbenes by exploiting the plant biosynthetic pathway in *Escherichia coli*'. *Chemistry and Biology*, **14**, 613–21.

KHAN, N. and MUKHTAR, H. (2007) 'Tea polyphenols for health promotion'. *Life Sciences*, **81**, 519–33.

KHAN, N., AFAQ, F. and MUKHTAR, H. (2008) 'Cancer chemoprevention through dietary antioxidants: progress and promise'. *Antioxidants & Redox Signaling*, **10**, 475–510.

KHAN, M. K., RAKOTOMANOMANA, N., LOONIS, M. and DANGLES, O. (2010) 'Chemical synthesis of citrus flavanone glucuronides'. *Journal of Agricultural and Food Chemistry*, **58**, 8437–43.

KIM, S. W. and KEASLING, J. D. (2001) 'Metabolic engineering of the nonmevalonate isopentenyl diphosphate synthesis pathway in *Escherichia coli* enhances lycopene production'. *Biotechnology and Bioengineering*, **72**, 408–15.

KIM, D. H., KIM, B. G., JUNG, N. R. and AHN, J. R. (2009) 'Production of genistein from naringenin using *Escherichia coli* containing isoflavone synthase-cytochrome P450 reductase fusion protein'. *Journal of Microbiology and Biotechnology*, **19**, 1612–16.

KIZER, L., PITERA, D. J., PFLEGER, B. F. and KEASLING, J. D. (2008) 'Application of functional genomics to pathway optimization for increased isoprenoid production'. *Applied and Environmental Microbiology*, **74**, 3229–41.

LANGE, B. M., RUJAN, T., MARTIN, W. and CROTEAU, R. (2000) 'Isoprenoid biosynthesis: The evolution of two ancient and distinct pathways across genomes'. *Proceedings of the National Academy of Sciences of the United States of America*, **97**, 13172–7.

LEONARD, E., YAN, Y. J., LIM, K. H. and KOFFAS, M. A. G. (2005) 'Investigation of two distinct flavone synthases for plant-specific flavone biosynthesis in *Saccharomyces cerevisiae*'. *Applied and Environmental Microbiology*, **71**, 8241–8.

LEONARD, E., CHEMLER, J., LIM, K. H. and KOFFAS, M. A. G. (2006a) 'Expression of a soluble flavone synthase allows the biosynthesis of phytoestrogen derivatives in *Escherichia coli*'. *Applied Microbiology and Biotechnology*, **70**, 85–91.

LEONARD, E., YAN, Y. J. and KOFFAS, M. A. G. (2006b) 'Functional expression of a P450 flavonoid hydroxylase for the biosynthesis of plant-specific hydroxylated flavonols in *Escherichia coli*'. *Metabolic Engineering*, **8**, 172–81.

LEONARD, E., LIM, K. H., SAW, P. N. and KOFFAS, M. A. G. (2007) 'Engineering central metabolic pathways for high-level flavonoid production in *Escherichia coli*'. *Applied and Environmental Microbiology*, **73**, 3877–86.

LEONARD, E., YAN, Y., FOWLER, Z. L., LI, Z., LIM, C. G., LIM, K. H. and KOFFAS, M. A. G. (2008) 'Strain improvement of recombinant *Escherichia coli* for efficient production of plant flavonoids'. *Molecular Pharmaceutics*, **5**, 257–65.

LEONARD, E., AJIKUMAR, P. K., THAYER, K., XIAO, W. H., MO, J. D., TIDOR, B., STEPHANOPOULOS, G. and PRATHER, K. L. J. (2010) 'Combining metabolic and protein engineering of a

© Woodhead Publishing Limited, 2013

terpenoid biosynthetic pathway for overproduction and selectivity control'. *Proceedings of the National Academy of Sciences of the United States of America*, **107**, 13654–9.

MA, S. M., GARCIA, D. E., REDDING-JOHANSON, A. M., FRIEDLAND, G. D., CHAN, R., BATTH, T. S., HALIBURTON, J. R., CHIVIAN, D., KEASLING, J. D., PETZOLD, C. J., LEE, T. S. and CHHABRA, S. R. (2011) 'Optimization of a heterologous mevalonate pathway through the use of variant HMG-CoA reductases'. *Metabolic Engineering*, **13**, 588–97.

MARTIN, V. J. J., PITERA, D. J., WITHERS, S. T., NEWMAN, J. D. and KEASLING, J. D. (2003) 'Engineering a mevalonate pathway in *Escherichia coli* for production of terpenoids'. *Nature Biotechnology*, **21**, 796–802.

MATSUI, T., UEDA, T., OKI, T., SUGITA, K., TERAHARA, N. and MATSUMOTO, K. (2001a) 'alpha-Glucosidase inhibitory action of natural acylated anthocyanins. 1. Survey of natural pigments with potent inhibitory activity'. *Journal of Agricultural and Food Chemistry*, **49**, 1948–51.

MATSUI, T., UEDA, T., OKI, T., SUGITA, K., TERAHARA, N. and MATSUMOTO, K. (2001b) 'alpha-Glucosidase inhibitory action of natural acylated anthocyanins. 2. alpha-Glucosidase inhibition by isolated acylated anthocyanins'. *Journal of Agricultural and Food Chemistry*, **49**, 1952–6.

MATSUI, T., EBUCHI, S., KOBAYASHI, M., FUKUI, K., SUGITA, K., TERAHARA, N. and MATSUMOTO, K. (2002) 'Anti-hyperglycemic effect of diacylated anthocyanin derived from *Ipomoea batatas* cultivar Ayamurasaki can be achieved through the alpha-glucosidase inhibitory action'. *Journal of Agricultural and Food Chemistry*, **50**, 7244–8.

MIDDLETON, E., KANDASWAMI, C. and THEOHARIDES, T. C. (2000) 'The effects of plant flavonoids on mammalian cells: Implications for inflammation, heart disease, and cancer'. *Pharmacological Reviews*, **52**, 673–751.

MIYAHISA, I., KANEKO, M., FUNA, N., KAWASAKI, H., KOJIMA, H., OHNISHI, Y. and HORINOUCHI, S. (2005) 'Efficient production of (2S)-flavanones by *Escherichia coli* containing an artificial biosynthetic gene cluster'. *Applied Microbiology and Biotechnology*, **68**, 498–504.

NAESBY, M., NIELSEN, S. V. S., NIELSEN, C. A. F., GREEN, T., TANGE, T. O., SIMON, E., KNECHTLE, P., HANSSON, A., SCHWAB, M. S., TITIZ, O., FOLLY, C., ARCHILA, R. E., MAVER, M., FIET, S. V. S., BOUSSEMGHOUNE, T., JANES, M., KUMAR, A. S. S., SONKAR, S. P., MITRA, P. P., BENJAMIN, V. A. K., KORRAPATI, N., SUMAN, I., HANSEN, E. H., THYBO, T., GOLDSMITH, N. and SORENSEN, A. S. (2009) 'Yeast artificial chromosomes employed for random assembly of biosynthetic pathways and production of diverse compounds in *Saccharomyces cerevisiae*'. *Microbial Cell Factories*, **8**, 45 (August).

NEWMAN, D. J. and CRAGG, G. M. (2007) 'Natural products as sources of new drugs over the last 25 years'. *Journal of Natural Products*, **70**, 461–77.

OHKATSU, Y., SAKURAI, T. and SATO, T. (2010) 'Relationship between chemical structure and antioxidant function of flavonoids'. *Journal of the Japan Petroleum Institute*, **53**, 213–21.

OKADA, K. (2011) 'The biosynthesis of isoprenoids and the mechanisms regulating it in plants'. *Bioscience Biotechnology and Biochemistry*, **75**, 1219–25.

OLDFIELD, E. (2010) 'Targeting isoprenoid biosynthesis for drug discovery: bench to bedside'. *Accounts of Chemical Research*, **43**, 1216–26.

OTUKI, M. F., VIEIRA-LIMA, F., MALHAIROS, K., YUNES, R. A. and CALIXTO, J. B. (2005) 'Topical antiinflammatory effects of the ether extract from *Protium kleinii* and alpha-amyrin pentacyclic triterpene'. *European Journal of Pharmacology*, **507**, 253–9.

PARADISE, E. M., KIRBY, J., CHAN, R. and KEASLING, J. D. (2008) 'Redirection of flux through the FPP branch-point in *Saccharomyces cerevisiae* by down-regulating squalene synthase'. *Biotechnology and Bioengineering*, **100**, 371–8.

© Woodhead Publishing Limited, 2013

PEANA, A. T., D'AQUILA, P. S., CHESSA, M. L., MORETTI, M. D. L., SERRA, G. and PIPPIA, P. (2003) '(-)-Linalool produces antinociception in two experimental models of pain'. *European Journal of Pharmacology*, **460**, 37–41.

PITERA, D. J., PADDON, C. J., NEWMAN, J. D. and KEASLING, J. D. (2007) 'Balancing a heterologous mevalonate pathway for improved isoprenoid production in *Escherichia coli*'. *Metabolic Engineering*, **9**, 193–207.

RAMIREZ-TORTOSA, C., ANDERSEN, O. M., GARDNER, P. T., MORRICE, P. C., WOOD, S. G., DUTHIE, S. J., COLLINS, A. R. and DUTHIE, G. G. (2001) 'Anthocyanin-rich extract decreases indices of lipid peroxidation and DNA damage in vitamin E-depleted rats'. *Free Radical Biology and Medicine*, **31**, 1033–7.

RANGANATHAN, S., SUTHERS, P. F. and MARANAS, C. D. (2010) 'OptForce: an optimization procedure for identifying all genetic manipulations leading to targeted overproductions'. *Plos Computational Biology*, **6**(4), e1000744.

REDDY, L. H. and COUVREUR, P. (2009) 'Squalene: A natural triterpene for use in disease management and therapy'. *Advanced Drug Delivery Reviews*, **61**, 1412–26.

REILING, K. K., YOSHIKUNI, Y., MARTIN, V. J. J., NEWMAN, J., BOHLMANN, J. and KEASLING, J. D. (2004) 'Mono and diterpene production in *Escherichia coli*'. *Biotechnology and Bioengineering*, **87**, 200–12.

RO, D. K., PARADISE, E. M., OUELLET, M., FISHER, K. J., NEWMAN, K. L., NDUNGU, J. M., HO, K. A., EACHUS, R. A., HAM, T. S., KIRBY, J., CHANG, M. C. Y., WITHERS, S. T., SHIBA, Y., SARPONG, R. and KEASLING, J. D. (2006) 'Production of the antimalarial drug precursor artemisinic acid in engineered yeast'. *Nature*, **440**, 940–3.

ROHDICH, F., KIS, K., BACHER, A. and EISENREICH, W. (2001) 'The non-mevalonate pathway of isoprenoids: genes, enzymes and intermediates'. *Current Opinion in Chemical Biology*, **5**, 535–40.

ROWINSKY, E. K. and DONEHOWER, R. C. (1995) 'Drug-therapy – paclitaxel (taxol)'. *New England Journal of Medicine*, **332**, 1004–14.

SCHMIDT, S., RAINIERI, S., WITTE, S., MATERN, U. and MARTENS, S. (2011) 'Identification of a *Saccharomyces cerevisiae* glucosidase that hydrolyzes flavonoid glucosides'. *Applied and Environmental Microbiology*, **77**, 1751–7.

SCHMIDT-DANNERT, C., UMENO, D. and ARNOLD, F. H. (2000) 'Molecular breeding of carotenoid biosynthetic pathways'. *Nature Biotechnology*, **18**, 750–3.

SHIBA, Y., PARADISE, E. M., KIRBY, J., RO, D. K. and KEASING, J. D. (2007) 'Engineering of the pyruvate dehydrogenase bypass in *Saccharomyces cerevisiae* for high-level production of isoprenoids'. *Metabolic Engineering*, **9**, 160–8.

TABART, J., KEVERS, C., PINCEMAIL, J., DEFRAIGNE, J. O. and DOMMES, J. (2009) 'Comparative antioxidant capacities of phenolic compounds measured by various tests'. *Food Chemistry*, **113**, 1226–33.

TANAKA, Y., SASAKI, N. and OHMIYA, A. (2008) 'Biosynthesis of plant pigments: anthocyanins, betalains and carotenoids'. *Plant Journal*, **54**, 733–49.

TRANTAS, E., PANOPOULOS, N. and VERVERIDIS, F. (2009) 'Metabolic engineering of the complete pathway leading to heterologous biosynthesis of various flavonoids and stilbenoids in *Saccharomyces cerevisiae*'. *Metabolic Engineering*, **11**, 355–66.

TYO, K. E. J., AJIKUMAR, P. K. and STEPHANOPOULOS, G. (2009) 'Stabilized gene duplication enables long-term selection-free heterologous pathway expression'. *Nature Biotechnology*, **27**, 760-U115.

VARMAN, A. M., XIAO, Y., LEONARD, E. and TANG, Y. J. J. (2011) 'Statistics-based model for prediction of chemical biosynthesis yield from *Saccharomyces cerevisiae*'. *Microbial Cell Factories*, **10**, 12.

VEITCH, N. C. and GRAYER, R. J. (2011) 'Flavonoids and their glycosides, including anthocyanins'. *Natural Product Reports*, **28**, 1626–95.

© Woodhead Publishing Limited, 2013

WAGNER, K. H. and ELMADFA, I. (2003) 'Biological relevance of terpenoids – Overview focusing on mono-, di- and tetraterpenes'. *Annals of Nutrition and Metabolism*, **47**, 95–106.

WANG, C. W., OH, M. K. and LIAO, J. C. (2000) 'Directed evolution of metabolically engineered *Escherichia coli* for carotenoid production'. *Biotechnology Progress*, **16**, 922–6.

WANG, Z. X., REN, Z. H., YAN, J., XU, X., SHI, X. X. and CHEN, G. R. (2006) 'Total synthesis of glycosylflavone Parkinsonin B'. *Chinese Journal of Organic Chemistry*, **26**, 1254–8.

WELCH, M., VILLALOBOS, A., GUSTAFSSON, C. and MINSHULL, J. (2009) 'You're one in a googol: optimizing genes for protein expression'. *Journal of the Royal Society Interface*, **6**, 10.

WERNER, S. R., CHEN, H., JIANG, H. X. and MORGAN, J. A. (2010) 'Synthesis of non-natural flavanones and dihydrochalcones in metabolically engineered yeast'. *Journal of Molecular Catalysis B-Enzymatic*, **66**, 257–63.

WINKEL-SHIRLEY, B. (1999) 'Evidence for enzyme complexes in the phenylpropanoid and flavonoid pathways'. *Physiologia Plantarum*, **107**, 142–9.

WINKEL-SHIRLEY, B. (2001) 'Flavonoid biosynthesis. A colorful model for genetics, biochemistry, cell biology, and biotechnology'. *Plant Physiology*, **126**, 485–93.

WITHERS, S. T. and KEASLING, J. D. (2007) 'Biosynthesis and engineering of isoprenoid small molecules'. *Applied Microbiology and Biotechnology*, **73**, 980–90.

XU, P., RANGANATHAN, S., FOWLER, Z. L., MARANAS, C. D. and KOFFAS, M. A. G. (2011) 'Genome-scale metabolic network modeling results in minimal interventions that cooperatively force carbon flux towards malonyl-CoA'. *Metabolic Engineering*, **13**, 578–87.

YAMANO, S., ISHII, T., NAKAGAWA, M., IKENAGA, H. and MISAWA, N. (1994) 'Metabolic engineering for production of beta-carotene and lycopene in *Saccharomyces cerevisiae*'. *Bioscience Biotechnology and Biochemistry*, **58**, 1112–14.

YAN, Y. J., KOHLI, A. and KOFFAS, M. A. G. (2005) 'Biosynthesis of natural flavanones in *Saccharomyces cerevisiae*'. *Applied and Environmental Microbiology*, **71**, 5610–13.

YAN, Y. J., LI, Z. and KOFFAS, M. A. G. (2008) 'High-yield anthocyanin biosynthesis in engineered *Escherichia coli*'. *Biotechnology and Bioengineering*, **100**, 126–40.

YAO, L. H., JIANG, Y. M., SHI, J., TOMAS-BARBERAN, F. A., DATTA, N., SINGANUSONG, R. and CHEN, S. S. (2004) 'Flavonoids in food and their health benefits'. *Plant Foods for Human Nutrition*, **59**, 113–22.

YOSHIKUNI, Y., FERRIN, T. E. and KEASLING, J. D. (2006) 'Designed divergent evolution of enzyme function'. *Nature*, **440**, 1078–82.

YOSHIKUNI, Y., DIETRICH, J. A., NOWROOZI, F. F., BABBITT, P. C. and KEASLING, J. D. (2008) 'Redesigning enzymes based on adaptive evolution for optimal function in synthetic metabolic pathways'. *Chemistry & Biology*, **15**, 607–18.

ZHA, W. J., RUBIN-PITEL, S. B., SHAO, Z. Y. and ZHAO, H. M. (2009) 'Improving cellular malonyl-CoA level in *Escherichia coli* via metabolic engineering'. *Metabolic Engineering*, **11**, 192–8.

© Woodhead Publishing Limited, 2013

11

Microbial production of enzymes used in food applications

K. Hellmuth, Chr Hansen Nienburg GmbH, Germany and
J. M. van den Brink, Chr Hansen A/S, Innovation, Denmark

DOI: 10.1533/9780857093547.2.262

Abstract: Enzymes are a major tool in the production of modern food. In this chapter enzymes used in the food industry are discussed. Production of enzymes by microorganisms in an industrial setup is discussed in more detail. Here production of the milk clotting enzyme chymosin is used as an example, since chymosin production systems in yeast, fungi and bacteria have been described in some detail in the literature. Although enzyme properties and enzyme production levels are important for the industry, an important hurdle for application of enzymes in the food industry is the approval of these enzymes. This is discussed briefly in the last part of the chapter.

Key words: Aspergillus, chymosin, enzyme production, Escherichia coli, food enzymes, Kluyveromyces lactis.

11.1 Introduction: microbial production of food enzymes

Enzymes have been used ever since mankind discovered ways to process food. Food processing steps like milk acidification, milk clotting, alcohol fermentation and soy bean fermentation are enzyme-mediated processes carried out by microorganisms. However, it was not until the late 19th century that purified enzymes were used for food processing. In 1907 Eduard Buchner was awarded the Nobel prize in biochemistry for his pioneering work on sugar fermentation by cell-free yeast extract, basically proving the existence of enzymes (Kohler 1972a,b). In 1946 three scientists (Northrop, Stanley and Sumner) were awarded the Nobel Prize for their work on enzymes. They proved that enzyme activity is carried out by a single pure protein and that the protein is not merely a carrier for the

© Woodhead Publishing Limited, 2013

enzyme activity, as was the general belief at that time (data obtained from Nobelprize.org, 2012).

Approximately 20 years before Buchner, a Danish pharmacist, Christian DA Hansen, invented and commercialized a procedure to purify rennet from calf stomachs. This invention – possibly the first commercial purified enzyme product – changed the dairy industry. Previously, cheese production was started by adding small pieces of calf stomachs to the milk, resulting in very inconsistent results. The availability of a purified and standardized enzyme revolutionized the dairy industry and can be considered to be the start of the global food enzyme industry.

Following Chr. Hansen's discovery it would take quite some time for the enzyme industry to develop to maturity. The major drawback was that enzymes had to be isolated from natural raw materials of very varying quality and typically high costs.

At the same time, industrialized fermentation processes, using microorganisms as cell factories, had been developed for the production of vitamins and citric acid, paving the way to use similar processes for enzyme production. The use of microbial production systems changed the enzyme industry. The first microbial enzyme production process was developed in 1894 by Dr Takamine, who used the mould *A. oryzae*, known from koji production, as a production host for takadiastase, a mixture of amylases and proteases, used in many starch processing steps (Collins, 1927; Lichtenstein, 1947; Suganuma *et al.*, 2007).

The use of microorganisms for enzyme production allowed the development of controlled and standardized processes, using safe and well-described raw materials. However, it was only after World War II that the enzyme industry became a major industry. Currently the enzyme industry is a large sector, having global sales of approximately US\$ 3.3 billion in 2010 (data obtained from Marketwire.com, 2011).

On the basis of application, industrial enzymes can be divided into four major categories, detergent enzymes, technical enzymes, food enzymes and feed enzymes. The technical enzymes segment can be further divided into textile enzymes, leather enzymes, pulp and paper enzymes, fuel ethanol enzymes and others. The total food enzymes market constitutes the largest market share of this large enzyme market. In 2008 the global sales of food enzymes was estimated to be around US\$ 1 billion (data obtained from PRWeb, 2012), while a more recent investigation estimates the global food enzyme market to be approx. US\$ 975 million, of a total enzyme market of US\$ 3.3 billion (Marketwire.com, 2011).

The industrial enzyme market is an oligopolistic market with the presence of three major suppliers, Novozymes A/S (headquartered in Denmark), Danisco A/S (headquartered in Denmark but recently taken over by US based DuPont) and DSM N.V (headquartered in the Netherlands). While the three main players are active in most or all of the enzyme market segments, a number of smaller players exist that focus on smaller segments of

© Woodhead Publishing Limited, 2013

the enzyme market. As an example, Chr. Hansen is focusing on enzymes for use in the dairy industry, while a company like Puratos specializes in enzymes for the bakery industry.

11.2   Requirements of a good food enzyme

A food enzyme needs to fulfil a range of important criteria. However, the primary criterion is low cost of use. The food industry in general operates with relatively low financial margins and high costs of basic raw materials. As an example, shown in Fig. 11.1, the production of 1 metric tonne of cheese requires the use of approximately 10,000 litres of milk, costing around €3500. This is roughly 80% of the total manufacturing costs. The cost of other ingredients is roughly €40 per tonne of cheese, with milk clotting enzymes amounting to around €10 (CDFA, 2012) (http://www.cdfa.ca.gov/dairy/uploader/postings/manufacturingcost/). The high costs of raw materials allow the producer only very limited financial space to purchase enzymes and use them to optimize food properties. This pushes enzyme producers to reduce production costs all the time. On the other hand it also makes clear that enzyme solutions resulting in reduced need for raw materials can have a major impact on manufacturing costs and are very valuable to the food producer.

Of course safety of the enzyme, both for consumers in the final product, but also for workers at the food producer as well as at the enzyme producing plants is very important. Enzymes, like most proteins, are allergenic and overexposure to allergens can cause serious problems (as reviewed by (Green and Beezhold, 2011). On the technical side, food enzymes need to fulfil a large range of criteria. The main criterion is that they should be able

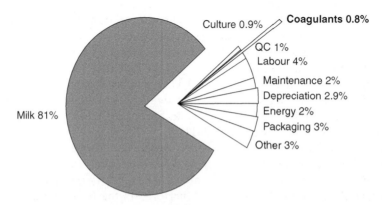

Fig. 11.1   Cost of production of cheese, divided in cost categories (for details see text).

© Woodhead Publishing Limited, 2013

to carry out their action in the food environment. This is not trivial. Some applications require the enzyme to be active at very high temperatures (bakery), other applications require the enzyme to be active at low temperatures and/or low pH (fermented milk products). Enzymes should preferably have no, or very limited, side activities. Side activities can result in unwanted properties of the food product. As an example, even very low levels of lipase activity in dairy enzymes can result in major off-taste of the final product. Side activities typically result from contaminating enzymes and can be strongly reduced, or even completely removed, by enzyme purification. However, this does compromise cost of use of the enzyme.

Some side activity is intrinsic to the enzyme. For example, milk clotting requires a very specific protease activity but most rennets, being proteases, can cleave other milk proteins to some extent, resulting in reduced cheese yield and negative sensory effects. Improved enzyme specificity can be achieved by either screening natural variation (Kappeler *et al.*, 2006) or by a dedicated protein engineering approach (protein engineering of food enzymes has been thoroughly reviewed by Aehle, 2004).

Enzyme properties are not only dependent on their intrinsic properties but are also to a large extent dependent on formulation. Most enzymes are highly unstable without the proper additives added to the formulation. For industrial use it is of great importance to formulate an enzyme such that it will have a reasonable shelf life, allowing constant product quality during distribution and during storage in the food plant. An optimal formulation of the enzyme will allow prolonged storage (months rather than weeks) at room temperature. The latter is important because of the global nature of the food enzyme industry; an intact cooling chain cannot be guaranteed in all cases.

The concentration is also important for formulation. The higher the enzyme concentration in a liquid formulation, the less water needs to be transported, saving both costs and the environment. However, solubility and crystallization of the enzyme product pose a limit to the maximal protein concentration in the product. For powder formulations, the allergenicity of the product is a major concern. Here granular products that produce a minimum amount of dust are preferred. Of course, components used in formulation are as much subject to food safety and worker safety evaluation as the actual enzymes.

11.3   Limitations of enzyme use in food applications

Despite the fact that enzymes offer huge advantages for improving food properties, there are some clear limits to their use. A major drawback of many enzymes is their need for co-factors and co-enzymes. Many enzymes have a need for a reducing or oxidizing power, provided by the NADP+/ NADPH or NAD+/NADH systems. In most of these cases alternative

© Woodhead Publishing Limited, 2013

solutions, like chemical processes or the use of living cells, are more appropriate for the producer.

A well-described example of such a system is malolactic fermentation in wine. In this process a bacterium, typically *Oenococcus oenii*, converts the harsh tasting malic acid into the smoother tasting lactic acid (Lafon-Lafourcade *et al.*, 1983; Liu, 2002). The key reaction in this process is decarboxylation of malate in the grape juice by malic enzyme (EC 1.1.1.39), which requires reduction of NADP+ to NADPH. Although in theory an enzymatic solution would be highly beneficial for the wine industry, this is not a practical option owing to the need for co-enzymes.

A second example is the conversion of lactose to lactate in fermented milk products. This conversion could, in theory, be done by enzymatic systems. However, the complexity of the systems that would be needed is so high that this is not a realistic option. In this case the use of living bacteria is also preferable.

Other factors limiting the use of enzymes in food applications are economy and legal factors. Enzymes will only be used to improve properties when the cost of use is minimal compared to the overall production costs. Often alternative solutions, for example chemical or physical treatment of raw materials can be used to obtain similar properties at lower costs.

Legal factors can be a major limitation for the use of enzymes in the food industry and will be discussed later. Use of genetically modified (GM)-derived enzymes in food products has been limited not only by legal aspects but also by the limited consumer acceptance of GM-derived products.

11.4   Enzymes currently used in the food industry

In this section an overview of some of the enzymes used in the food industry will be given. This overview is not meant to be complete, for a more complete overview please refer to Aehle (2004). A good overview of GM produced enzymes used in food applications today is found at www. gmo-compass.org, an objective, independent website sponsored by the EU.

11.4.1   Bakery
Amylases (EC3.1.1.x, EC3.2.1.x)
Amylases are the most widespread enzymes used in the bakery industry. The most commonly used amylase is the so called takaamylase from *A. oryzae*. Amylases are used to reduce dough viscosity by breaking down the starch in wheat flour. The result is an increased volume of the bread and a better crumb structure. Amylases are also important to increase shelf life of the bread, the anti-staling effect. Staling is directly related to crumb firmness. Amylases (especially maltogenic exo-amylases, EC 3.2.1.133, typically derived from *Bacillus* species) can help to reduce this crumb firmness.

© Woodhead Publishing Limited, 2013

Typically anti-staling amylases are added to the bread before baking and thus need to have a very high thermostability in order to be active after bread baking. Typically bacterial amylases, often improved by protein engineering, are used for this purpose.

Xylanases (a.o. EC3.2.1.8)

Xylanases are used to improve bread volume, by degrading arabinoxylan in the cereals. The typical result is an improved dough stability, improved crumb structure and increased volume of the bread. Different xylanases can show rather different specificities. Commercial xylanases are typically produced by *Aspergillus* species.

Glucose and hexose oxidase (EC1.1.3.4 and 1.1.3.5)

Glucose oxidase is an enzyme which in the presence of oxygen is capable of oxidizing D-glucose to its corresponding lactone. This process generates hydrogen peroxide which oxidizes thiol groups on flour proteins, thus creating disulfide bonds in the gluten network that contribute to a better dough quality. While glucose oxidase only uses glucose as a reducing sugar, hexose oxidase is able to use other reducing sugars, such as maltose, lactose and galactose, as well. Commercial glucose oxidase is produced by *Aspergillus* species while hexose oxidase is a GM derived product produced by the yeast *Hansenula polymorpha*. Both enzymes are also used in other food applications.

11.4.2   Juice and wines

Pectinases (Polygalacturonases)

Pectinases are a heterogeneous group of related enzymes involved in the degradation of pectin. They destabilize cell walls and improve extraction of colour and aroma. However, their prime use is to increase juice yield by maceration of the fruit cell wall. More often than not it is found that pectinase preparations are very impure, typically containing a large number of more or less related side activities (Arnous and Meyer, 2010). Pectinases are typically produced by *Aspergillus* species.

Glucanase (EC3.2.1.4)

Glucanase is used to clarify the juice and thus to increase filterability. The enzyme is also used in the beer industry to increase filterability.

11.4.3   Sugar

In the sugar and sweetener industry the use of enzymes is widespread. The use of sugar-modifying enzymes has received an extra boost in recent years owing to the focus on bioethanol. A major step in the bioethanol process is the conversion of plant materials to sugars.

© Woodhead Publishing Limited, 2013

Amylases (EC3.2.1.1 and 3.2.1.2)/*glucoamylases* (EC3.2.1.3)
Amylases and glucoamylases are widely used to convert starch-like products into sugar syrups. This class of enzymes was among the first to be produced in an industrial fermentation process (Hjort, 2007). Alpha-amylase (EC 3.1.1.1) cuts alpha-1,4 bonds in starch molecules at random positions while beta-amylase (EC 3.2.1.2) cuts off maltose molecules from the non-reducing end of the starch molecule. Glucoamylase (EC3.2.1.3) is also able to cleave alpha-1,6 bonds. The different types of amylases are usually used together to obtain full conversion of long chain starch molecules in monosaccharides.

Invertase (EC3.2.1.26)
Invertases can convert table sugar sucrose into fructose and glucose. This is advantageous as the resulting inverted sugar syrup has a higher sweetness than the starting product. Invertases can be obtained from fungi like *Aspergillus* as well as from yeast.

Glucose isomerase (EC 5.3.1.5)
Glucose isomarase is used to convert glucose into the fructose. The resulting high-fructose syrup is considerably sweeter than the original glucose syrup. Glucose isomerase is an expensive enzyme that is produced intracellularly by *Streptomyces* species. It is typically used as an immobilized enzyme in a flow-through reactor to reduce enzyme costs.

11.4.4   Food safety applications
Glucose oxidase (EC1.1.3.4)
Glucose oxidase (GO*x*) is used in many food applications as an oxygen scavenger (glucose + oxygen is converted in gluconolactone + hydrogen peroxide). As long as glucose is available, the enzyme will help to maintain a low oxygen level in the food product, thus reducing microbial growth and subsequently shelf life of the food product. In most cases it will be necessary to add a second enzyme, catalase, to the system to inactivate the hydrogen peroxide. GO*x* is also used to remove glucose from egg-white, thus preventing browning in the final application. Removal of glucose in grape juice has been described as an efficient way to control the final alcohol levels in wines (van den Brink and Bjerre 2009; Villettaz, 2011).

A more recent application of glucose oxidase is the reduction of acrylamide formation, as described in US patent 6989167 (Howie *et al.*, 2003). The key step in the formation of acrylamide is the formation of the so called Schiff base by the reaction of the alpha-amine group of free asparagine with a carbonyl source, typically a reduced sugar. Glucose oxidase will reduce the levels of reduced sugars and thus limit formation of acrylamide.

© Woodhead Publishing Limited, 2013

11.4.5 Dairy
Proteases and lipases
Proteases and lipases are used to digest milk proteins and milk fats, especially in cheese manufacturing. Although a controlled use of these enzymes can benefit taste and increase cheese maturation, the use of enzymes for this purpose is not widespread. This is mainly due to the fact that enzyme activity in a maturing cheese is difficult to control. Over-hydrolysis has a detrimental impact on taste and product quality. For this reason most dairies prefer to add living bacteria to the cheese milk. Upon cell-lysis these bacteria will release peptidases, proteases and lipases in moderate amounts, giving more controlled ripening (Fox *et al.*, 1996).

Phospholipase
A novel enzyme, called Yieldmax, was recently introduced into the market. This enzyme is a phospholipase and is used during the cheese making process. Phospholipases remove one of the fatty acids from a phospholipid, resulting in a free fatty acid and a lysophospholipid (Lilbæk *et al.*, 2007). During the cheese making process, the curd is subject to mechanical forces which damage the fat globules. The lysophospholipid molecule is a strong emulsifier, which helps to repair this damage, thus avoiding loss of fat into the whey. The use of YieldMax phospholipase results in up to 2% higher cheese yields, especially in cheese processes where large mechanical stress is applied to the cheese curd (i.e. mozzarella production).

Lactase (EC 3.2.1.108)
For many years milk was a major foodstock for adults in parts of the world only. As a consequence, a large part of the global adult population (estimated at around 70%) is unable to digest lactose, which is the major sugar in milk. For people who are lactose intolerant, consumption of milk and other dairy products can result in intestinal problems like diarrhoea and abdominal pain. Dairy products can be made digestable for lactose-intolerant people by treatment with lactase (beta-galactosidase EC3.2.1.108), converting one molecule of D-lactose into one molecule of D-glucose plus one molecule of D-galactose. Different lactases are currently on the market, isolated from either *Aspergillus* or *Kluyveromyces lactis*. The enzymes differ especially in their activity at different pH and temperature (Agrawal *et al.*, 1989; Asraf and Gunasekaran, 2011; Oliveira *et al.*, 2011). Lactase can be used to treat the dairy product before consumption but can also be obtained as OTC (over the counter) tablets that should be consumed prior to consumption of dairy products. OTC lactases are typically stable at lower pH.

Milk clotting enzymes
Milk clotting enzymes are the largest single group of food enzymes. Although many enzymes are applied as milk clotting enzymes, all of them

© Woodhead Publishing Limited, 2013

are aspartic proteases with a relative high specificity for kappa-casein in bovine milk.

In the field of milk clotting enzymes, the quality of a milk clotting enzyme is often indicated by its *C/P* ratio. The '*C*' stands for specific clotting activity, that is the ability of the milk clotting enzyme to cleave the Phe105-Met106 site in kappa-casein. The '*P*' stands for unspecific protease activity and refers to all other protein degradations. This aspecific protease activity is unwanted as it will reduce cheese yield and will have a negative effect on taste.

Animal rennet
Animal rennet is a milk clotting enzyme isolated from calf stomachs. The major component of rennet is chymosin (EC3.4.23.4) but in commercial preparations of rennet other proteases, typically bovine pepsin, are found in varying concentrations. The process of rennet isolation and formulation was commercialized more than a century ago, but even today animal rennet is still an important enzyme in the dairy industry. This is despite the fact that animal rennet prices swing a lot as a result of the swing in raw material prices and are considerably higher than alternative milk clotting enzymes.

Microbial rennets
Animal rennets are expensive and not vegetarian while the GM background of fermentation produced chymosin (see below) is unwanted in some markets. Based on these considerations there was a necessity for a third class of milk clotting enzymes that is vegetarian, not GM derived and preferably cheap. For this purpose some microbial proteases were developed as milk clotting enzymes. Currently the most important enzyme is the aspartic protease from *Rhizomucor miehei* (or *Mucor miehei*) (Harboe, 1998), while enzymes from *Mucor pussilus* and *Endothia parasitica* (Gagnaire *et al.* 2001) also are available. Typically microbial rennets are cheaper but have a poorer *C/P* ratio (seven-fold less than bovine chymosin), resulting in lower cheese yields and higher bitterness.

Fermentation-produced chymosin (EC3.4.23.4)
Production of animal rennet is dependent on raw materials of varying quality and with a limited supply. These problems could be circumvented by the use of GM technology. In a joint venture between Chr. Hansen and Genencor, a bovine chymosin-producing strain of *Aspergillus niger* was developed (Berka *et al.* 1991a,b; Dunn-Coleman *et al.*, 1991; Ward *et al.*, 1990; Ward, 1991), allowing for the production of chymosin using standard industrial fermentation technology. In parallel, other companies developed alternative expression systems. Pfizer developed a chymosin expression system in *Escherichia coli* (Emtage *et al.*, 1983; Marston *et al.*, 1985; McCaman, 1989; McCaman *et al.*, 1985), now owned by Chr. Hansen, while DSM developed a yeast expression system for chymosin in *K. lactis*

© Woodhead Publishing Limited, 2013

(Swinkels *et al.*, 1993; van den Berg *et al.*, 1990), which is still used for production of chymosin.

For many years GM produced bovine chymosin was the golden standard for milk clotting enzymes. However, recently an improved chymosin variant was isolated from camel stomachs. This camel chymosin, produced using the same *Aspergillus* production system as bovine chymosin shows improved cheese making properties (an increased *C/P* ratio) in bovine milk, resulting in higher cheese yields per liter of milk as well as less off-taste (Bansal *et al.*, 2009; Kappeler *et al.*, 2006).

Other milk clotting enzymes
Basically any aspartic protease will have some milk clotting activity. However, in general the high aspecfic protease activity will result in poor cheese yields, poor texture and bitter taste. Some alternative enzymes that are used in a small number of cheese applications are the protease from *Endothia parasitica* and a plant protease isolated from *Cynara cardunculus*.

11.5   Good production strain criteria for the food industry

The introduction of GM technology has made it possible to construct dedicated production strains, only producing the gene of interest. A good production strain needs to fulfil some important criteria. For approval purposes it is a major advantage if the host strain has a history of safe use (Olempska-Beer *et al.*, 2006) and of course it should be proven that the host strain does not produce any toxic compounds (Van Dijck *et al.*, 2003). A good food enzyme host organism should be able to grow in large fermentation vessels, using a cheap fermentation medium. Cheap fermentation media typically comprise undefined raw materials, preferably waste products from the food industry such as molasses and soy. From a genetic point of view it is important that the organism is genetically stable, while on the other hand, a GMO preferably should not be able to survive outside a bioreactor.

In order to make enzyme production economically attractive, it is important that the production costs are low. Production costs are controlled by reducing the costs of raw materials, but also the considerable costs of utilities like power, steam and water are important to consider. A main cost factor during enzyme production is downstream expenses. This typically implies that enzymes should be secreted, in large amounts, to the external medium.

It is of course essential that a food enzyme production host is a food grade organism, and thus will not produce toxic compounds. This is easier said than done; many of the known industrial work horses actually have the genetic setup to produce toxic compounds (Galagan *et al.*, 2005; Machida *et al.*, 2005; O'Callaghan and Dobson, 2006).

© Woodhead Publishing Limited, 2013

In order to introduce the gene of interest in the host it is essential that good molecular biological tools are available for the organism. Preferably it should be possible to introduce the gene of interest without introducing unwanted DNA sequences like antibiotic resistance genes or *E. coli*-derived plasmid sequences. Having the ability to use self-cloning (as described in EU Directive 90/219/EEC Annex II Part A) is even more preferable.

Currently the main work horses in the enzyme industry are fungi (like *Aspergillus* and *Trichoderma*), yeasts (*Saccharomyces, Hansenula, Kluyveromyces*) and bacteria (*E. coli, Bacillus*). Most food enzymes are currently being produced by *Aspergillus* species, *K. lactis* or *Bacillus* species (Olempska-Beer *et al.*, 2006).

Construction of a production strain can be a long process. It typically involves both classical mutagenesis and screening for improved production as well as the targeted deletion of unwanted genes. A well-described example is the construction of the chymosin production strain of *A. niger* var. *awamori*, by Genencor in collaboration with Chr. Hansen. This strain construction has been extensively described in the literature (Berka *et al.*, 1991a,b, 1992; Dunn-Coleman *et al.*, 1991; Lamsa and Bloebaum, 1990; Ward, 1991) and is summarized in Fig. 11.2. A wild type strain was selected for its ability to produce large amounts of fungal glucoamylase. This strain was used to remove a major protease (Berka *et al.*, 1990) before the expression plasmid was inserted in the genome. This primary strain was reported to produce only 300 mg l^{-1} of active chymosin. For this reason an extensive mutagenesis and screenings approach was performed resulting in a considerable increase in productivity (Bodie *et al.*, 1994). Production of chymosin was regulated by the glucoamylase promoter and thus subject to carbon catabolite repression. This repression was partially reduced by screening for a deoxyglucose mutant, resulting in a further yield increase. Final chymosin production yields were reported to be over 1 g l^{-1}.

11.6   Production processes

11.6.1   Introduction

Enzyme production with microorganisms can be grouped into two major process steps: first the propagation of microorganisms and generation of enzymes, the fermentation part, and second the purification of product from the broth, the downstream part.

Even though these two groups apply different technologies and are very often organized in different working groups or departments, fermentation and downstream processing strongly affect each other and therefore process development always has to be done as an integrated project.

In contrast to the 1970s when surface cultures (solid state fermentation, SSF) were mostly used, nowadays submerged cultivations in large scale bioreactors are mostly applied because of their higher productivity.

© Woodhead Publishing Limited, 2013

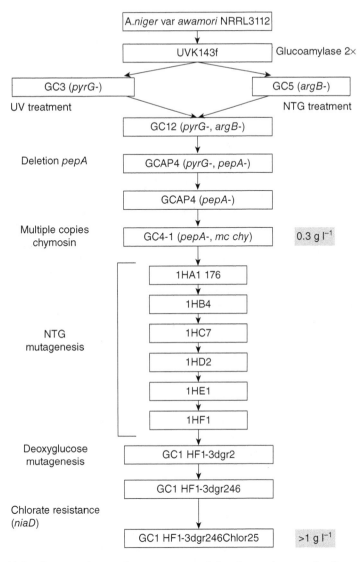

Fig. 11.2 Construction of a commercial chymosin production strain; NTG = *N*-methyl-*N*-nitro-*N*-nitrosoguanidine (Dunn-Coleman *et al.*, 1991; Lamsa and Bloebaum, 1990).

Depending on the amount of product that is needed, bioreactors with liquid volumes of 200 m³ and more are used in industry. However, the simple equation: 'the higher the fermentation volume, the lower the costs of production' is not always valid, because the financial loss in case of failure increases with the fermentation volume. Therefore the optimal scale for a

© Woodhead Publishing Limited, 2013

production process depends on technical and strain specific considerations, but also on the balance between economy and risk assessment.

11.6.2   Fermentation

An industrial fermentation is a process designed to maximize product yield and minimize the manufacturing costs. In order to achieve these two targets, product properties and physiology of the strain have to be adjusted to an optimal fermentation mode. Many strain-related factors have to be considered and the fermentation environment adapted to requirements of the microorganism. Important fermentation aspects are medium composition, pH, temperature, mixing and aeration.

The choice of medium components and carbon source determines the fermentation strategy. Many bacterial fermentation processes are still done as simple batch cultivations, but higher productivity is obtained by fed-batch processes. For fed-batch operations, strain specific knowledge is required in order to avoid critical inhibition or limitation. Nowadays this knowledge is available for most production strains and fed-batch cultivation is the typical process in industry. Continuous fermentations, as often used in academia for basic studies, are seldom used in industry because of the increased contamination risk. Bioreaction engineering principles, the fundamentals of mixing, power input, reactor design, mass- and heat-transfer, have been studied for decades and described in many publications and text books, for example recently by Villadsen et al. (2011).

As described in section 2 of Villadsen's book, the most crucial aspect of any fermentation process is mixing in the bioreactor. The agitation conditions require great attention, because efficient mixing is needed for optimal distribution of oxygen and nutrients (Diaz et al., 1996; Junker et al., 1998; Zlokarnik, 2000a,b).

Morphology and productivity

Morphology plays a very important role in fungal fermentation. Numerous authors have investigated the effect of environmental parameters on morphology. Especially the balance between pellet formation and filamentous growth which has a major effect on biomass and product formation (Carlsen et al., 1996; Trinci, 1974; van Suijdam et al., 1980; van Suijdam and Metz, 1981; Zetelaki, 1970). New methods for image analysis have been developed to classify and follow changes in morphology during cultivation (Barry et al., 2009; Cox and Thomas, 1992). The suitability of pellets or mycelia for product formation is discussed a lot in literature but conclusions are rather ambivalent (el Enshasy et al., 1999; Grimm et al., 2005; Papagianni et al., 2001). For an industrial scale fermentation (80 m^3) Li et al. (2000) report that fragmentation of hyphae had no effect on product formation.

The effect of viscosity on mass transfer is beyond dispute. High viscosity of fermentation broth leads to low oxygen transfer and, as a consequence,

© Woodhead Publishing Limited, 2013

a low dissolved oxygen concentration in the fermentation broth with a negative effect on product yield. The effect of agitation on morphology, viscosity, dissolved oxygen concentration and thereby productivity has been described in multiple studies (Amanullah *et al.*, 1999; Johansen *et al.*, 1998; Kelly *et al.*, 2004). The mechanical forces created by different impeller types are often key factors in process optimization (Justesen *et al.*, 2009; Olsvik *et al.*, 1993). Specific agitators were designed for mycelial fermentation and the traditional Rushton turbines are being replaced by new developments as described by Nienow (1990).

This topic was reviewed with particular focus on filamentous fungi by Wucherpfennig *et al.* (2010) who also modelled the complex relationship between productivity, biomass growth and rheology (Wucherpfennig *et al.*, 2010).

An alternative way to reduce viscosity is the directed screening of specific morphological mutants as described by Muller *et al.* (2003) to avoid these difficulties. Recently, new production organisms have been taken into use, which have been screened for a better morphology. An example of this is the so called C1 strain from Dyadic, a *Chrysosporium lucknowense* strain that is claimed to have improved enzyme production properties owing to the fact that it grows in small pellets (Burlingame and Verdoes, 2006).

11.6.3 Downstream

The design of downstream processes depends on two major aspects: is the enzyme being produced intracellular or extracellular and what is the required purification-grade for the product? Fungi usually secrete the product into the broth. Therefore cells and fermentation liquids have to be separated before the concentration and purification of product can start. The preferred technology is filtration because a dry filter cake can be obtained with a low residual content of product. Depending on the ratio of concentration and needed purification grade, different filtration (polish-, micro-, ultra filtration) and chromatographic technologies are used. It should be noted that many food enzymes are not purified but consist merely of spray dried fermentation broth.

In many of the bacterial processes the enzyme is located intracellularly, in the form of inclusion bodies or as soluble protein. Depending on the biomass concentration, a centrifugation step is needed before opening the cells with a homogenizer. Inclusion bodies have to be harvested by centrifugation. A refolding and enzyme reactivation procedure delivers a crude enzyme solution that can be purified by chromatography.

A major factor in industrial downstream processing is time. Owing to the large volumes most downstream processing steps take relatively long time. As an example, cooling down 50–200 m^3 of fermentation broth from 37°C to 4–10°C can take several hours. This long processing time poses special

© Woodhead Publishing Limited, 2013

challenges to enzyme stability but also to process design and even strain design (absence of proteases is a major advantage during downstream holding time).

11.7   Examples of heterologous enzyme production

In the 1980s several transformation systems were developed, but initially the choice of host organisms for production was limited to a few species. Today a wide variety of host organisms are used, but two microorganisms still play a major role in food enzyme production, *A. niger* and *K. lactis*. Both strains have a long history of safe use as food organisms and enzymes derived from these hosts have GRAS FDA status.

Chymosin is a good example of a microbially produced food enzyme because it is produced in both organisms. Considering chymosin, advantages and disadvantages of different production systems can be discussed. The first petition for production of chymosin was filed by Pfizer Inc. in 1988 and affirmed in 1990 for chymosin produced with *E. coli* as host organism. Therefore *E. coli* is also discussed as an alternative production system.

11.7.1   Chymosin production with fungi (*A. niger*)

The development of *A. niger* var. *awamori* as production system for chymosin has been well reviewed (Dunn-Coleman *et al.* 1992, 1991).

The process, as described in Fig. 11.3, starts with a stock culture of spores that is directly used to inoculate a seed fermentor. In many other fungal production processes longer inoculation trains, comprising sequential seed cultures, are used. A batch process, in a scale of several cubic metres, is performed to propagate biomass as the starting material for the main cultivation. Main cultivation is done in a large scale bioreactor. The cultivation is a fed-batch process using a complex medium and a choice of sugar (molasses, maltose, maltodextrin or glucose) as carbon sources. The choice of sugar depends on the genetic setup of the fungus. For example, maltose and maltodextrins are the preferred sugar source for genes for which expression is controlled by the maltose inducible glucoamylase promoter, as is the case in the chymosin production system. After about one week, just before cells stop their metabolic activity and at the time maximum concentration of the product is achieved, fermentation is finished by adding acids into the fermentor. At a pH below 2 the biomass (including DNA) is inactivated. This acid inactivation step, of course, cannot be used for production of acid-labile enzymes.

After a holding time of several hours, downstream processing starts with the addition of filter aid and the biomass is separated from the enzyme-containing liquid in a filter press or drum filter. At Chr. Hansen, enzymes

© Woodhead Publishing Limited, 2013

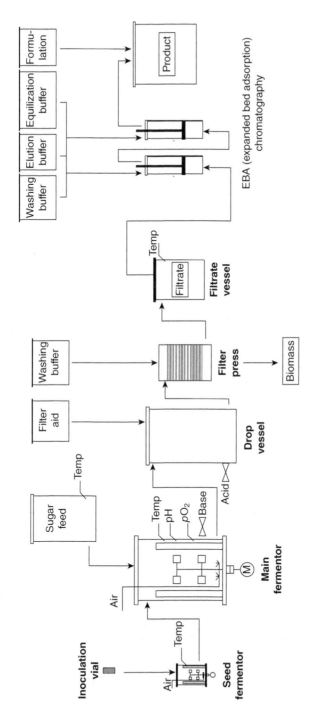

Fig. 11.3 Schematic overview of an industrial chymosin production processes with filamentous fungi (at Chr. Hansen).

© Woodhead Publishing Limited, 2013

are purified by chromatography, in contrast to many other suppliers that use technologies like spray drying or ultrafiltration to concentrate their enzymes. Owing to the fact that the EBA (expanded bed adsorption) chromatography allows the handling of turbid liquids, no polish filtration step is needed before initiating this process step. After loading, washing and elution, the final enzyme product can be formulated in order to achieve the desired specification.

11.7.2   Chymosin production with bacteria (*E. coli*)

E. coli is the only fully developed industrial host system that produces intracellular chymosin. The process was filed by Pfizer Inc. (Franke A.E., 1990) and this product was the first GM-derived food enzyme on the market. The production organism is an *E. coli* K-12 strain having the prochymosin gene under the control of the *trp* promoter in pBR322-derived vector system.

The process, as outlined in Fig. 11.4, is completely different to the process described for the *A. niger* process. A specific C-limited feeding strategy is required to gain a high biomass and product concentration. This tool is used in order to reduce the metabolic activity of the strain and avoid the accumulation of acetic acid. Under unlimited growth conditions *E. coli* forms acetic acid in a metabolic overflow which becomes toxic at higher concentrations and inhibits growth and product formation.

Chymosin accumulates in the cells as inclusion bodies (IBs). Particular procedures have been developed (Marston *et al.*, 1984) to renaturate the protein aggregates, refold and convert the protein into active chymosin. Therefore cells are disrupted by a homogenizer and IBs collected by centrifugation. After washing at pH 2, the inclusion bodies are dissolved by addition of urea (7–9 M) and the pH is increased to 10. After the renaturation step with NaCl and $Na_3(PO_4)$ buffers, the pH is stepwise adjusted to 5.5. Chymosin can be purified and concentrated by anion exchange chromatography.

11.7.3   Chymosin production with yeast (*K. lactis*)

K. lactis as host organism is used by the company DSM for the production of chymosin (Swinkels *et al.*, 1993; van den Berg *et al.*, 1990; van Ooyen *et al.* 2006). *K. lactis* has been well known in industry since the 1950s as a production strain for lactase, the enzyme that degrades the milk sugar lactose.

The bovine chymosin gene is controlled by the strong *lac4*-promoter. Efficient secretion of the product is ensured by the use of the yeast α-factor leader sequence. Like *E. coli*, *K. lactis* can be cultivated at high cell densities, using a specific feeding strategy and medium composition. In order to avoid the formation of critical concentrations of ethanol, a special fermentation

© Woodhead Publishing Limited, 2013

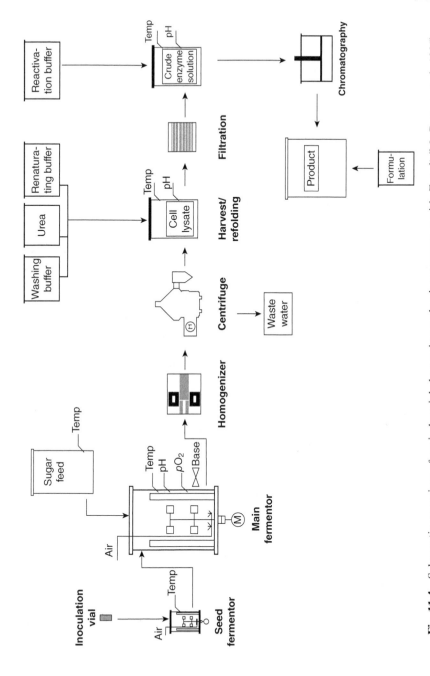

Fig. 11.4 Schematic overview of an industrial chymosin production processes with *E. coli* (McCaman *et al.*, 1985).

© Woodhead Publishing Limited, 2013

concept is applied. It has been reported that up-scaling to volumes of over 100 m^3 is possible without significant losses in productivity. The purification of product from the fermentation broth is comparable to the procedure described for *Aspergillus*, separation of biomass, concentration and purification of product.

11.7.4   Advantages and disadvantages of different fermentation processes

The main reason that fungi are considered to be one of the workhorses of the enzyme industry is their ability to secrete large amounts of product into the medium. This makes downstream processing relatively simple, recovery yields high and reduces process costs. However, the enzyme producer has to balance this with several disadvantages on the fermentation side, compared to bacterial cultivations. The long process time of 4–6 days means high costs of energy, high capacity utilisation, higher risk of contamination and the already described difficulties in morphology and rheology, resulting in challenges for up-scaling and reproducibility.

The main advantage of bacteria compared to fungi is their fast and relatively uncomplicated growth. In many cases a defined mineral salt medium can be used in fermentation, which improves the reproducibility of cultivation owing to the independence from variation in complex raw materials like soy or yeast extracts that are used in fungal fermentation processes. The fact that a bacterial fermentation takes less than half time of a fungal cultivation means a significant cost reduction for energy and a reduced need for expensive fermentation capacity. On the other hand, many bacteria do not secrete target proteins into the broth, they accumulate the product intracellularly, which is a major disadvantage for purification and compromises the overall yield. Besides the disadvantages with respect to yield, the downstream processes needed to isolate intracellular proteins from bacteria use large amounts of chemicals which pose a major waste problem.

The above-described disadvantages of bacterial systems are not true for all host systems. *K. lactis* and several *Bacillus* expression hosts combine the advantages of fungal systems (ie. high production levels and secretion in the external medium) with some of the advantages of *E. coli*-based systems like fast growth.

The choice of host systems is not one that usually can be made in advance. In most cases the production levels are tested in different host systems before making a final decision. Existing know-know, experience, available production capacity and set-up are all factors that will influence the final choice of a host system. The most commonly used host systems in industry at this time are *Aspergillus* and *Bacillus* species. However, the question of the best host/vector system for a specific product is continuously under revision. New technological or genetic developments, as well as

© Woodhead Publishing Limited, 2013

energy costs or environmental issues, move the preferences in the one or other direction.

11.8   Regulatory aspects of food enzymes

The use and production of food enzymes is controlled by many legal demands. These demands and rules differ from country to country, making introduction of new food enzymes to the market a complicated and expensive process. Even changes to the production process may require additional action in order to get approval, involving time and costs.

11.8.1   Europe

Approval of enzymes for food use in Europe is in a transition phase, going from different national approvals toward a generalized approval system Currently, only France and Denmark do not allow marketing of a food product without prior approval of the food product.

Food enzymes can either be labelled as food additives (requiring labelling of the final food product) or processing aids (no labelling requirement on the final food product). The difference is that food additives have a technological function in the final food product whereas processing aids have their technological function only during the processing of the food. As an example, milk clotting enzymes exert their technological function during the initial clotting of the milk (the processing) but have no technological function in the final cheese. Today, most enzymes are considered to be processing aids. This difference is important for the food industry owing to the different labelling requirements. Food producers prefer labels showing only 'natural' ingredients, without addition of other 'biotechnological' compounds. This wish gets even stronger when using GM-derived enzymes.

Approval of food enzymes in the EU is currently subject to some major changes, involving both existing enzymes and novel enzymes. National approval will be replaced by a European approval procedure administrated by the European Commission. Scientific support for the European Commission will come from the European Food Safety Authority, EFSA (www.efsa.europa.eu). In short, the EU will construct a positive list of enzymes whose use in food products is allowed. Enzymes not on this list may not be used in food products. Application for all enzymes, novel and currently marketed, will need to be made for inclusion on this positive list. Food safety will be judged based on an evaluation of both the enzyme (including products produced by the enzymatic reaction) and the production host being used as well as the exposure of the consumer (dietary uptake), as outlined in the expert opinion published by EFSA (The EFSA Journal, 2009). More often than not, such food safety evaluations will involve extensive, and thus expensive, animal toxicity tests.

11.8.2   Qualified presumption of safety (QPS)

According to the EFSA regulations species, and not production organisms, can qualify for QPS status in the EU (Barlow *et al.*, 2007). This requires that the species has a clear track record of safe use. Using species included on the QPS list makes approval procedures considerably easier, while the use of non-QPS production hosts will necessitate the application of a full safety study for each enzyme produced by such a host. The QPS list comprises some *Bacillus* species as well as some yeasts like *S. cerevisiae, K. lactis* and *Pichia pastoris*. Owing to the fact that the QPS system focuses on species, the QPS list does not contain well-known, well-documented and safe production organisms such as filamentous fungi. The QPS list is regularly updated and the most recent version is available at www.efsa.europe. eu. However, it will take considerable time, probably on the other side of 2020, before the final positive list will be in place.

11.8.3   Generally recognized as safe (GRAS)

In the USA, the FDA uses the term GRAS. The term GRAS is often falsely used to describe production organisms. However, the GRAS list refers to substances to be used in food. GRAS status can be obtained by either scientific evidence or by a long history of documented safe use. An example of the latter could be animal rennet that has been used for centuries in cheese processing (Olempska-Beer *et al.*, 2006). The rule of substantial equivalence is important for GM-derived products, meaning that GRAS status will be obtained for any product that is substantially equivalent to a GRAS approved product. A good example is the GRAS status for GM produced camel chymosin (Kappeler *et al.*, 2006), camel chymosin being functionally equivalent to bovine chymosin. The FDA allows introduction of enzymes to the market based on a so-called self-affirmed GRAS petition. Typically the producer will ask an external expert panel to evaluate a new product or process and document their findings in a statement.

11.8.4   Rest of the world

To make approval of food enzymes even more complicated, different approval guidelines exist all over the world. In countries like Japan and Korea, a major focus is directed towards GM-derived enzymes. For some countries (part of) the dossiers prepared for the EU or US approval can be used, for other countries a completely new file will have to be prepared, incurring considerable costs for the enzyme producers.

For more information please refer to Regulation (EC) No 1332/2008 of the European Parliament and of the Council of 16 December 2008 on food enzymes and amending Council Directive 83/417/EEC, Council Regulation (EC) No 1493/1999, Directive 2000/13/EC, Council Directive 2001/112/EC and Regulation (EC) No 258/97. A good introduction on how to perform

© Woodhead Publishing Limited, 2013

a safety evaluation of food processing enzymes can be found in Pariza and Johnson (2001).

11.9   References

AEHLE, W. (2004), *Enzymes in Industry, Production and Applications*, Wiley-VCH Verlag, Weinheim.

AGRAWAL, S., GARG, S. K. and DUTTA, S. M. (1989), 'Microbial beta-galactosidase:production, properties and industrial applicastions – a review', *Indian J Dairy Sci*, **42**(2), 251–62.

AMANULLAH, A., BLAIR, R., NIENOW, A. W. and THOMAS, C. R. (1999), 'Effects of agitation intensity on mycelial morphology and protein production in chemostat cultures of recombinant *Aspergillus oryzae*', *Biotechnol Bioeng*, **62**(4), 434–46.

ARNOUS, A. and MEYER, A. S. (2010), 'Discriminated release of phenolic substances from red wine grape skins (*Vitis vinifera* L.) by multicomponent enzymes treatment', *Biochem Eng J*, **49**(1), 68–77.

ASRAF, S. S. and GUNASEKARAN P. (2011), 'Current trends of β-galactosidase research and application'. *Current Research, Technology and Education Topics in Applied Microbiology and Microbial Biotechnology*. Microbiology book series, Number 2. Mendez-Vilas A. (ed.), Formatex Research Center, Badajoz, Spain, 880–90.

BANSAL, N., DRAKE, M. A., PIRAINO, P., BROE, K. L., HARBOE, M., FOX, P. F. and MCSWEENEY, P. L. H. (2009), 'Suitability of recombinant camel (*Camelus dromedarius*) chymosin as a coagulant for Cheddar Cheese', *Int Dairy J*, **19**, 510–17.

BARLOW, S., CHESSON, A., COLLINS, J. D., DYBING, E., FLYNN, A., FRUIJTIER-PÖLLOTH, C., HARDY, A., KNAAP, A., KUIPER, H., LE NEINDRE, P., SCHANS, J., SCHLATTER, J., SILANO, V., SKERFVING, S. and VANNIER, P. (2007), 'Introduction of a qualified presumption of safety (QPS) approach for assessment of selected microorganisms referred to EFSA', *EFSA J*, **587**, 1–16.

BARRY, D. J., CHAN, C. and WILLIAMS, G. A. (2009), 'Morphological quantification of filamentous fungal development using membrane immobilization and automatic image analysis', *J Ind Microbiol Biotechnol*, **36**(6), 787–800.

BERKA, R. M., WARD, M., WILSON, L. J., HAYENGA, K. J., KODAMA, K. H., CARLOMAGNO, L. P. and THOMPSON, S. A. (1990), 'Molecular cloning and deletion of the gene encoding aspergillopepsin A from *Aspergillus awamori*', *Gene*, **86**(2), 153–62.

BERKA, R. M., BAYLISS, F. T., BLOEBAUM, P., CULLEN, D., DUNN-COLEMAN, N., KODAMA, K. H., HAYENGA, K. J., HITZEMAN, R. A., LAMSA, M. H., PRZETAK, M., REY, M. W., WILSON, L. J. and WARD, M. (1991a), '*Aspergillus niger* var *awamori* as a host for the expression of heterologous genes,' in *Applications of Enzyme Biotechnology*, J. W. Kelly and T. O. Baldwin (eds), Plenum Press, New York, 273–92.

BERKA, R. M., KODAMA, K. H., REY, M. W., WILSON, L. J. and WARD, M. (1991b), 'The development of *Aspergillus niger* var. *awamori* as a host for the expression and secretion of heterologous gene products', *Biochem Soc Trans*, **19**(3), 681–5.

BERKA, R. M., DUNN-COLEMAN, N. and WARD, M. (1992), 'Industrial enzymes from *Aspergillus* species', *Biotechnology*, **23**, 155–202.

BODIE, E. A., ARMSTRONG, G. L. and DUNN-COLEMAN, N. S. (1994), 'Strain improvement of chymosin-producing strains of *Aspergillus niger* var. *awamori* using parasexual recombination', *Enzyme Microb Technol*, **16**(5), 376–82.

BURLINGAME, R. P. and VERDOES, J. C. (2006), 'Maximizing protein expression in filamentous fungi', *BioPharm Int*, 40–47.

CDFA (2012), http://www.cdfa.ca.gov/dairy/uploader/postings/manufacturingcost/

© Woodhead Publishing Limited, 2013

CARLSEN, M., SPOHR, A. B., NIELSEN, J. and VILLADSEN, J. (1996), 'Morphology and physiology of an alpha-amylase producing strain of *Aspergillus oryzae* during batch cultivations', *Biotechnol Bioeng*, **49**(3), 266–76.

COLLINS, I. D. (1927), 'Quantitative hydrolysis of starch by buffered taka-diastase', *Science*, **66**, 430–1.

COX, P. W. and THOMAS, C. R. (1992), 'Classification and measurement of fungal pellets by automated image analysis', *Biotechnol Bioeng*, **39**(9), 945–52.

DIAZ, M., GARCIA, A. I. and GARCIA L. A. (1996), 'Mixing power, external conversion and effectiveness in bioreactors', *Biotechnol and Bioeng*, **51**, 131–40.

DUNN-COLEMAN, N. S., BLOEBAUM, P., BERKA, R. M., BODIE, E., ROBINSON, N., ARMSTRONG, G., WARD, M., PRZETAK, M., CARTER, G. L., LACOST, R. WILSON, R. J., KODAMA, K. H., BALIU, E. F., BOWER, B., LASMA, M. and HEINSOHN, H. (1991), 'Commercial levels of chymosin production by Aspergillus', *Biotechnology (NY)*, **9**(10), 976–81.

DUNN-COLEMAN, N., BODIE, E. A., CARTER, G. L. and ARMSTRONG, G. L. (1992), 'Stability of recombinant strains under fermentation conditions,' in *Applied Molecular Genetics in Filamentous Fungi*, C. Charlier and Hall (eds), Blackie Academic & Professional, UK.

EL ENSHASY, H., HELLMUTH, K. and RINAS, U. (1999), 'Fungal morphology in submerged cultures and its relation to glucose oxidase excretion by recombinant *Aspergillus niger*', *Appl Biochem Biotechnol*, **81**(1), 1–11.

EMTAGE, J. S., ANGAL, S., DOEL, M. T., HARRIS, T. J., JENKINS, B., LILLEY, G. and LOWE, P. A. (1983), 'Synthesis of calf prochymosin (prorennin) in *Escherichia coli*', *Proc Natl Acad Sci USA*, **80**(12), 3671–5.

FOX, P. F., WALLACE, J. M., MORGAN, S., LYNCH, C. M., NILAND, E. J. and TOBIN, J. (1996), 'Acceleration of cheese ripening', *Antonie Van Leeuwenhoek*, **70**(2–4), 271-297.

FRANKE A. E. (1990), *Expression Plasmids for Improved Production of Heterologous Proteins in E. coli* US Patent 4935370.

GAGNAIRE, V., MOLLE, D., HERROUIN, M. and LEONIL, J. (2001), 'Peptides identified during Emmental cheese ripening: origin and proteolytic systems involved', *J Agric Food Chem*, **49**(9), 4402–13.

GALAGAN, J. E., CALVO, S. E., CUOMO, C., MA, L. J., WORTMAN, J. R., BATZOGLOU, S., LEE, S. I., BASTURKMEN, M., SPEVAK, C. C., CLUTTERBUCK, J., KAPITONOV, V., JURKA, J., SCAZZOCCHIO, C., FARMAN, M., BUTLER, J., PURCELL, S., HARRIS, S., BRAUS, G. H., DRAHT, O., BUSCH, S., D'ENFERT, C., BOUCHIER, C., GOLDMAN, G. H., BELL-PEDERSEN, D., GRIFFITHS-JONES, S., DOONAN, J. H., YU, J., VIENKEN, K., PAIN, A., FREITAG, M., SELKER, E. U., ARCHER, D. B., PENALVA, M. A., OAKLEY, B. R., MOMANY, M., TANAKA, T., KUMAGAI, T., ASAI, K., MACHIDA, M., NIERMAN, W. C., DENNING, D. W., CADDICK, M., HYNES, M., PAOLETTI, M., FISCHER, R., MILLER, B., DYER, P., SACHS, M. S., OSMANI, S. A. and BIRREN, B. W. (2005), 'Sequencing of *Aspergillus nidulans* and comparative analysis with *A. fumigatus* and *A. oryzae*', *Nature*, **438**(7071), 1105–15.

GREEN, B. J. and BEEZHOLD, D. H. (2011), 'Industrial fungal enzymes: an occupational allergen perspective', *J Allergy (Cairo)*, **2011**, 682574.

GRIMM, L. H., KELLY, S., KRULL, R. and HEMPEL, D. C. (2005), 'Morphology and productivity of filamentous fungi', *Appl Microbiol Biotechnol*, **69**(4), 375–84.

HARBOE, M. K. (1998), 'Rhizomucor miehei aspartic proteinases having improved properties', in *Aspartic Proteinases*, James, M. N. G. (ed.), Plenum Press, New York, 293–6.

HJORT, C. (2007), *Novel enzyme technology for food applications*, 150 edn, Woodhead Publishing, Cambridge.

HOWIE, J. K., LIN, P. Y. T., ZYZAK, D. V. and SCHAFERMEYER, R. G. (2003), *Method for Reducing Acrylamide in Foods Comprising Reducing the Level of Reducing Sugars, Foods Having Reduced Levels of Acrylamide, and Article of Commerce*, Procter + Gamble, US Patent 6989167.

© Woodhead Publishing Limited, 2013

JOHANSEN, C. L., COOLEN, L. and HUNIK, J. H. (1998), 'Influence of morphology on product formation in *Aspergillus awamori* during submerged fermentations', *Biotechnol Prog*, **14**(2), 233–40.

JUNKER, B. H., STANIK, M., BARNA, C., SALMON, P. and BUCKLAND, B. C. (1998), 'Influence of impeller type on mass transfer in fermentation vessels', *Bioprocess Eng*, **19**, 403–13.

JUSTESEN, S. F., LAMBERTH, K., NIELSEN, L. L., SCHAFER-NIELSEN, C. and BUUS, S. (2009), 'Recombinant chymosin used for exact and complete removal of a prochymosin derived fusion tag releasing intact native target protein', *Protein Sci*, **18**(5), 1023–32.

KAPPELER, S. R., VAN DEN BRINK, H. J., RAHBEK-NIELSEN, H., FARAH, Z., PUHAN, Z., HANSEN, E. B. and JOHANSEN, E. (2006), 'Characterization of recombinant camel chymosin reveals superior properties for the coagulation of bovine and camel milk', *Biochem Biophys Res Commun*, **342**(2), 647–54.

KELLY, S., GRIMM, L. H., HENGSTLER, J., SCHULTHEIS, E., KRULL, R. and HEMPEL, D. C. (2004), 'Agitation effects on submerged growth and product formation of *Aspergillus niger*', *Bioprocess Biosyst Eng*, **26**(5), 315–23.

KOHLER, R. (1972a), 'The background to Eduard Buchner's discovery of cell-free fermentation', *J History Biol*, **4**(1), 35–61.

KOHLER, R. (1972b), 'The reception of Eduard Buchner's discovery of cell-free fermentation', *J History Biol*, **5**(2), 327–53.

LAFON-LAFOURCADE, S., LONVAUD-FUNEL, A. and CARRE, E. (1983), 'Lactic acid bacteria of wines: stimulation of growth and malolactic fermentation', *Antonie Van Leeuwenhoek*, **49**(3), 349–52.

LAMSA, M. H. and BLOEBAUM, P. (1990), 'Mutation and screening to increase chymosin yield in a genetically-engineered strain of *Aspergillus awamori*', *J Ind Microbiol*, **5**, 229–38.

LI, Z. J., SHUKLA, V., FORDYCE, A. P., PEDERSEN, A. G., WENGER, K. S. and MARTEN, M. R. (2000), 'Fungal morphology and fragmentation behavior in a fed-batch *Aspergillus oryzae* fermentation at the production scale', *Biotechnol Bioeng*, **70**(3), 300–12.

LICHTENSTEIN, N. (1947), 'The proteolytic enzymes of taka-diastase', *Exp Med Surg*, **5**(2–3), 182–90.

LILBÆK, H. M., FATUM, T. M., IPSEN, R. and SOERENSEN, N. K. (2007), 'Modification of milk and whey surface properties by enzymatic hydrolysis of milk phospholipids', *J Agric Food Chem*, **55**(8), 2970–8.

LIU, S. Q. (2002), 'A review: malolactic fermentation in wine – beyond deacidification', *J Appl Microbiol*, **92**(4), 589–601.

MACHIDA, M., ASAI, K., SANO, M., TANAKA, T., KUMAGAI, T., TERAI, G., KUSUMOTO, K., ARIMA, T., AKITA, O., KASHIWAGI, Y., ABE, K., GOMI, K., HORIUCHI, H., KITAMOTO, K., KOBAYASHI, T., TAKEUCHI, M., DENNING, D. W., GALAGAN, J. E., NIERMAN, W. C., YU, J., ARCHER, D. B., BENNETT, J. W., BHATNAGAR, D., CLEVELAND, T. E., FEDOROVA, N. D., GOTOH, O., HORIKAWA, H., HOSOYAMA, A., ICHINOMIYA, M., IGARASHI, R., IWASHITA, K., JUVVADI, P. R., KATO, M., KATO, Y., KIN, T., KOKUBUN, A., MAEDA, H., MAEYAMA, N., MARUYAMA, J., NAGASAKI, H., NAKAJIMA, T., ODA, K., OKADA, K., PAULSEN, I., SAKAMOTO, K., SAWANO, T., TAKAHASHI, M., TAKASE, K., TERABAYASHI, Y., WORTMAN, J. R., YAMADA, O., YAMAGATA, Y., ANAZAWA, H., HATA, Y., KOIDE, Y., KOMORI, T., KOYAMA, Y., MINETOKI, T., SUHARNAN, S., TANAKA, A., ISONO, K., KUHARA, S., OGASAWARA, N. and KIKUCHI, H. (2005), 'Genome sequencing and analysis of *Aspergillus oryzae*', *Nature*, **438**(7071), 1157–61.

MARKETWIRE (2011), Available from Marketresearch.com (http://www.marketwire.com/press-release/industrial-enzymes-market-estimated-at-33-billion-in-2010-1395348.htm).

MARSTON, F. A., LOWER, P. A., DOEL, M. T., SCHOEMAKER, J. M., WHITE, S. and ANGAL, S. (1984), 'Purification of calf prochymosin (prorennin) synthesised in *E. coli.*', *Bio/technology*, **8**, 800–4.

© Woodhead Publishing Limited, 2013

MARSTON, F. A., ANGAL, S., WHITE, S. and LOWE, P. A. (1985), 'Solubilization and activation of recombinant calf prochymosin from *Escherichia coli*', *Biochem Soc Trans*, **13**(6), 1035.

MCCAMAN, M. T. (1989), 'Fragments of prochymosin produced in *Escherichia coli* form insoluble inclusion bodies', *J Bacteriol*, **171**(2), 1225–7.

MCCAMAN, M. T., ANDREWS, W. H. and FILES, J. G. (1985), 'Enzymatic properties of bovine prochymosin synthesized in *Escherichia coli*', *J Biotech*, **2**, 177–90.

MULLER, C., HANSEN, K., SZABO, P. and NIELSEN, J. (2003), 'Effect of deletion of chitin synthase genes on mycelial morphology and culture viscosity in *Aspergillus oryzae*', *Biotechnol Bioeng*, **81**(5), 525–34.

NIENOW, A. W. (1990), 'Agitators for mycelial fermentations', *TIBTECH*, **8**, 224–33.

NOBELPRIZE.ORG (2012), The official website of the Nobel prize. Available from http://nobelprize.org/nobel_prizes/chemistry/laureates/1946

O'CALLAGHAN, J. and DOBSON, A. D. (2006), 'Molecular characterization of ochratoxin A biosynthesis and producing fungi', *Adv Appl Microbiol*, **58**, 227–43.

OLEMPSKA-BEER, Z. S., MERKER, R. I., DITTO, M. D. and DINOVI, M. J. (2006), 'Food-processing enzymes from recombinant microorganisms – a review', *Regul Toxicol Pharmacol*, **45**(2), 144–58.

OLIVEIRA, C., GUIMARAES, P. M. and DOMINGUES, L. (2011), 'Recombinant microbial systems for improved beta-galactosidase production and biotechnological applications', *Biotechnol Adv*, **29**(6), 600–9.

OLSVIK, E., TUCKER, K. G., THOMAS, C. R. and KRISTIANSEN, B. (1993), 'Correlation of *Aspergillus niger* broth rheological properties with biomass concentration and the shape of mycelial aggregates', *Biotechnol Bioeng*, **42**(9), 1046–52.

PAPAGIANNI, M., NOKES, S. U. and FILER, K. (2001), 'Submerged and solid-state phytase fermentation by *A. niger*: effects of agitation and medium viscosity on phytase production, fungal morphology and inoculum performance', *Food Technol Biotechnol*, **39**, 319–26.

PARIZA, M. W. and JOHNSON, E. A. (2001), 'Evaluating the safety of microbial enzyme preparations used in food processing: update for a new century', *Regul Toxicol Pharmacol*, **33**(2), 173–86.

PRWEB (2012), Data available from Global Industry Analysts Inc. http://www.prweb.com/releases/industrial_enzymes/carbohydrases_proteases/prweb1569084.htm

SUGANUMA, T., FUJITA, K. and KITAHARA, K. (2007), 'Some distinguishable properties between acid-stable and neutral types of alpha-amylases from acid-producing koji', *J Biosci Bioeng*, **104**(5), 353–62.

SWINKELS, B. W., VAN OOYEN, A. J. and BONEKAMP, F. J. (1993), 'The yeast *Kluyveromyces lactis* as an efficient host for heterologous gene expression', *Antonie Van Leeuwenhoek*, **64**(2), 187–201.

THE EFSA JOURNAL (2009), 'Guidance of EFSA prepared by the Scientific Panel of Food Contact Material, Enzymes, Flavourings and Processing Aids on the Submission of a Dossier on Food Enzymes', *The EFSA Journal*, **1305**, 1–26.

TRINCI, A. P. (1974), 'A study of the kinetics of hyphal extension and branch initiation of fungal mycelia', *J Gen Microbiol*, **81**(1), 225–36.

VAN DEN BERG, J. A., VAN DER LAKEN, K. J., VAN OOYEN, A. J., RENNIERS, T. C., RIETVELD, K., SCHAAP, A., BRAKE, A. J., BISHOP, R. J., SCHULTZ, K., MOYER, D., RICHMAN M. and SHUSTER J. (1990), '*Kluyveromyces* as a host for heterologous gene expression: expression and secretion of prochymosin', *Biotechnology (NY)*, **8**(2), 135–9.

VAN DEN BRINK, J. M. and BJERRE, K. (2009), *A method for production of an alcoholic beverage with reduced content of alcohol*, Chr Hansen patent EP 20080101051.

VAN DIJCK, P. W., SELTEN, G. C. and HEMPENIUS, R. A. (2003), 'On the safety of a new generation of DSM *Aspergillus niger* enzyme production strains', *Regul Toxicol Pharmacol*, **38**, 27–35.

© Woodhead Publishing Limited, 2013

VAN OOYEN, A. J., DEKKER, P., HUANG, M., OLSTHOORN, M. M., JACOBS, D. I., COLUSSI, P. A. and TARON, C. H. (2006), 'Heterologous protein production in the yeast *Kluyveromyces lactis*', *FEMS Yeast Res*, **6**(3), 381–92.

VAN SUIJDAM, J. C. and METZ, B. (1981), 'Influence of engineering variables upon the morphology of filamentous moulds', *Biotechnol Bioeng*, **23**, 111–48.

VAN SUIJDAM, J. C., KOSSEN, N. W. F. and PAUL, P. G. (1980), 'An inoculum technique for the production of fungal pellets', *European J Appl Microbiol Biotechnol*, **10**, 211.

VILLADSEN, J., NIELSEN, J. and LIDEN, G. (2011), *Bioreaction Engineering Principles*, Springer Science+Business Media, Springer, New York.

VILLETTAZ, J. C. (2011), *Method for Production of Low Alcoholic Wine*, EP0194043 (A1), NOVO Industries.

WARD, M. (1991), 'Chymosin production in *Aspergillus*,' in *Molecular Industrial Mycology, systems and applications for filamentous fungi*, 8th edn, S. A. Leong & R. M. Berka (eds), 83–105.

WARD, M., WILSON, L. J., KODAMA, K. H., REY, M. W. and BERKA, R. M. (1990), 'Improved production of chymosin in *Aspergillus* by expression as a glucoamylase-chymosin fusion', *Biotechnology (NY)*, **8**(5), 435–40.

WUCHERPFENNIG, T., KIEP, K. A., DRIOUCH, H., WITTMANN, C. and KRULL, R. (2010), 'Morphology and rheology in filamentous cultivations', *Adv Appl Microbiol*, **72**, 89–136.

ZETELAKI, K. Z. (1970), 'The role of aeration and agitation in the production of glucose oxidase in submerged culture. II', *Biotechnol Bioeng*, **12**(3), 379–97.

ZLOKARNIK, M. (2000a), *Scale-Up*, Wiley-CH Verlag GmbH, Weinheim 3-527-29864-9.

ZLOKARNIK, M. (2000b), 'Typical problems and mistakes in application of modelling,' in *Scale-Up*, M. Zlokarnik (ed.), Wiley-CH Verlag GmbH, Weinheim, 83–94.

© Woodhead Publishing Limited, 2013

12

Microbial production of organic acids for use in food

M. Sauer, D. Mattanovich, and H. Marx, University of Natural Resources and Life Sciences (BOKU Wien-VIBT), Austria

DOI: 10.1533/9780857093547.2.288

Abstract: Organic acids are traditional products of food- and bio-technology. However, at the same time organic acids are among the most promising future products of industrial microbiology, owing to their possible use as building block chemicals. Considerable knowledge about microbial production processes on an industrial scale has been gained in producing organic acids. While strain and process development were previously based on classical methods, modern technologies like systems biotechnology and metabolic engineering are becoming crucial in this fast moving field of biotechnology. This chapter focuses on the respective microbial cell factories: natural producers and recombinant microorganisms, fungi and bacteria.

Key words: industrial microbiology, microbial cell factory, organic acid production.

12.1 Introduction

Organic acids are traditional products of food- and bio-technology. However, recently organic acids become more interesting for the chemical industry as building blocks. A wide variety of organic acids are of use for polymer and solvent production, but also as starting compounds or co-substrates for pharmaceuticals.

In line with the topic of this volume this chapter will focus on organic acids for food use. Furthermore, we will focus on biotechnology and not on traditional food technology. Consequently, the bulk products acetic and lactic acid will not be covered. Acetic acid for industrial use is produced chemically from petrol-derived resources. Acetic acid for food use (about 10% of the annual world production) is produced by microbial processes, as many food purity laws require that vinegar used in foods must be of biological origin (production processes reviewed by Raspor *et al.*, 2008). Lactic acid for food use is a traditional product, in most cases directly

© Woodhead Publishing Limited, 2013

produced by fermentation of the food with lactic acid bacteria. The industrially relevant production of purified lactic acid as a starting material for polymer production is predominantly based on yeast as production host (reviewed by Sauer *et al.*, 2010 and Miller *et al.*, 2011).

An interesting aspect of microbial organic acid production is that some of them, such as citric acid, are 'old' biotech products. Considerable knowledge about microbial production processes on an industrial scale has been gained in producing organic acids. For example, the fast establishment of mass production of antibiotics was only possible owing to experience with fungal bioprocesses from citric acid production. Clearly, strain and process development were based on classical methods, like mutagenesis and selection. However, modern technologies are becoming more important in this fast moving field of biotechnology and organic acids are among the most promising new products of industrial microbiology.

12.2 From filamentous fungi to genetically engineered bacteria and yeasts

Interestingly, filamentous fungi account for the majority of traditional processes for microbial organic acid production. First of all they are natural producers of various valuable acids and consequently also quite resistant to high concentrations of these substances. This has allowed establishment of efficient production processes by traditional means. However, many modern techniques based on genetic engineering are very difficult to apply to filamentous fungi. As a consequence, recent developments often focus on bacterial hosts and yeasts. The following part of this chapter exemplifies traditional processes and recent developments in the most important organic acids on the market. Figure 12.1 shows the chemical structures of the acids and Table 12.1 summarizes the world annual production and the fractions used for food applications.

12.3 Gluconic acid production

Gluconic acid is a mild organic acid with an annual market volume of around 100 000 metric tonnes per year ($t\ y^{-1}$). It is sold as δ-lactone or gluconate and serves in the food industry as a mild acidulant. It produces and improves a mild sour taste and complexes possible traces of heavy metals. Other uses of gluconic acid include the construction, textile and pharmaceutical industries, where gluconic acid in its various forms is useful because it forms water-soluble complexes with a variety of divalent and trivalent metal ions. The actual main product is sodium gluconate, which makes up about 80% of the gluconic acid market (Kirimura *et al.*, 2011). The conversion of glucose to gluconic acid is a site specific oxidation. The aldehyde

© Woodhead Publishing Limited, 2013

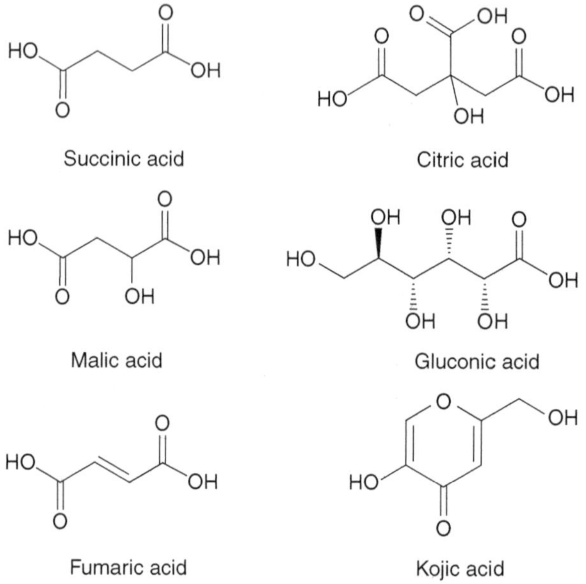

Fig. 12.1   Chemical structures of food-related organic acids.

group of glucose (C1) is oxidized to a carboxyl group. While this reaction can be achieved by chemical means, low yields and selectivity strongly favour microbial production.

Microbial processes for gluconic acid production are approved by the FDA. A variety of organisms have been described to accumulate this acid: fungal species like *Aspergillus niger* or *Penicillium luteum* or bacterial species of the genera *Gluconobacter*, *Pseudomonas* or *Acetobacter* among others (Ramachandran *et al.*, 2006). Most industrial production processes nowadays rely on *A. niger* or *Gluconobacter* (Kubicek *et al.*, 2010). Interestingly, the biochemical pathways for the conversion of glucose to gluconic acid are quite distinct in fungal as opposed to bacterial species.

12.3.1   Gluconic acid production with *Aspergillus niger*

Gluconic acid production by *A. niger* is an aerobic fermentation with a high oxygen demand. The process resembles more an enzymatic conversion rather than a microbial process. The key enzyme is glucose oxidase, a homodimeric flavoprotein, localized in the mycelial cell wall. Consequently, the conversion takes place entirely extracellularly with a net reaction: glucose + ½ O_2 → gluconic acid. Figure 12.2 outlines the involved enzymatic steps.

The production parameters are defined by the enzyme properties of glucose oxidase. The enzyme is rapidly inactivated at low pH, therefore the pH of the culture has to be kept between 4.5 and 6.5 for efficient gluconic

© Woodhead Publishing Limited, 2013

Table 12.1 Annual production of food-related organic acids

Organic acid	Annual production (t)	Fraction used for food applications (%)	Fraction produced by microbial processes (%)	Reference
Citric acid	2000 000[a]	70	100	Chemco (2010)
Acetic acid	6500 000	10	10	Continental (2011)
Lactic acid	370 000	<50[b]	100	Miller et al. (2011)
Gluconic acid	100 000	34.5	100	Singh et al. (2007)
				Kirimura et al. (2011)
Succinic acid	35 000	9	25	van den Tweel (2010)
	2000 000[c]	4[c]	100[c]	
Fumaric acid	90 000	22	0	Roa Engel et al. (2008)
		33[d]		
Malic acid	40 000	92[e]	0	Goldberg et al. (2006)
Kojic acid	no data available	no data available	no data available	

Note that data marked with footnotes indicate a different source from the Reference column.
[a]Personal information; [b]Malveda et al., 2009; [c]Prediction for 2020; [d]Bizarri and Blagoev, 2010a; [e]Bizarri and Blagoev, 2010b.

© Woodhead Publishing Limited, 2013

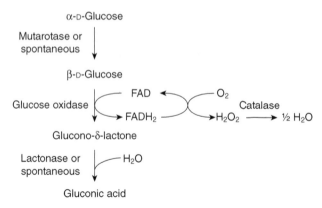

Fig. 12.2   Fungal gluconic acid biosynthetic pathway.

acid production. High glucose concentrations activate glucose oxidase. Typically, processes start with 110–250 g l^{-1} glucose. Oxygen as a substrate has to be kept at a high level by vigorous aeration throughout the process. In fact the (K_m) value Michaelis constant of glucose oxidase is in the range of air saturation in water. Hydrogen peroxide, formed by flavin adenin dinucleotide (FAD) recycling (see Fig. 12.2) is a strong inhibitor of glucose oxidase. Sufficient catalase activity is consequently of utmost importance for an efficient process. Nitrogen and phosphorus sources have to be kept at a very low level in order to limit growth (= loss of carbon source for biomass production).

The process of fungal gluconic acid production is extraordinarily efficient. Yields typically exceed 90%. Up to 98% have been reported (Singh and Kumar, 2007). Fermentation times are short, between 24 and 60 h. It is therefore not surprising that strain improvement did not play a significant role up to now. It has been suggested that the quick conversion of glucose to gluconic acid outside the fungal cell leads to an evolutionary advantage. *A. niger* lowers the pH of the environment quickly, thereby inhibiting competitive microorganisms. Furthermore, gluconic acid is a poor carbon source for competing microorganisms, but a good one for *A. niger*.

A future trend for process improvement could be the conversion of glucose by purified enzymes (Magnuson and Lasure, 2004; Wong *et al.*, 2008). This could ease the purification significantly, as all media components, which are only necessary to support growth of the fungus can be omitted. However, up to date the production of purified enzymes is still too expensive.

12.3.2   Gluconic acid production by bacteria

Gluconic acid production by bacteria relies on different enzyme systems than those for glucose oxidase (Ramachandran *et al.*, 2006) and their

© Woodhead Publishing Limited, 2013

metabolism is quite distinct from the fungal metabolism. *Gluconobacter oxydans*, for example, oxidizes glucose via two alternative pathways. The first pathway includes uptake and intracellular oxidation of glucose. The aim in this case is dissimilation of glucose by oxidation via the pentose phosphate pathway. The second route consists of the direct oxidation of glucose in the periplasmic space by membrane-bound glucose dehydrogenase. In fact both oxidizing enzymes are glucose dehydrogenases. However, the periplasmatic form depends on pyrroloquinoline quinone (PQQ) as coenzyme and the intracellular form is $NADP^+$ dependent.

The fate of extracellular gluconate under natural conditions is to be further oxidized to 2-keto-D-gluconate and 2,5-diketo-D-gluconate by the membrane-bound enzymes gluconate-dehydrogenase and 2-keto-D-gluconate dehydrogenase. However, under production conditions these ezymes are inhibited and gluconate is the major accumulated extracellular product. The suggested reason for this special metabolism of *G. oxydans* is an incomplete tricarboxylic acid (TCA) cycle, preventing the complete oxidation of the carbon source (De Muynck *et al.* 2007). Instead, *G. oxydans* posesses several membrane-bound dehydrogenases, which channel electrons into the respiratory chain. *G. oxydans* is therefore capable of oxidizing various compounds (like sugars and alcohols) incompletely, which leads to the accumulation of the products in the medium.

12.4   Oxidative branch of the citric acid cycle

12.4.1   Citric acid

Citric acid production with Aspergillus niger

Fungal citric acid production is an old industrial process. The first patents for this technology were issued before 1900 (reviewed by Berovič and Legiša, 2007). To date citric acid is one of the major biotech products on the market. The annual production of citric acid is approximately 2 million tonnes, therefore it is one of the most outstanding products in terms of volume. However, the process is also remarkable in terms of titer and yield, which is not surprising considering the long history of strain and process improvement.

The main use of citric acid is the food industry. It is used as a buffering, flavouring and preservative agent, particularly in soft drinks. It is the most versatile and widely used acidulant owing to its pleasant taste and its property of enhancing flavours.

In contrast to the majority of biotechnologically produced chemicals, citric acid was produced industrially before the development of a microbial process. Initially, the production was based on extraction of the acid from lemons until it was discovered that various filamentous fungi, but particularly *A. niger* accumulate citric acid in high amounts under certain conditions (reviewed by Berovič and Legiša, 2007 and Papagianni, 2007).

© Woodhead Publishing Limited, 2013

Nowadays the filamentous fungus *A. niger* is the major industrial production organism for citric acid.

Initially, industrial scale production relied on surface fermentation. The fungus was grown on static medium in trays, where it formed a mycelial mat. This mat was easy to separate from the liquid medium for purification of the acid. Another possibility for fungal production processes is solid fermentation, which was particularly widespread in Asia. In this case the substrate is solid acting as carrier and nutrient source. However, nowadays the majority of citric acid is produced in submerged production processes. Such processes are easily employed on a very large scale and provide several advantages such as higher productivity and yields and lower contamination risk. Most frequently, submerged citric acid fermentations are performed batchwise. Typically, the bioprocess is carried out around 30°C, lasting 5–10 days, with the cultivation media containing 150–220 g l^{-1} carbon sources such as sucrose (molasses) or glucose (from starch hydrolysate).

Efficient production of citric acid by *A. niger* is dependent on some parameters, which have been determined empirically: first of all the substrate concentration at the beginning of the process has to be high and the dissolved oxygen concentration needs to be high throughout the process. The nitrogen and phosphate concentrations in contrast have to be low and the concentrations of certain trace metals, such as manganese and iron need also to be kept low. Finally, the pH of the culture has to be low. The empirical definition of these parameters enabled the development of highly efficient fungal processes, which are currently used. The yield from glucose approaches 90% and productivity can exceed 2 g l^{-1} h^{-1}. The high substrate concentration is required to establish a state of fast substrate uptake (most probably by simple diffusion) and high glycolytic flux, which is essential for efficient citric acid production. It is believed that high glycolytic flux leads to unfavourably high intracellular concentrations of carboxylic acids, which in turn leads to citric acid production (Legiša and Mattey, 2007).

Peculiarly, citric acid is in principle a potent inhibitor of glycolysis. It has been shown that citric acid effectively inhibits phosphofructokinase I (*pfk I*), which is the most important enzyme controlling the flux through glycolysis. Initially, it was believed that a certain intracellular ammonium concentration would relieve this inhibition (Papagianni, 2007). However, Papagianni *et al.* (2005) showed that the intracellular ammonium concentration is low throughout the citric acid production process. Legiša and Mattey (2007) describe post-translational modifications of *pfk I* triggered by a drop in the intracellular pH and a cAMP (cyclic adenosine monophosphate) peak before the onset of citric acid accumulation. The enzyme is cleaved and phosphorylated leading to a fragment that is highly active and highly inducible by ammonium, but less susceptible to citrate inhibition, which explains why glycolysis is not repressed. However, this connects the high intracellular tricarboxylic acid concentration at the beginning of the process with a drop in intracellular pH and the onset of citric acid production.

© Woodhead Publishing Limited, 2013

The biochemical understanding of the high dissolved oxygen requirement is that citric acid overproduction depends on an efficient reoxidation of glycolytically produced nicotinamide adenine dinucleotide (NADH) by an alternative oxidase (Hattori *et al.*, 2009). This enzyme catalyzes the reduction of oxygen to water without direct coupling to adenosine triphosphate (ATP) production. This results in lower levels of ATP and consequently decreased anabolism, which in turn allows high fluxes through glycolysis and efficient conversion of glucose into citric acid with very limited cellular growth. Interestingly, the energy set free by this reaction leads to a remarkable heat production, requiring efficient cooling of the bioreactors. An unwanted drop in oxygen tension would lead to a shift to the main (cytochrome) respiratory pathway, resulting in a significantly decreased citric acid accumulation.

The physiological influence of the trace metals on citric acid production is less clear. Iron and manganese strongly affect cellular morphology and the correct morphological appearance of the cells is a crucial prerequisite for efficient organic acid production (Magnuson and Lasure, 2004; Grimm *et al.*, 2005). Only growth in the form of pellets of less than 1 mm in diameter is associated with high production rates and yields (Magnuson and Lasure, 2004). When the trace metal concentration is not tightly controlled, growth of unproductive filamentous mycelia is promoted instead of pellet formation. The conditions for obtaining such pellets have been determined empirically (Liao *et al.*, 2007; Liu *et al.*, 2008).

The correlation of fungal morphology and productivity is a general finding. While for efficient acid production pellet growth is required, protein secretion (e.g. enzymes) is optimal when the fungus grows as freely dispersed mycelium (Teng *et al.*, 2009; Driouch *et al.* 2010). The significance of the correct fungal morphology for good productivity has stimulated various attempts to manipulate the growth characteristics of the fungi. The bioprocess parameters are the first and obvious conditions, influencing morphology and productivity, as outlined before. However, desirable morphological characteristics were also achieved by random mutagenesis and selection. This has been successful, for example, for hemicellulolytic enzyme production (De Nicolas-Santiago *et al.*, 2006). A very different approach is the addition of inorganic microparticles to the fungal culture (Kaup *et al.*, 2007; Driouch *et al.* 2010). Driouch *et al.* could show that distinct morphological forms like pellets of different size, free dispersed mycelium, and short hyphae fragments could be reproducibly created by variation of the size and concentration of the micromaterial added. A final approach is to gain understanding of the biological regulation of the morphology development. By understanding the signalling networks it will be possible in the future to modulate the regulating signals in order to obtain the desired morphology.

A first molecular biology approach has been shown by Dai *et al.* (2004), who proved that antisense expression of a putative amino acid transporter

© Woodhead Publishing Limited, 2013

allows the formation of productive cell pellets even in the presence of normally inhibiting Mn^{2+} concentrations. Morphology engineering has been getting the attention which it deserves only recently (Grimm *et al.*, 2005; Driouch *et al.*, 2010; Wucherpfennig *et al.* 2011).

Finally, the production of organic acids by *A. niger* has been shown to be strictly dependent on the pH of the culture broth. Between pH 6 to 8 the fungus predominantly produces oxalic acid. Around pH 5.5 gluconic acid is the main product and citric acid production begins at pH 3.0 and is optimal below pH 2.0 (Andersen *et al.*, 2009). This suggests that an evolutionary process has been selected for production of a given acid at different pH values. This also implies that acid production in *A. niger* does not simply stem from a kind of overflow metabolism, but rather from an evolutionary objective (Andersen *et al.*, 2009). Consequently, the metabolism is regulated in an elaborate signal transduction pathway dependent on ambient pH (Andersen *et al.*, 2009).

Strategies for process improvement
Physiological understanding and rational strain design, rather than random mutagenesis and selection seem to be key to further improvements. Technologies like -omics analyses are used to understand the differences between a high producer and a low producer and metabolic modelling are applied. These technologies are outlined elsewhere in this book. Here, a few examples relating to *A. niger* and citric acid production shall be outlined.

Alvarez-Vasquez *et al.* (2000) presented an extensive mathematical model of the metabolism of *A. niger* as a rational basis for the optimization of citric acid production. The model predicts room for a five-fold increase in the citric acid production rate, even keeping the overall concentration of enzymes constant. The model predicts that a minimum of 13 enzymatic activities must be modulated before a significant increase in citric acid production rate can be observed. Transport processes, such as sugar uptake, citric acid excretion and transport between the cytosol and the mitochondrion, are prominent metabolic activities to be modulated. In fact, these transport processes are emerging as a crucial bottleneck for microbial organic acid production in general.

García and Torres (2011) integrated a set of new elements into the mathematical model, with the aim of relating the citrate excretion in *A. niger* to cytoplasmic pH regulation. They suggest that the fungus requires intracellular citric acid accumulation as well as secretion to regulate the cytoplasmic pH against external acidic conditions. The field of mathematical modelling is moving fast. So far the majority of models rely solely on stoichiometry. However, the kinetics of enzymatic reactions and the interdependency of the concentrations of various compounds and reaction rates are obviously essential parameters of metabolic networks. A first step in the right direction has been reported by Pozo *et al.* (2011) who suggest generalized mass action (GMA) models to describe citric acid production

© Woodhead Publishing Limited, 2013

with *A. niger*. Although detailed GMA genome-wide models are far in the future, the results published show that optimization results by this approach go beyond those possible with stoichiometric models.

It is indispensable for further modelling approaches to gain more information about the detailed peculiarities and metabolic functions of this fungus. Comparative metabolic genomics relating metabolic annotations of two *A. niger* strains with six other *Aspergilli* revealed more than 1100 enzyme-encoding genes, which are unique to *A. niger* (Sun *et al.*, 2007). However, most of these genes are additional copies of those genes that are present in other comparable fungi, indicating that it is genetic multiplicity which makes *A. niger* such a unique organism. In fact, only nine genes were identified to encode enzymes with EC numbers exclusively found in *A. niger*. These are mostly involved in the biosynthesis of complex secondary metabolites and degradation of aromatic compounds, so they are not involved in citric acid production. de Jongh and Nielsen (2008) describe how model predictions can be correlated with metabolic facts: in 1981 Röhr and Kubicek had already shown that a significant increase in cytosolic malate concentration precedes the onset of citrate production. Cytosolic malate concentration is linked to citric acid production because the export of citric acid from the mitochondria is linked to the import (antiport) of malate (Karaffa and Kubicek, 2003, see Fig. 12.3). A metabolic flux balance model has predicted that one of the rate-limiting steps for over-all citric acid production is citrate export from the mitochondrion (Guebel and Torres, 2001). de Jongh and Nielsen (2008) finally speculated that mitochondrial citrate export might be increased by increasing the cytosolic concentration of four carbon dicarboxylic acids (such as malate). To this end they

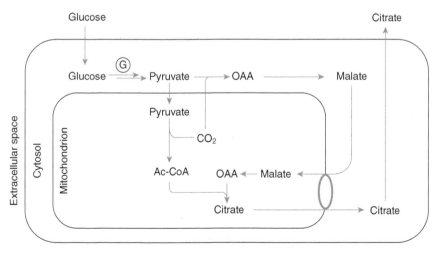

Fig. 12.3   Citric acid production by *A. niger*. OAA, oxaloacetic acid; Ac-CoA, acetyl-coenzyme A; G, glycolysis.

© Woodhead Publishing Limited, 2013

overexpressed several genes and found that yield and productivity increased as predicted.

Citric acid production with Yarrowia lipolytica
In addition to the well-established filamentous fungal species, the yeast *Y. lipolytica* has been suggested as a microbial cell factory for citric acid. It has been shown that *Y. lipolytica* naturally accumulates high amounts of organic acids, such as pyruvic, citric or isocitric acid in the presence of an excess of carbon source when growth is limited for example, by exhaustion of nitrogen, or mineral salts (Anastassiadis *et al.*, 2002; Förster *et al.*, 2007a). Advantages of *Y. lipolytica* compared to *A. niger* include less sensitivity to low oxygen concentrations and heavy metals in the medium (Stottmeister and Hoppe, 1991), but also a larger substrate range. For example n-paraffins (Crolla and Kennedy, 2004) and fatty acids (like animal fats) (Papanikolaou *et al.*, 2006), which are not converted by *A. niger* are efficiently metabolized by *Y. lipolytica*. Peculiarly, sucrose, a major carbon source for citric acid production by *A. niger*, cannot be converted by wild-type *Y. lipolytica* strains because they lack the enzyme invertase. However, recombinant introduction of the respective gene, allows this yeast to convert sucrose efficiently (Förster *et al.*, 2007b; Moeller *et al.*, 2011b).

The major disadvantage of *Y. lipolytica* as microbial host for citric acid production is the simultaneous secretion of isocitric acid, which disturbs the crystallization process of citric acid, which is required during purification (Förster *et al.*, 2007a). The ratio of accumulated citric acid compared with isocitric acid is on the one hand strain dependent and on the other hand significantly influenced by the growth conditions, such as the substrate. On plant oils or n-alkanes a wild-type strain typically forms about 35–45% isocitric acid. Grown on carbohydrates or glycerol, wild-type strains secrete 10–12% isocitric acid. Genetic engineering approaches have been suggested to influence the citric/isocitric ratio favourably. Examples include the overexpression of isocitrate lyase, which decreased isocitrate significantly (Förster *et al.*, 2007a) while overexpression of aconitase increased isocitrate accumulation (Holz *et al.*, 2009).

Citric acid concentrations in *Y. lipolytica* cultures growing on sucrose reach up to 140 g l^{-1} (Förster *et al.*, 2007b). However, yields are generally lower than with *A. niger*. Repeated fed-batch cultivation on glucose as the carbon source achieve citric acid yields between 0.39 and 0.69 g g^{-1} (Moeller *et al.*, 2010, 2011a). Productivities vary between 0.79 g l^{-1} h^{-1} and 1.41 g l^{-1} h^{-1}. However, the citric acid/isocitric acid ratio remains around 90%. Another important carbon source suggested for citric acid production with *Y. lipolytica* is glycerol (Rymowicz *et al.*, 2006; Makri *et al.*, 2010). Also in this case high titers for citric acid have been reported. Rywińska and Rymowicz (2010) achieved 197 g l^{-1} citric acid in repeated batch cultures. The average production was 154 g l^{-1}, corresponding to a yield of 0.78 g g^{-1} and a productivity of 1.05 g l^{-1} h^{-1}. Isocitric acid concentrations

© Woodhead Publishing Limited, 2013

were comparatively low, however polyols, such as erythritol, accumulated in significant amounts.

Interestingly, morphology also seems to play a crucial role in citric acid production in the case of *Y. lipolytica*. It has been shown that addition of Triton-X-100 to the culture leads to a significant suppression of pseudohyphae formation. The cells are smaller and thinner and citric acid accmuluation is increased up to 1.4-fold (Mirbagheri *et al.*, 2011). There are indications that *Y. lipolytica* is used on an industrial scale, although few details are known of actual production methods (Lopez-Garcia, 2002).

12.4.2   Succinic acid

Apart from the food industry, where succinic acid is used as acidulant, flavour and antimicrobial agent, this acid is currently used for pharmaceutical applications, in detergents and surfactants, as well as an ion chelator. The name derives from the Latin *succinum*, meaning amber, from which the acid was first obtained. The market for petrochemically produced succinic acid is currently about 30,000 t y^{-1}. However, significant efforts have been made to produce succinic acid microbially from renewable resources. With the price becoming competitive, succinic acid could replace petroleum-derived maleic anhydride, which has a market volume of 1,700,000 t y^{-1}. This market volume is conceivable for biogenic succinic acid as it is a versatile building-block chemical suitable for a wide variety of uses (Sauer *et al.*, 2008).

Succinic acid is one of the intermediates of the TCA cycle, but also an end-product of the mixed acid fermentation of various bacteria. It can therefore be produced oxidatively or reductively (Raab and Lang, 2011). See Fig. 12.4. for a summary of the involved metabolic pathways.

The fermentative biosynthesis of succinic acid via the reductive branch of the TCA cycle enables a theoretical yield of 2 mol mol^{-1} glucose because CO_2 is fixed. However, the energy yield of this pathway is very limited, not providing any excess ATP for maintenance or transport requirements and furthermore the redox balance is uneven. Unwanted by-product formation compromising yield and purification of the product or addition of reductants such as H_2 can close the redox balance, but are not favourable measures for an economic production process.

Redirecting the carbon flux into the glyoxylate shunt allows the oxidative production of succinic acid. The two decarboxylating steps of the oxidative branch of the TCA cycle are thereby by-passed. However, the maximal theoretical yield amounts to only 1 mol succinic acid mol^{-1} glucose. A third mode for succinic acid production is the simultaneous activity of the reductive branch of the TCA cycle and the glyoxylate shunt. This oxido/reductive route as suggested by Raab and Lang (2011) provides a positive energy balance and a closed redox balance with a succinic acid yield of glucose of 1.71 mol mol^{-1}.

© Woodhead Publishing Limited, 2013

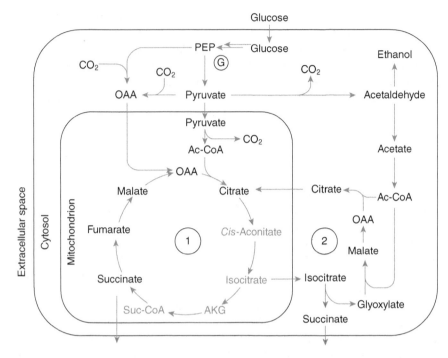

Fig. 12.4   Possible succinic acid production pathways. The reductive and the oxidative tricarboxylic acid cycle (1) and the glyoxylate shunt (2) are depicted. OAA, oxaloacetic acid; Ac-CoA, Acetyl-coenzyme A; Suc-CoA, succinyl-coenzyme A; AKG, alpha-ketoglutarate; G, glycolysis.

A variety of bacteria produce succinic acid as a natural product. First of all it is a minor product of the mixed acid fermentation of, for example, *Escherichia coli*. However, certain bacteria have been isolated from intestinal environments of different species, which accumulate significant amounts of succinic acid as one of their major products. In fact, succinate is an important metabolic intermediate in the rumen, where several bacteria obtain energy by decarboxylating succinate to propionate, which in turn serves as a nutrient for the ruminant (Leng *et al.*, 1967). Among the described succinic acid producing rumen bacteria are *Anaerobiospirillum succiniproducens*, *Actinobacillus succinogenes*, *Bacteroides fragilis* and *Mannheimia succiniproducens*.

Escherichia coli *as recombinant production host*
Figure 12.5 summarizes the metabolic pathways of the mixed acid fermentation of *E. coli*.

Strategies for producing succinate with *E. coli* centre around the abolition of by-products in order to increase production of the desired acid. Thus,

© Woodhead Publishing Limited, 2013

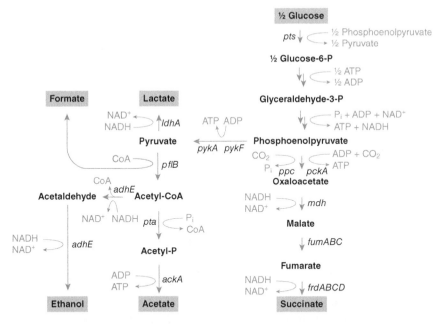

Fig. 12.5 Mixed acid fermentation. Major pathways of the mixed acid fermentation of *E. coli*. Corresponding pathways occur in other bacteria. However, the gene names are given only for *E. coli. ack*A, acetate kinase; *adh*E, alcohol dehydrogenase; *frd*ABCD, fumarate reductase; *fum*ABC, fumarase; *ldh*A, lactate dehydrogenase; *mdh*, malate dehydrogenase; *pck*A, phosphoenolpyruvate carboxykinase; *pff*B, pyruvate formate-lyase; *pta*, phosphate acetyltransferase; *pts*, sugar phosphotransferase system; *ppc*, phosphoenolpyruvate carboxylase; *pyk*A, pyruvate kinase II; *pyk*F, pyruvate kinase I; Pi, inorganic phosphate.

focus is given to two main pathways: the phosphoenolpyruvate (PEP)–pyruvate–oxaloacetate node and the TCA/glyoxylate route (Fig. 12.5). The PEP–pyruvate–oxaloacetate node forms a central point in the metabolism distributing carbon towards different biomass precursors, such as amino acids and fatty acids, and the energy metabolism (anaplerotic, catabolic and gluconeogenic route).

A key step in the metabolic pathway to succinate is the carboxylation of PEP to oxaloacetate. Two enzymes are present in *E. coli*: phosphoenolpyruvate carboxylase (*ppc*) and phosphoenolpyruvate carboxykinase (*pckA*). Succinic acid production from glucose by *E. coli* was significantly increased by overexpression of the endogenous *ppc*. However, overexpression of the endogenous *pckA*, which appears more promising at first glance because in contrast to *ppc* it links carboxylation with substrate level phosphorylation, had no effect (Millard *et al.*, 1996). The introduction of an *A. Succinogenes* phosphoenolpyruvate carboxykinase gene into *E. coli* resulted in a 6.5-fold increase of succinate production (Kim *et al.*, 2004). In *E. coli, pckA*

© Woodhead Publishing Limited, 2013

functions as a part of gluconeogenesis, which means that the enzyme has a high affinity for ATP and oxaloacetate, but a low affinity for ADP and carbon dioxide or carbonate. The corresponding genes from natural succinic acid producers, such as *A. succinogenes* have the opposite properties. Their K_m values favour the reaction from PEP and ADP to oxaloacetate and ATP, explaining these results.

A negative property of the natural metabolism of *E.coli* for succinic acid production is the link of PEP to the glucose uptake by the sugar phosphotransferase system (pts) system. From each molecule of glucose which is taken up, only one molecule of PEP is available for succinate production. Knocking the pts system out avoids the simultaneous conversion of PEP to pyruvate with glucose uptake. In this case, alternative sugar transporters, such as galactose permease or the glucokinase system take over. This increases the succinate yield significantly (Chatterjee *et al.*, 2001), however the growth rate of *E. coli* is negatively influenced since these sugar transporters are less efficient compared to the sugar pts system. Instead of knocking the pts system out, *ppc* and *pckA* can be overexpressed to draw PEP away from the pts system, aiming for the other sugar transporters to take over. Another important point is the decrease of by-product formation. The metabolic flux to succinic acid has to be increased while the metabolic flux to by-products has to be decreased.

Overexpression of *ppc* together with malate dehydrogenase (*mdh*) and inactivation of the pyruvate-formate lyase (*pflB*) and lactate dehydrogenase (*ldhA*) led to a strain producing 12.74 g l^{-1} of succinate with a molar yield of 0.98 mol mol^{-1} glucose (Millard *et al.*, 1996; Stols *et al.*, 1997). A very high yield of 1.6 mol succinate mol^{-1} glucose was reached with a strain overexpressing heterologous genes of citrate synthase (*citZ*) and pyruvate carboxylase (*pyc*) and inactivating the genes of isocitrate lyase repressor (*iclR*), alcohol dehydrogenase (*adhE*), lactate dehydrogenase (*ldhA*) and acetate kinase (*ackA*) (Sánchez *et al.*, 2005). Jantama *et al.* (2008) combined metabolic engineering with an evolutionary approach. Their strain accumulated a succinate concentration of 86.49 g l^{-1} with a yield of 1.41 mol mol^{-1}.

All of these processes are anaerobic, as the pathway leading to succinic acid production derives from the mixed acid fermentation. This poses a problem for the bioprocess: clearly, biomass concentration is of key importance for the desired high succinic acid productivities. However, growth of the cells is slow under anaerobic conditions, which is why usually dual-phase fermentation processes are employed. In a first aerobic phase biomass is formed efficiently, followed by a second anaerobic phase for succinic acid production (Vemuri *et al.*, 2002; Lu *et al.*, 2009a). With this process, the highest succinate productivity of 2.9 g l^{-1} h^{-1} was reported by Andersson *et al.* (2007).

Interestingly, a metabolic flux analysis using ^{13}C-labelled glucose showed that 61% of the PEP partitioned to oxaloacetate and 39% partitioned to pyruvate and 93% of the succinate was formed via the reductive arm of the

© Woodhead Publishing Limited, 2013

TCA cycle (Lu *et al.*, 2009b). To resolve the slow growth problem, Lin *et al.* (2005a, b) set out to construct *E. coli* strains producing succinic acid under aerobic conditions. Mutating a variety of genes (succinate dehydrogenase, *sdhAB*, phosphotransacetylase, *pta*, acetate kinase, *ackA*, pyruvate oxidase, *poxB*, isocitrate lyase repressor, *iclR* and *ptsG*), the *E. coli* strain produces 58.3 g l^{-1} succinic acid with a yield of 0.94 mol mol^{-1} aerobically in a fed-batch (Lin *et al.*, 2005b).

Corynebacterium glutamicum *for succinic acid production*
A similar approach to that used for *E. coli* has been chosen to construct a succinic acid producing *Corynebacterium glutamicum* strain (Okino *et al.*, 2005, 2008). This bacterium is currently used for industrial amino acid production and a wealth of knowledge, including genome sequence, metabolic models and well-developed genetic tools are available. *C. glutamicum* naturally produces succinic acid among other acids under oxygen-deprived non-growth conditions.

   Aerobic biomass formation followed by a fed-batch under anoxic conditions, with the addition of bicarbonate, led to the accumulation of 23 g l^{-1} succinic acid and 95 g l^{-1} lactic acid by a wild-type strain (Okino *et al.*, 2005). To abolish lactate production, the lactate dehydrogenase gene (*ldhA*) was deleted and succinic acid production was enhanced by overexpression of pyruvate carboxylase gene (*pyc*). This strain efficiently produced succinic acid with a very high volumetric productivity of 11.8 g l^{-1} h^{-1}. The product concentration reached 146 g l^{-1} within 46 h, with a yield of 1.40 mol mol^{-1} (Okino *et al.*, 2008).

Natural producers
While filamentous fungi, including *Penicillium simplicissimum*, have been shown to accumulate succinic acid naturally (Gallmetzer *et al.*, 2002) this has never led to titers and yields that were sufficiently economic for an industrial exploitation. In contrast to other organic acids, bacteria are efficient natural producers.

Actinobacillus succinogenes
A. succinogenes is an anaerobic bacterium producing relatively large amounts of succinic acid from a broad range of carbon sources, including glucose, fructose, galactose, mannose, arabinose or xylose (Guettler *et al.*, 1999). A further advantageous property of *A. succinogenes* is a good tolerance to high sugar concentrations owing to its moderate osmophily. However, succinic acid obviously is not the only fermentative product of the metabolism of *A. succinogenes* (McKinley and Vieille, 2008). Figure 12.5 shows the five branches of the pathway network, which lead from PEP to the production of succinate, acetate, formate, lactate and ethanol. Flux distribution between succinate and alternative fermentation products is affected by environmental conditions. Furthermore, *A. succinogenes*, like

© Woodhead Publishing Limited, 2013

all the other ruminal bacteria, is characterized by a variety of auxotrophies, which demand expensive media components and lead to problems in the purification of the product.

In any case, the key to a successful succinic acid production process is not only a process design for optimal productivity and titer of succinate but also to minimize the accumulation of the various by-products. Different strategies have been shown to be successful in this respect. Li *et al.* (2010) have shown that regulating the redox potential of the fermentation medium has a great impact on succinic acid production by *A. succinogenes*. They demonstrated that the redox potential has a close connection to metabolic flux partitioning, intracellular enzyme activity and the NADH/NAD ratio. By fixing the redox potential at −350 mV, the fermentation cycle was shortened, productivity of succinic acid enhanced and the formation of by-products minimized. The reported yield of succinic acid was 1.28 mol mol^{-1}, the productivity 1.18 g l^{-1} h^{-1} and the molar ratio of succinic acid to acetic acid 2.02.

CO_2 is one of the substrates for succinic acid production, which makes the microbial production of this acid particularly interesting from an environmental point of view. Increasing the available CO_2 also leads to higher succinate yields (van der Werf *et al.*, 1997). Zou *et al.* (2011) presented a quantitative mathematical model describing the dissolved CO_2 concentration in the fermentation broth. They showed that the highest succinic acid production (61.92 g l^{-1}) was obtained using $MgCO_3$ in combination with gaseous CO_2. The optimal dissolved CO_2 concentration is 159.22 mM and neither source alone was sufficient to reach this level. However, in the end optimizing the environmental conditions will not suffice to achieve homosuccinate fermentation. Metabolic engineering will be required. As a first step in characterizing the enzymes and mechanisms controlling flux distribution, the genome sequence of *A. succinogenes* has been deciphered (McKinlay *et al.*, 2010).

Mannheimia succiniproducens
Another rumen bacterium belonging to the family of *Pasteurellaceae*, producing large amounts of succinic acid from a wide variety of carbon sources is *M. succiniciproducens* (Lee *et al.*, 2002). Under CO_2-rich anaerobic conditions succinic acid is the major fermentation product, clearly along with various by-products, including acetic, formic and lactic acid. (Song *et al.*, 2007a, b). The genome has been completely sequenced and analyzed (Hong *et al.*, 2004) and used for the development of a genome-scale metabolic model, to allow for simulation of the metabolism of *M. succiniciproducens* (Kim *et al.*, 2007). Genetic tools for gene deletion and overexpression have also been developed (Jang *et al.*, 2007; Lee *et al.*, 2006). These efforts allowed the genome-based rational metabolic engineering of *M. succiniciproducens*, leading to the development of an engineered strain LPK7 that is capable of producing 52.43 g l^{-1} succinic acid with high productivity of up to

© Woodhead Publishing Limited, 2013

2.97 g l^{-1} h^{-1} and a yield of 1.16 mol mol^{-1} while producing small amounts of by-products (ratio succinic acid to by-products of 2.092) (Lee *et al.*, 2008).

Anaerobiospirillum succiniproducens
A. succiniproducens is another example of an intestinal bacterium naturally producing succinic acid from a wide variety of carbon sources (Davis *et al.*, 1976). A volumetric productivity of 1.8 g l^{-1} h^{-1} was reported for a wild-type strain, when CO_2 and H_2 was supplemented; 50.3 g l^{-1} of succinic acid were accumulated in 24 h in a batch process utilizing glucose and corn steep liquor (Glassner and Datta, 1992). An integrated membrane-bioreactor-electrodialysis process, which allows simultaneous production and separation of succinic acid in continuous fermentation to avoid the inhibitory effect of the end product led to a maximum productivity of 10.4 g l^{-1} h^{-1} (final succinic acid concentration: 83.0 g l^{-1}, Meynial-Salles *et al.*, 2008). However, problems including the accumulation of by-products still remain to be solved to reduce the costs of the purification of the acid.

Baker's yeast as a recombinant succinic acid producer
S. cerevisiae is a host organism that will be exploited industrially for succinic acid production. However, information about these processes is scarce in the literature. The most studies relating baker's yeast with succinate production were in the context of wine and liquor manufactoring (Arikawa *et al.*, 1999; Camarasa *et al.*, 2003).

Raab *et al.* (2010) reported a yeast deletion strain oxidatively producing 3.62 g l^{-1} succinic acid. This strain was constructed by disrupting the succinate and the isocitrate dehydrogenase, thereby redirecting the metabolic flux into the glyoxylate cycle. The strategy followed by DSM is the overexpression of a phosphoenolpyruvate carboxy kinase, which generates oxaloacetate from pyruvate and concomitantly ATP (Verwaal *et al.*, 2009a). Additionally, malate dehydrogenase and fumarase are overexpressed, enhancing the complete pathway from oxaloacetate to succinate. Decreasing the activity of alcohol dehydrogenase, glycerol-3-*P*-dehydrogenase and succinate dehydrogenase lead to a further increase in metabolic flux to the desired product and a decrease of by-product formation. The best disclosed strain accumulates about 12 g l^{-1} of succinic acid (Verwaal *et al.*, 2009b).

Biogenic succinic acid production: on the verge of industrialisation
Succinic acid is a success story in the establishment of industrial processes with new microbial cell factories for a new product. A variety of companies disclosed the startup of industrial production of biogenic succinic acid, based on different technologies: Bioamber is a joint venture of DNP Green Technology and Agro-industrie Recherches and Developpements (ARD). They built a pilot plant in France, with an annual capacity of 2000 t. Their process uses an *E. coli* strain producing succinic acid from wheat-derived glucose.

© Woodhead Publishing Limited, 2013

Another *E. coli*-based process was set up by Myriant in Louisiana, USA. Currently, the annual capacity is 15 000 t using unrefined sugar as the feedstock. However, an expansion to 70 000–80 000 t in 2012 is already planned.

DSM formed a partnership with Roquette, named Reverdia. They opened a demonstration plant in France that can produce 'hundreds of metric tonnes' per year, scale-up to a facility capable of producing 10 000–20 000 t y^{-1} is planned. The company believes it can achieve a scale of around 50 000 t. The production organism is *S. cerevisiae*, using a combination of reductive and oxidative TCA. Finally, BASF linked up with the Purac subsidiary CSM to make biosuccinic acid using *Basfia succiniprodu-cens*, a ruminal bacterium (Kuhnert *et al.*, 2010) and glycerol (Scholten *et al.*, 2009) or glucose as a feedstock. Commercial quality and volumes of succinic acid are produced at a Purac facility in Spain.

12.5   Reductive branch of the citric acid cycle

12.5.1   Fumaric acid

Fumaric acid is used as a food acidulent in beverages and baking powders. Furthermore, fumaric acid is a pharmaceutically active substance, used to treat psoriasis or multiple sclerosis (Gold *et al.*, 2011). The current world market is about 90,000 t y^{-1}. However, particular interest in fumaric acid – as in all dicarboxylic acids described in this chapter – is derived from its suitability as building block chemical. Use as a bulk chemical in the polymer industry is conceivable. Also its chemical relatedness to aspartic acid, a precursor for aspartam, a widely used sweetener, makes it an interesting product, provided that the production costs are low enough.

Originally, fumaric acid was isolated from plants belonging to the genus *Fumaria*, from which its name is derived. It is an old biotech product. In the 1940s fumaric acid was already being produced on a commercial scale (about 4000 t y^{-1}) by fermentation with the filamentous fungus *Rhizopus arrhizus* (Goldberg *et al.*, 2006). However, with the rapid development of the petrochemistry in the 1960s this process was replaced by chemical synthesis because conventional fumaric acid fermentation is more expensive owing to its low product yield and productivity. Nowadays, microbial production of chemicals from renewable resources is becoming interesting again.

Research has been focused on filamentous fungi, so far. Most strains used are from the genus *Rhizopus*. Typical published production values obtained with *Rhizopus arrhizus* grown on glucose are: product titer 107 g l^{-1}, yield 0.82 g g^{-1} and volumetric productivity 2.00 g l^{-1} h^{-1} (Ng *et al.*, 1986) and with *Rhizopus oryzae*: product titer 92 g l^{-1}, yield 0.85 g g^{-1} and volumetric productivity 4.25 g l^{-1} h^{-1} (Cao *et al.*, 1996).

Fumaric acid is an intermediate of the TCA cycle. Fungal production under aerobic conditions seems to be entirely via the reductive branch of the TCA, by cytosolic pyruvate carboxylase, malate dehydrogenase and

© Woodhead Publishing Limited, 2013

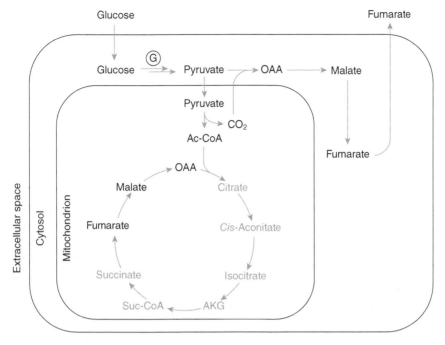

Fig. 12.6 Fumaric acid production. OAA, oxaloacetic acid; Ac-CoA, acetyl-coenzyme A; Suc-CoA, succinyl-coenzyme A; AKG, alpha-ketoglutarate; G, glycolysis.

fumarase (see Fig. 12.6). Pyruvate carboxylase catalyses the carboxylation of pyruvate to oxaloacetic acid. This acid is further converted to malic acid, which is finally converted to fumaric acid by fumarase. The exclusive cytosolic localization of pyruvate carboxylase is a trigger for the formation of fumaric acid in the cytosol since the respective enzymes are found in the mitochondrium but also in the cytosol. The cytosolic localization of the key enzyme pyruvate carboxylase in various filamentous fungal species is the reason for their ability to overproduce certain carboxylic acids. Pyruvate carboxylase is exclusively located in the mitochondria in mammalian cells and other eukaryotes. Malate dehydrogenase and fumarase are in any case located in the cytosol, most probably to convert 'unwanted' acids in this compartment (Goldberg *et al.*, 2006). The theoretical yield of fumaric acid production is 2 mol of fumaric acid per mole of glucose (corresponding to 1.29 g g^{-1}) owing to CO_2 fixation.

Similar to the other fungal acid fermentation processes, the culture conditions are crucial to obtain the correct cellular morphology, which promotes efficient production. Factors affecting fungal growth morphology and fumaric acid production by *R. oryzae* include the initial pH value and concentrations of trace metals such as zinc, magnesium, iron and manganese

© Woodhead Publishing Limited, 2013

(Zhou *et al.*, 2000). A lower initial pH value in the cultivation medium appears to be inhibitory to fungal growth and promotion of fumaric acid productivity. A higher initial pH value leads to fast growth and the formation of large pellets or filamentous forms, which remain unproductive. The importance of pellet morphology for fumarate productivity has also been shown for *R. delemar* (Zhou *et al.*, 2011).

Later in the culture the pH has to be kept high, at a value around 6 (Goldberg *et al.*, 2006). In contrast to citric acid, fumarate salts are produced, not the free acid. To this end neutralizing agents have to be used. $CaCO_3$ is a cheap and effective neutralizing agent with the benefit that it provides CO_2 for the biosynthesis of fumarate. However, calcium fumarate has a limited solubilty of 21 g l^{-1}, which leads to severe viscosity problems in the bioprocess. Higher viscosity leads to a lower oxygen transfer rate and more energy required for mixing. Na_2CO_3 or $NaHCO_3$, are possible replacements. An important prerequisite for entering the productive phase is growth limitation by nitrogen and phosphate depletion. A ratio of glucose to nitrogen (C : N) of 200 : 1 turned out to be optimal.

Riscaldati *et al.* (2000) showed that direct ammonium fumarate production is possible on a large scale, when fugal growth is limited solely by phosphate and $(NH_4)_2CO_3$ is used as neutralizing agent. Furthermore, optimal aeration is required throughout the process. This can be tricky, because *Rhizopus* species tend to clump, thereby limiting oxygen transfer. Addressing this problem, rotary biofilm contactors have been suggested instead of stirred tank reactors. These are characterized by high volumetric productivities. However, the scalability is limited. Du *et al.* (1997) suggest the use of airlift reactors, which are scalable and provide favourable mass transfer and high yields, thus appearing to be a reasonable choice for industrial purposes. While the processes so far have used wild-type strains or strains obtained by conventional methods for strain optimization, the genome sequence of *R. oryzae* is now available and will allow metabolic modelling and target prediction for strain improvement (Machida *et al.*, 2008). The use of bacteria like *E. coli* or *Lactobacillus* sp. for fumaric acid production, as for other acids discussed in this chapter, seems feasible. However, no effort in this direction has been disclosed so far.

12.5.2   Malic acid

Malic acid was first isolated from unripe apples (Latin: *malus*), from which its name is derived. Malic acid naturally occurs in the L-form, whereas chemical synthesis (hydration of maleic or fumaric acid) gives rise to a racemic mixture of DL-malic acid. Both are generally recognized as safe (GRAS) and conform to the FCC Food Chemicals Codex (Goldberg *et al.*, 2006). Malic acid is used in beverage and food production with a consumption of 55 000 t y^{-1}. It is used as an acidulant, improving the tartness or sweetness of fruit juices, carbonated soft drinks or candies. Furthermore,

© Woodhead Publishing Limited, 2013

malic acid is applied in the cosmetics industry to control the pH of creams (Lee *et al.*, 2011a) and it is used for metal cleaning and finishing, textile finishing and electroless plating (Goldberg *et al.*, 2006). Derivatives of malic acid have various medical applications (Lee *et al.*, 2011a).

While the annual production of malic acid is currently rather low (see Table 12.1), there is remarkable potential for malic acid to become a commodity chemical, serving as a feedstock for the production of poly-L-malic acid, a biodegradable polymer (Goldberg *et al.*, 2006). Besides chemical hydration of maleic or fumaric acid, which is the actual production mode for malic acid, enzymatic conversion of fumaric acid catalyzed by fumarase has been proposed. Fumarase is provided by immobilized bacterial cells of *Brevibacterium ammoniagenes*, obtaining a conversion yield of 70%, or from cells of *S. cerevisiae*, which have been genetically modified to overproduce fumarase, obtaining a conversion yield of 80–90%. Special bioreactor construction allowed even higher conversion yields of up to 100% (Goldberg *et al.*, 2006). Giorno *et al.* (2001) report the application of immobilized fumarase in a membrane reactor. However, since biogenic fumaric acid has not been available up to now at competitive costs, there is considerable interest in establishing a direct microbial production process for L-malic acid.

Theoretically, malic acid should be the most favourable of the C4-dicarboxylic acids for microbial production. Like for the others, fermentative production of malic acid includes CO_2 fixation, but additionally malate production is redox neutral (Abbott *et al.*, 2009). The formation of any C4 diacid is metabolically interconnected with the formation of other C4 acids, because the same arterial pathway is employed (Cao *et al.*, 2011). To enhance the production of the dicarboxylic acid of interest, the metabolic flux through the other dicarboxylic acids has to be increased (Cao *et al.*, 2011). This can be achieved by modification of enzyme activities, inactivation of competing by-product formation, enhancement of energy metabolism or provision of reducing power (Cao *et al.*, 2011). Additionally, the export of the carboxylic diacid from the cell can cause severe stress. Particularly, the energy that is required for the export imposes an additional burden on the cellular system. According to Abbott *et al.* (2009) the principles of mono-carboxylic acid transport also apply to dicarboxylic acids but the energetics of dicarboxylate export may be different.

Malic acid production with natural producers: Aspergillus flavus
The filamentous fungus *A. flavus* is a natural L-malic acid producer. Goldberg *et al.* (2006) report a production of 113 g l⁻¹ of L-malic acid from 120 g l⁻¹ glucose with a productivity of 0.59 g l⁻¹ h⁻¹ and a molar yield of 128%. Major by-products are succinic acid (20% molar yield from glucose) and fumaric acid (1–3% molar yield from glucose). Genetic engineering appears to be a solution to overcome the problem of by-product formation. The genome sequencing of several different *Aspergillus* strains is

© Woodhead Publishing Limited, 2013

a promising starting point for improved production of L-malic acid, by metabolic modelling and identifying knock-out or overexpression candidate genes. However, *A. flavus* remains a questionable microbial cell factory for food ingredients owing to possible aflatoxin formation by this fungus. Further natural produers of malic acid have been disclosed by Taing and Taing (2007) reporting L-malic acid production with *Zygosaccharomyces rouxii* and by Kawagoe *et al.* (1997) with *Schizophyllum commune*.

Recombinant malic acid producers: Saccharomyces cerevisiae
Zelle *et al.* (2008) suggest baker's yeast as production organism for malic acid. Three genetic modifications were introduced into a pyruvate decarboxylase-negative *S. cerevisiae* strain: overexpression of pyruvate carboxylase (*PYC2*), retargeting of malate dehydrogenase (*MDH3*) into the cytosol and functional expression of the *Schizosaccharomyces pombe* malate transporter (Sp*MAE*1). The resulting engineered strain produced up to 59 g l^{-1} of malic acid with a yield of 0.42 mol mol^{-1} in glucose-grown batch cultures. With the optimization of process parameters in bioreactor cultivations (pH and concentrations of CO_2, calcium and O_2) a malate yield of 0.48 mol per mol glucose was achieved and a 19% increase over yields compared to shake flask experiments (Zelle *et al.* 2010).

Recombinant malic acid producers: Escherichia coli
As mentioned above, the production of C4 dicarboxylic acids is interconnected. Zhang *et al.* (2011) used this fact in L-malic acid production. The starting point was an *E. coli* strain already engineered for the overproduction of succinic acid by overexpressing pyruvate carboxykinase in the genetic background of deleted lactate dehydrogenase, acetate kinase, alcohol dehydrogenase, pyruvate formate lyase, methylglyoxal synthase and pyruvate oxidase. Surprisingly, the initial rational pathway designs were ineffective and led to unexpected metabolite accumulations, giving deeper insight into the physiology of *E. coli*. Further modifications were required, namely deletions of fumarate reductase, fumarase and malic enzyme. With this strain a yield of 1.42 mol malic acid mol^{-1} glucose with a final product concentration of 34 g l^{-1} was obtained in a two step fermentation process (Zhang *et al.*, 2011).

12.6   Kojic acid

Kojic acid ($C_6H_6O_4$; 5-hydroxy-2-(hydroxymethyl)-4-pyrone, Fig. 12.1) has been identified as a by-product in the fermentation process of malting rice, for use in the manufacturing of sake. Several species of fungi produce this organic acid. A prominent producer is *Aspergillus oryzae*, called 'koji-kin' in Japanese, which is the origin of the common name of this acid. Kojic acid is a mild inhibitor of pigment formation in plant and animal tissues and is used in food and cosmetics to preserve or change colours of

© Woodhead Publishing Limited, 2013

substances. In the food industry it is also used as a precursor of flavour enhancers (Le Blanc and Akers, 1989) as well as on cut fruits and in seafood to prevent colour changes. Its main use, however, is in the cosmetics industry where it is used as a skin whitener (Ohyama and Mishima, 1990).

Kojic acid is produced microbiologically by a variety of fungi during aerobic fermentation. An early study included 19 *Aspergillus* species, five *Penicillium* species, and certain bacteria, such as *Bacterium xylinoides*, *Glucono-acetobacter opacus* var. *mobilis* and *Gyrinium roseum* (Wilson, 1971). The wide distribution of the ability to form kojic acid makes it a rather unusual secondary metabolite. At most, formation of secondary metabolites is restricted to a limited number of organisms.

Industrial production of kojic acid is restricted to filamentous fungi. High yields (0.456 g g^{-1} glucose) have been obtained with *A. flavus* (Ariff *et al.*, 1997; Rosfarizan and Ariff, 2007). However, the possible concomitant production of highly cancerogenic aflatoxins render this production host quite problematic for food or pharmaceutical use.

A better producer in this respect is *A. oryzae*. This fungus has been used in food production for centuries in the Far East and has received GRAS status by the US FDA (Machida *et al.*, 2008). Natural isolates have been improved by *N*-methyl-*N*-nitro-*N*-nitrosoguanidine (NTG) mutagenesis, UV irradiation and protoplastation; the academic literature reports titers of 41 g l^{-1} (Wan *et al.*, 2004). Carbon sources include starch, sucrose, maltose, glucose, fructose, mannose, galactose, xylose, arabinose, sorbitol, acetate, ethanol and glycerol. Glucose and sucrose are preferred.

The concentration of glucose has a big impact on the kojic acid production. The highest kojic acid titer (24.2 g l^{-1} yield 0.24 g g^{-1} glucose) was obtained using 100 g l^{-1} glucose. Using less glucose, the cabon source was entirely used for biomass formation, while more glucose was not converted at all. Organic nitrogen sources are better than inorganic nitrogen sources. Limitation of nitrogen supply is required to limit the growth. In fact a C/N ratio of about 100 was found to be optimal. More carbon did not increase the kojic acid titer. Less nitrogen lead to much less production (Rosfarizan, 2010). Low pH appears to be a requirement for efficient kojic acid production. Values as low as 1.9 can be optimal at the end of the process.

About 83 g l^{-1} kojic acid was produced from two rounds of repeated cultivation for 100 days using *A. oryzae* immobilized in Ca-alginate beads (Kwak and Rhee, 1992). Generally, kojic acid production is a slow process. Pilot-scale production of this acid using an improved strain of *A. oryzae* in repeated-batch fermentations with cell-retention achieved a productivity of 5.3 g l^{-1} day^{-1} (Wan *et al.*, 2005).

12.7   Conclusions

In conclusion it becomes clear that filamentous fungi are established natural producers of a variety of organic acids. First of all they evolved to overproduce these acids naturally under certain conditions. This includes the fact that they also tolerate these acids in high concentrations, which is a prerequisite for an industrial process. Particularly, this tolerance is what a lot of common laboratory host strains lack. The efficient production of organic acids is otherwise connected to precise culture conditions, such as pH, a defined C/N ratio, or concentration of metal ions. The productivity of filamentous fungi is furthermore strongly connected to morphology. While the parameters for obtaining a productive morphology have been defined empirically, the molecular basis has been unclear up to now. Studies in this respect are ongoing. It will be interesting to see how basic biologic findings and applied sciences work together.

Bacterial hosts for organic acid production are under construction. Acids for which no natural producer has been found are produced with metabolically engineered strains. Generally, the development of genetic tools seems to be easier for bacterial hosts compared to fungal hosts. Systems biology tools and modelling are key technologies for the development of new hosts (Sauer and Mattanovich, 2012). Importantly, the purification has to be seen as an integral part of the bioprocess and has to be taken into account for design of the microbial cell factories. In this chapter the technology for purification has not been outlined for space constraints. However, there are excellent papers describing this part of the process. The reader is kindly referred to Section 12.9.

12.8   Future trends

Without doubt microbial production of organic acids will be of major industrial importance in the future. A large number of these acids have been produced for food use in the past and the history of most of the production processes is closely connected to the food industry. However, a wide range of organic acids are useful building block chemicals, which can be produced from renewable resources and substitute for petrol-derived chemicals (Sauer et al., 2008; Lee et al., 2011b; Vennestrøm et al., 2011). Consequently, the societal urge for sustainability significantly fuels the development of new processes for microbial organic acid production.

Traditionally, sugar from sugar plants or starchy plants was the main carbon source for microbial production processes. However, production of sugar is in direct competition with food production. Furthermore, the overall efficiency of fixed carbon usage is low, when only sugar and starch are used. Consequently, it will be very important in the future to develop microbial production processes, using cheap and abundant substrates, such as lignocellulosic material. This poses however, a variety of problems, which have to be solved. First of all, hydrolysis of lignocellulosic material leads to a variety of carbohydrates, which are not natural substrates for most microbes (xylose or arabinose for example). Pathways allowing the use of

© Woodhead Publishing Limited, 2013

these substrates have therefore to be introduced by genetic engineering. Furthermore, the hydrolysis of lignocellulosic material leads to a substrate stream, which is highly contaminated with antimicrobial substances (such as acetic acid or furan derivates). The development of microbial cell factories, which are resistant to these contaminants, is therefore a very important step. One example, showing how worthwhile such efforts can be, was shown by Li *et al.* (2011) for succinic acid production. The total fermentation cost decreased by 55.9% using lignocellulosic hydrolysate and waste yeast hydrolysate as carbon and nitrogen sources instead of expensive glucose and yeast extract.

Last, but not least, many organic acids, which are interesting from the point of view of the chemical industry are not natural products of microbial metabolism. New metabolic pathways, perhaps relying on non-natural enzymatic activities, have to be created in order to produce these acids.

All of these endeavours profit from the modern technologies of analysis, modelling and engineering of microbial strains, as outlined elsewhere in this book. They will be of major importance for the development of microbial cell factories for organic acid production. An interesting point for the future development of microbial production processes is the question of oxygen supply. Aerobic processes are often limited by oxygen transfer rate or by heat transfer because oxidative processes produce heat which has to be dissipated (Weusthuis *et al.*, 2011). Consequently, anaerobic processes are more efficient and typically easier (= cheaper) to perform. However, growth is usually limited under anoxic conditions, which causes problems, because a certain biomass concentration is obviously required.

12.9   Sources of further information and advice

Further information about production of all the organic acids mentioned can be found in the book by Moo-Young (2011). The chapters in this book in particular also cover details of the downstream processes for the purification of the acids. This part of the process has been left out here owing to space constraints. However, it should be underlined that the purification is an integral part of any production process, often deciding about economic success or failure. Excellent review articles focusing on the microbial production of single organic acids include: Berovič and Legiša (2007) for citric acid, Beauprez *et al.* (2010) for succinic acid, Roa Engel *et al.* (2008) for fumaric acid, Singh and Kumar (2007) for gluconic acid and Bentley (2006) for kojic acid. Yu *et al.* (2011) and Cao *et al.* (2011) have published comprehensive comparisons of different production processes in the bacterial cell factory *E. coli*, outlining common lines with this platform host. Kubicek *et al.* (2010) review fungal production processes of organic acids, with a focus on citric and gluconic acid.

© Woodhead Publishing Limited, 2013

12.10   References

ABBOTT DA, ZELLE RM, PRONK JT and VAN MARIS AJA (2009) 'Metabolic engineering of *Saccharomyces cerevisiae* for production of carboxylic acids: current status and challenges'. *FEMS Yeast Res*, **9**, 1123–36.

ALVAREZ-VASQUEZ F, GONZÁLEZ-ALCÓN C and TORRES NV (2000) 'Metabolism of citric acid production by *Aspergillus niger*: model definition, steady-state analysis and constrained optimization of citric acid production rate'. *Biotechnol Bioeng*, **70**, 82–108.

ANASTASSIADIS S, AIVASIDIS A and WANDREY C (2002) 'Citric acid production by *Candida* strains under intracellular nitrogen limitation'. *Appl Microbiol Biotechnol*, **60**, 81–7.

ANDERSEN MR, LEHMANN L and NIELSEN, J (2009) 'Systemic analysis of the response of *Aspergillus niger* to ambient pH', *Genome Biol*, **10**, R47.

ANDERSSON C, HODGE D, BERGLUND KA and ROVA U (2007) 'Effect of different carbon sources on the production of succinic acid using metabolically engineered *Escherichia coli*'. *Biotechnol Prog*, **23**, 381–8.

ARIFF AB, ROSFARIZAN M, HERNG LS, MADIHAH S and KARIM MIA (1997) 'Kinetics and modelling of kojic acid production by *Aspergillus flavus*. Link in batch fermentation and resuspended cell mycelial'. *World J Microbiol Biotechnol*, **13**, 195–201.

ARIKAWA Y, KUROYANAGI T, SHIMOSAKA M, MURATSUBAKI H, ENOMOTO K, KODAIRA R and OKAZAKI M (1999) 'Effect of gene disruptions of the TCA cycle on production of succinic acid in *Saccharomyces cerevisiae*'. *J Biosci Bioeng*, **87**, 28–36.

BEAUPREZ JJ, DE MEY M and SOETART WK (2010) 'Microbial succinic acid production: Natural versus metabolic engineered producers'. *Process Biochem*, **45**, 1103–14.

BENTLEY (2006) 'From miso, saké and shoyu to cosmetics: a century of science for kojic acid'. *Nat Prod Rep*, **23**, 1046–62.

BEROVIČ M and LEGIŠA M (2007) 'Citric acid production'. *Biotechnol Annu Rev*, **13**, 303–43.

BIZZARI S and BLAGOEV M (2010a) *Fumaric Acid*. Available from: http://chemical.ihs.com/CEH/Public/Reports/659.5000/ (accessed 18 December 2011).

BIZZARI S and BLAGOEV M (2010b) *Malic Acid*. Available from: http://chemical.ihs.com/CEH/Public/Reports/672.8000/ (accessed 18 December 2011).

CAMARASA C, GRIVET JP and DEQUIN S (2003) 'Investigation by ^{13}C-NMR and tricarboxylic acid (TCA) deletion mutant analysis of pathways for succinate formation in *Saccharomyces cerevisiae* during anaerobic fermentation'. *Microbiology*, **149**, 2669–78.

CAO NJ, DU JX, GONG CS and TSAO GT (1996) 'Simultaneous production and recovery of fumaric acid from immobilized *Rhizopus oryzae* with a rotary biofilm contactor and an adsorption column'. *Appl Environ Microbiol*, **62**, 2926–31.

CAO Y, CAO Y and LIN X (2011) 'Metabolically engineered *Escherichia coli* for biotechnological production of four-carbon 1,4-dicarboxylic acids'. *J Ind Microbiol Biotechnol*, **38**, 649–56.

CHATTERJEE R, MILLARD CS, CHAMPION K, CLARK DP and DONNELLY MI (2001) 'Mutation of the ptsG gene results in increased production of succinate in fermentation of glucose by *Escherichia coli*'. *Appl Environ Microbiol*, **67**, 148–54.

CHEMCO (2010) *Citric Acid Overview*. Available from: http://www.thechemco.com/chemicals/Citric-Acid (accessed 18 December 2011).

CONTINENTAL (2011) Available from: http://www.continentalchemicalusa.com/bulk-chemicals/acetic-acid.php (accessed 20 December 2011).

CROLLA A and KENNEDY KJ (2004) 'Fed-batch production of citric acid by *Candida lipolytica* grown on n-paraffins'. *J Biotechnol*, **110**, 73–84.

DAI Z, MAO X, MAGNUSON JK and LASURE LL (2004) 'Identification of genes associated with morphology in *Aspergillus niger* by using suppression subtractive hybridization'. *Appl Environ Microbiol*, **70**, 2474–85.

© Woodhead Publishing Limited, 2013

DAVIS CP, CLEVEN D, BROWN J and BALISH E (1976) 'Anaerobiospirillum, a new genus of spiral-shaped bacteria'. Int J Syst Bacteriol. 26, 498–504.

DE JONGH WA and NIELSEN J (2008) 'Enhanced citrate production through gene insertion in Aspergillus niger'. Metab Eng, 10, 87–96.

DE MUYNCK C, PEREIRA CS, NAESSENS M, PARMENTIER S, SOETAERT W and VANDAMME EJ (2007) 'The genus Gluconobacter oxydans: comprehensive overview of biochemistry and biotechnological applications'. Crit Rev Biotechnol, 27, 147–71.

DE NICOLAS-SANTIAGO S, REGALADO-GONZALEZ C, GARCIA-ALMENDÁREZ B, FERNÁNDEZ FJ, TELLEZ-JURADO A and HUERTA-OCHOA S (2006) 'Physiological, morphological, and mannanase production studies on Aspergillus niger uam-gs1 mutants'. Electron J Biotechnol, 9, 50–60.

DRIOUCH H, SOMMER B and WITTMANN C (2010) 'Morphology engineering of Aspergillus niger for improved enzyme production'. Biotechnol Bioeng, 105, 1058–68.

DU JX, CAO NJ, GONG CS, TSAO GT and YUAN NJ (1997) 'Fumaric acid production in airlift loop reactor with porous sparger'. Appl Biochem Biotech, 63–5, 541–56.

FÖRSTER A, JACOBS K, JURETZEK T, MAUERSBERGER S and BARTH G (2007a) 'Overexpression of the ICL1 gene changes the product ratio of citric acid production by Yarrowia lipolytica'. Appl Microbiol Biotechnol, 77, 861–9.

FÖRSTER A, AURICH A, MAUERSBERGER S and BARTH G (2007b) 'Citric acid production from sucrose using a recombinant strain of the yeast Yarrowia lipolytica'. Appl Microbiol Biotechnol, 75, 1409–17.

GALLMETZER M, MERANER J and BURGSTALLER W (2002) 'Succinate synthesis and excretion by Penicillium simplicissimum under aerobic and anaerobic conditions'. FEMS Microbiol Lett, 210, 221–5.

GARCÍA J and TORRES N (2011) 'Mathematical modelling and assessment of the pH homeostasis mechanisms in Aspergillus niger while in citric acid producing conditions'. J Theor Biol, 282, 23–35.

GIORNO L, DRIOLI E, CARVOLI G, CASSANO A and DONATO L (2001) 'Study of an enzyme membrane reactor with immobilized fumarase for production of L-malic acid'. Biotechnol Bioeng, 72, 77–84.

GLASSNER DA and DATTA R (1992) Process for the Production and Purification of Succinic Acid. US Patent 5,143,834.

GOLD R, LINKER RA and STANGEL M (2011) 'Fumaric acid and its esters: An emerging treatment for multiple sclerosis with antioxidative mechanism of action'. Clin Immunol, 142, 44–8.

GOLDBERG I, ROKEM SJ and PINES O (2006) 'Organic acids: old metabolites, new themes'. J Chem Technol Biotechnol, 81, 1601–11.

GRIMM LH, KELLY S, KRULL R and HEMPEL DC (2005) 'Morphology and productivity of filamentous fungi'. Appl Microbiol Biotechnol, 69, 375–84.

GUEBEL DV and TORRES DNV (2001) 'Optimization of citric acid production by Aspergillus niger through a metabolic flux balance model'. Electron J Biotechnol, 4.

GUETTLER MV, RUMLER D and JAIN MK (1999) 'Actinobacillus succinogenes sp. Nov., a novel succinic-acid-producing strain from the bovine rumen'. Int J Syst Bacteriol, 49, 207–16.

HATTORI T, KINO K and KIRIMURA K (2009) 'Regulation of alternative oxidase at the transcription stage in Aspergillus niger under the conditions of citric acid production'. Curr Microbiol, 58, 321–5.

HOLZ M, FÖRSTER A, MAUERSBERGER S and BARTH G (2009) 'Aconitase overexpression changes the product ratio of citric acid production by Yarrowia lipolytica'. Appl Microbiol Biotechnol, 81, 1087–96.

HONG SH, KIM JS, LEE SY, IN YH, CHOI SS, RIH JK, KIM CH, JEONG H, HUR CG and KIM JJ (2004) 'The genome sequence of the capnophilic rumen bacterium Mannheimia succiniciproducens'. Nat Biotechnol, 22, 1275–81.

© Woodhead Publishing Limited, 2013

JANG YS, JUNG YR, LEE SY, KIM JM, LEE JW, OH DB, KANG HA, KWON O, JANG SH, SONG H, LEE SJ and KANG KY (2007) 'Construction and characterization of shuttle vectors for succinic acid-producing rumen bacteria'. *Appl Environ Microbiol*, **73**, 5411–20.

JANTAMA K, HAUPT MJ, SVORONOS SA, ZHANG X, MOORE JC, SHANMUGAM KT and INGRAM LO (2008) 'Combining metabolic engineering and metabolic evolution to develop nonrecombinant strains of *E. coli* C that produce succinate and malate'. *Biotechnol Bioeng*, **99**, 1140–53.

KARAFFA L and KUBICEK CP (2003) '*Aspergillus niger* citric acid accumulation: do we understand this well working black box?' *Appl Microbiol Biotechnol*, **61**, 189–96.

KAUP BA, EHRICH K, PESCHECK M and SCHRADER J (2007) 'Microparticle-enhanced cultivation of filamentous microorganisms: Increased chloroperoxidase formation by *Caldariomyces fumago* as an example'. *Biotech Bioeng*, **99**, 198–491.

KAWAGOE M, HYAKUMURA K, SUYE SI, MIKI K and NAOE K (1997) 'Application of bubble column fermentors to submerged culture of *Schizophyllum commune* for production of l-malic acid'. *J Ferment Bioeng*, **84**, 333–6.

KIM P, LAIVENIEKS M, VIEILLE C and ZEIKUS JG (2004) 'Effect of overexpression of *Actinobacillus succinogenes* phosphoenolpyruvate carboxykinase on succinate production in *Escherichia coli*'. *Appl Environ Microbiol*, **70**, 1238–41.

KIM TY, KIM HU, PARK JM, SONG H, KIM JS and LEE SY (2007) 'Genome-scale analysis of *Mannheimia succiniciproducens* metabolism'. *Biotechnol Bioeng*. **97**, 657–71.

KIRIMURA, K, HONDA Y and HATTORI T (2011) 'Gluconic and itaconic acids'. In: *Comprehensive Biotechnology*, 2nd edition, Moo-Young, M (ed.), Elsevier BV, Amsterdam, Volume 3, 143–7.

KUBICEK CP, PUNT P and VISSER J (2010) 'Production of organic acids by filamentous fungi'. In: *The Mycota X, Industrial Applications*, 2nd edition, Hofrichter M (ed.), Springer Verlag, Berlin Heidelberg.

KUHNERT P, SCHOLTEN E, HAEFNER S, MAYOR D and FREY J (2010) '*Basfia succiniciproducens* gen. nov., sp. nov., a new member of the family *Pasteurellaceae* isolated from bovine rumen'. *Int J Syst Evol Microbiol*, **60**, 44–50.

KWAK MY and RHEE JS (1992) 'Controlled mycelial growth for kojic acid production using Ca-alginate-immobilized fungal cells'. *Appl Microbiol Biotechnol*, **36**, 578–83.

LE BLANC DT and AKERS HA (1989) 'Maltol and ethyl maltol; from larch tree to successful food additives'. *Food Technol*, **26**, 78–87.

LEE PC, LEE SY, HONG SH and CHANG HN (2002) 'Isolation and characterization of a new succinic acid-producing bacterium, *Mannheimia succiniciproducens* MBEL 55E, from bovine rumen'. *Appl Microbiol Biotechnol*, **58**, 663–8.

LEE SJ, SONG H and LEE SY (2006) 'Genome-based metabolic engineering of *Mannheimia succiniciproducens* for succinic acid production'. *Appl Environ Microbiol*, **72**, 1939–48.

LEE SY, KIM JM, SONG H, LEE JW, KIM TY and JANG YS (2008) 'From genome sequence to integrated bioprocess for succinic acid production by *Mannheimia succiniciproducens*'. *Appl Microbiol Biotechnol*, **79**, 11–22.

LEE JW, HAN M-S, CHOI S, YI J, LEE TW and LEE SY (2011a) 'Succinic and malic acids'. In: *Comprehensive Biotechnology*, 2nd edition, Moo-Young, M (ed.), Elsevier BV, Amsterdam, Volume 3, 150–61.

LEE JW, KIM HU, CHOI S, YI J and LEE SY (2011b) 'Microbial production of building block chemicals and polymers'. *Curr Opin Biotechnol*, **22**, 758–67.

LEGIŠA M and MATTEY M (2007) 'Changes in primary metabolism leading to citric acid overflow in *Aspergillus niger*'. *Biotechnol Lett*, **29**, 181–90.

LENG RA, STEEL JW and LUICK JR (1967) 'Contribution of propionate to glucose synthesis in sheep'. *Biochem J*, **103**, 785–90.

© Woodhead Publishing Limited, 2013

LI J, JIANG M, CHEN KQ, YE Q, SHANG LA, WEI P, YING HJ and CHANG HN (2010) 'Effect of redox potential regulation on succinic acid production by *Actinobacillus succinogenes*'. *Bioprocess Biosyst Eng*, **33**, 911–20.

LI J, ZHENG XY, FANG XJ, LIU SW, CHEN KQ, JIANG M, WEI P and OUYANG PK (2011) 'A complete industrial system for economical succinic acid production by *Actinobacillus succinogenes*'. *Bioresour Technol*, **102**, 6147–52.

LIAO W, LIU Y, FREAR C and CHEN S (2007) 'A new approach of pellet formation of a filamentous fungus *Rhizopus oryzae*'. *Bioresour Technol*, **98**, 3415–23.

LIN H, BENNETT GN and SAN KY (2005a) 'Genetic reconstruction of the aerobic central metabolism in *Escherichia coli* for the absolute aerobic production of succinate'. *Biotechnol Bioeng*, **89**, 148–56.

LIN H, BENNETT GN and SAN KY (2005b) 'Fed-batch culture of a metabolically engineered *Escherichia coli* strain designed for high-level succinate production and yield under aerobic conditions'. *Biotechnol Bioeng*, **90**, 775–9.

LIU Y, LIAO W and CHEN S (2008) 'Study of pellet formation of filamentous fungi *Rhizopus oryzae* using a multiple logistic regression model'. *Biotechnol Bioeng*, **99**, 117–28.

LOPEZ-GARCIA R (2002) 'Citric acid'. In: *Kirk-Othmer Encyclopedia of Chemical Technology*, John Wiley & Sons, Inc, UK. (http://mrw.interscience.wiley.com/emrw/9780471238966/home/).

LU S, EITEMAN MA and ALTMAN E (2009a) 'Effect of CO_2 on succinate production in dual-phase *Escherichia coli* fermentations'. *J Biotechnol*, **143**, 213–23.

LU S, EITEMAN MA and ALTMAN E (2009b) 'pH and base counterion affect succinate production in dual-phase *Escherichia coli* fermentations'. *J Ind Microbiol Biotechnol*, **36**, 1101–9.

MACHIDA M, YAMADA O and GOMI, K (2008) 'Genomics of *Aspergillus oryzae*: learning from the history of koji mold and exploration of its future'. *DNA Res*, **15**, 173–83.

MAGNUSON JK and LASURE LL (2004) 'Organic acid production by filamentous fungi'. In: *Advances in Fungal Biotechnology for Industry, Agriculture, and Medicine*, Tkacz JS and Lange L (eds), Kluwer Academic/Plenum, Dordrecht, The Netherlands, 307–40.

MAKRI A, FAKAS S and AGGELIS G (2010) 'Metabolic activities of biotechnological interest in *Yarrowia lipolytica* grown on glycerol in repeated batch cultures'. *Bioresour Technol*, **101**, 2351–8.

MALVEDA M, BLAGOEV M and KUMAMOTO T (2009) *Lactic Acid, Its Salts and Esters*. Available from: http://chemical.ihs.com/CEH/Public/Reports/670.5000/ (accessed 18 December 2011).

MCKINLAY JB and VIEILLE C (2008) '13C-Metabolic flux analysis of *Actinobacillus succinogenes* fermentative metabolism at different $NaHCO_3$ and H_2 concentrations'. *Metab Eng*, **10**, 55–68.

MCKINLAY JB, LAIVENIEKS M, SCHINDLER BD, MCKINLAY AA, SIDDARAMAPPA S, CHALLACOMBE JF, LOWRY SR, CLUM A, LAPIDUS AL, BURKHART KB, HARKINS V and VIEILLE C (2010) 'A genomic perspective on the potential of *Actinobacillus succinogenes* for industrial succinate production'. *BMC Genomics*, **11**, 680.

MEYNIAL-SALLES I, DOROTYN S and SOUCAILLE P (2008) 'A new process for the continuous production of succinic acid from glucose at high yield, titer, and productivity'. *Biotechnol Bioeng*, **99**, 129–35.

MILLARD CS, CHAO Y-P, LIAO JC and DONNELLY MI (1996) 'Enhanced production of succinic acid by overexpression of phosphoenolpyruvate carboxylase in *Escherichia coli*'. *Appl Environ Microbiol*, **62**, 1808–10.

MILLER, FOSMER A, RUSH B, MCMULLIN T, BEACOM D, SUOMINEN P (2011) 'Industrial production of lactic acid'. *Comprehensive Biotechnology*, 2nd edition, Moo-Young M (ed), Elsevier BV, Amsterdam, Volume 3, 179–88.

© Woodhead Publishing Limited, 2013

MIRBAGHERI M, NAHVI I, EMTIAZI G and DARVISHI F (2011) 'Enhanced production of citric acid in *Yarrowia lipolytica* by triton X-100'. *Appl Biochem Biotechnol*, **165**, 1068–74.

MOELLER L, GRÜNBERG M, ZEHNSDORF A, STREHLITZ B and BLEY, T (2010) 'Biosensor online control of citric acid production from glucose by *Yarrowia lipolytica* using semincontinuous fermentation'. *Eng Life Sci*, **10**, 311–20.

MOELLER L, GRÜNBERG M, ZEHNSDORF A, AURICH A, BLEY T and STREHLITZ B (2011a) 'Repeated fed-batch fermentation using biosensor online control for citric acid production by *Yarrowia lipolytica*'. *J Biotechnol*, **153**, 133–7.

MOELLER L, ZEHNSDORF A, AURICH A, BLEY T and STREHLITZ B (2011b) 'Substrate utilization by recombinant *Yarrowia lipolytica* growing on sucrose'. *Appl Microbiol Biotechnol*, **93**, 1695–702.

MOO-YOUNG M (ed.) (2011) *Industrial Biotechnology and Commodity Products, Comprehensive Biotechnology*, 2nd edition, Volume 3, Elsevier B.V. Amsterdam.

NG TK, HESSER RJ, STIEGLITZ B, GRIFFITHS BS and LING LB (1986) 'Production of tetrahydrofuram-1,4 butanediol by a combined biological and chemical process'. *Biotech Bioeng Symp*, **17**, 344–63.

OHYAMA Y and MISHIMA Y (1990) 'Melanosis-inhibitory effect of Kojic acid and its action mechanism'. *Fragrance J*, **6**, 53–8 (in Japanese).

OKINO S, INUI M and YUKAWA H (2005) 'Production of organic acids by *Corynebacterium glutamicum* under oxygen deprivation'. *Appl Microbiol Biotechnol*, **68**, 475–80.

OKINO S, NOBURYU R, SUDA M, JOJIMA T, INUI M and YUKAWA H (2008) 'An efficient succinic acid production process in a metabolically engineered *Corynebacterium glutamicum* strain'. *Appl Microbiol Biotechnol*, **81**, 459–64.

PAPAGIANNI M (2007) 'Advances in citric acid fermentation by *Aspergillus niger*: biochemical aspects, membrane transport and modeling'. *Biotechnol Adv*, **25**, 244–63.

PAPAGIANNI, M *et al.* (2005) 'Fate and role of ammonium ions during fermentation of citric acid by *Aspergillus niger*'. *Appl Environ Microbiol*, **71**, 7178–86.

PAPANIKOLAOU, S *et al.* (2006) 'Influence of glucose and saturated free fatty acid mixtures on citric acid and lipid production by *Yarrowia lipolytica*'. *Curr Microbiol*, **52**, 134–42.

POZO C, GUILLÉN-GOSÁLBEZ G, SORRIBAS A and JIMÉNEZ L (2011) 'A spatial branch-and-bound framework for the global optimization of kinetic models of metabolic networks'. *Ind Eng Chem Res*, **50**, 5225–38.

RAAB AM and LANG C (2011) 'Oxidative versus reductive succinic acid production in the yeast *Saccharomyces cerevisiae*'. *Bioeng Bugs*, **2**, 120–3.

RAAB AM, GEBHARDT G, BOLOTINA N, WEUSTER-BOTZ D and LANG C (2010) 'Metabolic engineering of *Saccharomyces cerevisiae* for the biotechnological production of succinic acid'. *Metab Eng*, **12**, 518–25.

RAMACHANDRAN S, FONTANILLE P, PANDEY A and LARROCHE C (2006) 'Gluconic acid: properties, applications and microbial production'. *Food Technol Biotechnol*, **44**, 185–95.

RASPOR P and GORANOVIC D (2008) 'Biotechnological applications of acetic acid bacteria'. *Crit Rev Biotechnol*, **28**, 101–24.

RISCALDATI E, MORESI M, FEDERICI F and PETRUCCIOLI M (2000) 'Direct ammonium fumarate production by *Rhizopus arrhizus* under phosphorous limitation'. *Biotechnol Lett*, **22**, 1043–7.

ROA ENGEL CA, STRAATHOF AJ, ZIJLMANS TW, VAN GULIK WM and VAN DER WIELEN LA (2008) 'Fumaric acid production by fermentation'. *Appl Microbiol Biotechnol*, **78**, 379–89.

RÖHR GJG and KUBICEK CP (1981) 'Regulatory aspects of citric acid fermentation by *Aspergillus niger*'. *Process Biochem*, **16**, 3.

© Woodhead Publishing Limited, 2013

ROSFARIZAN M (2010) 'Kojic acid: Applications and development of fermentation process for production'. *Biotechnol Mol Biol Rev*, **5**, 24–37.

ROSFARIZAN M and ARIFF AB (2007) 'Biotransformation of various carbon sources to kojic acid by cell-bound enzyme system of *A. flavus* Link 44-1'. *Biochem Eng J*, **35**, 203–9.

RYMOWICZ W, RYWINSKA A, ZAROWSKA B and JUSZCZYK P (2006) 'Citric acid production from raw glycerol by acetate mutants of *Yarrowia lipolytica*'. *Chem Papers*, **60**, 391–4.

RYWIŃSKA A and RYMOWICZ W (2010) 'High-yield production of citric acid by *Yarrowia lipolytica* on glycerol in repeated-batch bioreactors'. *J Ind Microbiol Biotechnol*, **37**, 431–5.

SÁNCHEZ AM, BENNETT GN and SAN KY (2005) 'Novel pathway engineering design of the anaerobic central metabolic pathway in *Escherichia coli* to increase succinate yield'. *Metab Eng*, **7**, 229–39.

SAUER M and MATTANOVICH D (2012) 'Construction of microbial cell factories for industrial bioprocesses'. *J Chem Technol Biotechnol*, **87**, 445–500.

SAUER M, PORRO D, MATTANOVICH D and BRANDUARDI P (2008) 'Microbial production of organic acids: expanding the markets'. *Trends Biotechnol*, **26**, 100–8.

SAUER M, PORRO D, MATTANOVICH D and BRANDUARDI P (2010) '16 years research of lactic acid production with yeast – ready for the market?' *Biotechnol Genet Eng Rev*, **27**, 229–56.

SCHOLTEN E, RENZ T and THOMAS J (2009) 'Continuous cultivation approach for fermentative succinic acid production from crude glycerol by *Basfia succiniciproducens* DD1'. *Biotechnol Lett*, **31**, 1947–51.

SINGH OV and KUMAR R (2007) 'Biotechnological production of gluconic acid: future implications'. *Appl Microbiol Biotechnol*, **75**, 713–22.

SONG H, LEE JW, CHOI S, YOU JK, HONG WH and LEE SY (2007a) 'Effects of dissolved CO_2 levels on the growth of *Mannheimia succiniciproducens* and succinic acid production'. *Biotechnol Bioeng*, **98**, 1296–304.

SONG H, HUH YS, LEE SY, HONG WH and HONG YK (2007b) 'Recovery of succinic acid produced by fermentation of a metabolically engineered *Mannheimia succiniciproducens* strain'. *J Biotechnol*, **132**, 445–52.

STOLS L, KULKARNI G, HARRIS BG and DONNELLY MI (1997) 'Expression of *Ascaris suum* malic enzyme in a mutant *Escherichia coli* allows production of succinic acid from glucose'. *Appl Biochem Biotechnol*, **63–5**, 153–8.

STOTTMEISTER U and HOPPE K (1991) 'Organische Genußsäuren'. In: *Lebensmittelbiotechnologie*. Ruttloff H (ed.), Akademie-Verlag, Berlin, 516–47.

SUN J, LU X, RINAS U and ZENG AP (2007) 'Metabolic peculiarities of *Aspergillus niger* disclosed by comparative metabolic genomics'. *Genome Biol*, **8**, R182.

TAING O and TAING K (2007) 'Production of malic and succinic acids by sugar-tolerant yeast *Zygosaccharomyces rouxii*'. *Eur Food Res Technol*, **224**, 343–7.

TENG Y, XU Y and WANG D (2009) 'Changes in morphology of *Rhizopus chinensis* in submerged fermentation and their effect on production of mycelium-bound lipase'. *Bioprocess Biosyst Eng*, **32**, 397–405.

VAN DEN TWEEL W (2010) *Sustainable Succinic Acid*. Available from: http://www.dsm. com/en_US/cworld/public/sustainability/downloads/publications/oct_2010_dsm_ roquette_bio_based_sustainable_succinic_acid.pdf (accessed 18 December 2011).

VAN DER WERF MJ, GUETTLER MV, JAIN MK and ZEIKUS JG (1997) 'Environmental and physiological factors affecting the succinate product ratio during carbohydrate fermentation by *Actinobacillus* sp. 130Z'. *Arch Microbiol*, **167**, 332–42.

VEMURI GN, EITEMAN MA and ALTMAN E (2002) 'Succinate production in dualphase *Escherichia coli* fermentations depends on the time of transition from aerobic to anaerobic conditions'. *J Ind Microbiol Biotechnol*, **28**, 325–32.

© Woodhead Publishing Limited, 2013

VENNESTRØM PN, OSMUNDSEN CM, CHRISTENSEN CH and TAARNING E (2011) 'Beyond petrochemicals: the renewable chemicals industry'. *Angew Chem Int Ed*, **50**, 10502–9.

VERWAAL R, WU L, DAMVELD RA, SAGT C and MARIA J (2009a) *Dicarboxylic Acid Production in Eukaryotes*. International Patent Application WO2009065780A.

VERWAAL R, WU L, DAMVELD RA, SAGT C and MARIA J (2009b) *Succinic Acid Production in Eukaryotes*. International Patent Application WO2009065778A.

WAN HM, CHEN CC, CHANG TS, GIRIDHAR RN and WU WT (2004) 'Combining induced mutation and protoplasting for strain improvement of *Aspergillus oryzae* for kojic acid production'. *Biotechnol Lett*, **26**, 1163–6.

WAN HM, CHEN CC, GIRIDHAR R, CHANG TS and WU WT (2005) 'Repeated-batch production of kojic acid in a cell-retention fermenter using *Aspergillus oryzae* M3B9'. *J Ind Microbiol Biotechnol*, **32**, 227–33.

WEUSTHUIS RA, LAMOT I, VAN DER OOST J and SANDERS JP (2011) 'Microbial production of bulk chemicals: development of anaerobic processes'. *Trends Biotechnol*, **29**, 153–8.

WILSON BJ (1971) 'Fungal toxins.' In: *Microbial Toxins*, Ciegler A, Kadis S and Ajl SJ (eds), Academic Press, New York, Volume VI, 235–50.

WONG CM, WONG KH and CHEN XD (2008) 'Glucose oxidase: natural occurrence, function, properties and industrial applications'. *Appl Microbiol Biotechnol*, **78**, 927–38.

WUCHERPFENNIG T, HESTLER T and KRULL R (2011) 'Morphology engineering– osmolality and its effect on *Aspergillus niger* morphology and productivity'. *Microb Cell Fact*, **10**, 58.

XI YL, CHEN KQ, LI J, FANG XJ, ZHENG XY, SUI SS, JIANG M and WEI P (2011) 'Optimization of culture conditions in CO_2 fixation for succinic acid production using *Actinobacillus succinogenes*'. *J Ind Microbiol Biotechnol*, **38**, 1605–12.

YANG ST, ZHANG K, ZHANG B and HUANG H (2011) 'Fumaric acid'. *Comprehensive Biotechnology*, 2nd edition, Moo-Young M (ed), Elsevier BV, Amsterdam, Volume 3, 163–77.

YU C, CAO Y, ZOU H and XIAN M (2011) 'Metabolic engineering of *Escherichia coli* for biotechnological production of high-value organic acids and alcohols'. *Appl Microbiol Biotechnol*, **89**, 573–83.

ZELLE RM, DE HULSTER E, VAN WINDEN WA, DE WAARD P, DIJKEMA C, WINKLER AA, GEERTMAN JM, VAN DIJKEN JP, PRONK JT and VAN MARIS AJ (2008) 'Malic acid production by *Saccharomyces cerevisiae*: engineering of pyruvate carboxylation, oxaloacetate reduction, and malate export'. *Appl Environ Microbiol*, **74**, 2766–77.

ZELLE RM, DE HULSTER E, KLOEZEN W, PRONK JT and VAN MARIS AJ (2010) 'Key process conditions for production of C(4) dicarboxylic acids in bioreactor batch cultures of an engineered *Saccharomyces cerevisiae* strain'. *Appl Environ Microbiol*, **76**, 744–50.

ZHANG X, WANG X, SHANMUGAM KT and INGRAM LO (2011) 'L-malate production by metabolically engineered *Escherichia coli*'. *Appl Environ Microbiol*, **77**, 427–34.

ZOU W, ZHU LW, LI HM and TANG YJ (2011) 'Significance of CO_2 donor on the production of succinic acid by *Actinobacillus succinogenes* ATCC 55618'. *Microb Cell Fact*, **10**, 87.

ZHOU Y, DU JX and TSAO GT (2000) 'Mycelial pellet formation by *Rhizopus oryzae* ATCC 20344'. *Appl Biochem Biotechnol*, **84–6**, 779–89.

ZHOU Z, DU G, HUA Z, ZHOU J and CHEN J (2011) 'Optimization of fumaric acid production by *Rhizopus delemar* based on the morphology formation'. *Bioresour Technol*, **102**, 9345–9.

© Woodhead Publishing Limited, 2013

13

Production of viable probiotic cells

F. Grattepanche and C. Lacroix, ETH Zürich, Switzerland

DOI: 10.1533/9780857093547.2.321

Abstract: When considering the probiotic definition of FAO/WHO, products claiming probiotic effects should contain a sufficient number of viable cells to confer efficacy. Probiotic bacteria, being of intestinal origin, are sensitive to many stresses outside their natural habitat and therefore are difficult to propagate, store and deliver in a viable form. This chapter addresses different aspects of processes that aim to reach maximal cell yields with robust physiology and high resistance to conditions, from production to delivery to the gut. Novel culture and stabilization technologies aiming at the improvement of intrinsic tolerance of probiotic cells or conferring a physical protection against stresses are also discussed.

Key words: biomass production, drying technologies, encapsulation, probiotic, stress resistance.

13.1 Introduction

At the beginning of the probiotic story, the number of commercially available microorganisms with presumptive beneficial health effect was limited to a few species of lactic acid bacteria, and even fewer species of bifidobacteria that were well adapted to grow, or at least to maintain their viability, in fermented dairy products, which were the main carriers of probiotic cultures. Nowadays, a wide range of probiotic food, as well as feed, products and supplements are available on the market. Since 2000, the catalogue of probiotic cultures has significantly increased with about 40 different probiotic strains, mostly belonging to the *Lactobacillus* and *Bifidobacterium* genera, commercially available from the main manufacturers (Muller *et al.*, 2009). This development is closely related to the progress in production and stabilization technologies of bacterial cultures. In the literature though, many more bacterial strains have been reported to show beneficial health effects, some *in vivo* in animal studies and could be considered as probiotics based on functional and safety criteria. However they may exhibit very poor technological properties. Indeed, in respect of the probiotic definition of

© Woodhead Publishing Limited, 2013

FAO/WHO (Araya *et al.*, 2002), products claiming probiotic effects should contain a sufficient number of viable cells to confer efficacy (which is generally higher than 10^6–10^8 cfu g^{-1} or 10^8–10^{10} cfu/day, Champagne *et al.*, 2011). Probiotic bacteria, being of intestinal origin, are sensitive to many environmental stress conditions outside their natural habitat and are therefore difficult to propagate. As for all industrial processes, economic aspects have also to be considered in the manufacture of probiotic cultures.

This chapter addresses the process conditions, from biomass production to stabilization, which may positively or negatively affect viability of probiotic cells. Low cost alternatives to existing drying technologies will also be presented, as well as recent technological developments that aim to improve the intrinsic tolerance of probiotic cells or confer physical protection against stress conditions.

13.2   Biomass production

13.2.1   Culture media for large scale biomass production

Many commercial probiotic lactobacilli and bifidobacteria are of human or animal origin and possess complex nutritional requirements. For example, *in silico* analyses of the genome sequence of *Lactobacillus acidophilus* NCFM revealed auxotrophies for 14 amino acids, most vitamins and cofactors (Altermann *et al.*, 2005). In contrast to lactobacilli, *Bifidobacterium* genomics has underlined their relative broad autotrophy for amino acids, nucleotides, vitamins and cofactors (Kleerebezem and Vaughan, 2009). However, bifidobacteria can exhibit great heterogeneity in regard to vitamin requirements. Of 24 tested strains of bifidobacteria belonging to *Bifidobacterium bifidum*, *Bifidobacterium infantis*, *Bifidobacterium breve*, *Bifidobacterium adolescentis* and *Bifidobacterium longum* species, none was able to synthesize riboflavin which was either required or stimulatory for their normal growth in a synthetic medium (Deguchi *et al.*, 1985). Strains of *B. adolescentis* also exhibited growth requirements for thiamine and nicotinic acid (Deguchi *et al.*, 1985). The presence of amino acids in the culture medium can also be required to support growth of bifidobacteria (Biavati and Mattarelli, 2006). Other factors promoting growth of bifidobacteria, so-called bifidogenic factors, have been identified mainly from human milk (Coppa *et al.*, 2006). Among these, *N*-acetylglucosamine, an amino sugar mainly present in the peptidoglycan and lipopolysaccharides of bacterial cell walls, is essential to support growth of *B. bifidum* strains (Gyorgy *et al.*, 1954; Foley *et al.*, 2008).

Rich complex media such as de Man Rogosa and Sharpe (MRS) (de Man *et al.*, 1960), reinforced clostridium medium (RCM) (Hirsch and Grinsted, 1954) and trypticase phytone yeast extract (TPY) (Scardovi, 1986) broth are generally used to produce biomass of bifidobacteria and lactobacilli at small scale. However, the high cost of these media and/or presence

© Woodhead Publishing Limited, 2013

of compounds of animal origin (e.g. beef extract in MRS), allergens and off-flavours developed during the fermentations may considerably limit their use for large scale production in the feed and food industry.

Several media have been studied to counteract these limitations (Table 13.1). They are mainly based on milk, or milk components such as whey, a by-product of the cheese industry, soybean and cereals. Most probiotic bacteria exhibit poor growth in these media without supplementation. Yeast extract is therefore generally added at concentrations ranging from 1–2% to provide essential growth factors. Although milk contains all the essential nutrients for growth of probiotic bacteria, amino acids and small peptides are present in limited amounts or in a non-readily assimilated form to support effective growth of these non-dairy microorganisms (De Vuyst, 2000). Addition of milk hydrolysates to milk enhances biomass production of *Bifidobacterium lactis* Bo (Gomes *et al.*, 1998). Indeed, bifidobacteria exhibit low proteolytic activity compared to bacteria well adapted to grow in milk such as *Lactobacillus bulgaricus* or *Streptococcus thermophilus* (Shihata and Shah, 2000). Another alternative to provide peptides and free amino acids which support growth of bifidobacteria is to use co-culture containing high proteolytic bacteria. Pure cultures of different species of bifidobacteria grown in reconstituted skim milk yielded 1–3 log lower viable cell counts than in co-culture with *Lactococcus lactis* MCC857 (Yonezawa *et al.*, 2010).

Table 13.1 reports the composition of different media used to propagate probiotic bacteria bifidobacteria that are able to metabolize a wide variety of carbohydrates, including glucose, lactose, maltose, galactose and fructose as well as complex carbohydrates and polyols. The analysis of complete genome sequences of bifidobacteria reveals the presence of 17 to 38 genes, depending on the strain, predicted to encode many complex carbohydrate utilization enzymes (Lee and O'Sullivan, 2010). The ability to utilize host-indigestible complex carbohydrates including the so-called prebiotics (e.g. lactulose, fructooligosaccharides, manna-oligosaccharides, galactooligosaccharides, soybean oligosaccharides, lactosucrose, isomalto-oligosaccharides, gluco-oligosaccharides, xylo-oligosaccharides and palatinose) may confer to probiotic bacteria a selective advantage in a competitive environment like the human gut where simple carbohydrates are lacking. Prebiotics can also be used to enhance probiotic biomass production in fermented food products. Supplementation of milk with 4% fructooligosaccharides leads to a 1 log increase in cell concentration of *B. lactis* BL04 in co-culture with *S. thermophilus* TA040 compared to non-supplemented milk while no effect was observed on the population of the latter bacteria (Oliveira *et al.*, 2009).

13.2.2   Culture conditions
Redox potential/anaerobic conditions
Probiotic bifidobacteria and lactobacilli have been mainly isolated from the human gastrointestinal tract. Probiotic lactobacilli are facultative anaerobes

© Woodhead Publishing Limited, 2013

Table 13.1 Production of bifidobacteria and lactobacilli in different media using batch culture and free cells

Broth medium[a]	Bacterial strain	Initial cell concentration or inoculation rate	Maximum production	Reference
Soy peptone (25 g l^{-1}), glucose (25 g l^{-1}), YE (25 g l^{-1})	L. acidophilus La-5	10^4 and 10^5 cfu ml^{-1} for lactobacilli and bifidobacteria, resp.	8.7 × 10^6 cfu ml^{-1} (17 h)	Heenan et al. (2002)
	B. lactis Bb-12		1.3 × 10^8 cfu ml^{-1} (14 h)	
MRS broth	L. acidophilus La-5		5.4 × 10^8 cfu ml^{-1} (24 h)	
	B. lactis Bb-12		4.1 × 10^8 cfu ml^{-1} (16 h)	
Soybean milk, 0.05% L-cysteine, 0.05% threonine	B. longum	10^5–10^6 cfu ml^{-1}	3.2 × 10^9 cfu ml^{-1} (24 h)	Kamaly (1997)
	B. bifidum		5.0 × 10^9 cfu ml^{-1} (24 h)	
Reconstituted skim milk (12%), 2% glucose, 1% YE, 0.05% L-cysteine	B. longum		3.2 × 10^8 cfu ml^{-1} (24 h)	
	B. bifidum		4.0 × 10^8 cfu ml^{-1} (24 h)	
MRS broth, 0.05% L-cysteine	B. longum	1% with a 16-h preculture	7.9 × 10^8 cfu ml^{-1} (24 h)	Rozada et al. (2009)
	B. bifidum		1.3 × 10^9 cfu ml^{-1} (24 h)	
Malt extract (134 g l^{-1}), YE (10 g l^{-1})	B. breve NCIM 702257		2.2 × 10^9 cfu ml^{-1} (23 h)	
Malt extract (134 g l^{-1}), YE (20 g l^{-1})			2.6 × 10^9 cfu ml^{-1} (23 h)	
Malt extract (134 g l^{-1}), YE (10 g l^{-1}), L-cysteine (250 mg l^{-1})			2.8 × 10^9 cfu ml^{-1} (16 h)	
Malt extract	L. acidophilus NCIMB11951	10^7 cfu ml^{-1}	1.3 × 10^8 cfu ml^{-1} (10 h)	Charalampopoulos et al. (2002)
Barley extract			5.4 × 10^7 cfu ml^{-1} (8 h)	
Wheat extract			5.1 × 10^7 cfu ml^{-1} (8 h)	
Soy yogurt	Mixed cultures of S. thermophilus IM111 L. bulgaricus IM025 Lb. jonhsonii La-1	ca. 3 and 0.5 × 10^5 cfu ml^{-1} for the yogurt starter and probiotic bacteria, resp.	1.5 × 10^8 cfu ml^{-1} (6 h) 1.0 × 10^8 cfu ml^{-1} (12 h) 2.0 × 10^7 cfu ml^{-1} (12 h)	Farmworth et al. (2007)
	Mixed cultures of S. thermophilus IM111 L. bulgaricus IM025 Lb. rhamnosus GG		8.0 × 10^8 cfu ml^{-1} (12 h) 3.0 × 10^8 cfu ml^{-1} (12 h) 1.5 × 10^8 cfu ml^{-1} (12 h)	
	Mixed cultures of S. thermophilus IM111 L. bulgaricus IM025 Bifidobacterium sp.		3.0 × 10^8 cfu ml^{-1} (12 h) 1.0 × 10^8 cfu ml^{-1} (12 h) 2.0 × 10^7 cfu ml^{-1} (12 h)	

RBL00064

© Woodhead Publishing Limited, 2013

Growth medium	Bifidobacteria strain	Initial cell count	Final cell count (time)	Reference
Reconstituted skim milk (10%)	B. pseudocatenulatum G4	ca. 10^7 cfu ml^{-1}	6.3×10^5 cfu ml^{-1} (20 h)	Stephenie et al. (2007)
Reconstituted skim milk (2.8%), YE (2.2%)			1.3×10^9 cfu ml^{-1} (20 h)	
Reconstituted skim milk (10%)	In pure culture			Yonezawa et al. (2010)
	B. infantis ATCC15697	3.1×10^6 cfu ml^{-1}	4.1×10^7 cfu ml^{-1} (16 h)	
	B. breve ATCC15700	3.6×10^6 cfu ml^{-1}	4.4×10^7 cfu ml^{-1} (16 h)	
	B. longum ATCC15707	7.4×10^6 cfu ml^{-1}	3.3×10^7 cfu ml^{-1} (16 h)	
	B. adolescentis ATCC15703	6.5×10^6 cfu ml^{-1}	3.5×10^5 cfu ml^{-1} (16 h)	
	B. pseudocatenulatum ATCC27919	5.6×10^6 cfu ml^{-1}	2.3×10^4 cfu ml^{-1} (16 h)	
	B. longum BB536	7.9×10^6 cfu ml^{-1}	3.2×10^7 cfu ml^{-1} (16 h)	
	In co-culture with Lc. lactis MCC857			
	B. infantis ATCC15697	3.1×10^6 cfu ml^{-1}	4.4×10^8 cfu ml^{-1} (16 h)	
	B. breve ATCC15700	3.6×10^6 cfu ml^{-1}	3.4×10^8 cfu ml^{-1} (16 h)	
	B. longum ATCC15707	7.4×10^6 cfu ml^{-1}	4.5×10^8 cfu ml^{-1} (16 h)	
	B. adolescentis ATCC15703	6.5×10^6 cfu ml^{-1}	1.0×10^7 cfu ml^{-1} (16 h)	
	B. pseudocatenulatum ATCC27919	5.6×10^6 cfu ml^{-1}	4.7×10^7 cfu ml^{-1} (16 h)	
Whey, YE (1%), L-cysteine (0.05%)	B. longum BB536	7.9×10^6 cfu ml^{-1}	3.5×10^8 cfu ml^{-1} (16 h)	Corre et al. (1992)
Whey, YE (1%), ascorbic acid (0.1%)	B.bifidum TLR100	Not available	9.0×10^8 cfu ml^{-1} (16 h)	
MRS broth			9.0×10^8 cfu ml^{-1} (16 h)	
MRS broth, whey permeate (2.5%), Na$_2$CO$_3$ (0.02%), L-cysteine (0.05%)	B. longum ATCC15707	6.0×10^7 cfu ml^{-1}	1.4×10^9 cfu ml^{-1} (16 h)	Doleyres et al. (2002a)
			1.7×10^{10} cfu ml^{-1} (12 h)	
MRS broth, Na$_2$CO$_3$ (0.02%), L-cysteine (0.05%)			8.7×10^9 cfu ml^{-1} (12 h)	

[a]YE: yeast extract.

© Woodhead Publishing Limited, 2013

and predominate in the stomach and small intestine whereas their populations usually do not exceed 1% of the total microbiota in the colon or feces (Mueller *et al.*, 2006). In contrast to lactobacilli, bifidobacteria are strict anaerobes reaching high levels in the large intestine where the oxidoreduction (redox) potential of -415 ± 72 mV is lower than in the proximal and distal small intestines (-67 ± 90 and -196 ± 97 mV, respectively) (Sinha and Kumria, 2003).

Reducing agents such as L-cysteine or ascorbic acid are often added to culture media in order to decrease the redox potential and therefore to promote growth and survival of bifidobacteria (Table 13.1). Electrolysis can also be used to decrease the redox potential of media. Pasteurized milk has a redox potential in the region of *ca.* +200 mV which by applying an electroreduction treatment of 10 V for 15 min decreases to -520 mV (Bolduc *et al.*, 2006a). Bolduc *et al.* (2006b) reported an enhanced survival of bifidobacteria in electroreduced milk during extended storage at 7°C. Similar results were observed for survival of *B. bifidum* CIP56.7 which increased by 1.5 log after 28 days storage at 4°C in reduced milk degassed with a mixture of N_2 and H_2 and fermented by *S. thermophilus* and *L. bulgaricus* compared to control milk (Ebel *et al.*, 2011). In this study, nitrogen–hydrogen degassing decreased the redox potential of milk from +440 to -300 mV.

The intestines of all infants are initially colonized by large numbers of facultative anaerobic bacteria such as *Enterobacteriaceae*, *Enterococcus* and *Streptococcus* which lower the intestinal redox potential permitting the growth of strict anaerobic bacteria (Favier *et al.*, 2003). In fermented milk, the dissolved oxygen can be reduced and the viability of bifidobacteria may be improved by incorporating *S. thermophilus* with a high oxygen utilization ability (De Vuyst, 2000).

During biomass production, different gases, for example pure nitrogen, pure carbon dioxide or mixtures of nitrogen, carbon dioxide and hydrogen, are used to maintain anaerobic conditions (Ninomiya *et al.*, 2009). However, not all bifidobacteria are able to grow under pure nitrogen atmosphere (Kawasaki *et al.*, 2007). Absence of cell growth under anaerobic conditions in the absence of CO_2 could be attributed to limitations in metabolic pathways involving bicarbonate which is produced from CO_2 by carbonic anhydrase (Ninomiya *et al.*, 2009).

Temperature and pH
The optimum growth temperature of bifidobacteria ranges from 37–41°C, depending on the origin of the strains. Generally, no growth is observed below 20°C or above 46°C, with the exception of *Bifidobacterium psychroaerophilum*, *Bifidobacterium thermacidophilum* and *Bifidobacterium thermophilum* which are able to grow at 4°C and 49.5°C, respectively (Gavini *et al.*, 1991; Zhu *et al.*, 2003; Simpson *et al.*, 2004, von Ah *et al.*, 2007). Lactobacilli isolates from humans, animals and some dairy products are generally incubated at 37°C, from other habitats, at 30°C and from low

© Woodhead Publishing Limited, 2013

temperature sources, at 22°C (Hammes and Hertel, 2006). In contrast to lactobacilli, bifidobacteria are less acid tolerant, with a pH optimum between 6.5 and 7.0 and no growth below 5.0 or above 8.0 with some exceptions (Scardovi, 1986; Doleyres and Lacroix, 2005; von Ah *et al.*, 2007). Optimal conditions of temperature and pH result in high biomass production. However, the use of sub- or above optimal growth temperature or pH may be relevant for producing cells exhibiting a better survival rate during downstream processing, storage and digestion as discussed Section 13.7.2.

13.3   Fermentation technologies

Different fermentation technologies are discussed in the following section as illustrated in Fig. 13.1.

13.3.1   Batch cultures

The simplest conventional technology for biomass production is batch fermentation without pH control. However, growth of probiotic bacteria is rapidly inhibited by low pH resulting from carbohydrate metabolism and organic acids production. Up to ten times more biomass can be obtained when pH control is applied. As an example, cell counts of *B. longum* ATCC15707 grown in MRS supplemented with whey permeate increased from 3.3×10^9 to 1.7×10^{10} cfu ml^{-1} without and with pH controlled at 5.5, respectively (Doleyres *et al.*, 2002a). Control of pH can be done internally by adding into the medium soluble (e.g. phosphate buffer) or insoluble (e.g. calcium carbonate) buffers or compounds with a high buffering capacity. A five-fold concentration of skim milk by ultrafiltration leads to a two-fold higher cell concentration of bifidobacterial than using non-concentrated skim milk (Ventling and Mistry, 1993). Phosphate buffers used for internal pH control can exhibit inhibitory activity against some lactobacilli as reported for *Lactobacillus helveticus* 880 (Parente and Zottola, 1991). Indeed, phosphate may chelate divalent cations which play an important role in many biological functions of lactic acid bacteria (Boyaval, 1989; Parente and Zottola, 1991). External pH control is performed by addition of a base such as NaOH, KOH, NH$_4$OH or mixtures. Ammonium stimulates the growth of lactic acid bacteria compared to sodium hydroxide (Peebles *et al.*, 1969). Many strains of bifidobacteria are also able to use ammonium as a nitrogen source (Tamine *et al.*, 1995).

13.3.2   Continuous cultures

Free cells

Continuous culture technology with free cells has been little investigated for production of probiotics mainly because of the difficulty in controlling

© Woodhead Publishing Limited, 2013

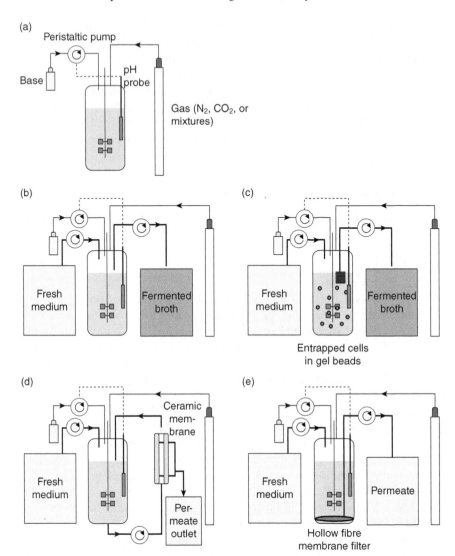

Fig. 13.1 Schematic representation of the different fermentation technologies presented in this book chapter. (a) Batch culture, (b) continuous culture, (c) continuous culture combined with immobilized cells, (d) membrane bioreactor and (e) submerged membrane bioreactor.

contaminants at the industrial scale and the lack of stability of strain characteristics such as bacteriocin production (Huang *et al.*, 1996; Desjardins *et al.*, 2001). However, productivity (i.e. biomass production per unit of volume and time) can largely be increased using continuous cultures instead of batch cultures. Kim *et al.* (2003) reported about a five-fold higher

© Woodhead Publishing Limited, 2013

productivity with a continuous culture of *B. longum* operating at a dilution rate of 0.33 h^{-1} compared with a 22-h batch fermentation.

Membrane bioreactors

Using free cells continuous culture, cells leave the fermentation vessel concomitantly with the effluent. The cells can be separated from the fermented broth by installing a filtration unit (submerged or not) (Fig. 13.1). Continuous removal of fermented broth and addition of fresh medium allows reduction of metabolite inhibition and continuous growth of the cells.

Corre *et al.* (1992) reported a 15-fold higher productivity of 2×10^{11} cfu l^{-1} h^{-1} for *B. bifidum* using a continuous stirred tank reactor coupled to an external ultrafiltation device operated under non-aerobic conditions compared to a batch culture with free cells. A seven-fold increase in *B. longum* cell productivity was obtained using a bioreactor with continuous cross-flow filtration compared to batch culture (Taniguchi *et al.*, 1987). However, cross-flow membrane bioreactors have some disadvantages such as poor viability and low metabolic activity of cells, the high power consumption of the system, and difficulty in controlling fouling (Bibal *et al.*, 1991; Kwon *et al.*, 2006). High productivity of 6.1×10^{11} cfu l^{-1} h^{-1} has been obtained for *B. bifidum* using a submerged membrane bioreactor which overcomes the limitations associated with the crossflow system (Kwon *et al.*, 2006).

Entrapped cells

Cell entrapment technology has many advantages compared to free cell systems, or membrane bioreactors in particular, when combined with continuous fermentation, such as high cell density, reuse of biocatalysts, improved resistance to contamination and bacteriophage attack, enhancement of plasmid stability, prevention from washing-out, physical and chemical protection of cells, and can positively affect the physiology of probiotic cells (Doleyres and Lacroix, 2005; Lacroix *et al.*, 2005; Lacroix and Yildirim, 2007). Cell entrapment in polysaccharide gel beads is the most widely investigated technique for biomass and metabolites production in food applications (Lacroix *et al.*, 2005; Grattepanche and Lacroix, 2010). Indeed, polymeric gels are among the few matrices fulfilling most of the criteria for food or feed applications (Margaritis and Kilonzo, 2005).

In addition to these criteria, the polymers should exhibit suitable thermal or ionotropic gelation properties under mild conditions in order to maintain high cell viability during the immobilization process. Spherical polymer beads containing probiotic cells are produced using extrusion or emulsification in a two-phase dispersion process (Lacroix *et al.*, 2005). In a culture medium, immobilized cells grow preferentially close to the bead surface owing to the diffusional limitation of both substrates and inhibitory end products of metabolism occurring in the catalyst (Arnaud *et al.*, 1992a; Cachon *et al.*, 1998; Doleyres *et al.*, 2002b; Lamboley *et al.*, 1997; Masson

© Woodhead Publishing Limited, 2013

et al., 1994). Spontaneous release into the culture medium of active cells from the surface layer of gel beads is favoured by shear forces resulting from mechanical agitation and bead contacts in the bioreactor (Arnaud *et al.*, 1992b; Sodini *et al.*, 1997). In contrast to traditional free cells systems, very high cell density typically ranging from 5×10^{10} to 5×10^{11} cfu ml^{-1} or g of matrix can be reached using entrapment technology (Champagne *et al.*, 1994; Lacroix and Yildirim, 2007).

Several studies reported very high productivities for biomass production using cell entrapment combined with continuous culture. During continuous culture of *B. longum* ATCC 15707, immobilized in gellan gum gel beads, in MRS broth supplemented with whey permeate, cell concentrations ranged from $3.5-4.9 \times 10^9$ cfu^{-1} ml^{-1} in the effluent for a dilution rate decreasing from $2-0.5$ h^{-1}, respectively, leading to a maximal productivity of 6.9×10^{12} cfu l^{-1} h^{-1} compared to 7.2×10^{11} cfu l^{-1} h^{-1} for free-cell batch cultures at an optimal pH of 5.5 (Doleyres *et al.*, 2002a).

Reimann *et al.* (2011) reported similar high biomass productivity for *B. longum* NCC2705 immobilized in gellan and xanthan gum gel beads and continuously cultured in MRS at a dilution rate of 2.25 h^{-1}. However, immobilization of *B. longum* NCC2705, in contrast to *B. longum* ATCC15707, led to the formation of macroscopic cell aggregates in the effluent, as already reported for an immobilized exopolysaccharide producing strain of *Lactobacillus rhamnosus* (Bergmaier *et al.*, 2005). Aggregate formation results in underestimation of viable cell concentrations measured by plate counting. This is an important limitation for probiotic products since defined viable cell counts should be guaranteed for product efficacy. On the other hand, aggregated cell separation from fermented broth could be facilitated and aggregates may provide physical and physiochemical protection for cells in products and during digestion. It has also been shown that the aggregation property of probiotic bacteria favours coaggregation with pathogens (Collado *et al.*, 2007; Schachtsiek *et al.*, 2004), as well as adhesion to epithelial cells (Del Re *et al.*, 2000; Kos *et al.*, 2003). Furthermore, cell entrapment may induce other physiological changes such as an increased resistance to environmental stressing conditions (Grattepanche and Lacroix, 2010).

Cell entrapment technology has successfully been tested for the production of mixed cultures containing competitive and non-competitive strains of lactic acid bacteria and bifidobacteria (Doleyres *et al.*, 2002b; Grattepanche *et al.*, 2007). In a continuous two-stage system, consisting of a first reactor containing a dominant *Lc. lactis* subsp. *lactis* biovar. *diacetylactis* strain and a less competitive *B. longum* strain immobilized separately in gel beads and a second reactor operating in series with free cells released from the first reactor, a stable and high production was measured for both strains over 17 days of culture (Doleyres *et al.*, 2004).

As explained in the two previous sections, very high cell densities can be reached using a suitable medium, operating fermentation conditions and

© Woodhead Publishing Limited, 2013

technology. However, growth under sub-optimal conditions may favour tolerance of the cells to environmental stressing conditions such as freezing (see Section 13.7.2 for an example) leading to a compromise between high cell density and high viability during downstream processing and digestion. In addition, the effects of the technology and process parameters on the probiotic functionality of the cells should be addressed.

13.4   Downstream processing of probiotic biomass

Probiotic cultures are mostly preserved in frozen state or, preferentially, in dried form to reduce costs associated with storage and distribution and for easier handling. Independently of the technologies and even if biomass concentration reached a high level during fermentation, a concentration step is often required. The following section deals with different environmental and processing conditions which may affect the viability of probiotic cells during concentration, freezing and drying of cultures.

13.4.1   Biomass concentration

Despite its important role, the effect of concentration technologies on cell viability has not been extensively investigated. Centrifugation is the main method used to concentrate cells before stabilization as well as to separate cells from the fermented broth which contains high concentrations of inhibitory metabolites such as organic acids. Streit *et al.* (2010) reported moderate effects of centrifugation speed and duration, but not temperature (i.e. 4°C or 15°C), on the specific acidification activity of *Lb. bulgaricus* during freezing and frozen storage. Surprisingly, a combination of centrifugation and acid adaptation treatment (pH 5.25 for 30 min) of the cells, enhanced significantly resistance to freezing and frozen storage. Cross-flow microfiltration has been proposed as an alternative to centrifugation for the production of lactic acid bacteria concentrates (Zokaee *et al.*, 1999; Streit *et al.*, 2011). Microfiltered cells of *Lb. bulgaricus* were less resistant to concentration than centrifuged cells. However, cryotolerance of the cells during freezing and frozen storage was improved in a range of 28–88%, depending on microfiltration conditions (Streit *et al.*, 2011).

13.4.2   Freezing and freeze drying

Freezing is commonly used to preserve the viability of lactic acid bacteria during storage and is also the first step of the lyophilization (freeze drying) process. The inactivation of freeze dried lactic acid bacteria is mostly attendant on the freezing step, 60–70% of cells that survived this step were reported to stay alive through the dehydration step (To and Etzel, 1997). Resistance to freezing varies among genera, species and, even strains, for

© Woodhead Publishing Limited, 2013

example streptococci generally survive better than lactobacilli (Fonseca *et al.*, 2000; Tsvetkov and Shishkova, 1982) and are strongly dependent on freezing conditions.

At a low freezing rate, cellular damage is caused by accumulation of extracellular ice which leads to concentration of extracellular solute and cell dehydration. In addition, mechanical forces that compress cells as the size of the extracellular ice crystals increases progressively with decreasing temperature may also induce cellular damage (Santivarangkna *et al.*, 2008). Simultaneous solidification of the unfrozen fraction of water and solutes (also called eutectic crystallization) occurring at low freezing rate can also directly damage the cell membrane, and/or the cell by propagating into the intracellular space across the membrane (Santivarangkna *et al.*, 2008).

At a fast cooling rate, cell water cannot migrate rapidly enough outside to maintain osmotic pressure, leading to intracellular ice formation. However, the question of whether intracellular ice formation is a cause rather than a result of cellular damage is still under debate. A wide range of cryoprotective agents such as skim milk supplemented or not with poly-ethylene glycol, dextran, bovine albumin, glycogen, sucrose, trehalose, glycerol; disaccharides such as sucrose, lactose, maltose and trehalose; dimethyl sulfoxide, betaine, sodium ascorbate and glutamate and maltodextrine have been investigated to improve the viability of lactic acid bacteria and bifidobacteria during freezing and frozen storage (Saarela *et al.*, 2005). The efficacy of these cryoprotectants is mainly based on their ability to penetrate the cells and also depends on the freezing rate. In contrast to penetrating cryoprotectants, addition of non-penetrating cryoprotectants partially dehydrates the cell prior to freezing, thereby reducing the chances that ice nucleates intracellularly, as occurs at a fast freezing rate (Fowler and Toner, 2005).

During drying, removal of bound water from bacterial cells leads to an osmotic shock, destabilization of structural integrity of important biological macromolecules (e.g. RNA, DNA, surface proteins) and damages the cell membrane, which is considered to be the primary target (Meng *et al.*, 2008). Intracellular accumulation of compatible solutes such as glycine betaine, which exhibits both osmo- and cryoprotective properties in prokaryotes (Cleland *et al.*, 2004), helps to counteract osmotic stress. A genetically modified strain of *Lactobacillus salivarius* able to accumulate glycine betaine was shown to be more resistant to freeze drying than its parent strain (Sheehan *et al.*, 2006). Nevertheless, the ability of glycine betaine to protect freeze-dried cells is genus dependent. For example, reduction in viability of freeze-dried *B. animalis* subsp. *lactis* was within 0.4 log cfu^{-1} g^{-1} using sucrose or reconstituted skim milk as protectants, whereas it ranged from 1.5–5 log with glycine betaine (Saarela *et al.*, 2005). Sucrose and reconstituted skim milk stabilize the cell membrane and protect the structure and the function of proteins during drying (Leslie *et al.*, 1995; Meng *et al.* 2008; Li *et al.*, 2010). High cell concentrations of *ca.* 10^{11} cfu g^{-1} of powder can be achieved for

© Woodhead Publishing Limited, 2013

bifidobacteria and lactic acid bacteria by fine tuning important process conditions and protectant formulations through empirical approaches supported by experimental design (Saxelin *et al.*, 1999; Saarela *et al.*, 2005; Kurtmann *et al.*, 2009a).

Other methods based on the exploitation of cellular stress response have also been investigated to improve the tolerance of probiotic bacteria to freezing and freeze drying (see Section 13.7).

13.4.3 Spray drying

Spray drying is widely used in the dairy industry for the production of milk powder and is considered, in regard to its cost and large scale potential, as a very interesting alternative to freeze drying for dehydration of microbial cultures including probiotics. Indeed, the fixed and operating costs of spray drying are 5–10 times lower than those of freeze drying (Santivarangkna *et al.*, 2007). Therefore, large scale production of starter and probiotic culture using this technology has been extensively investigated over the last 10 years (Corcoran *et al.*, 2004; Desmond *et al.*, 2002a; Fritzen-Freire *et al.*, 2012; Gardiner *et al.*, 2000; Lian *et al.*, 2002; Simpson *et al.*, 2005; Ying *et al.*, 2012). However, bacterial cells suffer much more lethal damage mainly attributed to heat exposure during spray rather than freeze drying (Ross *et al.*, 2005; Teixeira *et al.*, 1995).

The spray drying process can be divided into two steps (Santivarangkna *et al.*, 2007). During the first step, a suspension containing the probiotic cells is first atomized at high velocity and a spray of droplets is directed into a concurrent flow of hot air at temperatures up to 200°C, corresponding to the initial period of drying (constant drying rate period). In contrast to the first step, the temperature of the spray-dried particles increases in the next stage, the falling drying rate period, to reach the temperature of the outlet air. This latter stage is often considered as a critical factor for cell viability (Santivarangkna *et al.*, 2007). The residence time of particles in this stage can be controlled by drying tower size, as well as outlet temperature which in turn is influenced by inlet air temperature, air flow rate, product feed rate and atomized droplet size (Santivarangkna *et al.*, 2007). The survival rate of *Lactobacillus paracasei* decreased from 100–30% with an increase of outlet temperature from 70–75 to 90–95°C (Gardiner *et al.*, 2000). A decrease in cell viability as the outlet temperature increases is related to the degree of membrane disintegration (Ananta *et al.*, 2005). The effect of outlet temperature strongly depends on the carriers. An increase of the outlet temperature from 50–60°C had a less detrimental effects on the viability of *B. longum* spray dried using skim milk compared to gelatin or soluble starch (Lian *et al.*, 2002). However, the residual moisture content of the dried powder increases concomitantly with a decrease of the outlet temperature and may severely affect cell viability during long-term storage (see Section 13.5.2 on storage of dried product).

© Woodhead Publishing Limited, 2013

As for freeze-drying, a wide range of protective agents and carriers have been investigated to improve the viability of bacterial cells during spray drying (see Santivarangkna *et al.*, 2007, for a review). Reconstituted skim milk at concentrations ranging from 10–20% is often considered to be a suitable carrier owing to the stabilizing effect of milk protein on cell membrane constituents, the formation of a protective coating on the cell wall proteins and the presence of calcium which may increase cell survival after dehydration (Meng *et al.*, 2008).

Finally, cell resistance to spray drying varies between genera, species and even strains (Gardiner *et al.*, 2000; Lian *et al.*, 2002; Simpson *et al.*, 2005; Reimann *et al.*, 2010) leading to a labour-intensive, lengthy and costly strain-by-strain screening procedure for optimization of spray drying conditions and/or identification of the most robust strains. A fast screening method based on a randomly amplified polymorphic DNA method has been recently developed for identification of *B. longum* strains resistant to spray drying and storage in mixed strains bacterial preparations (Reimann *et al.*, 2010).

13.4.4   Alternative drying technologies

Freeze drying allows production of highly concentrated probiotic cultures in dried form but its relatively high operating costs may limit its use to high value food products. During both freeze- and spray-drying, bacterial cells are subjected to very low or high temperatures, respectively, which may severely affect their viability. Therefore, low cost alternative drying approaches, operated under mild conditions, have been investigated in order to improve cell viability during drying as well as to keep the moisture content of the dried powder below a critical value of 4% for long-term storage of the powder.

Fluidized bed drying

A fluidized bed is a bed of solid particles with a stream of air or gas passing upward through the particles at a rate great enough to set them in motion. As the air travels through the particle bed, it imparts unique properties to the bed (Santivarangkna *et al.*, 2007). The advantages of this technology include large-scale continuous production, rapid exchange of heat and mass between the gas and particles and use of relatively low air temperatures to minimize overheating (Mille *et al.*, 2004). In addition, fixed and operating costs of fluidized bed drying have been estimated to be 11.3 and 5.6-fold lower, respectively, than those of freeze drying and slightly less than those of spray drying (Santivarangkna *et al.*, 2007). Fluidized bed drying applied to microorganisms is a common technology for the production of active dry yeast for direct use in industrial fermentations (Jenkins *et al.*, 2011), but it has been little investigated for drying lactic acid bacteria. Mille *et al.* (2004) reported high cell concentrations of *ca.* 1.0×10^{10} cfu g^{-1} of dried powder for *Lactobacillus plantarum* and *Lb. bulgaricus* cells mixed with casein

© Woodhead Publishing Limited, 2013

powder prior dehydration using a fluidized bed dryer. In a study by Strasser *et al.* (2009), concentrated liquid cell suspensions were directly sprayed onto fluidized cellulose powder. In the presence of protectants, survival rates of *Enterococcus faecium* and *Lb. plantarum* after fluidized bed drying were similar to those observed using freeze drying, while in the absence of a protectant, the fluidized bed process caused more damage to the cells than lyophilization (Strasser *et al.*, 2009).

Vacuum drying
Vacuum drying is a particularly suitable technology for drying oxygen and heat sensitive compounds or microorganisms. Indeed, it can be operated at low temperature since the boiling point of water decreases with a decrease in pressure. As an example, in the study of Tymczyszyn *et al.* (2008) a vacuum below 10 Torr was used to dry *Lb. bulgaricus* cells. At this pressure, the boiling point of water is *ca.* 11.4°C meaning that the lowest temperature used in this study (i.e. 30°C) was high enough for efficient drying of the cell preparations. In contrast to spray drying, vacuum drying requires a longer time to reach equivalent moisture content although less than for freeze drying. Compared to 24 hours for freeze drying, controlled low-temperature vacuum dehydration of *Lb. acidophilus* took only four hours to dry to a final moisture content of 5% (King and Su, 1994). In addition, the fixed and manufacturing costs of vacuum drying are *ca.* 50% lower than those of freeze drying (Santivarangkna *et al.*, 2007). Nevertheless, dehydration of lactic acid bacteria and bifidobacteria using vacuum drying has not been extensively studied.

Using vacuum drying operating at a shelf temperature of 40–45°C, King and Su (1994) reported a survival rate of *Lb. acidophilus* ranging from 15.4–30.4% depending on the protectant used. Loss of viability is explained by long time exposure of cells to heat stress, the temperature of the cell concentrate in the vacuum chamber reaching 40 ± 5°C. The use of controlled low temperature vacuum drying (CLTVD) which operates at pressures just above the triple point of water and a product temperature close to 0°C minimizes exposure of the cells to heat stress. Under these conditions, survival rates of *Lb. acidophilus* increased to 50.0–74.6% according to the protectant used and were similar to those obtained using freeze drying (King and Su, 1994). Similar results were reported for *Lb. paracasei* which reached a cell concentration of 1.2×10^{11} cfu g^{-1} dried powder while a low survival rate was obtained for *B. lactis* after CLTVD compared to freeze drying (Foerst *et al.*, 2012; Bauer *et al.*, 2012).

Mixed drying technologies
Like all processes, the aforementioned technologies possess inherent advantages and disadvantages (Table 13.2). Mixed drying systems aim to combine the advantages of these technologies. Wolff *et al.* (1990) investigated the combination between freeze drying and fluidized bed for the dehydration

© Woodhead Publishing Limited, 2013

Table 13.2   Advantages and disadvantages of different drying technologies

Technology	Advantages	Disadvantages
Freeze drying	• High survival rate, generally considered as a reference • Equipment available for large scale production	– High fixed and operating costs – Duration of the process, however productivity can be increased using a continuous freeze dryer
Spray drying	• Very fast drying process • Low cost in comparison to freeze drying • Equipment available for large scale production • Continuous process	– Poor survival rate for very heat sensitive bacteria – Oxidative stress during drying
Fluidized bed drying	• Very low cost technology • High survival rate in presence of suitable protectants • Equipment available for large scale production • Faster than freeze drying	– Only granulatable materials can be dried – Process optimization and scale up based on empirical approach – Use of gas to maintain anaerobic condition for oxygen sensitive bacteria
Vacuum drying	• Good survival rate in particular for CLTVD process • Drying performed under low oxygen environment	– Long drying time, however continuous vacuum drying systems are now available and may improve the productivity

of *S. thermophilus*. This process, so-called atmospheric pressure freeze drying, is based on a decrease in the partial pressure water vapour, not by working under vacuum, but by adsorption in a fluidized adsorbent bed at atmospheric pressure. The adsorbing particles (starch) are mixed with the frozen mixture of cells and cryoprotectants. Water vapour emitted by sublimation or left in the apparatus is transferred to these particles. After drying, the viability of *S. thermophilus* reached 33% and 55% for vacuum and atmospheric freeze drying, respectively. Atmospheric freeze drying also required double the processing time (15 h vs. 8 h) compared to vacuum freeze drying. However, the energy costs can be decreased by one-third (Wolff *et al.*, 1990).

The spray drying process has many advantages over other technologies for industrial applications (Table 13.2). However, heat sensitive bacteria, like most probiotic bacteria, exhibit a poor survival rate during spray drying mainly due to the heat treatment required to reach suitable moisture content. A combination of spray drying and fluidized bed drying allows a reduction in the outlet temperature while the final moisture is reduced under moderate temperature in the fluidized bed. A high survival rate of

© Woodhead Publishing Limited, 2013

84.5% was reached for *Lb. paracasei* cells dehydrated using a spray dryer operated at inlet and outlet temperatures of 175°C and 68°C, respectively, and fitted with two external fluidized beds, set at air temperatures of 55°C and 25°C, respectively (Gardiner *et al.*, 2002). The same process was used to dehydrate *B. thermophilum*, with no improvement in cell viability observed in comparison to classic spray drying operated at inlet and outlet temperatures of 170°C and 85–90°C, respectively (22% vs. 26% respectively). However, fluidized bed spray drying led to a better resistance of cells during storage than with spray drying with a loss of 1.2 and 6.4 log cfu g^{-1} of powder, respectively, after 90 days of storage at 25°C (Simpson *et al.*, 2005). A high moisture content resulting from spray drying operating at a high outlet temperature can also be efficiently reduced using a vacuum dryer, as reported by Stadhouders *et al.* (1969).

Kitamura *et al.* (2009) have developed a vacuum spray dryer to dry lactic acid bacteria at lower temperatures, ranging from 35–120°C, than in conventional spray drying. However, the activity of dried cells was less than 5% at the highest operating temperature (i.e. 120°C) while the moisture content remained at high level of 6.4%. In a study by Semyonov *et al.* (2011), a spray dryer was fitted with a fluidized bed dryer operating both under vacuum. This system was applied to dehydrate *Lb. paracasei* cells. The highest survival rate of 70.6% was obtained under optimized conditions for the process (i.e. drying vacuum of 3.33 KPa) and carriers (i.e. maltodextrin DE 5 and trehalose at a ratio of 1:1 and concentration of 20 g/100 g solids).

13.5 Storage of frozen and dried probiotic concentrates

Probiotics are delivered as food supplements, generally in caplets or capsules containing dried cells, or in processed foods into which they are mainly incorporated by direct inoculation in frozen or dried forms for greater flexibility, and to standardize the delivery of the cultures better (Farnworth and Champagne, 2010). In general, probiotic concentrates are stored until incorporation in the final products. This storage of probiotic cells in frozen or dried forms may greatly affect their viability.

13.5.1 Storage of frozen concentrates

The main factor affecting the viability of bacterial cells during frozen storage is temperature. As an example, after 24 months storage, reduction in viability of *Lb. salivarius* and *Lb. acidophilus* increased from 0.3–1.4 and 2.5–4.2 log cfu ml^{-1}, respectively, with an increase in storage temperature from −70 to −20°C (Juárez Tomás *et al.*, 2004). The better preservation at −70°C can be ascribed to lower rates of enzymatic reactions which may take place up to −20°C and to a weaker recrystallization of the frozen matrix, limiting cell damage induced by ice crystals (Fonseca *et al.*, 2001). The

© Woodhead Publishing Limited, 2013

freezing rate also influences crystals as discussed previously (see Section 13.4.2). Addition of protectants such as glycerol may also attenuate the growth of intracellular and extracellular ice crystals (Fonseca *et al.*, 2003). These authors also showed that addition of sodium ascorbate exhibited a strong protective effect during frozen storage of *Lb. bulgaricus*. This effect was attributed to the preservation of the cell membrane from lipid oxidation (Fonseca *et al.*, 2003).

Maintenance of the cell membrane structure and fluidity that decreases because of the temperature downshift plays a key role in viability during frozen storage. Membrane fluidity is directly related to the fatty acid composition, in particular the ratio between unsaturated and saturated fatty acids (U/S). Factors modifying cellular fatty acid composition after an increase of membrane fluidity are therefore susceptible to improving resistance of cells to frozen storage. Streit *et al.* (2011) attributed the better tolerance of microfiltered cells of *Lb. bulgaricus* during frozen storage to an increase in the unsaturated to saturated and cyclic to saturated fatty acid ratios. Addition of oleic acid (or tween 80) to the culture medium also enhanced the concentration of dihydrosterculic acid and the U/S ratio and improved resistance of *S. thermophilus* to frozen storage (Béal *et al.*, 2001). Under stress conditions (e.g. acidic environment), bacterial cells are also able to modify the U/S ratio (see section 13.7.2).

13.5.2   Storage of dried cells

The control of two parameters related to water, namely water activity (a_w) and moisture content, is essential to preserve the viability of dried cells. Moisture content below 1% is desired to ensure good long-term shelf life (Nakamura, 1996). However, good stability of spray dried lactobacilli was reported during storage at a moisture content until 4% (Gardiner *et al.*, 2000; Corcoran *et al.*, 2004). With over 4% moisture content, lactose present in reconstituted skim milk, widely used as a suitable carrier for drying probiotic bacteria, may nevertheless crystallize during storage, depending on temperature, resulting in browning reactions and formation of radical species that are detrimental to cell viability (Jouppila and Roos, 1994; Thomsen *et al.*, 2005; Kurtmann *et al.*, 2009a,b). Miao *et al.* (2008) also reported the detrimental effects of crystallization of lactose, trehalose and sucrose that occurred at high relative vapour pressure on the viability of freeze dried cells of *Lb. rhamnosus* GG stored at room temperature.

The stability of dried lactic acid bacteria is generally considered optimal for a_w ranging from 0.1 and 0.2 (Champagne *et al.*, 1996; Teixeira *et al.*, 1995). During 30 months storage at 5°C, Abe *et al.* (2009) observed no significant decrease in the viability of freeze dried cells of *B. longum* at a_w of 0.04, 0.10, 0.16, 021, 0.32, 0.40 and 0.56. However, the lowest a_w resulted in better stability when storage temperature increased. At 37°C, inactivation rate constants were of 11.0 and 0.015 log cfu g^{-1} per month for dried cells with a_w

© Woodhead Publishing Limited, 2013

of 0.56 and 0.04, respectively. A similar observation has been reported for freeze- and vacuum-dried cells of *Lb. paracasei* (Foerst *et al.*, 2012).

Oxidation processes occurring during storage may cause damage to the cell wall, membrane and DNA leading to cell death. Oxidation reaction rates are strongly influenced by moisture content, water activity and temperature and, obviously, the presence of oxygen. Addition of antioxidants or free radical scavengers such as ascorbic acid to a certain extent may prevent damage resulting from oxidative stress (Teixeira *et al.*, 1995; Kurtmann *et al.*, 2009a). Diffusion of oxygen into the powder can also be prevented by packaging. New packages have recently been developed with regard to oxygen sensitivity of probiotic bacteria. ZerO$_2$® developed by Food Science Australia (CSIRO, Australia) is an oxygen scavenging additive containing a reducible organic compound, such as substituted anthraquinone, which is incorporated into a polymer for use as a layer in a laminated packaging film. This system is activated by UV light exposure before packaging (Lopez-Rubio *et al.*, 2008).

The stability of dried probiotic cells is affected by many factors which are strongly coupled. However, combination of the following conditions is necessary for optimal stability: $a_w < 0.3$; moisture content <4%, temperature <4°C and presence of antioxidants. Microencapsulation (see next section) may also help to preserve the viability of dried cells during storage by packaging cells in material layers exhibiting different barrier properties.

13.6   Microencapsulation

Encapsulation, also referred as microencapsulation with respect to the size of the capsules produced, aims to protect probiotic cells from further stress conditions encountered during their incorporation into foods or feeds, in the gastrointestinal tract following ingestion as well as to improve stability during storage. Microencapsulation is defined as 'the technology for packaging solid, liquid and gaseous materials in small capsules that release their contents at controlled rates over prolonged period of time' (Champagne and Fustier, 2007). The main encapsulation technologies used for probiotics are based on entrapment in gel particles, spray coating or emulsion followed by spray drying. Selection of the technology will depend on different criteria: degree of protection, cost and toxicity.

13.6.1   Gel particles

Gel particle technology consists of entrapment of probiotic cells within a polymeric matrix such as pectin, gellan gum, κ-carrageenan/locust bean gum and alginate by emulsion or extrusion techniques and various methodologies (Champagne and Fustier, 2007; Kailasapathy, 2002). Entrapment operating in general under mild conditions allows good preservation of cell

© Woodhead Publishing Limited, 2013

viability. In addition, the cost of this technology is particularly low in comparison to other microencapsulation methods. Plant polymers are often used as an encapsulation matrix and efficiently protect cells against oxygen, acidic environment, freezing and refrigerated storage in food (Grattepanche and Lacroix, 2010). Coating gel beads containing the cells with compounds such as chitosan, sodium alginate, poly-L-lysine or proteins confers additional protection against the harsh conditions encountered in the gastrointestinal tract (Krasaekoopt et al., 2004; Chen et al., 2006; Guerin et al., 2003; Gbassi et al., 2009).

Despite very promising results, entrapment of probiotic cells into gel particles possesses two major challenges when scaling up the technology, production of beads with diameters which do not affect food texture for the droplet extrusion methods and the large size dispersion in the emulsion techniques.

13.6.2 Spray coating
In contrast to entrapment of cells in gel particles, spray coating is a suitable encapsulation technology for large scale production. Spray coating consists of suspending, or fluidizing, particles of a core material in an upward stream of air and applying an atomized coating material to the fluidized particles (Augustin and Sanguansri, 2008). The spray-coating techniques have been recently reviewed by Champagne and Fustier (2007). Little information is available in the literature while spray coating is widely used in industry. A patent claimed that the high survival rate of bacteria and yeast can be achieved during spray coating through an appropriate selection of the operating parameters (i.e. the nature of the coating material, rotor, and air speed in the vessel, the pulverization pressure of the coating material, coating rate, temperature of pulverization and incoming air, the coating material and the product) (Durand and Panes, 2003). Spray-coated cells exhibit better resistance to simulated gastric conditions, high temperature, compression and, during storage in powdered milk, compared to uncoated freeze-dried cells as well as better resistance of cells to compression.

13.6.3 Emulsion technique combined with spray drying
Few studies dealing with this microencapsulation method applied to probiotic cells have been published, although this process is well established in the flavour, fragrance and pharmaceutical industries.

The technique is based on encapsulation of milk fat droplets containing freeze-dried probiotic bacteria in whey protein-based water-insoluble microcapsules, using emulsification and spray drying (Picot and Lacroix, 2003a,b). A stabilizing film of whey protein polymers is formed upon rehydration over milk fat globules containing dried cells. A large number of microcapsules with a low diameter can be produced in this two-step

© Woodhead Publishing Limited, 2013

continuous process at a competitive cost. However, micronization of the freeze-dried powder, which is required to control the distribution of the microcapsule size, severely affects the viability of the cells during spray drying (Picot and Lacroix, 2003c, 2004). Dispersion of fresh cells in heat-treated whey protein suspension followed by spray drying was a less destructive method, with survival rates of 26% for *B. breve* and 1.4% for the more heat sensitive *B. longum* (Picot and Lacroix, 2004). Viable counts of *B. breve* cells entrapped in whey protein microcapsules using this method were significantly higher than those of free cells after 28 days in yoghurt stored at 4°C (+2.6 log) and after sequential exposure to simulated gastric and intestinal juices (+2.7 log); in contrast, no protective effect of encapsulation was observed with *B. longum* (Picot and Lacroix, 2004).

13.7   Exploitation of adaptive stress response of bacteria

Bacteria, including lactic acid bacteria and bifidobacteria possess different defence mechanisms to cope with modifications, to a certain extent, of their natural environment (e.g. increase in temperature, pH drop, etc.). These mechanisms have been reviewed for probiotic bacteria by Corcoran *et al.* (2008), Grattepanche and Lacroix (2010) and Mills *et al.* (2011) and the following section will give an overview on how the adaptive stress response of bacteria can be manipulated to improve cell viability to a lethal stress.

13.7.1   Genetic engineering

The first approach consists of improvement of the efficiency of an existing defence mechanism. Molecular chaperones are proteins playing a key role in protein homeostasis in particular under stress conditions (Sugimoto *et al.*, 2008). Homologous overexpression of the *groESL* operon, encoding the molecular chaperone GroESL, in *Lb. paracasei* NFBC 338 improved resistance of the cells to butanol, salt and heat stresses as well as during spray- and freeze-drying (Desmond *et al.*, 2002b; Corcoran *et al.*, 2006). Homologous overproduction of the small heat shock proteins, Hsp 18.5, Hsp 18.55 and Hsp 19.3, in *Lb. plantarum* WCFS1 also alleviated the reduction in growth rate triggered by exposing exponentially growing cells to heat and cold shocks and enhanced survival in the presence of ethanol (Fiocco *et al.*, 2007).

The second approach, heterologous expression, is based on the transfer of an efficient defence mechanism from one bacterial strain to another one. For example, heterologous production by *Lb.* casei of a non-heme catalase from *Lb. plantarum* ATCC 14431 conferred enhanced oxidative stress resistance (Rochat *et al.*, 2006). In contrast to most lactobacilli, *S. thermophilus* possesses an antioxidant enzyme, superoxide dismutase (SOD), which was shown to be essential to support growth under aerobic

© Woodhead Publishing Limited, 2013

conditions. The expression of *sodA*, encoding a manganese SOD of *S. thermophilus*, in *Lactobacillus gasseri* and *Lb. acidophilus* provided protection against hydrogen peroxide stress (Bruno-Bárcena *et al.*, 2004). *Listeria monocytogenes* is able to survive in the presence of high bile concentration. This resistance is attributed to a bile exclusion mechanism, BilE. Heterologous expression of this system from *L. monocytogenes* into *B. breve* improved bile tolerance of the transformed cells as well as their persistence in a murine gastrointestinal tract (Watson *et al.*, 2008).

Targeted genetic modifications can efficiently be used to improve stress tolerance of probiotic cultures, as illustrated above. However, lack of acceptance of genetically modified organisms by consumers, together with technical limitations, for example, bifidobacteria cannot be easily transformed, hinder application. As an alternative to targeted genetic modifications, natural and stable bacterial variants exhibiting resistance to specific stress conditions can be obtained by evolution of strains under rational selective pressure, as illustrated by Mozzetti *et al.* (2010). In this study, *B. longum* NCC2705 was immobilized in gellan-xanthan gum gel beads and continuously cultivated in MRS broth containing an increasing concentration of hydrogen peroxide. After 18 days of culture, one isolate from the immobilized population exhibited stable H_2O_2 resistance phenotype for at least 70 generations as well as higher tolerance to oxygen than non-adapted wild type cells. This technology could be used to raise cells resistant to different and possibly combined stresses.

13.7.2   Application of sub-lethal stresses

Nutrient starvation, accumulation of inhibitory fermentation end products, in particular organic acids, and low pH for non-pH controlled cultures are the main factors leading to the entry of cells into a stationary phase of growth. During this period, bacterial cells develop a general stress response that can further improve their tolerance to lethal conditions (De Angelis and Gobbetti, 2004; Maus and Ingham, 2003; Saarela *et al.*, 2004). Starved cells of *Lb. acidophilus* exhibited an increase in synthesis of unsaturated cyclic and branched fatty acids, to the detriment of saturated fatty acids, as well as proteins involved in carbohydrate and energy metabolisms and in pH homeostasis, which can explain their better tolerance to freezing and frozen storage compared to non-starved cells (Wang *et al.*, 2011).

Application of a specific sub-lethal stress can also induce an adaptive response to protect cells against further homologous or heterologous stress, also called cross adaptation. For example, heat-adapted cells of *Lb. paracasei* NFBC 338 (52°C for 15 min) and *B. adolescentis* (47°C for 15 min) survived up to 300-fold and 128-fold better than non-adapted cells to a lethal heat stress of 60°C for 10 min and 55°C for 20 min, respectively (Desmond *et al.*, 2002b; Schmidt and Zink, 2000). Pretreatment of *Lb. paracasei* NFBC338 with a sub-lethal concentration of NaCl, H_2O_2, or bile

© Woodhead Publishing Limited, 2013

salts confers cross protection of cells against heat although less efficiently than the homologous sub-lethal heat treatment (Desmond *et al.*, 2002b).

However, some hurdles are inherent in the application of sub-lethal stress to improve the robustness of probiotic bacteria. Some treatments, such as heat shock, may be difficult to perform at large scale. Adaptation of a specific stress can detrimentally affect the resistance to another stress. For example, a decrease in membrane fluidity through a shift in fatty acid composition enhances resistance to bile salts (Ruiz *et al.*, 2007; Reimann *et al.*, 2011) but could negatively affect viability during freezing and frozen storage. Moreover, the effects of sub-lethal stress depend on the growing phase of the cells, the species or even strains, and can result in reduced cell yields, cell activity, and/or process volumetric productivity (Doleyres and Lacroix, 2005). Therefore, a labour intensive and costly screening procedure has to be done for each bacterium to optimize sub-lethal stressing conditions (Lacroix and Yildirim, 2007). A two stage continuous culture has recently been proposed as a high throughput alternative to batch for screening sub-lethal stresses (Mozzetti *et al.*, 2012 and 2013).

13.8   Conclusion

For a long time, probiotic strains have been mainly selected based on their technological properties for large scale production. Recent advances in technology and increasing knowledge of bacterial physiology have extended the number of probiotic cultures with more stress sensitive strains for applications in food, feed and supplements. However, fine tuning of process conditions is labour intensive and has to be carried out independently for each strain. Efficient screening methods such as continuous culture could be helpful in the near future in selecting the optimal parameters yielding high cell production with robust physiology and high resistance to stresses. The actual costs of some technologies (e.g. freeze drying, microencapsulation using spray coating) are also relatively high in regard to some large scale applications like feed additives and low cost alternatives need to be developed.

Until now, the focus has been on maximal biomass production while the dose effects of probiotics have not been extensively studied. The physiological state of the cells has also to be considered to maximize health benefits. In this regard, technologies and operating conditions may be of critical importance and should be assessed.

13.9   References

ABE F, MIYAUCHI H, UCHIJIMA A, YAESHIMA T and IWATSUKI K (2009), 'Effects of storage temperature and water activity on the survival of bifidobacteria in powder form', *Int. J. Dairy Technol.*, **62**, 234–9.

© Woodhead Publishing Limited, 2013

ALTERMANN E, RUSSELL WM, AZCARATE-PERIL MA, BARRANGOU R, BUCKL BL, MCAULIFFE O, SOUTHER N, DOBSON A, CALLANAN M, LICK S, HAMRICK A, CANO R and KLAENHAMMER TR (2005), 'Complete genome sequence of the probiotic lactic acid bacterium *Lactobacillus acidophilus* NCFM', *Proc. Natl. Acad. Sci. USA*, **102**, 3906–12.

ANANTA E, VOLKERT M and KNORR D (2005), 'Cellular injuries and storage stability of spray-dried *Lactobacillus rhamnosus* GG', *Int. Dairy J.*, **15**, 399–409.

ARAYA M, MORELLI L, REID G, SANDERS ME, STANTON C, PINEIRO M and BEN EMBAREK P (2002), 'Guidelines for the evaluation of probiotics in food', *Joint FAO/WHO Working Group Report on Drafting Guidelines for the Evaluation of Probiotics in Food*, London (ON, Canada) April 30 and May 1 (2002) ftp://ftp.fao.org/es/esn/food/wgreport2.pdf.

ARNAUD JP, LACROIX C and CASTAIGNE F (1992a), 'Counter-diffusion of lactose and lactic acid in kappa-carrageenan/locust bean gum gel beads with or without entrapped lactic acid bacteria', *Enz. Microb. Technol.*, **14**, 715–24.

ARNAUD JP, LACROIX C and CHOPLIN L (1992b), 'Effect of agitation rate on cell release rate and metabolism during continuous fermentation with entrapped growing *Lactobacillus casei* subsp. *casei*', *Biotechnol. Tech.*, **6**, 265–70.

AUGUSTIN MA and SANGUANSRI L (2008) 'Encapsulation of bioactives', *Food Materials Science, Principles and Practice*, Aguilera JM and Lillfor PJ (eds), Springer, New York, 577–601.

BAUER SA, SCHNEIDER S, BEHR J, KULOZIK U and FOERST P (2012), 'Combined influence of fermentation and drying conditions on survival and metabolic activity of starter and probiotic cultures after low-temperature vacuum drying', *J. Biotechnol.*, **159**, 351–7.

BÉAL C, FONSECA F and CORRIEU G (2001), 'Resistance to freezing and frozen storage of *Streptococcus thermophilus* is related to membrane fatty acid composition', *J. Dairy Sci.*, **84**, 2347–56.

BERGMAIER F, CHAMPAGNE CP and LACROIX C (2005), 'Growth and exopolysaccharide production during free and immobilized cell chemostat culture of *Lactobacillus rhamnosus* RW-9595M', *J. Appl. Microbiol.*, **98**, 272–84.

BIAVATI B and MATTERELLI P (2006), 'The family Bifidobacteriaceae', *The Prokaryotes Vol.3*, Dworkin M, Falkow S, Rosenberg E, Schleifer KH and Stackebrandt E (eds), Springer, New York, 322–82.

BIBAL B, VAYSSIER Y, GOMA G and PAREILLEUX A (1991), 'High-concentration cultivation of *Lactococcus cremoris* in a cell-recycle reactor', *Biotechnol. Bioeng.*, **37**, 746–54.

BOLDUC MP, BAZINET L, LESSARD J, CHAPUZET JM and VUILLEMARD JC (2006a), 'Electrochemical modification of the redox potential of pasteurized milk and its evolution during storage', *J. Agric. Food Chem.*, **54**, 4651–7.

BOLDUC MP, RAYMOND Y, FUSTIER P, CHAMPAGNE CP and VUILLEMARD JC (2006b), 'Sensitivity of bifidobacteria to oxygen and redox potential in non-fermented pasteurized milk', *Int. Dairy J.*, **16**, 1038–48.

BOYAVAL P (1989), 'Lactic acid bacteria and metal ions', *Lait*, **69**, 87–113.

BRUNO-BÁRCENA JM, ANDRUS JM, LIBBY SL, KLAENHAMMER TR and HASSAN HM (2004), 'Expression of a heterologous manganese superoxide dismutase gene in intestinal lactobacilli provides protection against hydrogen peroxide toxicity', *Appl. Environ. Microbiol.*, **70**, 4702–10.

CACHON R, ANTERIEUX P and DIVIES C (1998), 'The comparative behavior of *Lactococcus lactis* in free and immobilized culture process', *J. Biotechnol.*, **63**, 211–18.

CHAMPAGNE CP and FUSTIER P (2007), 'Microencapsulation for the improved delivery of bioactive compounds into foods', *Curr. Opin. Biotechnol.*, **18**, 184–90.

CHAMPAGNE CP, LACROIX C and SODINI-GALLOT I (1994), 'Immobilized cell technologies for the dairy industry', *Crit. Rev. Biotechnol.*, **14**, 109–34.

© Woodhead Publishing Limited, 2013

CHAMPAGNE CP, MONDOU F, RAYMOND Y and ROY D (1996), 'Effect of polymers and storage temperature in the stability of freeze-dried lactic acid bacteria', *Food Res. Int.*, **29**, 555–62.

CHAMPAGNE CP, ROSS RP, SAARELA M, HANSEN KF and CHARALAMPOPOULOS D (2011), 'Recommendations for the viability assessment of probiotics as concentrated cultures and in food matrices', *Int. J. Food Microbiol.*, **149**, 185–93.

CHARALAMPOPOULOS D, PANDIELLA SS and WEBB C (2002). 'Growth studies of potentially probiotic lactic acid bacteria in cereal-based substrates', *J. Appl. Microbiol.*, **92**, 851–9.

CHEN L, REMONDETTO GE and SUBIRADE M (2006), 'Food protein-based material as nutraceuticals delivery systems', *Trends Food Sci. Technol.*, **17**, 272–83.

CLELAND D, KRADER P, MCCREE C, TANG J and EMERSON D (2004), 'Glycine betaine as a cryoprotectant for prokaryotes', *J. Microbiol. Methods*, **58**, 31–8.

COLLADO MC, MERILUTO J and SALMINEN S (2007), 'Measurement of aggregation properties between probiotics and pathogens: *in vitro* evaluation of different methods', *J. Microbiol. Methods*, **71**, 71–4.

COPPA GV, ZAMPINI L, GALEAZZI T and GABRIELLI O (2006), 'Prebiotics in human milk: a review', *Dig. Liver Dis.*, **38**, S291–4.

CORCORAN BM, ROSS RP, FITZGERALD GF and STANTON C (2004), 'Comparative survival of probiotic lactobacilli spray-dried in the presence of prebiotic substances', *J. Appl. Microbiol.*, **96**, 1024–39.

CORCORAN BM, ROSS RP, FITZGERALD GF, DOCKERY P and STANTON C (2006), 'Enhanced survival of GroESL-overproducing *Lactobacillus paracasei* NFBC 338 under stressful conditions induced by drying', *Appl. Environ. Microbiol.*, **72**, 5104–7.

CORCORAN BM, STANTON C, FITZGERALD G and ROSS RP (2008), 'Life under stress: the probiotic stress response and how it may be manipulated', *Curr. Pharm. Des.*, **14**, 1382–99.

CORRE C, MADEC MN and BOYAVAL P (1992), 'Production of concentrated *Bifidobacterium bifidum*', *J. Chem. Tech. Biotechnol.*, **53**, 189–94.

DE ANGELIS M and GOBBETTI M (2004), 'Environmental stress responses in *Lactobacillus*: a review', *Proteomics*, **4**, 106–22.

DEGUCHI Y, MORISHITA T and MUTAI M (1985), 'Comparative studies on synthesis of water-soluble vitamins among human species of bifidobacteria', *Agric. Biol. Chem.*, **49**, 13–19.

DEL RE B, SGORBATI B, MIGLIOLI M and PALENZONA D (2000), 'Adhesion, autoaggregation and hydrophobicity of 13 strains of *Bifidobacterium longum*', *Lett. Appl. Microbiol.*, **31**, 438–42.

DE MAN JD, ROGOSA M and SHARPE ME (1960), 'A medium for the cultivation of lactobacilli', *J. Appl. Bact.*, **23**, 130–5.

DE VUYST L (2000), 'Technology aspects related to the application of functional starter cultures', *Food Technol. Biotechnol.*, **38**, 105–12.

DESJARDINS P, MEGHROUS J and LACROIX C (2001), 'Effect of aeration and dilution rate on nisin Z production during continuous fermentation with free and immobilized *Lactococcus lactis* UL719 in supplemented whey permeate', *Int. Dairy J.*, **11**, 943–51.

DESMOND C, ROSS RP, O'CALLAGHAN E, FITZGERALD G and STANTON C (2002a) 'Improved survival of *Lactobacillus paracasei* NFBC 338 in spray-dried powders containing gum acacia', *J. Appl. Microbiol.*, **93**, 1003–11.

DESMOND C, FITZGERALD GF, STANTON C and ROSS RP (2002b), 'Improved stress tolerance of GroESL-overproducing *Lactococcus lactis* and probiotic *Lactobacillus paracasei* NFBC 338', *Appl. Environ. Microbiol.*, **70**, 5929–36.

DOLEYRES Y and LACROIX C (2005), 'Technologies with free and immobilized cells for probiotic bifidobacteria production and protection', *Int. Dairy J.*, **15**, 973–88.

© Woodhead Publishing Limited, 2013

DOLEYRES Y, PAQUIN C, LEROY M and LACROIX C (2002a), '*Bifidobacterium longum* ATCC 15707 cell production during free- and immobilized-cell cultures in MRS-whey permeate medium', *Appl. Microbiol. Biotechnol.*, **60**, 168–73.

DOLEYRES Y, FLISS I and LACROIX C (2002b), 'Quantitative determination of the spatial distribution of pure- and mixed-strain immobilized cells in gel beads by immunofluorescence', *Appl. Microbiol. Biotechnol.*, **59**, 297–302.

DOLEYRES Y, FLISS I and LACROIX C (2004), 'Continuous production of mixed lactic starters containing probiotics using immobilized cell technology', *Biotechnol. Prog.*, **20**, 145–50.

DURAND H and PANES J (2003), *Particles Containing Coated Living Microorganisms, and Method for Producing Same*, US patent 2003/0109025.

EBEL B, MARTIN F, LE LD, GERVAIS P and CACHON R (2011), 'Use of gases to improve survival of *Bifidobacterium bifidum* by modifying redox potential in fermented milk', *J. Dairy Sci.*, **94**, 2185–91.

FARNWORTH ER and CHAMPAGNE CP (2010), 'Production of probiotic cultures and their incorporation into foods', *Bioactive Foods in Promoting Health*, Watson R and Preedy VR (eds), Academic Press, London, 3–17.

FARNWORTH ER, MAINVILLE I, DESJARDINS MP, GARDNER N, FLISS I and CHAMPAGNE CP (2007), 'Growth of probiotic bacteria and bifidobacteria in a soy yogurt formulation', *Int. J. Food Microbiol.*, **116**, 174–81.

FAVIER CF, DE VOS WM and AKKERMANS AD (2003), 'Development of bacterial and bifidobacterial communities in feces of newborn babies', *Anaerobes*, **9**, 219–29.

FIOCCO D, CAPOZZI V, GOFFIN P, HOLS P and SPANO G (2007), 'Improved adaptation to heat, cold, and solvent tolerance in *Lactobacillus plantarum*', *Appl. Microbiol. Biotechnol.*, **77**, 909–15.

FOERST P, KULOZIK U, SCHMITT M, BAUER S and SANTIVARANGKNA C (2012) 'Storage stability of vacuum-dried probiotic bacterium *Lactobacillus paracasei* F19', *Food Bioprod. Process*, **90**, 295–300.

FOLEY S, STOLARCZYK E, MOUNI F, BRASSART C, VIDAL O, AÏSSI E, BOUQUELET S and KRZEWINSKI F (2008), 'Characterisation of glutamine fructose-6-phosphate amidotransferase (EC 2.6.1.16) and N-acetylglucosamine metabolism in *Bifidobacterium*', *Arch. Microbiol.*, **189**, 157–67.

FONSECA F, BÉAL C and CORRIEU G (2000), 'Method of quantifying the loss of acidification activity of lactic acid starters during freezing and frozen storage', *J. Dairy Sci.*, **67**, 83–90.

FONSECA F, BÉAL C and CORRIEU G (2001), 'Operating conditions that affect the resistance of lactic acid bacteria to freezing and frozen storage', *Cryobiology*, **43**, 189–98.

FONSECA F, BÉAL C, MIHOUB F, MARIN M and CORRIEU G (2003), 'Improvement of cryopreservation of *Lactobacillus delbrueckii* subsp. *bulgaricus* CF1 with additives displaying different protective effects', *Int. Dairy J.*, **13**, 917–26.

FOWLER A and TONER M (2005), 'Cryo-injury and biopreservation', *Ann. N.Y. Acad. Sci.*, **1066**, 119–35.

FRITZEN-FREIRE CB, PRUDÊNCIO ES, AMBONI RDMC, PINTO SS, NEGRÃO-MURAKAMI AN and MURAKAMI FS (2012), 'Microencaspulation of bifidobacteria by spray drying in the presence of prebiotics', *Food Res. Int.*, **45**, 306–12.

GARDINER GE, O'SULLIVAN E, KELLY J, AUTY MA, FITZGERALD GF, COLLINS JK, ROSS RP and STANTON C (2000), 'Comparative survival rates of human-derived probiotic *Lactobacillus paracasei* and *L. salivarius* strains during heat treatment and spray drying', *Appl. Environ. Microbiol.*, **66**, 2605–12.

GARDINER GE, BOUCHIER P, O'SULLIVAN E, KELLY J, COLLINS JK, FITZGERALD G, ROSS RP and STANTON C (2002), 'A spray-dried culture for probiotic Cheddar cheese manufacture', *Int. Dairy J.*, **12**, 749–56.

© Woodhead Publishing Limited, 2013

GAVINI F, POURCHER AM, NEUT C, MONGET D, ROMOND C, OGER C and IZARD D (1991), 'Phenotypic differentiation of bifidobacteria of human and animal origins', *Int. J. Syst. Bacteriol.*, **41**, 548–57.

GBASSI GK, VANDAMME T, ENNAHAR S and MARCHIONI E (2009), 'Microencapsulation of *Lactobacillus plantarum* spp. in alginate matrix coated with whey proteins', *Int. J. Food Microbiol.*, **129**, 103–5.

GOMES AMP, MALCATA FX and KLAVER FAM (1998), 'Growth enhancement of *Bifidobacterium lactis* Bo and *Lactobacillus acidophilus* Ki by milk hydrolyzates', *J. Dairy Sci.*, **81**, 2817–25.

GRATTEPANCHE F and LACROIX C (2010), 'Production of high-quality probiotics using novel fermentation and stabilization technologies', *Biotechnology in Functional Foods and Nutraceuticals*, Bagchi D, Lau FC and Gosh D (eds), Taylor and Francis, Boca Raton, 361–87.

GRATTEPANCHE F, AUDET P and LACROIX C (2007), 'Enhancement of functional characteristics of mixed lactic culture producing nisin Z and exopolysaccharides during continuous prefermentation of milk with immobilized cells', *J. Dairy Sci.*, **90**, 5361–73.

GUERIN D, VUILLEMARD JC and SUBIRADE M (2003), 'Protection of bifidobacteria encapsulated in polysaccharide-protein gel beads against gastric juice and bile', *J. Food Prot.*, **66**, 2076–84.

GYORGY P, KUHN R, ROSE CS and ZILLIKEN F (1954), 'Bifidus factor II. Its occurrence in milk from different species and in other natural products', *Arch. Biochem. Biophys.*, **48**, 202–8.

HAMMES WP and HERTEL C (2006), 'The genera *Lactobacillus* and *Carnobacterium*', *The Prokaryotes Vol.4*, Dworkin M, Falkow S, Rosenberg E, Schleifer KH and Stackebrandt E (eds), Springer, New York, 320–403.

HEENAN CN, ADAMS MC, HOSKEN RW and FLEET GH (2002), 'Growth medium for culturing probiotic bacteria for applications in vegetarian food products', *LWT Food Sci. Technol.*, **35**, 171–6.

HIRSCH A and GRINSTEAD E (1954), 'Methods for the growth and enumeration of anaerobic spore-formers from cheese, with observations on the effect of nisin', *J. Dairy Res.*, **21**, 101–10.

HUANG J, LACROIX C, DABA H and SIMARD RE (1996), 'Pediocin 5 production and plasmid stability during continuous free and immobilized cell cultures of *Pediococcus acidilactici* UL5', *J. Appl. Bacteriol.*, **80**, 635–44.

JENKINS DM, POWELL CD, FISCHBORN T and SMART KA (2011), 'Rehydration of active dry brewing yeast and its effect on cell viability', *J. Inst. Brew.*, **117**, 377–82.

JOUPPILA K and ROOS YH (1994), 'Glass transitions and crystallization in milk powders', *J. Dairy Sci.*, **77**, 2907–15.

JUÁREZ TOMÁS MS, OCÃNA VS and NADER-MACIÁS ME (2004), 'Viability of vaginal probiotic lactobacilli during refrigerated and frozen storage', *Anaerobe*, **10**, 1–5.

KAILASAPATHY K (2002), 'Microencapsulation of probiotic bacteria: technology and potential applications', *Curr. Issues Intest. Microbiol.*, **3**, 39–48.

KAMALY KM (1997). 'Bifidobacteria fermentation of soybean milk', *Food Res. Int.*, **30**, 675–82.

KAWASAKI S, NAGASAKU M, MIMURA T, KATASHIMA H, IJYUIN S, SATOH T and NIIMURA Y (2007), 'Effect of CO2 on colony development by *Bifidobacterium* species', *Appl. Environ. Microbiol.*, **73**, 7796–8.

KIM TB, SONG SH, KANG SC and OH DK (2003), 'Quantitative comparison of lactose and glucose utilization in *Bifidobacterium longum* cultures', *Biotechnol. Prog.*, **19**, 672–5.

KING VAE and SU JT (1994), 'Dehydration of *Lactobacillus acidophilus*', *Process Biochem.*, **28**, 47–52.

KITAMURA Y, ITOH H, ECHIZEN H and SATAKE T (2009), 'Experimental vacuum spray drying of probiotic foods included lactic acid bacteria', *J. Food Process. Pres.*, **33**, 714–26.

KLEEREBEZEM M and VAUGHAN EE (2009), 'Probiotic and gut lactobacilli and bifidobacteria: molecular approaches to study diversity and activity', *Ann. Rev. Microbiol.*, **63**, 269–90.

KOS B, SUSKOVIĆ J, VUKOVIĆ S, SIMPRAGA M, FRECE J and MATOSIĆ S (2003), 'Adhesion and aggregation ability of probiotic strain *Lactobacillus acidophilus* M92', *J. Appl. Microbiol.*, **94**, 981–7.

KRASAEKOOPT W, BHANDARI B and DEETH H (2004), 'The influence of coating materials on some properties of alginate beads and survivability of microencapsulated probiotic bacteria', *Int. Dairy J.*, **14**, 737–43.

KURTMANN L, CARLSEN CU, RISBO J and SKIBSTED LH (2009a), 'Storage stability of freeze-dried *Lactobacillus acidophilus* (La-5) in relation to water activity and presence of oxygen and ascorbate', *Cryobiology*, **58**, 175–80.

KURTMANN L, SKIBSTED LH and CARLSEN CU (2009b), 'Browning of freeze-dried probiotic bacteria cultures in relation to loss of viability during storage', *J. Agric. Food Chem.*, **57**, 6736–41.

KWON SG, SON JW, KIM HJ, PARK CS, LEE JK, JI GE and OH DK (2006), 'High concentration cultivation of *Bifidobacterium bifidum* in a submerged membrane bioreactor', *Biotechnol. Prog.*, **22**, 1591–7.

LACROIX C and YILDIRIM S (2007), 'Fermentation technologies for the production of probiotics with high viability and functionality', *Curr. Opin. Biotechnol.*, **18**, 176–83.

LACROIX C, GRATTEPANCHE F, DOLEYRES Y and BERGMAIER D (2005), 'Immobilised cell technologies for the dairy industry', *Applications of Cell Immobilisation Biotechnology*, Nedovic V and Willaert R (eds), Springer, Dordrecht, 295–319.

LAMBOLEY L, LACROIX C, CHAMPAGNE CP and VUILLEMARD JC (1997), 'Continuous mixed strain mesophilic lactic starter production in supplemented whey permeate medium using immobilized cell technology', *Biotechnol. Bioeng.*, **5**, 502–16.

LEE JH and O'SULLIVAN DJ (2010), 'Genomic insights into bifidobacteria', *Microbiol. Mol. Biol. Rev.*, **74**, 378–416.

LESLIE SB, ISREALI E, LIGHTHART B, CROWE JH and CROWE LM (1995), 'Trehalose and sucrose protect both membranes and proteins in intact bacteria during drying', *Appl. Environ. Microbiol.*, **61**, 3592–7.

LI H, LU M, GUO H, LI W and ZHANG H (2010), 'Protective effect of sucrose on the membrane properties of *Lactobacillus casei* Zhang subjected to freeze-drying', *J. Food Prot.*, **73**, 715–19.

LIAN WC, HSIAO HC and CHOU CC (2002), 'Survival of bifidobacteria after spray-drying', *Int. J. Food Microbiol.*, **74**, 79–86.

LINDERS LJM, DE JONG GIW, MEERDINK G and VAN'T RIET K (1997), 'Carbohydrates and the dehydration inactivation of *Lactobacillus plantarum*: the role of moisture distribution and water activity', *J. Food Eng.*, **31**, 237–50.

LÓPEZ-RUBIO A, LAGARÓN JM and OCIO MJ (2008), 'Active polymer packaging of non-meat food products', *Smart Packaging Technologies for Fast Moving Consumers Goods*, Kerry J and Butler P (eds), Wiley, Chichester, 19–32.

MARGARITIS A and KILONZO PM (2005), 'Production of ethanol using immobilised cell bioreactor systems', *Applications of Cell Immobilisation Biotechnology*, Nedovic V and Willaert R (eds), Springer, Dordrecht, 375–405.

MASSON F, LACROIX C and PAQUIN C (1994) 'Direct measurement of pH profiles in gel beads immobilizing *Lactobacillus helveticus* using a pH sensitive microelectrode', *Biotechnol. Tech.*, **8**, 551–6.

© Woodhead Publishing Limited, 2013

MAUS, JE and INGHAM SC (2003), 'Employment of stressful conditions during culture production to enhance subsequent cold- and acid-tolerance of bifidobacteria', *J. Appl. Microbiol.*, **95**, 146–54.

MENG XC, STANTON C, FITZGERALD GF, DALY C and ROSS RP (2008), 'Anhydrobiotics: the challenges of drying probiotic cultures', *Food Chem.*, **106**, 1406–16.

MIAO S, MILLS S, STANTON C, FITZGERALD GF, ROOS Y and ROSS RP (2008), 'Effect of disaccharides on survival during storage of freeze dried probiotics', *Dairy Sci. Technol.*, **88**, 19–30.

MILLE Y, OBERT JP, BENEY L and GERVAIS P (2004), 'New drying process for lactic acid bacteria based on their dehydration behavior in liquid medium', *Biotechnol. Bioeng.*, **88**, 71–6.

MILLS S, STANTON C, FITZGERALD GF and ROSS RP (2011), 'Enhancing the stress response of probiotics for a lifestyle from gut to product and back again', *Microb. Cell Fact.*, **10**, S19.

MOZZETTI V, GRATTEPANCHE F, MOINE D, BERGER B, REZZONICO E, MEILE L, ARIGONI F and LACROIX C (2010), 'New method for selection of hydrogen peroxide adapted bifidobacteria cells using continuous culture and immobilized cell technology', *Microb. Cell Fact.*, **9**, 60.

MOZZETTI V, GRATTEPANCHE F, MOINE D, BERGER B, REZZONICO E, REIMANN S, MEILE L, ARIGONI A, LACROIX C (2013), 'High-throughput screening of sublethal stress conditions for optimal fitness of *B. longum* NCC2705 in a two stage continuous culture system', *Benef. Microbes*, in press.

MOZZETTI V, GRATTEPANCHE F, MOINE D, BERGER B, REZZONICO E, REIMANN S, MEILE L, ARIGONI A, LACROIX C (2012), 'Physiological stability of *Bifidobacterium longum* NCC2705 under continuous culture conditions', *Benef. Microbes*, **3**, 261–72.

MUELLER S, SAUNIER K, HANISCH C, NORIN E, ALM L, MIDTVEDT T, CRESCI A, SILVI S, ORPIANESI C, VERDENELLI MC, CLAVEL T, KOEBNICK C, ZUNFT HJ, DORÉ J and BLAUT M (2006), 'Differences in fecal microbiota in different European study populations in relation to age, gender, and country: a cross-sectional study', *Appl. Environ. Microbiol.*, **72**, 1027–33.

MULLER JA, ROSS RP, FITZGERALD GF and STANTON C (2009), 'Manufacture of probiotic bacteria', *Prebiotics and Probiotics Science and Technology*, Charalampopoulos and Rastall RA (eds), Springer-Verlag, New York, 725–59.

NAKAMURA LK (1996), 'Preservation and maintenance of eubacteria', *Maintaining Cultures for Biotechnology and Industry*, Hunter-Cervera JC and Belt A (eds), Academic Press, London, 65–84.

NINOMIYA K, MATSUDA K, KAWAHATA T, KANAYA T, KOHNO M, KATAKURA Y, ASADA M and SHIOYA S (2009), 'Effect of CO_2 concentration on the growth and exopolysaccharide production of *Bifidobacterium longum* cultivated under anaerobic conditions', *J. Biosci. Bioeng.*, **107**, 535–7.

OLIVEIRA RP, FLORENCE AC, SILVA RC, PEREGO P, CONVERTI A, GIOIELLI LA and OLIVEIRA MN (2009), 'Effect of different prebiotics on the fermentation kinetics, probiotic survival and fatty acids profiles in nonfat symbiotic fermented milk', *Int. J. Food Microbiol.*, **128**, 467–72.

PARENTE E and ZOTTOLA EA (1991), 'Growth of thermophilic starters in whey permeate media', *J. Dairy Sci.*, **74**, 20–8.

PEEBLES MM, GILLILAND SE and SPECK ML (1969), 'Preparation of concentrated lactic streptococcus starters', *Appl. Microbiol.*, **17**, 805–10.

PICOT A and LACROIX C (2003a), 'Optimization of dynamic loop mixer operating conditions for production of o/w emulsion for cell microencapsulation', *Lait*, **83**, 237–50.

PICOT A and LACROIX C (2003b), 'Production of multiphase water-insoluble microcapsules for cell microencapsulation using an emulsification/spray-drying technology', *J. Food Sci.*, **68**, 2693–700.

© Woodhead Publishing Limited, 2013

PICOT A and LACROIX C (2003c), 'Effects of micronization on viability and thermotolerance of probiotic freeze dried cultures', *Int. Dairy J.*, **13**, 455–62.

PICOT A and LACROIX C (2004), 'Encapsulation of bifidobacteria in whey protein-based microcapsules and survival in simulated gastrointestinal conditions and in yoghurt', *Int. Dairy J.*, **14**, 505–15.

REIMANN S, GRATTEPANCHE F, BAGGENSTOS C, REZZONICO E, BERGER B, ARIGONI F and LACROIX C (2010), 'Development of a rapid screening protocol for selection of strains resistant to spray drying and storage in dry powder', *Benef. Microbes*, **1**, 165–74.

REIMANN S, GRATTEPANCHE F, BENZ R, MOZZETTI V, REZZONICO E, BERGER B and LACROIX C (2011), 'Improved tolerance to bile salts of aggregated *Bifidobacterium longum* produced during continuous culture with immobilized cells', *Bioresour. Technol.*, **102**, 4559–67.

ROCHAT T, GATADOUX JJ, GRUSS A, CORTHIER G, MAGUIN E, LANGELLA P and VAN DE GUCHTE M (2006), 'Production of a heterologous nonheme catalase by *Lactobacillus casei*: an efficient tool for removal of H_2O_2 and protection of *Lactobacillus bulgaricus* from oxidative stress in milk', *Appl. Environ. Microbiol.*, **72**, 5143–9.

ROSS RP, DESMOND C, FITZGERALD GF and STANTON C (2005), 'Overcoming the technological hurdles in the development of probiotic foods', *J. Appl. Microbiol.*, **98**, 1410–17.

ROZADA R, VÁZQUEZ JA, CHARALAMPOPOULOS D, THOMAS K and PANDIELLA SS (2009), 'Effect of storage temperature and media composition on the survivability of *Bifidobacterium breve* NCIMB 702257 in a malt hydrolisate', *Int. J. Food Microbiol.*, **133**, 14–21.

RUIZ L, SÁNCHEZ B, RUAS-MADIEDO P, DE LOS REYES-GAVILÁN CG and MARGOLLES A (2007), 'Cell envelope changes in *Bifidobacterium animalis* ssp. *lactis* as a response to bile', *FEMS Microbiol. Lett.*, **274**, 316–22.

SAARELA M, RANTALA M, HALLAMAA K, NOHYNEK L, VIRKAJARVI I and MATTO J (2004), 'Stationary-phase acid and heat treatments for improvement of the viability of probiotic lactobacilli and bifidobacteria', *J. Appl. Microbiol.*, **96**, 1205–14.

SAARELA M, VIRKAJÄRVI I, ALAKOMI HL, MATILLA-SANDHOLM T, VAARI A, SUOMALAINEN T and MATTÖ J (2005), 'Influence of fermentation time, cryoprotectant and neutralization of cell concentrate on freeze-drying survival, storage stability, and acid and bile exposure of *Bifidobacterium animalis* ssp. *Lactis* cells produced without milk-based ingredients', *J. Appl. Microbiol.*, **99**, 1330–9.

SANTIVARANGKNA C, KULOZIK U and FOERST P (2007), 'Alternative drying processes for the industrial preservation of lactic acid starter cultures', *Biotechnol. Prog.*, **23**, 302–15.

SANTIVARANGKNA C, KULOZIK U and FOERST P (2008), 'Inactivation mechanisms of lactic acid starter cultures preserved by drying processes', *J. Appl. Microbiol.*, **105**, 1–13.

SAXELIN M, GRENOV B, SVENSSON U, FONDÉN R, RENIERO R and MATILLA-SANDHOLM T (1999), 'The technology of probiotics', *Trends Food Sci. Technol.*, **10**, 387–92.

SCARDOVI V (1986), 'Genus *Bifidobacterium* Orla-Jensen', *Bergey's Manual of Systematic Bacteriology Vol.2*, Sneath PHA, Mair NS, Sharpe ME and Holt JG (eds), Williams and Wilkins, Baltimore, 1418–34.

SCHACHTSIEK M, HAMMES WP and HERTEL C (2004), 'Characterization of *Lactobacillus coryneformis* DSM 20001T surface protein Cpf mediating coaggregation with and aggregation among pathogens', *Appl. Environ. Microbiol.*, **70**, 7078–85.

SCHMIDT G and ZINK R (2000), 'Basic features of the stress response in three species of bifidobacteria: *B.longum*, *B. adolescentis*, and *B. breve*', *Int. J. Food Microbiol.*, **55**, 41–5.

SEMYONOV D, RAMON O and SHIMONI E (2011), 'Using ultrasonic vacuum spray dryer to produce highly viable dry probiotics', *LWT-Food Sci. Technol.*, **44**, 1844–52.

© Woodhead Publishing Limited, 2013

SHEEHAN VM, SLEATOR RD, FITZGERALD GF and HILL C (2006), 'Heterologous expression of BetL, a betaine uptake system, enhances the stress tolerance of *Lactobacillus salivarius* UCC118', *Appl. Environ. Microbiol.*, **72**, 2170–7.

SHIHATA A and SHAH NP (2000), 'Proteolytic profiles of yogurt and probiotic bacteria', *Int. Dairy J.*, **10**, 401–8.

SIMPSON PJ, ROSS RP, FITZGERALD GF and STANTON C (2004), '*Bifidobacterium psychraerophilum* sp. nov. and *Aeriscardovia aeriphila* gen. nov., sp. nov., isolated from a porcine caecum', *Int. J. Syst. Evol. Microbiol.*, **54**, 401–6.

SIMPSON PJ, STANTON C, FITZGERALD GF and ROSS RP (2005), 'Intrinsic tolerance of *Bifidobacterium* species to heat and oxygen and survival following spray drying and storage', *J. Appl. Microbiol.*, **99**, 493–501.

SINHA VR and KUMRIA R (2003), 'Microbially triggered drug delivery to the colon', *Eur. J. Pharm. Sci.*, **18**, 3–18.

SODINI I, BOQUIEN CY, CORRIEU G and LACROIX C (1997), 'Microbial dynamics of co- and separately entrapped mixed cultures of mesophilic lactic acid bacteria during the continuous prefermentation of milk', *Enz. Microb. Technol.*, **20**, 381–8.

STADHOUDERS J, JANSEN LA and HUP G (1969), 'Preservation of starters and mass production of starter bacteria', *Neth. Milk Dairy J.*, **23**, 182–9.

STEPHENIE W, KABEIR BM, SHUHAIMI M, ROSFARIZAN M and YAZID AM (2007), 'Growth optimization of a probiotic candidate, *Bifidobacterium pseudocatenulatum* G4, in milk medium using response surface methodology', *Biotechnol. Bioprocess Eng.*, **12**, 106–13.

STRASSER S, NEUREITER M, GEPPL M, BRAUN R and DANNER H (2009), 'Influence of lyophilization, fluidized bed drying, addition of protectants, and storage on the viability of lactic acid bacteria', *J. Appl. Microbiol.*, **107**, 167–177.

STREIT F, CORRIEU G and BÉAL C (2010), 'Effect of centrifugation conditions on the cryotolerance of *Lactobacillus bulgaricus* CFL1', *Food Bioprocess Technol.*, **3**, 36–42.

STREIT F, ATHÈS V, BCHIR A, CORRIEU G and BÉAL C (2011), 'Microfiltration conditions modify *Lactobacillus bulgaricus* cryotolerance in response to physiological changes', *Bioprocess Biosyst. Eng.*, **34**, 197–204.

SUGIMOTO S, AL-MAHIN A and SONOMOTO K (2008), 'Molecular chaperones in lactic acid bacteria: physiological consequences and biochemical properties', *J. Biosci. Bioeng.*, **106**, 324–36.

TAMINE AY, MARSHALL VM and ROBINSON RK (1995), 'Microbiological and technological aspects of milks fermented by bifidobacteria', *J. Dairy Res.*, **62**, 151–87.

TANIGUCHI M, KOTANI N and KOBAYASHI T (1987), 'High concentration cultivation of *Bifidobacterium longum* in fermenter with cross-flow filtration', *Appl. Microbiol. Biotechnol.*, **25**, 438–41.

TEIXEIRA PC, CASTRO MH, MALCATA FX and KIRBY RM (1995), 'Survival of *Lactobacillus delbrueckii* spp. *bulgaricus* following spray-drying', *J. Dairy Sci.*, **78**, 1025–31.

THOMSEN MK, LAURIDSEN L, SKIBSTED LH and RISBO J (2005), 'Two types of radicals in whole milk powder. Effect of lactose crystallization, lipid oxidation, and browning reactions', *J. Agric. Food Chem.*, **53**, 1805–11.

TO BCS and ETZEL MR (1997), 'Spray drying, freeze drying, or freezing of three different lactic acid bacteria species', *J. Food Sci.* **62**, 576–85.

TSVETKOV T and SHISHKOVA I (1982), 'Studies on the effects of low temperatures on lactic acid bacteria', *Cryobiology*, **19**, 211–14.

TYMCZYSZYN EE, DIAZ R, PATARO A, SANDONATO N, GOMEZ-ZAVAGLIA A and DISALVO EA (2008), 'Critical water activity for the preservation of *Lactobacillus bulgaricus* by vacuum drying', *Int. J. Food Microbiol.*, **128**, 342–7.

VENTLING BL and MISTRY W (1993), 'Growth characteristics of bifidobacteria in ultrafiltered milk', *J. Dairy Sci.*, **76**, 962–71.

© Woodhead Publishing Limited, 2013

VON AH U, MOZZETTI V, LACROIX C, KHEADR EE, FLISS I and MEILE L (2007), 'Classification of a moderately oxygen-tolerant isolate from baby faeces as *Bifidobacterium thermophilum*', *BMC Microbiol.*, **7**, 79.

WANG Y, DELETTRE J, CORRIEU G and BÉAL C (2011), 'Starvation induces physiological changes that act on the cryotolerance of *Lactobacillus acidophilus* RD758', *Biotechnol. Prog.*, **27**, 342–50.

WATSON D, SLEATOR RD, HILL C and GAHAN CG (2008), 'Enhancing bile tolerance improves survival and persistence of *Bifidobacterium* and *Lactococcus* in the murine gastrointestinal tract', *BMC Microbiol.*, **8**, 176.

WOLFF E, DELISLE B, CORRIEU G and GIBERT H (1990), 'Freeze-drying of *Streptococcus thermophilus*: a comparison between the vacuum and the atmospheric method', *Cryobiology*, **27**, 569–75.

YING DY, SUN J, SANGUANSRI L, WEERAKKODY R, AUGUSTIN MA (2012), 'Enhanced survival of spray-dried microencapsulated *Lactobacillus rhamnosus* GG in the presence of glucose', *J. Food Eng.*, **109**, 6.

YONEZAWA S, XIAO JZ, ODAMAKI T, ISHIDA T, MIYAJI K, YAMADA A, YAESHIMA T and IWATSUKI K (2010), 'Improved growth of bifidobacteria by cocultivation with *Lactococcus lactis* subspecies *lactis*', *J. Dairy Sci.*, **93**, 1815–23.

ZHU L, LI W and DONG X (2003), 'Species identification of genus *Bifidobacterium* based on partial HSP60 gene sequences and proposal of *Bifidobacterium thermacidophilum* subsp. *porcinum* subsp. nov.', *Int. J. Syst. Evol. Microbiol.*, **53**, 1619–23.

ZOKAEE F, KAGHAZCHI T and ZARE A (1999), 'Cell harvesting by microfiltration in a deadend system', *Process Biochem.*, **34**, 803–10.

© Woodhead Publishing Limited, 2013

14

Microbial production of bacteriocins for use in foods

D. G. Burke, P. D. Cotter, R. P. Ross, Teagasc Food Research Centre, Ireland and C. Hill, University College Cork, Ireland

DOI: 10.1533/9780857093547.2.353

Abstract: Bacteriocins are ribosomally synthesised, antimicrobial peptides produced by bacteria. Many bacteriocins produced by food grade lactic acid bacteria exhibit the potential to control spoilage and pathogenic bacteria in food. Here we review the means by which these bacteriocins are employed by the food industry, that is through the production of bacteriocins by bacteriocinogenic bacteria from within the food or by the addition of a bacteriocin to the food in the form of an ingredient or preservative.

Key words: bacterial fermentation, bacteriocin, bacteriocin purification, biopreservatives, lactic acid bacteria.

14.1 Introduction

Bacteriocins are ribosomally synthesised, small heat stable, antimicrobial peptides produced by bacteria. They are typically active against closely related species but can also have a broad spectrum of activity across genera. Bacteriocinogenic bacteria are protected from the bacteriocins which they produce as a consequence of the production of dedicated immunity (self-protective) proteins (Cotter *et al.*, 2005; Rea *et al.*, 2011). Gram positive bacteriocins can be divided into two classes: Class I, the post-translationally modified bacteriocins and Class II, the unmodified bacteriocins (Cotter *et al.*, 2005; Rea *et al.*, 2011). It is estimated that between 30 and 99% of bacteria produce at least one bacteriocin (Klaenhammer, 1988; Riley, 1998). Indeed, the frequency with which bacteriocin encoding gene clusters occur has been borne out by genome sequencing studies (Begley *et al.*, 2009; Marsh *et al.*, 2010). This fact indicates that there continue to be great opportunities for the discovery and development of new bacteriocins for commercial applications. To date such commercial applications have most

© Woodhead Publishing Limited, 2013

frequently involved the use of bacteriocins to control spoilage or pathogenic bacteria in food. The lactic acid bacteria (LAB) are the most important bacteriocin producers with regard to such applications. LAB have been
used for millennia for the preservation and microbial safety of fermented
foods by inhibiting the growth of pathogenic and spoilage bacteria (Caplice
and Fitzgerald, 1999). Food preservation is mediated by the production of
a number of end products of LAB fermentation such as organic acids,
ethanol, hydrogen peroxide and, of course, bacteriocins (Jack *et al.*, 1995).

Many bacteria isolated from food fermentations have been found to
produce bacteriocins. Examples include nisin and lacticin 3147, which are
Class I bacteriocins (or lantibiotics). These bacteriocins are produced by
members of the genus *Lactococcus* and the benefits of using producers of
such bacteriocins in cheese manufacture have been highlighted (Roberts
et al., 1992, Ryan *et al.*, 1996). Similarly Class II pediocin PA-1 producing
Pediococcus acidilactici have been used in the fermentation of dry fermented sausage (Foegeding *et al.*, 1992) and several bacteriocin producing
lactobacilli have been employed in the fermentation of sausages and olives
(Dicks *et al.*, 2004; Messens *et al.*, 2003; Ruiz-Barba *et al.*, 1994). As a consequence of their long history of safe use, LAB have attained a generally
regarded as safe (GRAS) status. Despite this, nisin remains the only bacteriocin approved for use as a preservative in foods. There are however,
products such as the pediocin PA-1-containing Alta 2341® (Kerry Bioscience, Carrigaline, Co. Cork, Ireland) and other fermentates which are
employed by the food industry. As a consequence of its approved status,
nisin has been the focus of much attention. It was first discovered in 1928
(Rogers and Whittier, 1928) and was first marketed in England in 1953. In
1969, nisin was assessed as safe for use in food by the Joint Food and Agriculture Organization/World Health Organization (FAO/WHO) and was
later added to the European food additive list and given the number E234
by the EEC (EEC, 1983). It has also been approved for use in food by the
US Food and Drug Administration. While nisin is widely available as Nisa-
plin™ (Danisco, Copenhagen, Denmark), other commercial preparations
are also available.

There are numerous benefits to the use of bacteriocins to preserve and
increase the microbial safety of food. First, as bacteriocins are metabolites
of bacteria they are seen as 'natural' products that can be used in place of
chemical preservatives in foods, therefore gaining wider acceptance by
consumers. Bacteriocins can also be applied in combination with other
treatments such as high pressure or temperature to improve food preservation. This is exemplified by lacticin 3147 which, when combined with hydrostatic pressure, has an increased killing effect on *Staphylococcus aureus* and
Listeria innocua than was observed with either treatment alone (Morgan *et
al.*, 1999). Similarly, a combination of the *Enterococcus*-produced enterocin
AS-48 with a mild heat treatment (80–95°C for 5 min) caused a considerable reduction in the viability of *Bacillus coagulans* CECT 12 endospores

© Woodhead Publishing Limited, 2013

when compared to the heat treatment alone, which did not impact significantly on the endospores (Lucas *et al.*, 2006). Thus the effective use of bacteriocins in hurdle technology can result in increased food safety and nutritional quality by the improved killing of contaminating bacteria and a reduction in the need for harsh processing procedures, respectively.

This chapter will focus on the use of bacteriocins for the preservation of food through their production by microbes *in situ* in foods and their incorporation as preservatives/fermentates, and will review the different approaches which have been taken to optimise their yield.

14.2   *In situ* production of bacteriocins in food

As previously stated, LAB have been used for millennia in food fermentations. However, defined starter cultures have replaced traditional undefined mixed culture starters in modern commercial fermentations. Use of defined starter culture systems allows for improved control over fermentations as well as the selection of strains that possess specific beneficial traits, such as bacteriocin production. In the last few decades bacteriocinogenic strains have been studied extensively with regard to their potential as starter cultures, starter adjuncts and protective cultures to control the growth of pathogenic and food spoilage bacteria. This section will explore the topic of *in situ* bacteriocin production for food preservation.

14.2.1   Bacteriocinogenic starter cultures

Unbeknownst to us, mankind may have been reaping the benefits of *in situ* bacteriocin production in fermented foods since the practice of fermentation first began. Ancient fermentations were the result of the outgrowth of microflora naturally present on the raw material from which the fermented product was made. Backslopping, that is the practice of inoculating a new fermentation with a small quantity of fermented product from a successful fermentation was later used as a means of replicating successful fermentations. In effect this process lead to the selection of the best starters, although the underlying science was not understood at the time (Leroy and De Vuyst, 2004). Modern industrial fermentations utilise well-characterised and defined starter systems with properties which are beneficial from the perspective of the end product. Such properties include rapid acid production, phage resistance, the formation of aromatic compounds and, of course, the production of bacteriocins. Bacteriocin production by starter cultures can contribute to food safety and preservation thereby limiting the need for chemical preservatives.

Lactococci are commonly used as starter cultures in the dairy industry, most notably as starters in the manufacture of cheese. As previously stated, nisin is produced by a number of *Lactococcus lactis* strains, some of which

© Woodhead Publishing Limited, 2013

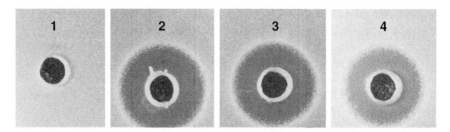

Fig. 14.1   Residual lactcin 3147 activities in Cheddar cheeses made with bacterio-cinogenic strains. Wells: 1, control cheese manufactured by using *L. lactis* DPC4268 (non-bacteriocin-producing starter); 2, 3 and 4 cheeses made with bacteriocin producers (DPC3147, DPC 3256 and DPC 3204) sampled after 2 weeks and 4 and 6 months, respectively (adapted from Ryan *et al.*, 1996).

have been investigated with a view to their use as starter cultures for the control of pathogenic and spoilage bacteria in various food systems. Rodríguez *et al.* (1998) used a nisin producing *L. lactis* strain, ESI 515, in the production of Manchego style cheese made from raw ewe's milk. The strain displayed desirable properties in terms of making this cheese but, importantly, also produced sufficient nisin to reduce counts of *L. innocua* by 4.08 log units relative to a control cheese which was produced with a non-bacteriocinogenic starter culture. *L. lactis* DPC 3147 and a transconjugant, *L. lactis* DPC 4275, both producing the broad spectrum two-component bacteriocin lacticin 3147, have also been successfully used to manufacture Cheddar cheese. Levels of the bacteriocin remained constant over a six month ripening period (see Fig. 14.1) and were sufficient to control non-starter LAB (Ryan *et al.*, 1996), which can lead to inconsistencies and off-flavours. This strain was also used as a starter for cottage cheese manufacture, where it produced 2560 activity units (AU) ml^{-1} of lacticin 3147. There was a 99.9% reduction in *L. monocytogenes* Scott A numbers in cheese made with the lacticin 3147 producing starter after five days whereas no change in pathogen levels was seen in the control cheese (McAuliffe *et al.*, 1999).

In addition to inhibiting pathogens, bacteriocins can be of great use in controlling spoilage bacteria. Bacterial spores that survive milk pasteurisation is a significant problem as they can contaminate cheese and sporulate during ripening causing the formation of off odours and a late blowing defect caused by butyric acid formation. The latter phenomenon is caused by the outgrowth of clostridial spores and is a major cause of spoilage in hard and semi-hard cheeses (McSweeney and Fox, 2004). Garde *et al.*, (2011) successfully prevented the outgrowth of *Clostridium beijerinckii* spores in ovine milk cheese by using a bacteriocinogenic *L. lactis* starter culture. Production of nisin and another Class I bacteriocin, lacticin 481, by the starter culture during fermentation prevented outgrowth of the spores

© Woodhead Publishing Limited, 2013

and hence late blowing. After 120 days ripening this defect occurred in the control cheese but not in cheese made with the bacteriocinogenic starter. As well as preventing spoilage, the starter also generated the desired sensory characteristics for this type of cheese and thus was deemed a suitable replacement starter. The use of bacteriocinogenic cultures in the production of cheese to prevent late blowing offers a natural alternative to lysozyme, which is commonly added to cheese to prevent this defect (Crawford, 1987). This method of prevention is not only more cost effective, but lysozyme has also become an increasingly less attractive preservative in recent years as a consequence of the fact that it is purified from eggs and thus there are fears associated with its potential allergenicity (Carmen Martínez-Cuesta et al., 2010).

Moving away from the topic of dairy products temporarily, the production of dry fermented sausages is also worthy of discussion. This practice traditionally involves fermentation followed by drying without a heat treatment. This minimal processing can potentially lead to contamination and the proliferation of spoilage and pathogenic bacteria, such as *L. monocytogenes*. Pediococci are commonly used as starters in the fermentation of sausages and many *Pediococcus*-produced bacteriocins (also known as pediocins), are active against important pathogenic bacteria, such as those of *Listeria* and *Clostridium* sp. (Christensen and Hutkins, 1992; Luchansky et al., 1992; Nieto-Lozano et al., 2010).

In one such study, a pediocin PA-1 producer, *Pediococcus pentosaceus* BCC 3772 was selected as a consequence of its anti-listerial activity and was evaluated as a starter culture for the fermentation of Nham (Thai traditional pork sausage). The strain performed agreeably when used as a starter culture in the fermentation of Nham in that it did not significantly alter the sensory characteristics of the sausage. *In situ* production of PA-1 was sufficient to reduce spiked *L. monocytogenes* numbers by 3.2 logs compared to initial counts, within 18–24 h (Kingcha et al., 2011).

Pe. acidilactici MCH14, a starter culture commonly used in the Spanish meat industry, also produces pediocin PA-1 and was tested to assess its ability to control the growth of *L. monocytogenes* in Spanish dry-fermented sausage. Sausages were made using either the bacteriocinogenic starter, or a non-bacteriocin producing *P. acidilactici* as a control and were spiked with *L. monocytogenes* (10^5 CFU g^{-1}) before being vacuum stuffed and stored for four weeks. After ripening, the numbers of *L. monocytogenes* were reduced by 2 logs (2×10^1 CFU g^{-1}) in the MCH14-containing sausage compared to the control (7×10^3 CFU g^{-1}) but, importantly, no significant difference was observed between the two starters with regard to A_w and lactic acid production (Nieto-Lozano et al., 2010).

Members of the genus *Lactobacillus* are also commonly used as starter cultures for the fermentation of meat (Hammes et al., 1990). Bacteriocin production appears to provide these lactobacilli with a competitive advantage in such environments, as a number of bacteriocinogenic lactobacilli

© Woodhead Publishing Limited, 2013

have been isolated from fermented sausages (Aymerich *et al.*, 2000; Schillinger and Lucke, 1989; Vignolo *et al.*, 1993; Vogel *et al.*, 1993). For example, *Lactobacillus sake* CTC 494, which was isolated from dry fermented sausage, has been found to have excellent starter capabilities, as well as the ability to produce the *Listeria*-active bacteriocin sakacin K (Hugas *et al.*, 1996), under conditions (pH and temperature) similar to that found during the fermentation of dry fermented sausage (pH 5.0–5.5, 20–25°C) (Leroy and De Vuyst, 1999b). This strain was tested for its ability to control the growth of *L. monocytogenes* in three types of fermented sausage; Belgin-type, Cacciatore-type and Italian salami. Notably, *L. monocytogenes* numbers were reduced by between 0.6 and 1.0 log CFU g^{-1} when compared to bacteriocinogenic starter free controls after the completion of sausage production (Ravyts *et al.*, 2008).

Another example relates to *Lactobacillus pentosus* 31-1. This strain was isolated from a Chinese meat product (Xuanwei ham) and produces the bacteriocin pentocin 31-1. This strain was evaluated as a starter culture for the production of fermented sausage. The strain performed very well, producing a product with desirable organoleptic properties. During challenge tests with *L. innocua* and *S. aureus*, pathogen cell numbers in the *Lb. pentosus* 31-1 sausage were reduced by between 4.4 and 5.1 log units compared to the control. Antimicrobial activity was detected up to 7 days post production in homogenised sausages whereas, as expected, no activity was observed in the controls. Activity was lost during the ripening period suggesting that the bacteriocin was inactivated over time (Liu *et al.*, 2010).

14.2.2   Bacteriocinogenic starter culture adjuncts.

Bacteriocinogenic strains are not always suitable as starter cultures as they may not be capable of the requisite acidification rates or may lack the proteolytic activity required. Such problems have previously been associated with some nisin-producing strains (Lipinska, 1973, 1977). This problem can be solved by introducing the bacteriocin encoding genes into a suitable starter by conjugation or genetic manipulation or, more simply, by combining the bacteriocinogenic culture with a suitable starter culture that is resistant to the bacteriocin being produced.

A number of authors have described the use of bacteriocinogenic enterococci as starter-culture adjuncts (Arantxa *et al.*, 2004; Giraffa *et al.*, 1995; Nascimento *et al.*, 2008; Oumer *et al.*, 2001). However, owing to the ongoing debate about the safety of enterococci in food, these will not be discussed in this section.

The combination of bacteriocinogenic adjunct cultures with resistant/ insensitive commercial starter cultures is an economical alternative to the application of either chemical preservatives or purified bacteriocin preparations for controlling spoilage and pathogenic bacteria. Rilla *et al.* (2003) successfully controlled the growth of *Clostridium tyrobutyricum* CECT4011,

© Woodhead Publishing Limited, 2013

which has been associated with the previously described butyric acid forma-tion/late blowing defect in cheese, through the use of a mesophilic mixed starter, IPLA-001, in combination with the nisin Z-producing strain, *L. lactis* subsp. lactis IPLA 729. Nisin Z is a variant of nisin (nisin A) which differs by one amino acid. Nisin Z levels reached 1600 AU ml^{-1} in this cheese after day 1, this level of activity was retained for 15 days. The experimental cheese, along with a control and a commercial cheese made with the anti-blowing agent potassium nitrate, were spiked with the CECT4011 strain. During ripening, *Cl. tyrobutyricum* numbers were reduced from 1.2×10^5 CFU g^{-1} to 1.3×10^3 CFU g^{-1} in the nisin Z cheese, but increased to 1.99×10^9 CFU g^{-1} and 3.5×10^7 CFU g^{-1} in the control and commercial cheeses, respectively. The nisin Z producing starter *L. lactis* subsp. lactis biovar diacetylactis UL719 was also successfully used as a starter adjunct by Bouksaim *et al.* (2000) in the production of Gouda cheese. Nisin Z levels reached a maximum of 512 AU g^{-1} after 6 weeks ripening but decreased to 128 and 32 AU g^{-1} after 27 and 45 weeks, respectively. The level of nisin present is significant as it has been shown that 40 Au g^{-1} is sufficient to prevent butyric acid formation (Hugenholtz and De Veer, 1991).

As previously stated, bacteriocins can potentially be used in combination with other treatments (or hurdles) to maximise food quality and safety. By utilising two bacteriocinogenic cultures producing different bacteriocins, with differing mechanisms of action but with coinciding spectrums of activ-ity, it is possible to affect a synergistic killing of a target bacterium. This approach can be particularly useful in preventing the emergence of resist-ance (Vignolo *et al.*, 2000). *Lb. plantarum* LMG P-26358, which produces an anti-listerial Class II bacteriocin with 100% homology to plantaricin 423 was used as a starter adjunct in combination with a nisin-producer, *L. lactis* CSK65 in laboratory scale cheese production. In challenge trials, cheese produced with the *Lb. plantarum* LMG P-26358 starter adjunct had a greater killing effect against *L. innocua* than observed in cheese produced with the nisin producer alone. It was also noted that when used in combina-tion, no viable *L. innocua* were recovered after 28 days ripening. Mass spectrometry also revealed that both bacteriocins were present after 18 weeks ripening (Mills *et al.*, 2011). Despite the success of this approach, it is important to ensure the bacteriocins produced are not overly antagonistic against the other bacteriocinogenic culture(s) present. Mills *et al.* (2011) did observe the inhibition of *Lb. plantarum* LMG P-26358 by nisin, which slowed its growth during the first eight hours of fermentation. However, despite this, the *Lb. plantarum* strain did reach the optimal cell density required for bacteriocin production (i.e. 10^8 CFU ml^{-1}) on day one of cheese production in this case.

In addition to inhibiting spoilage and pathogenic microbes, bacteriocins can also be exploited to lyse starter and/or adjunct cultures partially during cheese production which enhances cheese maturation. This is due to the release of intracellular enzymes, such as proteinases, peptidases, amino acid

© Woodhead Publishing Limited, 2013

catabolic enzymes and esterases, all of which have an impact upon flavour formation (Lortal and Chapot-Chartier, 2005). A number of authors have described the use of bacteriocinogenic adjunct starter cultures to increase starter cell lysis (Ávila *et al.*, 2005; Garde *et al.*, 2002, 2005, 2006; Lortal and Chapot-Chartier, 2005; Morgan *et al.*, 1997; O'Sullivan *et al.*, 2002, 2003).

14.2.3   Bacteriocinogenic protective cultures

In situ bacteriocin production for the protection of foods from pathogenic and spoilage bacteria is not an option which is exclusively reserved for fermented foods. Bacteriocinogenic bacteria can also be applied to the surface of non-fermented foods, allowing *in situ* bacteriocin production. A suitable bacteriocinogenic protective culture must not be capable of causing spoilage, must grow at the intended storage temperature (usually refrigeration temperatures), must not affect the organoleptic properties of the food during storage and must produce sufficient bacteriocin to have a protective role under these conditions.

Two lactobacilli producing bacteriocin-like inhibitory substances were assayed for their ability to enhance the preservation of vacuum-packed sliced beef meat stored at 4°C over 28 days by Katikou *et al.* (2005). The two lactobacilli, *Lactobacillus sakei* CECT 4808 and *Lactobacillus curvatus* CECT 904^T, were applied to the sliced beef meat either individually and incombination. Counts of Enterobacteriaceae, *Pseudomonas* spp., LAB, *Brochothrix thermosphacta*, yeasts and moulds as well as the organoleptic properties of the beef were assayed over 28 days. Beef inoculated with either the *Lb. sakei* or the *Lb. sakei* in combination with *Lb. curvatus* had significantly ($P < 0.05$) lower counts of spoilage microbes than beef inoculated with *Lb. curvatus* alone or controls. The use of *Lb. sakei* in isolation was more effective in controlling spoilage microbes and the associated product attained greater organoleptic scores than was achieved by the other treatments. The sakacin P producer *Lb. curvatus* CWBI-B28 has also been shown to be effective as a protective culture, in this instance when employed on the surface of cold-smoked salmon under vacuum-packed and refrigerated conditions. Counts of *L. monocytogenes* on this product were reduced to below the detectable limit (0.7 CFU cm^{-2}) within the first week of storage after the application of *Lb. curvatus* as a protective culture. However a 1.3-log increase in colony numbers was observed after 14 days (Ghalfi *et al.*, 2006). *Lb. curvatus* CWBI-B28 has also been shown to be effective in to controlling *L. monocytogenes* in raw beef and in raw chicken meat when combined with the sakacin G producing *Lb. sakei* CWBI-B1365 (Dortu *et al.*, 2008).

When considering commercial applications for storage of bacteriocinogenic protective cultures, the retention of antimicrobial activity may become an issue. Spray drying offers an inexpensive method for industrial scale production of cultures containing high levels of viable bacteriocinogenic

© Woodhead Publishing Limited, 2013

bacteria suitable for storage and transport. Silva *et al.* (2002) assessed the antimicrobial activity of three bacteriocinogenic protective cultures for their antimicrobial activity before and after spray drying. All three strains, *Lb. sakei*, *Lb. salivarius* and *Carnobacterium divergens*, exhibited antimicrobial activity against the target strains: *Staph. aureus*, *L. innocua* and *L. monocytogenes*. After spray drying, the cultures were stored at either 4°C or 18°C and at a relative humidity of 0.3%. Although *Carn. divergens* lost its antimicrobial activity against *S. aureus* immediately after spray drying, it retained activity against *L. innocua* and *L. monocytogenes*. The two lactobacilli retained antimicrobial activity against all target bacteria after spray drying. Storage temperature had a significant impact on survival of the bacteriocinogenic spray-dried cultures. Numbers of *Carn. divergens* and *Lb. sakei* decreased during storage at 4°C and 18°C, with a greater decrease being seen at 18°C. Numbers of *Lb. salivarius* did not significantly decrease until the third month of storage. Rodgers *et al.* (2002) also demonstrated the retention of nisin and pediocin A production by the protective cultures, *L. lactis* CSCC 146 and *P. pentosaceus* ATCC 43,200, respectively, after freeze-drying. On the basis of these results spray- and freeze-drying could become important processes in the production of bacteriocinogenic cultures for use in food production.

14.2.4 *In situ* bacteriocin production: conclusion

The use of physical processes (heating, hydrostatic pressure, gamma radiation etc.) to kill undesirable microbes in foods can be expensive and can have a negative impact on the sensory characteristics of certain foods and therefore are not always suitable. The addition of chemical preservatives is highly regulated and is negatively perceived by consumers. For these reasons bacteriocin production *in situ* is a very attractive alternative for maintaining food safety and quality. Also, because the bacteriocin is produced *in situ* by GRAS cultures this provides an alternative to the use of concentrated bacteriocin preparations as food preservatives and the associated regulatory restrictions. It is important to note, however, that the bacterial strains to be used as starter, adjunct or protective cultures for *in situ* bacteriocin production need to be extensively tested within the food environment as both their growth rate and levels of bacteriocin production can be affected by environmental factors that prevail within the food environment (Leroy and de Vuyst, 1999a; Neysens *et al.*, 2003; Sarantinopoulos *et al.*, 2002).

14.3 *Ex situ* production of bacteriocins

Bacteriocins for use as preservatives and food additives can be produced as either purified preparations or as fermentates (see Table 14.1). Like many industrial fermentations, maximal product formation is only achieved

© Woodhead Publishing Limited, 2013

Table 14.1 Table of bioactive powders containing bacteriocins

Bacteriocin	Strain	Indicator strain	Growth medium	Method of powder preparation	Activity	Author
Alta 2341™ (Pediocin PA-1)	N/A	N/A	N/A	N/A	3 000 AU g⁻¹	Drider et al. (2006)
AS-48	Enterococcus faecalis A-48-32	Enterococcus faecalis S-47	E-300ᵃ	Spray dried	600 AU ml⁻¹	Ananou et al. (2008)
Bavaricin MN	Lactobacillus bavaricus MN	Lactobacillus sakei ATCC15521	DWPᵇ	Freeze dried	500 AU ml⁻¹	Dimitrieva-Moats et al. (2011)
Brevicin 286	Lactobacillus brevis VB 286	Listeria monocytogenes 4A	YEGᶜ medium	Spray dried	512 000 AU g⁻¹	Coventry et al. (1996)
Lacticin 3147	Lactoccous lactis DPC 3147	Lactococcus lactis HP	DWᵈ	Spray dried	102 400 AU g⁻¹	Morgan et al. (1999)
Lactocin G13	Lactococcus lactis G13	Lactobacillus sakei ATCC15521	CCWᵉ	Freeze dried	500 AU ml⁻¹	Dimitrieva-Moats et al. (2011)
Microgard™	Propinonibacterium freundenreichii subsp. shermanii	N/A	Skim milk	N/A	N/A	Morgan et al. (1999)
Nisaplin™ (Nisin A)	Lactococcus lactis	N/A	Skim milk	N/A	N/A	Morgan et al. (1999)
Nisin A	Lactococcus lactis subsp. cremoris ATCC11454	Lactobacillus sakei ATCC15521	CCW	Freeze dried	500 AU ml⁻¹	Dimitrieva-Moats et al. (2011)
Nisin-like bacteriocin	Lactococcus lactis BFE 920	Lactobacillus sakei ATCC15521	CCW	Freeze dried	500 AU ml⁻¹	Dimitrieva-Moats et al. (2011)
Pediocin PA-1	Pediococcus acidilactici	Pediococcus pentosaceus FBB63	MRSᶠ	Freeze dried	16 000 AU g⁻¹	Pucci et al. (1988)
Varicin	Kocuria varians NCC1482	Lactobacillus helveticus NCC639	RSM + YEᵍ	Spray dried	20 000 AU ml⁻¹	O'Mahony et al. (2001)

ᵃEspiron-300 whey derived substrate; ᵇDemineralised whey protein; ᶜYeast-extract & glucose medium; ᵈDemineralised whey; ᵉCheddar cheese whey; ᶠMRS with 2% yeast extract, 10% (wt/vol) dry milk powder was added to cell free supernatant prior to drying; ᵍReconstituted skim milk + 1.5% yeast extract. * Activity is given as described in the literature.

© Woodhead Publishing Limited, 2013

through tight control of fermentation conditions. For this reason there have been many studies into the effects of medium composition, temperature, pH and fermentation design on bacteriocin production.

14.3.1   Media composition

Production of bacteriocins on an industrial scale is an expensive process. This is due in part to the high cost of commercial media required for the cultivation of the bacteriocin producing LAB, examples include deMan, Rogosa, Sharpe media (MRS), tryptone, glucose, yeast extract media (TGY) and all purpose tween media (APT) (Daba *et al.*, 1993; De Vuyst, 1995; Jensen and Hammer, 1993). While these general media may be ideal for the growth of LAB in the laboratory, they may not be optimal for bacteriocin production.

MRS is commonly used in the production of sakacin A (180 AU ml^{-1}) from *Lb. sakei*. Trinetta *et al.* (2008) used a 'one variable at a time' approach to develop an alternative culture medium for the production of sakacin A. Using this approach an optimal medium for sakacin A production was developed which cost 50% less. In this alternative medium consisting of bactopeptone, meat peptone, milk whey, yeast autolysate, glucose and calcium carbonate, sakacin A production increased to 480 AU ml^{-1}. Juárez Tomás *et al.* (2010) designed an optimal medium for growth and production of the Class II bacteriocin salivaricin CRL 1328 by *Lb. salivarius* CRL 1328. Growth and bacteriocin production were assessed under a range of culture conditions. Using a desirability function, an optimal medium composition was predicted which cost between 25 and 40% less than commercial alternatives, that is MRS.

Cultivation of bacteriocinogenic bacteria in medium based on industrial waste products has also been assessed in a bid to reduce the cost of bacteriocin production. LAB require diverse peptic sources owing to their fastidious nature. Peptones from fish, while uncommon, are a quite good alternative to peptone sources commonly used such as bactopeptone, tryptone, yeast extract and meat extract (Vázquez *et al.*, 2004). Octopus peptone (OP) media made from octopus tissue (as a peptone source) was compared to MRS and a medium (medium B) made with commercial peptones (bactopeptone), for its ability to support growth and the production of nisin and pediocin by strains of *L. lactis* and *P. acidilactici*, respectively. Increased nisin and biomass production was observed in OP relative to that achieved in either MRS or medium B. OP outperformed medium B for biomass and pediocin production but not MRS (Vázquez *et al.*, 2004). A similar study was performed using peptones derived from fish visceral and muscle residues. Nisin and pediocin PA-1 production was assessed in medium prepared with fish peptones (FP) or bactopeptone (D medium) as well as in MRS. FP compared favourably to D medium and MRS for biomass and nisin production by *L. lactis*. Pediocin and biomass production in *P. acidilactici*

© Woodhead Publishing Limited, 2013

improved 500% with FP compared to medium D or MRS (Vázquez *et al.*, 2006). Snow crab hepatopancreas (a by-product of crustacean processing) medium has also been shown to be an effective substrate for the production of bacteriocins. A maximum activity of 3.7×10^4 AU ml^{-1} of divergicin M35 was produced by *Carnobacterium divergens* M35 in snow crab hepatopancreas medium after 10 h (Tahiri *et al.*, 2009).

Cereal-based by-products from various industrial processes have also been investigated for their application as a low-cost medium for large scale bacteriocin production. Condensed corn soluble (CSS) is a by product of fuel ethanol production. CSS was found to be considerably cheaper than the commercial media laurel-tryptose broth (LT broth) for the production of nisin when buffered by NaHCO$_3$. Costs were reduced from US $600 kg^{-1} nisin (LT broth) to US $35–40 kg^{-1} nisin (Wolf-Hall *et al.*, 2009). Malt sprout extract (MSE) medium is another low cost cereal-based medium derived from food industry by-products. MSE medium supplemented with glucose (G) and yeast extract (YE) was investigated as an alternative to MRS for the growth and production of antimicrobial activity by *Lb. plantarum* VTT E-79098. MSE(GYE) compared favourably with MRS for biomass production and, interestingly, was superior for the production of antimicrobial activity. The cost of MSE medium was estimated to be 20% that of MRS (Laitila *et al.*, 2004).

A number of studies have demonstrated that milk and by-products of the dairy industry (whey) are cost effective food grade substrates for commercial bacteriocin production (Arakawa *et al.*, 2008; Daba *et al.*, 1993; Dimov, 2007; Goulhen *et al.*, 1999; Guerra and Pastrana, 2001; Liu *et al.*, 2005; Morgan *et al.*, 1999). Ananou *et al.* (2008) found production of the bacteriocin AS-48 to be significantly cheaper in a commercial whey (supplemented with 1% glucose) when compared to the cost of production in BHI (€1.53 kg^{-1} Vs €140.00 kg^{-1}). As well as being cost-effective food grade substrates, milk and whey have also been used to produce bacteriocin-containing bioactive powders by lyophilising the fermentate of bacteriocinogenic cultures, that is Nisaplin^{TM} and Alta 2341®. Owing to the fastidious nature of LAB, the low concentration or absence of certain growth factors (vitamins and amino acids) in whey may be limiting for bacteriocin production (Guerra and Pastrana, 2001). The addition of a complex nitrogen source can facilitate increased bacteriocin production through the availability of free amino acids, short peptides as well as additional growth factors (Aasen *et al.*, 2000; Cheigh *et al.*, 2002). This is exemplified by addition of yeast extract to whey, which has been shown to increase bacteriocin production in a number of studies (Anthony *et al.*, 2009; Avonts *et al.*, 2004; Cladera-Olivera *et al.*, 2004; Enan and Al Amri, 2006; Liao *et al.*, 1993; Pérez Guerra *et al.*, 2005).

The availability of carbon can also have an impact upon bacteriocin production. At low concentrations, carbon availability can limit biomass production and hence final bacteriocin titres (De Vuyst and Vandamme,

© Woodhead Publishing Limited, 2013

1992). However, it has been shown that increasing the initial carbon concentration does not result in a proportional increase in bacteriocin titre and above a certain concentration can result in a reduction in biomass and bacteriocin formation (Parente *et al.*, 1997). Maximum bacteriocin titre is usually reached at the end of the log phase or early stationary phase, after which time a decrease in bacteriocin activity is commonly seen (Callewaert *et al.*, 2000; Leroy and De Vuyst, 2002; Mataragas *et al.*, 2003). The depletion of the energy source has been proposed for this decline in bacteriocin activity (Callewaert *et al.*, 2000; De Vuyst *et al.*, 1996). Parente *et al.* (1997) observed that at low glucose concentrations (<25 g l^{-1}) bacteriocin activity dropped after it had reached its peak. At higher glucose concentrations, no decline in activity was observed.

14.3.2   Environmental conditions

Bacteriocin production has frequently been reported as being greatest during the active growth phase, displaying primary metabolite kinetics (De Vuyst *et al.*, 1996). While bacteriocin production is growth associated, conditions required for maximum growth rate may not be optimal for maximum bacteriocin production (Nel *et al.*, 2001). A number of studies have shown that temperatures favouring bacteriocin production are lower than those required for optimum growth (Bizani and Brandelli, 2004; Cheigh *et al.*, 2002; Delgado *et al.*, 2005; 2007; Mataragas *et al.*, 2003; Messens *et al.*, 2003). Aasen *et al.* (2000) observed that sakacin P production ceased earlier when grown at 30°C (optimum for growth) than at 20°C and that lower quantities of glucose were consumed at 30°C than was observed when grown at 20°C. The authors proposed that the greater bacteriocin activity observed at 20°C was due to rate limiting reactions dependent on temperature, resulting in more efficient carbohydrate utilisation at lower growth rates. It has also been proposed that a decrease in bacteriocin activity at higher temperatures could be due to the action of proteases (Bizani and Brandelli, 2004; Messens *et al.*, 2003).

The pH of the growth medium during fermentation can have a profound effect on the yield of bacteriocin. As with temperature, sub-optimal pH has frequently been reported as being required for optimal bacteriocin production (Drosinos *et al.*, 2006; Herranz *et al.*, 2001; Mataragas *et al.*, 2003; Matsusaki *et al.*, 1996; Parente and Ricciardi, 1994). Once an optimal pH for bacteriocin production has been established, pH controlled fermentations can be used to maintain maximum bacteriocin production throughout the fermentation (Abriouel *et al.*, 2003; Liu *et al.*, 2010; Mataragas *et al.*, 2003; Naghmouchi *et al.*, 2008; Wolf-Hall *et al.*, 2009). In contrast to this, the natural fall in pH resulting from the production of lactic acid by LAB has also been shown to be beneficial in increasing bacteriocin yields. Guerra and Pastrana (2003) investigated the effect of pH drop on nisin and pediocin PA-1 production in whey. They found that nisin and pediocin

© Woodhead Publishing Limited, 2013

production rates were higher in unbuffered fermentations than in those buffered with different concentrations of potassium hydrogen phthalate-NaOH. Specific bacteriocin production rates increased with decreasing pH, until a final pH inhibitory to the producer was reached. Similar results were reported by Cabo *et al.* (2001) and Yang and Ray (1994). The adsorption of bacteriocins by producer cells at high pH has also frequently been described (Bhunia *et al.*, 1991; D'Angelis *et al.*, 2009; Klaenhammer, 1988; Ray, 1992; Wu *et al.*, 2008; Zhang *et al.*, 2009) and occurs once growth-associated bacteriocin production has ceased (De Vuyst *et al.*, 1996). In such situations, a reduction of the culture pH to around 2.0 once maximum bacteriocin titre has been reached can be employed to prevent adsorption by producer cells and release any bacteriocin already bound to the producer cells (Yang *et al.*, 1992).

14.3.3   Effect of fermentation processes on bacteriocin production

Batch fermentation is what is described as a 'closed system', whereby the substrate and producing microorganism are added to the system at time zero and are not removed until the fermentation is complete. This represents the simplest and most commonly employed method of fermentation for the production of bacteriocins. Optimisation of bacteriocin production in batch fermentation is achieved by the manipulation of growth conditions (pH, temperature etc.) and medium composition. Batch fermentation is ideal for studying bacteriocin production in the laboratory or in small scale trials, but is not economically viable on a commercial scale (De Vuyst and Vandamme, 1991; Liao *et al.*, 1993). Fed-batch fermentation is a modified form of batch fermentation whereby growth limiting substrates are fed into the fermenter at a controlled rate. This allows tight control over the growth rate and can alleviate problems such as catabolite repression. By controlling the sucrose feeding rate in a fed-batch system, Lv *et al.* (2005) increased the maximum nisin titre from 2658 IU ml^{-1} in batch fermentation to 4185 IU ml^{-1}. Controlled carbohydrate feeding facilitated a maximum growth rate without substrate inhibition having an impact upon the rate of bacteriocin production. A number of studies have demonstrated that bacteriocin production can be improved through the use of fed batch, rather than batch fermentations (Ekinci and Barefoot, 2006; Guerra *et al.*, 2005; Lv *et al.*, 2005; Paik and Glatz, 1997). While promising, these experiments were performed on a small scale and thus it is difficult to determine how scaling up of the process would impact upon bacteriocin yields. In one study, bacteriocin production by *Propionibacterium thoenii* in small and large scale fed-batch fermentations has been assessed. Paik and Glatz (1997) found that scaling up of the process led to a reduction in bacteriocin activity. Even with the reduced activity, the authors concluded that fed-batch fermentations have the potential to facilitate the production of high concentrations of bacteriocins by propionibacteria.

© Woodhead Publishing Limited, 2013

While fed-batch fermentation is an improvement over batch fermenta-tion, it has been proposed that continuous culture techniques are superior to both. Continuous fermentation is preferred to the batch process owing to its high productivity, reduced product inhibition, no batch to batch vari-ation and ultimately reduced production costs (Tejayadi and Cheryan, 1995). Continuous fermentation is an 'open system' whereby sterile sub-strate is added to the system at a specific dilution rate (D, volume of vessel h^{-1}), while an equal volume of converted substrate containing the product together with the producer is simultaneously removed. The continuous flow through the system maintains fermentation conditions at the optimum required for bacteriocin production. This results in the producer being maintained at the growth rate/phase which results in the specific bacteriocin production rate being optimal (Bhugaloo-Vial *et al.*, 1997). The dilution rate (D, vessel volume h^{-1}) in a continuous fermentation can greatly affect the productivity of the process. The maximum dilution rate in a continuous fermentation must not exceed the specific growth rate of the bacterial strain being used to prevent cells being washed out of the system.

Kaiser and Montville (1993) compared the production of the Class II bacteriocin Bavaricin MN by *Lb. bavaricus* MN in batch and continuous culture. In a pH and temperature controlled batch fermentation a maximum titer of 3200 AU ml^{-1} was reached. However this fell to 800 AU ml^{-1} after 76 h. In a continuous culture at the same temperature and pH conditions, a maximum titer of 6400 AU ml^{-1} was obtained, which was subsequently maintained for 345 h.

The effect of carbon sources as well as dilution rates on the production of the Class I bacteriocin, Plantaricin C, by *Lb. plantarum* LL441 was inves-tigated in a continuous culture (Bárcena *et al.*, 1998). The carbon sources, glucose, sucrose and fructose were tested at different dilution rates to deter-mine the optimum conditions for bacteriocin production. Plantaricin C was only detected at low D (0.05 h^{-1}) when glucose was used as the carbon source. From this observation it was postulated that glucose may mediate catabolite repression and therefore the use of sucrose and fructose as carbon sources became a priority. With these carbon sources plantaricin C production was detected at D (0.1–0.12 h^{-1}) that is double that of glucose. Under optimum conditions, similar bacteriocin titres (~3200 AU ml^{-1}) in culture supernatants were achieved for all three carbon sources, however bacteriocin yield was doubled in media containing sucrose and fructose owing to the higher dilution rates.

It has also been noted that immobilisation of bacteriocin producing cells onto a solid matrix facilitates the application of dilution rates far in excess of the maximum specific growth rate, without cell washout (Lamboley *et al.*, 1997) and can also provide increased cell density (Dervakos and Webb, 1991) and plasmid stability (Huang *et al.*, 1996). In one instance nisin Z production by *L. lactis* UL179 immobilised in κ-carrageenan/locust bean gum gel (IC) and in free cell culture (FC) was compared in a continuous

© Woodhead Publishing Limited, 2013

fermentation. It was noted that nisin Z production increased with increasing D in IC fermentations relative to FC fermentations. Sonomoto *et al.* (2000) compared nisin Z production by FC and IC adsorbed on to ENTG-3800 in continuous fermentation. FC displayed good nisin Z production at a dilution rate of 0.1 h^{-1} but, at a dilution rate of 0.2 h^{-1}, cell wash out resulted in reduced production. In contrast an increase in productivity at higher dilution rates was observed in IC fermentations. This phenomenon was also reported by Liu *et al.* (2005) during the production of nisin by *L. lactis* subsp. *lactis* ATCC 11454, in other words the authors observed that nisin activity increased with increased dilution rates up to a D of 0.31 h^{-1}, although a further increase in D resulted in a decrease in nisin activity.

The immobilisation of bacteriocinogenic bacteria in calcium alginate gel beads has also been frequently reported (Bhugaloo-Vial *et al.*, 1997; Ivanova *et al.*, 2002; Naghmouchi *et al.*, 2008; Scannell *et al.*, 2000; Wan *et al.*, 1995). Calcium alginate is an appropriate scaffold for cell immobilisation owing to its low cost, food grade status and relative ease in encapsulation of cells (Bhugaloo-Vial *et al.*, 1997). The production of nisin and lacticin 3147 by two *L. lactis* strains either immobilised in calcium alginate beads (IC) or in free culture (FC) was assessed in continuous fermentation (Scannell *et al.*, 2000). The beads were found to be quite stable over 180 h, although some cell leakage from the beads was reported. While both nisin and lacticin 3147 were detected earlier in FC bioreactors, a sharp decline in production occurred mid way through the fermentation. Bacteriocin production took longer to reach its maximum in IC bioreactors but, once achieved, maximum production was maintained for the remainder of the fermentation. It should be noted, however, that Mg^{2+}, MgSO$_4$ acetate, citrate and phosphate ions can cause instability in calcium alginate beads (Vignolo *et al.*, 1995; Yang and Ray, 1994) and that the stability is also affected by the strain that is being encapsulated (Scannell *et al.*, 2000).

The natural attachment of cells to fibrous surfaces in packed bed bioreactors has also been employed when immobilising bacteriocinogenic bacteria (Cho *et al.*, 1996; Liu *et al.*, 2005). Using this design, Liu *et al.* (2005) immobilised the nisin producer, *L. lactis* subsp. *lactis* ATCC 11454, in a packed bed bioreactor. This bioreactor was run continuously for six months without encountering any problems. Under optimal conditions a maximum nisin titre of 5.1×10^4 AU ml^{-1} was achieved.

As well as being used in continuous fermentations, immobilised cells have also been investigated in repeated cycle batch (RCB) fermentations. In RCB fermentation, immobilised bacteriocinogenic bacteria are used to perform consecutive batch fermentations. After each fermentation cycle, the fermentate is removed and replaced with fresh media and the process is repeated. During one hour cycle RCB cultures the nisin Z producer *L. lactis* UL719, when immobilised in calcium alginate beads, was capable of producing up to 8200 IU ml^{-1} of the bacteriocin. This corresponds to a volumetric productivity of 5730 IU ml^{-1}. An aerated continuous IC culture with

© Woodhead Publishing Limited, 2013

the same strain run at D 2.0 h^{-1}, resulted in a maximum volumetric productivity of only 1760 IU ml^{-1} (Bertrand *et al.*, 2001). RCB fermentations have also been assessed with respect to pediocin PA-1 production by *P. acidilactici* UL5 immobilised in κ-carrageenan/locust bean gum gel. Maximum pediocin PA-1 activity, 4096 AU ml^{-1}, was obtained after 0.45 and 2 h of incubation in MRS and supplemented whey permeate, respectively. This corresponded to volumetric productivities of 5461 and 2048 AU ml^{-1} h^{-1}. In contrast, pediocin activity in pH-controlled batch fermentations with free cells yielded only 4096 AU ml^{-1} after a 12 h incubation, resulting in a much lower volumetric productivity of just 342 ml^{-1} h^{-1} (Naghmouchi *et al.*, 2008).

14.3.4 Purification of bacteriocins

The purification of bacteriocins from culture media is necessary to study their composition and mode of action. It is also required for the production of pure peptides for application as natural biopreservatives in food. The purification of bacteriocins from culture media is a notoriously difficult process and so many methods have been developed (Saavedra and Sesma, 2011). This section will give a brief overview of the methods employed for the purification of bacteriocins. For a more detailed review, please refer to Saavedra and Sesma (2011).

When purifying bacteriocins, the first step is to concentrate the peptides. This process is compounded by the presence of a substantial quantity of small peptides of similar size to that of most bacteriocins (3000–6000 Da) in the complex media required for the cultivation of bacteriocinogenic LAB (Parente and Ricciardi, 1999). Therefore, it is not possible simply to concentrate the peptides by water removal. The two most widely reported methods for the concentration of bacteriocins are acid extraction (Yang *et al.*, 1992) and salt precipitation (Muriana and Klaenhammer, 1991). Acid extraction relies on the adsorption and desorption of a bacteriocin to its producing cells at different pHs. Yang *et al.* (1992) studied this phenomenon with the bacteriocins nisin, pediocin PA-1, sakacin A and leuconocin Lcm1. It was discovered that by adjusting the pH to ~ 6.0 (bacteriocin specific), it was possible to adsorb between 93 and 100% of bacteriocin molecules on to the surface of the bacteriocin-producing cells. Cells were removed from the culture supernatant by centrifuging and released from the cells by adjusting the pH to ~ 2.0 (bacteriocin specific) in the presence of 100 mM NaCl. Over 90% recovery was reported for nisin, pediocin PA-1 and leuconocin Lcm1. However, sakacin A recovery was lower at 44%. This is a relatively simple method that provides good yields of bacteriocin and so could be suitable for the concentration of bacteriocin peptides on an industrial scale.

The other most common method for concentrating bacteriocins from the culture media is salt precipitation with ammonium sulphate. Ammonium sulphate is added to the culture supernatant to the point of saturation. The

© Woodhead Publishing Limited, 2013

saturated culture supernatant is then gently mixed overnight and the peptides are allowed to precipitate out of solution. The precipitate is subsequently removed from the culture by centrifuging and is washed in buffer
(Muriana and Klaenhammer, 1991). While this is the more commonly
applied method for peptide bacteriocin concentration, the percentage
recovery can vary greatly, from below 10% to nearly 100% (Bayoub et al.,
2011; Kamoun et al., 2005; Kumar et al., 2010; Liu et al., 2008a).

After concentration, the bacteriocin peptides still need to be separated
from the contaminants present in the media. Generally this is achieved by
applying concentrated peptides to a cation exchange column (Callewaert
and De Vuyst, 1999; Uteng et al., 2002). Anionic compounds can pass freely
through the column, while bacteriocins, which are frequently cationic, are
retained in the columns matrix. Bacteriocins are subsequently eluted from
the column using a NaCl solution (Uteng et al., 2002). Active fractions
containing bacteriocin peptides are then applied to a reverse-phase high-
performance liquid chromatography (RP-HPLC) column for final concentration. Bacteriocins are eluted from the column using a gradient of water
miscible organic solvents (Saavedra and Sesma, 2011).

14.3.5   Partially purified bioactive powders

Purification of highly purified bacteriocins on an industrial scale is not
viable owing to the long processing time and low percentage recovery. For
these reasons, bacteriocins are more commonly produced as partially purified bioactive powders such as Nisaplin™ and Alta 2341™ (Danisco, Kerry
bioscience). The long term stability and ease of transport associated with
powdered preparations also make them a more attractive option (Gardiner
et al., 2000). Spray dried bioactive powders have also been described in the
literature (Ananou et al., 2008; Morgan et al., 1999). A lacticin 3147-contain-
ing powder has been produced by Morgan et al. (1999). Briefly, L. lactis
DPC 3147 was inoculated into reconstituted demineralised whey (10%) and
fermented for 24 h under constant pH. The resulting fermentate was pas-
teurised (72°C for 15s) and concentrated by evaporation to 40% solids. The
concentrate was then spray dried to produce a bioactive lacticin 3147
powder. The addition of 10% total product weight lacticin 3147 powder was
required to reduce numbers of Bacillus cereus and L. monocytogenes sufficiently in a range of food trials (Morgan et al., 2001). While this exhibits
the potential of a lacticin 3147 based bioactive powder, the addition of 10%
total product weight of such a powder is neither feasible nor economic.

A spray-dried bioactive powder containing the enterocin AS-48 has also
been produced. Enterococcus faecalis A-48-32 was cultivated in the whey-
derived substrate Esprion-300 (E-300), at 28°C for 18–20 h under controlled a pH of 6.5. After fermentation AS-48 was recovered from the E-300
by cation exchange chromatography on a carboxymethyl Sephadex CM-25.
The recovered fractions were either heat treated (80°C, 20 min) or UV light

© Woodhead Publishing Limited, 2013

irradiated (5 min), to inactivate the producer, and were subsequently spray dried (Ananou *et al.*, 2010). However, yet again, 5% and 10% wt/vol was still required to control the growth of *L. monocytogenes* and *S. aureus* in skim milk, respectively (Ananou *et al.*, 2010). Also while both the lacticin 3147 and the AS-48 powders retain full activity for at least 4 months at refrigeration temperatures (4–5°C), however both undergo a 50% reduction in activity after 9 months at room temperature (Ananou *et al.*, 2010; Morgan *et al.*, 2001). The stability of these powders clearly does not match that of the commercial nisin preparation Nisaplin™ which is stable between 4°C and 25°C over 2 years from the date of manufacture (Morgan *et al.*, 2001). For these powders to reach their true commercial potential, the issues of stability and total activity will have to be addressed.

14.4   Improvement of bacteriocinogenic bacteria

Genetic manipulation of bacteriocinogenic bacteria can be exploited to address many of the problems associated with bacteriocin production for food applications, such as low production, production in suitable bacteria and spectrum of activity. A number of authors have demonstrated the effectiveness of genetic manipulation in improving bacteriocin production (Cheigh *et al.*, 2005; Cotter *et al.*, 2006; Heinzmann *et al.*, 2006). While the genetically modified nature of the strains in question usually precludes their use in food, such studies have often provided valuable insights and may be of relevance to the food industry in the future.

Cotter *et al.* (2006) investigated the provision of additional copies of the lacticin 3147 genes on the production of lacticin 3147 in *L. lactis* MG1363 (pMRC01), a strain expressing the parental lacticin 3147 encoding plasmid (pMRC01). A high copy number plasmid containing the entire lacticin 3147 encoding region (pOM02) was introduced into MG1363 by electroporation and the resulting strain, MG1363(pMRC01, pOM02) was found to produce four-fold more lacticin 3147 than strains containing either plasmid alone. Further investigation revealed that additional copies of the two lacticin 3147 structural genes (*ltnA1A2*) were not necessary, as addition of a plasmid containing all other lacticin 3147 encoding genes resulted in a 3.5-fold increase in production relative to the strain carrying pMRC01 alone. In another instance, an increase in the production of nisin Z by *L. lactis* A163 was achieved by introducing a multiple copy plasmid harbouring the nisin regulatory genes, *nisR* and *nisK*. This resulted in an increase in production from 16 000 AU ml^{-1}, observed in the control to 25 000 AU ml^{-1} in the strain over-expressing *nisRK*. This was as a consequence of the increased transcription of the *nisZ* gene (Cheigh *et al.*, 2005).

As enterocin-producing *Enterococci* can potentially carry virulence genes and therefore are not considered GRAS organisms (Eaton and Gasson, 2001; Franz *et al.*, 2001; Shankar *et al.*, 2002), there has been a

© Woodhead Publishing Limited, 2013

growing interest in producing enterocins in GRAS hosts. An example of this is the heterologous expression of the Class II bacteriocin enterocin P (EntP) from *Enterococcus faecium* P13 by a strain of *L. lactis*. The EntP structural gene, *entP*, and its immunity gene, *entiP* were cloned into *L. lactis* NZ9000 under a nisin-inducible expression system. The resulting strain, *L. lactis* NZ9000(pJR199), exhibited a higher specific activity than was produced by any other *L. lactis* hosts or the parent strain *E. faecium* P13 (Gutiérrez *et al.*, 2006). Liu *et al.* (2008b) heterologously expressed the Class II bacteriocin, enterocin A (EntA) in *L. lactis* MG1614(pLP712). The EntA encoding plasmid pEnt02 was introduced into *L. lactis* MG1614(pLP712) by electroporation and although the resulting strain, *L. lactis*$_{Ent+}$ produced four-fold less EntA than the parent strain *E. faecium* DPC 1146, however this was deemed acceptable owing to the strong anti-listerial properties of EntA.

The natural transfer of bacteriocin-encoding plasmids by conjugation has also been used to confer a bacteriocin-producing phenotype on strains that are better adapted to specific food environments. Over 30 food grade starter cultures producing lacticin 3147 have been constructed by exploiting the conjugal nature of the lacticin 3147 encoding plasmid pMRC01 (Coakley *et al.*, 1997; Trotter *et al.*, 2004), some of which have also been used as protective cultures in food fermentations (Coffey *et al.*, 1998; McAuliffe *et al.*, 1999). One such transconjugant, *L. lactis* DPC 4275 was found to produce variable titres of lacticin 3147. In this strain it was found that an 80 kb cointegrate plasmid, pMRC02 had formed from the incorporation of the lacticin 3147 genes into the resident plasmid pMT60. It was revealed that when pMRC02 was present at a high copy number, lacticin 3147 titre was roughly double that of a low copy variant (Trotter *et al.*, 2004).

The limited antimicrobial potency and spectrum of inhibition of some bacteriocins restrict their value with respect to food-related applications. One solution to this problem is the creation of multi-bacteriocinogenic bacteria producing two or more bacteriocins to enhance or broaden their activity. By combining unrelated bacteriocins with different modes of action, it is also possible to prevent the emergence of resistance to either bacteriocin (Horn *et al.*, 1999). Gutiérrez *et al.* (2006) described the heterologous production of enterocin P in the nisin-producing strain *L. lactis* DPC 5598. The resulting transformant was capable of simultaneous production of both bacteriocins, although it should be noted that the levels of EntP activity of this strain were lower than those observed among other *L. lactis* host strains. Lower bacteriocin production relative to the parent strain was also observed when enterocin A and pediocin PA-1 were heterologously co-produced in *L. lactis* IL1403, although this could be attributed to an inefficient host (Martínez *et al.*, 2000).

Genetic manipulation has also been employed to modify bacteriocins with a view to improving their spectrum of inhibition, antimicrobial activity and the solubility and stability of the bacteriocins for applications in food

© Woodhead Publishing Limited, 2013

environments. This is generally achieved either by mutagenesis of the bacteriocin encoding genes or by fusing genes from different species to create chimeric bacteriocins (Gillor *et al.*, 2005). Yuan *et al.* (2004) created a number of mutants producing derivatives of nisin Z by site-directed mutagenesis at the hinge region of the nisin Z gene, *nisZ*. The resulting mutants had decreased activity versus *Micrococcus flavus* NCIB8166 and *Streptococcus thermophilus*. However, peptides with enhanced activity against Gram negative, that is *Shigella*, *Salmonella* and *Pseudomonas* species, or with increased solubility and stability compared to that of nisin Z were identified. Random mutagenesis has also been employed to create the largest bank of randomly mutated nisin derivates (Field *et al.*, 2008). Use of this approach resulted in the identification of derivatives with enhanced activity against *L. monocytogenes, S. aureus, Streptococcus agalactiae* (Field *et al.*, 2008) and various mycobacteria (Carroll *et al.*, 2010) and others with apparently enhanced ability to diffuse through complex matrices (Rouse *et al.*, 2012). Notably, although the initial producers of these variants were genetically modified microorganisms (GMM), strategies exist which can facilitate the generation of corresponding strains through self cloning, meaning that such producers would fall outside the scope of directives regarding the contained use of GMMs. This fact may ultimately facilitate the application of these nisin derivatives for food applications.

An alternative strategy, DNA shuffling, was employed to develop an 'improved' derivative of pediocin PA-1. A DNA shuffling library was created by shuffling four specific regions of the N-terminal half of pediocin PA-1 with 10 other class IIa bacteriocins. A library of 280 shuffled DNA mutants was created, 63 of which displayed antimicrobial activity. Shuffled mutants displayed increased activity against various species of *Lactobacillus, Pediococcus* and *Carnobacterium*. One of the mutants identified was also active against *L. lactis*, which was immune to the parent pediocin PA-1 (Tominaga and Hatakeyama, 2007).

14.5   Conclusions

The *in situ* production of bacteriocins by bacteriocinogenic starter, adjunct or protective cultures has been demonstrated to be an effective delivery system with respect to the incorporation of bacteriocins into the food environment. As a consequence of the diverse array of bacteriocinogenic LAB that are available and the fact that they can be produced as lyophilised bacteriocinogenic starter, adjunct or protective cultures, there is great potential for the use of cultures as biopreservatives in food. As there are no regulatory issues that limit the use of bacteriocinogenic LAB in food, this approach may be an economical alternative to the application of chemical preservatives or purified bacteriocin preparations for controlling spoilage and pathogenic bacteria.

© Woodhead Publishing Limited, 2013

The optimisation of fermentation processes, as well as the development of food grade media from industrial waste products, has greatly reduced the cost of producing bacteriocins by large scale fermentation. However, the production of purified bacteriocin peptides can still be a difficult and expensive process and peptides other than nisin have not been approved for use as food preservatives. Therefore, the production of bioactive fermentates containing bacteriocins is preferred. Such bioactive fermentates can be added to foods as food ingredients rather than as preservatives and thus are not subject to the same regulatory scrutiny as a purified peptide preservative. While the production of bacteriocins as bioactive fermentates is relatively cheaper and less complex than the production of purified bacteriocin peptides, in some instances the antibacterial activity and stability of such preparations would need to be improved first before commercial applications could be considered.

Finally, many strategies have been employed to improve bacteriocinogenic bacteria and the bacteriocins they produce, as current regulations prohibit the use of GMMs in foods and consumer resistance to GMMs also hinders their application in food production, for the short to medium term there will continue to be an emphasis on the use of food grade strategies to generate new and improved bacteriocinogenic strains.

14.6   Acknowledgements

The Alimentary Pharmabiotic Centre is a research centre funded by Science Foundation Ireland (SFI). The authors and their work were supported by SFI CSET grant APC CSET 2 grant 07/CE/B1368.

14.7   References

AASEN, I. M., MØRETRØ, T., KATLA, T., AXELSSON L. and STORRØ, I. (2000) 'Influence of complex nutrients, temperature and pH on bacteriocin production by *Lactobacillus sakei* CCUG 42687'. *Applied Microbiology and Biotechnology*, **53**, 159–66.

ABRIOUEL, H., VALDIVIA, E., MARTÍNEZ-BUENO, M., MAQUEDA, M. and GÁLVEZ, A. (2003) 'A simple method for semi-preparative-scale production and recovery of enterocin AS-48 derived from *Enterococcus faecalis* subsp. *liquefaciens* A-48-32'. *Journal of Microbiological Methods*, **55**, 599–605.

ANANOU, S., MUÑOZ, A., GÁLVEZ, A., MARTÍNEZ-BUENO, M., MAQUEDA, M. and VALDIVIA, E. (2008) 'Optimization of enterocin AS-48 production on a whey-based substrate'. *International Dairy Journal*, **18**, 923–7.

ANANOU, S., MUÑOZ, A., MARTÍNEZ-BUENO, M., GONZÁLEZ-TELLO, P., GÁLVEZ, A., MAQUEDA, M. and VALDIVIA, E. (2010) 'Evaluation of an enterocin AS-48 enriched bioactive powder obtained by spray drying'. *Food Microbiology*, **27**, 58–63.

ANTHONY, T., RAJESH, T., KAYALVIZHI, N. and GUNASEKARAN, P. (2009) 'Influence of medium components and fermentation conditions on the production of bacteriocin(s) by *Bacillus licheniformis* AnBa9'. *Bioresource Technology*, **100**, 872–7.

© Woodhead Publishing Limited, 2013

ARAKAWA, K., KAWAI, Y., FUJITANI, K., NISHIMURA, J., KITAZAWA, H., KOMINE, K.-I., KAI, K. and SAITO, T. (2008) 'Bacteriocin production of probiotic *Lactobacillus gasseri* LA39 isolated from human feces in milk-based media'. *Animal Science Journal*, **79**, 634–40.

ARANTXA, MUÑOZ, MERCEDES, M., ANTONIO, G., MANUEL, M., ANA, R. and EVA, V. (2004) 'Biocontrol of psychrotrophic enterotoxigenic *Bacillus cereus* in a nonfat hard cheese by an enterococcal strain-producing enterocin AS-48'. *Journal of Food Protection*, **67**, 1517–21.

ÁVILA, M., GARDE, S., GAYA, P., MEDINA, M. and NUÑEZ, M. (2005) 'Influence of a bacteriocin-producing lactic culture on proteolysis and texture of Hispánico cheese'. *International Dairy Journal*, **15**, 145–53.

AVONTS, L., UYTVEN E. V., and VUYST, L. D. (2004) 'Cell growth and bacteriocin production of probiotic *Lactobacillus* strains in different media'. *International Dairy Journal*, **14**, 947–55.

AYMERICH, M. T., GARRIGA, M., MONFORT, J. M., NES I. and HUGAS, M. (2000) 'Bacteriocin-producing lactobacilli in Spanish-style fermented sausages: characterization of bacteriocins'. *Food Microbiology*, **17**, 33–45.

BÁRCENA, J. M. B., SIÑERIZ, F., DE LLANO, D. G., RODRÍGUEZ A. and SUÁREZ, J. E. (1998) 'Chemostat production of plantaricin C by *Lactobacillus plantarum* LL441'. *Applied and Environmental Microbiology*, **64**, 3512–4.

BAYOUB, K., MARDAD, I., AMMAR, E., SERRANO A. and SOUKRI, A. (2011) 'Isolation and purification of two bacteriocins 3D produced by *Enterococcus faecium* with inhibitory activity against *Listeria monocytogenes*'. *Current Microbiology*, **62**, 479–85.

BEGLEY, M., COTTER, P. D., HILL, C. and ROSS, R. P. (2009) 'Identification of a novel two-peptide lantibiotic, lichenicidin, following rational genome mining for LanM proteins'. *Applied Environmental Microbiology*, **75**, 5451–60.

BERTRAND, N., FLISS, I. and LACROIX, C. (2001) 'High nisin-Z production during repeated-cycle batch cultures in supplemented whey permeate using immobilized *Lactococcus lactis* UL719'. *International Dairy Journal*, **11**, 953–60.

BHUGALOO-VIAL, P., GRAJEK, W., DOUSSET, X. and BOYAVAL, P. (1997) 'Continuous bacteriocin production with high cell density bioreactors'. *Enzyme and Microbial Technology*, **21**, 450–7.

BHUNIA, A. K., JOHNSON, M. C., RAY, B. and KALCHAYANAND, N. (1991) 'Mode of action of pediocin AcH from *Pediococcus acidilactici* H on sensitive bacterial strains'. *Journal of Applied Microbiology*, **70**, 25–33.

BIZANI, D. and BRANDELLI, A. (2004) 'Influence of media and temperature on bacteriocin production by *Bacillus cereus* 8A during batch cultivation'. *Applied Microbiology and Biotechnology*, **65**, 158–62.

BOUKSAIM, M., LACROIX, C., AUDET, P. and SIMARD, R. E. (2000) 'Effects of mixed starter composition on nisin Z production by *Lactococcus lactis* subsp. *lactis* biovar. diacetylactis UL 719 during production and ripening of Gouda cheese'. *International Journal of Food Microbiology*, **59**, 141–56.

CABO, M. L., MURADO, M. A., GONZÁLEZ, M. P. and PASTORIZA, L. (2001) 'Effects of aeration and pH gradient on nisin production. A mathematical model'. *Enzyme and Microbial Technology*, **29**, 264–73.

CALLEWAERT, R. and DE VUYST, L. (1999) 'Expanded bed adsorption as a unique unit operation for the isolation of bacteriocins from fermentation media'. *Bioseparation*, **8**, 159–68.

CALLEWAERT, R., HUGAS, M. and VUYST, L. D. (2000) 'Competitiveness and bacteriocin production of Enterococci in the production of Spanish-style dry fermented sausages'. *International Journal of Food Microbiology*, **57**, 33–42.

© Woodhead Publishing Limited, 2013

CAPLICE, E. and FITZGERALD, G. F. (1999) 'Food fermentations: role of microorganisms in food production and preservation'. *International Journal of Food Microbiology*, **50**, 131–49.

CARMEN MARTÍNEZ-CUESTA, M., BENGOECHEA, J., BUSTOS, I., RODRÍGUEZ, B., REQUENA, T. and PELÁEZ, C. (2010) 'Control of late blowing in cheese by adding lacticin 3147-producing *Lactococcus lactis* IFPL 3593 to the starter'. *International Dairy Journal*, **20**, 18–24.

CARROLL, J., FIELD, D., O'CONNOR, P. M., COTTER, P. D., COFFEY, A., HILL, C., ROSS, R. P. and O'MAHONY, J. (2010) 'Gene encoded antimicrobial peptides, a template for the design of novel anti-mycobacterial drugs'. *Bioengineered Bugs*, **1**, 408.

CHEIGH, C.-I., CHOI, H.-J., PARK, H., KIM, S.-B., KOOK, M.-C., KIM, T.-S., HWANG, J.-K. and PYUN, Y.-R. (2002) 'Influence of growth conditions on the production of a nisin-like bacteriocin by *Lactococcus lactis* subsp. *lactis* A164 isolated from kimchi'. *Journal of Biotechnology*, **95**, 225–35.

CHEIGH, C. I., PARK, H., CHOI, H. J. and PYUN, Y. R. (2005) 'Enhanced nisin production by increasing genes involved in nisin Z biosynthesis in *Lactococcus lactis* subsp. *lactis* A164'. *Biotechnology Letters*, **27**, 155–60.

CHO, H. Y., YOUSEF, A. E. and YANG, S. T. (1996) 'Continuous production of pediocin by immobilized *Pediococcus acidilactici* PO_2 in a packed-bed bioreactor'. *Applied Microbiology and Biotechnology*, **45**, 589–94.

CHRISTENSEN, D. P. and HUTKINS, R. W. (1992) 'Collapse of the proton motive force in *Listeria monocytogenes* caused by a bacteriocin produced by *Pediococcus acidilactici*'. *Appl. Environ. Microbiol.*, **58**, 3312–5.

CLADERA-OLIVERA, F., CARON, G. R. and BRANDELLI, A. (2004) 'Bacteriocin production by *Bacillus licheniformis* strain P40 in cheese whey using response surface methodology'. *Biochemical Engineering Journal*, **21**, 53–8.

COAKLEY, M., FITZGERALD, G. and ROS, R. P. (1997) 'Application and evaluation of the phage resistance-and bacteriocin-encoding plasmid pMRC01 for the improvement of dairy starter cultures'. *Applied and Environmental Microbiology*, **63**, 1434.

COFFEY, A., RYAN, M., ROSS, R. P., HILL, C., ARENDT, E. and SCHWARZ, G. (1998) Use of a broad-host-range bacteriocin-producing *Lactococcus lactis* transconjugant as an alternative starter for salami manufacture'. *International Journal of Food Microbiology*, **43**, 231–5.

COTTER, P. D., HILL, C. and ROSS, R. P. (2005) 'Bacteriocins: developing innate immunity for food'. *Nature Reviews Microbiology*, **3**, 777–88.

COTTER, P. D., DRAPER, L. A., LAWTON, E. M., MCAULIFFE, O., HILL, C. and ROSS, R. P. (2006) 'Overproduction of wild-type and bioengineered derivatives of the lantibiotic lacticin 3147'. *Applied and Environmental Microbiology*, **72**, 4492.

COVENTRY, M. J., WAN, J., GORDON, J. B., MAWSON, R. F. and HICKEY, M. W. (1996) 'Production of brevicin 286 by *Lactobacillus brevis* VB286 and partial characterization'. *Journal of Applied Microbiology*, **80**, 91–8.

CRAWFORD, R. J. M. (1987) 'The use of lysozyme in the prevention of late blowing in cheese'. *International Dairy Federation*, **216**, 16.

D'ANGELIS, C. E. M., POLIZELLO, A. C. M., NONATO, M. C., SPADARO, A. C. C. and DE MARTINIS, E. C. P. (2009) 'Purification, characterization and N-terminal amino acid sequencing of sakacin 1, a bacteriocin produced by *Lactobacillus sakei* 1'. *Journal of Food Safety*, **29**, 636–49.

DABA, H., LACROIX, C., HUANG, J. and SIMARD, R. E. (1993) 'Influence of growth conditions on production and activity of mesenterocin 5 by a strain of *Leuconostoc mesenteroides*'. *Applied Microbiology and Biotechnology*, **39**, 166–73.

DE VUYST, L. (1995) 'Nutritional factors affecting nisin production by *Lactococcus lactis* subsp. *lactis* NIZO 22186 in a synthetic medium'. *Journal of Applied Microbiology*, **78**, 28–33.

© Woodhead Publishing Limited, 2013

DE VUYST, L. and VANDAMME, E. J. (1991) 'Microbial manipulation of nisin biosynthesis and fermentation'. *Nisin and Novel Lantibiotics.* ESCOM Science Publishers, Leiden, The Netherlands, 397–409.

DE VUYST, L. and VANDAMME, E. J. (1992) 'Influence of the carbon source on nisin production in *Lactococcus lactis* subsp. *lactis* batch fermentations'. *Journal of General Microbiology*, **138**, 571.

DE VUYST, L., CALLEWAERT, R. and CRABBÉ, K. (1996) 'Primary metabolite kinetics of bacteriocin biosynthesis by *Lactobacillus amylovorus* and evidence for stimulation of bacteriocin production under unfavourable growth conditions'. *Microbiology*, **142**, 817.

DELGADO, A., BRITO, D., PERES, C., NOÉ-ARROYO, F. and GARRIDO-FERNÁNDEZ, A. (2005) 'Bacteriocin production by *Lactobacillus pentosus* B96 can be expressed as a function of temperature and NaCl concentration'. *Food Microbiology*, **22**, 521–8.

DELGADO, A., ARROYO LÓPEZ, F. N., BRITO, D., PERES, C., FEVEREIRO, P. and GARRIDO-FERNÁNDEZ, A. (2007) 'Optimum bacteriocin production by *Lactobacillus plantarum* 17.2 b requires absence of NaCl and apparently follows a mixed metabolite kinetics'. *Journal of Biotechnology*, **130**, 193–201.

DERVAKOS, G. A. and WEBB, C. (1991) 'On the merits of viable-cell immobilisation'. *Biotechnology Advances*, **9**, 559–612.

DICKS, L. M. T., MELLETT, F. D. and HOFFMAN, L. C. (2004) 'Use of bacteriocin-producing starter cultures of *Lactobacillus plantarum* and *Lactobacillus curvatus* in production of ostrich meat salami'. *Meat Science*, **66**, 703–8.

DIMITRIEVA-MOATS, G. Y. and ÜNLÜ, G. (2011) 'Development of freeze-dried bacteriocin-containing preparations from lactic acid bacteria to inhibit *Listeria monocytogenes* and *Staphylococcus aureus*'. *Probiotics and Antimicrobial Proteins*, 1–12.

DIMOV, S. (2007) 'A novel bacteriocin-like substance produced by *Enterococcus faecium*'. *Current Microbiology*, **55**, 323–7.

DORTU, C., HUCH, M., HOLZAPFEL, W. H., FRANZ, C. M. A. P. and THONART, P. (2008) 'Anti-listerial activity of bacteriocin-producing *Lactobacillus curvatus* CWBI-B28 and *Lactobacillus sakei* CWBI-B1365 on raw beef and poultry meat'. *Letters in Applied Microbiology*, **47**, 581–6.

DRIDER, D., FIMLAND, G., HECHARD, Y., MCMULLEN, L. M. and PREVOST, H. (2006) 'The continuing story of class IIa bacteriocins'. *Microbiology and Molecular Biology Reviews*, **70**, 564.

DROSINOS, E. H., MATARAGAS, M. and METAXOPOULOS, J. (2006) 'Modeling of growth and bacteriocin production by *Leuconostoc mesenteroides* E131'. *Meat Science*, **74**, 690–6.

EATON, T. J. and GASSON, M. J. (2001) 'Molecular screening of enterococcus virulence determinants and potential for genetic exchange between food and medical isolates'. *Applied and Environmental Microbiology*, **67**, 1628–35.

EEC (1983) EEC Commission Directive 83/463/EEC. 'Introducing temporary measures for the designation of certain ingredients in the labelling of foodstuffs for sale to the ultimate consumer'. *Official Journal of the European Communities*, **26**, 255.

EKINCI, F. Y. and BAREFOOT, S. F. (2006) 'Fed-batch enhancement of jenseniin G, a bacteriocin produced by *Propionibacterium thoenii* (jensenii) P126'. *Food Microbiology*, **23**, 325–30.

ENAN, G. and AL AMRI, A. A. (2006) 'Novel plantaricin UG 1 production by *Lactobacillus plantarum* UG 1 in enriched whey permeate in batch fermentation processes'. *International Journal of Food, Agriculture and Environment*, **4**, 85–8.

FIELD, D., CONNOR, P. M. O., COTTER, P. D., HILL C. and ROSS, R. P. (2008) 'The generation of nisin variants with enhanced activity against specific Gram-positive pathogens'. *Molecular Microbiology*, **69**, 218–30.

© Woodhead Publishing Limited, 2013

FOEGEDING, P. M., THOMAS, A. B., PILKINGTON D. H. and KLAENHAMMER, T. R. (1992) 'Enhanced control of *Listeria monocytogenes* by *in situ*-produced pediocin during dry fermented sausage production'. *Appl. Environ. Microbiol.*, **58**, 884–90.

FRANZ, C. M. A. P., MUSCHOLL-SILBERHORN, A. B., YOUSIF, N. M. K., VANCANNEYT, M., SWINGS J. and HOLZAPFEL, W. H. (2001) 'Incidence of virulence factors and antibiotic resistance among enterococci isolated from food'. *Applied and Environmental Microbiology*, **67**, 4385–9.

GARDE, S., CARBONELL, M. A., FERNÁNDEZ-GARCÍA, E., MEDINA M. and NUÑEZ, M. (2002) 'Volatile compounds in Hispánico cheese manufactured using a mesophilic starter, a thermophilic starter, and bacteriocin-producing *Lactococcus lactis* subsp. *lactis* INIA 415'. *Journal of Agricultural and Food Chemistry*, **50**, 6752–7.

GARDE, S., ÁVILA, M., MEDINA M. and NUÑEZ, M. (2005) 'Influence of a bacteriocin-producing lactic culture on the volatile compounds, odour and aroma of Hispánico cheese'. *International Dairy Journal*, **15**, 1034–43.

GARDE, S., ÁVILA, M., GAYA, P., MEDINA M. and NUÑEZ, M. (2006) 'Proteolysis of Hispánico cheese manufactured using lacticin 481-producing *Lactococcus lactis* ssp. *lactis* INIA 639'. *Journal of Dairy Science*, **89**, 840–9.

GARDE, S., ÁVILA, M., ARIAS, R. N., GAYA P. and NUÑEZ, M. (2011) 'Outgrowth inhibition of *Clostridium beijerinckii* spores by a bacteriocin-producing lactic culture in ovine milk cheese'. *International Journal of Food Microbiology*, **150**, 59–65.

GARDINER, G. E., O'SULLIVAN, E., KELLY, J., AUTY, M. A. E., FITZGERALD, G. F., COLLINS, J. K., ROSS R. P. and STANTON, C. (2000) 'Comparative survival rates of human-derived probiotic *Lactobacillus paracasei* and *L. salivarius* strains during heat treatment and spray drying'. *Applied and Environmental Microbiology*, **66**, 2605.

GHALFI, H., ALLAOUI, A., DESTAIN, J., BENKERROUM N. and THONART, P. (2006) 'Bacteriocin activity by *Lactobacillus curvatus* CWBI-B28 to inactivate *Listeria monocytogenes* in cold-smoked salmon during 4C storage'. *Journal of Food Protection*, **69**, 1066–71.

GILLOR, O., NIGRO, L. M. and RILEY, M. A. (2005) 'Genetically engineered bacteriocins and their potential as the next generation of antimicrobials'. *Current Pharmaceutical Design*, **11**, 1067–75.

GIRAFFA, G., CARMINATI, D. and TARELLI, G. T. (1995) 'Inhibition of *Listeria innocua* in milk by bacteriocin-producing *Enterococcus faecium* 7C5'. *Journal of Food Protection*, **58**, 621–3.

GOULHEN, F., MEGHROUS, J. and LACROIX, C. (1999) 'Production of a nisin Z/pediocin mixture by pH-controlled mixed-strain batch cultures in supplemented whey permeate'. *Journal of Applied Microbiology*, **86**, 399–406.

GUERRA, N. P. and PASTRANA, L. (2001) 'Enhanced nisin and pediocin production on whey supplemented with different nitrogen sources'. *Biotechnology Letters*, **23**, 609–12.

GUERRA, N. P. and PASTRANA, L. (2003) 'Influence of pH drop on both nisin and pediocin production by *Lactococcus lactis* and *Pediococcus acidilactici*'. *Letters in Applied Microbiology*, **37**, 51–5.

GUERRA, N. P., AGRASAR, A. T., MACI'AS, C. L. and PASTRANA, L. (2005) 'Modelling the fed-batch production of pediocin using mussel processing wastes'. *Process Biochemistry*, **40**, 1071–83.

GUTIÉRREZ, J., LARSEN, R., CINTAS, L., KOK, J. and HERNÁNDEZ, P. (2006) 'High-level heterologous production and functional expression of the sec-dependent enterocin P from *Enterococcus faecium* P13 in *Lactococcus lactis*'. *Applied Microbiology and Biotechnology*, **72**, 41–51.

HAMMES, W. P., BANTLEON, A. and MIN, S. (1990) 'Lactic acid bacteria in meat fermentation'. *FEMS Microbiology Letters*, **87**, 165–73.

HEINZMANN, S., ENTIAN, K. D. and STEIN, T. (2006) 'Engineering *Bacillus subtilis* ATCC 6633 for improved production of the lantibiotic subtilin'. *Applied Microbiology and Biotechnology*, **69**, 532–6.

HERRANZ, C., MARTÍNEZ, J. M., RODRÍGUEZ, J. M., HERNANDEZ, P. E. and CINTAS, L. M. (2001) 'Optimization of enterocin P production by batch fermentation of *Enterococcus faecium* P13 at constant pH'. *Applied Microbiology and Biotechnology*, **56**, 378–83.

HORN, N., MARTÍNEZ, M. I., MARTÍNEZ, J. M., HERNÁNDEZ, P. E., GASSON, M. J., RODRÍGUEZ, J. M. and DODD, H. M. (1999) 'Enhanced production of pediocin PA-1 and coproduction of nisin and pediocin PA-1 by *Lactococcus lactis*'. *Applied and Environmental Microbiology*, **65**, 4443.

HUANG, J., LACROIX, C., DABA, H. and SIMARD, R. E. (1996) 'Pediocin 5 production and plasmid stability during continuous free and immobilized cell cultures of *Pediococcus acidilactici* UL5'. *Journal of Applied Microbiology*, **80**, 635–44.

HUGAS, M., NEUMEYER, B., PAGES, F., GARRIGA, M. and HAMMES, W. P. (1996) 'Die antimikrobielle Wirkung von Bakteriozin bildenden Kulturen in Fleischwaren: 2. Vergleich des Effektes unterschiedlicher Bakteriozin bildender Laktobazillen auf Listerien in Rohwurst'. *Fleischwirtschaft*, **76**, 649–52.

HUGENHOLTZ, J. and DE VEER, G. J. M. (1991) 'Application of nisin A and nisin Z in dairy technology'. *Nisin and Novel Lantibiotics.* ESCOM Science Publishers, Leiden, 440–7.

IVANOVA, E., CHIPEVA, V., IVANOVA, I., DOUSSET, X. and PONCELET, D. (2002) 'Encapsulation of lactic acid bacteria in calcium alginate beads for bacteriocin production'.

JACK, R. W., TAGG, J. R. and RAY, B. (1995) 'Bacteriocins of gram-positive bacteria'. *Microbiological Reviews*, **59**, 171–200.

JENSEN, P. R. and HAMMER, K. (1993) 'Minimal requirements for exponential growth of *Lactococcus lactis*'. *Applied and Environmental Microbiology*, **59**, 4363.

JUÁREZ TOMÁS, M., BRU, E., WIESE, B. and NADER-MACÍAS, M. (2010) 'Optimization of low-cost culture media for the production of biomass and bacteriocin by a urogenital *Lactobacillus salivarius* strain'. *Probiotics and Antimicrobial Proteins*, **2**, 2–11.

KAISER, A. L. and MONTVILLE, T. J. (1993) 'The influence of pH and growth rate on production of the bacteriocin, bavaricin MN, in batch and continuous fermentations'. *Journal of Applied Microbiology*, **75**, 536–40.

KAMOUN, F., FGUIRA, I., HASSEN, N., MEJDOUB, H., LERECLUS, D. and JAOUA, S. (2005) 'Purification and characterization of a new *Bacillus thuringiensis* bacteriocin active against *Listeria monocytogenes*, *Bacillus cereus*, and *Agrobacterium tumefaciens*'. *Applied Biochemistry and Biotechnology*, **165**, 300–14.

KATIKOU, P., AMBROSIADIS, I., GEORGANTELIS, D., KOIDIS, P. and GEORGAKIS, S. A. (2005) 'Effect of Lactobacillus-protective cultures with bacteriocin-like inhibitory substances' producing ability on microbiological, chemical and sensory changes during storage of refrigerated vacuum-packaged sliced beef'. *Journal of Applied Microbiology*, **99**, 1303–13.

KINGCHA, U., VISESSANGUAN, W., TOSUKHOWONG, A., ZENDO, T., ROYTRAKUL, S., LUXANANIL, P., CHAREONPORNSOOK, K., VALYASEVI, R. and SONOMOTO, K. (2011) 'Anti-listeria activity of *Pediococcus pentosaceus* BCC 3772 and application as starter culture for Nham, a traditional fermented pork sausage'. *Food Control*, **25**, 190–6.

KLAENHAMMER, T. R. (1988) 'Bacteriocins of lactic acid bacteria'. *Biochimie*, **70**, 337–49.

KUMAR, M., TIWARI, S. and SRIVASTAVA, S. (2010) 'Purification and characterization of enterocin LR/6, a bacteriocin from *Enterococcus faecium* LR/6'. *Applied Biochemistry and Biotechnology*, **160**, 40–9.

LAITILA, A., SAARELA, M., KIRK, L., SIIKA-AHO, M., HAIKARA, A., MATTILA-SANDHOLM, T. and VIRKAJÄRVI, I. (2004) 'Malt sprout extract medium for cultivation of *Lactobacillus plantarum* protective cultures'. *Letters in Applied Microbiology*, **39**, 336–40.

© Woodhead Publishing Limited, 2013

LAMBOLEY, L., LACROIX, C., CHAMPAGNE, C. P. and VUILLEMARD, J. C. (1997) 'Continuous mixed strain mesophilic lactic starter production in supplemented whey permeate medium using immobilized cell technology'. *Biotechnology and Bioengineering*, **56**, 502–16.

LEROY, F. and DE VUYST, L. (1999a) 'The presence of salt and a curing agent reduces bacteriocin production by *Lactobacillus sakei* CTC 494, a potential starter culture for sausage fermentation'. *Applied Environmental Microbiology*, **65**, 5350–6.

LEROY, F. and DE VUYST, L. (1999b) 'Temperature and pH conditions that prevail during fermentation of sausages are optimal for production of the antilisterial bacteriocin sakacin K'. *Applied and Environmental Microbiology*, **65**, 974.

LEROY, F. and DE VUYST, L. (2002) 'Bacteriocin production by *Enterococcus faecium* RZS C5 is cell density limited and occurs in the very early growth phase'. *International Journal of Food Microbiology*, **72**, 155–64.

LEROY, F. and DE VUYST, L. (2004) 'Lactic acid bacteria as functional starter cultures for the food fermentation industry'. *Trends in Food Science and Technology*, **15**, 67–78.

LIAO, C.-C., YOUSEF, A. E., RICHTER, E. R. and CHISM, G. W. (1993) '*Pediococcus acidilactici* PO₂ bacteriocin production in whey permeate and inhibition of *Listeria monocytogenes* in foods'. *Journal of Food Science*, **58**, 430–4.

LIPINSKA, E. (1973) 'Use of nisin-producing lactic streptococci in cheesemaking'. In: *Annual Bulletin of International Dairy Federation*, **73**, 1–24.

LIPINSKA, E. (1977) 'Nisin and its applications'. In: *Antimicrobials and Antibiosis in Agriculture*. M. Woodbine (ed). Butterworths, London, 107–30.

LIU, X., CHUNG, Y.-K. YANG, S.-T. and YOUSEF, A. E. (2005) 'Continuous nisin production in laboratory media and whey permeate by immobilized *Lactococcus lactis*'. *Process Biochemistry*, **40**, 13–24.

LIU, G., LV, Y., LI, P., ZHOU, K. and ZHANG, J. (2008a) 'Pentocin 31-1, an anti-Listeria bacteriocin produced by *Lactobacillus pentosus* 31-1 isolated from Xuan-Wei Ham, a traditional China fermented meat product'. *Food Control*, **19**, 353–9.

LIU, L., O'CONNER, P., COTTER, P. D., HILL, C. and ROSS, R. P. (2008b) 'Controlling *Listeria monocytogenes* in cottage cheese through heterologous production of enterocin A by *Lactococcus lactis*'. *Journal of Applied Microbiology*, **104**, 1059–66.

LIU, G., GRIFFITHS, M. W., SHANG, N., CHEN, S. and LI, P. (2010) 'Applicability of bacteriocinogenic *Lactobacillus pentosus* 31-1 as a novel functional starter culture or coculture for fermented sausage manufacture'. *Journal of Food Protection*, **73**, 292–8.

LORTAL, S. and CHAPOT-CHARTIER, M. P. (2005) 'Role, mechanisms and control of lactic acid bacteria lysis in cheese'. *International Dairy Journal*, **15**, 857–71.

LUCAS, R., GRANDE, M. J., ABRIOUEL, H., MAQUEDA, M., BEN OMAR, N., VALDIVIA, E., MARTÍNEZ-CAÑAMERO, M. and GÁLVEZ, A. (2006) 'Application of the broad-spectrum bacteriocin enterocin AS-48 to inhibit *Bacillus coagulans* in canned fruit and vegetable foods'. *Food and Chemical Toxicology*, **44**, 1774–81.

LUCHANSKY, J. B., GLASS, K. A., HARSONO, K. D., DEGNAN, A. J., FAITH, N. G., CAUVIN, B., BACCUS-TAYLOR, G., ARIHARA, K., BATER, B. and MAURER, A. J. (1992) 'Genomic analysis of *Pediococcus* starter cultures used to control *Listeria monocytogenes* in turkey summer sausage'. *Applied Environmental Microbiology*, **58**, 3053–9.

LV, W., ZHANG, X. and CONG, W. (2005) 'Modelling the production of nisin by *Lactococcus lactis* in fed-batch culture'. *Applied Microbiology and Biotechnology*, **68**, 322–6.

MARSH, A., O'SULLIVAN, O., ROSS, R. P., COTTER, P. and HILL, C. (2010) '*In silico* analysis highlights the frequency and diversity of type 1 lantibiotic gene clusters in genome sequenced bacteria'. *BMC Genomics*, **11**, 679.

MARTÍNEZ, J. M., KOK, J., SANDERS, J. W. and HERNÁNDEZ, P. E. (2000) 'Heterologous coproduction of enterocin A and pediocin PA-1 by *Lactococcus lactis*: detection

© Woodhead Publishing Limited, 2013

by specific peptide-directed antibodies'. *Applied and Environmental Microbiology*, **66**, 3543.

MATARAGAS, M., METAXOPOULOS, J., GALIOTOU, M. and DROSINOS, E. H. (2003) 'Influence of pH and temperature on growth and bacteriocin production by *Leuconostoc mesenteroides* L124 and *Lactobacillus curvatus* L442'. *Meat Science*, **64**, 265–71.

MATSUSAKI, H., ENDO, N., SONOMOTO, K. and ISHIZAKI, A. (1996) 'Lantibiotic nisin Z fermentative production by *Lactococcus lactis* IO-1: relationship between production of the lantibiotic and lactate and cell growth'. *Applied Microbiology and Biotechnology*, **45**, 36–40.

MCAULIFFE, O., HILL, C. and ROSS, R. P. (1999) 'Inhibition of *Listeria monocytogenes* in cottage cheese manufactured with a lacticin 3147-producing starter culture'. *Journal of Applied Microbiology*, **86**, 251–6.

MCSWEENEY, P. L. H., FOX, P. F. (2004) 'Metabolism of residual lactose and of lactate and citrate'. In: *Cheese: Chemistry, Physics and Microbiology,* Springer, New York, 361–71.

MESSENS, W., VERLUYTEN, J., LEROY, F. and DE VUYST, L. (2003) 'Modelling growth and bacteriocin production by *Lactobacillus curvatus* LTH 1174 in response to temperature and pH values used for European sausage fermentation processes'. *International Journal of Food Microbiology*, **81**, 41–52.

MILLS, S., SERRANO, L. M., GRIFFIN, C., O'CONNOR, P., SCHAAD, G., BRUINING, C., HILL, C., ROSS, R. P. and MEIJER, W. (2011) 'Inhibitory activity of *Lactobacillus plantarum* LMG P-26358 against *Listeria innocua* when used as an adjunct starter in the manufacture of cheese'. *Microbial Cell Factories*, **10**, S7.

MORGAN, S., ROSS, R. P. and HILL, C. (1997) 'Increasing starter cell lysis in Cheddar cheese using a bacteriocin-producing adjunct'. *Journal of Dairy Science*, **80**, 1–10.

MORGAN, S. M., GALVIN, M., KELLY, J., ROSS, R. P. and HILL, C. (1999) 'Development of a lacticin 3147-enriched whey powder with inhibitory activity against foodborne pathogens'. *Journal of Food Protection*, **62**, 1011–6.

MORGAN, S. M., GALVIN, M., ROSS, R. P. and HILL, C. (2001) 'Evaluation of a spray-dried lacticin 3147 powder for the control of *Listeria monocytogenes* and *Bacillus cereus* in a range of food systems'. *Letters in Applied Microbiology*, **33**, 387–91.

MURIANA, P. M. and KLAENHAMMER, T. R. (1991) 'Purification and partial characterization of lactacin F, a bacteriocin produced by *Lactobacillus acidophilus* 11088'. *Applied and Environmental Microbiology*, **57**, 114.

NAGHMOUCHI, K., FLISS, I., DRIDER, D. and LACROIX, C. (2008) 'Pediocin PA-1 production during repeated-cycle batch culture of immobilized *Pediococcus acidilactici* UL5 cells'. *Journal of Bioscience and Bioengineering*, **105**, 513–7.

NASCIMENTO, M. S., MORENO, I. and KUAYE, A. Y. (2008) 'Applicability of bacteriocin-producing *Lactobacillus plantarum*, *Enterococcus faecium* and *Lactococcus lactis* ssp. *lactis* as adjunct starter in Minas Frescal cheesemaking'. *International Journal of Dairy Technology*, **61**, 352–7.

NEL, H. A., BAUER, R., VANDAMME, E. J. and DICKS, L. M. T. (2001) 'Growth optimization of *Pediococcus damnosus* NCFB 1832 and the influence of pH and nutrients on the production of pediocin PD-1'. *Journal of Applied Microbiology*, **91**, 1131–8.

NEYSENS, P., MESSENS, W. and DE VUYST, L. (2003) 'Effect of sodium chloride on growth and bacteriocin production by *Lactobacillus amylovorus* DCE 471'. *International Journal of Food Microbiology*, **88**, 29–39.

NIETO-LOZANO, J. C., REGUERA-USEROS, J. I., PELÁEZ-MARTÍNEZ, M. D. C., SACRISTÁN-PÉREZ-MINAYO, G., GUTIÉRREZ-FERNÁNDEZ, Á. J. and DE LA TORRE, A. H. (2010) 'The effect of the pediocin PA-1 produced by *Pediococcus acidilactici* against *Listeria monocytogenes* and *Clostridium perfringens* in Spanish dry-fermented sausages and frankfurters'. *Food Control*, **21**, 679–85.

O'MAHONY, T., REKHIF, N., CAVADINI, C. and FITZGERALD, G. F. (2001) 'The application of a fermented food ingredient containing 'variacin', a novel antimicrobial

© Woodhead Publishing Limited, 2013

produced by *Kocuria varians*, to control the growth of *Bacillus cereus* in chilled dairy products'. *Journal of Applied Microbiology*, **90**, 106–14.

O'SULLIVAN, L., MORGAN, S. M., ROSS, R. P. and HILL, C. (2002) 'Elevated enzyme release from lactococcal starter cultures on exposure to the lantibiotic lacticin 481, produced by *Lactococcus lactis* DPC5552'. *Journal of Dairy Science*, **85**, 2130–40.

O'SULLIVAN, L., ROSS, R. P. and HILL, C. (2003) 'A lacticin 481-producing adjunct culture increases starter lysis while inhibiting nonstarter lactic acid bacteria proliferation during Cheddar cheese ripening'. *Journal of Applied Microbiology*, **95**, 1235–41.

OUMER, A., GAYA, P., FERNÁNDEZ-GARCÍA, E., MARIACA, R. A., UACUTE, GARDE, S., MEDINA, M. and NUÑEZ, M. (2001) 'Proteolysis and formation of volatile compounds in cheese manufactured with a bacteriocin-producing adjunct culture'. *Journal of Dairy Research*, **68**, 117–29.

PAIK, H. D. and GLATZ, B. A. (1997) 'Enhanced bacteriocin production by *Propionibacterium thoenii* in fed-batch fermentation'. *Journal of Food Protection*, **60**, 1529–33.

PARENTE, E. and RICCIARDI, A. (1994) 'Influence of pH on the production of enterocin 1146 during batch fermentation'. *Letters in Applied Microbiology*, **19**, 12–5.

PARENTE, E. and RICCIARDI, A. (1999) 'Production, recovery and purification of bacteriocins from lactic acid bacteria'. *Applied Microbiology and Biotechnology*, **52**, 628–38.

PARENTE, E., BRIENZA, C., RICCIARDI, A. and ADDARIO, G. (1997) 'Growth and bacteriocin production by *Enterococcus faecium* DPC1146 in batch and continuous culture'. *Journal of Industrial Microbiology & Biotechnology*, **18**, 62–7.

PÉREZ GUERRA, N., BERNÁRDEZ, P. F., AGRASAR, A. T., LÓPEZ MACÍAS, C. and CASTRO, L. P. (2005) 'Fed-batch pediocin production by *Pediococcus acidilactici* NRRL B-5627 on whey'. *Biotechnology and Applied Biochemistry*, **42**, 17–23.

PUCCI, M. J., VEDAMUTHU, E. R., KUNKA, B. S. and VANDENBERGH, P. A. (1988) 'Inhibition of *Listeria monocytogenes* by using bacteriocin PA-1 produced by *Pediococcus acidilactici* PAC 1.0'. *Applied and Environmental Microbiology*, **54**, 2349.

RAVYTS, F., BARBUTI, S., FRUSTOLI, M. A., PAROLARI, G., SACCANI, G., DE VUYST, L. and LEROY, F. (2008) 'Competitiveness and antibacterial potential of bacteriocin-producing starter cultures in different types of fermented sausages'. *Journal of Food Protection*, **71**, 1817–27.

RAY, B. (1992) 'Bacteriocins of starter culture bacteria as food biopreservatives: an overview'. *Food Biopreservatives of Microbial Origin,* Ray, B. and M. Daeschel (eds). CRC Press, Boca Raton, 177–205.

REA, M. C., ROSS, R. P., COTTER, P. D., HILL, C., DRIDER, D. and REBUFFAT, S. (2011) 'Classification of bacteriocins from Gram-positive bacteria'. *Prokaryotic Antimicrobial Peptides*. Drider, D. and Rebuffat, S. (eds). Springer, New York, 29–53.

RILEY, M. A. (1998) 'Molecular mechanisms of bacteriocin evolution'. *Annual Reviews Genetics*, **32**, 255–78.

RILLA, N., DELGADO, S., RODRÍGUEZ, MARTÍNEZ, T., (2003) 'Inhibition of *Clostridium tyrobutyricum* in Vidiago cheese by *Lactococcus lactis* ssp. lactis IPLA 729, a nisin Z producer'. *International Journal of Food Microbiology*, **85**, 23–33.

ROBERTS, R. F., ZOTTOLA, E. A. and MCKAY, L. L. (1992) 'Use of a nisin-producing starter culture suitable for cheddar cheese manufacture'. *Journal of Dairy Science*, **75**, 2353–63.

RODGERS, S., KAILASAPATHY, K., COX, J. and PEIRIS, P. (2002) 'Bacteriocin production by protective cultures'. *Food Service Technology*, **2**, 59–68.

RODRÍGUEZ, E., GAYA, P., NUÑEZ, M. and MEDINA, M. (1998) 'Inhibitory activity of a nisin-producing starter culture on *Listeria innocua* in raw ewes milk Manchego cheese'. *International Journal of Food Microbiology*, **39**, 129–32.

© Woodhead Publishing Limited, 2013

ROGERS, L. A. and WHITTIER, E. O. (1928) 'Limiting factors in the lactic fermentation'. *Journal of Bacteriology*, **16**, 211.

ROUSE, S., FIELD, D., DALY, K. M., O'CONNOR, P. M., COTTER, P. D., HILL, C. and ROSS, R. P. (2012) 'Bioengineered nisin derivatives with enhanced activity in complex matrices'. *Microbial Biotechnology*, **5**, 501–8.

RUIZ-BARBA, J. L., CATHCART, D. P., WARNER, P. J. and JIMÉNEZ-DÍAZ, R. (1994) 'Use of *Lactobacillus plantarum* LPCO10, a bacteriocin producer, as a starter culture in Spanish-style green olive fermentations'. *Applied Environmental Microbiology*, **60**, 2059–64.

RYAN, M. P., REA, M. C., HILL, C. and ROSS, R. P. (1996) 'An application in Cheddar cheese manufacture for a strain of *Lactococcus lactis* producing a novel broad-spectrum bacteriocin, lacticin 3147'. *Applied Environmental Microbiology*, **62**, 612–9.

SAAVEDRA, L. and SESMA, F. (2011) 'Purification techniques of bacteriocins from lactic acid bacteria and other Gram-positive bacteria'. *Prokaryotic Antimicrobial Peptides*, 99–113.

SARANTINOPOULOS, P., LEROY, F. D. R., LEONTOPOULOU, E., GEORGALAKI, M. D., KALANTZOPOULOS, G., TSAKALIDOU, E. and VUYST, L. D. (2002) 'Bacteriocin production by *Enterococcus faecium* FAIR-E 198 in view of its application as adjunct starter in Greek feta cheese making'. *International Journal of Food Microbiology*, **72**, 125–36.

SCANNELL, A. G. M., HILL, C., ROSS, R. P., MARX, S., HARTMEIER, W. and ARENDT, E. K. (2000) 'Continuous production of lacticin 3147 and nisin using cells immobilized in calcium alginate'. *Journal of Applied Microbiology*, **89**, 573–9.

SCHILLINGER, U. and LUCKE, F. K. (1989) 'Antibacterial activity of *Lactobacillus sake* isolated from meat'. *Applied Environmental Microbiology*, **55**, 1901–6.

SHANKAR, N., BAGHDAYAN, A. S. and GILMORE, M. S. (2002) 'Modulation of virulence within a pathogenicity island in vancomycin-resistant *Enterococcus faecalis*'. *Nature*, **417**, 746–50.

SILVA, J., CARVALHO, A. S., TEIXEIRA, P. and GIBBS, P. A. (2002) 'Bacteriocin production by spray-dried lactic acid bacteria'. *Letters in Applied Microbiology*, **34**, 77–81.

SONOMOTO, K., CHINACHOTI, N., ENDO, N. and ISHIZAKI, A. (2000) 'Biosynthetic production of nisin Z by immobilized *Lactococcus lactis* IO-1'. *Journal of Molecular Catalysis B: Enzymatic*, **10**, 325–34.

TAHIRI, I., DESBIENS, M., LACROIX, C., KHEADR, E. and FLISS, I. (2009) 'Growth of *Carnobacterium divergens* M35 and production of Divergicin M35 in snow crab by-product, a natural-grade medium'. *LWT–Food Science and Technology*, **42**, 624–32.

TEJAYADI, S. and CHERYAN, M. (1995) 'Lactic acid from cheese whey permeate. Productivity and economics of a continuous membrane bioreactor'. *Applied Microbiology and Biotechnology*, **43**, 242–8.

TOMINAGA, T. and HATAKEYAMA, Y. (2007) 'Development of innovative pediocin PA-1 by DNA shuffling among class IIa bacteriocins'. *Applied and Environmental Microbiology*, **73**, 5292.

TRINETTA, V., ROLLINI, M. and MANZONI, M. (2008) 'Development of a low cost culture medium for sakacin A production by *L. sakei*'. *Process Biochemistry*, **43**, 1275–80.

TROTTER, M., MCAULIFFE, O. E., FITZGERALD, G. F., HILL, C., ROSS, R. P. and COFFEY, A. (2004) 'Variable bacteriocin production in the commercial starter *Lactococcus lactis* DPC4275 is linked to the formation of the cointegrate plasmid pMRC02'. *Applied and Environmental Microbiology*, **70**, 34.

UTENG, M., HAUGE, H. H., BRONDZ, I., NISSEN-MEYER, J. and FIMLAND, G. (2002) 'Rapid two-step procedure for large-scale purification of pediocin-like bacteriocins and other cationic antimicrobial peptides from complex culture medium'. *Applied and Environmental Microbiology*, **68**, 952–6.

© Woodhead Publishing Limited, 2013

VÁZQUEZ, J. A., GONZÁLEZ, M. P. and MURADO, M. A. (2004) 'Nisin and pediocin production by *Lactococcus lactis* and *Pediococcus acidilactici* using waste protein sources from octopus'. *Electronic Journal Environmental Agricultural Food Chemistry*, **3**, 648–57.

VÁZQUEZ, J. A., GONZÁLEZ, M. P. and MURADO, M. A. (2006) 'Preliminary tests on nisin and pediocin production using waste protein sources: Factorial and kinetic studies'. *Bioresource Technology*, **97**, 605–13.

VIGNOLO, G. M., SURIANI, F., HOLGADO A. P. D. R., and OLIVER, G. (1993) 'Antibacterial activity of *Lactobacillus* strains isolated from dry fermented sausages'. *Journal of Applied Microbiology*, **75**, 344–9.

VIGNOLO, G. M., DE KAIRUZ, M. N., DE RUIZ HOLGADO, A. A. P. and OLIVER, G. (1995) 'Influence of growth conditions on the production of lactocin 705, a bacteriocin produced by *Lactobacillus casei* CRL 705'. *Journal of Applied Microbiology*, **78**, 5–10.

VIGNOLO, G., PALACIOS, J., FARÍAS, M. A. E., SESMA, F., SCHILLINGER, U., HOLZAPFEL, W. and OLIVER, G. (2000) 'Combined effect of bacteriocins on the survival of various listeria species in broth and meat system'. *Current Microbiology*, **41**, 410–6.

VOGEL, R. F., POHLE, B. S., TICHACZEK, P. S. and HAMMES, W. P. (1993) 'The competitive advantage of *Lactobacillus curvatus* LTH 1174 in sausage fermentations is caused by formation of curvacin A'. *Systematic and Applied Microbiology*, **16**, 457–62.

WAN, J., HICKEY, M. W. and COVENTRY, M. J. (1995) 'Continuous production of bacteriocins, brevicin, nisin and pediocin, using calcium alginate-immobilized bacteria'. *Journal of Applied Microbiology*, **79**, 671–6.

WOLF-HALL, C., GIBBONS, W. and BAUER, N. (2009) 'Development of a low-cost medium for production of nisin from *Lactococcus lactis* subsp. *Lactis*'. *World Journal of Microbiology and Biotechnology*, **25**, 2013–9.

WU, Z., JI, Y., GUO, Y., and HU, J. (2008) 'Application of ceramic membrane filtration to remove the solid in nisin fermentation broth'. *International Journal of Food Engineering*, **4**, 2.

YANG, R. and RAY, B. (1994) 'Factors influencing production of bacteriocins by lactic acid bacteria'. *Food Microbiology*, **11**, 281–91.

YANG, R., JOHNSON, M. C. and RAY, B. (1992) 'Novel method to extract large amounts of bacteriocins from lactic acid bacteria'. *Applied and Environmental Microbiology*, **58**, 3355.

YUAN, J., ZHANG, Z. Z., CHEN, X. Z., YANG, W. and HUAN, L. D. (2004) 'Site-directed mutagenesis of the hinge region of nisinZ and properties of nisinZ mutants'. *Applied Microbiology and Biotechnology*, **64**, 806–15.

ZHANG, J., LIU, G., SHANG, N., CHENG, W., CHEN, S. and LI, P. (2009) 'Purification and partial amino acid sequence of pentocin 31-1, an anti-listeria bacteriocin produced by *Lactobacillus pentosus* 31-1'. *Journal of Food Protection*, **72**, 2524–9.

© Woodhead Publishing Limited, 2013

15

Microbial production of amino acids and their derivatives for use in foods, nutraceuticals and medications

H. Suzuki, Kyoto Institute of Technology, Japan

DOI: 10.1533/9780857093547.2.385

Abstract: More than a century has passed since Kikunae Ikeda identified sodium L-glutamate as the umami component of 'kombu', a kind of seaweed. Since then, many novel physiological functions and new applications of amino acids have been revealed. The biosynthetic pathways of amino acids are tightly controlled by several regulation mechanisms that ensure specific amino acids do not accumulate and homeostasis is maintained. Therefore, the attempt to overproduce a specific amino acid was challenging. It has been a competition between cellular regulation mechanisms and the sophisticated innovations of humans. The cellular regulation mechanisms have been solved along with the development of molecular biology and methods for the production of various amino acids by fermentation and enzymes have been developed. In this chapter, typical regulation mechanisms of the biosynthetic pathways of amino acids are illustrated and how they have been overcome are described.

Key words: feedback inhibition, analog-resistant mutant, auxotroph, α-dipeptide, γ-glutamyl amino acid

15.1 Introduction

In 1909, Kikunae Ikeda discovered that the major umami component of 'kombu', a kind of seaweed, is sodium glutamate (Ikeda, 1909). Subsequently, Ajinomoto industrialized the production of sodium glutamate by the hydrolysis of gluten for use as a food additive. In 1957, Shukuo Kinoshita and his colleagues invented its direct fermentation method using *Corynebacterium glutamicum* (originally named *Micrococcus glutamicus*) (Kinoshita *et al.*, 1957b). Since then, methods for producing various amino acids and their derivatives using either bacteria or bacterial enzymes have been developed. Since L-amino acids are the building units of proteins, their biosyntheses are regulated to prevent the accumulation of a specific amino

© Woodhead Publishing Limited, 2013

acid and homeostasis is maintained. Therefore, the discovery of L-glutamate accumulation by a wild-type bacteria was a breakthrough of the amino acid industry. Since then, methods of amino acid fermentation have been developed so as to avoid metabolic regulations. It is reported that more than 2.3 million tonnes of amino acids per year were produced in 2000 by using bacteria or bacterial enzymes (estimated by the Essential Amino Acid Association, Japan). In this chapter, the methods of microbial amino acid production are discussed, in addition to description of the production and usage of amino acid derivatives.

15.2 Microbial production of amino acids

15.2.1 Bacterial biosynthetic pathways of amino acids

The major 20 L-amino acids are synthesized via several groups of biosynthetic pathways in bacteria (Fig. 15.1). L-Lysine, L-methionine, L-threonine, L-isoleucine and L-asparagine are synthesized via L-aspartic acid. L-Proline, L-glutamine and L-arginine are synthesized via L-glutamic acid. L-Alanine, L-valine and L-leucine are synthesized from pyruvic acid. L-Cysteine and glycine are synthesized via L-serine. And L-tryptophan, L-phenylalanine and

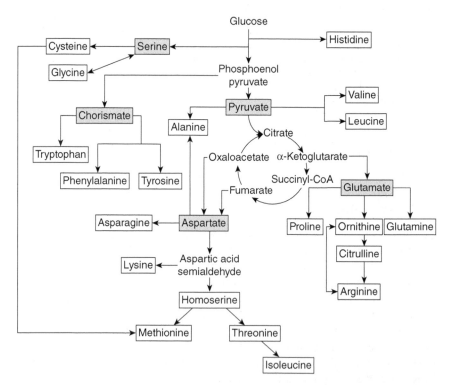

Fig. 15.1 Typical microbial metabolic pathways of amino acids.

© Woodhead Publishing Limited, 2013

L-tyrosine are synthesized from chorismic acid. Although there are slight differences in the biosynthetic pathways between bacteria, they are fundamentally the same. It is important to understand their biosynthetic pathways and their regulation mechanisms because amino acids are overproduced industrially by avoiding and/or blocking the regulatory mechanisms of bacteria.

15.2.2   Fermentation of amino acids by wild-type strains

L-Glutamate fermentation is an example of a typical amino acid fermentation using a wild-type strain. It was long believed that the direct fermentation of an amino acid was impossible because of the tightly controlled bacterial homeostasis. However, researchers at Kyowa Hakko Kogyo succeeded in discovering a bacterial strain, which produced and excreted L-glutamate into the medium, by a very elegant method (Udaka, 1960). They grew a lawn of lactic acid bacteria, which require several amino acids including L-glutamate for growth, on a minimal medium agar plate without L-glutamate. Bacteria isolated from the environment were then inoculated and growth halos of the lactic acid bacterium were observed around the bacterial strains that excrete L-glutamate into the medium. The strains were further tested in liquid medium to confirm the synthesis of L-glutamate and its excretion into the medium; in this way *Corynebacterium glutamicum* (originally named *Micrococcus glutamicus*) was isolated.

The excretion of L-glutamate was observed under biotin-limited conditions which cause a defect in phospholipid biosynthesis. However, it is difficult to limit the concentration of biotin using molasses, which is the preferred carbon source for industrial applications because of its low price. The addition of penicillin (Sommerson and Phillips, 1965) or some detergents (Takinami *et al.*, 1963; Duperray *et al.*, 1992) to the medium solved this problem, by inhibiting the activity of α-ketoglutarate (2-oxoglutarate) dehydrogenase complex and leading to the overproduction of L-glutamate (Kawahara *et al.*, 1997; Asakura *et al.*, 2007). It was also found that perturbations of the membrane or its synthesis lead to the secretion of L-glutamate into the medium. Recently, Nakamura and his colleagues indicated that alterations to the membrane tension caused by limited biotin or the addition of detergent or penicillin led to a structural transformation in the mechanosensitive channel that eventually caused L-glutamate excretion (Nakamura *et al.*, 2007).

15.2.3   Fermentation of amino acids by auxotrophic and analog-resistant mutant strains

It is not simple to overproduce an amino acid by using wild-type bacteria because of the tightly controlled amino acid homeostasis. Therefore, many methods using auxotrophic and analog-resistant mutant strains have been developed. In this section, some typical examples are described.

© Woodhead Publishing Limited, 2013

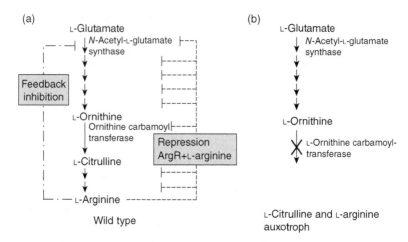

Fig. 15.2   L-Ornithine fermentation. (a) L-Arginine biosynthetic pathway and its regulations. (b) Metabolism of L-citrulline and L-arginine auxotroph.

Examples of utilization of auxotrophic strains to produce amino acids
L-Ornithine fermentation was the first successful example of amino acid fermentation using an auxotrophic strain. Kinoshita and his colleagues found that some arginine or citrulline auxotrophs accumulated L-ornithine in a medium supplemented with limited amounts of L-arginine for growth (Kinoshita *et al.*, 1957a). The L-arginine biosynthetic pathway is regulated by feedback inhibition of the first enzyme of the pathway, *N*-acetylglutamate synthase, as well as repression of the genes coding for the enzymes of the pathway (Fig. 15.2). Overproduction of L-ornithine cannot be explained simply as an accumulation of the intermediate owing to the blockage of the biosynthetic pathway, but it can be explained as the result of elimination of the feedback inhibition of *N*-acetylglutamate synthase and de-repression of the genes coding for the enzymes of the pathway caused by the limited amounts of the effector, L-arginine.

   Another good example of utilization of an auxotrophic strain for amino acid production is L-lysine fermentation. As shown in Fig. 15.3, L-lysine, L-threonine, L-isoleucine and L-methionine are synthesized from L-aspartate *via* aspartic acid semi-aldehyde. Aspartokinase (aspartate kinase) of *C. glutamicum*, which catalyzes the conversion of L-aspartate to L-aspartyl phosphate, is synergistically inhibited by L-threonine and L-lysine. A mutant strain that is deficient in homoserine dehydrogenase cannot synthesize homoserine, so that it cannot synthesize L-threonine, L-isoleucine and L-methionine. When this strain is grown with limited amounts of L-threonine and L-methionine, or homoserine, synergistic (concerted) inhibition of aspartokinase does not occur (Nakayama *et al.*, 1966) and a large amount of aspartic acid semi-aldehyde is synthesized. Since this strain is deficient in

© Woodhead Publishing Limited, 2013

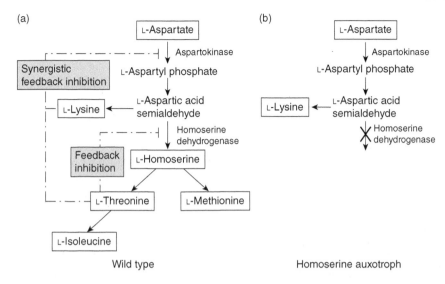

Fig. 15.3 L-Lysine fermentation. (a) L-Lysine, L-methionine, L-threonine and L-isoleucine biosynthetic pathway and its regulations in *Corynebacterium glutamicum*. (b) Metabolism of homoserine dehydrogenase deficient mutant.

homoserine synthesis, all aspartic acid semialdehyde synthesized is converted into L-lysine and accumulated as a consequence (Kinoshita *et al.*, 1958; Nakayama *et al.*, 1961). Recently, the molecular mechanism of synergistic inhibition of aspartokinase of *C. glutamicum* by L-threonine and L-lysine was revealed (Yoshida *et al.*, 2010). There are two effector-binding sites in aspartokinase, one for L-threonine and the other for L-lysine. Binding of two L-threonine molecules and one L-lysine molecule for each α-subunit stabilizes the closed inhibitory form, which leads to conformational change in the active center of the enzyme. As a result, the substrate, L-aspartate, cannot bind to the catalytic residue and instead L-lysine occupies its space.

In the case of *Escherichia coli*, the situation is more complicated because it has three aspartokinases and two homoserine dehydrogenases and each has a different mode of feedback inhibition and repression. Furthermore, one peptide codes for both aspartokinase I and homoserine dehydrogenase I, and another for aspartokinase II and homoserine dehydrogenase II.

Example of utilization of analog-resistant mutant strains to produce amino acids
Utilization of an auxotroph for amino acid production requires the addition of a minimal amount of effector (L-threonine and L-methionine, or homoserine in the case of L-lysine fermentation in section 15.2.3) to support the growth of the bacterium, but it should not exceed the amount that induces feedback inhibition. Since it is difficult to provide a limited amount of an effector, not many amino acids are produced industrially using auxotrophs.

© Woodhead Publishing Limited, 2013

Instead, several amino acids are produced using analog-resistant mutant strains.

α-Amino-β-hydroxyvaleric acid (AHV) is an analog of L-threonine. Although this compound cannot substitute for L-threonine in protein biosynthesis, homoserine dehydrogenase of *C. glutamicum* is subjected to feedback inhibition by this compound (Fig. 15.3a). Therefore, a wild-type strain cannot grow on a minimal medium agar plate containing AHV and L-methionine but without L-threonine. A mutant strain which was able to grow on this medium plate was isolated and it was found that its homoserine dehydrogenase was desensitized to AHV. Since the homoserine dehydrogenase of this strain was also desensitized to L-threonine, this mutant over-produced L-threonine (Shiio and Nakamori, 1970). Fermentation of L-threonine was first industrialized by Ajinomoto by using an AHV-resistant mutant of *C. glutamicum* (originally named *Brevibacterium flavum*). Later, it was found that the C-terminal region of homoserine dehydrogenase had an amino acid substitution (Sugimoto *et al.*, 1997).

They also developed a threonine-producing *E. coli* strain by genetic engineering (Miwa *et al.*, 1983): an AHV-resistant desensitized mutant gene for aspartokinase-homoserine dehydrogenase (*thrA*) and other genes responsible for L-threonine biosynthesis were amplified on a plasmid and a mutant strain with an AHV-desensitized aspartokinase-homoserine dehydrogenase gene was transformed by this plasmid. After further improvement, L-threonine production using the recombinant strain was industrialized by Eurolysine. Later, Okamoto and his colleagues at Kyowa Hakko Kogyo identified an *E. coli* strain with impaired L-threonine uptake that maintained a lower intracellular threonine concentration and overproduced L-threonine up to 100 g l^{-1} (Okamoto *et al.*, 1997). There are many reports of analog-resistant amino acid producing strains, such as L-arginine, L-glutamine, L-histidine, L-isoleucine, L-leucine, L-phenylalanine, L-proline, L-serine, L-tryptophan and L-valine producers.

15.2.4   Enzymatic production of amino acids

Since some amino acids cannot be produced efficiently by fermentation, enzymatic methods using microbial enzymes were developed. Microorganisms, especially bacteria, are the preferred sources of such enzymes because they can grow fast in inexpensive media and there are potential possibilities to find new enzymes from bacteria.

Enzymatic production of amino acids from naturally occurring intermediates

Tosa and his colleagues of Tanabe Seiyaku (which has merged into Mitsubishi Tanabe Pharma) developed a continuous enzymatic production method to produce L-aspartic acid from fumaric acid using immobilized aspartase from *E. coli* (Sato and Tosa, 1993). They also found L-aspartic acid

© Woodhead Publishing Limited, 2013

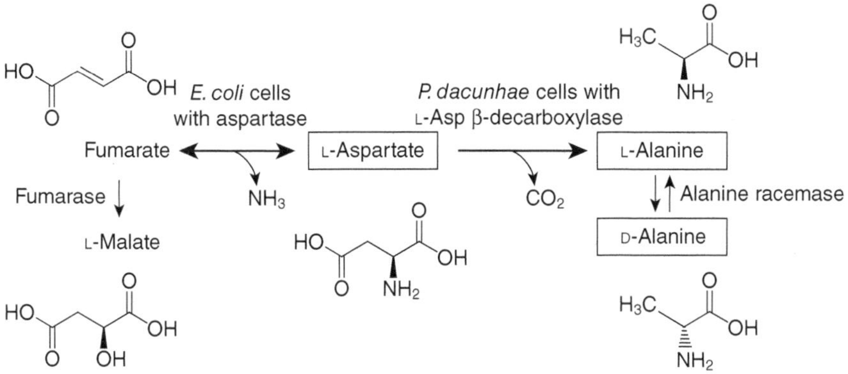

Fig. 15.4   Enzymatic production of L-aspartic acid and L-alanine from fumarate.

β-decarboxylase from *Pseudomonas dacunhae* and developed the enzymatic synthesis of L-alanine from L-aspartate using this enzyme (Chibata *et al.*, 1965). Generation of CO_2 gas in this enzymatic reaction, however, prevented the use of a conventional column packed with immobilized *P. dacunhae* cells. This problem was overcome by packing a closed column with the immobilized cells and performing the enzyme reaction under high pressure to prevent the generation of CO_2 gas (Furui and Yamashita, 1993).

In the next step, they tried to produce L-alanine directly from fumaric acid by the sequential reactions of immobilized *E. coli* cells with aspartase and *P. dacunhae* cells with L-aspartic acid β-decarboxylase (Fig. 15.4). However, the yield of L-alanine was not high because these strains contain fumarase and alanine racemase, respectively, and L-malic acid and D-alanine were produced as by-products. This was overcome by separately pretreating these two bacterial cells at low pH and high temperature to inactivate fumarase and alanine racemase, and then immobilizing cells of each bacterium respectively (Furui and Yamashita, 1993). These immobilized cells had high aspartase and L-aspartic acid β-decarboxylase activities, respectively, but practically no fumarase and alanine racemase activities. A column of immobilized *E. coli* cells and another of immobilized *P. dacunhae* cells were connected in tandem, and L-alanine was successfully produced by sequential reactions. This method was industrialized by Tanabe Seiyaku in 1982 (Chibata *et al.*, 1984). Tanabe Seiyaku was the pioneer in immobilization of enzymes and bacterial cells, and their applications of immobilized enzymes and bacterial cells were developed in parallel with their enzymatic productions of L-aspartic acid and L-alanine.

Enzymatic production of amino acids from chemically
synthesized intermediates
Large portions of L-cysteine are extracted from raw materials of animal origin with high L-cysteine content. DL-2-Amino-Δ^2-thiazoline-4-carboxylic

© Woodhead Publishing Limited, 2013

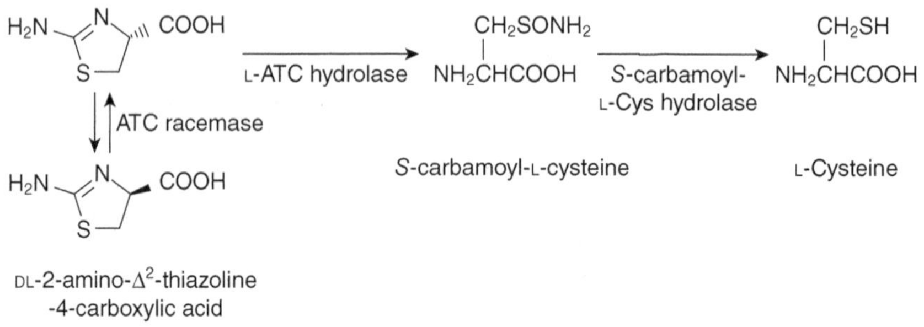

H₂N $\diagup$ N $\diagdown$ ‖ COOH
S $\diagdown$
L-ATC hydrolase → CH₂SONH₂ | NH₂CHCOOH
S-carbamoyl- L-Cys hydrolase →
CH₂SH | NH₂CHCOOH

↑↓ ATC racemase

H₂N $\diagup$ N $\diagdown$ COOH
S $\diagdown$

S-carbamoyl-L-cysteine

L-Cysteine

DL-2-amino-Δ²-thiazoline
-4-carboxylic acid

Fig. 15.5   Enzymatic production of L-cysteine.

acid (DL-ATC) is an intermediate in the chemical synthesis of DL-cysteine. Sano and his colleagues isolated several bacteria from soil that can grow in the medium containing DL-ATC as the sole nitrogen source. They further searched for the optimum conditions for the conversion of DL-ATC to L-cysteine using intact cells of *Pseudomonas thiazolinophilum* (Sano and Mitsugi, 1978); this enzymatic production of L-cysteine was industrialized by Ajinomoto. It was suggested that this strain converts DL-ATC to L-cysteine by the sequential reactions of ATC racemase, L-specific ATC hydrolase and *S*-carbamoyl-L-cysteine hydrolase (Sano *et al.*, 1977) (Fig. 15.5).

15.3   Amino acid derivatives of interest

15.3.1   L-Dihydroxylphenylalanine (L-DOPA)

Dopamine is a neurotransmitter in the brain. Patients with Parkinson's disease are deficient in L-tyrosine hydroxylase and cannot convert L-tyrosine to L-DOPA in the brain. They are deficient in dopamine as a result. Since dopamine cannot pass through the blood–brain barrier, L-DOPA is administered as a symptomatic therapeutic medication for patients. As W. S. Knowles was eager to synthesize the L-form of DOPA specifically, he invented a chiral specific production method for chemicals and this chemical synthetic method was industrialized. Indeed, L-DOPA was originally made according to his chemical method and this led to him being awarded the Nobel Prize. Kumagai and his colleagues at Kyoto University discovered tyrosine phenol-lyase (originally named β-tyrosinase) from *Citrobacter intermedius* (originally named *Escherichia intermedia*) (Kumagai *et al.*, 1970) and *Ervinia herbicola* (Kumagai *et al.*, 1972). This enzyme degrades L-tyrosine into phenol, ammonia and pyruvic acid (Fig. 15.6a) and these strains utilize ammonia as the nitrogen source and pyruvic acid as the carbon source for growth. They later discovered that this enzyme can catalyze the reverse reaction and synthesize L-tyrosine from the three substrates

© Woodhead Publishing Limited, 2013

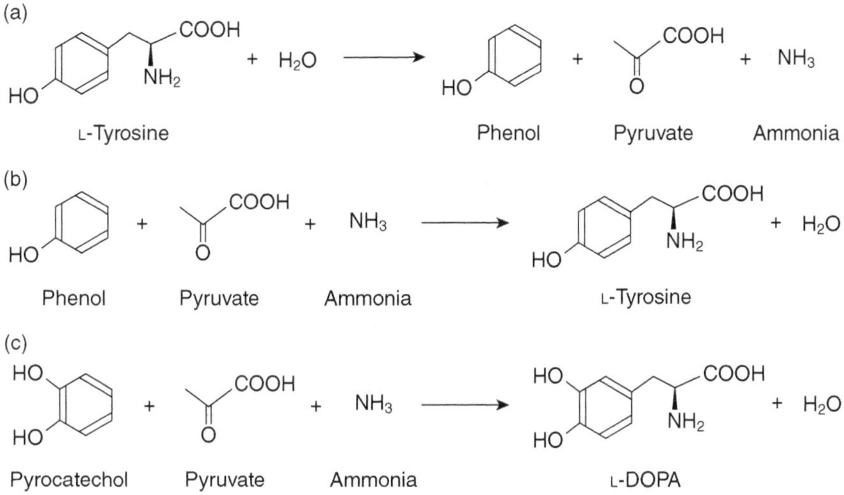

(a)

L-Tyrosine   + H₂O —→   Phenol   +   Pyruvate   + NH₃   Ammonia

(b)

Phenol   +   Pyruvate   +   NH₃   —→   L-Tyrosine   + H₂O

Ammonia

(c)

Pyrocatechol   Pyruvate   +   NH₃   —→   L-DOPA   + H₂O

Ammonia

Fig. 15.6   Reactions catalyzed by tyrosine phenol-lyase.

by a single-step reaction (Fig. 15.6b). In addition, it was determined that pyrocatechol can be used instead of phenol when L-DOPA was synthesized (Fig. 15.6c) (Enei *et al.*, 1973b). This method of L-DOPA production by *Er. herbicola* cells was industrialized by Ajinomoto in 1992 (Enei *et al.*, 1996) and the growth conditions were optimized to overexpress tyrosine phenol-lyase and allow the use of whole bacterial cells (Enei *et al.*, 1973a). Since this reverse reaction is physiologically insignificant and the solubility of L-DOPA in aqueous solution is very low, the reaction is inclined toward L-DOPA synthesis. The proportion of enzymatic production of L-DOPA worldwide is increasing because it is a simple, environmentally friendly, efficient and L-form specific reaction.

The addition of L-tyrosine to the medium is required for the induction of tyrosine phenol-lyase (Enei *et al.*, 1973a). Since the solubility of L-tyro-sine is extremely low, L-tyrosine added to the medium remains as a precipi-tate in the medium. When the cells are harvested and transferred into the reactor, precipitated L-tryrosine is also carried into the reactor with the cells. Although this amount of L-tyrosine carry-over is very little, it remains even after L-DOPA production. For use as a therapeutic agent, the required level of purity of L-DOPA is quite high and unfortunately it is difficult to separate these two compounds chemically because the only structural dif-ference between them is an OH group. This is the defect of this method.

The induction mechanism of tyrosine phenol-lyase gene (*tpl*) of *Er. her-bicola* was studied by Katayama and his collegues at Kyoto University. This work revealed that the gene is induced by L-tyrosine and repressed by glucose (Suzuki *et al.*, 1995), while further studies demonstrated that the

© Woodhead Publishing Limited, 2013

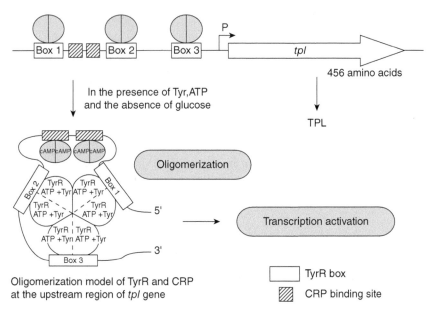

Fig. 15.7 Induction mechanism of *tpl* gene of *Er. herbicola* by TyrR protein and cAMP receptor protein. Oligomerization of TyrR protein in the presence of ATP and L-tyrosine with the aid of cAMP receptor protein induces the transcription of the *tpl* gene.

expression of the *tpl* gene of *Er. herbicola* is regulated by TyrR and the cAMP receptor protein. Three TyrR dimers bind to the three TyrR boxes located in the upstream region of the *tpl* gene, and its hexamerization mediated with tyrosine by the aid of the cAMP receptor protein, activates its transcription (Katayama *et al.*, 1999). In addition, an *in vitro* study showed that oligomerization of TyrR requires adenosine triphosphate (ATP) (Koyanagi *et al.*, 2008) (Fig. 15.7).

Overexpression of tyrosine phenol-lyase without the addition of L-tyrosine to the medium was attempted by isolating suppressor mutants of TyrR^{E275Q} that cannot oligomerize, and N324D and A503T mutations were isolated (Koyanagi *et al.*, 2008) (Fig. 15.8). TyrR$^{V67A\,Y72C\,E201G\,N324D\,A503T}$ mutant, which contains these mutations with other effective mutations (Katayama *et al.*, 2000) was constructed and the *Er. herbicola* strain with this mutant TyrR showed hyperproduction of L-DOPA without the addition of L-tyrosine to the medium (Koyanagi *et al.*, 2009).

15.3.2   D-*p*-Hydroxyphenylglycine

D-*p*-Hydroxyphenylglycine is a side-chain precursor for semi-synthetic β-lactam antibiotics, such as amoxicillin and cefadroxil. Yamada and his

© Woodhead Publishing Limited, 2013

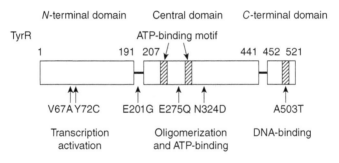

Fig. 15.8 Domain structure of TyrR protein of *Erwinia herbicola*.

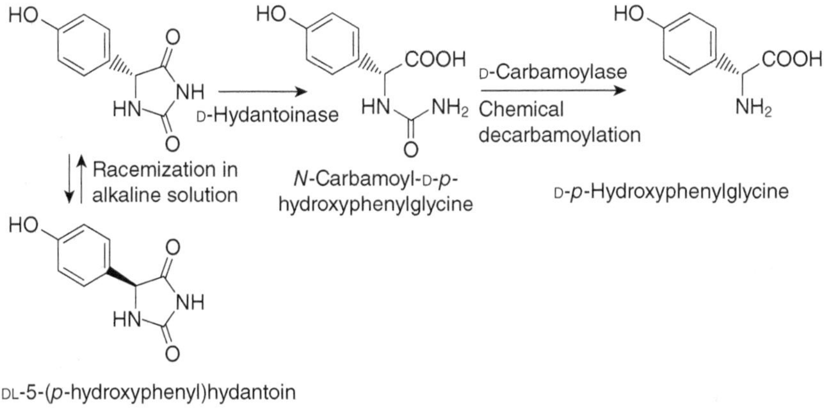

HO, HN, NH, O, D-Hydantoinase

HO, COOH, HN, NH₂, O — *N*-Carbamoyl-D-*p*-hydroxyphenylglycine

D-Carbamoylase / Chemical decarbomylation

HO, COOH, NH₂ — D-*p*-Hydroxyphenylglycine

Racemization in alkaline solution

HO, HN, NH, O

DL-5-(*p*-hydroxyphenyl)hydantoin

Fig. 15.9 Enzymatic production of D-*p*-hydroxyphenylglycine.

colleagues found several bacteria capable of synthesizing hydantoinase, which stereospecifically hydrolyzes 5-substituted D-hydantoins to release D-amino acids (Yamada *et al.*, 1978). They developed a method involving the cleavage of chemically synthesized DL-5-(*p*-hydroxyphenyl)hydantoin with D-hydantoinase to release *N*-carbamoyl-D-*p*-hydroxyphenylglycine. Then, 5-(*p*-hydroxyphenyl)hydantoin which remained was racemized under mild alkaline conditions, but *N*-carbamoyl-D-*p*-hydroxyphenylglycine liberated was not racemized under these mild conditions (Takahashi *et al.*, 1979). Eventually, all DL-5-(*p*-hydroxyphenyl)hydantoin is converted to *N*-carbamoyl-D-*p*-hydroxyphenylglycine. In the next step, *N*-carbamoyl-D-*p*-hydroxyphenylglycine was subjected to chemical decarbamoylation with nitrous acid under acidic conditions and D-*p*-hydroxyphenylglycine was produced (Fig. 15.9) (Takahashi, 1986). This D-*p*-hydroxyphenylglycine production was industrialized by Kaneka (originally Kanegafuchi Chemical Industry) (Takahashi *et al.*, 2000). The side reaction tended to take place in

© Woodhead Publishing Limited, 2013

the second chemical step and a large amount of acid and alkaline waste was generated. Takahashi and his colleagues at Kaneka substituted the enzymatic decarbamoylation of *N*-carbamoyl-D-*p*-hydroxyphenylglycine for a chemical decarbamoylation step. Nanba and his colleagues found a D-carbamoylase with high specific activity from *Agrobacterium* sp., cloned its gene and overexpressed it in *E. coli* (Nanba *et al.*, 1998b). They successfully increased the thermostability of this enzyme by random mutagenesis (Ikenaka *et al.*, 1999). The mutant enzyme was immobilized with Duolite A-568 (Nanba *et al.*, 1998a) and it was proved that the immobilized enzyme could be used stably and repeatedly with batchwise reactions (Nanba *et al.*, 1999). The sequential use of immobilized D-hydantoinase followed by immobilized mutant D-carbamoylase was industrialized by Kanaka Singapore in 1995. They are expanding the method to produce D-amino acids.

15.3.3 Hydroxy-L-proline

Hydroxy-L-proline is synthesized by the post-translational modification of L-proline residues of collagen by procollagen-proline dioxygenase (prolyl hydroxylase) (Kivirikko and Prockop, 1967). In collagen molecules, three polypeptide strands, called α peptides, are twisted together into right-handed 'super helices'. This structure is stabilized by water-mediated hydrogen bonds in which hydroxy-L-proline residues are involved (Shoulders and Raines, 2009).

There are eight stereoisomers and regioisomers of hydroxyproline. Among them, *trans*-4-hydroxy-L-proline is a precursor for various pharmaceuticals such as antiphlogistics, some angiotensin-converting enzyme inhibitors and some carbapenem antibiotics (Remuzon, 1996). However, mammalian procollagen-proline dioxygenase can only add hydroxyl groups to the proline residues of proteins and cannot use free proline as a substrate. Therefore, *trans*-4-hydroxy-L-proline was purified from the hydrolysate of collagen.

Ozaki and his co-workers developed a detection method for *trans*-4-hydroxy-L-proline by separating it from other isomers and proline (Ozaki *et al.*, 1995). Then, they screened microorganisms for proline 4-hydroxylase activity and chose *Dactylosporangium* sp. strain RH1, an actinomycete strain, since it exhibited the highest enzymatic activity (Shibasaki *et al.*, 1999). They then cloned the gene coding for proline 4-hydroxylase from *Dactylosporangium* sp. strain RH1 (Shibasaki *et al.*, 1999). Its promoter was substituted by the *trp* tandem promoter and the codon usage of the 5′ end of the gene was optimized for overexpression of this enzyme in *E. coli* (Shibasaki *et al.*, 2000).

2-Oxoglutamic acid, which is required for the hydroxylation of L-proline by proline 4-hydroxylase, was supplied by *E. coli* and the proline utilization pathway was blocked by introducing *putA* (Shibasaki *et al.*, 2000) (see Fig. 15.10). The *E. coli* strain generated for L-proline fermentation was used as

© Woodhead Publishing Limited, 2013

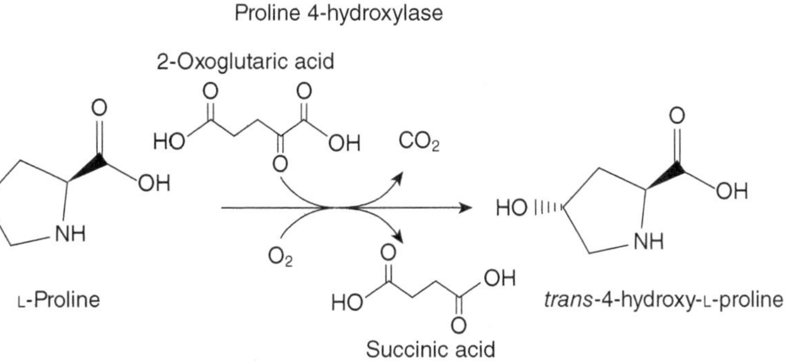

Fig. 15.10   Enzymatic production of *trans*-4-hydroxy-L-proline.

a host cell for the cloned proline 4-hydroxylase gene. This method of *trans*-4-hydroxy-L-proline production was industrialized by Kyowa Hakko Kogyo in 1997. Because of mad cow disease, demands for products from non-animal sources have increased and the introduction of this method was well-timed. Recently, Kino and his colleagues at Waseda University found a novel proline 4-hydroxylase, which can produce *cis*-4-hydroxy-L-proline (Hara and Kino, 2009). The production of *cis*-4-hydroxy-L-proline was industrialized by Kyowa Hakko Bio in 2011 according to their press release.

15.4   Short peptides

R. B. Merrifield developed a solid-phase peptide synthesis method and showed that several peptide hormones can be synthesized by his method (Merrifield, 1969). This method, however, is costly and it is only affordable for highly value-added peptides; it is not feasible for large-scale industrial production of short peptides.

15.4.1   Short peptides with α-peptide linkages
*Utilization of reverse reactions of proteases to make short peptides:
aspartame; N-(L-α-aspartyl)-L-phenylalanine methylester*
James M. Schlatter, a chemist working for G. D. Searle & Company, discovered aspartame in 1965 in the course of developing an antiulcer medication. When he licked his finger to pick up a piece of paper, he realized its unexpected strong sweetness. He retraced his study and identified aspartame as the source of that sweetness. Although the calories per gram of aspartame are about the same of those of sucrose, it is about 200 times sweeter than sucrose. That is, the amount of aspartame needed to produce the same sweetness is 1/200 of sugar and, therefore, 1/200th the calories are required

© Woodhead Publishing Limited, 2013

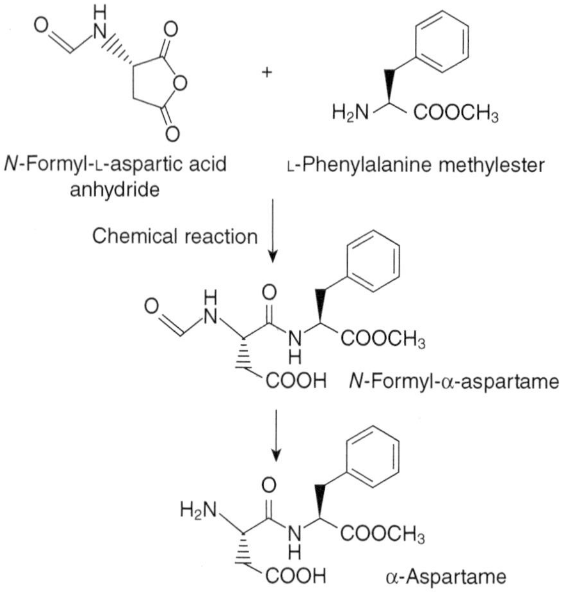

N-Formyl-L-aspartic acid anhydride

L-Phenylalanine methylester

Chemical reaction

N-Formyl-α-aspartame

α-Aspartame

Fig. 15.11   Chemical production of α-aspartame.

to attain the same sweetness. This makes aspartame useful as an artificial sweetener with low calories.

Ajinomoto developed a synthetic chemical method for aspartame (Fig. 15.11). In this method, N-protected L-aspartic acid anhydride is condensed with L-phenylalanine-methylester to obtain N-protected aspartame, followed by deprotection (Ariyoshi et al., 1974a,b). The defect of this method is that N-(L-β-aspartyl)-L-phenylalanine-methylester, which has bitter taste, is formed as a by-product. Formyl groups are used to protect the amino group of L-aspartic acid and HCl is used to deprotect it. During the deprotection step by HCl, four analogs of aspartame are generated. A purification method was developed in which only aspartame, not the other three analogs, can crystallize. According to a news release, Ajinomoto will begin the biotechnological production of aspartame in 2012, but currently the details have not been announced (https://bio.nikkeibp.co.jp/article/oc/2007/5036/).

Enzymatic production of aspartame using thermolysin, a protease from *Bacillus thermolyticus*, was developed by Tosoh (previously Toyo Soda) (O'Donnell, 2006) (Fig. 15.12). When thermolysin is used for stereospecific production of aspartame, the regiospecificity of the enzyme eliminates the need to protect the β-carboxyl group of L-aspartic acid, but its α-amino group must still be protected. Chemically synthesized DL-phenylalanine methylester and N-protected L-aspartic acid with a benzyloxycarbonyl group were subjected to the condensing reaction of thermolysin. Only

© Woodhead Publishing Limited, 2013

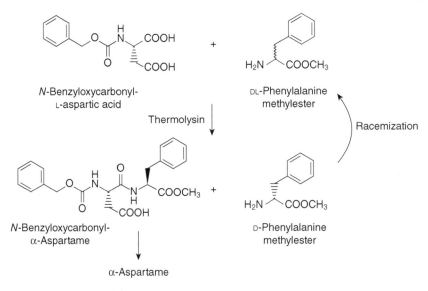

Fig. 15.12 Enzymatic production of α-aspartame using thermolysin.

L-phenylalanine methylester binds with N-benzyloxycarbonyl-L-aspartic acid to form N-benzyloxycarbonylaspartame, and D-phenylalanine methylester was racemized to DL-phenylalanine methylester. The protection group of benzyloxycarbonylaspartame was then deprotected to release aspartame. This is one of the successful examples utilizing the reverse reaction of a protease. This enzymatic production was industrialized by Holland Sweetener (a joint venture of Royal DSM and Tosoh) in 1988. This method had the advantage that chemically synthesized DL-phenylalanine was much cheaper than L-phenylalanine. When the price of L-phenylalanine produced by fermentation became overwhelmingly less expensive by technical innovations, this method lost its competitiveness because they used a costly enzyme and protective group. Holland Sweetener stopped supplying aspartame and closed the company in 2006.

Advantame (N-[N-[3-(3-hydroxy-4-methoxyphenyl) propyl]-α-aspartyl]-L-phenylalanine 1-methyl ester, monohydrate), an N-substituted analogue of aspartame, has been developed as a next-generation high-intensity sweetener. It is approximately 100 times sweeter than aspartame; Ajinomoto began its industrial scale production in the United States in 2011.

Utilization of newly discovered enzymes for dipeptide production:
L-alanyl-L-glutamine
L-Glutamine is a non-essential amino acid in healthy tissues, however it can be an essential one in patients receiving enteric nutrition for recovery from certain pathological conditions (Lacey and Wilmore, 1990) or injuries

© Woodhead Publishing Limited, 2013

(Houdijk *et al.*, 1998). Enteral preoperative supplementation of L-glutamine attenuates intestinal and lung damage during surgical manipulation (Prabhu *et al.*, 2003) and oral L-glutamine administration after intense exercise has been shown to decrease the risk of athletes developing infections caused by immunosuppression (Castell and Newsholme, 1997). Despite these favorable effects, L-glutamine cannot be distributed as an aqueous solution, such as an infusion solution and as aqueous amino acid supplements, because of its instability in aqueous solution and its relatively low solubility in water (Gandini *et al.*, 1993; Shih, 1985). On the contrary, L-alanyl-L-glutamine is quite stable in aqueous solutions and its solubility in water is dramatically higher than that of L-glutamine. Fürst showed its usability for infusion solution as it can be readily cleaved in serum and utilized as its constituent amino acids (Fürst, 2001). Although this dipeptide is a promising compound, an efficient production method for dipeptides has been studied only recently. Two enzymatic methods, involving different enzymes, were separately developed by Ajinomoto and Kyowa Hakko Kogyo (Yagasaki and Hashimoto, 2008).

Yokozeki and Hara of Ajinomoto screened for microorganisms that can synthesize L-alanyl-L-glutamine (L-ala-L-gln) from L-alanine methylester (L-ala-*O*-Me) and L-glutamine (L-gln) (Fig. 15.13a) (Yokozeki and Hara, 2005; Yokozeki, 2005). Among the several microorganisms that produced L-alanyl-L-glutamine, *Empedobacter brevis* was chosen. The purified enzyme was shown to produce 83 mM of L-alanyl-L-glutamine from 100 mM L-alanine methylester and 200 mM L-glutamine at 18°C in 2 h. In this method, a methyl-esterified amino acid attaches to the *N*-terminal side and methanol is released when L-alanyl-L-glutamine is formed. As by-products, small amounts of L-alanine and scarce amounts of L-alanyl-L-alanyl-L-glutamine were found, but L-glutamyl-L-alanine was never formed. They also showed the use of this enzyme for oligopeptide formation, although not all amino acids can be used efficiently as *N*-terminal side residues.

In contrast, Tabata and his colleagues at Kyowa Hakko Kogyo screened enzymes *in silico* (Tabata *et al.*, 2005). Their idea was the following; the enzyme they were looking for should have an ATP-binding motif, should

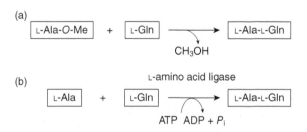

Fig. 15.13   Enzymatic production of L-alanyl-L-glutamine.

© Woodhead Publishing Limited, 2013

be unassigned and should have slight homology with D-alanine:D-alanine ligase. The only gene that satisfied these three conditions was *ywfE* of *Bacillus subtilis*. His-tagged YwfE was expressed in *E. coli* and the protein was purified on a nickel column. Purified YwfE was incubated with L-alanine, L-glutamine and ATP, and L-alanyl-L-glutamine was formed. This enzyme is an L-amino acid ligase. The purified enzyme was shown to produce 19 mM of L-alanyl-L-glutamine from 30 mM L-alanine, L-glutamine and 60 mM ATP at 37°C in 12 h. L-Alanyl-L-alanine (1.8 mM) was formed as a by-product, but L-glutamyl-L-alanine and tripeptides were not found.

The possibility of using this enzyme for the formation of other dipeptides was also shown although not all amino acid combinations can be condensed efficiently. Since this enzyme is a ligase, it must be coupled with an ATP-regenerating system to produce L-alanyl-L-glutamine. The gene for polyphosphate kinase and the *ywfE* gene were overexpressed in *E. coli*. The detergent-treated cells were incubated with L-alanine, L-glutamine and polyphosphate, and L-alanyl-L-glutamine was produced (Hashimoto *et al.*, 2005). In addition, an *E. coli* strain for the direct fermentation of L-alanyl-L-glutamine into the culture medium was constructed (Tabata and Hashimoto, 2007). The genes encoding aminopeptidases and dipeptidases, namely *pepA, B, D, N* genes, and *dpp* gene, which encodes a dipeptide transporter, were deleted to increase the yield of dipeptides. *glnE* and *glnB* genes, whose products negatively regulate glutamine synthesis, were deleted to increase the synthesis of L-glutamine, and the *ald* gene, which encodes alanine dehydrogenase, was introduced to increase the synthesis of L-alanine. Then, the *ywfE* gene, which encodes L-amino acid ligase, was introduced to produce L-alanyl-L-glutamine. After 47 h of fed-batch culture, 100 mM L-alanyl-L-glutamine was produced with 26 mM L-alanyl-L-alanine as a by-product. Therefore, separation of L-alanyl-L-alanine from L-alanyl-L-glutamine is critical for this method. Kyowa Hakko Kogyo industrialized the fermentation method in 2009 and supplies L-alanyl-L-glutamine for infusion solutions.

15.4.2 γ-Glutamylated derivatives of amino acids

γ-Glutamyl linkages are a relatively common natural amide bond between the γ-carboxyl group of glutamic acid and the amino group of another compound. If the C-terminus is an amino acid, the bond is a γ-peptide linkage. Glutamine is the most abundant free amino acid in living organisms (Curi *et al.*, 2005) and is a kind of γ-glutamyl compound. Glutathione (γ-glutamylcysteinylglycine), which is a cellular antioxidant, is the most abundant free thiol compound in living cells (Meister and Anderson, 1983). Folic acid usually exists as a γ-glutamyl derivative in food and is cleaved by γ-glutamyl hydrolase (Green *et al.*, 1996). Peptidoglycan, the major component of bacterial cell walls, contains γ-glutamyl linkages (Park, 1996). Some bacteria have capsules made of poly-γ-glutamic acid and the relationship

© Woodhead Publishing Limited, 2013

between virulence and the capsule is well known for *Bacillus anthracis* (Makino *et al.*, 1989).

Increase in solubility by γ-glutamylation

Some compounds with low water solubility become dramatically more soluble by γ-glutamylation. For instance, the solubility of cystine in water is extremely low at around 0.1 g l^{-1}. When cystine is γ-glutamylated, the solubility of γ-glutamylcystine dramatically increases to >742 g l^{-1}, that is a more than 7000-fold increase (Hara *et al.*, 1992). Nonetheless the solubility of alanylcystine is very high compared with that of cystine, as described above; it is less soluble than γ-glutamylcystine. A medication that cannot be administered orally must be administered either by injection or by intravenous drip. In addition, medications with low water solubility must be taken intravenously in large volumes, which is a burden for patients. Therefore, water solubility is critical for therapeutic design.

Increase in stability in serum by γ-glutamylation

Since normal peptidases in serum cannot cleave the γ-glutamyl linkage, the half lives of compounds are greatly increased by γ-glutamylation (Hara *et al.*, 1992). γ-Glutamyl linkages are first cleaved in organs that express γ-glutamyltranspeptidase, such as the kidney. Hence, some γ-glutamyl compounds are appropriate as pro-drugs specific for organs that express γ-glutamyltranspeptidase. For example, the concentration of dopamine in the brain markedly increased after the administration of γ-L-glutamyl-L-DOPA to animals (Wilk *et al.*, 1978; Ichinose *et al.*, 1987), indicating that γ-L-glutamyl-L-DOPA is a promising pro-drug for Parkinson's disease.

Preferable taste of γ-glutamyl compounds

L-Theanine, also known as γ-L-glutamylethylamide, is the most abundant free amino acid in tea leaves. Sakato showed that L-theanine is the major umami component of tea (Sakato, 1949) and a positive correlation between the grade of Japanese green tea and the amount of L-theanine has been observed (Goto *et al.*, 1996).

High concentrations of branched-chain amino acids, aromatic amino acids and basic amino acids are added to nutritional supplements in Japan. However, the bitter taste of these L-amino acids is a crucial problem when designing oral supplements. It was found that γ-glutamylation of these amino acids dramatically reduces their bitterness and increases sourness (Suzuki *et al.*, 2002b). Importantly, this sourness is a refreshing lemon-like taste and, therefore, the preference for the γ-glutamyl amino acids increased. Bacterial γ-glutamyltranspeptidases prefer aromatic amino acids and basic amino acids as γ-glutamyl acceptors (Suzuki *et al.*, 1986; Minami *et al.*, 2003), and therefore, it is easy to γ-glutamylate these amino acids using bacterial γ-glutamyltranspeptidase. Although branched-chain amino acids are not

© Woodhead Publishing Limited, 2013

good γ-glutamyl acceptors, a method has been improved to γ-glutamylate these amino acids fairly efficiently (Suzuki *et al.*, 2004b).

Physiological effects of γ-glutamyl compounds
Several γ-glutamyl compounds have various preferable physiological effects on mammals. For example, some γ-D-glutamyl compounds, such as γ-D-glutamyltaurine, have antagonistic effects on the excitatory nervous system (Davies *et al.*, 1982).

Methods for production of γ-glutamylated derivatives of amino acids
γ-Glutamylpeptides were synthesized by bacterial γ-glutamylcysteine synthetase, which catalyzes the first reaction in the glutathione synthesis (Nakayama *et al.*, 1981). γ-Glutamylmethylamide and L-theanine were synthesized by bacterial glutamine synthetase (Tachiki *et al.*, 1986; Yamamoto *et al.*, 2005). These enzymes are ligases and require ATP to generate γ-glutamyl linkages, but they are also inhibited by the by-product adenosine diphosphate (ADP). An ATP-regenerating system has to be coupled to the process in order to use ligases for the synthesis of γ-glutamyl compounds. L-Theanine can be synthesized by bacterial glutaminase; this method was industrialized by Taiyo Kagaku (Abelian *et al.*, 1993a,b). However, because of its substrate specificity, it is difficult to synthesize γ-glutamyl amino acids by glutaminase.

In contrast, substrate specificities of bacterial γ-glutamyltranspeptidases are very broad and they can utilize various compounds with amino groups, such as amino acids, dipeptides, taurine and ethylamine as their γ-glutamyl acceptor. It is also an advantage that bacterial γ-glutamyltranspeptidases can utilize less expensive glutamine as well as glutathione as their γ-glutamyl donor (Suzuki *et al.*, 1986; Minami *et al.*, 2003). It can not only use L-glutamine but also D-glutamine as a γ-glutamyl donor and various γ-D-glutamyl compounds can be produced efficiently (Suzuki *et al.*, 2003). It is difficult to synthesize γ-D-glutamyl compounds by other enzymes, such as glutamine synthetase, because most enzymes are L-amino acid specific. Moreover, since γ-glutamyltranspeptidase is a transferase, not a synthetase, it does not require ATP to generate γ-glutamyl linkages (Fig. 15.14). A large amount of bacterial γ-glutamyltranspeptidase is readily available because it can be easily purified from strains overproducing γ-glutamyltranspeptidase (Suzuki *et al.*, 1988; Claudio *et al.*, 1991; Minami *et al.*, 2003). Altogether, the use of bacterial γ-glutamyltranspeptidase to produce γ-glutamyl compounds is advantageous.

15.4.3   Examples of γ-glutamylated derivatives of amino acids
L-*Theanine (γ-L-glutamylethylamide)*
L-Theanine is the most abundant free amino acid in tea leaves and the major umami compound of tea. This amino acid reduces the blood pressure

© Woodhead Publishing Limited, 2013

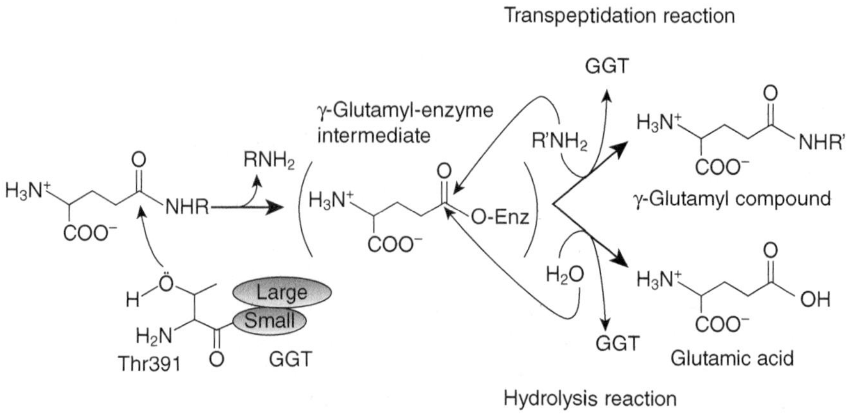

Fig. 15.14   Two reactions catalyzed by γ-glutamyltranspeptidase (GGT).

of spontaneously hypertensive rats (Yokogoshi *et al.*, 1995) and the excitation caused by caffeine (Kimura *et al.*, 1975), and has relaxing effects on humans (Juneja *et al.*, 1999). When orally administered, L-theanine can be absorbed into the blood stream through the intestinal tract (Kitaoka *et al.*, 1996; Unno *et al.*, 1999). It crosses the blood–brain barrier and may affect the metabolism and/or release of some neurotransmitters in the brain (Yokogoshi *et al.*, 1998). As described previously (see Section 15.4.2; 'Methods for production of γ-glamylated derivatives of amino acids' above), Taiyo Kagaku industrialized L-theanine production from L-glutamine and ethylamine using glutaminase from *Pseudomonas nitroreducens* (Abelian *et al.*, 1993a,b) and it is now sold as a supplement in the United States. It was also shown that L-theanine can be produced by either γ-glutamyl-transpeptidase (Suzuki *et al.*, 2002a) or γ-glutamylcysteine synthetase (Miyake and Kakita, 2009) from *E. coli*.

γ-D-Glutamyl-L-tryptophan
γ-D-Glutamyl-L-tryptophan, in combination with standard chemotherapy, was very effective in the treatment of tuberculosis in phase 2 clinical trials (Orellana, 2002). Oral administration of γ-D-glutamyl-L-tryptophan may also be effective as a treatment and thus is a promising medication for tuberculosis, considering the expected future demands of the pharmaceutical industry. Its chemical synthesis, however, is complicated because γ-D-glutamyl-L-tryptophan has several reactive groups and consists of D- and L-amino acids, connected by a γ-glutamyl linkage. In contrast, γ-D-glutamyl-L-tryptophan is an ideal compound to produce from bacterial γ-glutamyl-transpeptidase (Suzuki *et al.*, 2004a). γ-Glutamyltranspeptidase can utilize

© Woodhead Publishing Limited, 2013

a γ-D-glutamyl compound as a γ-glutamyl donor, but not as a γ-glutamyl acceptor (it cannot utilize D-amino acids as γ-glutamyl acceptors). Therefore, γ-D-glutamyl-D-glutamine is never synthesized as a by-product when D-glutamine is utilized as a γ-glutamyl donor. Moreover, it prefers aromatic amino acids as γ-glutamyl acceptors. Since L-tryptophan is an aromatic amino acid, it is a good substrate.

γ-L-*Glutamyl*-L-*DOPA*

L-DOPA is a medication for symptomatic therapy for Parkinson's disease. However, its solubility in water is low. Since its water solubility is increased by γ-glutamylation and the γ-glutamyl linkages can be cleaved at the blood–brain barrier where γ-glutamyltranspeptidase is expressed (see Section 15.4.2; 'Increase in stability in serum by γ-glutamylation' above), γ-L-glutamyl-L-DOPA is a promising pro-drug for Parkinson's disease. This is supported by the finding in mice that the amount of dopamine in the brain increased markedly after the administration of γ-L-glutamyl-L-DOPA synthesized by bacterial γ-glutamyltranspeptidase (Ichinose *et al.*, 1987).

γ-L-*Glutamyl*-L-*glutamine*

As previously described, despite its favorable effects, L-glutamine cannot be distributed as an aqueous solution such as an infusion solution and aqueous supplement, because it is unstable in aqueous solution (see Section 15.4.1; 'Utilization of newly discovered enzymes for dipeptide production: L-alanyl-L-glutamine' above). Therefore, L-alanyl-L-glutamine is used instead of L-glutamine and two major industrialized production methods are utilized. γ-L-Glutamyl-L-glutamine is quite stable in aqueous solution and is an alternative to L-alanyl-L-glutamine. As such, an enzymatic method for its production using bacterial γ-glutamyltranspeptidase was developed (El Sayed *et al.*, 2010).

15.5   Future trends in amino acid production

In the 20th century, many fermentation and enzymatic methods using microbial enzymes were developed to produce amino acids. Along with these, the regulation mechanisms of their metabolism were revealed. Many new physiological functions of amino acids have been identified and their applications are expanding. However, the possibility of finding new functions and applications are relatively limited because there are only 20 amino acids that make up proteins. In contrast, for dipeptides arising from 20 amino acids, $20 \times 20 = 400$ kinds of peptides are possible and, therefore, there are more possibilities of finding new functions and applications. Previously, most dipeptides were not available at large scale from commercial sources. Since two companies have recently developed the enzymatic and

© Woodhead Publishing Limited, 2013

fermentation methods to produce dipeptides, the 21st century may become a century focused on dipeptides.

In addition to the 20 L-amino acids that constitute proteins, there are a number of other amino acids, including D-amino acids and their derivatives of interest in nature. Compared with the 20 L-amino acids, very few studies have been performed on these minor amino acids and their derivatives. Therefore, applications of these amino acids may be developed in the near future.

15.6   Sources of further information and advice

15.6.1   Key books and reviews

Amino Acids Biosynthesis and Genetic Regulation (1983), edited by K. M. Herrmann and R. L. Somerville, Addison-Wesley, Reading, MA, USA. ISBN 1-201-10520-9: This is rather old, but still a great review book on microbial amino acid metabolism. Unfortunately, this book covers only part of the modern genetic and molecular biological progress of amino acid metabolisms.

Biotechnology of Amino Acid Production (1986), edited by K. Aida, I. Chibata, K. Nakayama, K. Takinami and H. Yamada, *Progress in Industrial Microbiology, Volume 24*. Kodansha and Elsevier, Tokyo. ISBN 4-06-201742-3: This is also rather old, but still a great review book to learn about industrial amino acid production.

Kumagai, H. (2000). Microbial production of amino acids in Japan. *Advances in Biochemical Engineering/Biotechnology*, **69**, 71–85: This review illustrates the successful industrial production of several amino acids in Japan.

Ikeda, M. (2003). Amino acid production processes. *Advances in Biochemical Engineering/Biotechnology,* **79**, 1–35: This article illustrates well ideas about strain construction and practical strategies for amino acid production.

Nakamori, S. (2008). 'Systematic survey of the technical development of fermentative production of amino acids'. In: *National Science Museum: Report of Systematic Survey of the Technology.* Center of the History of Japanese Industrial Technology (ed.) National Science Museum, Tokyo, Japan, pp 55–91 (in Japanese): This is a great review article of the technical development of amino acid fermentation in Japan. Unfortunately, it is written in Japanese.

15.6.2   Websites

Ajinomoto: http://www.ajinomoto.com/about/rd/index.html
Kaneka: http://www.fcd.kaneka.co.jp/tech/index.html
Kyowa Hakko Kogyo: http://www.kyowahakko-bio.co.jp/english/r_d/

© Woodhead Publishing Limited, 2013

15.7   References

ABELIAN, V. H., OKUBO, T., MUTOH, K., CHU, D.-C., KIM, M. and YAMAMOTO, T. (1993a). 'A continuous production method for theanine by immobilized *Pseudomonas nitroreducens* cells'. *J Ferment Bioeng*, **76**, 195–8.

ABELIAN, V. H., OKUBO, T., SHAMTSIAN, M. M., MUTOH, K., CHU, D.-C., KIM, M. and YAMAMOTO, T. (1993b).'A novel method of production of theanine by immobilized *Pseudomonas nitroreducens* cells'. *Biosci Biotechnol Biochem*, **57**, 481–3.

ARIYOSHI, Y., NAGAO, M., SATO, N., SHIMIZU, A. and KIRIMURA, J. (1974a). *Method of producing* α-L-*aspartyl-*L-*phenylalanine Lower Alkyl Esters*. US Patent 1974/3786039.

ARIYOSHI, Y., YAMATANI, T., UCHIYAMA, N., YASUDA, N., TOI, K. and SATO, N. (1974b). *Method of Producing* α-L-*aspartyl-*L-*phenylalanine Alkyl Esters*. US Patent 1974/3833553.

ASAKURA, Y., KIMURA, E., USUDA, Y., KAWAHARA, Y., MATSUI, K., OSUMI, T. and NAKAMATSU, T. (2007). 'Altered metabolic flux due to deletion of *odhA* causes L-glutamate overproduction in *Corynebacterium glutamicum*'. *Appl Environ Microbiol*, **73**, 1308–19.

CASTELL, L. M. and NEWSHOLME, E. A. (1997). 'The effects of oral glutamine supplementation on athletes after prolonged, exhaustive exercise'. *Nutrition*, **13**, 738–42.

CHIBATA, I., KAKIMOTO, T. and KATO, J. (1965). 'Enzymatic production of L-alanine by *Pseudomonas dacunhae*'. *Appl Microbiol*, **13**, 638–45.

CHIBATA, I., TOSA, T. and TAKAMATSU, S. (1984). 'Industrial production of L-alanine using immobilized *Escherichia coli* and *Pseudomonas dacunhae*'. *Microbiol Sci*, **1**, 58–62.

CLAUDIO, J. O., SUZUKI, H., KUMAGAI, H. and TOCHIKURA, T. (1991). 'Excretion and rapid purification of γ-glutamyltranspeptidase from *Escherichia coli* K-12'. *J Ferment Bioeng*, **72**, 125–7.

CURI, R., LAGRANHA, C. J., DOI, S. Q., SELLITTI, D. F., PROCOPIO, J., PITHON-CURI, T. C., CORLESS, M. and NEWSHOLME, P. (2005). 'Molecular mechanisms of glutamine action'. *J Cell Physiol*, **204**, 392–401.

DAVIES, J., EVANS, R. H., JONES, A. W., SMITH, D. A. and WATKINS, J. C. (1982). 'Differential activation and blockade of excitatory amino acid receptors in the mammalian and amphibian central nervous systems'. *Comp Biochem Physiol C*, **72**, 211–24.

DUPERRAY, F., JEZEQUEL, D., GHAZI, A., LETELLIER, L. and SHECHTER, E. (1992). 'Excretion of glutamate from *Corynebacterium glutamicum* triggered by amine surfactants'. *Biochim Biophys Acta*, **1103**, 250–8.

EL SAYED, A. S. A. F., FUJIMOTO, S., YAMADA, C. and SUZUKI, H. (2010).'Enzymatic synthesis of γ-glutamylglutamine, a stable glutamine analogue, by γ-glutamyltranspeptidase from *Escherichia coli* K-12'. *Biotechnol Lett*, **32**, 1877–81.

ENEI, H., NAKAZAWA, H., OKUMURA, S. and YAMADA, H. (1973a). 'Synthesis of L-tyrosine or 3, 4-dihydroxyphenyl-L-alanine from pyruvic acid, ammonia and phenol or pyrocatechol'. *Agric Biol Chem,* **37**, 725–35.

ENEI, H., YAMASHITI, K., OKUMURA, S. and YAMADA, H. (1973b). 'Culture conditions for the preparation of cells containing high tyrosine phenol lyase activity'. *Agric Biol Chem*, **37**, 485–92.

ENEI, H., NAKAZAWA, H., TSUCHIDA, T., NAMEKAWA, T. and KUMAGAI, H. (1996). 'Development of L-DOPA production by enzymatic synthesis'. *Biosci Technol*, **54**, 11–15. (in Japanese)

FÜRST, P. (2001). 'New developments in glutamine delivery'. *J Nutr*, **131**, 2562S–8S.

FURUI, M. and YAMASHITA, Y. (1993). 'Pressurized reaction method for continuous production of L-alanine by immobilized *Pseudomonas dacunhae* cells'. *J Ferment Technol*, **61**, 587–91.

© Woodhead Publishing Limited, 2013

GANDINI, C., DE LORENZI, D., KITSOS, M., MASSOLINI, G. and CACCIALANZA, G. (1993). 'HPLC determination of pyroglutamic acid as a degradation product in parenteral amino acid formulations'. *Chromatographia*, **36**, 75–8.

GOTO, T., YOSHIDA, Y., AMANO, I. and HORIE, H. (1996). 'Chemical composition of commercially available Japanese green tea'. *Foods Food Ingredients J Jpn*, **170**, 46–51.

GREEN, J. M., NICHOLS, B. P. and MATTHEWS, R. G. (1996). 'Folate biosynthesis, regulation, and polyglutamylation'. In: Escherichia coli *and* Salmonella: *Cellular and Molecular Biology*, Neidhardt, F. C., Curtiss, R. I., Ingraham, J. L., Lin, E. C. C., Low, K. B., Magasanik, B., Reznikoff, W. S., Riley, M., Schaechter, M. and Umbarger, H. E. (eds), The American Society for Microbiology, Washington, DC. 665–73.

HARA, R. and KINO, K. (2009). 'Characterization of novel 2-oxoglutarate dependent dioxygenases converting L-proline to cis-4-hydroxy-1-proline'. *Biochem Biophys Res Commun*, **379**, 882–6.

HARA, T., YOKOO, Y. and FURUKAWA, T. (1992). 'Potential of γ-L-glutamyl-L-cystine and bis-γ-L-glutamyl-L-cystine as a cystine-containing peptide for parental nutrition'. In: *Frontiers and New Horizons in Amino Acid Research*, Takai, K. (ed.), Elsevier, Amsterdam.

HASHIMOTO, S., IKEDA, H. and YAGASAKI, M. (2005). *Process for Producing Dipeptides or Dipeptide Derivatives*. US Patent 2005/0287627.

HOUDIJK, A. P., RIJNSBURGER, E. R., JANSEN, J., WESDORP, R. I., WEISS, J. K., MCCAMISH, M. A., TEERLINK, T., MEUWISSEN, S. G., HAARMAN, H. J., THIJS, L. G. and VAN LEEUWEN, P. A. (1998). 'Randomised trial of glutamine-enriched enteral nutrition on infectious morbidity in patients with multiple trauma'. *Lancet*, **352**, 772–6.

ICHINOSE, H., TOGARI, A., SUZUKI, H., KUMAGAI, H. and NAGATSU, T. (1987). 'Increase of catecholamines in mouse brain by systemic administration of γ-glutamyl L-3,4-dihydroxyphenylalanine'. *J Neurochem*, **49**, 928–32.

IKEDA, K. (1909). 'About the new seasoning'. *Tokyo Kagaku Kaishi*, **30**, 820–36.

IKENAKA, Y., NANBA, H., YAJIMA, K., YAMADA, Y., TAKANO, M. and TAKAHASHI, S. (1999). 'Thermostability reinforcement through a combination of thermostability-related mutations of N-carbamyl-D-amino acid amidohydrolase'. *Biosci Biotechnol Biochem*, **63**, 91–5.

JUNEJA, L. R., CHU, D.-C., OKUBO, T., NAGATO, Y. and YOKOGOSHI, H. (1999). 'L-Theanine-A unique amino acid of green tea and its relaxation effect in humans'. *Trends Food Sci Technol*, **10**, 199–204.

KATAYAMA, T., SUZUKI, H., YAMAMOTO, K. and KUMAGAI, H. (1999). 'Transcriptional regulation of tyrosine phenol-lyase gene mediated through TyrR and cAMP receptor protein'. *Biosci Biotechnol Biochem*, **63**, 1823–7.

KATAYAMA, T., SUZUKI, H., KOYANAGI, T. and KUMAGAI, H. (2000). 'Cloning and random mutagenesis of the *Erwinia herbicola tyrR* gene for high-level expression of tyrosine phenol-lyase'. *Appl Environ Microbiol*, **66**, 4764–71.

KAWAHARA, Y., TAKAHASHI-FUKE, K., SHIMIZU, E., NAKAMATSU, T. and NAKAMORI, S. (1997). 'Relationship between the glutamate production and the activity of 2-oxoglutarate dehydrogenase in *Brevibacterium lactofermentum*'. *Biosci Biotechnol Biochem*, **61**, 1109–12.

KIMURA, R., KURITA, M. and MURATA, T. (1975). 'Influence of alkylamides of glutamic acid and related compounds on the central nervous system. III. Effect of theanine on spontaneous activity of mice'. *Yakugaku Zasshi*, **957**, 892–5. (in Japanese)

KINOSHITA, S., NAKAYAMA, K. and UDAKA, S. (1957a). 'The fermentative production of L-ornithine (preliminary report)'. *J Gen Appl Microbiol*, **3**, 276–7.

KINOSHITA, S., UDAKA, S. and SHIMONO, M. (1957b). 'Studies on the amino acid fermentation. Part 1. Production of L-glutamic acid by various microoganism'. *J Gen Appl Microbiol*, **3**, 193–205.

© Woodhead Publishing Limited, 2013

KINOSHITA, S., NAKAYAMA, K. and KITADA, S. (1958). 'L-Lysine production using microbial auxotroph (preliminary report)'. *J Gen Appl Microbiol*, **4**, 128–9.

KITAOKA, S., HAYASHI, H., YOKOGOSHI, H. and SUZUKI, Y. (1996). 'Transmural potential changes associated with the in vitro absorption of theanine in the guinea pig intestine'. *Biosci Biotechnol Biochem*, **60**, 1768–71.

KIVIRIKKO, K. I. and PROCKOP, D. J. (1967). 'Purification and partial characterization of the enzyme for the hydroxylation of proline in protocollogen'. *Arch Biochem Biophys*, **118**, 611–8.

KOYANAGI, T., KATAYAMA, T., SUZUKI, H. and KUMAGAI, H. (2008). 'Altered oligomerization properties of N316 mutants of *Escherichia coli* TyrR'. *J Bacteriol*, **190**, 8238–43.

KOYANAGI, T., KATAYAMA, T., SUZUKI, H., ONISHI, A., YOKOZEKI, K. and KUMAGAI, H. (2009). 'Hyperproduction of 3,4-dihydroxyphenyl-L-alanine (L-Dopa) using *Erwinia herbicola* cells carrying a mutant transcriptional regulator TyrR'. *Biosci Biotechnol Biochem*, **73**, 1221–3.

KUMAGAI, H., MATSUI, H. and YAMADA, H. (1970). 'Formation of tyrosine phenol-lyase by bacteria'. *Agric Biol Chem*, **34**, 1259–61.

KUMAGAI, H., KASHIMA, N., TORII, H., YAMADA, H., ENEI, H. and OKUMURA, S. (1972). 'Purification, crystallization and properties of tyrosine phenol lyase from *Erwinia herbicola*'. *Agric Biol Chem*, **36**, 472–82.

LACEY, J. M. and WILMORE, D. W. (1990). 'Is glutamine a conditionally essential amino acid?' *Nutr Rev*, **48**, 297–309.

MAKINO, S., UCHIDA, I., TERAKADO, N., SASAKAWA, C. and YOSHIKAWA, M. (1989). 'Molecular characterization and protein analysis of the *cap* region, which is essential for encapsulation in *Bacillus anthracis*'. *J Bacteriol*, **171**, 722–30.

MEISTER, A. and ANDERSON, M. E. (1983). 'Glutathione'. *Annu Rev Biochem*, **52**, 711–60.

MERRIFIELD, R. B. (1969). 'Solid-phase peptide synthesis;. *Adv Enzymol Relat Areas Mol Biol*, **32**, 221–96.

MINAMI, H., SUZUKI, H. and KUMAGAI, H. (2003). 'Salt-tolerant γ-glutamyltranspeptidase from *Bacillus subtilis* 168 with glutaminase activity'. *Enzyme Microb Technol*, **32**, 431–8.

MIWA, K., TSUCHIDA, T., KURAHASHI, O., NAKAMORI, S., SANO, K. and MONOSE, H. (1983). 'Construction of L-threonine ovrproducing strains of *Escherichia coli* K-12 using recombinant DNA techniques'. *Agric Biol Chem*, **47**, 2339–4.

MIYAKE, K. and KAKITA, S. (2009). 'A novel catalytic ability of γ-glutamylcysteine synthetase of *Escherichia coli* and its application in theanine production'. *Biosci Biotechnol Biochem*, **73**, 2677–83.

NAKAMURA, J., HIRANO, S., ITO, H. and WACHI, M. (2007). 'Mutations of the *Corynebacterium glutamicum* NCgl1221 gene, encoding a mechanosensitive channel homolog, induce L-glutamic acid production'. *Appl Environ Microbiol*, **73**, 4491–8.

NAKAYAMA, K., KITADA, S. and KINOSHITA, S. (1961). 'Studies on lysine fermentation I. The control mechanism on lysine accumulation by homoserine and threonine'. *J Gen Appl Microbiol*, **7**, 145–54.

NAKAYAMA, K., TANAKA, H., HAGINO, H. and KINOSHITA, S. (1966). 'Studies on lysine fermentation. Part V. Concerted feedback inhibition of aspartokinase and the absence of lysine inhibition on aspartic semialdehyde-pyruvate condensation in *Micrococcus glutamicus*'. *Agric Biol Chem*, **30**, 611–6.

NAKAYAMA, R., KUMAGAI, H., MARUYAMA, T., TOCHIKURA, T., UENO, T. and FUKAMI, H. (1981). 'Synthesis of γ-glutamylpeptides by γ-glutamylcysteine synthetase from *Proteus mirabilis*'. *Agric Biol Chem*, 2839–45.

NANBA, H., IKENAKA, Y., YAMADA, Y., YAJIMA, K., TAKANO, M., OHKUBO, K., HIRAISHI, Y., YAMADA, K. and TAKAHASHI, S. (1998a). 'Immobilization of *N*-carbamyl-D-amino acid amidohydrolase'. *Biosci Biotechnol Biochem*, **62**, 1839–44.

© Woodhead Publishing Limited, 2013

NANBA, H., IKENAKA, Y., YAMADA, Y., YAJIMA, K., TAKANO, M. and TAKAHASHI, S. (1998b). 'Isolation of *Agrobacterium sp.* strain KNK712 that produces *N*-carbamyl-D-amino acid amidohydrolase, cloning of the gene for this enzyme, and properties of the enzyme'. *Biosci Biotechnol Biochem*, **62**, 875–81.

NANBA, H., IKENAKA, Y., YAMADA, Y., YAJIMA, Y., TAKANO, M., OHKUBO, K., HIRAISHI, H., YAMADA, K. and TAKAHASHI, S. (1999). 'Immobillization of thermotolerant *N*-carbamyl-D-amino acid amidohydrolase'. *J Mol Catal B Enzymatic*, **6**, 257–63.

O'DONNELL, K. (2006). 'Aspartame and neotame'. In: *Sweeteners and Sugar Alternatives in Food technology.* Mitchell, H. L. (ed.) Blackwell Publishing, Oxford, pp 86–102.

OKAMOTO, K., KINO, K. and IKEDA, M. (1997). 'Hyperproduction of L-threonine by an *Escherichia coli* mutant with impaired L-threonine uptake'. *Biosci Biotechnol Biochem*, **61**, 1877–82.

ORELLANA, C. (2002). 'Immune system stimulator shows promise against tuberculosis'. *Lancet Infect Dis*, **2**, 711.

OZAKI, A., SHIBASAKI, T. and MORI, H. (1995). 'Specific proline and hydroxyproline detection method by post-column derivatization for high-performance liquid chromatography'. *Biosci Biotechnol Biochem*, **59**, 1764–5.

PARK, J. T. (1996). 'The murein sacculus'. In: Escherichia coli *and* Salmonella: Cellular and Molecular Biology. Neidhardt, F. C., Curtiss, R. I., Ingraham, J. L., Lin, E. C. C., Low, K. B., Magasanik, B., Reznikoff, W. S., Riley, M., Schaechter, M. and Umbarger, H. E. (eds), The American Society for Microbiology, Washington, DC, pp 48–57.

PRABHU, R., THOMAS, S. and BALASUBRAMANIAN, K. A. (2003). 'Oral glutamine attenuates surgical manipulation-induced alterations in the intestinal brush border membrane'. *J Surg Res*, **115**, 148–56.

REMUZON, P. (1996). '*Trans*-4-hydroxy-L-proline, a useful and versatile chiral starting block'. *Tetrahedron*, **52**, 13803–35.

SAKATO, Y. (1949). 'Studies on the chemical constituents of tea. Part III. On a new amide theanine'. *Nippon Nogeikagaku Kaishi*, **23**, 262–7.

SANO, K. and MITSUGI, K. (1978). 'Enzymatic production of L-cysteine from DL-2-amino-Δ^2-thiazoline-4-carboxylic acid by *Pseudomonas thiazolinophilum*: optimal conditions for the enzyme formation and enzymatic reaction'. *Agric Biol Chem*, **42**, 2315–21.

SANO, K., YOKOZEKI, K., TAMURA, F., YASUDA, N., NODA, I. and MITSUGI, K. (1977). 'Microbial conversion of DL-2-amino-Δ^2-thiazoline-4-carboxylic acid to L-cysteine and L-cystine: screening of microorganisms and identification of products'. *Appl Environ Microbiol*, **34**, 806–10.

SATO, T. and TOSA, T. (1993). 'Production of L-aspartic acid'. *Bioprocess Technol*, **16**, 15–24.

SHIBASAKI, T., MORI, H., CHIBA, S. and OZAKI, A. (1999). 'Microbial proline 4-hydroxylase screening and gene cloning'. *Appl Environ Microbiol*, **65**, 4028–31.

SHIBASAKI, T., MORI, H. and OZAKI, A. (2000). 'Enzymatic production of *trans*-4-hydroxy-L-proline by regio- and stereospecific hydroxylation of L-proline'. *Biosci Biotechnol Biochem*, **64**, 746–50.

SHIH, F. F. (1985). 'Analysis of glutamine, glutamic acid and pyroglutamic acid in protein hydrolysates by high-performance liquid chromatography'. *J Chromatogr*, **322**, 248–56.

SHIIO, I. and NAKAMORI, S. (1970). 'Microbial production of L-threonine. Part II. Production by α-amino-β-hydroxyvaleric acid resistant mutants of glutamate producing bacteria'. *Agric Biol Chem*, **34**, 448–56.

SHOULDERS, M. D. and RAINES, R. T. (2009). 'Collagen structure and stability'. *Annu Rev Biochem*, **78**, 929–58.

SOMMERSON, N. L. and PHILLIPS, T. (1965). 'Production method of glutamic acid'. JP patent S37-1695.

© Woodhead Publishing Limited, 2013

SUGIMOTO, M., TANAKA, A., SUZUKI, T., MATSUI, H., NAKAMORI, S. and TAKAGI, H. (1997). 'Sequence analysis of functional regions of homoserine dehydrogenase genes from L-lysine and L-threonine-producing mutants of *Brevibacterium lactofermentum*'. *Biosci Biotechnol Biochem*, **61**, 1760–2.

SUZUKI, H., KUMAGAI, H. and TOCHIKURA, T. (1986). 'γ-Glutamyltranspeptidase from *Escherichia coli* K-12: purification and properties'. *J Bacteriol*, **168**, 1325–31.

SUZUKI, H., KUMAGAI, H., ECHIGO, T. and TOCHIKURA, T. (1988). 'Molecular cloning of *Escherichia coli* K-12 *ggt* and rapid isolation of γ-glutamyltranspeptidase'. *Biochem Biophys Res Commun*, **150**, 33–8.

SUZUKI, H., KATAYAMA, T., YAMAMOTO, K. and KUMAGAI, H. (1995). 'Transcriptional regulation of tyrosine phenol-lyase gene of *Erwinia herbicola* AJ2985'. *Biosci Biotechnol Biochem*, **59**, 2339–41.

SUZUKI, H., IZUKA, S., MIYAKAWA, N. and KUMAGAI, H. (2002a). 'Enzymatic production of theanine, an "umami" component of tea, from glutamine and ethylamine with bacterial γ-glutamyltranspeptidase, *Euzyme Microb Technol*, **31**, 884–9.

SUZUKI, H., KAJIMOTO, Y. and KUMAGAI, H. (2002b). 'Improvement of the bitter taste of amino acids through the transpeptidation reaction of bacterial γ-glutamyltranspeptidase'. *J Agric Food Chem*, **50**, 313–8.

SUZUKI, H., IZUKA, S., MINAMI, H., MIYAKAWA, N., ISHIHARA, S. and KUMAGAI, H. (2003). 'Use of bacterial γ-glutamyltranspeptidase for enzymatic synthesis of γ-D-glutamyl compounds'. *Appl Environ Microbiol*, **69**, 6399–404.

SUZUKI, H., KATO, K. and KUMAGAI, H. (2004a). 'Development of an efficient enzymatic production of γ-D-glutamyl-L-tryptophan (SCV-07), a prospective medicine for tuberculosis, with bacterial γ-glutamyltranspeptidase'. *J Biotechnol*, **111**, 291–5.

SUZUKI, H., KATO, K. and KUMAGAI, H. (2004b). 'Enzymatic synthesis of γ-glutamylvaline to improve the bitter taste of valine'. *J Agric Food Chem*, **52**, 577–80.

TABATA, K. and HASHIMOTO, S. (2007). 'Fermentative production of L-alanyl-L-glutamine by a metabolically engineered *Escherichia coli* strain expressing L-amino acid α-ligase. *Appl Environ Microbiol*, **73**, 6378–85.

TABATA, K., IKEDA, H. and HASHIMOTO, S. (2005). '*ywfE* in *Bacillus subtilis* codes for a novel enzyme, L-amino acid ligase'. *J Bacteriol*, **187**, 5195–202.

TACHIKI, T., SUZUKI, H., WAKISAKA, S., YANO, T. and TOCHIKURA, T. (1986). 'Production of γ-glutamylmethylamide and γ-glutamylethylamide by coupling of baker's yeast preparations and bacterial glutamine synthetase'. *J Gen Appl Microbiol*, **32**, 545–8.

TAKAHASHI, S. (1986). 'Microbial production of D-*p*-hydroxyphenylglycine. In: *Biotechnology of Amino Acid Production*. Aida, K., Chibata, I., Nakayama, K., Takinami, K. and Yamada, H. (eds) Kodansha-Elsevier, Tokyo, pp 269–79.

TAKAHASHI, S., OHASHI, T., KII, Y., KUMAGAI, H. and YAMADA, H. (1979). 'Microbial transformation of hydantoins to N-carbamyl-D-amino acids'. *J Ferment Technol*, **57**, 328–32.

TAKAHASHI, S., NANBA, H., IKENAKA, Y. and YAJIMA, K. (2000). 'Application of a bioreactor to the D-amino acid production process'. *Nippon Nogeikagaku Kaishi*, **74**, 961–6. (in Japanese)

TAKINAMI, K., OKADA, H. and TSUNODA, T. (1963). 'Biochemical effects of fatty acid and its derivatives on L-glutamic acid fermentation. Part 1. Accumulation of L-glutamic acid in the presence of sucrose fatty acid ester.' *Agric Biol Chem*, **27**, 858–63.

UDAKA, S. (1960). 'Screening method for microorganisms accumulating metabolites and its use in the isolation of *Micrococcus glutamicus*'. *J Bacteriol*, **79**, 754–5.

UNNO, T., SUZUKI, Y., KAKUDA, T., HAYAKAWA, T. and TSUGE, H. (1999). 'Metabolism of theanine, γ-glutamylethylamide, in rats'. *J Agric Food Chem*, **47**, 1593–6.

WILK, S., MIZOGUCHI, H. and ORLOWSKI, M. (1978). 'γ-Glutamyl dopa: a kidney-specific dopamine precursor'. *J Pharmacol Exp Ther*, **206**, 227–32.

© Woodhead Publishing Limited, 2013

YAGASAKI, M. and HASHIMOTO, S. (2008). 'Synthesis and application of dipeptides; current status and perspectives'. *Appl Microbiol Biotechnol*, **81**, 13–22.

YAMADA, H., TAKAHASHI, S., KII, Y. and KUMAGAI, H. (1978). 'Distribution of hydantoin hydrolyzing activity in microorganisms'. *J. Ferment. Technol.*, **56**, 484–91.

YAMAMOTO, S., WAKAYAMA, M. and TACHIKI, T. (2005). 'Theanine production by coupled fermentation with energy transfer employing *Pseudomonas taetrolens* Y-30 glutamine synthetase and baker's yeast cells'. *Biosci Biotechnol Biochem*, **69**, 784–9.

YOKOGOSHI, H., KATO, Y., SAGESAKA, Y. M., TAKIHARA-MATSUURA, T., KAKUDA, T. and TAKEUCHI, N. (1995). 'Reduction effect of theanine on blood pressure and brain 5-hydroxyindoles in spontaneously hypertensive rats'. *Biosci Biotechnol Biochem*, **59**, 615–8.

YOKOGOSHI, H., KOBAYASHI, M., MOCHIZUKI, M. and TERASHIMA, T. (1998). 'Effect of theanine, γ-glutamylethylamide, on brain monoamines and striatal dopamine release in conscious rats'. *Neurochem Res*, **23**, 667–73.

YOKOZEKI, K. (2005). 'A powerful new tool for peptide production'. *Speciality Chemicals Magazine,* March, 42.

YOKOZEKI, K. and HARA, S. (2005). 'A novel and efficient enzymatic method for the production of peptides from unprotected starting materials'. *J Biotechnol*, **115**, 211–20.

YOSHIDA, A., TOMITA, T., KUZUYAMA, T. and NISHIYAMA, M. (2010). 'Mechanism of concerted inhibition of α2β2-type hetero-oligomeric aspartate kinase from *Corynebacterium glutamicum*'. *J Biol Chem*, **285**, 27477–86.

© Woodhead Publishing Limited, 2013

16

Production of microbial polysaccharides for use in food

I. Giavasis, Technological Educational Institute of Larissa, Greece

DOI: 10.1533/9780857093547.2.413

Abstract: Microbial polysaccharides comprise a large number of versatile biopolymers produced by several bacteria, yeast and fungi. Microbial fermentation has enabled the use of these ingredients in modern food and delivered polysaccharides with controlled and modifiable properties, which can be utilized as thickeners/viscosifiers, gelling agents, encapsulation and film-making agents or stabilizers. Recently, some of these biopolymers have gained special interest owing to their immunostimulating/therapeutic properties and may lead to the formation of novel functional foods and nutraceuticals. This chapter describes the origin and chemical identity, the biosynthesis and production process, and the properties and applications of the most important microbial polysaccharides.

Key words: biosynthesis, food biopolymers, functional foods and nutraceuticals, microbial polysaccharides, structure–function relationships.

16.1 Introduction

Microbial polysaccharides form a large group of biopolymers synthesized by many microorganisms, as they serve different purposes including cell defence, attachment to surfaces and other cells, virulence expression, energy reserves, or they are simply part of a complex cell wall (mainly in fungi). Many of them have been used for many years in the food industry and in human diet, either as an ingredient naturally present in food (e.g. in edible mushrooms or brewer's/baker's yeast) or mainly as a purified food additive recovered from microbial fermentation processes, as well as in pharmaceuticals (as bioactive compounds, or media for encapsulation and controlled drug release), cosmetics and other industrial applications, such as oil drilling and recovery, film formation, biodegradable plastic, tissue culture substrate, and other applications which go beyond the scope of this chapter (Sutherland, 1998). Their broad spectrum of applications is due to their diverse and modifiable properties as viscosifiers/thickeners, gelling and film-forming

© Woodhead Publishing Limited, 2013

agents, stabilizers, texturizers and emulsifiers. In addition, research in recent years has revealed that some microbial polysaccharides possess significant immunomodulating properties (anti-tumour, anti-inflammatory, antimicrobial), or hypocholesterolaemic and hypoglycaemic properties, thus making them perfect candidates for use in 'functional foods' or 'nutraceuticals' (Giavasis and Biliaderis, 2006). The world market for this type of foods is currently expanding and scientific interest in this field is growing, as consumers realize the importance of food to the quality of life (Hardy, 2000).

In comparison with polysaccharides isolated from plant sources (carrageenan, guar gum, modified starch, cereal glucans, etc.), which are also used for similar purposes, microbial polysaccharides have the advantages of well-controlled production processes in a large scale within a comparatively limited space and production time, stable chemical characteristics and unhindered availability in the market, as opposed to plant derivatives whose availability, yearly production and chemical characteristics often vary (Reshetnikov et al., 2001). However, in some cases, high production costs, low polysaccharide yields, and tedious downstream processing needed for isolation and purification are still a matter of concern for microbial processes, and appropriate strategies for bioprocess optimization have to be adopted (Kumar et al., 2007).

Apart from well-established microbial polysaccharides, such as xanthan, gellan, curdlan, pullulan or scleroglucan, many new polysaccharides from fungi, yeasts or bacteria emerge, as research on polysaccharide-producing strains continues and the properties and functionality of these biopolymers become better elucidated. The present chapter discusses the types and sources, the physicochemical and biological properties, and the applications of a number of well-established, commercial microbial polysaccharides, such as xanthan, gellan, alginate, curdlan, pullulan, scleroglucan and some less industrialized or less studied biopolymers such as elsinan, levan, alternan, microbial dextrans, lactic acid bacteria (LAB) polysaccharides and last but not least, mushrooms polysaccharides, such as lentinan, ganoderan, grifolan, zymosan, and so on.

16.2   Types, sources and applications of microbial polysaccharides

Microbial polysaccharides are found in many microorganisms, being part of the cell wall (such as fungal β-glucans), or serving as an energy reserve for the cell (such as polyhydroxybutyrate), or as a protective capsule or a slime-facilitating attachment to other surfaces (such as xanthan and gellan), the latter being characteristic of pathogens, especially plant pathogens (Giavasis et al., 2000). Cell wall polysaccharides are generally difficult to isolate and purify, as cell lysis and fractionation are needed to remove other cell impurities prior to alcohol precipitation, while extracellular polysaccharides

© Woodhead Publishing Limited, 2013

(EPS), which are excreted out of the cell, can generally be separated by filtration or centrifugation which removes cells, followed by precipitation. The main producers of microbial polysaccharides are fungi of the Basidiomycetes family, and several Gram negative (*Xanthomonas, Pseudomonas, Alcaligenes*, etc) and Gram positive (LAB) bacteria. Some yeasts may also synthesize polysaccharides in significant quantities, mostly belonging to the *Saccharomyces* genus (Giavasis and Biliaderis, 2006).

16.2.1   Bacterial polysaccharides

Xanthan is probably the most common bacterial polysaccharide used as a food additive owing to its viscosifying and stabilizing properties. It is produced by *Xanthmonas campestris*, a Gram negative plant pathogen which yields xanthan as a means of attachment to plant surfaces (Kennedy and Bradshaw, 1984). It was discovered in 1963 at Northern Regional Research Center of the United States Department of Agriculture (USDA) and commercial production for use in the food industry started soon after. Xanthan was approved by the United States Food and Drug Administration (FDA) for use in food additive without any quantity limitations, as it is non-toxic (Kennedy and Bradshaw, 1984). Xanthan comprises a linear (1,4) linked β-D-glucose backbone with a trisaccharide side chain on every other glucose at C-3, containing a glucuronic acid residue (1,4)-linked to a terminal mannose unit and (1,2)-linked to a second mannose of the backbone (Jansson *et al.*, 1975; Casas *et al.*, 2000). Its chemical structure is shown in Fig. 16.1. Its molecular weight ranges from 2000,000–20,000,000 Da (Daltons), depending on bioprocess conditions and the level of aggregation of individual chains (Casas *et al.*, 2000). Native xanthan is pyruvylated by 50% at the terminal mannose and acetylated at non-terminal mannose residues at C-6.

Xanthan has found multiple uses as a viscosifier and stabilizer in syrups, sauces, dressings, bakery products, soft cheese, restructured meat, and so on, where it is characterized by thermal stability even under acidic conditions, good freeze–thaw stability, and excellent suspending properties (Casas *et al.*, 2000; Sharma *et al.*, 2006; Palaniraj and Jayaraman, 2011). In bakery products xanthan gum is used to improve volume and texture (especially of gluten-free breads), water binding during baking and shelf life of baked foods, freeze–thaw stability of refrigerated doughs, to replace egg white in low calorie cakes and to increase flavour release and reduce synersis in creams and fruit fillings (Sharma *et al.*, 2006). In dressings, sauces and syrups xanthan gum facilitates emulsion stability to acid and salt and a stable viscosity over a wide temperature range; it imparts desirable body, texture and pourability and improved flavour release. In buttered syrups and chocolate toppings xanthan offers excellent consistency and viscosity and freeze–thaw stability (Sharma *et al.*, 2006; Rosalam and England, 2006). Xanthan is also an effective stabilizer and bodying agent in cream cheese where it improves

© Woodhead Publishing Limited, 2013

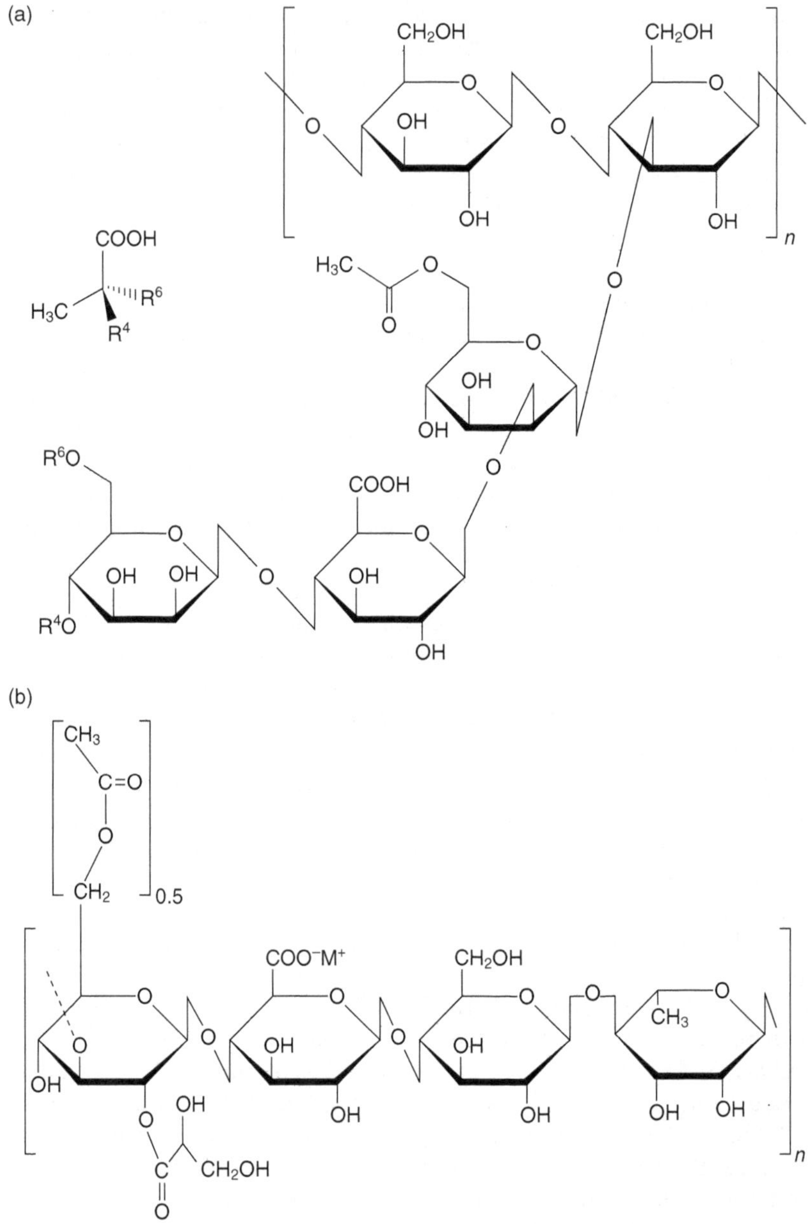

(a)

(b)

Fig. 16.1 Structures of some important bacterial polysaccharides. (a) xanthan repeating unit, (b) native gellan repeating unit (acetylated), (c) dextran repeating unit, (d) levan repeating unit.

© Woodhead Publishing Limited, 2013

(c)

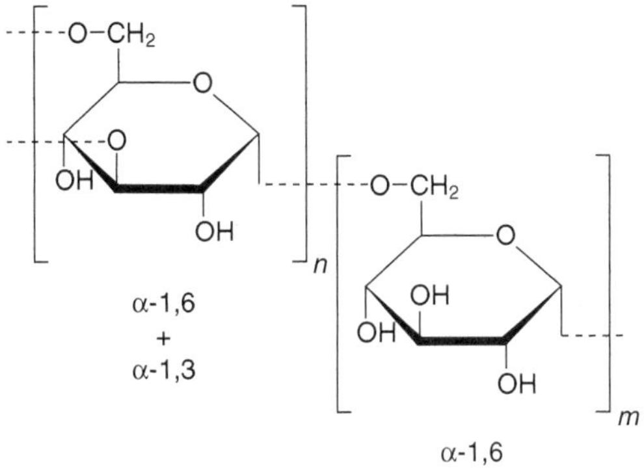

α-1,6
+
α-1,3

α-1,6

(d)

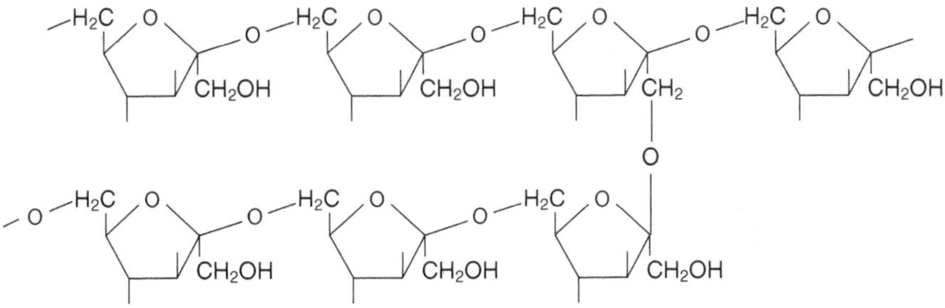

Fig. 16.1   *Continued*

flavour, shelf life, heat-shock protection and reduces syneresis, and is also suitable for beverages as it is soluble and stable at low pH and improves the suspension of insoluble partices (e.g. in fruit juices) and the body and mouthfeel of the products (Sharma *et al.*, 2006; Palaniraj and Jayaraman, 2011).

Acetan (also known as xylinan) is another EPS structurally related to xanthan and is produced by *Acetobacter xylinum*, a strain that is used in the food industry for the production of a sweet confectionery and vinegar (van Kranenburg *et al.*, 1999). It is an anionic heteropolysaccharide with a MW of approximately 1000,000 Da, consisting of a pentasacchride main chain where the (1, 2)-D-mannose residue of the main chain and the (1,3,4)-D glucose residue are *O*-acetylated (Ridout *et al.*, 1994, 1998; Ojinnaka *et al.*, 1996).

The same microorganism is the best industrial producer of microbial cellulose, a β-(1,4)-linked glucopyranose biopolymer with a low degree of branching or no branching at all, which lacks the hemicellulose, pectin and lignin moieties of plant-derived cellulose. It was granted a 'GRAS' (generally recognized as safe) status by FDA in 1992 for food applications (Khan

© Woodhead Publishing Limited, 2013

et al., 2007). *Acetobacter xylinum* cellulose also differs from plant-derived cellulose in that it has high purity and crystallinity, gel strength, moldability and increased water-holding capacity (Jonas and Farah, 1998; Iguchi *et al.*, 2000; Khan *et al.*, 2007). It is used mainly in Asian speciality food '*nata*', for instance '*Nata de Coco*', a jelly food with coconut water used in confectionery and desserts (Iguchi *et al.*, 2000; Khan *et al.*, 2007). Other potential food applications of microbial cellulose include dressings, sauces, icings, whipped toppings and aerated desserts, frozen dairy products where it functions as a low-calorie additive, thickener, stabilizer and texture modifier (Okiyama *et al.*, 1993; Khan *et al.*, 2007).

Another plant pathogen, *Sphingomonas paucimobilis* (formerly *Pseudomonas elodea*), produces gellan, an EPS of approximately 500,000 Da on average, which facilitates cell attachment to plant surfaces, such as water lilies, the plants from which it was first isolated (Kang *et al.*, 1982; Pollock, 1993; Giavasis *et al.*, 2006). Native gellan is composed of a linear anionic tetrasaccharide repeat unit containing two molecules of D-glucose, one of D-glucuronic acid and one of L-rhamnose, as well as glucose-bound acyl substituents (one L-glycerate and two O-acetate substituents per two repeat units on average) (Jay *et al.*, 1998). Its structure is depicted in Fig. 16.1. In its industrial form, gellan gum is usually deacylated after an alkaline thermal treatment, which transforms the soft elastic gels of native gellan to hard, brittle, thermoreversible, acid-tolerant, transparent gels, especially after addition of divalent cations (Jay *et al.* 1998; Giavasis *et al.*, 2000; Rinaudo and Milas, 2000; Rinaudo, 2004). Commercial gellan is available in three forms with distinct degree of acetylation: no, low and high acyl content corresponding to the brand names of Gelrite®, Kelcogel® F and Kelcogel® LT100 (Fialho *et al.*, 2008). Gellan has found several food applications as viscosifier, stabilizer, gelling agent in dessert gels, icings, sauces, puddings and restructured foods, as a bodying agent in beverages, or as an edible film and coating agent when blended with other gums (Giavasis *et al.*, 2000; Fialho *et al.*, 2008; Stalberg *et al.*, 2011). Other species of the genus *Sphingomonas* produce other biopolymers structurally related to gellan, such as wellan, rhamsan, diutan and gums S-88 and S-657 (all with different acylation patterns compared to gellan), which lack the strong gelling properties of deacylated gellan, but perform well as suspension agents with high resistance to shear stress and have found several applications in the food industry (Kang and Pettitt, 1993; Rinaudo, 2004; Fialho *et al.*, 2008).

Dextrans are some of the most common bacterial polysaccharides, and some of the first to be produced on industrial scale, with applications in foods, as well as pharmaceuticals, separation technology and so on (Glicksman, 1982; Alsop, 1983; Leathers, 2002a). Although many bacterial strains belongining to the genera *Leuconostoc, Lactobacillus, Streptococcus, Acetobacter,* and *Gluconobacter* are capable of synthesizing dextrans, dextran is industrially produced by *Leuconostoc mesenteroides* strains grown on a

© Woodhead Publishing Limited, 2013

sucrose medium via the action of dextran sucrase (a glucosyltransferase) which catalyses sucrose to form D-fructose and D-glucose and transfers the latter to an acceptor molecule where polymerization takes place. Purely enzymatic (bioconversion) processes, involving polymerization via dextran-sucrases, have also been developed (Jeanes *et al.*, 1954; Brown and McAvoy, 1990; Khalikova *et al.*, 2005; Khan *et al.*, 2007). Microbial dextran was initially identified and characterized after attempts to solve the problem of thickening or ropeyness that occurred in sugar juices and wines in the 1980s, but soon its water-binding properties led to its utilization in several applications as a food thickener and viscosifier (Glicksman, 1982; Vandamme and Soetaert, 1995).

Commercial dextran is produced by the lactic acid bacterium *L. mesenteroides* strain NRRL-B512 and consists of a α-(1,6)-D-glucan backbone (by 95% or less) and α-(1,3)- branches (by 5% or more) (Leathers, 2002a). Its chemical structure is shown in Fig. 16.1. Crude dextran has relatively high MW, around or above 1000,000 Da, although much higher MW values have also been reported, probably caused by the tendency of dextran molecules to aggregate (Khalikova *et al.*, 2005). In industrial processes, dextran is partly hydrolysed (by acid or enzymatic hydrolysis) and fractionated, yielding a wide range of dextrans with different MW values (Khalikova *et al.*, 2005). When used in food applications MW ranges from 15,000 to 90,000 Da (Glicksman, 1982; Kumar *et al.*, 2007). Food applications of dextrans include confectionery products where they act as stabilizers and bodying agents (e.g. in puddings), as crystallization inhibitors (e.g. in ice cream), or as moisture retention agents and viscosifiers in food pastes (Khan *et al.*, 2007). Dextrans from *L. mesenteroides* or other lactic bacteria (e.g. *Lactobacillus curvatus*) have also been used as texturizers in bread, especially gluten-free bread, where they enhance water-holding capacity, elasticity and specific volume of bread (Ruhmkorf *et al.*, 2012). The α-(1,6) linkages of the molecule are resistant to depolymerization, which results in the slow digestion of dextran in humans (Glicksman, 1982).

Alternan is another glucan similar to dextran, yet with unusual structure. It is synthesized mainly by *L. mesenteroides* strain NRRL B-1355, which is grown in a complex sucrose-based medium, in a process that resembles that of dextran production and is mediated through alternan sucrases (Cote and Robyt, 1982; Raemaekers and Vandamme, 1997). Although several *L. mesenteroides* strains that produce alternan also synthesize dextrans as undesirable contaminants, genetically engineered strains producing only alternan have been isolated (Kim and Robyt, 1994; Monsan *et al.*, 2001). The unique characteristic of alternan is the alternating structure of α-(1,6) and α-(1,3) linkages, with approximately 10% branching through 3,6-di-substituted D-glucosyl units (Seymour and Knapp 1980; Leathers *et al.*, 2003).

Several *Agrobacterium* and *Rhizobium* species, can each produce exopolysaccharides such as curdlan, a neutral 1,3-β-D-glucan with a low MW

© Woodhead Publishing Limited, 2013

(around 74,000 Da) (Sutherland, 1998). Curdlan, along with xanthan and gellan, has been approved for use in food by FDA and it is industrially produced from *Agrobacterium sp.* ATCC 31749, or *sp.* NTK-u, or *Agrobacterium radiobacter* (Jezequel 1998; Zhan *et al.*, 2012). Curdlan is a homopolysaccharide formed in the stationary phase following depletion of nitrogen and is insoluble in cold water but can be dissolved in hot water or in dimethylsulphoxide, forming stable gels. Many food applications of curdlan utilize its thermo-irreversible gel form, its stability during freeze–thawing cycles or during deep-fat frying, its lipid-mimicking properties and the fact that it provides a pleasant mouthfeel compared to other biopolymers (Lo *et al.*, 2003; McIntosh *et al.*, 2005). Curdlan has been used in various food products, mainly freezable and low-calorie foods, since it is not degraded in the gastrointestinal tract (McIntosh *et al.*, 2005). In Japan, curdlan is commonly used in food as a texturizer and water-holding agent in pasta, tofu, jellies, fish pastes, and reconstituted food and confectionery (Sutherland, 1998; Laroche and Michaud, 2007). In addition to the above properties and applications, the sulphated derivatives of curdlan have shown important immunostimulatory, antitumour and antiviral properties which have been reported extensively (Goodridge *et al.*, 2009; Zhan *et al.*, 2012) and might be exploited in the formulation of novel neutraceuticals.

Rhizobium and *Agrobacterium* species, as well as microorganisms such as *Alcaligenes faecalis* var. *myxogenes* and *Pseudomonas* sp. also produce succinoglycan, an acidic biopolymer which is commercialized and used mainly in oil recovery, but is also suitable for food applications for its thickening and stabilizing properties, even under extreme process conditions (Freitas *et al.*, 2011; Moosavi-Nasab *et al.*, 2012). It comprises large (octasaccharide) repeating units of D-glucose and D-galactose and carries *O*-acetyl groups, *O*-succinyl half-esters and pyruvate ketals as substituents, which form a molecule of relatively high MW (in the order of 10^6 Da) (Ridout *et al.*, 1997; Sutherland, 2001). Natural and chemically modified succinoglycans show high stability under high temperature and pressure, high/low pH and high shear stress (Moosavi-Nasab *et al.*, 2012).

Many other extracellular polysaccharides (EPS) have been isolated from a large number of LAB, namely *Lactobacillus, Streptococcus, Lactococcus, Pediococcus,* as well as *Bifidobacterium* sp. and *Weissella* strains found in fermented dairy products (De Vuyst and Degeest, 1999, Notararigo *et al.*, 2012). They excrete linear or branched biopolymers of galactopyranose, glucopyranose, fructopyranose, rhamnopyranose or other residues (e.g. *N*-acetylglucosamine, *N*-acetylgalactosamine, or glucuronic acid), characterized by a large range of MW values (10^4–10^6 kDa); for instance, homopolysaccharides (α-glucans or β-glucans) such as reuteran from *Lactocccbacillus reuteri,* mutan from *Streptococcus mutans,* polygalactan from *Lactococcus lactis* H414, and heteropolysaccharides such as kefiran from *Lactobacillus hilgardii,* and several other EPS from *Lactobacillus bulgaricus, Lactobacillus helveticus, Lactobacillus rhamnosus, Lactococcus lactis* NIZO-B39 or

© Woodhead Publishing Limited, 2013

NIZO-B891 and *Streptococcus thermophillus* (Ruas-Madiedo *et al.*, 2002; Tieking *et al.*, 2005; Patel *et al.*, 2010). Most LAB produce polysaccharides extracellularly from sucrose by glycansucrases or intracellularly by glycosyltransferases from sugar nucleotide precursors (Ruas-Madiedo *et al.*, 2002). These molecules and the producer strains have been thoroughly studied, since they can improve rheological and textural properties in dairy and other food products where LAB are already used (Laws *et al.*, 2001). Also, EPS from LAB such as kefiran have been used in the formulation of edible films with various plasticizers (Ghasemlou *et al.*, 2011). Moreover, some of these slimy homo/heteropolymers are associated with anticarcinogenic and immunomodulating properties, or reported to act as prebiotics promoting the growth of the producer strain or other LAB (Oda *et al.*, 1983; Adachi, 1992; Nakajima *et al.*, 1995; Sreekumar and Hosono, 1998; Ruas-Madiedo *et al.*, 2002), which could be great assets in formulating novel foods with bioactivity and health-promoting properties.

In spite of the fact that LAB and their products are considered GRAS and acceptance and incorporation of these polysaccharides in traditional and new functional food products should be easy, the substantially low production yields of these biopolymers, especially of the heteropolysaccharides (i.e. 50–1000 mg l^{-1} compared to 15–25 g l^{-1} of xanthan gum), remain a serious drawback for their broad commercial application in foods, which could be overcome with the aid of genetic engineering and better understanding of microbial physiology (Laws *et al.*, 2001). An exception to these low yields are two types of homobiopolymers, a glucan and a fructan synthesized by *Lb. reuteri* strain LB 121 which can reach a concentration of nearly 10 g l^{-1} during fermentation on a sucrose-based medium (van Geel-Schutten *et al.*, 1999). Most applications of LAB polysaccharides are related to fermented dairy food, beverages and sour doughs where the specific LAB are either part of the natural fermenting microflora or inoculated in purified form in order to contribute to the improvement in texture and viscosity, owing to the synthesis of the above biopolymers (Elizaquível *et al.*, 2011; Notararigo *et al.*, 2012). Also in another application, the *in situ* production of EPS from LAB cultures was useful in the production of low-fat Mozzarella cheese where they improved moisture retention (Bhaskaracharya and Shah, 2000).

Another class of bacterial polysaccharides is levans, extracellular homopolysaccharides of D-fructose (fructans). These biopolymers are characterized by β-(2,6)-fructofuranosidic bonds in their main chain and β-(2,1)-linked side chains (Huber *et al.*, 1994). A typical levan structure is illustrated in Fig. 16.1. They are produced by several bacteria, such as *Streptococcus salivarius* (a bacterium of the oral flora*), Lactobacillus sanfranciscensis, Bacillus subtilis* and *Bacillus polymyxa, Acetobacter xylinum, Gluconoacetobacter xylinus, Microbacterium levaniformans, Zymomonas mobilis* and a few more microorganisms which express the biosynthetic enzyme levan sucrase in sucrose-rich culture media (Newbrun and Baker, 1967; Han, 1990;

© Woodhead Publishing Limited, 2013

Keith *et al.*, 1991; Yoo *et al.* 2004; Notararigo *et al.*, 2012). Alternatively, they can be synthesized enzymatically by levan sucrases using sucrose as subtrate (Jang *et al.*, 2001; Castillo and Lopez-Munguia, 2004). Levans often reach a very high MW value (over 10^6–10^7 Da), while low MW levans can also be produced, depending on the microorganism used and the fermentation/ biocatalysis conditions (Newbrun and Baker, 1967; Calazans *et al.*, 2000; Shih *et al.* 2005).

The properties and potential applications of levan in food resemble those of dextrans, but levans from *Aerobacter levanicum* and *Z. mobilis* (an industrial ethanol-producing strain) have also exhibited immunostimulating and anti-tumour properties (Calazans *et al.*, 2000; Bekers *et al.*, 2002; Yoo *et al.*, 2004), as well as hypolipidaemic and hypocholesterolaemic effects (Kang *et al.*, 2004). Most food applications of levans utilize their texturizing, and water and air retention properties in doughs and breads, as well as their ability to act as a stabilizer, thickener, osmoregulator, cryoprotector, sweetener and a carrier of flavours and fragrances (Han, 1990; Bekers *et al.*, 2005; Tieking *et al.*, 2005; Kang *et al.*, 2009). Levan from *Lactobacillus sanfranciscensis* was reported to affect dough rheology and texture positively (Waldherr and Vogel, 2009). Also, Huber *et al.* (1994) proposed the use of levan as an ingredient for forming edible films. These are too brittle when levan is the sole ingredient, but when blended with other polymers, such as glycerol, elastic and extrudable films can be formed (Barone and Medynets, 2007). Furthermore, levan has exhibited anti-obesity and hypolipidaemic effects as well as antitumour and anti-radiation protective properties (Han, 1990; Kang *et al.*, 2004; Yoo *et al.*, 2004; Bekers *et al.*, 2005; Combie, 2006) which could be exploited in novel nutraceuticals.

Bacterial alginate is another biopolymer with food applications. It is currently produced from the marine brown algae on the industrial scale thanks to the comparatively low cost of this process, but can also be produced by liquid cultures of bacteria such as *Azotobacter vinelandii*, *Azotobacter chroococcum* and *Pseudomonas aeruginosa*, with *Azotobacter* being preferable for microbial alginate production, owing to the potential pathogenicity of *P. aeruginosa*. (Sabra *et al.*, 2001; Remminghorst and Rehm, 2006). Alginate is an acidic copolymer of β-D-mannuronic acid (M) and α-L-guluronic acid (G), with varying content of G and M and chain length (although alginate from *P. aeruginosa* lacks the G blocks). Its molecular weight is in the order of 10^6 Da (Sabra *et al.*, 2001; Celik *et al.*, 2008; Freitas *et al.*, 2011). Homopolymeric M and G groups are normally interconnected with alternating residues of both acids (MG groups) in *Azotobacter* and brown algae. Microbial alginates are acetylated on some mannuronic acid residues, which is a main difference from alginate derived from algae (Sabra *et al.*, 2001).

Alginic acid and its sodium calcium and, potassium salts are safe for use in food (GRAS) as thickeners, stabilizers, or gelling agents. They are usually added to jams, confectionery (candies, ice cream, milk shakes), beverages,

© Woodhead Publishing Limited, 2013

soups and sauces, margarine, liquors, structured meat and fish, as well as dairy products (Sabra *et al.*, 2001; Giavasis and Biliaderis, 2006). Calcium alginate is also a common medium for cell and enzyme immobilization and microencapsulation of bioactive molecules and can be used as an edible film coating (Freitas *et al.*, 2011). Recently, several physiological effects of alginate have been disclosed, including dietary fibre effects, anti-inflammatory (anti-ulcer) and immunostimulating properties, as well detoxifying properties (Khotimchenko *et al.*, 2001) which may establish this biopolymer as a functional ingredient in the manufacture of functional foods or nutraceuticals. In fact, a bioactive food additive ('Detoxal') containing calcium alginate can reduce lipid peroxidation products and normalize the concentrations of lipids and glycogen in the liver, while it has also shown antitoxic effects, for example against tetrachlorometan-induced hepatitis in mice, or via adsorption and elimination of heavy metals in humans (Khotimchenko *et al.*, 2001).

16.2.2 Fungal polysaccharides

One of the most common and well-studied fungal polysaccharides is pullulan. It was back in 1958 when Bernier (1958) observed that *Pullularia* (now *Aureobasidium*) *pullulans*, a yeast-like fungus, can synthesize an extracellular polysaccharide, a neutral glucan which was called pullulan a year later (Bender *et al.*, 1959). It was first commercialized by Hayashibara Biochemical Laboratories (Japan) and protected by patents for several years (Sugimoto, 1978; Singh *et al.*, 2008). Pullulan is a white, tasteless, water-soluble homopolymer of glucose consisting of repeating units of maltotriose with a regular alternation of two α-(1,4) linkages, and one α-(1,6) linkage on the outer glucosyl unit (ratio 2 : 1) as periodate oxidation, permethylation and infra-red spectrum analysis suggest (Bender *et al.*, 1959; Catley, 1970; Taguchi *et al.*, 1973a; Sandford, 1982; Le Duy *et al.*, 1988, Leathers, 2002b), although other structures comprising α-maltotetraose units and (1,3)-linked residues have also been proposed (Ueda *et al.*, 1963; Taguchi *et al.*, 1973b). This variance is not surprising since several extracellular polysaccharides have been isolated from the same microorganism (Sandford, 1982).

Figure 16.2 depicts a typical pullulan structure. The MW of pullulan is generally in the range 10,000–1000,000 Da with a average MW of 360–480 KDa, depending on process conditions and the strain used (McNeil and Harvey, 1993; Cheng *et al.* 2011), but the two main industrial products from Hayashibara Company Ltd, a food grade pullulan (PF) and a deionized pullulan (PI), have a mean molecular weight of 100,000 Da (PF-10 and PI-10), or 200,000 Da (PF-20 and PI-20) (Singh *et al.*, 2008). Pullulan can also form oil-resistant, water-soluble, odourless, thin and transparent films with low oxygen permeability which can act as edible food coatings that improve self life (e.g. of fruits and nuts) (Leathers, 2003;

© Woodhead Publishing Limited, 2013

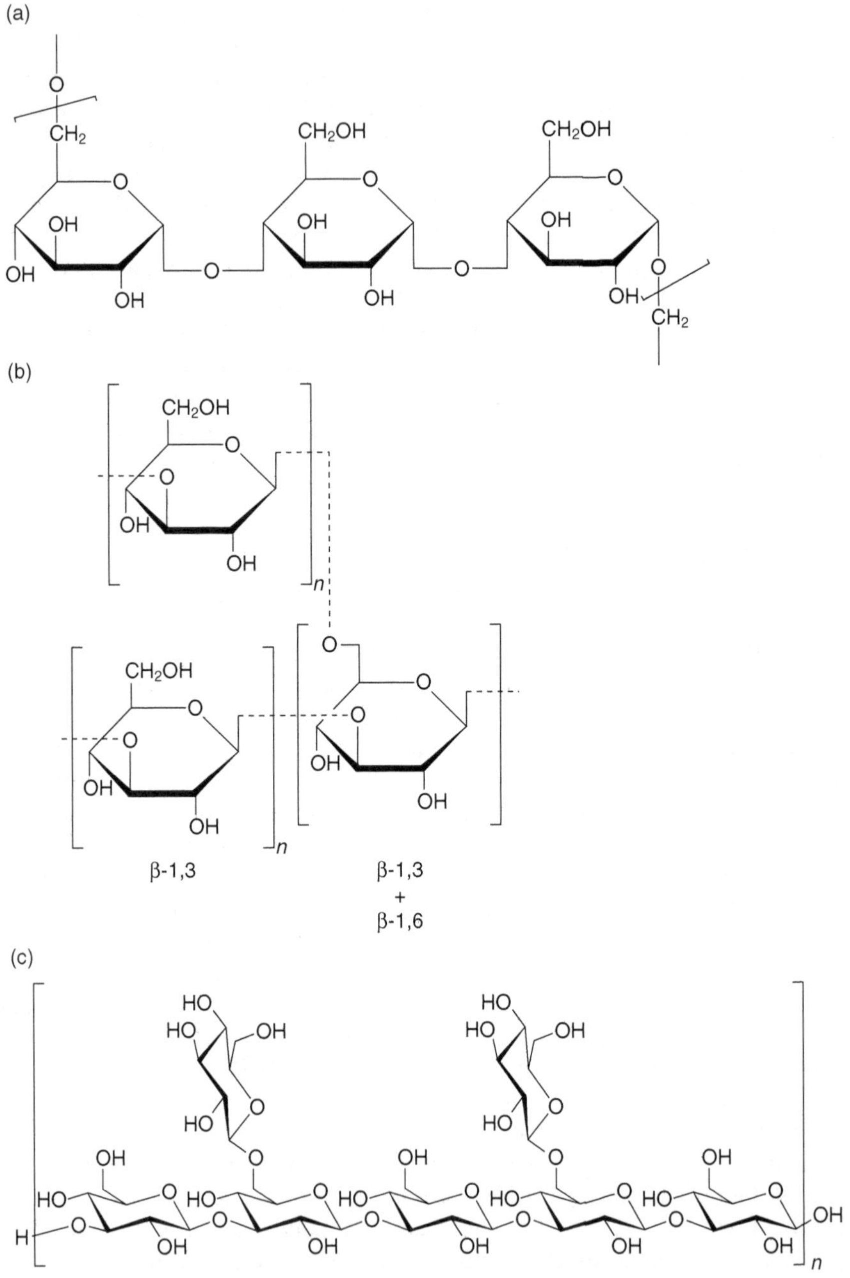

(a)

(b)

β-1,3

β-1,3
+
β-1,6

(c)

Fig. 16.2   Structures of some important fungal polysaccharides. (a) Pullulan repeating unit, (b) schizophyllan (sizofiran) repeating unit, (c) lentinan repeating unit.

© Woodhead Publishing Limited, 2013

Gounga *et al.*, 2008; Cheng *et al.*, 2011). These applications have been marketed in Japan, but are apparently limited elsewhere (Sutherland, 1998; Leathers, 2003).

Pullulan has been proposed as a replacement for starch in solid and liquid food, especially pastas and baked products, where it strengthens food consistency, moisture and gas retention and dispersibility. In addition, it can be used as a stabilizer/viscosifier in sauces and beverages, offering low but stable viscosity with temperature and pH changes, or as a binder in food pastes and confectionery products where its adhesive properties may be exploited (e.g. for adhesion of nuts on cookies). It has also been applied as a dietary fibre and as a prebiotic to promote growth of *Bifidobacterium* spp. owing to its partial degradation to indigestible short-chain oligomers by human salivary α-amylase (Okada *et al.*, 1990; Singh *et al.*, 2008; Cheng *et al.*, 2011). In food packaging, pullulan–polyethylene films could be used to offer high water and oxygen resistance, and better rigidity and strength comparable to expanded polystyrene films (Paul *et al.*, 1986).

The fungus *Elsinoe leucospila*, isolated from a white spot of tea leaves, produces elsinan, an extracellular, linear α-D-glucan composed of glucose units linked by approximately 70% (1,4)-linkages (maltotriose) and 30% (1,3)-linkages (maltotetraose) (Sandford, 1982; Misaki *et al.*, 1978, 1982). The proposed structure of elsinan, as determined by methylation and periodate oxidation studies, as well as partial acid hydrolysis, acetolysis and enzymic degradation by glucanases, is similar to that of pullulan, which has (1,6)-links instead of the (1,3)-links in elsinan (Misaki *et al.*, 1978, 1982; Misaki, 2004). Like pullulan, elsinan was manufactured by Hayashibara Biochemical Laboratories (Japan), but despite its viscosifying and film-forming properties it has found little application as a food additive so far (Misaki, 2004). However, there is a significant potential for food applications of elsinan owing to its dietary fibre properties (i.e. reduction of serum cholesterol in hypercholesterolemic rats) and its ability to form oxygen impermeable edible films and coatings, and viscous solutions which are stable over a wide range of pH (3–11), temperature (30–70°C) and salt concentrations (Misaki, 2004). It can also be used in food packaging as a biodegradable film (Yokobayashi and Sugimoto, 1979; Sandford, 1982). In experiments with oleic acid and fresh fish packed with elsinan films, no colorization caused by self-oxidation occurred over 3 and 4 months, respectively, while acidic conditions (pH 1 to 4) did not affect the stability of these films (Sandford, 1982). Moreover, its cholesterol-lowering and anti-tumour properties can be utilized in the formulation of novel functional foods (Shirasugi and Misaki, 1992; Misaki, 2004). Additionally, Shirasugi and Misaki (1992) have isolated a cell wall polysaccharide from *Elsinoe leucospila*, which exhibited antitumour activity. This polymer, obtained from cold alkali cell wall extract, was a β-D-glucan with a main chain of eight (1,3)-glucose residues and single β-D-glucosyl side groups at the O-6 position.

© Woodhead Publishing Limited, 2013

Scleroglucan is another extracellular glucose homopolysaccharide with a high MW (about or over 1000 000 Da) with a β-(1,3) linked backbone, where a single D-glucosyl side group is bound via a β-(1,6) linkage to every third or fourth unit of glucose in the main chain (Holzwarth, 1985; Giavasis et al., 2002). The main producer microorganisms are the filamentous phytopathogenic fungi *Sclerotium glucanicum* and *Sclerotium rolfsii*. Scleroglucan was first brought into the market by Pillsbury Co (Minneapolis, USA), followed by CECA S.E. (France) and Satia S.A. (France), serving mainly as a viscosifier in chemically enhanced oil recovery, where it performs better than xanthan (Holzworth, 1985; McNeil and Harvey, 1993). In the food industry, scleroglucan would be ideal for the stabilization of dressings, sauces, ice creams and other desserts, as well as low calorie or thermally processed and acidic products (sterilization, salts and acids do not affect its stabilizing capacity), but its use in food is not yet approved in Europe and the USA (Survase, 2007; Schmid et al., 2011). Nevertheless, there are several Japanese patents on the application of scleroglucan as a stabilizer and thickener in frozen or heat-treated food, such as steamed foods and bakery products (Schmid et al., 2011), showing the interest that exists for such applications.

Vinarta et al. (2006) investigated the stabilizing properties of scleroglucan in cooked starch pastes and showed that scleroglucan offered high water retention and significantly reduced syneresis during refrigeration, and this effect was even more pronounced when scleroglucan was blended with corn starch before being added. Scleroglucan could also be utilized in the formation of edible films and tablets for neutraceuticals, owing to its chemical stability, biocompatibility and biodegradability (Grassi et al., 1996; Coviello et al., 1999). Although it does not act as a surfactant, it can stabilize oil-in-water emulsions, by preventing coalescence (Sandford, 1982). Additionally, this β-glucan has shown significant antitumour and antiviral activity (Jong and Donovick, 1989; Pretus et al., 1991; Mastromarino et al., 1997), which could be a great asset in designing functional foods.

Two similar polysaccharides (only of lower MW than scleroglucan), namely schizophyllan (also called sizofiran) and lentinan, are produced by the edible mushrooms *Schizophyllum commune* and *Lentinus edodes*, respectively (Giavasis et al., 2002). They are two of the most well-studied immunostimulating microbial β-(1,3)-D-glucans, while *L. elodes*, is the most common edible mushroom in Japan (Maeda et al., 1998). Their chemical structure is illustrated in Fig. 16.2. Both lentinan and schizophyllan are characterized by a main chain of β-(1,3)-D-glucose residues to which β-(1,6)-D-glucose side groups are attached (one branch to every third main chain unit), and an average molecular weight of about 500 000 Da (Misaki et al., 1993). Their addition to food in purified form has not been commercialized yet, in contrast to several pharmaceutical applications where they are used (Giavasis and Biliaderis, 2006), but as they both come from edible

© Woodhead Publishing Limited, 2013

mushrooms, they have a great potential for use in novel foods and nutraceuticals.

(1,3)(1,6)-β-D glucans from *L. edodes* were used as a replacement for a portion of the wheat flour in baked foods such as cakes, in an attempt to produce a novel functional food with low calories and high fibre content (Kim *et al.*, 2011). In this application *L. edodes* glucans from mushroom powder which was incorporated in batter improved pasting parameters and increased batter viscosity and elasticity, without having any adverse effects on air holding capacity (volume index) or hardness compared to the control, when used at concentrations of 1 g pure glucan per 100 g of cake. Reduced volume and increased hardness were only observed when glucan concentration was 2% or more (Kim *et al.*, 2011). In similar studies, *L. edodes* glucan from unmarketable mushrooms was added to noodles as a partial wheat flour replacement and resulted in a fibre-rich functional food with antioxidant and hypocholesterolaemic effects and improved quality charasterictics (Kim *et al.*, 2008, 2009). In another study (Kim *et al.*, 2010) *L. edodes* mushroom powders (LMP) rich in β-glucans were utilized effectively as oil barriers and texture-enhancing ingredients in frying batters.

Several other mushrooms, many of which are part of the traditional diet in East Asian (especially Chinese and Japanese) or South American populations, contain a number of polysaccharides, mainly β-D-glucans, which have been associated with healthy diet have fortified the immune system of the consumers (Hobbs, 1995; Wasser, 2002; Giavasis and Biliaderis, 2006; He *et al.*, 2012) and could find novel applications as functional food ingredients. *Agaricus blazei*, for instance, is a well-known edible and medicinal mushroom originating from Brazil, containing several antitumour polysaccharides in its fruit body (Mizuno *et al.*, 1990). The water-soluble fraction of these polysaccharides includes a β-(1,6);β-(1,3) glucan an acidic β-(1,6);α-(1,3) glucan, and an acidic β-(1,6);α-(1,4) glucan. Unlike most known glucans, *A. blazei* glucans have a main chain of β-(1,6) glycopyranose, instead of the common β-(1,3) linked main chain (Mizuno *et al.*, 1990). The fruit body also contains a water-soluble proteoglucan with a α-(1,4) glucan backbone and β-(1,6) branches at a ratio of 4 : 1. It has a MW of 380,000 Da and it consists mainly of glucose (Fujimiya *et al.*, 1998). Moreover, the water-insoluble fraction of *A. blazei* fruit body, which has also shown immunostimulating activity, includes two heteroglucans consisting of glucose, galactose and mannose, one consisting of glucose and ribose, a xyloglucan and a proteoglucan (Cho *et al.*, 1999; Mizuno, 2002). Notably, submerged cultures of *A. blazei* synthesize somewhat different (medicinal) polysaccharides compared to those from the mushroom fruit body (Mizuno, 2002). Among these biopolymers, some of which are covered by patents (Hikichi *et al.*, 1999; Tsuchida *et al.*, 2001), an extracellular protein–polysaccharide polymer with significant antitumour properties and a high MW (1000 000–10 000 000 Da) has been isolated. The sugar components of this

© Woodhead Publishing Limited, 2013

biomolecule include mainly mannose, as well as glucose, galactose and ribose (Mizuno, 2002).

Ganoderma lucidum is another medicinal mushroom belonging to the *Basiomycetes* family, which has been used for many years in traditional East Asian medicine as a dry powder, or consumed as a hot water extract (in a type of bitter mushroom tea). The bioactive component of the fungi, termed 'ganoderan', is a typical β-(1,3) glucan branching at C-6 with β-(1,6) glucose units and with a high (Bao *et al.*, 2002) or low (Misaki *et al.*, 1993) degree of branching, which can be isolated either from the water-soluble fraction of the fruit body (Misaki *et al.*, 1993; Bao *et al.*, 2002), or from the filtrates of liquid cultures of *G. lucidum* mycelia. The latter is a water-soluble β-D-glucan with a MW of $1.2–4.4 \times 10^6$ Da, degradable by pectinases and dextranases (Lee *et al.*, 1996; Xie *et al.*, 2012). Apart from the above glucans, a few more heteroglucans and proteoglucans are also present in fruit bodies of *G. lucidum* (Eo *et al.*, 2000). Kozarski *et al.* (2011, 2012) studied the antioxidant and immnomodulatory properties of glucans from *G. lucidum* and *Ganoderma applanatum* with respect to their potential application in food, and reported a significant free radical scavenging activity and protective action against lipid peroxidation, as well as significant enhancement of interferone synthesis in human blood cells.

Other antioxidant and immunostimulating basidiomycetal polysaccharides from edible mushrooms include krestin, a commercialized proteoglucan synthesized by the mushroom *Coriolous versicolor* (also called *Trametes versicolor*) which has a β-(1,3)-D-glucan moiety (Ooi and Liu, 2000) and grifolan, a gel-forming β-(1,3)-D glucan with β-(1,6) branches at every third glucopyranosyl residue, elaborated by the edible fungus *Grifola frondosa* (Adachi *et al.*, 1998; Laroche and Michaud, 2007), which could also be utilized as a food grade functional ingredient.

Kozarksi *et al.* (2012) also reported significant antioxidant properties of polysaccharides extracted from *T. versicolor* and *L. edodes* mushrooms, which exhibited chelating properties and inhibited lipid oxidation. The latter were correlated with the presence of an α-glucan and a phenolic (mainly tyrosine and ferrulic acid) moiety linked to the main β-glucan backbone by covalent bonds. He *et al.* (2012) studied the antioxidant properties of edible mushroom glucans, namely the water soluble β-glucans of *Agaricus bisporus* (one of the most popular edible mushrooms in Europe), *Auricularia auricula, Flammulina velutipes* and *L. edodes*. The glucans of the first three mushrooms were composed of D-mannose, D-galactose and D-glucose, while the glucan from *F. velutipes* contained L-arabinose, D-mannose, D-galactose and D-glucose. Based on their reducing capacity and their hydroxyl, superoxide ion and DPPH radical scavenging ability, the use of these biopolymers in food applications was suggested owing to their significant antioxidant properties (especially those of *A. bisporus* glucans which showed the highest antioxidant acitivity). Although commercial applications of the above glucans in the food industry are not available so far, there

© Woodhead Publishing Limited, 2013

are patents (especially in Japan) related to the use of Ganoderma, Agaricus and other mushroom glucans in edible film coatings and water-soluble capsules, for example inclusion of pickling liquids in soups and sauces (Laroche and Michaud, 2007) and a great potential exists for future food applications.

16.2.3   Yeast polysaccharides

Although most microbial polysaccharides derive from fungi and bacteria, *Saccharomyces cerevisiae*, probably the most common food grade yeast in fermented food and drinks, is known for the production of a food-related glycan which is extracted from yeast cells walls. Cell wall polysaccharides are usually insoluble in water, but their solubility and properties can readily be altered by chemical or enzymatic derivatization and facilitate their use in foods or pharmaceuticals. BYG is the general term for commercialized 'brewer's yeast glucan' (or more precisely glycan), which may also contain non-carbohydrate moieties, produced from *S. cerevisiae*. BYG is efficient in improving the physical properties of foods as a thickening and water-holding agent, or as a fat replacer giving a rich mouthfeel, and it also enhances gel strength in solutions, when used alone or in combination with other food grade polymers, such as carrageenan (Reed and Nagodawithana, 1991; Xu *et al.*, 2009). Firm gels of BYG can be formed after heating and subsequent cooling of solutions above 5–10% concentration. The glycan also has emulsifying properties and is reported to improve the organoleptic characteristics of the foods where it is added (Sandford, 1982). Thammakiti *et al.* (2004) studied the production of such a β-glucan with a β-(1,3)-glucose backbone chain and a minor branch (about 3%) of β-(1,6)-glucose with an additional 4.5–6.5% protein content from spent brewer's yeast after alkali extraction from homogenized cell walls, which had potential applications in food as an emulsion stabilizing agent, as it exhibited high viscosity and water holding and oil binding capacities.

Baker's yeast glycan is a similar product composed of D-glucose and D-mannose in 3 : 2 ratio and used mainly as stabilizer/emulsifier in dressings and desserts (Robbins and Seeley, 1977, 1978; Sandford, 1982). The same yeast has also been studied and utilized for the production of therapeutic glucans (Williams *et al.*, 1992). The wild type strain of *S. cerevisiae* excretes an extracellular β-(1,3)-D-glucan with a degree of branching (DB) of 0.2, and a genetically modified strain produces PGG (also known as Betafectin), a commercial bioactive (1,6)-β-D-glucopyranosyl-(1,3)-β-D-glucopyranose glucan with DB of 0.5 which has several pharmaceutical properties (Jamas *et al.*, 1991; Wakshull *et al.*, 1999; Kim *et al.*, 2006). In addition, *S. cerevisiae* is the industrial producer of zymosan, a complex immunoactive and anti-inflamatory glycan (proteoglucan) comprising a cell wall β-glucan with long (1,3)- and (1,6)-glucosyl groups, in conjunction with mannan, protein and nucleic acid (Ohno *et al.*, 2001; Goodridge *et al.*, 2009). These health-

© Woodhead Publishing Limited, 2013

promoting effects of glucans from edible yeast cell walls could find new applications in novel functional foods.

16.3   Production of microbial polysaccharides

A brief look at the literature on microbial polysaccharides shows that despite the numerous biopolymers that have been discovered and studied in the laboratory and the interesting properties and miscellaneous proposed applications, only a handful of these have made their way into industry and the market. The reasons for this vary, but it is principally the production process on a large scale and the problems related to it, which may make such an application economically unfeasible. High production costs, low polysaccharide yields, by-product formation and laborious downstream processing (separation and purification of the final product) are therefore issues that have to be resolved (Freitas *et al.*, 2011; Donot *et al.*, 2012). In this direction, the understanding of microbial physiology, polysaccharide biosynthesis and genetics, bioprocess (fermentation) conditions and separation/purification steps, are valuable tools. In addition, as can be deduced from the above description of microbial biopolymers, there is sometimes a diversity in structure and composition of polysachharides produced by the same microorganism, which can be a problem when commercializing these polymers. This is attributed partly to the cultivation/fermentation process conditions adopted, the composition of nutrients in the cultivation medium, and the fractionation and purification steps that are followed, which can altogether influence polysaccharide composition, branching and molecular weight. Besides this, fruit bodies of fungi generally contain more biopolymers than cultured mycelia (Wasser, 2002; Lee *et al.*, 2004; Giavasis and Biliaderis, 2006; Donot *et al.*, 2012). All these parameters have to be taken into account in the standardization of commercial products and will be briefly discussed below.

16.3.1   Biosynthesis

Microbial polysaccharides are either a part of the cell wall or excreted from the cell (extracellular polysaccharides) and are characterized as primary (e.g. several cell wall biopolymers) or secondary (e.g. several bacterial capsular biopolymers) metabolites. Their role in the cell can be to form an external slimy layer as a means of attachment to other cells and cell-to-cell interaction (a characteristic of many pathogenic speices) or a more rigid capsule or glycocalyx closely attached to the cell wall offering protection from unfavourable conditions (such as high acid or alkali concentrations, desiccation, oxygen stress, antibiotics, phagocytes, etc.), the mechanical stability of the cell wall, the control of the diffusion of molecules into the cell and the export of other metabolites, or the formation of an energy reserve,

© Woodhead Publishing Limited, 2013

as some polysaccharide-producing microorganisms also possess degrading enzymes (polysaccharide lyases) in order to hydrolyse these biopolymers to sugar monomers (Sutherland, 1990; Herrera, 1991; Sharon and Lis, 1993; Whitfield and Valvano, 1993; McNeil, 1996; Sutherland, 1997, Kumar *et al.*, 2007).

The biosynthetic steps in polysaccharide production generally include the import and assimilation of sugar monomers inside the cell by passive or active transport, their conversion to activated sugar-phospho-nucleotides after intracellular phosporylation (e.g. uridine diphospate, UDP, and thimidine diphosphate, TDP) which act as sugar donors, the transfer of sugars to lipid carriers (located in the cytoplasmic membrane) by specific glycosyltranferases, and subsequent polymerization by polymerases (Whitfield and Valvano, 1993; Stephanopoulos *et al.*, 1998; Laws *et al.*, 2001; Sutherland, 2001; Freitas *et al.*, 2011). A key step in this process is the interconversion of glucose-6-phosphate (a glycolysis intermediate) into glucose-1-phosphate (which acts as sugar nucleotide precursor), which is catalysed by phosphoglucomutase (PGM), a key enzyme in polysaccharide biosynthesis (Patel *et al.*, 2010). From this point onwards, the biosynthesis of sugar nucleotides begins, which is the other crucial step in the assembly of the main repeat unit.

The biosynthetic route of sugar nucleotides involved in gellan formation is depicted in Fig. 16.3. Cell wall polysaccharides (e.g. mushroom polysaccharides) and many exopolysaccharides are synthesized totally intracellularly, but in the case of some exopolysaccharides, such as dextran, levan, alternan, mutan and reuteran, a simpler and partially extracellular process takes place, involving lipoprotein biosynthetic enzymes excreted at the cell surface (Vanhooren and Vandamme, 1998; Sutherland 2001; Patel

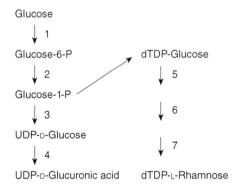

Fig. 16.3 Proposed pathway for biosynthesis of nucleotide precursors for gellan formation (adapted from Fialho *et al.*, 2008). (1) Phosphoglucomutase, (2) UDP-glucose pyrophosphorylase, (3) UDP-glucose dehydrogenase, (4) TDP-glucose pyrophosphorylase, (5) TDP-D-glucose-4,6-dehydratase, (6) TDP-6-deoxy-D-glucose-3,5-epimerase, (7) TDP-6-deoxy-L-mannose dehydrogenase.

© Woodhead Publishing Limited, 2013

et al., 2010). For instance, biosynthesis of levan is carried out via the dual action of levansucrases, which possess hydrolase activity to break down sucrose to fructose and glucose, and transferase activity, which is responsible for the transfer of the fructose moiety to a fructosyl-acceptor molecule (Han, 1990; Patel *et al.*, 2010). Similarly, in dextran synthesis by *Leuconostoc sp.* the major enzyme involved is a dextransucrase or D-glycosyl transferase which transfers glucose molecules to a monosaccharide or oligosaccharide acceptor, and polymerization takes place by the addition of D-glucose to the reducing end of the growing chain. Notably, these acceptors do not act as primers for dextran synthesis and their synthesis is competitive with dextran synthesis (Robyt *et al.*, 2008; Donot *et al.*, 2012). Dextran, as well as levan can also be synthesized by a purely enzymatic process, after isolation of the sucrases from cell cultures and mixing with sucrose. In the enzymatic process of dextran and levan synthesis it was observed that although biopolymer concentration increases at high enzyme concentration, the molecular weight of the polysaccharide is not proportional to sucrase concentration (Abdel-Fattah *et al.*, 2005; Robyt *et al.*, 2008).

In exopolysaccharide synthesis, apart from the biosynthetic enzymes, lipid transporters play a significant role in biosynthesis. They are long-chain phosphate esters and isoprenoide alcohols, similar to those involved in the biosynthesis of lipopolysaccharides, *O*-antigen and peptidoglycans (Sutherland, 1990). In EPS synthesis, lipid carriers are attached to the inner side of the cell membrane and are the anchor molecules on which the carbohydrate chain is orderly assembled. The chain is then transfered to the outer membrane where it is polymerized by a polymerase, although in some cases polymerization takes place on the inner side of the membrane and the whole chain is transferred out of the cell by exporter proteins linked to the lipid carrier (De Vuyst *et al.*, 2001; Donot *et al.*, 2012).

The biosynthetic route of heteropolysaccharides such as xanthan, gellan and LAB EPS are generally more complex than those of homopolysaccharides like fungal β-glucans. Xanthan is built up from cytoplasmic sugar nucleotides, acetyl-CoA and phosphoenolpyruvate with an inner-membrane polyisoprenol phosphate as an acceptor (Becker *et al.*, 1998). In xanthan synthesis, the repeating unit is formed by the sequential addition of glycosyl-1-phosphate from an UDP-glucose molecule to a polyisoprenol phosphate of a lipid carrier, followed by the transfer of D-mannose and D-glucuronic acid from GDP-mannose and UDP-glucuronic acid, while the acetyl groups attach to the internal mannose residue and pyruvate groups to the terminal mannose (Rosalam and England, 2006; Donot *et al.*, 2012).

The biosynthetic pathway of EPS production from LAB, although relatively complex, can be separated into four reaction sequences, one involved in sugar transport into the cytoplasm, one regulating the synthesis of sugar-1-phosphates, one responsible for activation of and coupling of sugars (formation of sugar nucleotides) and one regulating the export processes of the

© Woodhead Publishing Limited, 2013

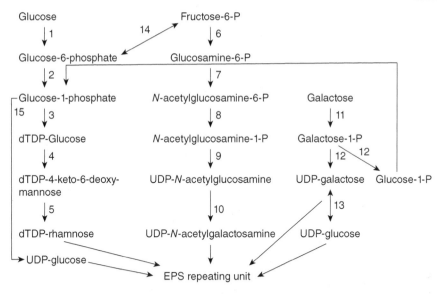

Fig. 16.4 Schematic representation of metabolic pathways for sugar nucleotide and heteropolysaccharide synthesis in LAB: (1) glucokinase, (2) phosphoglucomutase, (3) dTDP-glucose pyrophosphorylase, (4) dehydratase, (5) epimerase reductase, (6) glutamine-fructose-6-phosphate transaminase, (7) glucosamine-phosphate acetyltransferase, (8) acetylglucosamine-phosphate mutase, (9) UDP-glucosamine pyrophosphorylase, (10) UDP-N-acetylglucosamine-4-epimerase, (11) galactokinase, (12) galactose-1-phosphate uridylyl transferase, (13) UDP-galactose 4-epimerase, (14) phosphoglucose isomerase, (15) UDP-glucose pyrophosphorylase (adapted from DeVuyst *et al.*, 2001).

EPS (Laws *et al.*, 2001). The heteropolysaccharide biosynthetic route in LAB is described in Fig. 16.4.

Fungal glucans are in most cases not well studied at a biochemical and genetic level and identification of some enzymes involved in biosynthesis is still missing. However general postulated pathways have been described. Scleroglucan formation starts with the assimilation of glucose by glucose transporter(s) and its phosphorylation to glucose-6-phosphate via a hexokinase reaction. After isomerization to glucose-1-phosphate via the action of a phosphoglucomutase, UDP-glucose is formed by an UTPglucose-1-phosphate uridylyltransferase. A (1,3)-β-glucan synthase uses UDP-glucose for the synthesis of the main chain, while a (1 ~ 3);(1 ~ 6)-β-glucosyltransferase is postulated to mediate the addition of the (1,6)-β-linked glucosyl side chain into the (1,3)-β-glucan backbone (Schmid *et al.* 2011). Although the β-glucan synthase activity of *S. rolfsii* involved in the assembly of the (1,3)-β-glucan has been studied in membrane and protoplast fractions, the branching activity has not been assigned to a specific enzyme yet (Kottutz

© Woodhead Publishing Limited, 2013

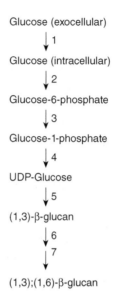

Glucose (exocellular)

↓ 1

Glucose (intracellular)

↓ 2

Glucose-6-phosphate

↓ 3

Glucose-1-phosphate

↓ 4

UDP-Glucose

↓ 5

(1,3)-β-glucan

↓ 6
| 7
↓

(1,3);(1,6)-β-glucan

Fig. 16.5 Postulated pathway for biosynthesis of scleroglucan by *S. rolfsii*; (1) glucose transporter, (2) hexocinase, (3) phosphoglumutase, (4) UTP-glucose-1-phosphate-uridyltransferase, (5) (1,3)-β-glucansynthase, (6) glycosyltransferase, (7) glucosidase (adapted from Schmid *et al.*, 2011).

and Rapp, 1990; Schmid *et al.*, 2011). Fig. 16.5 summarizes a general proposed pathway for scleroglucan synthesis.

In pullulan biosynthesis, three key enzymes are nessesary for glucose to be converted into pullulan, namely α-phosphoglucomutase, uridine diphosphoglucose pyrophosphorylase (UDPG-pyrophosphorylase) and glucosyltransferase. Hexokinases and isomerases are needed for the convertion of sugars other than glucose to the key sugar nucleotide UDPG, which acts as the pullulan precursor by transferring a D-glucose residue to the lipid carriers (lipid hydroperoxides with a phosphoester bridge) to form a lipid-linked isomaltosyl and subsequently an isopanosyl residue. The latter is finally polymerized into the pullulan chain (Simon *et al.*, 1998; Cheng *et al.*, 2011). Notably, a somewhat distinct process has been proposed concerning the sugar utilization in pullulan biosynthesis, where it has been observed that *A. pullulans* cells are able to store sugars in the form of an intracellular storage polysaccharide (glycogen) which is broken down to monosaccharides from which pullulan is formed (Simon *et al.*, 1998; Cheng *et al.*, 2011).

The activity of these biosynthetic enzymes, the availability of lipid carrier or acceptor molecules (usually mono- or oligosaccharides) and the number of phosphorylated sugars and sugar nucleotides strongly influence the

© Woodhead Publishing Limited, 2013

degree of polymerization, molecular weight and total yield of polysaccharides. The (over)expression of these molecules and the regulation of the corresponding genes are the targets of metabolic and genetic engineering efforts for improved biopolymer processes (Stephanopoulos *et al.*, 1998; Van Kranenburg *et al.*, 1999; Ruffing and Chen, 2006).

For most exopolysaccharides synthesized intracellularly, a typical gene sequence of the order of 12–17 kb may be required for biosynthesis. One gene cluster usually regulates the synthesis of sugar nucleotides and acyl groups if required. A different gene cluster may control the assembly of sugar precursors on lipid carriers, and a separate cluster seems to be responsible for polymerization and export (Kumar *et al.*, 2007). Gene size and complexity depend on the complexity of the polysaccharide structure and significant similarities in gene clusters have been observed among structurally similar polysaccharides (Sutherland *et al.*, 2001). Fig. 16.6 shows a proposed sequence of genes involved in the biosynthesis of LAB EPS, xanthan and gellan, which are some of the most well-studied biopolymers at a genetic level. In contrast, information on the gene cassette required for glucan synthesis in fungi is scarce.

During or after the biosynthetic process, polysaccharide lyases are activated in many microorganisms. The action of these enzymes is often triggered by glucose or carbon source depletion (i.e. in a prolonged fermentation process), or by the need to break down the extracullar slime or capsule in order to improve mass and oxygen transfer into the cell, which may be hindered otherwise. In addition, many of these degrading enzymes are necessary during the polymerization process, to control the size of the biopolymer and cleave parts of it, if necessary, and deletion of the genes encoding polysaccharide lyases may be detrimental to the synthesis of the biopolymer (Mattysse *et al.*, 1995). For instance, several hydrolases may appear upon a prolonged process of LAB EPS production (Degeest *et al.*, 2001), or during the stationary or death phase of *S. paucimobilis* in gellan production (for instance under high aeration rate conditions) (Giavasis *et al.*, 2006), or during alginate formation by *Azotobacter* or *Pseudomonas* sp. (Sutherland, 2001), while a β-1,3-endoglucanase and a β-glucosidase may hydrolyse scleroglucan to glucose molecules owing to carbon exhaustion (Rapp, 1989).

Interestingly, acetyl groups of gellan, alginate and acetan seem to have an inhibitory effect on the corresponding lyases, while xanthan lyases are unaffected by the presence of acyl groups (Sutherland, 1995). The susceptibility of several other glucans, such as lentinan, and *S. scerevisiae* and *Candida albicans* glucans has also been exhibited (Cutfield *et al.*, 1999; Minato *et al.*, 1999; Fernandez *et al.*, 2003). In pullulan production the decline in MW with processing time is attributed to the action of α-amylases (Manitchotpisit *et al.*, 2010). The reduction in MW and DB as a result of polysaccharases is usually undesirable since it may deteriorate the rheological properties of the biopolymers and thus process conditions have to aim

© Woodhead Publishing Limited, 2013

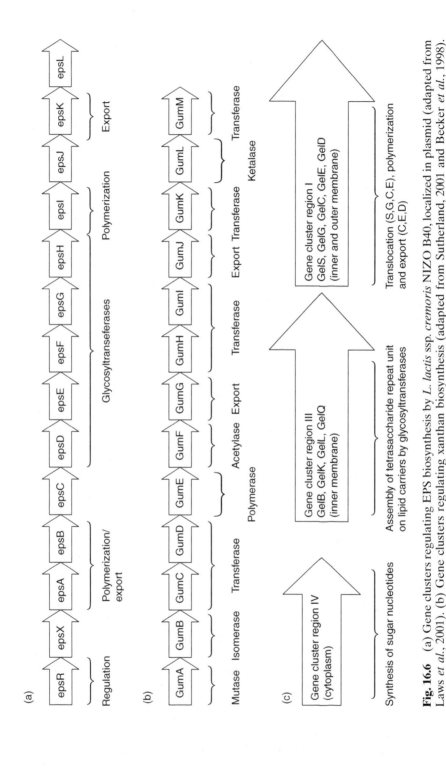

Fig. 16.6 (a) Gene clusters regulating EPS biosynthesis by *L. lactis* ssp. *cremoris* NIZO B40, localized in plasmid (adapted from Laws *et al.*, 2001). (b) Gene clusters regulating xanthan biosynthesis (adapted from Sutherland, 2001 and Becker *et al.*, 1998). (c) Gene clusters regulating gellan biosynthesis (adapted from Fialho *et al.*, 2008).

at minimal activity of these enzymes, especially after the biopolymer is formed at high concentrations. Having said this, the complete elimination of polysaccharide lyases in mutant strains may result in reduced biopolymer production as they may also be involved in biosynthesis (Sutherland, 2001; Giavasis *et al.*, 2006).

16.3.2 Industrial production

The industrial production of most microbial polysaccharides involves the batch or fed-batch cultivation (so-called 'fermentation') of the selected industrial strain in a bioreactor (or 'fermentor') under controlled conditions of agitation, aeration, pH, temperature, dissolved oxygen and process medium composition. The latter is usually a liquid synthetic medium with a standard composition based on a glucose or sucrose carbon source as the main ingredient, or a complex, carbohydrate-rich, non-synthetic medium derived from agricultural by-products (such as molasses, fruit pulp, potato pulp, corn syrup, deproteinized whey, etc.). The exception to this 'fermentation' process is the biocatalytic synthesis of some polysaccharides using cell-free biosynthetic enzymes (Lopez-Romero and Ruiz-Herrera, 1977; Finkelman and Vardanis, 1987, Abdel-Fattah *et al.*, 2005), or the cultivation of mushrooms on solid media (soil, manure, sawdust, straw, etc.), to produce fruit bodies from which the polysaccharides are extracted.

The control and modulation of bioprocess conditions normally has a significant effect on the quantity and the quality of the biopolymer and is a key parameter for process optimization, along with metabolic and genetic engineering, as well as downstream process efficiency. The latter is the process of isolation and purification of the end product in its final form via centrifugation or filtration, chromatographic separation, precipitation and drying, all of which play a significant role in the total cost and efficacy of the production process.

In terms of bioprocesss optimization, although plenty of data exist for polysaccharides such as xanthan, gellan, pullulan and scleroglucan, there is limited research on other microbial polysaccharide processes. However, some general rules apply to most processes. The way process parameters like nutrient composition affect polysaccharide production depends largely on whether the products are growth associated (primary metabolites) or non-growth associated (secondary metabolites). For example, a high carbon (e.g. glucose)/nitrogen (e.g. ammonium nitrate) ratio in the process medium, and subsequent nitrogen limitation, enhances scleroglucan (Farina *et al.*, 1998) and pullulan (Harvey, 1993) synthesis, by reducing the utilization of glucose for biomass (cell) synthesis. Taurhesia and McNeil (1994) reported that scleroglucan was produced at a higher concentration in a phosphate-limited medium (18.9 g l^{-1}) than in a nitrogen-limited medium (11.4 g l^{-1}). In xanthan and gellan synthesis, a high C/N ratio is required. Carbon sources are preferably used at 20–40 g l^{-1} and nitrogen sources, despite being

© Woodhead Publishing Limited, 2013

essential nutrients for growth, should be exhausted before polysaccharide formation arises (De Vuyst *et al.*, 1987; Garcia-Ochoa *et al.*, 2000; Giavasis *et al.*, 2006). Also, in the case of xanthan, a high concentration of nitrogen sources (low C/N ratio) has led to a low pyruvilation degree of the polymer (Casas *et al.*, 2000).

Culdlan synthesis is also boosted by nitrogen limitation and in this case it was observed that this was linked to enhanced levels of nucleotide precursors under these conditions (Kim *et al.*, 1999; McIntosh *et al.*, 2005). Conversely, cell wall polysaccharides from *S. cerevisiae* which depend on the rate of cell growth are accumulated at higher concentrations when a low carbon/nitrogen (C/N) ratio exists (Aguilar-Uscanga and François, 2003), probably due to the necessity of nitrogen for cell growh and proliferation. Carbon and nitrogen demands in exopolysaccharide-producing lactic acid bacteria differ between mesophilic species (*Lactococcus cremoris, Lactobacillus casei*) where EPS are not growth associated, and thermophilic species (*L. bulgaricus, L. helveticus*), where the synthesis of EPS is usually growth related (Cerning *et al.*, 1992; DeVuyst *et al.*, 2001).

The type of carbon and nitrogen source, also influences biopolymer production via fermentation. For instance levan from *B. subtilis* and *Z. mobilis* is elaborated at high amounts only in a sucrose medium, as opposed to glucose-based media or more complex carbon sources such as molasses and corn syrup (Senthilkumar and Gunasekaran, 2005; Shih *et al.*, 2005; De Oliveira *et al.*, 2007). Also, the composition of the carbon source(s) in the fermentation medium may influence the ratio of sugar monomers in heteropolysaccharides from LAB (Grobben *et al.*, 1996). In homopolymers from LAB, sucrose is an inducer of dextran synthesis by *L. mesenteroides*, while fructose represses dextran sucrase in the same microorganism and stimulates levan sucrase activity (Dols *et al.*, 1998; Leathers, 2002a). In gellan production, when lactose was used as the main carbon source instead of the commonly used glucose, acyl levels increased, having a negative effect on the rheological properties of lactose-derived gellan (Fialho *et al.*, 1999). Xanthan can be produced by various carbon sources and the best production yields in declining order were obtained by glucose, sucrose, maltose and soluble starch (Leela and Sharma, 2000).

The feeding strategy is another parameter influencing the efficacy of a biopolymer producing process. Although batch cultures are usually adopted on an industrial scale, in many cases a fed-batch process with a stepwise addition of the carbon source (and other nutrients) can improve the concentration of the final product, by eliminating potential substrate inhibition. For instance, optimization via a fed-batch approach has been reported for curdlan (Lee *et al.*, 1997), gellan (Wang *et al.*, 2006), scleroglucan (Survase *et al.*, 2007), ganoderan (Tang and Zhong, 2002) and *S. cerevisiae* glucan (Kim *et al.*, 2007).

Temperature and pH also influence these polysaccharide processes. For most bacterial EPS an optimal tempertature around 28–30°C is chosen

© Woodhead Publishing Limited, 2013

(e.g. in the case of gellan and xanthan) (Giavasis *et al.*, 2000; Casas *et al.*, 2000), while an optimal temperature for fungal polysaccharide synthesis is usually somewhat lower (around 25–28°C) (Cheng *et al.*, 2011; Fosmer and Gibbons, 2011). The chosen process temperature is often a compromise between optimal temperature for cell growth and optimal temperature for polysaccharide synthesis, as in the case of pullulan synthesis where optimum temperature for pullulan formation (27.4 g l^{-1}) was 26°C, whereas 32° C was optimal for cell growth (10.0 g l^{-1}). In such cases a bi-staged process can be adopted, in order to achieve high biomass at a first stage and then stimulate polysaccharide synthesis at a second stage (Wu *et al.*, 2010).

In the production of EPS from LAB, the optimal temperature varies significantly. Mesophiles such as *L. cremoris* and *L. casei* strains produce higher amounts of polysaccharides at 18–20°C, conditions which are sub-optimal for growth, while high process temperatures (37–42°C) are used for EPS production by thermophiles, such as *L. bulgaricus* and *L. helveticus* (Cerning *et al.*, 1992; Mozzi *et al.*, 1995; De Vuyst and Degeest, 1999). Temperature (and other process condition) may also affect the composition of the biopolymer, as in the case of xanthan where a relatively low temperature (25°C) caused an increase in the MW and a decrease in acetate and pyruvate groups (Casas *et al.*, 2000).

For many bacterial EPS such as xanthan and gellan, synthesis as well as growth is optimal at neutral pH, and the potential acidification of the process fluid (if pH is not controlled) owing to the formation of organic acidic (such as acetate, pyruvate) is detrimental to the production process (Giavasis *et al.*, 2000; Palaniraj and Jayaraman, 2011). However, in the curdlan process, a downshift of pH from 7 to 5.5 at the exponential growth phase increased the metabolic flux for the formation of sugar nucleotides, leading to increased curdlan production compared to processes with a stable pH of 7 (Zhan *et al.*, 2012). For *Zymomonas* levan an optimal concentration (27.2 g l^{-1} for a genetically engineered strain and 15.4 g l^{-1} for its natural parent strain) has been achieved at pH 5 and 25°C, while an increase in process temperature up to 35°C stimulates ethanol synthesis instead of levan formation (Senthilkumar and Gunasekaran, 2005).

In fungal fermentation processes pH values are preferably low, for instance pH 3.5–4.5 for scleroglucan and pH 4.5–5.5 for pullulan (Harvey, 1993; Giavasis *et al.*, 2002). The pH value of the process fluid may also determine cell morphology in fungal polysaccharide processes. At low pH *A. pullulans* cells acquire a yeast-like (unicellular) morphology which is essential for glucan formation, while at higher pH most cells are in mycelial form, which elaborates little or no pullulan (Reeslev *et al.*, 1997). Other fungal polysaccharides are optimally synthesized at low pH, such as grifolan at pH 5.5 (Lee *et al.*, 2004) and ganoderan at pH 4–4.5 (Yang and Liau, 1998) or during an uncontrolled process with a pH drop from 5 to 4 (Kim *et al.*, 2002). Shu *et al.* (2004) tested the effect of different pH levels (from 4 to 7.0) upon polysaccharide formation by *A. blazei*. They reported an

© Woodhead Publishing Limited, 2013

increase in biopolymer yield with increased pH, but at low pH, MW (and biological activity of the glucans) was maximal.

Aeration and agitation are also important factors for microbial polysaccharide production. As almost all microorganisms used are aerobic, oxygen is necessary for cell growth and, in the case of pullulan and xanthan, it is also stimulatory for polysaccharide synthesis (Rho *et al.*, 1988). Nevertheless, it has been reported that low dissolved oxygen (DO) in the bioreactor, and even DO limitation, can enhance scleroglucan and schizophyllan synthesis, in contrast to cell growth (Rau *et al.*, 1992). A possible explanation for this may be the presence of oxygen-sensitive biosynthetic enzymes, or the utilization of the carbon source primarily for glucan synthesis under growth-limiting conditions. Further, high (DO) and oxidative condition in the fermentor may cause a radical-induced degradation of scleroglucan (Hjerde *et al.*, 1998). Low DO levels also favour exopolysaccharide (ganoderan) production by *Ganoderma lucidum*, in contrast to cell growth (Tang and Zhong, 2003). In gellan production, some oxygen limitation may favour EPS synthesis and although good mixing (e.g. 500 rpm) is essential in the viscous gellan fermentation broth, high aeration rate (above 1 vvm, volume per volume minute) reduces gellan yield and MW (Giavasis *et al.*, 2006).

Conversely, the synthesis of cell wall polyssacchrides by *S. cerevisiae* is restricted at low aeration rate and DO (in the range 0–50%), following a decrease in biomass concentration (Aguilar-Uscanga and François, 2003). Process conditions for EPS production by lactic acid bacteria differ compared to the above processes. Here, little or no aeration is needed as most strains are microaerophilic and low agitation (e.g. 100 rpm) is usually applied, since the low EPS content of the fermentation broth (around or below 1 g l^{-1}) does not affect broth viscosity and bulk mixing (De Vuyst and Degeest, 1999). This is also the case for dextran-producing strains of *L. mesenteroides*, which require no aeration (Leathers, 2002a), while the alternan-producing *L. mesentoroides* NRRL B-1355 showed optimal levansucrase activity at a moderate DO level (controlled at 75%), as opposed to the non-aerated processes (Raemaekers and Vandamme, 1997).

Oxygen regulation is critical for the production of bacterial alginate which is optimal under microaerophilic conditions. This has been ascribed to the inactivation of some alginate biosynthetic enzymes through higher oxygenation, inactivation of the metabolically important nitrogenases and the fact that at high DO alginate forms a very rigid capsule which hinders nutrient transfer into the cell (Leitão and Sá-Correia 1997; Sabra *et al.*, 2001). The latter, that is the compromise of mass and oxygen transfer by the high viscosity that gradually develops in most microbial polysaccharide processes is a universal problem and may become more intense in fungal fermentation where the floating mycelia intensify this problem. In this sense, high agitation is usually required for adequate mixing. For example, the amount and quality (molecular weight) of pullulan is improved at high

© Woodhead Publishing Limited, 2013

agitation rates, as this promotes the formation of yeast-type (unicellular) cells (McNeil and Kristiansen, 1987). However, in the case of scleroglucan, high agitation stimulates cell growth at the expense of polysaccharide concentration (Schilling *et al.*, 1999). The same authors suggest that moderate agitation has to be applied for production of high molecular weight scleroglucan.

Apart from the above-mentioned process parameters, variance in polysaccharide concentration is also observed among different microbial strains. For example, levan production by *B. subtilis* and *Z. mobilis* is much higher $(40–50 \text{ g l}^{-1})$ than that achieved with *Streptococcus salivarus* (Newbrun and Baker, 1967; Viikari, 1984; Shih *et al.*, 2005). Variations between scleroglucan yields obtained from *S. glucanicum* and *S. rolfsii* have also been reported (Giavasis *et al.*, 2002).

Another step in the polysaccharide production process that is crucial for biopolymer yield and quality is the isolation and purification process, in other words the downstream processing. Downstream processing may affect the molecular size or molecular weight distribution, DB, composition (presence of side groups) and of course the total product yield, and accounts for a large part of the total production costs (Lee *et al.*, 1996; Wasser, 2002; Wang *et al.*, 2010). Therefore optimization of this process is a challenging task for chemical engineers. There is no simple method for extracting all microbial polysaccharide. For each biopolymer an appropriate extraction technique must be chosen and optimized based on the structural and physicochemical characteristics of the molecule to be extracted and the desired purity and intended use. For food applications, biopolymers should be free of biomass (cell debris and other intracellular components) and of the reagents used for extraction.

In the case of extracellular slimy polysaccharides, this processing generally involves sterilization or pasteurization of the fermentation broth to kill cells, inactivate undesirable enzymes (e.g. lyases) and facilitate separation of cells from the EPS, subsequent removal of cells by filtration or centrifugation, alchohol pecipitation of the polysaccharide in the cell-free filtrate or centrifugate, followed by further purification (if required) by ultrafiltration, gel permeation/ion-exchange chromatography or diafiltration. The end product in powder form is obtained after drying with air/inert gas/under vacuum, or spray drying, or lyophilization, and final milling to the desired mesh size (Giavasis and Biliaderis, 2006; Singh *et al.*, 2008; Donot *et al.*, 2012).

When the exopolysaccharide is firmly attached to the cell wall (capsular EPS), an initial step of hot alkali treatment or sonication is usually employed to facilitate disengagement of the biopolymer from the cell (Morin, 1998; Wang *et al.*, 2010). This treatment can also result in decolorization or deacylation of the gum, which is sometimes needed to improve its sensory and rheological characteristics (for example in the case of gellan or pullulan) (Giavasis *et al.*, 2000; Kumar *et al.*, 2007). Moreover, in some capsular EPS

© Woodhead Publishing Limited, 2013

such as xanthan, enzymatic cell lysis for cell removal may be more effective and preferable than hot alkali treatment, as it does not affect the composition and physicochemical properties of the polymer (Shastry and Prasad, 2005).

In the case of viscous biopolymers such as xanthan and gellan, thermal pretreatment reduces the high viscosity of the fermentation broth which facilitates further chemical and physical treatment and improves xanthan removal from the cells. However, care must be taken so that pasteurization/sterilization of the fermentation broth at elevated temperatures does not cause thermal degradation of the polysaccharide (Smith and Pace, 1982). Dilution in water or dilute solutions of salts or alcohols can by used alternatively to reduce the viscosity of the process fluid and facilitate filtration and removal of impurities (Garcia-Ochoa et al., 1993). Fosmer and Gibbons (2011) suggested a 50% dilution for optimal extraction of scleroglucan from the fermentation broth via filtration. Cross-flow filtration is a common technique for membrane separation of fungal EPS, where a series of connected filtration cassettes with different MW cutoffs separated cells, proteins, sugars and salts from the EPS. However, the use of ceramic membranes may prove a better alternative, as they provide sterility, stability (against fouling) and reusability (Schmid et al., 2011).

Precipitation of capsular gums like xanthan and gellan is usually achieved by mixing with double (or triple) volume of alcohol or acetone (which represent a significant part of the total processing cost), or divalent, trivalent or tetravalent salts. For the precipitation of food grade xanthan, FDA prescribes the use of isopropanol. Using a combined alcohol and salt precipitation has led to improved xanthan precipitation and yield and reduced the amount of alcohol required (Garcia-Ochoa et al., 1993). In order to reduce the use of solvents, ultrafiltration can prove useful, as it can concentrate xanthan broth by at least five times before the solvent is added (Palaniraj and Jayaraman, 2011). Lo et al. (1996) have suggested ultrafiltration of dilute xanthan broths as a complete alternative to alchohol precipitation, which did not impair the physicochemical properties and molecular weight of xanthan. For charged polysaccharides an ion exchange resin may be well suited for purifying the final product without use of undesirable chemical reagents. However, physical extraction (via sonication and separation in resins) may have a lower yield of extracted polysaccharides compared to chemical methods (treatments with alkali, alcohols and acids) (Donot et al., 2012).

To isolate (fungal) cell-wall polysaccharides, a somewhat different process is followed compared to extracellular biopolymers, which includes extraction of polysaccharides from the cell mass with hot alkali or hot water, followed by filtration, centrifugation or dialysis of the extracted material, in order to remove insoluble impurities and finally drying of the purified polysaccharide (Cutfield et al., 1999; Wasser, 2002; Hromadkova et al., 2003). Insoluble polysaccharides are usually isolated by washing the debris of both

© Woodhead Publishing Limited, 2013

cells and biopolymer with dilute acid solution, solubilization of the polysaccharide with alkali solution, centrifugation and collection of the polysaccharide from the supernatant after addition of a new, denser acid solution (Lee *et al.*, 1997).

16.4   Properties and structure–function relationships

The physicochemical properties of microbial polysaccharides such as solubility, viscosity, gelation, water binding, dispersability, emulsifying and stabilizing ability, film formation, interaction with other molecules, chemical stability and degradation are crucial for their application in food and other materials, as has been illustrated in Section 16.2 by several examples of food applications. Similarly, their dietary, immunostimulating and other health effects are important for designing functional foods and nutraceuticals. These properties are influenced principally by the composition, the chemical structure and the conformational order (random coil, single, double or triple helix), the degree of branching of biopolymer, the presence of salts or other molecules, and the processing or treatments that they undergo before being used. Below, the relationship between structure and function of these biopolymers will be highlighted.

16.4.1   Physicochemical properties

Xanthan gum is highly soluble biopolymer in both cold and hot water, thanks to its polyelectrolyte nature (presence of acyl groups) and can produce highly viscous solutions, but does not form gels (Garcia-Ochoa *et al.*, 2000). The xanthan molecule seems to have two conformations, helix and random coil, depending on the dissolution temperature and the presence of cations. The conformation shifts from disordered state (at high dissolution temperature) to an ordered state (at low-dissolution temperature and high ionic strength), thus solutions formed at 40–60°C are more viscous than those formed above 60°C (Garcia-Ochoa and Casas, 1994; Casas *et al.*, 2000). X-ray diffraction showed that in the molecular conformation of a (right-handed) helix the trisaccharide side chain is aligned with the backbone and stabilizes the overall structure, principally by hydrogen bonds. In solutions this side chain folds around the backbone protecting it from enzymic or acidic hydrolysis (Palaniraj and Jayaraman, 2011).

The pyruvyl and acetyl groups of xanthan can be removed at low (pH 3) or high pH (pH 9), but this does not have a significant effect on rheological properties (Garcia-Ochoa *et al.*, 2000). In contrast, the acyl content of gellan (acetate/glycerate) plays a crucial role in the gelling properties of the molecule. Thus, a high-acyl gellan produces a soft gel and can be useful in increasing viscosity and improving body and mouthfeel of

© Woodhead Publishing Limited, 2013

foods, while a low-acyl gellan forms rigid, brittle and stable gels which can withstand high temperatures and low pH (Giavasis et al., 2000; Fialho et al., 2008). This phenomenon has been ascribed primarily to the removal of the glycerate groups during deacylation by hot alkali treatment, which decide the level of rigidity or elasticity of gellan gells, along with the presence of divalent cations (Jay et al. 1998; Giavasis et al., 2000). Gellan molecules acquire a disordered coil conformation at high temperatures which transform to a double helix upon cooling. At high concentrations, the double helices become compact rod-like aggregates and form gels (Khan et al., 2007).

The effect of deacylation and the removal of side groups of succinoglycan depends on the molecule that is removed. Deacetylation of succinoglycan gels decreased the transition temperature, while the desuccinylated biopolymer had increased the transition temperature and greatly improved pseudoplasticity (Ridout et al., 1997). In the case of acetan aqueous solutions a thermoreversible helix-coil formation is obtained, where acetylation does not play a crucial role (triple helix is not prevented by the presence of acetyl groups) (Ojinnaka et al., 1996; Ridout et al., 1998).

Dextran physicochemical properties vary significantly, depending mainly on the microbial strain used for production, but even in a single strain there is heterogeneity in the molecular size and the proportion of α-(1,6) linkages of dextrans, thus the dextrans resulting from fermentation are fractionated according to molecular size and the intended application (Alsop, 1983; Naessens et al., 2005). Commercial dextran is a white fine powder that dissolves easily in hot or cold water to produce moderately viscous, clear solutions. It does not form gels and remains dissolved even at high concentrations (50%). However, high molecular weight dextrans tend to have a higher degree of branching which in turn has an adverse effect on dextrans solubility in water (Khalikova et al., 2005; Khan et al., 2007). Dextrans with more than 43% of α-(1,3) linked branches, such as mutan are considered insoluble in water (Mehvar, 2000; Monsan et al., 2001). Interestingly, dextrans of medium and low MW (< 500 kDa) exhibit Newtonian behaviour in water solutions at concentrations below 30% (w/v) (Khan et al., 2007). Purified dextran with a MW of 500 000 Da exhibits Newtonian behaviour below a 30% concentration, while 'native' dextran with a higher MW shows a slight pseudoplasticity above 1.5% concentration (McCurdy et al., 1994). The conformation of the molecule also depends on the molecular mass and polymer concentration. At low concentrations dextran (MW 500 000) appears as a random coil, but high concentrations in solutions lead to a more compact coil. The non-Newtonian native dextran is characterized by inter-chain interactions. In addition, concentration positively affects the melting point of dextran (T_g) (McCurdy et al., 1994).

Alternan acquires an extended and tightly coiled conformation in solutions, owing to its unique alternating structure of α-(1,6) and α-(1,3)

© Woodhead Publishing Limited, 2013

linkages (with approximately 10% DB) (Seymour and Knapp, 1980; Leathers *et al.*, 2003). This structure imparts properties of high solubility and low viscosity, despite its high MW (10^6–10^7 Da), as well as resistance to hydrolysis by most microbial and mammalian enzymes (Cote, 1992). In a test period of seven days, alternan was shown to form stable solutions in a pH range of 3–9, and a temperature range of 4–70°C (Leathers *et al.*, 2003). Despite its relatively low viscosity, dense solutions of native alternan (above 12–15%) are difficult to attain, owing to its high molecular weight and relatively high viscosity at that concentration. Thus, attempts have been made to reduce the MW of native alternan. Sonication of alternan yields a modified alternan of MW below 10^6 Da, which is able to give solutions of 50% concentration or higher, with novel rheological properties resembling those of gum arabic (Cote, 1992). Alternatively, native alternan can be modified after hydrolysis by *Penicillium* sp. to yield a polymer with similar viscosifying and bulking properties to gum arabic (Leathers, 2003). More extensive hydrolysis with isomaltodextranase gives a very low MW polymer (MW around 3500 D), with similar rheological characteristics to maltodextrins (Cote, 1992).

Curdlan hydrogels are comparatively weak when prepared by heating at 55–70°C, followed by cooling. Further heating to 80–100°C increases the gel strength and produces an impressively firm and resilient gel, which will not melt upon reheating below 140°C. Autoclaving at 120°C converts the molecular structure to a triple helix which is resistant to microbial degradation (Sutherland, 1998). The curdlan gels are firmer and less elastic than gelatine gels and less brittle than agar and gellan gels. One drawback is that they are susceptible to shrinkage and syneresis. Gels are also formed when alkaline solutions of curdlan are dialysed against distilled water or are neutralized. Interestingly, curdlan gel strength depends on the level and length of heating and on curdlan concentration, and is stable over a wide pH range (2–10), or in the presence of other sugars, salt and lipids (McIntosh *et al.*, 2005; Laroche and Michaud, 2007).

Bacterial levans exhibit high solubility in water, despite their high molecular weight. This is attributed to the highly branched structure of levan (Newbrun and Baker, 1967; Stivala *et al.*, 1975). On the other hand, levan generally has a much lower intrinsic viscosity compared to other polysaccharides of similar molecular weight (Arvidson *et al.*, 2006), which is assigned to the compact spherical or ellipsoidal conformation of the molecule, from which the branches extend radially (Stivala *et al.*, 1975; Rhee *et al.*, 2002). In fact, the viscosity of levan solutions from *Pseudomonas syringae* pv. *phaseolica* remain Newtonian up to a polymer concentration of 20% (Kasapis *et al.*, 1994). *B. subtilis* levans may be fractionated to low molecular weight levan and high molecular weight levan, the former having a lower viscosity than the latter (Shih *et al.*, 2005). Kasapis *et al.* (1994) indicate that levans from *P. syringae* pv. *phaseolica* exhibit solution properties similar to those of disordered linear polysaccharides and no detectable

© Woodhead Publishing Limited, 2013

conformational change with temperature. The same authors observed thermodynamic incompatibility and subsequent reduction of viscosity in mixed solutions of levan–pectin and levan–locust bean gum, which may be related to the role of levan in microbial phytopathogenesis, where levan may act as a barrier protecting bacterial cells from plant cell defence. In addition, levan is non-gelling and non-swelling in aqueous solutions at ambient temperature (Kasapis and Morris, 1994), biocompatible and may form turbid lyotropic crystalline solutions (liquid crystals) at low concentrations (Huber et al., 1994). Relative structural stability of Z. mobilis levan is achieved in purified levan–water solutions at pH 4 to 5 and temperatures between 25 and 70°C (Bekers et al., 2005). Bacterial levans are susceptible to enzymic or acid hydrolysis, thus their use as low-calorie ingredient in food is doubtful, in contrast to plant levans which are resistant to digestion in humans (Bekers et al., 2005; Izydorczyk et al., 2005).

Alginate gels have an adjustable strength. This depends on the number of intermolecular cross-links that can be formed between chains, on the type of ions facilitating cross-linking and on the length of blocks between the links. Alginates with high polyguluronate content can form rigid gels in the presence of calcium, especially after deacetylation, while viscosity is mainly a function of the molecular weight of alginate (Moe et al. 1995; Sutherland, 2001).

Pullulan is distinguished by its structural flexibility and high solubility, owing to its unique alternating (1,4) and (1,6) structure (Leathers, 2002b). It is readily soluble in water (except for esterified or etherified pullulan), insoluble in organic solvents and non-hydroscopic (Le Duy et al., 1988; Leathers, 2002b). Pullulan forms solutions that are stable in the presence of cations, but does not form gels, probably owing to its linear structure (Izydorczyk et al., 2005). Upon drying, pullulan solutions can form films, the elasticity of which can be improved by addition of plasticers like glycerol, or sorbitol (Singh et al., 2008).

Elsinan is readily dissolved in hot water and gives stable and very viscous solutions, but at high concentrations (approximately ten times that of pullulan). Unlike the structurally similar pullulan, it forms gels at concentrations of 2% when the temperature is lowered to around 4°C, or 5% at higher temperatures (Tsumuraya et al., 1978). Elsinan solutions are highly pseudoplastic and at high shear rates viscosity falls rapidly. Temperature affects the viscosity of these solutions in the following manner: when a 2% solution is gradually heated from 30–45°C a slight decrease in viscosity is observed, after which viscosity increases to reach a maximum value at 60°C. Beyond this temperature viscosity is again reduced and preheating the solution at 90°C for 30 min results overall in lower viscosity (Sandford, 1982). In contrast, viscosity is unaffected by pH in range from 3 to 11 and by several electrolytes at various concentrations, probably owing to its neutral, non-ionic nature (Misaki, 2004). Salivary and

© Woodhead Publishing Limited, 2013

pancreatic amylases as well as *B. subtilis* amylases are capable of hydrolysing elsinan-releasing maltotriose units, while *Aspergillus oryzae* amylases only cleave maltotetraose units (Tsumuraya *et al.*, 1978; Misaki *et al.*, 1982; Misaki, 2004).

Scleroglucan, in a refined form, is also soluble in water at ambient temperatures, owing to the presence of β-D-glucopyranosyl residues that enhance solubility and reduce gelation, resulting in stable viscous solutions over a temperature range of 10–90°C and a pH range of 1–11 (Wang and McNeil, 1996, Giavasis *et al.*, 2002). Loss of viscosity occurs only above pH 12, or after addition of DMSO (dimethylsulphoxide) which results in the disruption of the triple helix of the polymer (Nardin and Vincendon, 1989). Both refined scleroglucan and crude scleroglucan (60–75% gum) solutions (either hot or cold) are pseudoplastic (shear thinning) in concentrations above 0.2–0.5% and compatible with many salts (for example no change in solubility is observed in a 5% NaCl, 20% $CaCl_2$, 5% $NaSO_4$ or a 10% Na_2HPO_4 solution), as well as other polymers, such as gelatine, gellan, xanthan, carrageenan, guar gum and locust bean gum, although no synergism with other viscosifiers is observed (Brigand, 1993). In aqueous solutions of 1.2–1.5% concentration, purified scleroglucan forms sliceable thermoreversible hydrogels at approximately 25°C, caused by cross-linking of the triple helix. At low temperatures around 7°C, the swollen gels are softer and more diffused (Bluhm *et al.*, 1982). The stability of scleroglucan solutions is distinctively better than that of xanthan and other biopolymers, owing to its high molecular weight (Giavasis *et al.*, 2002; Schmid *et al.*, 2011). Apart from the microbial hydrolysis observed in *S. glucanicum* cultures, scleroglucan is not easily degradable and is considered undigestible for humans (Rapp, 1989).

Like many other glucans, schizophyllan and lentinan acquire a single or triple helix conformation with short side groups in their soluble form, stabilized by hydrogen bonds. The degree of branching (DB) and MW strongly influence the solubility and functionality of these molecules as will be discussed later (Bluhm *et al.*, 1982; Toshifumi and Ogawa, 1998; Falch *et al.*, 2000). Fang *et al.* (2004) studied the ability of schizophyllan to form co-gels with gelatin, where the concentration of schizophyllan affected the elasticity or rigidness of the gels (higher elasticity at low gelatin/schizophyllan ratio). Their findings could be useful in food applications where gelatin is already used as a thickener.

Brewer's yeast glycan (BYG) dissolves readily in water, at low or high temperatures, the latter leading to higher solution viscosity. Measurable viscosity is attained only above a 4.5% concentration of the glycan and it is unaffected by pH over the range 2–7 or by the addition of sodium chloride up to a 5% concentration. Viscous solutions of BYG become thinner upon heating and thicker upon cooling and have good freeze–thaw stability (Sandford, 1982).

© Woodhead Publishing Limited, 2013

16.4.2   Nutritional and health effects

Even though not all of the microbial polysaccharides mentioned above are used in commercial food applications, most of them are potentially useful as indigestible dietary fibres, as described in Section 16.2. Toxocilogical tests for those already used in food have revealed that they impose no danger to human health, while many others come from edible mushrooms, so no concerns over potential toxicity should arise. On the contrary, over the last few years their beneficial effects on human health have been widely researched and are already or expected to be utilized in the design of novel functional foods or nutraceuticals. More specifically, many of these polysaccharides possess antitumour, immunostimulatory, antimicrobial or hypolipidaemic and hypocholesterolaemic properties, which are influenced by the structural properties of the biopolymers.

Pullulan is only partially degraded by human amylases and thus has no nutritional value, neither is it toxic or mutagenic (Okada *et al.*, 1990). In studies with mice fed with pullulan as a 40% replacement for starch, organ-to-body weight was normal (equivalent to the control group fed with starch only), but hypertrophies in the large intestine were found (Sandford, 1982). Interestingly, pullulan has been proposed as a probiotic for humans, since it proved to promote the growth of bifidobacteria (Mitsuhashi *et al.*, 1990). As for scleroglucan, no toxicity, tissue pathology or blood abnormalities have been observed in studies on rats and dogs, and no eye or skin irritation in pigs, rabbits and humans, while its role as an undigestible dietary fibre for humans has been suggested (Rodgers, 1973). Elsinan is gradually digested by human amylases, but more slowly compared to starch (Sandford, 1982).

Feeding tests with dextran have shown that it is slowly but totally metabolized in humans (Halleck, 1972). Despite this, it has been reported that a diet rich in dextrans with high proportion of α-(1,6) linkages contributed to body weight loss (probably owing to slow absorption in the intestine), as these type of linkages are resistant to enzymic attack in the gastrointestinal tract, as opposed to α-(1,4) glycosidic linkages of starch and glycogen which are readily hydrolysed by human α-amylases (Naessens *et al.*, 2005). Dextran, as well as the non-digestible alternan, are biocompatible and non-toxic for humans and the producer strains are safe for use in food (GRAS), as are all LAB and their products. Nutritionally, the susceptibility of levans to acid and enzymatic hydrolysis, suggests that their use as dietary fibre in food is doubtful, in contrast to their plant equivalents (Bekers *et al.*, 2005; Izydorczyk *et al.*, 2005). However, levan-producing strains of *Z. mobilis, B. subtilis* and some Lactobacilli, concomitantly synthesize fructo-oligosaccharides (FOS). FOS, or the mixture of levan and FOS (referred to as 'fructan syrup') are considered new prebiotic substances, as they induce the growth of *Bifidobacterium* sp. and other beneficial microflora in the intestine, improve intestinal functionality, as well as having a low energetic value and a pleasant honey-like taste (Yun, 1996; Dal Bello *et al.*, 2001; Bekers *et al.*, 2004).

© Woodhead Publishing Limited, 2013

For *S. cerevisiae*, another GRAS microorganism, no adverse toxicological data exists for the polysaccharides it elaborates and the use of the organism and its products in food is considered safe.

The multiple health effects of microbial polysaccharides have been studied extensively in the last few years as a response to consumers' demand for healthier food and the need to develop new, milder and more natural bioactive or pharmaceutical ingredients for 'preventive medicine'. These effects can be summarized as immunostimulating-antitumour, antimicrobial and hypocholesterolaemic–hypoglycaemic. The relationship between structure and function of these biopolymers is of great importance and has been reviewed previously (Sutherland, 1994; Bohn and BeMiller, 1995; Wasser 2002; Giavasis and Biliaderis, 2006). Generally, the primary characteristics that affect the bioactivity of biopolymers are the MW, the type of glycosidic linkage and DB, the conformational state (random coil or helix) and the chemical composition and potential derivatization (Giavasis and Biliaderis, 2006). It appears that high MW and medium or high DB and a (triple) helix conformation, as well as the presence of β-(1,3) linkages play an important role in the expression of antitumour or immunomodulating activity of glucans, possibly owing to the presence of a β-(1,3)-glucan receptor in macrophages, which enables macrophage activation by β-(1,3)-glucans (however there are also reports on immunopotentiating β-(1,6) glucans or bioactive glucans lacking a triple helix, which probably have a distinct immunostimulatory mechanism).

Anti-cholesterol and anti-glycaemic effects are also linked to medium or high MW of the biopolymer, even though some controversial reports are available in the literature (Wasser 2002; Giavasis and Biliaderis, 2006; Laroche and Michaud, 2007; Zhang *et al.*, 2011). Trying to shed light on the impact of structural conformation on the immunostimulating effects of several β-glucans, Suzuki *et al.* (1992) concluded that the triple helix can enhance the alternative pathways of complement (APC) in the immune system more effectively than the single helix, whereas a single helix is a better stimulant of the classical pathways of complement (CPS), which explains some controversial reports on structure–function relationships, since different modes of bioactivity (all contributing to the overall performance of the immune system) are triggered by structurally different biopolymers.

Many fungal β-glucans are associated with a fortified immune system and treatment of cancer, especially in traditional oriental medicine. They act either directly against the tumour cells, or indirectly by boosting the immune system. For example, scleroglucan has very important attributes as a biological response modifier. Sinofilan, the immunopharmacological form of scleroglucan, is used effectively in the clinical treatment of cancer. Scleroglucan has a reportedly high affinity for human monocytes and stimulates phagocytic cells and monocyte, neutrophil and platelet haemopoietic activity (Jamas *et al.*, 1996; Giavasis *et al.*, 2002). In addition, *Sclerotinia*

© Woodhead Publishing Limited, 2013

sclerotiorum glucan (SSG), the pharmacological form of a similar glucan from *Sclerotinia sclerotiorum*, exhibits antitumour and immunostimulating properties when administered parenterally, or orally (Suzuki *et al.*, 1991; Bohn and BeMiller, 1995). Interestingly, oral administration which is important for avoiding pain and side effects and indicative of the potential for incorporation of the biopolymer in nutraceuticals, is a distinct advantage of scleroglucan compared to other immunoactive glucans.

Schizophyllan (a biopolymer chemically and structurally similar to scleroglucan) and lentinan are another two fungal immunotherapeutic glucans used clinically for cancer treatment since 1986, usually in combination with conventional cancer therapies. Schizophyllan, administered along with antineoplastic drugs, seemed to prolong the life of patients with lung, gastric or cervical cancer (Furue, 1987; Wasser, 2002). Further, it can restore mitosis in bone marrow cells suppressed by antitumour drugs (Zhu, 1987). Lentinan from the mushroom *L. edodes* is also effective against gastric, colorectal or breast cancer and in the prevention of metastasis. Notably, both lentinan and schizophyllan, as well as scleroglucan, have low or no toxicity even at high doses and are more effective when administered at the early stages of a treatment (Bohn and BeMiller, 1995; Ikekawa, 2001). In terms of structure–function relationships, it has been clarified that high MW schizophyllan (100 000–200 000 Da or higher) which acquires a triple helix, exhibits significant antitumour activity, while lower MW fractions (5 000–50 000 Da) or denatured polymers lacking a triple helix are biologically inactive (Kojima *et al.*, 1986). Similarly, high molecular weight lentinan with a triple helix conformation had the highest bioactivity compared to low MW or partially denatured polymers (Zhang *et al.*, 2011). In contrast, Kulicke *et al.* (1997) concluded that bioactivity of scleroglucan was not dependent on or favoured by an ordered helix structure, in fact random coil conformations of scleroglucan were better activators of human blood monocytes.

Other immunomodulating glucans include Krestin or PSK (a commercialized antitumour glucan with broad immunostimulating and antineoplastic activity) and the proteglucans from *A. blazei* fruit bodies (Zhu, 1987; Hobbs, 1995; Fujimiya *et al.*, 1998).

The antitumour activity of fungal β-D-glucans is probably due to the mitogenic activity of soluble glucan molecules (especially when they adopt a triple helix conformation), which provoke a number of immune responses, such as activation of natural killer (NK) cell and T-cell mediated cytotoxicity, stimulation of monocytes, increased synthesis of immunoglobulins and cytokines, for example interferons (IFN), interleukines (IL) and tumour necrosis factor-α (TNF-α), activation of peripheral mononuclear cells (PMNC), and enhancement of phagocytosis by neutrophils and macrophages, which can destroy malignant cells (Reizenstein and Mathe, 1984; Arinaga *et al.*, 1992; Falch *et al.*, 2000; Giavasis *et al.*, 2002; Wasser, 2002). It is interesting that interactions of lentinan and other bioactive polymers with

© Woodhead Publishing Limited, 2013

carrageenans may restrict their antitumour activity (Hamuro and Chihara, 1985). This underlines the need for *in vivo* studies to be carried out in a realistic food matrix, rather than polysaccharides in pure form, before these polysaccharides are applied in potential functional foods and nutraceuticals.

The insoluble glucans synthesized by *S. cerevisiae* exhibit various immunostimulating effects. PGG and Zymosan have mitogenic activity and increase the production of cytokines and monocyte and neutrophil phagocytosis. Also, solubilized *S. cerevisiae* glucans after chemical derivatization (e.g. glucan sulphate, sulphoethyl glucan, carboxymethyl glucan and oxidized glucan) show equal or higher antitumour activity compared to the native insoluble glucans (Jamas *et al.*, 1991; Bohn and BeMiller, 1995; Sandula *et al.*, 1999). With regard to structure, *S. cerevisiae* β-D-glucans with a DB of 0.2 had higher immunostimulating activity compared to glucans with a DB of 0.05, owing to the higher affinity of the branched glucan for the β-glucan receptor of human macrophages. Also, immunomodulatory activity of *S. cerevisiae* particulate β-D-glucans was associated with high molecular weight among glucans of 500 000–4000 000 Da (Cleary *et al.*, 1999).

Levans from *Z. mobilis, B. subtilis* (natto), *Aerobacter* sp., *Microbacterium laevaniformans* and *Rahnella aquatilis* also exhibit anticarcinogenic, radioprotective and immunomodulating properties, including the prevention of allergic disorders. These are mediated by generation of mononuclear cells, increase in peripheral leucocytes and spleen cell antibodies, and the stimulation of macrophages, the induction of interleukin and the control of immunoglobulin levels in the serum (Calazans *et al.*, 2000; Yoo *et al.*, 2004; Yoon *et al.*, 2004; Xu *et al.*, 2006). The high molecular weight and increased degree of branching is reported to play a decisive role in the expression of such properties by levans (Yoo *et al.*, 2004; Yoon *et al.*, 2004).

Although native, unbranched and insoluble curdlan does not posses immunostimulating properties, chemically modified and branched curdlans are biological response modifiers and exhibited significant antitumor activity, which may not involve the typical stimulation of macrophages and phagocytosis observed in other bioactive glucans (Bohn and BeMiller, 1995; McIntosh *et al.*, 2005).

Moreover, exopolysaccharides from lactic acid bacteria (other than dextran or alternan) have been isolated, which have antitumor or immunostimulatory activity *in vitro* and *in vivo*. Specifically, a phosphopolysaccharide from *L. lactis* ssp. *cremoris* stimulates lymphocyte mitogenicity, macrophage cytostaticity, cytokine synthesis in macrophages and antigen-specific antibody production (Nakajima *et al.*, 1995). A similar phosphopolysaccharide from *Lactobacillus delbrueckii* spp. *bulgaricus* injected intraperitoneally (solutions of 10–100 µg ml^{-1} administered in a 100 mg kg^{-1} dose in mice), caused an increase in the number and the tumouricidal activity of intraperitoneal macrophages (Kitazawa *et al.*, 2000). The phosphate

© Woodhead Publishing Limited, 2013

group of these polymers seems to be essential for the expression of these properties (Kitazawa *et al.*, 1998, 2000).

Apart from the anticarcinogenic–immunomodulatory effects, the antimicrobial activity of several microbial glucans may find novel applications in food. Sizofiran (a commercialized pharmacological schizophyllan product) was used successfully in order to stimulate immune responses of patients with hepatitis B virus, via the increased excretion of interferon-gamma and the proliferation of peripheral blood mononuclear cells (PBMC) (Kakumu *et al.*, 1991).

Lentinan is also active against bacterial infections, such as tuberculosis and *Listeria monocytogenes* infection. The antimicrobial activity of lentinan is reportedly accomplished by an improvement in phagocytosis of microbial cells by neutrophils and macrophages (Furue, 1987). The immunomodulating and microbiocidal activity of lentinan against *Salmonella enteritis* and *Staphylococcus aureus* was also shown in immunological studies (Mattila *et al.*, 2000).

Insoluble glucan from baker's yeast, as well as SSG glucan from *S. sclerotiorum* helped control the growth of *Mycobacterium tuberculosis* (*in vitro*) (Hetland and Sandven, 2002), while the supply of a relatively low dose of PGG glucan from *S. cerevisiae* inhibited the growth of antibiotic-resistant *S. aureus* in the blood of contaminated rats, which was linked to elevated activity of monocytes and neutrophils (Liang *et al.*, 1998). In addition, oral immunization with levan from *A. levanicum* was tested successfully against pneumonia caused by *P. aeruginosa* and proved to induce levan-specific titres of serum immunoglobulin A, especially when supplied at the beginning of the infection (Abraham and Robinson, 1991).

The indigestible or slowly degraded biopolymers also have potentially hypocholesterolaemic and hypoglycaemic properties, although these have not been extensively studied in clinical experiments. Few reports are available with regard to the potential anticholesterol and antiglycaemic effects of commercial microbial polysaccharides, with some exceptions, as in the case of alginates (Khotimchenko *et al.*, 2001). Elsinan and levan, for example, have exhibited remarkable cholesterol-lowering effects in hypercholesterolaemic rats (Yamamoto *et al.*, 1999; Misaki, 2004).

Additionally, the documented contribution of dextrans to body weight loss is probably due to the slow and gradual hydrolysis of the molecule, which suppresses blood glucose levels (Naessens *et al.*, 2005). The ability of sodium alginate to lower blood glucose and increase faecal excretion of cholesterol has been reported to depend on the MW, with polymers of 100 000 Da being more effective than polymers of 50 000 or 10 000 Da (Kimura *et al.*, 1996) Although there are no reports on biological activity of native pullunan, chemical derivatization of pullulan may infer anticholesterol activity, as has been achieved with diethyl-amino-ethyl (DEAE) derivatized gellan which obtained negative and positive charges and acquired novel bile acid binding and anticholesterolaemic capacity (Yoo

© Woodhead Publishing Limited, 2013

et al., 2005). Lentinan from shiitake mushrooms can also be used for the treatment of cholesterol in humans. It works by reducing overall levels of lipoproteins (both high dersity lipoprotein (HDL) and low density lipoprotein (LDL) in blood (Breene, 1990). It is generally proposed that the reduction of cholesterol levels is due to the interruption of enterohepatic circulation of bile acids, which leads to higher liver cholesterol and bile acid excretion in the feces (Seal and Mathers, 2001), while the regulation in blood glucose levels results from the attachment of undigestible polysaccharides to the intestinal surface, which decelerating glucose absorption (Hikino *et al.*, 1985; Kimura *et al.*, 1996). The incorporation of these biopolymers in novel foods could lead to the production of innovative products which might help regulate the cholesterol blood levels of consumers.

16.5   Future trends

Microbial polysaccharides are complex molecules with versatile properties and numerous applications in foods; the search for new biopolymers with attractive properties continues. However, only a handful of experimentally studied polysaccharides have been commercialized owing to problems related to low production yields, costly manufacture or regulatory restraints, which may be overcome in the future. Future research at a biosynthetic and genetic level and optimization of bioprocesses and extraction methods is necessary and is expected to broaden their use in the food industry and allow the commercialization of novel biopolymers. For instance, the regulation of genes and the overexpression or downregulation of key biosynthetic enzymes involved in the synthesis of LAB polysaccharides or mushroom biopolymers (where genetic and metabolic engineering studies are still scarce), coupled with cost-effective downstream processing (e.g. using robust filtration systems and a single-step extraction of exopolysaccharides) may lead to economically viable production of food grade biopolymers from GRAS microorganisms, which could be readily adopted by the food industry. More research is needed with regard to the nutritional and health effects of microbial polysaccharides, especially at the level of clinical trials, in order to test their performance as ingredients in functional foods and consolidate potential health claims of novel food products. The latter is expected to boost the adoption of several biopolymers as bioactive molecules in novel foods and nutraceuticals, as many of them have shown impressive medicinal properties *in vitro* or in clinical trials in their purified form. Since these biopolymers are to be incorporated in a complex food matrix, the interactions with other food components and the impact of food processing (i.e. effect of high/low temperature, high pressure or vacuum, drying, acidic environment and presence of cations, presence of other polysaccharides or proteins which may alter their functional properties) on these molecules should be thoroughly studied. For established food

© Woodhead Publishing Limited, 2013

polysaccharides, novel uses, for example in edible films or coatings that improve shelf life and stability of food (alone or in combination with other ingredients) and modification of their structure, composition and properties are other interesting areas of research. All the above ideas are expected to bring about innovative food applications with high consumer acceptance and commercial success for these exciting products of microbial metabolism.

16.6   References

ABDEL-FATTAH AF, MAHMOOD DAR and ESAWY MAT (2005), 'Production of levansucrase from *Bacillus subtilis* NRC 33a and enzymic synthesis of levan and fructo-oligosaccharides', *Current Microbiol*, **51**, 402–7.

ABRAHAM E and ROBINSON A (1991), 'Oral immunization with bacterial polysaccharide and adjuvant enhances antigen-specific pulmonary secretory antibody response and resistance to pneumonia', *Vaccine*, **9**, 757–64.

ADACHI S (1992), 'Lactic acid bacteria and the control of tumours', in *The Lactic Acid Bacteria in Health and Disease*', Wood JB (ed.). Elsevier, London, 233–61.

ADACHI Y, SUZUKI Y, OHNO N and YADOMAE T (1998), 'Adjuvant effect of grifolan on antibody production in mice', *Biol Pharm Bull*, **21**, 974–7.

AGUILAR-USCANGA B and FRANÇOIS JM (2003), 'A study of the yeast cell wall composition and structure in response to growth conditions and mode of cultivation', *Lett Appl Microbiol*, **37**, 268–74.

ALSOP RM (1983), 'Industrial production of dextrans', in *Microbial Polysaccharides*, Bushell ME (ed.), Elsevier, Amsterdam, 1–44.

ARINAGA S, KARIMINE N, TAKAMUKU K, NANBARA S, NAGAMATSU M, UEO H and AKIYOSHI T (1992), 'Enhanced production of interleukin 1 and tumor necrosis factor by peripheral monocytes after lentinan administration in patients with gastric carcinoma', *Int J Immunopharm*, **14**, 43–7.

ARVIDSON SA, RINEHART BT and GADALA-MARIA F (2006), 'Concentration regimes of solutions of levan polysaccharides from *Bacillus* sp.', *Carbohydr Polym*, **65**, 144–9.

BARONE JR and MEDYNETS M (2007), 'Thermally processed levan polymers', *Carbohydr Polym*, **69**, 554–61.

BAO XF, WANG XS, DONG Q, FANG JN and LI XY (2002), 'Structural features of immunologically active polysaccharides from *Ganoderma lucidum*', *Phytochem*, **59**, 175–81.

BECKER A, KATZEN F, PUEHLER A and IELPI L (1998), 'Xanthan gum biosynthesis and application: a biochemical/genetic perspective', *Appl Microbiol Biotechnol*, **50**, 145–52.

BEKERS M, LAUKEVICH J, UPITE D, KAMINSKA E,VIGANTS A, VIESTURS U, PANKOVA L and DANILEVICH A (2002), 'Fructooloigosaccharides and levan producing activity of *Zymomonas mobilis* extracellular levansucrase', *Process Biochem*, **38**, 701–6.

BEKERS M, MARAUSKA M, GRUBE M, KARKLINA D and DUMA M (2004), 'New prebiotics for functional food', *Acta Alimentaria*, **33**, 31–7.

BEKERS M, UPITE D, KAMINSKA E, LAUKEVICS J, GRUBE M, VIGANTS A and LINDE R (2005), 'Stability of levan produced by *Zymomonas mobilis*', *Process Biochem*, **40**, 1535–9.

BENDER H, LEHMANN J and WALLENFELS K (1959), 'Pullulan, ein extracellulaeres glucan von *Pullularia pullulans*', *Biochim Biophys Acta*, **36**, 309–16.

BERNIER B (1958), 'The production of polysaccharides by fungi active in the decomposition of wood and forest litter', *Can J Microbiol*, **4**, 195–204.

© Woodhead Publishing Limited, 2013

BHASKARACHARYA RK and SHAH NP (2000), 'Texture characteristics and microstructure of skim milk Mozzarella cheese made using exopolysaccharide and non-exopolysaccharide producing starter cultures', *Austral J Dairy Technol*, **55**, 132–8.

BLUHM C, DESLANDS Y, MARCHESSAULT R, PERZ S and RINAUDO M (1982), 'Solid-state and solution conformations of scleroglucan', *Carbohydr Res*, **100**, 117–30.

BOHN JA and BEMILLER JN (1995), '(1→3)-β-D-glucans as biological response modifiers: a review of structure-functional activity relationships', *Carbohydr Polym*, **28**, 3–14.

BREENE WM (1990), 'Nutritional and medicinal value of specialty mushrooms', *J Food Protect*, **53**, 883–94.

BRIGAND G (1993), 'Scleroglucan', in *Industrial Gums*, Whistler RL and Be Miller JN (eds), 3rd edition, Academic Press, San Diego, 461–72.

BROWN DE and MCAVOY A (1990), 'A pH-controlled fed-batch process for dextransucrase production', *J Chem Technol Biotechnol*, **48**, 405–14.

CALAZANS GMT, LIMA RC, DE FRANCA FP and LOPES CE (2000), 'Molecular weight and antitumour activity of *Zymomonas mobilis* levans', *Int J Biol Macromol*, **27**, 245–7.

CASAS JA, SANTOS VE and GARCIA-OCHOA FG (2000), 'Xanthan gum production under several operational conditions: molecular structure and rheological properties', *Enz Microb Technol*, **26**, 282–91.

CASTILLO E and LOPEZ-MUNGUIA A (2004) 'Synthesis of levan in water-miscible organic solvents', *J Biotechnol*, **114**, 209–17.

CATLEY BJ (1970), 'Pullulan, a relationship between molecular weight and fine structure', *FEBS Lett*, **10**, 190–3.

CELIK GY, ASLIM B and BEYATLI Y (2008). 'Characterization and production of the exopolysaccharide (EPS) from *Pseudomonas aeruginosa* G1 and *Pseudomonas putida* G12 strains', *Carbohydr Polym*, **73**(1), 178–82.

CERNING J, BOUILLANNE C, LANDON M and DESMAZEAUD MJ (1992), 'Isolation and characterization of exopolysaccharides from slime-forming mesophilic lactic acid bacteria', *J Dairy Sci*, **75**, 692–9.

CHENG KC, DEMIRCI A and CATCHMARK JM (2011), 'Pullulan: biosynthesis, production, and applications', *Appl Microbiol Biotechnol*, **92**, 29–44.

CHO SM, PARK JS, KIM KP, CHA DY, KIM HM and YOO ID (1999), 'Chemical features and purification of immunostimulating polysaccharides from the fruit bodies of *Agaricus blazei*', *Korean J Mycol*, **27**, 170–4.

CLEARY JA, KELLY, GE and HUSBAND AJ (1999), 'The effect of molecular weight and β-1,6-linkages on priming of macrophage function in mice by (1,3)-β-D-glucan', *Immunol Cell Biol*, **77**, 395–403.

COMBIE J (2006), 'Properties of levan and potential medical uses', in *Polysaccharides for Drug Delivery and Pharmaceutical Applications*, ACS Symposium Series, American Chemical Society, **934**, 263–9.

COTE GL (1992), 'Low-viscosity α-D-glucan fractions derived from sucrose which are resistant to enzymatic digestion', *Carbohydr Polym*, **19**, 249–52.

COTE GL and ROBYT JF (1982), 'Isolation and partial characterization of an extracellular glucan sucrase from *L. mesenteroides* NRRL B1355 that synthesizes an alternating (1 ~ 6), (1 ~ 3) α-D-glucan', *Carbohydr Res*, **101**, 57–74.

COVIELLO T, GRASSI M, RAMBONE G, SANTUCCI E, CARAFA M, EVELINA MURTAS, RICCIERI F and ALHAIQUE F (1999), 'Novel hydrogel systems from scleroglucan synthesis and characterization', *J Control Release*, **60**, 367–78.

CUTFIELD SM, DAVIES GJ, MURSHUDOV G, ANDERSON BF, MOODY PCE, SULLIVAN PA and CUTFIELD JF (1999), 'The structure of the exo-β-(1,3)-glucanase from *Candida albicans* in native and bound forms: relationship between a pocket and groove in family 5 glycosyl hydrolases', *J Mol Biol*, **294**, 771–83.

DAL BELLO F, WALTER J, HERTEL C and HAMMES WP (2001), '*In vitro* study of prebiotic properties of levan-type exopolysaccharides from lactobacilli and non-digestible

© Woodhead Publishing Limited, 2013

carbohydrates using denaturating gradient gel elecrophoresis', *Syst Appl Microbiol*, **24**, 1–6.

DE OLIVEIRA MR, DA SILVA RSSF, BUZATO JB and CELLIGOI MAPC (2007), 'Study of levan production by *Zymomonas mobilis* using regional low-cost carbohydrate sources', **37**, 177–83.

DE VUYST L and DEGEEST B (1999), 'Heteropolysaccharides from lactic acid bacteria', *FEMS Microbiol Rev*, **23**, 153–77.

DE VUYST L, VAN LOO J and VANDAMME EJ (1987), 'Two-step fermentation process for improved xanthan production by *Xanthomonas campestris* NRRL B-1459', *J Chem Technol Biotechnol* **39**, 263–73.

DE VUYST L, DE VIN F, VANINGELGEM F and DEGEEST B (2001), 'Recent developments in the biosynthesis and applications of heteropolysaccharides from lactic acid bacteria', *Int Dairy J*, **11**, 687–707.

DEGEEST B, VANINGELGEM F and DE VUYST L (2001), 'Microbial physiology, fermentation kinetics, and process engineering of heteropolysaccharide production by lactic acid bacteria', *Int Dairy J*, **11**, 747–57.

DOLS M, REMAUD-SIMEON M and MONSAN R (1998), 'Optimisation of the production of dextransucrase from *Leuconostoc mesenteroides* NRRL B-1299 and its application to the synthesis of non digestible glucooligosaccharides', in *Proceedings of the Second European Symposium on Biochemical Science*, de Azevedo SF, Ferreira EC, Luyben AM and Ossenweijer P (eds), European Federation of Biotechnology, Porto, 86–92.

DONOT F, FONTANA A, BACCOU JC AND SCHORR-GALINDO S, (2012), 'Microbial exopolysaccharides: Main examples of synthesis, excretion, genetics and extraction', *Carbohydr Polym*, **87**, 951–62.

ELIZAQUÍVEL P, SÁNCHEZ G, SALVADOR A, FISZMAN S, DUENAS MT, LÓPEZ P, FERNÁNDEZ DE PALENCIA P and AZNAR R (2011), 'Evaluation of yogurt and various beverages as carriers of lactic acid bacteria producing 2-branched (1,3)-beta-D-glucan', *J Dairy Sci*, **94**, 3271–8.

EO SK, KIM YS, LEE CK and HAN SS (2000), 'Possible mode of antiviral acivity of acidic protein bound polysaccharide isolated from *Ganoderma lucidum* on herpes simplex viruses', *J Ethnopharm*, **72**, 475–81.

FALCH BH, ESPEVIK T, RYAN L and STOKKE BT (2000), 'The cytokine stimulating activity of (1→3)-β-D-glucans is dependent on the triple helix conformation', *Carbohydr Res*, **329**, 587–96.

FANG Y, TAKAHASHI R and NISHINARI K (2004). 'Protein/polysaccharide co-gel formation based on gelatin and chemically modified schizophyllan', *Biomacromolecules*, **5**, 126–36.

FARINA J, SINERIZ F, MOLINA O and PEROTTI N (1998), 'High scleroglucan production by *Sclerotium rolfsii*: influence of medium composition', *Biotechnol. Lett*, **20**, 825–96.

FERNANDEZ LF, ESPINOSA JC, FERNANDEZ-GONZALEZ M and BRIONES A (2003), 'β-Glucosidase activity in a *Saccharomyces cerevisiae* wine strain', *Int J Food Microbiol*, **80**, 171–6.

FIALHO AM, MARTINS LO, DONVAL ML, LEITÃO JH, RIDOUT MJ, JAY AJ, MORRIS VJ and SÁ-CORREIA I (1999), 'Structures and properties of gellan polymers produced by *Sphingomonas paucimobilis* ATCC 31461 from lactose compared with those produced from glucose and from cheese whey', *Appl Environ Microbiol*, **65**, 2485–91.

FIALHO AM, MOREIRA LM, GRANJA AT, POPESCU AO, HOFFMANN K and SÁ-CORREIA I (2008), 'Occurrence, production, and applications of gellan: current state and perspectives', *Appl Microbiol Biotechnol*, **79**, 889–900.

FINKELMAN MAJ and VARDANIS A (1987), 'Synthesis of β-glucan by cell-free extracts of *Aureobasidium pullulans*', *Can J Microbiol*, **33**, 123–7.

© Woodhead Publishing Limited, 2013

FOSMER A and GIBBONS W (2011), 'Separation of scleroglucan and cell biomass from *Sclerotium glucanicum* grown in an inexpensive, by-product based medium', *Int J Agric Biol Eng*, **4**, 52–60.

FREITAS F, ALVES VD and REIS MAM (2011), 'Advances in bacterial exopolysaccharides: from production to biotechnological applications', *Trends Biotechnol*, **29**, 388–98.

FUJIMIYA Y, SUZUKI Y, OSHIMAN K, KOBORI H, MORIGUCHI K, NAKASHIMA H, MATUMOTO Y, TAKAHARA S, EBINA T and KATAKURA R (1998), 'Selective tumoricidal effect of soluble proteoglucan extracted from the basidiomycete *Agaricus blazei* Murill, mediated via the natural killer cell activation and apoptosis', *Cancer Immunol Immunother*, **46**, 147–59.

FURUE H (1987), 'Biological characteristics and clinical effects of sizofilan (SPG)', *Drugs Today*, **23**, 335–46.

GARCIA-OCHOA F, CASAS JA and MOHEDANO AF (1993), 'Precipitation of xanthan gum', *Separ Sci Technol*, **28**, 1303–13.

GARCIA-OCHOA F and CASAS JA (1994), 'Apparent yield stress in xanthan gum solution at low concentration', *Chem Eng J*, **53**, B41–6.

GARCIA-OCHOA F, SANTOS VE, CASAS JA and GOMEZ E (2000), 'Xanthan gum: production, recovery, and properties', *Biotechnol Advan*, **18**, 549–79.

GHASEMLOU M, KHODAIYAN F, OROMIEHIE A (2011), 'Rheological and structural characterisation of film-forming solutions and biodegradable edible film made from kefiran as affected by various plasticizer types', *Int J Biol Macromol*, **49**, 814–21.

GIAVASIS I and BILIADERIS C (2006), 'Microbial polysaccharides', in *Functional Food Carbohydrates*, Biliaderis C and Izydorczyk M (eds), CRC Press, New York, 167–214.

GIAVASIS I, HARVEY LM and MCNEIL B (2000), 'Gellan gum', *Crit Rev Biotechnol*, **20**, 177–211.

GIAVASIS I, HARVEY LM and MCNEIL B (2002), 'Scleroglucan', in *Biopolymers*, Steinbuchel A (ed), Vol. 8, Wiley-VCH, Munster, chap. 2, p 37.

GIAVASIS I, HARVEY LM and MCNEIL B (2006), 'The effect of agitation and aeration on the synthesis and molecular weight of gellan in batch cultures of *Sphingomonas paucimobilis*', *Enz Microb Technol*, **38**, 101–8.

GLICKSMAN M (1982), 'Dextran', in *Food Hydrocolloids*, Glicksman M (ed), CRC Press, Florida.

GOODRIDGE HS, WOLF AJ and UNDERHILL DM (2009), 'β-glucan recognition by the innate immune system', *Immunol Rev*, **230**, 38–50.

GOUNGA ME, XU SY, WANG Z and YANG WG (2008), 'Effect of whey protein isolate–pullulan edible coatings on the quality and shelf life of freshly roasted and freeze-dried Chinese chestnut', *J Food Sci*, **73**, E155–E161.

GRASSI M, LAPASIN R, PRICL S and COLOMBO I (1996), 'Apparent non-Fickian release from a scleroglucan gel matrix', *Chem Eng Commun*, **155**, 89–112.

GROBBEN GJ, SMITH MR, SIKKEMA J and DE BONT JAM (1996), 'Influence of fructose and glucose on the production of exopolysaccharides and the activities of enzymes involved in the sugar metabolism and the synthesis of sugar nucleotides in *Lactobacillus delbrueckii* subsp. *bulgaricus* NCFB 2772', *Appl Microbiol Biotechnol*, **46**, 279.

HALLECK FE (1972), *Cosmetic Composition Employing Water-soluble Polysaccharide*, US Patent, 3,659,025.

HAMURO J and CHIHARA G (1985), 'Lentinan, a T-cell orientated immunopotentiator: its experimental and clinical applications and possible mechanism of immune modulation', in *Immunomodulation Agents and their Mechanisms*, Fenichel RL and Chirigos MA (eds), Dekker, New York, 409–36.

HAN YW (1990), 'Microbial levan', *Adv Appl Microbiol*, **35**, 171–94.

© Woodhead Publishing Limited, 2013

HARDY G (2000), 'Nutraceuticals and functional foods: Introduction and meaning', *Nutrition*, **16**, 688–9.

HARVEY LM (1993), 'Viscous fermentation products' *Crit Rev Biotechnol*, **13**, 275–304.

HE JZ, RU QM, DONG DD and SUN PL (2012), 'Chemical characteristics and antioxidant properties of crude water soluble polysaccharides from four common edible mushrooms', *Molecules*, **17**, 4373–87.

HERRERA JR (1991), 'Biosynthesis of β-glucans in fungi', *Antonie van Leeuwenhoek*, **60**, 73–81.

HETLAND G and SANDVEN P (2002), 'β-1,3-glucan reduces growth of *Mycobacterium tuberculosis* in macrophage cultures', *FEMS Immunol Med Microbiol*, **33**, 41–5.

HIKICHI M, HIROE E and OKUBO S (1999), *Protein Polysaccharide 0041*, European Patent 0939082.

HIKINO H, KONNO C, MIRIN Y and HAYASHI T (1985), 'Isolation and hypoglycemic activity of ganoderans A and B, glycans of *Ganoderma lucidum* fruit bodies', *Planta Med*, **4**, 339–40.

HJERDE T, STOKKE BT, SMIDSROD O and CHRISTENSEN BE (1998), 'Free-radical degradation of triple-stranded scleroglucan by hydrogen peroxide and ferrous ions', *Carbohydr Polym*, **37**, 41–8.

HOBBS C (1995), *Medicinal Mushrooms: an Exploration of Tradition, Healing and Culture*. Botanica Press, Santa Cruz, California.

HOLZWORTH G (1985), 'Xanthan and scleroglucan: structure and use in enhanced oil recovery', *Dev Ind Microbiol* **26**, 271–80.

HROMADKOVA Z, EBRINGEROVA A, SASINKOVA V, SANDULA J, HRIBALOVA V and OMELKOVA J (2003), 'Influence of the drying method on the physicochemical properties and immunomodulatory activity of the particulate (1→3)-β-D-glucan from *Saccharomyces cerevisiae*', *Carbohydr Polym*, **51**, 9–15.

HUBER AE, STAYTON PS, VINEY C and KAPLAN DL (1994), 'Liquid crystallinity of a biological polysaccharide: the levan/water phase diagram', *Macromol*, **27**, 953–7.

IGUCHI M, YAMANAKA S, BUDHIONO A (2000), 'Bacterial cellulose a masterpiece of nature's arts', *J Material Sci*, **35**, 261–70.

IKEKAWA T (2001), 'Beneficial effects of edible and medicinal mushrooms in health care', *Int J Med Mushrooms*, **3**, 291–8.

IZYDORCZYK M, CUI SW and WANG Q (2005), 'Polysaccharide gums: Structures, functional properties and applications', in *Food Carbohydrates*, Cui SW (ed), CRC Press, Florida.

JAMAS S, EASSON DD, OSTROFF GR and ONDERDONK AB (1991), 'PGG-glucans. A novel class of macrophage-activating immunomodulators', *ACS Symp Ser*, **469**, 44–51.

JAMAS S, EASSON J, DAVIDSON D, OSTROFF G (1996), *Use of Aqueous Soluble Glucan Preparations to Stimulate Platelet Production*, US Patent 5,532,223.

JANG KH, SONG KB, KIM CH, CHUNG BH, KANG SA, CHUN UH, CHOUE RW and RHEE SK (2001), 'Comparison of characteristics of levan produced by different preparations of levansucrases from *Zymomonas mobilis*', *Biotechnol Lett*, **23**, 339–44.

JANSSON PE, KENNE L and LINDBERG B (1975), 'Structure of the extracellular polysaccharide from *Xanthomonas campestris*', *Carbohydr Res*, **45**, 275–82.

JAY AJ, COLQUHOUN IJ, RIDOUT MJ, BROWNSEY GJ, MORRIS VJ, FIALHO AM, LEITÃO JH and SÁ-CORREIA I (1998), 'Analysis of structure and function of gellans with different substitution patterns', *Carbohydr Polym*, **35**, 179–88.

JEANES A, HAYNES WC, WILHAM CA, RANKIN JC, MELVIN EH, AUSTIN MJ, CLUSKEY JE, FISCHER BE, TSUCHIYA HM and RIST CE (1954), 'Characterization and classification of dextrans from ninety-six strains of bacteria', *J Am Chem Soc*, **76**, 5041–52.

JEZEQUEL V (1998), 'Curdlan: a new functional β-glucan', *Cereal Foods World*, **43**, 361–4.

© Woodhead Publishing Limited, 2013

JONAS R and FARAH, LF (1998), 'Production and application of microbial cellulose', *Polym Degrad Stabil*, **59**, 101–6.

JONG SC and DONOVICK R (1989), 'Anti-tumor and anti-viral ssubstances from fungi', *Adv Appl Microbiol*, **34**, 183–262.

KAKUMU S, ISHIKAWA T, WAKITA T, YOSHIOKA K, ITO Y and SHINAGAWA T (1991), 'Effect of sizofiran, a polysaccharide, on interferon gamma, antibody production and lymphocyte proliferation specific for hepatitis-B virus antigen in patients with chronic hepatitis-B', *Int J Immunopharm*, **13**, 969–75.

KANG KS and PETTITT, DJ (1993), 'Xanthan, gellan, wellan and rhamsan', in *Industrial Gums: Polysaccharides and Their Derivates*, Whistler RL and Bemiller JN (eds), Academic Press, San Diego, CA, 341–97.

KANG KS, VEEDER GT, MIRRASOUL PJ, KANEKO T and COTTRELL W (1982), 'Agar-like polysaccharide produced by a *Pseudomonas* species: production and basic properties', *Appl Environ Microbiol*, **43**, 1086–9.

KANG SA, HONG K, JANG KH, KIM S, LEE KH, CHANG BI, KIM CH and CHOUE R (2004), 'Anti-obesity and hypolipidemic effects of dietary levan in high fat diet-induced obese rats', *J Microb Biotechnol*, **14**, 796–804.

KANG SA, JANG KH, SEO JW, KIM KH, KIM YH, RAIRAKHWADA D, SEO MH, LEE JO, HA SD, KIM CH and RHEE SK (2009), 'Levan: applications and perspectives', in *Microbial Production of Biopolymers and Polymer Precursors*, Rehm BHA (ed), Caister Academic Press, Norfolk, 145–62.

KASAPIS S and MORRIS E (1994), 'Conformation and physical properties of two unusual microbial polysaccharides: *Rhizobium* CPS and levan', in *Food Hydrocolloids: Structures, Properties and Functions*, Nishinari K and Doi E (eds), Plenum Press, New York, 97–103.

KASAPIS S, MORRIS E, GROSS M and RUDOLPH K (1994), 'Solution properties of levan polysaccharide from *Pseudomonas syringae* pv. *phaseolica*, and its possible primary role as a blocker of recognition during pathogenesis', *Carbohydr Polym*, **23**, 55–64.

KEITH J, WILEY B, BALL D, ARCIDIACONO S, ZORFASS D, MAYER J and KAPLAN D (1991), 'Continuous culture systems for production of biopolymer levan using *Erwinia herbicola*', *Biotechnol Bioeng*, **38**, 557–60.

KENNEDY JF and BRADSHAW IJ (1984). 'Production, properties and applications of xanthan', *Prog Ind Microbiol*, **19**, 319–71.

KHALIKOVA E, SUSI P and KORPELA T (2005), 'Microbial dextran-hydrolyzing enzymes: Fundamentals and applications', *Microbiol Molec Biol Rev*, 306–25.

KHAN T, PARK JK and KWON KH (2007), 'Functional biopolymers produced by biochemical technology considering applications in food engineering', *Korean J Chem Eng*, **24**, 816–26.

KHOTIMCHENKO YS, KOVALEV VV, SAVCHENKO UV and ZIGANSHINA OA (2001), 'Physical-chemical properties, physiological activity, and usage of alginates, the polysaccharides of brown algae', *Russian J Marine Biol*, **27**, S53–S64.

KIM D and ROBYT JF (1994), 'Production and selection of mutants of *Leuconostoc mesenteroides* constitutivefor glucansucrases', *Enzyme Microb Technol*, **16**, 659–64.

KIM MK, LEE IY, KO JH, RHEE YH, PARK YH (1999), 'Higher intracellular levels of uridinemonophosphate under nitrogen limited conditions enhance metabolic flux of curdlan synthesis in Agrobacterium species', *Biotechnol Bioeng*, **62**, 317–23.

KIM SW, HWANG HJ, PARK JP, CHO YJ, SONG CH and YUN JW (2002), 'Mycelial growth and exo-biopolymer production by submerged culture of various edible mushrooms under different media', *Lett Appl Microbiol*, 2002, **34**, 56–61.

KIM SY, SONG HJ, LEE YY, CHO KH and ROH YK (2006), 'Biomedical issues of dietary fibre β-glucan', *J Korean Med Sci*, **21**, 781–9.

© Woodhead Publishing Limited, 2013

KIM YH, KANG SW, LEE JH, CHANG HI, YUN CW, PAIK HD, KANG CW and KIM SW (2007), 'High cell density fermentation of *Saccharomyces cerevisiae* JUL3 in fed-batch culture for the production of β-glucan', *J Ind Eng Chem*, **13**, 153–8.

KIM SY, KANG MY and KIM MH (2008). 'Quality characteristics of noodle added with browned oak mushroom (*Lentinus edodes*)', *Korean J Food Cook Sci*, **24**, 665–71.

KIM SY, CHUNG SI, NAM SH and KANG MY (2009). 'Cholesterol lowering action and antioxidant status improving efficacy of noodles made from unmarketable oak mushroom (*Lentinus edodes*) in high cholesterol fed rats', *J Korean Soc Appl Biol Chem*, **52**, 207–12.

KIM J, LIM J, BAE IY, PARK HG, LEE HG, LEE S (2010), 'Particle size effect of *Lentinus edodes* mushroom (Chamsong-I) powder on the physicochemical, rheological and oil-resisting properties of frying batters', *J Texture Stud*, **41**, 381–95.

KIM J, LEE S, BAE IY, PARK HG, LEE HG and LEE S (2011). '(1–3)(1–6)-β-Glucan-enriched materials from *Lentinus edodes* mushroom as a high-fibre and low-calorie flour substitute for baked foods', *J Sci Food Agric*, **91**, 1915–9.

KIMURA Y, WATANABE K and OKUDA H (1996), 'Effects of soluble sodium alginate on cholesterol excretion and glucose tolerance in rats', *J Ethnopharmacol*, **54**, 47–54.

KITAZAWA H, HARATA T, UEMURA J, SAITO T, KANEKO T and ITOH T (1998), 'Phosphate group requirement for mitogenic activation of lymphocytes by an extracellular phosphopolysaccharide from *Lactobacillus dulbreeckii* sp. *bulgaricus*', *Int J Food Microbiol*, **40**, 169–75.

KITAZAWA H, ISHII Y, UEMURA J, KAWAI Y, SAITO T, KANEKO T, NODA K and ITOH T (2000), 'Augmentation of macrophage functions by an extracellular phosphopolysaccharide from *Lactobacillus delbrueckii* sp. *bulgaricus*', *Food Microbiol*, **17**, 109–18.

KOJIMA T, TABATA K, ITOH W and YANAKI T (1986), 'Molecular weight dependence of the antitumor activity of schizophyllan', *Agric Biol Chem*, **50**, 231–2.

KOTTUTZ E and RAPP P (1990), '1,3-β-Glucan synthase in cell-free extracts from mycelium and protoplasts of *Sclerotium glucanicum*', *J Gen Microbiol*, **136**, 1517–23.

KOZARSKI M, KLAUS A, NIKSIC M, JAKOVLJEVIC D, HELSPER JPFG and VAN GRIENSVEN LJLD (2011), 'Antioxidative and immunomodulating activities of polysaccharide extracts of the medicinal mushrooms *Agaricus bisporus, Agaricus brasiliensis, Ganoderma lucidum* and *Phellinus linteus*', *Food Chem*, **129**, 1667–75.

KOZARSKI M, KLAUS A, NIKSIĆ M, VRVIĆ MM, TODOROVIĆ N, JAKOVLJEVIĆ D, and VAN GRIENSVEN LJLD (2012), 'Antioxidative activities and chemical characterization of polysaccharide extracts from the widely used mushrooms *Ganoderma applanatum, Ganoderma lucidum, Lentinus edodes* and *Trametes versicolor*', *J Food Composit and Anal*, **26**, 144–53.

KULICKE WM and HEINZE T (2005), 'Improvements in polysaccharides for use as blood plasma expanders', *Macromol Symposia*, **231**, 47–59.

KULICKE WM, LETTAU AI and THIELKING H (1997), 'Correlation between immunological activity, molar mass, and molecular structure of different (1→3)-β-D-glucans', *Carbohydr Res*, **297**, 135–43.

KUMAR AS, MODY K and JHA B (2007), 'Bacterial expolysaccharides – a perception', *J Basic Microbiol*, **47**, 103–17.

LAROCHE C and MICHAUD P (2007), 'New developments and prospective applications for β(1,3) glucans', *Recent Patents Biotechnol*, **1**, 59–73.

LAWS A, YUCHENG G and VALERIE M (2001), 'Biosynthesis, characterization and design of bacterial exopolysaccharides from lactic acid bacteria', *Biotechnol Adv*, **19**, 597–625.

LE DUY A, CHOPLIN L, ZAJIC JE and LUONG JHT (1988), 'Pullulan' in *Encyclopedia of Polymer Science and Engineering*, Mark HF and Bikales NM (eds), 2nd edition, John Wiley & Sons, New York.

© Woodhead Publishing Limited, 2013

LEATHERS TD (2002a), 'Dextran' in *Biopolymers, Vol 5. Polysaccharides I:Polysaccharides from Procaryotes*, Vandamme EJ, De Baets S and Steinbuechel A (eds), Wiley-VCH, Weinheim, 299–321.

LEATHERS TD (2002b), 'Pullulan', in *Biopolymers, Vol 6. Polysaccharides II: Polysaccharides from Eucaryotes*, Vandamme EJ, De Baets S and Steinbuechel A (eds), Wiley-VCH, Weinheim, 1–35.

LEATHERS TD (2003), 'Biotechnological production and applications of pullulan', *Appl Microbiol Biotechnol*, **62**, 468–73.

LEATHERS TD, NUNNALLY MS, AHLGREN JS and COTE GL (2003), 'Characterization of a novel modified alternan', *Carbohydr Polym*, **54**, 107–13.

LEE IY, SEO WT, KIM GJ, KIM MK, PARK CS and PARK YH (1997), 'Production of curdlan using sucrose or sugar cane molasses by two-step fed-batch cultivation of *Agrobacterium* species', *J Ind Microbiol Biotechnol*, **18**, 255–9.

LEE KH, KANG TS, MOON SO, LEW ID and LEE MY (1996), 'Fractionation and antitumor activity of the water soluble exo-polysaccharide by submerged cultivation of *Ganoderma lucidum* mycelium', *Korean J Appl Microbiol Biotechnol*, **24**, 459–64.

LEE C, BAE JT, PYO HB, CHOE TB, KIM SW, HWANG HJ and YUN JW (2004), 'Submerged culture conditions for the production of mycelial biomass and exopolysaccharides by the edible Basidiomycete *Grifola frondosa*', *Enz Microb Technol*, **35**, 369–76.

LEELA JK and SHARMA G (2000), 'Studies on xanthan production from *Xanthomonas Compestris*', *Bioproc Engineer*, **23**, 687–9.

LEITÃO JH and SÁ-CORREIA I (1997), 'Oxygen dependent upregulation of transcription of alginate genes *algA*, *algC* and *algD* in *Pseudomonas aeruginosa*', *Res Microbiol*, **148**, 37–43.

LIANG J, MELICAN D, CAFRO L, PALACE G, FISETTE L, ARMSTRONG R and PATCHEN ML (1998), 'Enhanced clearance of a multiple antibiotic resistant *Staphylococcus aureus* in rats treated with PGG-glucan is associated with increased leukocyte counts and increased neutrophil oxidative burst activity', *Int J Immunopharm*, **20**, 595–614.

LO YM, YANG ST and MIN DB (1996), 'Kinetic and feasibility studies of ultra-filtration of viscous xanthan gum fermentation broth', *J Membr Sci* **117**, 237–49.

LO YM, ROBBINS KL, ARGIN-SOYSAL S and SADAR LN (2003), 'Viscoelastic effects on the diffusion properties of curdlan gels', *J Food Sci*, **68**, 2057–63.

LOPEZ-ROMERO E and RUIZ-HERRERA J (1977), 'Biosynthesis of β-glucans by cell free extracts from *Saccharomces cerevisiae*', *Biochim Biophys Acta*, **500**, 372–84.

MAEDA YY, TAKAHAMA S and YONEKAWA H (1998), 'Four dominant loci for the vascular responses by the antitumor polysaccharide lentinan', *Immunogenet*, **47**, 159–65.

MANITCHOTPISIT P, SKORY CD, LEATHERS TD, LOTRAKUL P, EVELEIGH DE, PRASONGSUK S and PUNNAPAYAK H (2010), 'α-Amylase activity during pullulan production and α-amylase gene analyses of *Aureobasidium pullulans*', *J Ind Microbiol Biotechnol*, **38**, 1211–8.

MASTROMARINO P, PETRUZZIELLO R, MACCHIA S, RIETI S, NICOLETTI R and ORSI N (1997), 'Antiviral activity of natural and semisynthetic polysaccharides on early steps of rubella virus infection', *J Antimicrob Chemother*, **39**, 339–45.

MATTILA P, SUONPAA K and PIIRONEN V (2000), 'Functional properties of edible mushrooms', *Nutrition*, **16**, 694–6.

MATTYSSE AG, WHITE S and LIGHTFOOT R (1995), 'Genes required for cellulose synthesis in *Agrobacterium tumefaciens*', *J Bacteriol*, **177**, 1069–75.

MCCURDY RD, GOFF HD, STANLEY DW and STONE AP (1994), 'Rheological properties of dextran related to food applications', *Food Hydrocol*, **8**, 609–23.

MCINTOSH M, STONE BA and STANISICH VA (2005), 'Curdlan and other bacterial (1→3)-β-D-glucans', *Appl Microbiol Biotechnol*, **68**, 163–73.

MCNEIL B (1996), 'Fungal biotechnology', in *Encyclopedia of Molecular Biology and Molecular Medicine*, Meyers R (ed), VCH, New York.

© Woodhead Publishing Limited, 2013

MCNEIL B and HARVEY LM (1993), 'Viscous fermentation products', *Crit Rev Biotechnol*, **13**, 275–304.

MCNEIL B and KRISTIANSEN B (1987), 'Influence of impeller speed upon the pullulan fermentation', *Biotechnol Lett*, **9**, 101–4.

MEHVAR R (2000), 'Dextrans for targeted and sustained delivery for therapeutic and imaging agents', *J Control Release*, **69**, 1–25.

MINATO K, MIZUNO M, TERAI H and TSUCHIDA H (1999), 'Autolysis of lentinan, an antitumor polysaccharide, during storage of *Lentinus elodes*, Shiitake mushroom', *J Agric Food Chem*, **47**, 1530–2.

MISAKI A (2004), 'Elsinan, an extracellular α-1,3 : 1,4 glucan produced by *Elsinoe leucospila*: Production, structure, properties and potential food utilization', *Foods Food Ingred J Japn*, **209**(4), 286–97.

MISAKI A, TSUMURAYA Y and TAKAYA S (1978), 'A new fungal a-D-glucan, Elsinan, elaborated by *Elsinoe leucospila*', *Agric Biol Chem*, **42**, 491–3.

MISAKI A, NISHI H and TSUMURAYA Y (1982), 'Degradation of elsinan by alpha amylases and elucidation of its fine structure', *Carbohydr Res*, **109**, 207–19.

MISAKI A, KISHIDA E, KAKUTA M and TABATA K (1993), 'Antitumor fungal β-(1→3)-D-glucans: structural diversity and effects of chemical modification', in *Carbohydrates and Carbohydrate Polymers*, Yalpani M (ed), ATL Press, Illinois.

MITSUHASHI M, YONEYAMA M and SAKAI S (1990), *Growth Promoting Agent for Bacteria Containing Pullulan with or without Dextran*, Canadian Patent 1,049,245.

MIZUNO T (2002), 'Medicinal properties and clinical effects of culimary-medicinal mushroom *Agaricus blazei Murill (Agaricomycetidae)*', *Int J Med Mushrooms*, **4**, 32.

MIZUNO T, HAGIWARA T, NAKAMURA T, ITO, H, SHIMURA K, SUMIYA T and ASAKURA A (1990), 'Antitumor activity and some properties of water-soluble polysaccharides from 'Himematsutake', the fruiting body of *Agaricus blazei* Murrill', *Agric Biol Chem*, **54**, 2889–96.

MOE ST, DRAGET KI, SKJAK-BRAEK G and SMIDSROD O (1995), 'Alginates', *Food Polysaccharide and Application*, New York, Marcel Dekker, 245–86.

MONSAN P, BOZONET S, ALBENNE C, JOUCLA G, WILLEMONT RM and REMAUD-SIMEON M (2001), 'Homopolysaccharides from lactic acid bacteria', *Int Dairy J*, **11**, 675–85.

MOOSAVI-NASAB M, TAHERIAN AR, BAKHTIYARI M, FARAHNAKY A and ASKARI H (2012), 'Structural and rheological properties of succinoglycan biogums made from low-quality date syrup or sucrose using *Agrobacterium radiobacter* inoculation', *Food Bioprocess Technol*, **5**, 638–47.

MORIN A (1998), 'Screening of polysaccharide-producing microorganisms, factors influencing the production, and recovery of microbial polysaccharides', in *Polysaccharides: Structural Diversity and Functional Versatility*, Dumitriu S (ed), Marcel Dekker, New York.

MOZZI F, OLIVER G, SAVOY DE GIORI G and FONT DE VALDEZ G (1995), 'Influence of temperature on the production of exopolysaccharides by thermophilic lactic acid bacteria', *Milchwissenshaft*, **50**, 80–2.

NAESSENS M, CERDOBBEL A, SOETAERT W and VANDAMME EJ (2005), 'Leuconostoc dextransucrase and dextran: production, properties and applications', *J Chem Technol Biotechnol*, **80**, 845–60.

NAKAJIMA H, TOBA T and TOYODA S (1995), 'Enhancement of antigen-specific antibody production by extracellular slime products from slime-forming *Lactococcus lactis* subspecies *cremoris* SBT 0495 in mice', *Int J Food Microbiol*, **25**, 153–8.

NARDIN P and VINCENDON M (1989), 'Isotopic exchange study of the scleroglucan chain in solution' *Macromol*, **22**, 3551–4.

NEWBRUN E and BAKER S (1967), 'Physico-chemical characteristics of the levan produced by *Sreptococcus salivarious*', *Carbohydr Res*, **6**, 165–70.

© Woodhead Publishing Limited, 2013

NOTARARIGO S, NÁCHER-VÁZQUEZ M, IBARBURU I, WERNING ML, DE PALENCIA PF, DUENAS MT, AZNAR R, LÓPEZ P and PRIETO A (2012), 'Comparative analysis of production and purification of homo- and hetero-polysaccharides produced by lactic acid bacteria', *Carbohyd Polym*, in press.

ODA M, HASEGAWA H, KOMATSU S and TSUCHIYA F (1983), 'Anti-tumor polysaccharide from *Lactobacillus* sp.', *Agric Biol Chem*, **47**, 1623–25.

OHNO N, MIURA T, MIURA NN, ADACHI Y and YADOMAE T (2001), 'Structure and biological activities of hypochlorite oxidized zymosan', *Carbohydr Polym*, **44**, 339–49.

OJINNAKA C, JAY AJ, COLQUHOUN IJ, BROWNSEY GJ, MORRIS ER and MORRIS VJ (1996), 'Structure and conformation of acetan polysaccharide', *Int J Biol Macromol*, **19**, 149–56.

OKADA K, YONEYAMA M, MANDAI T, AGA H, SAKAI S and ICHIKAWA T (1990), 'Digestion and fermentation of pullulan', *J Japn Soc Nutr Food Sci*, **43**, 23–9.

OKIYAMA A, MOTOKI M and YAMANAKA S (1993), 'Bacterial cellulose IV. Application to processed foods', *Food Hydrocol*, **6**, 503–11.

OOI VEC and LIU F (2000), 'Immunomodulation and anti-cancer activity of polysaccharide-protein complexes', *Curr Med Chem* **7**, 715–29.

PALANIRAJ A and JAYARAMAN V (2011), 'Production, recovery and applications of xanthan gum by *Xanthomonas campestris*', *J Food Eng*, **106**, 1–12.

PATEL AK, MICHAUD P, SINGHANIA RR, SOCCOL CR and PANDEY A (2010), 'Polysaccharides from probiotics: new developments as food additives', *Food Technol Biotechnol*, **48**, 451–63.

PAUL F, MORIN A and MONSAN P (1986), 'Microbial polysaccharides with actual potential for industrial applications', *Biotechnol Adv*, **4**, 245–9.

POLLOCK TJ (1993), 'Gellan-related polysaccharides and the genus *Sphingomonas*', *J Gen Microbiol*, **139**, 1939–45.

PRETUS H, EUSLEY H, MCNAMEE R, JONES E, BROWDER I and WILLIAMS D (1991), 'Isolation, physicochemical characterisation and pre-clinical efficacy evaluation of a soluble scleroglucan', *J Pharmacol Exp Ther*, **257**, 500–10.

RAEMAEKERS MHM and VANDAMME EJ (1997), 'Production of levansucrase by *Leuconostoc mesenteroides* NRRL B-1355 in batch fermentation with controlled pH and dissolved oxygen', *J Chem Technol Biotechnol*, **69**, 470–8.

RAPP P (1989), '1,3-β-glucanase, 1,6-β-glucanase, and β-glucosidase activities of *Sclerotium glucanicum*: synthesis and properties' *J Gen Microbiol* **135**, 2847–55.

RAU U, GURA E, OLZEWSKI E and WAGNER F (1992), 'Enhanced glucan formation of filamentous fungi by effective mixing, oxygen limitation and fed-batch processing', *J Industr Microbiol*, **9**, 19–26.

REED G and NAGODAWITHANA TW (1991), 'Yeast-derived products and food and feed yeast', in *Yeast Technology*, Rose AH (ed), Van Nostrand Reinhold, New York, 369–440.

REESLEV M, STORM T, JENSEN B and OLSEN J (1997), 'The ability of yeast form of *Aureobasidium pullulans* to elaborate exopolysaccharide in chemostat culture at various pH values', *Mycol Res*, **101**, 650–2.

REIZENSTEIN P and MATHE G (1984), 'Immunomodulating agents', *Immunol Ser*, **35**, 347.

REMMINGHORST U and REHM BHA (2006), 'Bacterial alginates: from biosynthesis to applications', *Biotechnol Lett*, **28**, 1701–12.

RESHETNIKOV SV, WASSER SP, TAN KK (2001), 'Higher Basidiomycota as a source of antitumor and immunostimulating polysaccharides', *Int. J. Med. Mushrooms*, **3**, 361–94.

RHEE S, SONG K, KIM C, PARK B, JANG E and JANG K (2002), 'Levan', in *Biopolymers : Polysaccharides I*, Steinbuchel A (ed.), Wiley-VCH, Weinheim, 351–77.

RHO D, MULCHANDANI A, LUONG JHT and LEDUY A (1988), 'Oxygen requirement in pullulan fermentation', *Appl Microbiol Biotechnol*, **28**, 361–6.

© Woodhead Publishing Limited, 2013

RIDOUT MJ, BROWNSEY GJ, MORRIS VJ and CAIRNS P (1994), 'Physicochemical characterization of an acetan variant secreted by *Acetobacter xylinurn* strain CR1/4', *Int J Biol Macromol*, **16**, 324–30.

RIDOUT, MJ, BROWNSEY, GJ, YORK, GM,WALKER, GC and MORRIS, VJ (1997), 'Effect of O-acyl substituents on the functional behaviour of *Rhizobium meliloti* succinoglycan', *Int J Biol Macromol*, **20**, 1–7.

RIDOUT MJ, BROWNSEY GJ and MORRIS VJ (1998), 'Synergistic interactions of acetan with carob or konjac mannan', *Macromolecules*, **31**, 2539–44.

RINAUDO M (2004), 'Role of substituents on the properties of some polysaccharides', *Biomacromolecules*, **5**, 1155–65.

RINAUDO M and MILAS M (2000), 'Gellan gum, a bacterial gelling polymer', in *Novel Macromolecules in Food Systems*, Doxastakis G and Kiosseoglou V (eds), Elsevier, Amsterdam, 239–63.

ROBBINS EA and SEELEY RD (1977), 'Cholesterol lowering effect of dietary yeast and yeast fractions', *J Food Sci*, **42**, 694–8.

ROBBINS EA and SEELEY RD (1978), '*Process for the Manufacture of Yeast Glycan*', US Patent 4,122,196.

ROBYT JF, YOON SH and MUKERJEA R (2008), 'Dextransucrase and the mechanism for dextran biosynthesis', *Carbohydr Res*, **343**, 3039–48.

RODGERS N (1973), 'Scleroglucan', in *Industrial Gums*, Whistler R and Bemiller J (eds), 2nd ed, Academic Press, New York, 499–511.

ROSALAM S and ENGLAND R (2006), 'Review of xanthan gum production from unmodified starches by *Xanthomonas campestris* sp.', *Enzym Microb Technol*, **39**, 197–207.

RUAS-MADIEDO P, HUGENHOLTZ J, ZOON P (2002), 'An overview of the functionality of exopolysaccharides produced by lactic acid bacteria', *Int Dairy J*, **12**, 163–71.

RUFFING A and CHEN RR (2006), 'Metabolic engineering of microbes for oligosaccharide and polysaccharide synthesis' *Microb Cell Factories*, **5**, 25–33.

RUHMKORF C, RUBSAM H, BECKER T, BORK C, VOIGES K, MISCHNICK P, BRANDT MJ and VOGEL RF (2012), 'Effect of structurally different microbial homoexopolysaccharides on the quality of gluten-free bread', *Europ Food Res and Technol*, **235**, 139–46.

SABRA W, ZENG AP and DECKWER WD (2001), 'Bacterial alginate: physiology, product quality and process aspects', *Appl Microbiol Biotechnol*, **56**, 315–25.

SANDFORD, PA (1982), 'Potentially important microbial gums' in *Food Hydrocolloids*, Glicksman M, CRC Press, Florida, 167–202.

SANDULA J, KOGAN G, KACURACOVA M and MACHOVA E (1999), 'Microbial (1→3)-β-D-glucans, their preparation, physichochemical characterization and immuno-modulatory activity', *Carbohydr Polym*, **38**, 247–53.

SCHILLING BM, RAU U, MAIER T and FANKHAUSER P (1999), 'Modelling and scale up of unsterile scleoglucan production process with *Sclerotium rolfsii* 15205', *Bioprocess Biosyst Eng*, **20**, 195–201.

SCHMID J, MEYER V and SIEBER V (2011), 'Scleroglucan: Biosynthesis, production and application of a versatile hydrocolloid', *Appl Microbiol Biotechnol*, **91**, 937–47.

SEAL CJ and MATHERS JC (2001), 'Comparative gastrointestinal and plasma cholesterol responses of rats fed on cholesterol-free diets supplemented with guar gum and sodium alginate', *Brit J Nutr*, **85**, 317–24.

SENTHILKUMAR V and GUNASEKARAN P (2005), 'Influence of fermentation conditions on levan production by *Zymomonas mobilis* CT2', *Ind J Biotechnol*, **4**, 491–6.

SEYMOUR FR and KNAPP RD (1980), 'Unusual dextrans: 13. Structural analysis of dextrans from strains of *Leuconostoc mesenteroides* and related genera, that contain 3-*O*-α-D-glucosylated α-D-glucopyranosyl residues at the branch points, or in consecutive linear position', *Carbohydr Res*, **81**, 105–129.

SHARMA BR, NARESH L, DHULDHOYA NC, MERCHANT SU and MERCHANT UC (2006). 'Xanthan gum – a boon to food industry', *Food Promot Chronic*, **1** (5), 27–30.

© Woodhead Publishing Limited, 2013

SHARON N and LIS H (1993), 'Carbohydrates in cell recognition', *Sci Am*, **268**, 82–9.

SHASTRY S and PRASAD MS (2005), 'Technological application of an extracellular cell lytic enzyme in xanthan gum clarification', *Braz J Microbiol*, **36**, 57–62.

SHIH IL, YU YT, SHIEH CJ and HSIEH CY (2005), 'Selective production and characterization of levan by *Bacillus subtilis* (Natto) Takahashi', *Agric Food Chem*, **53**, 8211–5.

SHIRASUGI N and MISAKI A (1992), 'Isolation, characterization, and antitumor activities of the wall polysaccharides from *Elsinoe leucospila*', *Biosci Biotechnol Biochem*, **56**, 29–33.

SHU CH, LIN KJ and WEN BJ (2004), 'Effects of culture pH on the production of bioactive polysaccharides by *Agaricus blazei* in batch cultures', *J Chem Technol Biotechnol*, **79**, 998–1002.

SIMON L, BOUCHET B, BREMOND K, GALLANT DJ and BOUCHONNEAU M (1998), 'Studies on pullulan extracellular production and glycogen intracellular content in Aureobasidium pullulans', *Can J Microbiol*, **44**, 1193–9.

SINGH RS, GAGANPREET KS and KENNEDY J (2008), 'Pullulan: Microbial sources, production and applications', *Carbohydr Polym*, **73**, 515–31.

SMITH JH and PACE GW (1982), 'Recovery of microbial polysaccharides', *J Chem Technol Biotechnol*, **32**, 119–29.

SREEKUMAR O and HOSONO A (1998), 'The antimutagenic properties of a polysaccharide produced by *Bifidobacetrium longum* and its cultured milk against some heterocyclic amines', *Can J Microbiol*, **44**, 1029.

STALBERG S, DELIUS U and FERON B (2011). *Smoke-and steam-permeable Food Skin made from a Thermoplastic Mixture with a Natural Appearance*. US Patent 7,976,942, 2011-July-12.

STEPHANOPOULOS G, ARISTIDOU A and NIELSEN J (1998), 'Review of cellular metabolism', in *Metabolic Engineering: Principles and Methodologies*, Stephanopoulos G, Aristidou A and Nielsen J (eds), Academic Press, New York.

STIVALA SS, BAHARY WS, LONG LW, EHRLICH J and NEWBRUN E (1975), 'Levans II. Light scattering and sedimentation data of *Streptococcus salivarious* levan in water', *Biopolymers*, **14**, 1283–92.

SUGIMOTO K (1978), 'Pullulan: production and applications', *J Ferm Ind Japn*, **36**, 98–108.

SURVASE SA, SAUDAGAR PS, BAJAJ IB and SINGHAL RS (2007), 'Scleroglucan: fermentative production, downstream processing and applications', *Food Technol Biotechnol*, **45**, 107–18.

SUTHERLAND IW (1990), Sutherland IW (ed), *Biotechnology of Microbial Exopoly-saccharides*, Cambridge University Press, Cambridge.

SUTHERLAND IW (1994), 'Structure-function relationships in microbial exopoly-saccharides', *Biotechnol Adv*, **12**, 393–448.

SUTHERLAND IW (1995), 'Polysaccharide lyases', *FEMS Microbiol Rev*, **16**, 323–47.

SUTHERLAND IW (1997), 'Bacterial exopolysaccharides-their nature and production', in *Surface Carbohydrates of the Procaryotic Cell*, Sutherland IW (ed), Academic Press, London.

SUTHERLAND IW (1998), 'Novel and established applications of microbial polysaccharides', *Trends Biotechnol*, **16**, 41–6.

SUTHERLAND IW, (2001), 'Microbial polysaccharides from gram-negative bacteria', *Int Dairy J*, **11**, 663–74.

SUZUKI T, SAKURAI T, HASHIMOTO K, OIKAWA S, MASUDA A, OHSAWA M and YADOMAE T (1991), 'Inhibition of experimental pulmonary metastasis of Lewis lung carcinoma by orally administered β-glucan in mice', *Chem Pharm Bull*, **39**, 1606–8.

SUZUKI T, OHNO N, SAITO K, YADOMAE T (1992), 'Activation of the complement system by (1-3)-beta-D-glucans having different degrees of branching and different ultrastructures', *J Pharmacobiodyn*, **15**, 277–85.

TAGUCHI R, SANAKO Y, KIKUCHI Y, SAKUMA M and KOBAYASHI T (1973a), 'Synthesis of pullulan by acetone dried cells and cell free enzyme from *Pullularia pullulans* and the participation of lipid intermediates', *Agric Biol Chem*, **37**, 1635–41.

TAGUCHI R, SANAKO Y, KIKUCHI Y, SAKANO Y and KOBAYASHI T (1973b), 'Polysaccharide production by *Pullularia pullulans*. Part I. Structural uniformity of pullulan produced by several strains of *Pullularia pullulans*', *Agr Biol Chem*, **37**, 1583–8.

TANG YJ AND ZHONG JJ (2002), 'Fed-batch fermentation of *Ganoderma lucidum* for hyperproduction of polysaccharide and ganoderic acid', *Enz Microb Technol*, **31**, 20–8.

TANG YJ and ZHONG JJ (2003), 'Role of oxygen in submerged fermentation of *Ganoderma lucidum* for production of *Ganoderma* polysaccharide and ganoderic acid', *Enz Microb Technol*, **32**, 478–84.

TAURHESIA S and MCNEIL B (1994), 'Physicochemical factors affecting the formation of the biological response modifier scleroglucan', *J Chem Technol Biotechnol*, **59**, 157–63.

TIEKING M, KADITZKY S, VALCHEVA R, KORAKLI M, VOGEL RF and GANZLE MG (2005), 'Extracellular homopolysaccharides and oligosaccharides from intestinal lactobacilli', *J. Appl. Microbiol*, **99**, 692–702.

THAMMAKITI S, SUPHANTHARIKA M, PHAESUWAN T and VERDUYN C (2004), 'Preparation of spent brewer's yeast β-glucans for potential applications in the food industry', *Internat J Food Sci Technol*, **39**, 21–9.

TOSHIFUMI Y and OGAWA K (1998), 'X-ray diffraction of polysaccharides', in *Polysaccharides: Structural Diversity and Functional Versatility*, Dumitriu S (ed), Marcel Dekker, New York, 99–122.

TSUCHIDA H, MIZUNO M, TANIGUCHI Y, ITO H, KAWADE M and AKASAKA K (2001), *Glucomannan Separated from* Agaricus blazei *Mushroom Culture and Antitumor Agent Containing as Active Ingredient*, Japanese Patent 11-080206.

TSUMURAYA Y, MISAKI A, TAKAYA S and TORII M (1978), 'A new fungal α-D-glucan, elsinan, elaborated by *Elsinoe leucospila*', *Carbohydr Res*, **66**, 53–65.

UEDA S, FUJITA K, KOMATSU K and NAKASHIMA Z (1963), 'Polysaccharide produced by the genus *Pullularia*. I. Production of pullulan by growing cells', *J Appl Microbiol*, **11**, 211–5.

VAN GEEL-SCHUTTEN GH, FABER EJ, SMIT E, BONTING K, SMITH MR, TEN-BRINK BB, KAMERLING, JP, VLIEGENTHART JFG and DIJKHUIZEN L (1999), 'Biochemical and structural characterization of the glucan and fructan exopolysaccharides synthetized by the *Lactobacillus reuteri* wild-type strain and by mutant strains', *Appl Environ Microbiol*, **65**, 3008–14.

VAN KRANENBURG R, BOELS IC, KLEEREBEZEM M and DE VOS WM (1999), 'Genetics and engineering of microbial exopolysaccharides for food: approaches for the production of existing and novel polysaccharides', *Curr Opin Biotechnol*, **10**, 498–504.

VANDAMME EJ and SOETAERT W (1995), 'Biotechnological modification of carbohydrates', *FEMS Microbiol Rev*, **16**, 163–86.

VANHOOREN P and VANDAMME EJ (1998), 'Biosynthesis, physiological role, use and fermentation process characteristics of bacterial polysaccharides', *Recent Res Dev Ferment Bioeng*, **1**, 253–99.

VIIKARI L (1984), 'Formation of levan and sorbitol from sucrose by *Zymomonas mobilis*', *Appl Microbiol Biotechnol*, **19**, 252–5.

VINARTA SC, MOLINA OE, FIGUEROA LIC and FARINA JI (2006), 'A further insight into the practical applications of exopolysaccharides from *Sclerotium rolfsii*', *Food Hydrocol*, **20**, 619–29.

WAKSHULL E, BRUNKE-REESE D, LINDERMUTH J, FISETTE L, NATHANS RS, CROWLEY JJ, TUFTS JC, ZIMMERMAN J, MACKIN W and ADAMS DS (1999), 'PGG-glucan, a soluble beta-

© Woodhead Publishing Limited, 2013

(1,3)-glucan, enhances the oxidative burst response, microbicidal activity, and activates an NF-kappa B-like factor in human PMN: evidence for a glycosphingolipid beta-(1,3)-glucan receptor', *Immunopharmacol*, **41**, 89–107.

WALDHERR FW and VOGEL RF (2009), 'Commercial exploitation of homo-exopolysaccharides in non-dairy food systems', in *Bacterial polysaccharides: Current innovations and future trends*, Ullrich M (ed), Caister Academic Press, Norfolk, 313–32.

WANG Y and MCNEIL B (1996), 'Scleroglucan', *Crit Rev Biotechnol*, **16**, 185–215.

WANG X, XU P, YUAN Y, LIU C, ZHANG D, YANG Z, YANG C and MA C (2006), 'Modeling for gellan gum production by *Sphingomonas paucimobilis* ATCC 31461 in a simplified medium', *Appl Environ Microbiol*, **72**, 3367–74.

WANG ZM, CHEUNG YC, LEUNG PH and WU JY (2010), 'Ultrasonic treatment for improved solution properties of a high-molecular weight exopolysaccharide produced by a medicinal fungus', *Bioresource Technol*, **101**, 5517–22.

WASSER SP (2002), 'Medicinal mushrooms as a source of antitumor and immunomodulating polysaccharides', *Appl Microbiol Biotechnol*, **60**, 258–74.

WHITFIELD C and VALVANO MA (1993), 'Biosynthesis and expression of cell-surface polysaccharides', *Adv Microb Physiol*, **35**, 135–246.

WILLIAMS DL, PRETUS HA, MCNAMEE RB, JONES EL, ENSLEY HE and BROWDER IW (1992), 'Development of a water-soluble, sulfated (1→3)-β-D-glucan biological response modifier derived from *Saccharomyces cerevisiae*', *Carbohydr Res*, **235**, 247–57.

WU S, CHEN H, JIN Z, TONG Q (2010), 'Effect of two-stage temperature on pullulan production by *Aureobasidium pullulans*', *World J Microbiol Biotechnol*, **26**, 737–41.

XIE J, ZHAO J, HU DJ, DUAN JA, TANG YP and LI SP (2012), 'Comparison of polysaccharides from two species of *Ganoderma*', *Molecule*, **17**, 740–52.

XU Q, TAJIMA T, LI W, SAITO K, OHSHIMA Y and YOSHIKAI Y (2006), 'Levan (β-2,6-fructan), a major fraction of fermented soybean mucilage, displays immunostimulating properties via Toll-like receptor 4 signalling: induction of interleukin-12 production and suppression of T-helper type 2 response and immunoglobulin E production', *Clin Exper Allergy*, **36**, 94–101.

XU X, PU Q, HE L, NA Y, WU F and JIN Z (2009), 'Rheological and SEM studies on the interaction between spent brewer's yeast β-glucans and k-carrageenan', *J Texture Stud*, **40**, 482–96.

YAMAMOTO Y, TAKAHASHI Y, KAWANO M, IIZUKA M, MATSUMOTO T, SAEKI S and YAMAGUCHI H (1999), '*In vitro* digestibility and fermentability of levan and its hypocholesterolemic effects in rats', *J Nutr Biochem*, **10**, 13–8.

YANG FC and LIAU CB (1998), 'The influence of environmental conditions on polysaccharide formation by *Ganoderma lucidum* in submerged cultures', *Process Biochem*, **33**, 547–53.

YOKOBAYASHI K and SUGIMOTO T (1979), *Molded Body Consisting of or with Content of Glucan*, German Patent 2,842,855.

YOO SH, YOON EJ, CHA J and LEE HG (2004), 'Antitumor activity of levan polysaccharides from selected microorganisms', *Biol Macromol*, **34**, 37–41.

YOO SH, KYUNG HL, LEE JS, CHA J, PARK CS and LEE HG (2005), 'Physicochemical properties and biological activities of DEAE-derivatised *Sphingomonas paucimobilis* gellan', *J Agric Food Chem*, **53**, 6235–9.

YOON EJ, YOO SH, CHA J and LEE HG (2004), 'Effect of levan's branching structure on antitumor activity', *Int J Biol Macromol*, **34**, 191–4.

YUN JW (1996), 'Fructo-oligosaccharides: occurrence, preparation and application', *Enz Microb Technol*, **19**, 107–17.

© Woodhead Publishing Limited, 2013

ZHAN XB, LIN CC and ZHANG HT (2012), 'Recent advances in curdlan biosynthesis, biotechnological production, and applications', *Appl Microbiol Biotechnol*, **93**, 525–31.

ZHANG Y, LI S, WANG X, ZHANG L and CHEUNG PCK (2011), 'Advances in lentinan: Isolation, structure, chain conformation and bioactivities', *Food Hydrocol*, **25**, 196–206.

ZHU D (1987), 'Recent advances on the active components in Chinese medicines', *Abstr Chin Med*, **1**, 251–86.

© Woodhead Publishing Limited, 2013

17

Microbial production of xylitol and other polyols

T. Granström and M. Leisola, Aalto University, Finland

DOI: 10.1533/9780857093547.2.469

Abstract: D-glucose is the prevalent carbohydrate in human and microbial metabolism. Other common monomeric sugars are D-fructose, D-mannose, D-galactose, D-xylose and L-arabinose. In large quantities these sugars are found in bound polymeric forms in plants. When sugars are reduced they are converted to sugar alcohols (polyols) which are rare in nature. The history of sugar chemistry goes back to 18th century when the first sugar alcohols were detected. All sugars and their reduced forms can be consumed by at least some living organisms and the metabolism of the most common ones is discussed in detail. Sugar alcohols such as sorbitol or xylitol have found use in human consumption owing to their low calorie content and their positive physiological effects such as prevention of tooth decay. The common sugar alcohols can be produced in reductive reactions by microorganisms and in this chapter we discuss the present status of these microbial production technologies, including the use of genetically modified organisms. Sugar alcohols can also be produced by chemical catalytic reduction which is in most cases the preferred method for their large scale production. Sugar alcohols are key intermediates owing to their non-chiral nature in biosynthesis of several rare sugars and their derivatives.

Key words: biotechnical production, chemical dehydrogenation, rare sugars, remineralization, sugar alcohols.

17.1 Introduction

All biomass on earth is produced via sugars obtained by photosynthesis. Glucose is the most common sugar in the metabolism but it is found in large quantities in structural plant polymer cellulose and the storage carbohydrate polymer starch. Hemicelluloses are other important structural polymers in plants and their basic monosaccharide components include D-mannose, D-xylose, L-arabinose and D-galactose. Chemically, altogether 59 different tetrose, pentose and hexose sugars can be synthesized, most of them being rare or very rare in nature. The global annual synthesis of plant

© Woodhead Publishing Limited, 2013

biomass is estimated to be 170 billion metric tonnes per year, from which 6 billion tonnes are used by humans for agriculture and forestry and 3.7 billion tonnes are used for food production (Vandamme *et al.*, 2004).

The most important carbohydrate feedstocks are starch and sucrose in terms of human consumption. Starch comes from different sources, but mostly from corn, wheat, rice and potatoes. They all have different starch content ranging from 70% in rice to 20% in potatoes. Starch is composed of straight chain α-glucose polymer i.e. amylose linked together with α-1,4 glucosidic bonds. The branched chain polymer amylopectin is linked together by an α-1,4 glucosidic bond and α-1,6 side chains. Sucrose is composed of glucose and fructose units. The main sources of sucrose are sugar beet and sugar cane, where the typical sucrose content is 15–20%. The world annual production of starch was 66 million metric tonnes in 2008 and sucrose production was 168 million tonnes in 2011 (Foreign Agricultural Service, USDA).

Rare sugars are defined by the International Society of Rare Sugars (ISRS, 2012) as monosaccharide sugars and their derivatives that rarely exist in nature. Rare sugars include all hexoses, pentoses and tetroses and their different isomeric and reduced forms. Starch and sucrose serve as feedstocks for rare sugar production. Starch is hydrolyzed to glucose by a well-known saccharification process using enzymes. The bond between glucose and fructose units in sucrose can be hydrolyzed by the enzyme invertase, but many microbes can use sucrose as such as a carbon source. Sorbitol is the biggest industrial product of all sugar alcohols which are also quite rare in nature. It is produced on a large industrial scale by catalytic hydrogenation of glucose with a production volume of 1.1 million tonnes per year (Patel *et al.*, 2006). The total sugar alcohol production is estimated to be 1.5 million tonnes in 2013. The remaining market share after sorbitol is divided between xylitol, maltitol, mannitol, erythritol and lactitol.

17.2   History of sugars and sugar alcohols

Anselme Payen (1795–1878) was a French chemist known for discovering the carbohydrate cellulose and developing processes for refining sugar and starch. Another chemist who greatly contributed to carbohydrate chemistry was Hermann Emil Fischer (1852–1919). In 1902 he was awarded the Nobel Prize in chemistry for his work on sugar and purine chemistry. He also established the stereochemical configuration of sugars, systematic structure of tetroses, pentoses and hexoses and synthesized glucose, fructose and mannose from glycerol. Emil Fischer and his assistant Rudolf Stahel were the first to manufacture xylitol from beech chips (Fischer and Stahel, 1891) using an acid treatment. Simultaneously with the work of Fischer and Stahel, a French chemist M.G. Bertrand prepared xylitol syrup from wheat and oat straw (Bertrand, 1891).

© Woodhead Publishing Limited, 2013

Sorbitol was discovered and isolated from the *Sorbus aucuparia* tree for the first time in 1872 by a French chemist J. B. Boussingault. Sorbitol production from glucose with a platinum catalyst was patented in Germany in 1926 by Müller, Eppstein and Hofmann from I.G. Farbenindustrie AG, Germany and later on in the USA in 1935. Industrial production began in 1950s using enzymatic hydrolysis of starch to produce glucose followed by catalytic hydrogenation of the glucose. Sorbitol has about 60% of the sweetness of sucrose and is considered to have 2.6 calories per gram (sucrose has 4.0 calories per g). Sorbitol is used in candies and baked goods and mixes, chewing gums, cough drops, cosmetics and in vitamin C production.

Xylose was first discovered and isolated by a Finnish chemist F. Koch in 1881 (Koch, 1886). The isolation method from cottonseed hulls was further improved by Schreiber *et al.* (1930). Xylitol is produced by hydrogenation of xylose. Interest in xylitol production was originally created during the war as an alternative sweetener and later by its applicability as a sweetener for diabetic people (Mellinghoff 1961; Lang 1964). Chemical and microbial production of xylitol was reported by Lohman (1957) and Onishi and Suzuki (1966, 1969). At that time, however, xylitol production from D-glucose was studied, as D-xylose was an expensive substrate. In 1970, an industrial scale chromatographic method for separating different wood hemicellulose sugars was developed in Finland. This enabled the mass production of pure D-xylose. The method was developed where D-xylose was subsequently reduced to xylitol by a metal catalyst under high hydrogen pressure (Melaja *et al.* 1981; Härkönen and Nuojua 1979).

Mannitol was originally isolated from the *Fraxinus ornus* plant. D-Mannose sugar and the sugar alcohol mannitol both derive their name from the extract obtained from the plant. The sap of the manna ash tree contains 30–50% mannitol on a dry weight basis (Schiweck *et al.*, 1994; Lawson, 1997). Commercially mannitol is manufactured from fructose derived from enzymatically hydrolyzed starch or sucrose. The theoretical yield of mannitol from 1 : 1 glucose-fructose syrup is 25% mannitol and 75% sorbitol. A mannitol yield of 50% is in theory possible if sucrose is first inverted to a glucose-fructose mixture; fructose is separated by chromatographic techniques and hydrogenated to mannitol. Mannitol can be separated from sorbitol by fractional crystallization owing to its lower solubility (Soetaert, 1992; Albert *et al.*, 1980; Makkee *et al.*, 1985). Mannitol has about 65% of the sweetness of sucrose (Johnson, 1976) and is considered to have 1.6 calories per gram. It is widely used in the pharmaceutical and food industry mainly for the same reasons as the other sugar alcohols: low caloric value and suitability for diabetic people, but also for its hygroscopic properties as a tableting agent.

Maltitol is manufactured by hydrogenating maltose which is the α-1,4-linked glucose disaccharide derived from starch. Its energy value is 2.1 calories per gram. The use of maltitol in food is approved in Europe and Japan and it has GRAS (generally recognised as safe) status in USA. Maltitol is

© Woodhead Publishing Limited, 2013

about 90% as sweet as sucrose. It is available in crystalline, powdered and syrup forms. It can also be used as a plasticizer in gelatine capsules and as a humectant.

The discovery and extraction of the 4-carbon erythritol from lichen was originally described by John Stenhouse in 1848 in his work 'Examination of the proximate principles of some of the lichens'. He called the substance pseudo-orcin. Erythritol has a caloric value of 0.2 calories per gram and is approximately 70% as sweet as sucrose. It has a mild cooling effect in the mouth and it is used mainly in confectionery and baked products, chewing gums and some beverages.

17.3   Physiological effects of sugar alcohols

Sugar alcohols are used as alternative sweeteners owing to their sweet taste and low caloric content compared to sucrose. In addition, insulin is not involved in their metabolism and therefore they are suitable for daily consumptions by diabetics. Most of the sugar alcohols are laxative when consumed in excess, owing to their poor digestibility in human intestine. The only exception is erythritol which can be absorbed by the gut before entering the colon. Other sugar alcohols accumulate in the colon and when they are metabolized they absorb water causing a laxative effect. Other physiological effects have been discovered for sugar alcohols such as the beneficial effect of xylitol against tooth decay. All sugar alcohols have similar effects on dental caries, but certain properties raise xylitol above the others in this respect (Table 17.1).

The first indication of the beneficial effects of xylitol on tooth decay was shown in the so-called Turku Sugar Studies in 1972–1973 in Finland (Scheinin and Mäkinen, 1975). The purpose of the study was to find out the differences in the caries increment rate as influenced by various sugars. The trial involved almost complete substitution of sucrose (S) by fructose (F) or xylitol (X) during a period of two years. There were no significant initial differences in caries status between the respective sugar groups; 35 subjects in the S-group, 38 in the F-group and 52 in the X-group. After two years the mean increment of decayed, missed and filled tooth surfaces was 7.2 in the S-group, 3.8 in the F-group and 0.0 in the X-group. The results showed a massive reduction of the caries increment in relation to xylitol consumption. Fructose was found to be less cariogenic than sucrose.

It was suggested that the non- and anti-cariogenic properties of xylitol principally depend on its lack of suitability for microbial metabolism and physicochemical effects on plaque and saliva (Scheinin et al., 1976). Shyu and Hsu (1980) studied the cariogenicity of xylitol, mannitol, sorbitol and sucrose on weanling rats. They reported that in comparison to sucrose, xylitol, mannitol, and sorbitol reduced dental caries by 86, 70, and 48%, respectively. All the sugar alcohols studied had lower cariogenicity than

© Woodhead Publishing Limited, 2013

Table 17.1 Properties and physiological effects of sugar alcohols

Sugar alcohol	Low calorie	Suitable for diabetics	Prevention of tooth decay	Chemical hydrogenation	Fermentation	Medical applications
Erythritol	Yes	Yes	Yes	No	Yes	Yes
Xylitol	Yes	Yes	Yes	Yes	Intermediate	Yes
Sorbitol	Yes	Yes	Intermediate	Yes	No	No
Mannitol	Yes	Yes	Intermediate	Yes	No	Yes

© Woodhead Publishing Limited, 2013

sucrose, especially, xylitol, which is a good substitute for sucrose to prevent dental caries. All sugar alcohols can form complexes with metals particularly with Ca-ions. Xylitol–calcium complex accumulates in plaque on the surface of the tooth in a soluble form, which is a prerequisite for tooth remineralization (i.e. minerals are used to reharden the acid softened tooth surface), whereas sorbitol supports the growth of dental plaque and *Streptococcus mutans*, the main cause of tooth decay (Mäkinen, 2010). Scheinin and Mäkinen (1975) showed that sucrose and xylitol gums differed significantly from each other in their tooth decay prevention ability in a situation where the salivary involvement through gum chewing was regarded as similar in both studies. Partly owing to these consumption levels, the Finnish health authorities have recommended the daily use of xylitol in the prevention of caries. This recommendation has been widely followed in the instructions issued by national dental associations in several European and Asian countries.

In certain respects erythritol seems to be even more interesting than xylitol. It is regarded as non-cariogenic in humans. The effects of a six-month use of erythritol, xylitol and sorbitol were investigated in a cohort of 136 teenage subjects assigned to the respective sugar alcohol groups or to an untreated control group ($n = 30$–36 per group). The daily use of sugar alcohols was 7.0 g in the form of chewable tablets, supplemented by twice-a-day use of a dentifrice containing those sugar alcohols. The use of erythritol and xylitol was associated with a statistically significant reduction in plaque and saliva levels. These effects were not observed in other experimental groups. Chemical analyses showed sorbitol to be a normal finding in dental plaque while xylitol was less consistently detected. Erythritol was detected in measurable amounts only in the plaque of subjects receiving this compound. Erythritol and xylitol may exert similar effects on dental caries, although the biochemical mechanism of the effects may differ. These *in vivo* studies were supported by cultivation experiments in which xylitol, and especially erythritol, inhibited the growth of several strains of *S. mutans* (Mäkinen *et al.*, 2005).

Erythritol has been studied for its antioxidant capacity and attenuation of oxidative stress in diabetic rats. Yokozawa *et al.* (2002) reported that oral administration of erythritol [100, 200, or 400 mg/kg body wt/ day for 10 days] to rats with streptozotocin (STZ) -induced diabetes resulted in significant decreases in the glucose levels of serum, liver and kidney. Erythritol also reduced the elevated serum 5-hydroxymethylfurfural level which is an indicator of glycosylated protein and consequently may stimulate autoxidation reactions of sugars. These results suggest that erythritol affects glucose metabolism and reduces lipid peroxidation, thereby lowering the damage caused by oxidative stress involved in the pathogenesis of diabetes. Den Hartog *et al.* (2010) studied the antioxidant properties of erythritol *in vitro* and subsequently detected its antioxidant activity and its vasoprotective effect in the STZ diabetic rat. Erythritol was shown to be an excellent HO•

© Woodhead Publishing Limited, 2013

radical scavenger and an inhibitor of 2,2'-azobis-2-amidinopropane dihydrochloride-induced haemolysis but inert toward superoxide radicals. They concluded that erythritol acts as an antioxidant *in vivo* and may help protect against hyperglycaemia-induced vascular damage (Den Hartog *et al.*, 2010).

Both xylitol and erythritol have been reported to have beneficial effects on the human body. In the case of xylitol, the following physiological effects have been reported: (1) prevention of acute middle ear infections in children (Uhari *et al.*, 1996, 1998); (2) prevention of experimental osteoporosis and improvement in the properties of bones and collagen molecules (Svanberg and Knuuttila, 1994; Mattila *et al.*, 1995; Mäkinen, 2000); (3) increasing the levels of retinol-binding proteins (Georgieff *et al.*, 1985); (4) beneficial in adenosine deaminase deficiency (Bruyland and Ebinger, 1994) and (5) prevention of adrenocortical suppression during steroid therapy (Georgieff *et al.*, 1985).

According to Nissenson *et al.* (1979) mannitol has multiple uses in human medical applications as a diuretic and an obligate extracellular solute. As a diuretic it can be used to treat patients with intractable oedema states, to increase urine flow and flush out debris from the renal tubules in patients with acute tubular necrosis, and to increase toxin excretion in patients with barbiturate, salicylate or bromide intoxication. As an obligate extracellular solute it may be useful to ameliorate symptoms of the dialysis disequilibrium syndrome, to decrease cerebral oedema following trauma or cerebrovascular accident and to prevent cell swelling related to renal ischemia following cross-clamping of the aorta.

17.4   Biochemistry of sugar alcohol metabolism

17.4.1   Erythritol

Erythritol is the only sugar alcohol that is produced predominantly by fermentation. Erythritol production has been reviewed by Moon *et al.* (2010). Many microorganisms and especially yeasts such as *Zygosaccharomyces*, *Debaryomyces*, *Hansenula* and *Pichia* produce erythritol. Erythritol is a metabolite in the pentose phosphate pathway, explaining why microbes are able to produce it. In yeasts glucose-6-phosphate is converted by glucose-6-phosphate dehydrogenase into ribulose-5-phosphate generating 2 moles of nicotinamide adenine dinucleotide phosphate (NADPH) and CO_2. Ribulose-5-phosphate is converted to ribose-5-phosphate and xylulose-5-phosphate by enzyme action of ribose-5-isomerase and ribulose-5-phosphate 3-epimerase, respectively. By the action of transaldolase, glyceraldehyde-3-phosphate and sedoheptulose-7-phosphate are converted to fructose-6-phosphate and erythrose-4-phosphate. Finally, erythrose-4-phosphate and xylulose-5-phosphate are converted to glyceraldehyde-3-phosphate and fructose-6-phosphate by the enzyme action of transketolase.

© Woodhead Publishing Limited, 2013

In microbial metabolism, erythrose-4-phosphate is first dephosphorylated by erythrose-4-phosphokinase and then reduced to erythritol by NADPH-dependent aldose reductase (Lee *et al.* 2003). In bacteria, ribulose-5-phosphate is converted to xylulose-5-phosphate and then split into acetyl phosphate and glyceraldehyde-3-phosphate by pentose phosphate phosphoketolase present in heterofermentative lactic acid bacteria. Erythritol production is reported, for example, with *Pichia, Zygopichia, Candida, Torulopsis, Trigonopsis* and *Moniliella* (Onishi 1960; Hajny *et al.* 1964). The erythritol yield and productivity with *Aureobasidium* sp. SN-G42 was 47% and 2.0 g l^{-1} h^{-1} respectively in 96 h in a 100 m^3 bioreactor (Sawada *et al.* 2009).

17.4.2 Xylitol

Xylose derivatives are intermediates in the pentose phosphate pathway. Microbes that have been reported to carry out the reduction of xylose to xylitol include yeasts and fungi. So far the research on converting xylose to xylitol has focused on using yeasts as the production organisms. The most efficient producers belong to the genus *Candida*. These yeasts are able to reduce xylose to xylitol using a one-step enzymatic reaction. First xylose is taken up by the cell and then reduced to xylitol by the action of xylose reductase (XR) enzyme. The xylitol formed is either used for growth in aerobic conditions or accumulates in oxygen limited conditions. Most of the accumulated xylitol leaves the cell by diffusion. The remaining xylitol inside the cell is oxidized to xylulose by xylitol dehydrogenase (XDH) and then phosphorylated to xylulose 5-phosphate by xylulokinase (XK), before it enters the pentose phosphate pathway and then the central metabolic pathways. Yeasts can produce xylitol from xylose with up to 70–85% yield, but only if its oxygen availability is restricted (Granström, 2002; Kwon *et al.*, 2006). Part of the xylose has to be used for cell metabolism and maintenance energy and the remaining part goes to carbon dioxide and biomass.

Xylitol accumulation is a response to the reduction/oxidation imbalance inside the cell caused by oxygen limited conditions. This redox imbalance takes place because xylose reductase uses predominantly NADPH as a cofactor, whereas xylitol dehydrogenase enzyme uses NAD^+. These yeasts, which are natural xylose users like *Candida* spp., can generate a sufficient amount of NADPH through the pentose phosphate pathway (PPP) for high uptake rates of xylose and xylitol excretion. However, many strains of *Candida* spp. are opportunistic or even pathogenic microorganisms, which limit their applicability to the food industry (Fridkin and Javis, 1996).

Considerable time and effort has been invested in metabolic engineering of common baker's yeast (*Saccharomyces cerevisiae*) in order to make it an efficient xylitol producer from xylose since the initial work by Kötter *et al.*

© Woodhead Publishing Limited, 2013

(1990). This is due to the fact that *S. cerevisiae* would make an ideal host organism for xylitol production since it has a GRAS status. Furthermore, regarding ethanol production from wood biomass it can produce and tolerate high amounts of ethanol. However, *S. cerevisiae* is not a natural xylose user like *Candida* yeasts. This defect has been attributed to the inability to regenerate sufficient amounts of NADPH for the activity of xylose reductase (van Dijken and Scheffers, 1986; Rizzi *et al.* 1989). In the case of natural xylose users, the flux rate through the pentose phosphate pathway (PPP) is increased according to the requirement of NADPH in the cell metabolism.

Natural lactic acid bacteria (LAB) are not known to produce xylitol, although it has been reported that some strains of *Streptococcus avium* and *Lactobacillus casei* are able to metabolize it (London, 1990). The production of xylitol by LAB has been studied using *Lactococcus lactis* as the production host. A yeast xylose reductase catalyzing the reduction of xylose to xylitol has been expressed in this strain (Nyyssölä et al., 2005). Xylitol production was investigated using resting cells in high cell density with a glucose–xylose mixture as substrate. With a high initial xylose concentration ($160 \ g \ l^{-1}$) the ratio of xylitol produced per glucose consumed was 2.5 mol mol^{-1} and the volumetric productivity was 2.7 g l h^{-1} at 20 h. At this point 34% of the xylose initially present was consumed. This volumetric productivity reported for *L. lactis* is not far from the highest reported for yeasts of 4–5 g l h^{-1}. An obvious downside was that the xylose was not fully converted to xylitol.

Considerable efforts have been made in order to produce xylitol from D-glucose which is a cheaper raw material than D-xylose. Xylitol phosphate dehydrogenase (XPDH) genes from several Gram-positive bacteria were isolated and expressed in *Bacillus subtilis*. Expression of XPDH enzyme in D-ribulose and D-xylulose producing *B. subtilis* strain resulted in D-glucose conversion into xylitol at around 23% yield (Povelainen and Miasnikov, 2006). Genetically engineered *S. cerevisiae* was able to produce only 3.6% sugar alcohols from glucose of which 50% were xylitol (Toivari *et al.*, 2007).

17.4.3   Sorbitol

In the cell metabolism sorbitol is produced from glucose by aldehyde reductase using NADPH as a cofactor. Sorbitol can be converted to fructose by L-iditol dehydrogenase using NAD$^+$ as a cofactor. Fructose can be converted to glucose directly by the enzyme glucose isomerase. The difference between these two metabolic pathways is that the first one includes reduction and oxidation reactions whereas the second one is an isomerization reaction. The other approach would be to use sorbitol dehydrogenase enzyme, which converts fructose 6-phosphate to sorbitol 6-phosphate and remove the phosphoryl group using the enzyme, phosphatase. Successful

© Woodhead Publishing Limited, 2013

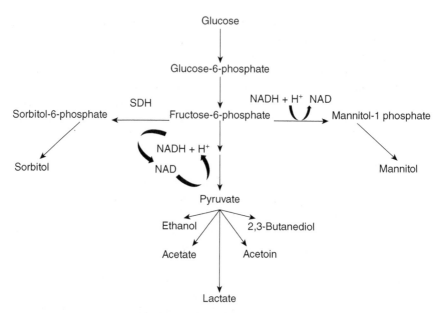

Fig. 17.1   Metabolism of sorbitol production in *Lactobacillus plantarum* (Nyyssölä and Leisola, 2005).

engineering of sorbitol production into *Lactobacillus. plantarum* has been reported (Fig. 17.1).

The reduction of fructose-6-phosphate to sorbitol-6-phosphate is catalyzed by a sorbitol dehydrogenase. Since sorbitol is a relatively cheap bulk product, it would at first hand seem difficult to develop a biotechnological process that could compete with catalytic hydrogenation of glucose. The production of sorbitol by *Zymomonas mobilis* has been described by Viikari (1984). The results indicated that sorbitol was not produced from glucose or fructose alone. When 75+75 g l^{-1} of sugar mixture was used 17 g l^{-1} of sorbitol was produced along with ethanol. The conclusion was that when fructokinase is inhibited by glucose, fructose is utilized through reduction to sorbitol by sorbitol dehydrogenase. In the reduction of fructose, nicotinamide adenine dinucleotide (NADH) is oxidized to NAD$^+$, which is normally obtained through the reduction of acetaldehyde to ethanol. In *Z. mobilis* the enzyme responsible for sorbitol production is glucose-fructose oxidoreductase (GFOR). It was originally found by Zachariou and Scopes (1986) and it converts glucose to glucono-δ-lactone and reduces fructose to sorbitol. Gluconolactonase enzyme (GL) involved in glucono-δ-lactone conversion to gluconic acid and eventually to ethanol. Consequently, ethanol and sorbitol are the main products of *Z. mobilis* fermentation (Viikari, 1984).

© Woodhead Publishing Limited, 2013

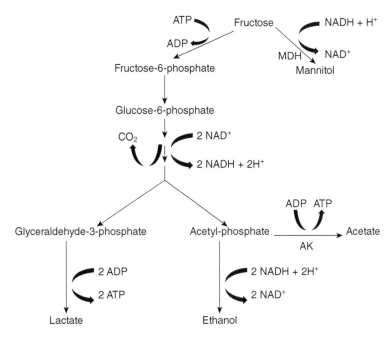

Fig. 17.2   Overview of hexose metabolism and mannitol synthesis in heterofermentative LAB (Nyyssölä and Leisola, 2005).

17.4.4   Mannitol

The ability of heterofermentative LAB to produce mannitol has been known since the 1920s, when it was shown that *Lactobacillus pentoaceticus* was able to reduce fructose to mannitol (Peterson and Fred, 1920). This is based on the use of the pentose phosphate pathway for glucose metabolism instead of Embden–Meyerhoff pathway as in the case of homofermentative LAB. In heterofermentative LAB this reaction is catalyzed by a mannitol dehydrogenase (Martinez *et al.*, 1963; Aarnikunnas *et al.*, 2002). As discussed above, reduction of fructose to mannitol allows the heterofermentative mannitol producers to regenerate NAD$^+$ in this reaction. This enables the cells to divert the acetate–phosphate formed under the heterolactic fermentation to adenosine triphosphate (ATP) production instead of ethanol production (Fig. 17.2). When only fructose is available, it is used as a substrate for growth as well as an electron acceptor. Part of the fructose is phosphorylated to fructose-6-phosphate, which is then isomerized to glucose-6-phosphate and metabolized further via the normal heterofermentative pathway (Axelsson, 1998; Eltz and Vandermark, 1960). Unlike homofermentative LAB, heterofermentative LAB do not appear to convert fructose-6-phosphate to mannitol (Wisselink *et al.*, 2002).

It has been shown in a *Lb. plantarum* strain deficient in lactate dehydrogenase that pyruvate is converted mainly to acetoin with organic acids,

© Woodhead Publishing Limited, 2013

2,3-butanediol, ethanol and mannitol as the secondary end products (Ferain *et al.*, 1996). A similar phenomenon has been shown also for a *L. lactis* strain in which the lactate dehydrogenase gene was disrupted. Since NAD^+ was no longer regenerated in the reaction in which pyruvate is reduced to lactate, the mannitol 1-phosphate dehydrogenase catalyzed reaction was used for this purpose instead. After the depletion of glucose, mannitol was transiently produced and metabolized in the lactate dehydrogenase deficient *L. lactis* cells (Neves *et al.*, 2000). Many natural heterofermentative isolates have also been shown to be very efficient mannitol producers.

Heterofermentative strains that have been studied for mannitol production belong mainly to *Lactobacillus* and *Leuconostoc* species. Mannitol can accumulate in very high concentrations in the medium of these LAB. With the most efficient processes, mannitol concentrations close to the solubility limit (180 g l^{-1} at 25 °C) can be reached (Soetaert *et al.*, 1995). The problem with using LAB for industrial production of commercially interesting bulk metabolites is that they have complex nutritional requirements for amino acids, peptides, nucleotide bases, vitamins, minerals and fatty acids. Therefore, complex media components have to be used, which increases the costs of the fermentation processes. The study of mannitol production by LAB has mainly been carried out using pure sugar substrates. The use of pure sugars is, however, not economically feasible on the industrial scale owing to the high cost of these substrates and to the relatively low price of the products. Cheaper raw materials such as hydrolysates of plant materials or side-streams of the sugar industry would have to be considered as the first choice of raw material.

17.5   Biotechnological production strategies

Xylitol is particularly in demand from chewing gum manufacturers in Asia. It has been estimated that 80–90% of the chewing gum sold in the region now have xylitol in their formulations. Danisco Sweeteners is the world leader in xylitol production, but some other manufacturers have moved into the market, for example Futaste from China. In 2007 Danisco Sweeteners invested €23 million in its factory located in Lenzing, Austria. Lenzing AG provides Danisco with xylose containing liquor from its dissolving pulp process. The xylitol manufacturing process helps to achieve sustainability since only wood biomass is used as a feedstock. More capacity has also come from Chinese suppliers, although this is based on corn cobs and is not considered to be as sustainable as lignocellulosic biomass. Despite high expectations a few years ago, xylitol markets and production have remained more or less static. There are numerous reasons for the slow development of xylitol markets. The main reason is that xylitol has not gained the same kind of recognition in US markets as it has in Scandinavian and Asian countries. Furthermore, xylitol has without doubt beneficial effects on tooth

© Woodhead Publishing Limited, 2013

decay, but people are increasingly more receptive to natural food without any supplements particularly in European countries. It should, however, been pointed out that xylitol is present in plums, raspberries and cauliflower and can thus be considered to be natural. The recent acquirement of Danisco Ltd. by DuPont Ltd. may change the future of xylitol in US markets since they have the expertise and experience in working in the North American continent.

Biotechnological production of xylitol from xylose containing industrial waste streams has been studied extensively to date. However, very few production plants, if any, have been built to produce xylitol by a fermentation route and none of those are in Europe or the USA. The key issue is that the chemical hydrogenation method is very cost effective and the yield is almost 100% from a pure xylose stream. The corresponding yield from a fermentation process under industrial conditions is below 50% even with natural xylose utilising yeasts like *Candida* or with metabolically engineered *S. cerevisiae*. High xylitol yields from D-xylose have been frequently reported under laboratory conditions. The research has mainly focused on optimizing production conditions and the effect of oxygen (Table 17.2). Recent efforts include using hydrolysates (Chiung-Fang *et al.*, 2011) and, cell recycling (Kwon *et al.*, 2006; Granström, 2002) to prolong the production phase and thus to increase the xylitol productivity or screening for new strains (Guo *et al.* 2006; Lopez *et al.* 2004), or using metabolically engineered *S. cerevisiae* (Bae *et al.*, 2004).

The oxygen availability is the most important factor in terms of xylitol production from D-xylose with *Candida* yeasts. Under oxygen limited conditions, oxidative phosphorylation is not able to reoxidize all the generated NADH. Therefore, the intracellular concentration of NADH increases, resulting in xylitol accumulation. This mechanism has been confirmed in controlled continuous culture conditions. We established a descending agitation gradient in the bioreactor to lower the oxygen availability from fully aerobic to anaerobic conditions in a continuous chemostat culture. This resulted in a gradual decrease in dissolved oxygen concentration and onset of xylitol accumulation in *Candida guilliermondii* (Granström *et al.* 2001) and *Candida tropicalis* (Granström and Leisola, 2002). The time of onset of xylitol accumulation was controlled by increasing the dilution rate (i.e. flow rate of xylose substrate feed) in the chemostat simultaneously with decreasing oxygen availability. These two strains differ in their cofactor dependency in the xylose metabolic pathway. *C. guilliermondii*, has an exclusively NADPH-dependent xylose reductase enzyme. Consequently it accumulates acetate to regenerate NADPH since acetaldehyde dehydrogenase enzyme converting acetate into ethanol is also NADPH dependent. In the case of *C. tropicalis*, the xylose reductase enzyme has dual dependence on NADH and NADPH. Therefore, it does not accumulate acetate under the same conditions as *C. guillermondii* even though it produces ethanol, indicating that it has an acetaldehyde dehydrogenase enzyme. Both strains regenerate

© Woodhead Publishing Limited, 2013

Table 17.2 Summary of microbiological production of xylitol by *Candida tropicalis*, *Candida guilliermondii* and metabolically engineered *S. cerevisiae*

Strain	Yield (g g^{-1})	Xylose (g l^{-1})	Productivity (g l^{-1} h^{-1})	Process strategy	Ref.
C. guilliermondii	0.78	250	Not given	Fed batch O$_2$ limit	Ojamo (1994)
C. guilliermondii	0.73	62	0.52	Batch O$_2$ limit	Roberto et al. (1999)
C. tropicalis	0.82	750	4.94	Cell recycling yeast extract Glu-xyl feed O$_2$ limit	Choi et al. (2000)
C. tropicalis	0.69	100	5.7	Cell recycling mineral medium O$_2$ limit	Granström (2002)
C. tropicalis	0.85	214	12.0	Cell recycling O$_2$ limit	Kwon et al. (2006)
C. tropicalis	0.71	45	0.51 (calc)	Non detoxified rice straw hydrolysate	Chiung-Fang et al. (2011)
S. cerevisiae (recombinant)	0.95	190	0.40 (calc)	Fed batch Glu-xyl feed	Hallborn et al.(1991)
S. cerevisiae (recombinant)	0.81	10	0.19	Shakeflask Glu-xyl O$_2$ limit	Hallborn et al.(1994)
S. cerevisiae (recombinant)		80 + 18 (glu)	2.34	Cell recycle Xyl-glu grown cells	Bae et al. (2004)

© Woodhead Publishing Limited, 2013

NAD^+ by accumulation of glycerol, which is a known redox sink in *S. cerevisiae* (Oura, 1997). The NAD^+ regeneration is even more pronounced in *C. tropicalis* because it produces both glycerol and ethanol in these conditions.

The dual dependency of NADH/NADPH of xylose reductase increases the demand of NADH for xylose flux. The different responses to oxidative stress caused by increased intracellular NADH concentration have been studied by formate co-feeding. A fully aerobic steady state was established with no xylitol accumulation. Consequently, formate was fed as a co-substrate, resulting in increased carbon dioxide production and intracellular NADH concentration by the action of formate dehydrogenase enzyme. *C. guilliermondii* produced only glycerol, indicating the most prominent pathway for regenerating NAD^+. *C. tropicalis* produced glycerol, ethanol and xylitol, indicating several possibilities for regenerating NAD^+ (Granström and Leisola, 2002).

The conclusions of the above experiments are that *Candida* yeasts require some oxygen since they cannot grow or produce xylitol in either fully anaerobic nor aerobic conditions. This might be one of the bottlenecks in producing xylitol via a fermentation route since the accurate control of the oxygen level in industrial size bioreactors is difficult and thus expensive. The only way to make biotechnological production of xylitol from xylose more competitive than by the chemical hydrogenation process is to use lignocellulosic hydrolysates or industrial side streams without the need of xylose purification.

Chiung-Fang *et al.* (2011) isolated a yeast strain, *C. tropicalis* JH030 that was shown to have a capacity for xylitol production from hemicellulosic hydrolysate without detoxification. The yeast gives a promising xylitol yield of 0.71 g g^{-1} from non-detoxified rice straw hydrolysate that had been prepared by diluted acid pretreatment under severe conditions. The yeast's capacity was also found to be practicable with various other raw materials, such as sugarcane bagasse, silvergrass, napiergrass and pineapple peel. Bae *et al.* (2004) produced xylitol from xylose in repeated fed-batch and cell-recycle fermentations using a recombinant *S. cerevisiae* BJ3505/δXR harbouring the xylose reductase gene from *Pichia stipitis*. Batch fermentations with 20 g l^{-1} xylose and 18 g l^{-1} glucose resulted in 9.52 g l^{-1} dry cell mass and 20.1 g l^{-1} xylitol concentrations. Repeated fed-batch operation to remove 10% of the culture broth and to supplement an equal volume of 200 g l^{-1} led to a high accumulation of xylitol at 48.7 g l^{-1}. To overcome the loss of xylitol-producing biocatalysts in repeated fed-batch fermentations, cell-recycle equipment with a hollow fibre membrane was implemented into a xylitol production system. A final dry cell mass of 22 g l^{-1}, and 116 g l^{-1} xylitol concentration with a 2.34 g l^{-1} h^{-1} overall xylitol productivity was obtained in cell-recycle fermentations.

A new bioprocess concept for production of mannitol was described by von Weymarn *et al.* (2002). Initially, the ability of ten heterofermentative

© Woodhead Publishing Limited, 2013

LAB to produce D-mannitol from D-fructose in a resting state were compared. *Leuconostoc mesenteroides* ATCC-9135 was selected for further examination in high cell density membrane cell-recycle cultures. High volumetric mannitol productivity (26.2 g l^{-1} h^{-1}) and mannitol yield (97%) were achieved. Furthermore, using the same initial biomass, a stable high-level production of mannitol was maintained for 14 successive bioconversion batches. Using a simple purification protocol, the crystallization yield was 85%. Hence, the total yield for the process was 0.77 g crystalline mannitol per gram of fructose and 0.52 g crystalline mannitol per gram initial sugar (fructose and glucose). Furthermore, in the novel bioprocess, only 0.67 g by-products are formed for each gram of crystalline mannitol obtained. According to Devos (1995) the respective values for a typical chemical mannitol production process are 0.39 g crystalline mannitol per gram initial sugar and 1.58 g by-products for each gram of crystalline mannitol. The by-products of the bioprocess are mainly acetate and lactate. Hence, to improve the economy of the process, applications for these compounds should be found (von Weymarn *et al.*, 2002).

17.5.1 Chemical dehydrogenation

The first patent for xylitol manufacturing was issued in 1975 by Melaja and Hämäläinen (Melaja *et al.*, 1975, 1978) working for the Finnish sugar refining company, Suomen Sokeri Oy (Table 17.3). The patent describes a method for recovering xylitol from xylan containing raw materials. The process includes hydrolyzing the raw material, purifying the hydrolysate by ion exclusion and colour removal. The purified solution is subjected to chromatographic fractionation to produce a solution containing a high level of xylose. The xylose solution is hydrogenated and a xylitol-rich solution is recovered by chromatographic fractionation using ion exchange resins. Xylitol is produced industrially by chemical hydrogenation from xylose with Raney nickel type or Ruthenium-based catalysts at high temperature and pressurized hydrogen atmosphere. Hydrogen reacts with xylose and the resulting xylitol yield is over 90%. Xylose and xylitol are easily crystallized and therefore this production method is favoured by the industry because of its cost effectiveness and straightforward procedure.

Suomen Sokeri was issued a patent for recovery of xylitol (Melaja, Virtanen and Heikkilä, 1978). The patent describes a method of preparing pharmaceutical grade xylitol from an aqueous solution containing mixtures of sugar alcohols including xylitol. In this method the solution is subjected to crude crystallization and recrystallization of xylitol followed by the recovery of residual xylitol from the mother liquor by fractionating the solution using at least two columns of ion-exchange resin in two different metal forms. In the next generation patents, a method for fractionating a solution by a chromatographic simulated moving bed method was filed by Heikkilä *et al.* (1996). The patent describes a method where the liquid flow

© Woodhead Publishing Limited, 2013

Table 17.3 Patented xylitol production processes

Patent	Inventors	Assignee	Filed
Process for large scale chromatography	Melaja, Hämäläinen, Rantanen	Suomen Sokeri Oy, Finland	February 14, 1975
Process for making xylitol	Melaja, Hämäläinen	Suomen Sokeri Oy	June 18, 1975
Process of making xylose	Melaja, Hämäläinen	Suomen Sokeri Oy	August 28, 1975
Method of recovering xylitol	Melaja, Virtanen, Heikkilä	Suomen Sokeri Oy	January 3, 1978
Production of pure sugars and lignosulphonates from sulphite spent liquor	Heikkilä	Suomen Sokeri Oy	October 4, 1985
Method for fractionation of a solution	Heikkilä, Kuisma, Paananen	Suomen Sokeri Oy	April 19, 1995

© Woodhead Publishing Limited, 2013

moves in a system comprising at least two sectional beds in different ionic forms. The fractions enriched with different components are recovered during a multi-step sequence including the following operations, (i.e. phases): feeding phase, eluting phase and recycling phase. The liquid present in the sectional packing material beds with its dry solids concentration profile is recycled during the recycling phase in a loop comprising one, two, or more sectional packing material beds. The method can be employed for fractionating, for example sulphite cooking liquor, molasses and vinasse.

Müller *et al.* (1935) from I.G. Farbenindustrie AG, Germany filed a US patent to produce polyhydric alcohols from aldoses in 1935 (intially the patent was applied in Germany in 1926) with a platinum catalyst in the presence of alkali hydrogen. In this way they were able to convert glucose or mannose into sorbitol and mannitol. The platinum catalyst was improved and patented in 1935 by Rothrock from E.I du Pont de Nemours & Company from Delaware USA by increasing the nickel portion of the catalyst to 70%. According to Silveira and Jonas (2002) the reaction temperature is in the range 120–150°C, the pressure is *ca.* 70 bar and time of hydrogenation is 2–4 h. Approximately 70% sorbitol solution is obtained and the catalyst is eliminated by precipitation and filtration. Purification of the sorbitol solution is performed by ion exchange chromatography and activated charcoal filter. *Z. mobilis* strain could be considered as a production host for microbiological production of sorbitol. The fermentation process would produce ethanol and gluconic acids as end products in addition to sorbitol. All these products have markets, which would increase the cost effectiveness of the process. In the case where a sorbitol fermentation plant could be integrated into a glucose–fructose conversion plant this might bring some advantages over the chemical catalysis process in terms of cost effectiveness. Otherwise, it is very difficult for the biotechnological method to compete with the chemical hydrogenation method in terms of cost effectiveness.

17.5.2   Rare sugars via sugar alcohols from readily available sugar molecules

In recent years new enzymatic and chemical methods have been developed which allow mass production of all monosaccharides; tetroses, pentoses and hexoses from readily available feedstock. This is achieved by using D-tagatose 3-epimerase, aldose isomerase, aldose reductase and oxidoreductase enzymes or whole cells as biocatalysts. Bioproduction strategies for all rare sugars are illustrated using ring form structures given the name Izumoring according to its inventor Professor Izumori (Fig. 17.3). In this production strategy the sugar alcohols are key intermediates. For example the ketopentulose sugars are very reactive in the C-2 position of the molecule and they can be used as skeletons for new rare sugar molecules. The deoxy forms of L-xylulose and L-ribulose can be produced from xylitol and ribitol by respective polyol dehydrogenases (Granström *et al.*, 2005; Kylmä *et al.*, 2004;

© Woodhead Publishing Limited, 2013

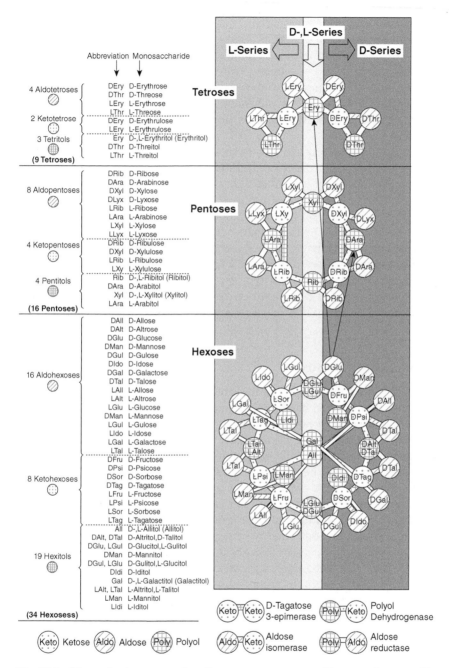

Fig. 17.3 Biosynthesis strategy for all tetroses, pentoses and hexoses from readily available raw materials starch, wood and lactose. Sugar alcohols have a key role in this strategy (Granström *et al.*, 2004).

© Woodhead Publishing Limited, 2013

Poonperm *et al.*, 2007; Aarnikunnas *et al.*, 2006). The main principle that allows the production of both D- and L-configuration sugars from polyols is that, with the exception of D-mannitol, they are optically inactive. All the D- and L-sugars can be obtained from readily available D-glucose. These conversions are achieved with specific enzymes, that is D-tagatose 3-epimerase, aldose isomerase, aldose reductase and oxidoreductase enzymes or whole cells as biocatalysts according to the Izumoring strategy (Granström *et al.*, 2004).

A chemical method has also been developed for producing L-ribulose from L-arabinose (Ekeberg *et al.*, 2002). L-xylulose and L-ribulose can be used as precursors for a number of interconversions to aldopentoses, keto-pentoses or pentitols (Granström *et al.*, 2004). The effect of these kinds of rare sugar molecules on human metabolism is largely unknown. However, it can be assumed to be fundamental since DNA and RNA molecules also have deoxysugar skeletons. Rare sugar molecules and sugar molecules with different chiral centres can be used for screening new enzyme reactions from various sources such as microbes, plants and crops. Rare sugar molecules are studied as potential new drugs for cancer treatment and viral infections as well as potential sweeteners.

17.6   Future trends

Most of the commercial sugar alcohols have already established markets based on their proven benefits to human metabolism. The markets will not probably grow much owing to the costs of production unless new health-promoting properties are found. Their use as an alternative sweetener, for example for diabetics, will be the main driver for their consumption. However, there will be an ever-increasing competition from natural low calorie products such as stevia or even D-xylose. These kinds of products may be better accepted by consumer groups which emphasize the requirement for healthy and natural food products, particularly in Europe. In the future it would be important to integrate sugar alcohol production into existing industrial processes in order to keep the production costs at a reasonable level. Biotechnological processes that can be replaced by chemical dehydrogenation are not viable in terms of process economy unless the product is significantly more expensive than the raw material. A good example is erythritol which has promising properties compared to xylitol, but can only be produced by fermentation. In order to expand the markets for erythritol, a biotechnological process using lignocellulosic feedstock should be developed. One interesting future trend for sugar alcohols is to act as a carbon skeleton for building branched rare sugar molecules. The human metabolism is based on sugars and their effect is therefore elemental. They have the possibility to act as new drugs for a number of diseases that have no cure currently or to replace the weakening effect of antibiotics.

For example, a patent has been issued to Fletre *et al.* (2011) Roquette Freres AS for an efficacy of branched maltodextrin in combination with insoluble fibres to protect colon mucosa.

17.7   References

AARNIKUNNAS, J., RÖNNHOLM, K. and PALVA, A. (2002) 'The mannitol dehydrogenase gene (mdh) from *Leuconostoc mesenteroides* is distinct from other known bacterial mdh genes'. *Appl Microbiol Biotechnol*, **59**, 665–71.

AARNIKUNNAS, J.S., PIHLAJANIEMI, A., PALVA, A., LEISOLA, M. and NYYSSOLÄ, A. (2006) 'Cloning and expression of a xylitol-4-dehydrogenase gene from *Pantoea ananatis*'. *Appl Environ Microbiol*, **72**, 368–77.

ALBERT, R., STRÄTZ, A. and VOLLHEIM, G. (1980) 'Die katalytische Herstellung von Zuckeralkoholen und deren Verwendung'. *Chem Ing Tech*, **52**, 582.

AXELSSON, L. (1998) '*Lactic Acid Bacteria: Microbiology and Functional Aspects*, Salminen, S. and von Wright, A. (eds), 2nd edn., Marcel Decker, New York, 1.

BAE, S-M., PARK, Y-C., LEE, T-H., KWEON, D-H., CHOI, J-H., KIM, S-K., RYU, Y-W. and SEO, J-H. (2004) 'Production of xylitol by recombinant *Saccharomyces cerevisiae* containing xylose reductase gene in repeated fed-batch and cell-recycle fermentations'. *Enzyme Microb Technol*, **35**, 545–9.

BERTRAND, M.G. (1891) 'Recherches zur quelques derives du xylose'. *Bull Soc Chim Fra*, **5**, 554–7.

BRUYLAND, M. and EBINGER, G. (1994) 'Beneficial effect of a treatment with xylitol in a patient with myoadenylate deaminase deficiency'. *Clin Neuropharmacol*, **17**, 492–3.

CHIUNG-FANG, H., YI-FENG, J., GIA-LUEN, G. and WEN-SONG, H. (2011) 'Development of a yeast strain for xylitol production without hydrolysate detoxification as part of the integration of co-product generation within the lignocellulosic ethanol process'. *Biores Technol*, **102**, 3, 3322–9.

CHOI, J.H., MOON, K.H., RYU, Y.W. and SEO, J.H. (2000) 'Production of xylitol in cell recycle fermentations of *Candido tropicalis*', *Biotechnol Lett*, **22**, 1625–8.

DEN HARTOG, G., BOOTS, A., ADAM-PERROT, A., BROUNS, F., VERKOOIJEN, I., WESELER, A., HAENEN, G. and BAST, A. (2010) 'Erythritol is a sweet antioxidant'. *Nutrition*, **26**, 449–58.

DEVOS F. (1995) '*Process for the Production of Mannitol*'. US Patent US 5466795.

EKEBERG, D., MORGENLIE, S. and STENSTRØM, Y. (2002) 'Base catalysed isomerisation of aldoses of the arabino and lyxo series in the presence of aluminate'. *Carbohydr Res*, **337**, 779–86.

ELTZ, R.W. and VANDEMARK, P.J. (1960) 'Fructose dissimilation by *Lactobacillus brevis*'. *J Bacteriol*, **79**, 763–76.

FERAIN, T., SCHANK, A.N. and DELCOUR, J. (1996) '13C Nuclear magnetic resonance analysis of glucose and citrate end products in an ldhL-ldhD double-knockout strain of *Lactobacillus plantarum*'. *J Bacteriol*, **178**, 7311–15.

FISCHER, E. and STAHEL, R. (1891) 'Zur Kenntniss der Xylose'. *Ber Chem Ges*, **24**, 528–39.

FLETRE, O.H., MORBECQUE, D.W., SIEPMANN, PHALEMPIN, J.S. and LILLE, Y.K. (2011) '*Water Insoluble Polymer: Indigestible Water-Soluble Polysaccharide Film Coatings for Colon Targeting*'. Roquette Freres, France, US 20110256230A1.

FRIDKIN, S.K. and JARVIS, W.R. (1996) 'Epidemiology of nosocomial fungal infections'. *Clin Microbiol Rev*, **9**, 499–511.

© Woodhead Publishing Limited, 2013

GEORGIEFF, M., MOLDAWER, L.L., BISTRIAN, B.R. and BLACKBURN, G.L. (1985) 'Xylitol, an energy source for intravenous nutrition after trauma'. *J Parenter Enteral Nutr*, **9**, 199–209.

GRANSTRÖM, T. (2002) *'Biotechnological production of xylitol with* Candida *yeasts'*. PhD Thesis, Helsinki University of Technology, Finland.

GRANSTRÖM, T. and LEISOLA, M. (2002) 'Controlled transient changes reveal differences in metabolite production in two *Candida* yeasts'. *Appl Microbiol Biotechnol*, **58**, 511–17.

GRANSTRÖM, T., OJAMO, H. and LEISOLA, M. (2001) 'Chemostat study of xylitol production by *Candida guilliermondii'*. *Appl Microbiol Biotechnol*, **55**, 36–42.

GRANSTRÖM, T.B., TAKATA, G., TOKUDA, M. and IZUMORI, K. (2004) 'A novel and complete strategy for bioproduction of rare sugars'. *J Biosci Bioeng*, **97**, 89–94.

GRANSTRÖM, T.B., TAKATA, G., MORIMOTO, K., LEISOLA, M. and IZUMORI, K. (2005) 'L-Lyxose and L-xylose production from xylitol using Alcaligenes 701B strain and immobilized L-rhamnose isomerase enzyme'. *Enzyme Microb Tech*, **36**, 976–81.

GUO, C., ZHAO, C., HE, P., LU, D., SHEN, A. and JIANG, N. (2006) 'Screening and characterization of yeasts for xylitol production'. *J Appl Microbiol*, **101**, 1096–104.

HAJNY, G., SMITH, J. and GARVER, J. (1964) 'Erythritol production by a yeast like fungus'. *Appl Environ Microbiol*, **12**, 240–6.

HALLBORN, J., WALFRIDSSON, M., AIRAKSINEN, U., OJAMO, H., HAHN-HÄGERDAL, B., PENTTILÄ, M. and KERÄNEN, S. (1991) 'Xylitol production by recombinant *Saccharomyces cerevisiae'*. *Biotechnology (NY)*, **9**, 1090–5.

HALLBORN, J., GORWA, M.F., MEINANDER, N., PENTTILÄ, M., KERÄNEN, S. and HAHN-HÄGERDAHL, B. (1994) 'The influence of cosubstrate and aeration on xylitol formation by recombinant *Saccharomyces cerevisiae* expressing XYL1 gene'. *Appl Microbiol Biotechnol*, **42**, 326–33.

HÄRKÖNEN, M. and NUOJUA, P. (1979) 'Eri tekijöiden vaikutus ksyloosin katalyyttiseen hydraukseen ksylitoliksi'. *Kemia-Kemi*, **6**, 445–7.

HEIKKILÄ, H.O., KUISMA, J. and PAANANEN, H. (1996) *'Method for Fractionation of a Solution'*. Xyrofin Oy Finland, US 5730877A.

ISRS (2012) *Activities*, International Society of Rare Sugars. can be found at http://isrs.kagawa-u.ac.jp/activities.html.

JOHNSON, J.C. (1976) *'Specialized Sugars for the Food Industry.'* Park Ridge, Better World Books, New Jersey, Noyes Data Corp., 360 pp.

KOCH, F. (1886) 'Experimentelle Prüfung des Holzgummi und dessen Verbreitung im Pflanzenreiche'. *Pharm Z Russland*, **26**, 657.

KÖTTER, P., AMORE, R., HOLLENBERG, C.P. and CIRIACY, M. (1990) 'Isolation and characterization of the *Pichia stipitis* xylitol dehydrogenase gene, XYL2, and construction of a xylose-utilizing *Saccharomyces cerevisiae* transformant'. *Curr Genet*, **18**, 493–500.

KWON, S.G., PARK, S.W., OH, D.K. (2006) 'Increase of xylitol productivity by cell-recycle fermentation of *Candida tropicalis* using submerged membrane bioreactor'. *J Biosci Bioeng*, **101**, 13–18.

KYLMÄ, A.K., GRANSTRÖM, T.B. and LEISOLA, M. (2004) 'Growth characteristics and oxidative capacity of *Acetobacter acetii* IFO 3281: implications for L-ribulose production'. *Appl Microbiol Technol*, **63**, 584–91.

LANG, K. (1964) 'Die ernährungsphysiologischen Eigenschaften von Xylit'. *Int Z Vitamforsch*, **34**, 117–22.

LAWSON, M.E. (1997) *Encyclopedia of Chemical Technology*, Vol. 23, R.E. Kirk, D.F. Othmer, J. Kroschwitz and M. Howe-Grant (eds), 4th edn., John Wiley & Sons, New York, p. 93.

© Woodhead Publishing Limited, 2013

LEE, J.K., KIM, S.Y., RYU, Y.W., SEO, J.H. and KIM, J.H. (2003) 'Purification and characterization of a novel erythrose reductase from *Candida magnoliae*'. *Appl Environ Microbiol*, **69**, 3710–18.

LOHMAN, R.L. (1957) 'The polyols'. In: *The Carbohydrates: Chemistry, Biochemistry and Physiology*. Pigman, W. (ed). Academic Press, New York, pp 245–6.

LONDON, J. (1990) 'Uncommon pathways of metabolism among lactic acid bacteria'. *FEMS Microbiol Rev*, **87**, 103.

LOPEZ, F., DELGADO, O.D., MARTINEZ, M.A., SPENCER, J.F. and FIGUEROA, L.I. (2004) 'Characterization of a new xylitol-producer *Candida tropicalis* strain'. *Antonie van Leeuwenhoek*, **85**, 281–6.

MÄKINEN, K.K. (2000) 'Can the pentitol-hexitol theory explain the clinical observations made with xylitol?' *Med Hypotheses*, **54**, 603–13.

MÄKINEN, K.K. (2010) 'Sugar alcohols, caries incidence, and remineralization of caries lesions: a literature review'. *Int J Dent*, 1–23.

MÄKINEN, K.K., SAAG, M., ISOTUPA, K.P. J. OLAKB, J., NÖMMELAB R., SÖDERLINGA, E. and MÄKINEN, P.-L. (2005) 'Similarity of the effects of erythritol and xylitol on some risk factors of dental caries'. *Caries Research*, **39**, 207–15.

MAKKEE, M., KIEBOOM, A.P.G. and VAN BEKKUM, H. (1985) 'Production methods of D-mannitol', *Starch*, **37**,136–41.

MARTINEZ, G., BARKER, H.A. and HORECKER, B.L. (1963) 'A specific mannitol dehydrogenase from *Lactobacillus brevis*'. *J Biol Chem*, **238**, 1598–603.

MATTILA, P., SVANBERG, M. and KNUUTTILA, M. (1995) 'Diminished bone resorption in rats after oral xylitol administration: a dose-response study'. *Calcif Tissue Int*, **56**, 232–5.

MELAJA, A.J., HÄMÄLÄINEN, L. and RANTANEN, L. (1975) '*Process for Large Scale Chromatography*'. *Suomen Sokeri Osakeyhtio*, US 3928193.

MELAJA, A.J., VIRTANEN, J.J. and HEIKKILA, H.O. (1978) '*Method for Recovering Xylitol*'. *Suomen Sokeri Osakeyhtio*, US 4066711.

MELAJA, A., HÄMÄLÄINEN, L. and HEIKKILÄ, H.O. (1981) '*Menetelmä ksylitolin suhteen rikastuneen polyolin vesiliuoksen valmistamiseksi*'. FI589388 (Finnish patent).

MELLINGHOFF, C.H. (1961) 'Über die Verwendbarkeit des Xylit als Ersatzzukker bei Diabetikern'. *Klin Wochenschr*, **39**, 447–8.

MOON, H-J., JEYA, MARIMUTHU, KIM, I-W. and LEE, J-K. (2010) 'Biotechnological production of erythritol and its applications'. *Appl Microbiol Biotechnol*, **86**, 1017–25.

MÜLLER, J., HOFFMANN, E. and HOFFMANN U. (1935) 'Production of polyhydric alcohols'. US 1990245.

NEVES, A.R., RAMOS, A., SHEARMAN, C., GASSON, M.J., ALMEIDA, J.S. and SANTOS, H. (2000) 'Metabolic characterization of *Lactococcus lactis* deficient in lactate dehydrogenase using H/ *vivo* ^{13}C-NMR'. *Eur J Biochem*, **267**, 3859.

NISSENSON, A.R., WESTON, R.E. and KLEEMAN, C.R,. (1979) 'Mannitol'. *West J Med*, **131**, 277–84.

NYYSSÖLÄ, A., PIHLAJANIEMI, A., PALVA, A., VON WEYMARN, N. and LEISOLA, M. (2005) 'Production of xylitol from D-xylose by recombinant *Lactococcus lactis*'. *J Biotechnol*, **118**, 55.

NYYSSÖLÄ, A. and LEISOLA, M. (2005) 'Production of sugar alcohols by lactic acid bacteria'. *Recent Res Devel Biotech Bioeng*, **7**, 1–21.

OJAMO, H. (1994) '*Yeast Xylose Metabolism and Xylitol Production*', PhD thesis, Helsinki University of Technology, Finland.

ONISHI, H. (1960) 'Studies on osmophilic yeasts. Part IX. Isolation of a new obligate halophilic yeast and some consideration on halophilism'. *Bull Agric Chem Soc Jpn*, **24**, 226–30.

© Woodhead Publishing Limited, 2013

ONISHI, H. and SUZUKI, T. (1966) 'The production of xylitol, L-arabinitol and ribitol by yeasts'. *Agric Biol Chem*, **30**, 1139–44.

ONISHI, H. and SUZUKI, T. (1969) 'Microbial production of xylitol from glucose'. *Appl Environ Microbiol*, **18**, 1031–5.

OURA, E. (1997) 'Reaction products of yeast fermentation'. *Process Biochem*, **12**, 19–21.

PATEL, M., CRANK, M., DORNBURG, V., HERMAN, B., ROES, L., HÜSING, B., OVERBEEK, L., TERRAGNI, F. and RECHIA, E. (2006) '*Medium and Long-term Opportunities and Risks of the Biotechnological Production of Bulk Chemicals from Renewable Resources*'. BREW Report, Utrecht University.

PETERSON, W.H. and FRED, E.B. (1920) 'Fermentation of fructose by *Lactobacillus pentoaceticus*'. *J Biol Chem*, **41**, 431–50.

POONPERM, W., TAKATA, G., MORIMOTO, K., GRANSTRÖM, T.B. and IZUMORI, K. (2007) 'Production of l-xylulose from xylitol by a newly isolated strain of *Bacillus pallidus* Y25 and characterization of its relevant enzyme xylitol dehydrogenase'. *Enzyme Microb Technol*, **40**, 1206–12.

POVELAINEN, M. and MIASNIKOV, A.N. (2006) 'Production of D-arabitol by a metabolic engineered strain of *Bacillus subtilis*', *J Biotechnol*, **1**, 214–219.

ROBERTO, I.C., DE MANCILHA, I.M. and SATO, S. (1999) 'Influence of kLa on bioconversion of rice straw hemicellulose hydrolysate to xylitol'. *Bioprocess Eng*, **21**, 505–8.

RIZZI, M., HARWART, K., BUI THANH, N.A. and DELLWEG, H. (1989) 'A kinetic study of the NAD$^+$-xylitol-dehydrogenase from the yeast *Pichia stipitis*'. *J Ferment Bioeng*, **67**, 25–30.

SAWADA, K., TAKI, A., NAKANO, S., ASABA, E. and MAEHARA, T. (2009) 'Scale-up of erythritol continuous culture'. *Program and Abstract for the Annual Meeting of the Japan Society for Bioscience, Biochemistry and Agrochemistry*, 3-2 Da-05.

SCHEININ, A. and MÄKINEN, K.K. (1975) 'Turku sugar studies I-XXI'. *Acta Odontol Scand*, **33**, 70, 1–351.

SCHEININ A, MÄKINEN KK and YLITALO K. (1976) 'Turku sugar studies. V. Final report on the effect of sucrose, fructose and xylitol diets on the caries incidence in man'. *Acta Odontol Scand*, **34** (4), 179–216.

SCHIWECK, H., BÄR, A., VOGEL, R., SCHWARZ, E. and KUNZ, M. (1994) '*Ullmann's Encyclopedia of Industrial Chemistry*'. Vol. A 25, B. Elvers, S. Hawkins, W. Russey (eds) 5th edn., VCH Verlagsgesellschaft, Weinheim, 413.

SCHREIBER, W. T., GEIB, N. V., WINGFIELD, B. and ACREE, S. F. (1930) 'Semi-commercial production of xylose', *Ind Eng Chem*, **22**, 497–501.

SHYU, K. W. and HSU, M. Y. (1980) 'The cariogenicity of xylitol, mannitol, sorbitol, and sucrose', *Proceedings of the National Science Council. Republic of China*, **4**, 21–6.

SILVEIRA, M.M. and JONAS, R.(2002) 'The biotechnological production of sorbitol'. *Appl Microbiol Biotechnol*, **59**, 400–8.

SOETAERT, W. (1992) '*Synthesis of D-mannitol and L-sorbose by Microbial Hydrogenation and Dehydrogenation of Monosaccharides*'. Doctoral Thesis, University of Ghent, Faculteit van de Landbouwwetenschappen, Belgium.

SOETAERT, W., BUCHHOLZ, K. and VANDAMME, E.J. (1995) 'Production on D-mannitol and D-lactic acid by fermentation with *Leuconostoc mesenteroides*'. *Agro-Food-Industry Hi-Tech.*, **6**, 41–4.

STENHOUSE, J. (1848) 'Examination of the proximate principles of some of the lichens'. *Phil Trans R Soc*, **138**, 63–89.

SVANBERG M. and KNUUTTILA M (1994) 'Dietary xylitol retards bone resorption in rats'. *Miner Electrolyte Metab*, **20**, 153–7.

TOIVARI, M.H., RUOHONEN, L., MIASNIKOV, A.N., RICHARD, P. and PENTTILÄ, M. (2007) 'Metabolic engineering of *Saccharomyces cerevisiae* for conversion of D-glucose to xylitol and other five-carbon sugars and sugar alcohols'. *Appl Environ Microbiol*, **73**, 5471–6.

© Woodhead Publishing Limited, 2013

UHARI, M., KONTIOKARI, T., KOSKELA, M. and NIEMELÄ, M. (1996) 'Xylitol chewing gum in prevention of acute otitis media: double blind randomised trial'. *Br Med J*, **313**, 1180–4.

UHARI, M., KONTIOKARI, T. and NIEMELÄ, M. (1998) 'A novel use of xylitol sugar in preventing acute otitis media'. *Pediatrics*, **102**, 879–84.

VANDAMME, E., BIENFAIT, C.G. and SOETAERT, W. (2004) *Industrial Biotechnology and Sustainable Chemistry*. Royal Belgian Academy Council of Applied Science, Brussels.

VIIKARI, L. (1984) 'Formation of sorbitol by *Zymomonas mobilis*'. *Appl Microbiol Biotechnol*, **20**, 118–23.

VAN DIJKEN, J.P. and SCHEFFERS, W.A. (1986) 'Redox balances in the metabolism of sugars by yeasts'. *FEMS Microbiol Rev*, **32**, 199–224.

VON WEYMARN, N., KIVIHARJU, K. and LEISOLA, M. (2002) 'High-level production of D-mannitol with membrane cell-recycle bioreactor'. *J Ind Microbiol Biotech*, **29**, 44–9.

WISSELINK, H.W., WEUSTHUIS, R.A., EGGINK, G., HUGENHOLTZ, J., and GROBBEN, G.J. (2002) 'Mannitol production by lactic acid bacteria: a review'. *Int Dairy J*, **12**, 151.

YOKOZAWA, T., KIM, H. and CHO, E. (2002) 'Erythritol attenuates the diabetic oxidative stress through modulating glucose metabolism and lipid peroxidation in streptozotocin-induced diabetic rats'. *J Agric FoodChem*, **50**, 5485–9.

ZACHARIOU, M. and SCOPES, R.K. (1986) 'Glucose-fructose oxidoreductase, a new enzyme isolated from *Zymomonas mobilis* that is responsible for sorbitol production'. *J Bacteriol*, **167**, 863–9.

© Woodhead Publishing Limited, 2013

18

Microbial production of prebiotic oligosaccharides

T.-H. Nguyen and D. Haltrich, University of Natural Resources and
Life Sciences, (BOKU Wien), Austria

DOI: 10.1533/9780857093547.2.494

Abstract: Prebiotic oligosaccharides have attracted an increasing amount of
attention owing to their physiological importance and functional effects on
human health, as well as physicochemical properties, which are of interest for
various applications in the food industry. This chapter reviews the production of
prebiotic (fructo-oligosaccharides, galacto-oligosaccharides) and some candidate
prebiotic oligosaccharides by transglycosylation or hydrolysis using glycosidases
or glycosyl transferases. The chapter also includes emerging trends in the
production of novel oligosaccharides.

Key words: enzymatic preparation methods, functional oligosaccharides, microbial
enzymes, prebiotics.

18.1 Introduction

18.1.1 Prebiotic concept

Since its first introduction (Gibson and Roberfroid, 1995), the concept of
prebiotics has attracted an increasing amount of attention and stimulated
both scientific and industrial interest. This concept was later revised (Gibson
et al., 2004; ISAPP, 2008), and the development of the definition of the
prebiotic concept has recently been summarized (Roberfroid *et al.*, 2010).
According to an updated definition of the prebiotic concept, 'a dietary
prebiotic is a selectively fermented ingredient that results in specific changes,
in the composition and/or activity of the gastrointestinal microbiota, thus
conferring benefit(s) upon host health' (ISAPP, 2008). Based on the criteria
(Gibson *et al.*, 2004; Roberfroid, 2007a) (i) resistance to gastric acidity, to
hydrolysis by mammalian enzymes, and to gastrointestinal absorption;
(ii) fermentation by intestinal microflora; and (iii) selective stimulation
of growth and/or activity of intestinal bacteria associated with health/
well-being, only inulin/fructo-oligosaccharides, galacto-oligosaccharides

© Woodhead Publishing Limited, 2013

and lactulose fulfil these requirements for prebiotics, as documented and proven in several studies, although promise exists for several other dietary oligosaccharides (Gibson *et al.*, 2004; Roberfroid, 2007a, *et al.,* 2010; Barreteau *et al.*, 2006; Barclay *et al.*, 2010). At present, commercially important oligosaccharides with prebiotic status, including fructo-oligosaccharides (FOS), inulin, galacto-oligosaccharides (GOS) and lactulose are available mainly in the Japanese, European and USA markets (Cummings *et al.*, 2001; Swennen *et al.*, 2006; Nakakuki, 2002; Patel and Goyal, 2011; Figueroa-González *et al.*, 2011). Candidate prebiotics include lactosucrose, isomalto-oligosaccharides, soybean oligosaccharides, xylo-oligosaccharides, and gentio-oligosaccharides (Roberfroid, 2007a; Crittenden and Playne, 1996).

18.1.2   Properties of prebiotic oligosaccharides

An oligosaccharide is 'a molecule containing a small number (2 to about 10) of monosaccharide residues connected by glycosidic linkages' (IUB-IUPAC, 1982). Other authorities classify saccharides with 3 to 19 monosaccharide units in this group (Mussatto and Mancilha, 2007). However, there is no rational physiological or chemical reason for setting these limits (Voragen, 1998). Consequently, oligosaccharides can be defined as low molecular weight carbohydrates or 'short-chain carbohydrates' (Englyst and Hudson, 1996). Carbohydrates can be classified as digestible or non-digestible based on their physiological properties. The concept of non-digestible oligosaccharides originates from the observation that the anomeric C atom (C_1 or C_2) of the monosaccharide units of some dietary oligosaccharides has a configuration that makes their osidic bonds non-cleavable by the hydrolytic activity of the human digestive enzymes (Roberfroid and Slavin, 2000). The main categories of non-digestible oligosaccharides presently available or in development as food ingredients include carbohydrates in which the monosaccharide unit is fructose, galactose, glucose and/or xylose (Crittenden and Playne, 1996; Roberfroid and Slavin, 2000).

Prebiotic oligosaccharides can serve as fermentable substrates for certain members of the gut microbiota and have been found to modulate the colonic flora (Holzapfel and Schillinger, 2002; Rastall *et al.*, 2005; Macfarlane *et al.*, 2008). The physiological importance and health benefits of prebiotic oligosaccharides have been reported extensively in several recent reviews on prebiotics and functional oligosaccharides (Patel and Goyal, 2011; Eiwegger *et al.*, 2010; Broekaert *et al.*, 2011; De Preter *et al.*, 2011; Roberfroid *et al.*, 2010; Qiang *et al.*, 2009). Prebiotic oligosaccharides have been found to modulate the colonic flora by stimulation of beneficial bacteria (such as bifidobacteria and lactobacilli) as well as inhibit 'undesirable' bacteria (Holzapfel and Schillinger, 2002; Blaut, 2002; Swennen *et al.*, 2006; Rastall *et al.*, 2005; Macfarlane *et al.*, 2008), affect bowel functions beneficially (Cummings *et al.*, 2001; Macfarlane *et al.*, 2006; Steed *et al.*, 2008;

© Woodhead Publishing Limited, 2013

Roberfroid *et al.*, 2010), reduce intestinal disturbances (constipation and diarrhoea), cardiovascular disease and intestinal cancer (Ziemer and Gibson, 1998; Scheppach *et al.*, 2001; Macfarlane *et al.*, 2006, 2008), stimulate absorption and retention of several minerals, particularly magnesium, calcium and iron (Scholz-Ahrens *et al.*, 2001; Macfarlane *et al.*, 2008; Abrams *et al.*, 2005, 2007; Barclay *et al.*, 2010; De Preter *et al.*, 2011) and reduce the risk of colon cancer (Wollowski *et al.*, 2001; Marteau and Boutron-Ruault, 2002; Swennen *et al.*, 2006; Roberfroid *et al.*, 2010). In addition to these confirmed functional effects, postulated effects include modulation of the immune response (Swennen *et al.*, 2006; Eiwegger *et al.*, 2010; Roberfroid *et al.*, 2010), reduction of the serum cholesterol level, and production or improved bio-availability of certain nutrients (Holzapfel and Schillinger, 2002). Because of these benefits, prebiotic oligosaccharides are of great interest for both human and animal nutrition.

In addition to these functional effects on human health, the physico-chemical properties of prebiotic oligosaccharides are also of significant interest for their application in the food industry. Prebiotic oligosaccharides are low-calorie sweeteners because they pass through the human small intestine without being digested. Typically, the oligosaccharides are about 0.3–0.6 times as sweet as sucrose, and sweetness decreases with longer chain length. Certain prebiotic oligosaccharides are stable over a wide range of pH and temperature, with stability of the oligosaccharides depending on the sugar moiety content, ring form, anomeric configuration and linkage types (Patel and Goyal, 2011). Generally β-glycosidic linkages are more stable than α-linkages and hexoses are more strongly linked than pentoses. These properties enable prebiotic oligosaccharides to be applied in a wide variety of food products. Apart from being used as sweeteners, oligosaccharides are nowadays incorporated in a wide range of products such as fermented milk products, breads, jams, confectionery, beverages and infant milk formula. Prebiotic oligosaccharides have also been used extensively as prebiotic supplements, for drug delivery and as immunostimulators in either cosmetics, animal feed and agrochemicals (Qiang *et al.*, 2009).

18.2　Microbial production of prebiotic oligosaccharides

Food-grade prebiotic oligosaccharides are produced using enzymatic processes with the exception of soybean oligosaccharides, which are obtained by direct extraction processes (Kim *et al.*, 2003), and lactulose, which is produced using an alkali-catalysed reaction (Schuster-Wolff-Bühring *et al.*, 2010). They can be formed by controlled enzymatic hydrolysis of polysaccharides such as starch (Sako *et al.*, 1999), inulin (Singh and Singh, 2010; Kango and Jains, 2011) or xylan (Vázquez *et al.*, 2001; Aachary and Prapulla, 2011). In addition, enzymatic synthesis of prebiotic oligosaccharides from simple sugars such as lactose or sucrose using glycosidases, glycosynthases

© Woodhead Publishing Limited, 2013

or glycosyl transferases (Crittenden and Playne, 1996; Nakakuki, 2002; Swennen *et al.*, 2006) has attracted an increasing amount of attention. The recent advent of glycosynthases, which are specifically mutated glycosidases that efficiently synthesize oligosaccharides but do not hydrolyse them, provides new tools for oligosaccharides synthesis (Perugino *et al.*, 2004). The products from these processes typically are mixtures of oligosaccharides which differ in their degree of polymerization and glycosidic linkages, unless highly specific glycosyl transferases are used. The chemical structures and composition of these oligosaccharide mixtures greatly depend on the enzyme source and process conditions. Here, we review the production of prebiotic and candidate prebiotic oligosaccharides by transglycosylation using glycosidases or glycosyl transferases.

18.2.1   Fructo-oligosaccharides and oligofructose

Fructo-oligosaccharides (FOS) and oligofructose are referred to as inulin-type prebiotics (Kelly, 2008). Roberfroid (Roberfroid, 2007b) considers that FOS and oligofructose are both synonyms for mixtures of small inulin oligomers with a maximum degree of polymerization DP_{max} or total number of fructosyl units of less than ten. Other authors use the term short-chain FOS for inulin-type prebiotics synthesized from sucrose using β-fructofuranosidases (Bouhnik *et al.*, 2007; Giacco *et al.*, 2004; Linde *et al.*, 2009). In this review, the term FOS is used to describe short-chain fructans, in which one to three fructosyl moieties are linked to the basic sucrose skeleton by different glycosidic linkages. These fructans are the products of transfructosylation reactions catalysed by β-fructofuranosidases when using sucrose as the substrate (Fig. 18.1(a)). Furthermore, the term oligofructose is used to refer to inulin-type fructans with $DP_{max} < 10$ that are produced by partial enzymatic hydrolysis of inulin using endoinulinases (Fig. 18.1(b), (c)).

Transfructosylation of sucrose using β-fructofuranosidases

Fructo-oligosaccharides (FOS) are produced from the disaccharide sucrose exploiting the transfructosylation activity of the enzyme β-fructofuranosidase. A high concentration of the starting material is required for efficient transglycosylation (Crittenden and Playne, 1996). Currently, *Aspergillus* fructosyltransferases are the main enzymes used in production of FOS (Zeng *et al.*, 2011; Kurakake *et al.*, 2010; Sangeetha *et al.*, 2005a), and these product mixtures contain β-(2→1)-linked fructosyl units ([1]F-FOS: 1-kestose Glu-Fru$_2$, nystose Glu-Fru$_3$ or [1]F-fructofuranosylnystose Glu-Fru$_4$). Even though the FOS with an inulin-type-structure are available on the market today, there is great interest in the development of novel products. Conceivably, these may possess improved and desirable attributes such as improved prebiotic and physiological properties that are not present in the substances currently available. β-(2→6)-linked FOS ([6]F-FOS: 6-kestose, with a β-(2→6)-linkage between the two fructosyl units; or [6]G-FOS: neokestose, neonystose,

© Woodhead Publishing Limited, 2013

(a)

(b)   (c)

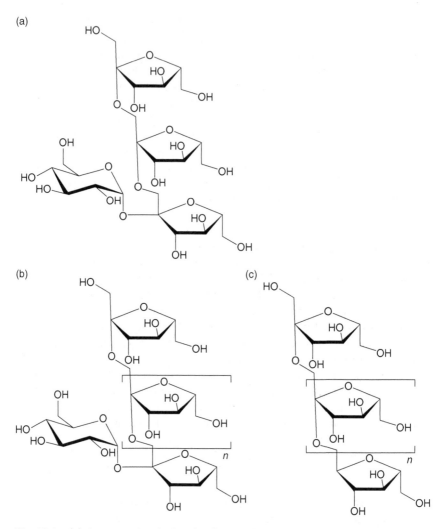

Fig. 18.1   (a) An example of a fructo-oligosaccharide or 1-nystose, containing three β-(2→1)-linked fructosyl units linked to a terminal α-D-glucose residue; (b) structure of inulin-type oligofructoses (where $n < 8$) and inulin (where $n \geq 8$); and (c) structure of inulin-type oligofructoses without a terminal α-D-glucose residue (where $n < 8$).

and neo-fructofuranosylnystose, with glycosidic linkages between a fructosyl unit and a terminal α-D-glucose residue of sucrose) may have enhanced prebiotic properties compared to commercially available FOS (Kilian *et al.*, 2002; Linde *et al.*, 2009).

The enzymatic synthesis of β-(2→6)-linked FOS using β-fructofuranosidases has been reported in yeasts or fungi such as *Saccharomyces cerevisiae*

© Woodhead Publishing Limited, 2013

(De Abreu *et al.*, 2011), *Schwanniomyces occidentalis* (De Abreu *et al.*, 2011; Álvaro-Benito *et al.*, 2007), *Xanthophyllomyces dendrorhous* (Linde *et al.*, 2009; Ning *et al.*, 2010), *Rhodotorula dairenensis* (Gutiérrez-Alonso *et al.*, 2009) or *Penicillium citrinum* (Lim *et al.*, 2007). In addition, different bifidobacteria including *Bifidobacterium adolescentis*, *Bifidobacterium longum*, *Bifidobacterium breve*, and *Bifidobacterium pseudocatenulatum* were reported for their ability to metabolize β-(2→6)-linked FOS (Marx *et al.*, 2000), but there is no report of their ability to form these FOS in transfructosylation mode. Different enzyme preparations such as purified native enzymes (Gutiérrez-Alonso *et al.*, 2009; Linde *et al.*, 2009), recombinant enzymes (De Abreu *et al.*, 2011), immobilized enzymes (Mussatto *et al.*, 2009; Kurakake *et al.*, 2010), whole-cell preparations (Ning *et al.*, 2010), immobilized and co-immobilized whole-cell preparations (Lim *et al.*, 2007), or recycling cultures (Sangeetha *et al.*, 2005a) are used for the synthesis of FOS. The initial concentration of sucrose varied from 40% (Ning *et al.*, 2010) to 60% (Sangeetha *et al.*, 2005a), and depending on the enzyme sources and preparations, the FOS yields obtained ranged from 49.4 g l^{-1} for whole-cell immobilisates of *P. citrinum* (Lim *et al.*, 2007), 87.9 g l^{-1} for homogenous native β-fructofuranosidase from *R. dairenensis* (Gutiérrez-Alonso *et al.*, 2009), 116 g l^{-1} for *Aspergillus japonicus* cells immobilized on vegetal fibres (Mussatto *et al.*, 2009) up to 227.7 g l^{-1} for free whole-cell biotransformations using *X. dendrorhous* (Ning *et al.*, 2010).

Major manufacturers of FOS from sucrose are Meiji Seika Kaisha Ltd. (Japan) whose products are marketed under the trade name Meioligo, Beghin-Meiji Industries (France) with Actilight, Cheil Foods and Chemicals Inc. (Korea) with Oligo-Sugar, GTC Nutrition (USA) with NutraFlora, and Victory Biology Engineering Co. Ltd. (China) with Prebiovis scFOS (Singh and Singh, 2010; Crittenden and Playne, 1996).

Enzymatic hydrolysis of inulin using endoinulinases
Inulin, after starch, is the most abundant storage carbohydrate in a wide range of plants (Franck and Bosscher, 2006). Inulin consists of β-(2→1)-linked fructosyl units linked to a terminal α-D-glucose residue (Fig. 18.1(b)). Inulin is a potent substrate for the production of inulinases (Chi *et al.*, 2009; Vijayaraghavan *et al.*, 2009; Singh and Singh, 2010), high-fructose syrup (Ricca *et al.*, 2007; Singh and Singh, 2010; Kango and Jains, 2011) and fructo-oligosaccharides and oligofructoses (Sangeetha *et al.*, 2005b; Singh and Singh, 2010). Inulinases, which can be produced by growing various microorganisms in an inulin-based medium (Vijayaraghavan *et al.*, 2009), are fructofuranosyl hydrolases produced by a wide range of organisms including plants, bacteria, molds and yeasts. The microbial production of inulinases has been summarized in recent reviews (Kango and Jains, 2011; Singh and Singh, 2010). Inulinases are classified as exoinulinases and endoinulinases based on the cleavage action of the β-(2→1) linkage in inulin. Exoinulinases cleave the terminal fructose units from inulin and are used

© Woodhead Publishing Limited, 2013

for the production of high-fructose syrup, whereas endoinulinases break down internal linkages of inulin. FOS and oligofructose can be produced by controlled enzymatic hydrolysis of inulin using endoinulinases. Microbial sources of endoinulinases include fungi, yeasts and bacteria (Kango and Jains, 2011; Chi *et al.*, 2009; Vijayaraghavan *et al.*, 2009). Among the fungi, *Aspergillus* and *Penicillium* are the most prominent inulinase producers, while among the yeasts this is *Klyveromyces* (Singh and Singh, 2010). Well-known and studied producers of inulinase among bacteria include some species of *Xanthomonas*, *Pseudomonas*, *Bacillus* and *Streptomyces* (Kango and Jains, 2011; Singh and Singh, 2010). In two recent reviews (Singh and Singh, 2010; Kango and Jains, 2011), the authors report a temperature range of 37–55°C and pH optima of 6.0–7.0 for endoinulinase activity, depending on the enzyme sources used for the hydrolysis of inulin (Mutanda *et al.*, 2008; Kim *et al.*, 1997; Yun *et al.*, 1997; Park *et al.*, 1999; Yokota *et al.*, 1995; Cho *et al.*, 2001; Naidoo *et al.*, 2009; Zhengyu *et al.*, 2005). In all the cases, FOS and oligofructose yields ranged from 60–86 % under optimal hydrolysis conditions.

The major manufacturer of FOS and oligofructose from inulin is Orafti Active Food Ingredients (USA) and the product, available under the trade name Raftilose, ranges from two to nine monosaccharide units in length, with an average length of four sugar moieties (Singh and Singh, 2010; Crittenden and Playne, 1996). Raftiline, another inulin-derived product marketed by Orafti Active Food Ingredients, contains longer fructose chains, ranging from 10 to 60 sugar moieties with an average DP of 25 (Singh and Singh, 2010). Products similar to Raftilose that are commercially available on the markets include Fibrulose by Cosucra Groupe Warcoing (Belgium) and Inulin FOS by Jarrow Formulas (USA).

18.2.2   Galacto-oligosaccharides (GOS)
Transgalactosylation of lactose using β-galactosidases
β-Galactosidases (β-gal; EC 3.2.1.23) catalyse the hydrolysis and transgalactosylation of β-D-galactopyranosides (such as lactose) (Prenosil *et al.*, 1987; Nakayama and Amachi, 1999; Pivarnik *et al.*, 1995) and are widespread in nature. They catalyse the cleavage of lactose (or related compounds) in hydrolysis mode and are thus used in the dairy industry to remove lactose from various products. An attractive biocatalytic application is found in the transgalactosylation potential of these enzymes, which is based on their catalytic mechanism (Nakayama and Amachi, 1999; Petzelbauer *et al.*, 2000). The products of transgalactosylation, galacto-oligosaccharides (GOS), are non-digestible carbohydrates that meet the criteria of 'prebiotics' and therefore have attracted increasing attention.

Microbial β-galactosidases have been isolated and characterized from yeasts, fungi and bacteria (Nakayama and Amachi, 1999; Husain, 2010; Park and Oh, 2010). The major industrial enzymes are obtained from *Aspergillus*

© Woodhead Publishing Limited, 2013

spp. and *Kluyveromyces* spp. (Husain, 2010) where *Kluyveromyces lactis* is probably the most widely used enzyme (Husain, 2010; Zhou and Chen, 2002, 2001; Kim and Ji, 2004; Maugard *et al.*, 2003). The use of lactic acid bacteria (LAB) as the sources of β-galactosidases offers substantial potential for food applications of these enzymes (Gänzle *et al.*, 2008). LAB, which constitute a diverse group of lactococci, streptococci and lactobacilli, have been studied intensively with respect to their enzymes for various different reasons, one of which is their 'generally recognized as safe' (GRAS) status. Hence, enzymes derived from these organisms can be used without extensive purification in various food-related applications (Vasiljevic and Jelen, 2002; Somkuti *et al.*, 1998). An additional attractive application of β-galactosidases from probiotic bacteria such as lactobacilli or bifidobacteria has been proposed, namely their use in the production of tailor-made prebiotics targeting specifically advantageous and beneficial intestinal microorganisms (Rabiu *et al.*, 2001; Rastall and Maitin, 2002). β-Galactosidases, which efficiently hydrolyse galacto-oligosaccharide structures, presumably can also form these galacto-oligosaccharides when they act in their transgalactosylation mode. It is conceivable that β-galactosidases from probiotic microorganisms produce galacto-oligosaccharide structures that have a strong prebiotic potential since they will be preferentially utilized by this group of microorganisms.

Galacto-oligosaccharides are the products of transgalactosylation reactions catalysed by β-galactosidases when lactose or other structurally related galactosides are used as the substrate. β-Galactosidases are generally classified as hydrolases, in fact, hydrolysis of the glycosidic bond is a special case of transgalactosylation in which the galactosyl acceptor is water (Gosling *et al.*, 2010). Scheme 18.1 illustrates the possible lactose conversion reactions catalysed by β-galactosidases, and structures of some galacto-oligosaccharides are given in Fig 18.2.

Transgalactosylation is described to involve intermolecular as well as intramolecular reactions. Intramolecular or direct galactosyl transfer to D-glucose yields regio-isomers of lactose. The glycosidic bond of lactose (β-D-Gal$p$-(1→4)-D-Glc) is cleaved and immediately formed again at a different position in the glucose molecule before it diffuses out of the active site. This is how allolactose (β-D-Gal$p$-(1→6)-D-Glc), the presumed natural inducer of β-galactosidases in certain microorganisms, can be formed even in the absence of significant amounts of free D-glucose (Mahoney, 1998; Splechtna *et al.*, 2006). Various di-, tri-, tetrasaccharides and eventually higher oligosaccharides are produced by intermolecular transgalactosylation. Any sugar molecule in the reaction mixture can be the nucleophile accepting the galactosyl moiety from the galactosyl–enzyme complex, which is formed as an intermediate in the reaction. The GOS produced are not the product of an equilibrium reaction, but must be regarded as kinetic intermediates as they are also substrates for hydrolysis, and hence transgalactosylation reactions are kinetically controlled (Splechtna *et al.*, 2006;

© Woodhead Publishing Limited, 2013

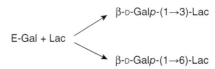

If Nu is Lac, trisaccharides are formed such as:

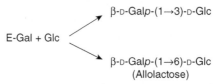

If Nu is Glc, disaccharides are formed such as:

E-Gal + Glc
⟶ β-ᴅ-Galp-(1→3)-ᴅ-Glc
⟶ β-ᴅ-Galp-(1→6)-ᴅ-Glc
     (Allolactose)

Scheme 18.1   Hydrolysis and galactosyl transfer reactions, both intra- and intermolecular, during the conversion of lactose catalyzed by β-galactosidases. E, Enzyme; Lac, lactose; Gal, galactose; Glc, glucose; Nu, nucleophile (Splechtna *et al.*, 2006).

Boon *et al.*, 2000). For these reasons GOS yield and composition change dramatically with reaction time and the GOS mixture thus obtained is very complex and can hardly be predicted.

It is well-known that β-galactosidases from different species possess very different specificities for building glycosidic linkages and therefore produce different GOS mixtures. For example, the β-galactosidase from *K. lactis* produced predominantly β-(1→6)-linked GOS (Asp *et al.*, 1980), the β-galactosidase from *Aspergillus oryzae* produced predominantly β-(1→3) and β-(1→6) linkages in their GOS (Toba *et al.*, 1985), *Bacillus circulans* β-galactosidase forms β-(1→2), β-(1→3), β-(1→4), β-(1→6) linked GOS (Yanahira *et al.*, 1995), whereas β-galactosidases from *Lactobacillus* spp. showed a preference to form β-(1→3) and β-(1→6) linkages in transgalactosylation mode (Splechtna *et al.*, 2006; Iqbal *et al.*, 2010, 2011; Nguyen *et al.*, 2007; Maischberger *et al.*, 2010).

Production of galacto-oligosaccharides
Galacto-oligosaccharides are one of the main prebiotics produced commercially (Sako *et al.*, 1999; Gosling *et al.*, 2010; Torres *et al.*, 2010). Galacto-oligosaccharides have attracted increasing attention because of the presence of structurally related oligosaccharides together with different complex structures in human breast milk (see below) and therefore the use of GOS in infant formula is nowadays of great interest (Crittenden and Playne, 1996;

© Woodhead Publishing Limited, 2013

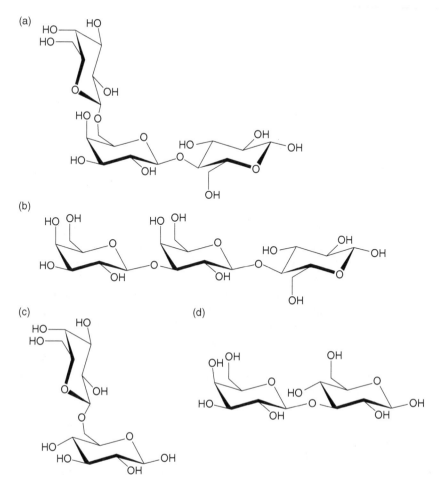

Fig. 18.2 Structures of some galacto-oligosaccharides: (a) β-D-Gal$p$-(1→6)-Lac, (b) β-D-Gal$p$-(1→3)-Lac, (c) β-D-Gal$p$-(1→6)-D-Glc (allolactose) and (d) β-D-Gal$p$-(1→3)-D-Glc.

Gopal *et al.*, 2001; Sangwan *et al.*, 2011). In addition, GOS are also incorporated in a wide range of products including fermented milk products, breads, jams, confectionery and beverages, to name a few (Sangwan *et al.*, 2011; Sako *et al.*, 1999; Otieno, 2010).

GOS are produced from lactose by microbial β-galactosidases employing different enzyme sources and preparations including crude enzymes, purified enzymes, recombinant enzymes, immobilized enzymes, whole-cell biotransformations, toluene-treated cells and immobilized cells. The enzyme sources, the process parameters as well as the yield and the productivity of these processes for GOS production are summarized in detail in recent

© Woodhead Publishing Limited, 2013

reviews (Otieno, 2010; Park and Oh, 2010; Torres *et al.*, 2010; Sangwan *et al.*, 2011). The highest GOS productivity, 106 g l^{-1} h^{-1}, was observed when β-galactosidase from *Aspergillus oryzae* immobilized on cotton cloth was used for GOS production in a packed-bed reactor (Albayrak and Yang, 2002).

Despite the importance of LAB, in particular *Lactobacillus* spp. and bifidobacteria for food technology and dairy applications, and despite numerous studies on the gene clusters involved in lactose utilization by these bacteria, only recently β-galactosidases from *Lactobacillus* and *Bifidobacterium* spp. have been characterized in detail with regard to their biochemical properties or investigated for their ability to produce GOS in biocatalytic processes. Previous studies reported the presence of multiple β-galactosidases in *B. infantis, B. adolescentis*, and *B. bifidum, B. breve* (Hung *et al.*, 2001; Hung and Lee, 2002; Moller *et al.*, 2001; Van Laere *et al.*, 2000a; Hinz *et al.*, 2004; Goulas *et al.*, 2007, 2009; Yi *et al.*, 2011) and revealed that these enzymes are very different with respect to substrate specificity and regulation of gene expression. Furthermore, these reports described the cloning and characterization of these enzymes and studied their transgalactosylation activity in detail, for example β-galactosidase BgbII from *B. adolescentis* showed high preference towards the formation of β-(1→4) linkages while no β-(1→6) linkages were formed (Hinz *et al.*, 2004). In contrast, the β-galactosidase BgbII from *B. bifidum* showed a clear preference for the synthesis of β-(1→6) linkages over β-(1→4) linkages (Goulas *et al.*, 2009). A recombinant β-galactosidase from *B. infantis* (Hung and Lee, 2002) was found to be an excellent biocatalyst for GOS production giving the highest GOS yield of 63% (mass of GOS of the total sugars in the reaction mixture).

β-Galactosidases of lactobacilli play an important role in a number of commercial processes, for example, milk processing or cheese making (Hébert *et al.*, 2000; Adam *et al.*, 2004). Recent studies of β-galactosidases, especially with respect to their enzymatic and molecular properties, from *Lactobacillus reuteri, Lactobacillus acidophilus, Lactobacillus plantarum, Lactobacillus sakei*, and *Lactobacillus pentosus* (Nguyen *et al.*, 2006, 2007; Splechtna *et al.*, 2006, 2007a, 2007b; Iqbal *et al.*, 2010, 2011; Maischberger *et al.*, 2010) revealed that these enzymes were found to be very well suited to the production of galacto-oligosaccharides. Maximum GOS yields at 30°C were 38% when using purified β-galactosidases from *L. reuteri* and *L. acidophilus* with initial an lactose concentration of 205 g l^{-1} and at ~ 80% lactose conversion (Splechtna *et al.*, 2006; Nguyen *et al.*, 2007). When using purified β-galactosidases from *L. plantarum* and *L. sakei*, the yields at 30°C were 41% with similar initial lactose concentrations and at 77% and 85% lactose conversion, respectively (Iqbal *et al.*, 2010, 2011). Purified β-galactosidase from *L. pentosus* gave the lowest yields, 31%, compared to the above-mentioned β-galactosidases from lactobacilli (Maischberger *et al.*, 2010).

© Woodhead Publishing Limited, 2013

In order to reduce enzyme costs by avoiding laborious and expensive chromatographic steps for the purification of the biocatalyst, crude β-galactosidase extract from *Lactobacillus* sp. directly obtained after cell disruption and separation of cell debris by centrifugation was used in lactose conversion for GOS production (Splechtna *et al.*, 2007b). It was reported that there was no obvious difference in the GOS yields obtained using either purified or crude β-galactosidase at 37°C, and in addition the crude enzyme was found to be equally stable as the purified one. Therefore, crude β-galactosidase extracts are suitable for a convenient and simple process of GOS production. Because of the GRAS status of *Lactobacillus* spp. it is also safe to use these crude extracts in food and feed applications. The reduction of reaction temperature to 17°C to limit microbial growth and the use of a cheap lactose source, such as whey permeate powder, did not have significant effects on GOS yield (Splechtna *et al.*, 2007b). For the described process about 100 kU of enzyme activity are required to produce 1 kg of GOS from 3.85 kg lactose within 15 h (Splechtna *et al.*, 2007b).

Since enzyme recovery methods such as immobilization or ultrafiltration can be problematic for mesophilic enzymes such as enzymes from lactic acid bacteria owing to microbial safety issues, especially when using technical substrate solutions, a strategy to reduce enzyme costs further could be the optimization of enzyme production or the enzyme itself. Over-expression of a β-galactosidase in another food-grade organism and enzyme engineering to increase transgalactosylation activity are valid strategies to improve this process further. Different approaches to engineering β-galactosidases and glycosidases were quite successful in producing improved enzymes with high transglycosylation activity that could be efficiently used in the synthesis of oligosaccharides (Hancock *et al.*, 2006; Feng *et al.*, 2005). Recently, a new food-grade host/vector system for *L. plantarum* based on the pSIP expression vectors (Sørvig *et al.*, 2003, 2005; Mathiesen *et al.*, 2004a, 2004b) was constructed using the homologous alanine racemase gene (*alr*) as selection marker (Nguyen *et al.*, 2011). The vectors were applied for intracellular overexpression of the β-galactosidase genes from various *Lactobacillus* spp. in an *alr*-deletion mutant of *L. plantarum* WCFS1 used as host organism (Nguyen *et al.*, 2011).

The main GOS structures formed from lactose by lactobacillal β-galactosidases are β-D-Gal*p*-(1→6)-D-Glc (allolactose), β-D-Gal*p*-(1→6)-D-Gal, β-D-Gal*p*-(1→3)-D-Gal, β-D-Gal*p*-(1→3)-D-Glc, β-D-Gal*p*-(1→6)-Lac and β-D-Gal*p*-(1→3)-Lac. β-Galactosidases from *Lactobacillus* spp. show a high specificity for the formation of β-(1→6) and β-(1→3) linkages in transgalactosylation mode and no major products are formed with β-(1→4) linkages which are found in commercially available GOS such as Vivinal GOS (Friesland Foods Domo, the Netherlands), Cup-Oligo (Nissin Sugar Manufacturing Company, Japan) and Purimune (GTC, Nutrition, United States) (Torres *et al.*, 2010; Sangwan *et al.*, 2011). A strong prebiotic effect was attributed to some of these former sugars in a study comparing a number

© Woodhead Publishing Limited, 2013

of disaccharides with respect to *in vitro* prebiotic selectivity (Sanz *et al.*, 2005).

Most galacto-oligosaccharide preparations currently available on the market contain a significant amount of monosaccharides and lactose (Tzortzis and Vulevic, 2009, Playne and Crittenden, 2009). To remove efficiently both monosaccharides and lactose from GOS mixtures containing significant amounts of prebiotic non-lactose disaccharides, a biocatalytic approach coupled with subsequent chromatographic steps was used (Splechtna *et al.*, 2001; Maischberger *et al.*, 2008). Lactose was first oxidized to lactobionic acid using fungal cellobiose dehydrogenases and then lactobionic acid and monosaccharides were removed by ion-exchange and gel permeation chromatography. This process was used to purify a GOS mixture produced from lactose using recombinant β-galactosidase from *L. reuteri* containing 48% monosaccharides, 26.5% lactose and 25.5% GOS (Maischberger *et al.*, 2008). Two different fungal cellobiose dehydrogenases (CDH), originating from *Sclerotium rolfsii* and *Myriococcum thermophilum*, were compared with respect to their applicability for this process. CDH from *S. rolfsii* showed higher specificity for the substrate lactose and only a few other components of the GOS mixture were oxidized during prolonged incubation. Removal of ions by ion-exchange chromatography using a strong cation resin (Lewatit S 2528) and removal of monosaccharides using a strong cation-exchange material (UBK-530) gave an essentially pure GOS product, containing less than 0.3% lactose and monosaccharides (Maischberger *et al.*, 2008). Separation of monosaccharides from a mixture of saccharides based on chromatography using high-throughput polyacrylamide gel chromatography (Bystrický *et al.*, 2011) or a reversed-phase high-performance liquid chromatography (Lv *et al.*, 2009) have been reported recently. Alternatively, digestible sugars in GOS mixtures can be depleted after fermentation with yeasts such as *S. cerevisiae* or *K. lactis* resulting in a monosaccharide-free and high-purity GOS mixture (Li *et al.*, 2008).

GOS have been manufactured and commercialized mainly in Japan, the United States and Europe. The major manufacturers are Yakult Honsha (Japan) with their product Oligomate, Nissin Sugar Manufacturing (Japan) with Cup-Oligo, Snow Brand Milk Products (Japan) with P7L, GTC Nutrition (United States) with Purimune, Friesland Foods Domo (the Netherlands) with Vivinal GOS, Clasado Ltd. (UK) with Bimuno, and Dairy Food Ingredients (Ireland) with Dairygold GOS (Crittenden and Playne, 1996; Coulier *et al.*, 2009; Tzortzis and Vulevic, 2009; Playne and Crittenden, 2009; Torres *et al.*, 2010; Sangwan *et al.*, 2011). Commercial GOS preparations typically are transparent syrups or white powders containing oligosaccharides with different DP, non-converted lactose and the monosacharides glucose and galactose. They differ in the purity of GOS product and also in the linkages of the oligosaccharide chains, which depend on the enzymes used for GOS production. Oligomate contains mainly β-(1→6) linked GOS,

© Woodhead Publishing Limited, 2013

Vivinal GOS, Cup-Oligo and Purimune contain mainly β-(1→4) linkages, whilst Bimuno contains mainly β-(1→3) linked GOS (Torres *et al.*, 2010).

18.2.3   Other oligosaccharides

Xylo-oligosaccharides

Xylo-oligosaccharides (XOS) are sugar oligomers made up of xylose units, which are found naturally, for example, in bamboo shoots, some fruits, vegetables, milk and honey (Vázquez *et al.*, 2001). Generally, XOS are mixtures of oligosaccharides formed by xylose residues linked through β-(1→4) linkages (Aachary and Prapulla, 2008) (Fig. 18.3). The number of xylose residues can vary from 2 to 10 and they are known as xylobiose, xylotriose, and so on. In addition to xylose residues, xylan is usually found in combination with arabinofuranosyl, α-D-glucopyranosyl uronic acid or its 4-O-methyl derivative, acetyl or phenolic substituents at positions C2 or C3 resulting in branched XOS with diverse biological properties (Vázquez *et al.*, 2001; Moure *et al.*, 2006; Aachary and Prapulla, 2011).

XOS are produced from xylan, which is extracted mainly from corn cobs (Crittenden and Playne, 1996) or other lignocellulosic materials by chemical methods, direct enzymatic hydrolysis of a susceptible xylan-containing substrate, or a combination of chemical and enzymatic treatments (Aachary and Prapulla, 2011). In chemical methods, XOS are produced through hydrolytic degradation of xylan by steam, water, diluted solutions of mineral acids or alkaline solutions (Vázquez *et al.*, 2001). The direct production of XOS by enzymatic methods from xylan-containing materials must be carried out with materials susceptible to direct enzymatic hydrolysis such as citrus peels (Alonso *et al.*, 2003; Aachary and Prapulla, 2011). To produce XOS using chemical and enzymatic methods, xylan is first extracted from lignocellulosic materials with alkalis such as NaOH, KOH, $Ca(OH)_2$, ammonia or a mixture of these compounds (Vázquez *et al.*, 2001; Akpinar *et al.*, 2007). Once the xylan is extracted in a soluble form, degradation of xylan can be accomplished by hydrolysis with suitable xylanases (Vázquez *et al.*, 2001). For production of XOS, enzyme complexes with low exo-xylanase and/or β-xylosidase activity are used (Vázquez *et al.*, 2001; Akpinar *et al.*, 2009). Enzyme preparations containing high exo-xylanase and/or β-xylosidase activities produce high amounts of xylose, which causes inhibition effects in XOS production (Vázquez *et al.*, 2002).

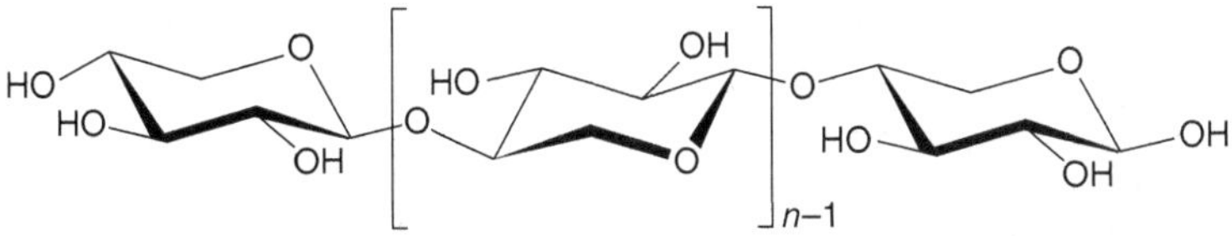

Fig. 18.3   General structure of xylo-oligosaccharides (where $n = 1$–9).

© Woodhead Publishing Limited, 2013

A refining step is necessary for the production of food-grade XOS. Vacuum evaporation increases the XOS concentration and removes volatile compounds such as acetic acid and the flavours of their precursors. Solvent extraction with organic solvents such as ethanol, acetone and 2-propanol is useful for removing non-saccharide components that are also the products of the xylan-degradation step. Adsorption using adsorbents such as activated charcoal, acid clay, bentonite, diatomaceous earth, aluminium hydroxide or oxide, titanium, silica and porous synthetic materials has been used to remove non-saccharide compounds and monosaccharides. Membrane techniques, of which ultrafiltration and nanofiltration are the most promising methods, are used for refining and separation of XOS within a given DP range. Ion-exchange and chromatographic separation are useful for desalting and for colour removal, respectively. Chromatographic separation is used for XOS purification on an analytical level yielding high-purity fractions, however, this is not suitable for economic reasons for large-scale production of GOS intended to be used in the food industry (Vázquez et al., 2001; Alonso et al., 2003; Aachary and Prapulla, 2011).

XOS have a considerable prebiotic potential and seem to exert their nutritional benefits in various animal species by having an intestinal tract populated by a complex, bacterial intestinal ecosystem (Aachary and Prapulla, 2011). Besides the potential prebiotic effect, immunostimulating effects, growth regulatory activity in aquaculture and poultry, antioxidant activity, anti-allergy, anti-infection and anti-inflammatory properties were reported for XOS. These biological properties are given in detail in a recent review of XOS (Aachary and Prapulla, 2011). In addition to the beneficial health effects, XOS have interesting physicochemical properties; they are only moderately sweet and stable over a wide range of pH and temperatures which make XOS suitable for incorporation into many food products such as in combination with soy milk, soft drinks, dairy products, sweets and confectionaries.

Malto-oligosaccharides

Malto-oligosaccharides contain 2–10 units of α-D-glucopyranose linked by α-$(1{\rightarrow}4)$ glycosidic linkages (Fig. 18.4). These compounds are used in the food industry because of their properties such as low sweetness, anti-hygroscopic agent, truncating agent, humectant and prevention of sucrose crystallization (Duedahl-Olesen et al., 2000; Lee et al., 2004). Malto-oligosaccharides are hydrolysed and absorbed in the small intestine and do not reach the colon intact, therefore, they are not generally claimed to increase the number of beneficial bacteria in the human colon and hence they are not considered to be prebiotics. They are, however, claimed to be effective in improving colonic conditions by reducing the levels of intestinal putrefactive bacteria such as Clostridium perfringens and members of the Enterobacteriaceae family (Crittenden and Playne, 1996).

© Woodhead Publishing Limited, 2013

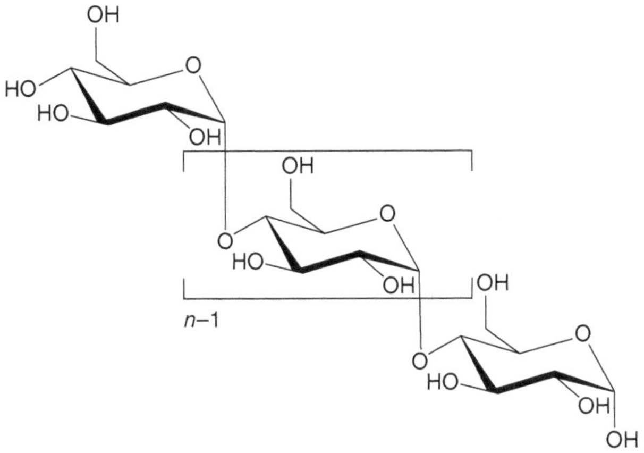

Fig. 18.4   Structure of malto-oligosaccharides (where $n = 1$–9).

Malto-oligosaccharides are produced from starch by the action of de-branching enzymes such as pullulanase and isoamylase, combined with hydrolysis catalysed by α-amylases. Depending on the source of α-amylases, product mixtures that are rich in malto-oligosaccharides differ in their chain lengths of these oligosaccharides. The preparation of malto-oligosaccharides that are larger than maltotriose in large amounts previously had been expensive and difficult, however, the discovery of microbial enzymes that produce malto-oligosaccharides of a specific length allows the production of these oligosaccharides in large amounts (Min *et al.*, 1998; Duedahl-Olesen *et al.*, 2000). Several amylases producing specific malto-oligosaccharides were reported including maltohexaose-producing amylases from *Bacillus circulans* (Takasaki, 1982), *Bacillus caldovelox* (Fogarty *et al.*, 1991), maltopentaose-producing amylases from *Pseudomonas* spp. (Shida *et al.*, 1992), maltotetraose-producing amylases from *Bacillus* sp. MG-4 (Takasaki *et al.*, 1991) and *Pseudomonas* sp. IMD 353 (Fogarty *et al.*, 1994), maltotriose-producing amylases from *Bacillus subtilis* (Takasaki, 1985) and maltose-producing amylases from *Bacillus licheniformis* (Kim *et al.*, 1992) and *Bacillus megaterium* G-2 (Takasaki, 1989). In particular, maltopentaose, which has five glucopyranose units, is in high demand as a high-value-added material in the medical field since pure maltopentaose can be used as a diagnostic reagent to examine the activity of α-amylase in serum (Min *et al.*, 1998; Lee *et al.*, 2004). Therefore, separation and purification processes to obtain pure maltopentaose from the reaction mixtures, including raw materials in enzymatic hydrolysis of starch, is crucial. For this purpose, various separation techniques such as gel filtration, solvent precipitation, chromatographic separation and adsorption by activated carbon (Lee *et al.*, 2003, 2004) have been applied.

© Woodhead Publishing Limited, 2013

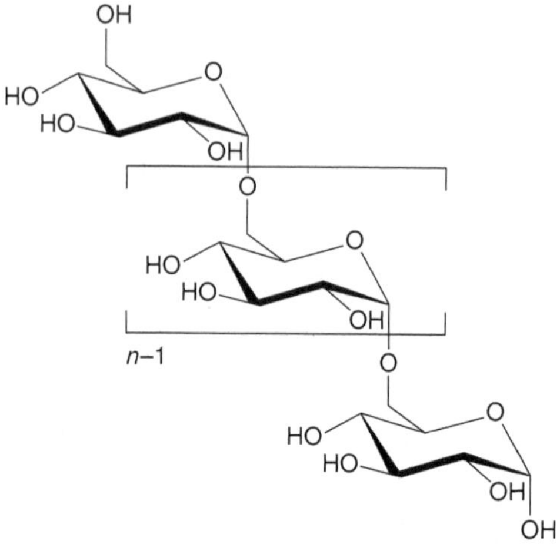

Fig. 18.5   Structure of isomalto-oligosaccharides (where $n = 1$–4).

Isomalto-oligosaccharides

Isomalto-oligosaccharides (IMO) consist of α-D-glucopyranose units linked by α-(1→6) glycosidic linkages (Fig. 18.5). IMO show promise as prebiotics and IMO have been used for treatment of chronic constipation and hyperlipidemia occurring as complications of maintenance haemodialysis (Goulas *et al.*, 2004). Among other oligosaccharides, which are widely used as food ingredients or additives (Qiang *et al.*, 2009) based on their nutritional and health benefits (Mussatto and Mancilha, 2007), IMO are of interest because of their availability, high stability and low cost (Zhang *et al.*, 2010). IMO are produced through sequential reactions catalysed by α-amylase, β-amylase and α-glucosidase. In the first stage, α-amylase liquefies starch, and then in the second stage, liquefied starch is hydrolysed to maltose using β-amylase. Finally, the transglucosidase activity of α-glucosidase is exploited in the production of IMO (Crittenden and Playne, 1996). The resulting IMO mixtures contain oligosaccharides with both α-(1→6)- and α-(1→4)-linked glucose residues. In recent years, much research had been focused on improvement of the efficiency of IMO production by screening for new and better enzymes for high yield IMO synthesis (Cho *et al.*, 2007; Fernández-Arrojo *et al.*, 2007; Zhou *et al.*, 2009). Efforts also have been made to develop novel processes such as synthesis of IMO from sucrose using free dextransucrase and dextranase (Goulas *et al.*, 2004; Lee *et al.*, 2008b), production of IMO by immobilized dextranase and glucosidase (Aslan and Tanriseven, 2007) and efficient conversion of maltose into IMO using transglucosidase immobilized in biomimetic polymer-inorganic hybrid capsules (Zhang *et al.*, 2009), or using an enzyme membrane reactor (Zhang *et al.*, 2010).

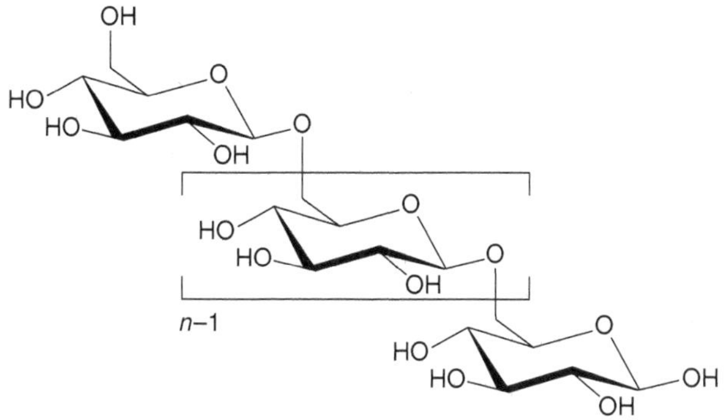

Fig. 18.6   Structure of gentio-oligosaccharides (where $n = 1$–4).

Gentio-oligosaccharides

Gentio-oligosaccharides (GenOS) consist of 2–5 glucose residues linked by β-(1→6) glycosidic linkages (Fig. 18.6). These oligosaccharides are not hydrolysed in the stomach or small intestine and therefore reach the colon intact, thus fulfilling a criterion of a prebiotic (Crittenden and Playne, 1996). GenOS were further reported to possess bifidogenic activity (Rycroft *et al.*, 2001; Gibson *et al.*, 2004). GenOS are usually produced from glucose syrup by enzymatic transglucosylation. Despite the prebiotic potential of GenOS, research on the novel production of GenOS is sparse. Fujimoto and colleagues have reported the use of a crude enzyme preparation from *Penicillium multicolor* for the efficient production of mainly gentiotriose to gentiopentaose by transglycosylation using a high concentration of gentiobiose as the starting substrate (Fujimoto *et al.*, 2009). This study reported that the transglycosylation involved two stages by a combination of β-glucosidase and β-(1→6)-glucanase. In the initial stage, which was the rate-limiting step in the overall process, β-glucosidase converted gentiobiose mainly to gentiotriose. In the second step, β-(1→6)-glucanase acted on the resulting gentiotriose, which served as both donor and acceptor, to produce higher GenOS by transglycosylation (Fujimoto *et al.*, 2009). Recently, GenOS of genistein and glycitein, the soy isoflavonoids, were successfully produced by biocatalytic glycosylation with cultured cells of *Eucalyptus perriniana* (Shimoda *et al.*, 2011).

Chito-oligosaccharides

Chitin is a linear polysaccharide consisting of β-(1→4)-linked *N*-acetyl-D-glucosamine (GlcNAc) residues. It is considered the second most abundant polysaccharide in nature, after cellulose, and it is a structural component of the exoskeletons of insects and arthropods (e.g. crabs, lobsters and shrimps), as well as of the cell walls of many fungi and yeasts (Kim and Rajapakse, 2005; Aam *et al.*, 2010). Chitin is insoluble in water and exists mainly in two

© Woodhead Publishing Limited, 2013

crystalline polymorphic forms, α and β. Chitosan is a non-toxic biopolymer of GlcNAc and D-glucosamine (GlcN). Chitosan can be prepared from chitin by partial deacetylation and it is soluble in dilute aqueous acid solutions (Aam et al., 2010; Xia et al., 2011).

Chito-oligosaccharides (CHOS) are the degraded products of chitosan and have recently received much attention because of their biological properties including their hypocholesterolaemic effect, antimicrobial effect, antioxidant effect, immunostimulating effect, calcium and iron absorption, anti-inflammatory effect and so on (Kim and Rajapakse, 2005; Xia et al., 2011). CHOS can be prepared by several methods (Mourya et al., 2011), of which acidic hydrolysis and enzymatic hydrolysis are widely used. Chemical hydrolysis is used more commonly in the industrial-scale production, however, its main drawback is the formation of some undesirable, toxic compounds during hydrolysis, higher risk associated with environmental pollution and lower production yields (Kim and Rajapakse, 2005). Enzymatic preparation methods using glycosyl hydrolases like chitinases or chitosanases (De Assis et al., 2010; Mourya et al., 2011) are nowadays drawing more interest because of their safety and ease of control (Xia et al., 2011). The fraction of N-acetylated residues, degree of polymerization, the molecular weight distribution and the pattern of N-acetylation of the resulting CHOS mixtures greatly depend on the chitosan preparation employed as a starting material and the specificity of the enzyme used (Aam et al., 2010). Some structures of CHOS are shown in Fig. 18.7. Furthermore, several nonspecific enzymes such as lipases, pectinases, proteases and cellulases have also been used for the preparation of CHOS (Lee et al., 2008a; Kittur et al., 2005; Lin et al., 2009).

18.3   Future trends

18.3.1   Human milk oligosaccharides

Human milk is special since it is the only food an infant may take in during the first months of its life and it contains all the essential nutrients needed by the infant to develop and grow. Moreover, it contains ingredients that go beyond traditional nutrients in that they provide certain health benefits to the infant. Human milk oligosaccharides (HMO) are prominent among these functional components of human milk. Mature human milk contains approximately 7–12 g l^{-1} of free oligosaccharides in addition to lactose, which typically is present in concentrations of 55–70 g l^{-1} (Boehm and Stahl, 2007; Kunz et al., 2000; Wu et al., 2010, 2011). Considerable efforts have been put into the structural elucidation of these different oligosaccharides and currently up to 200 unique oligosaccharide structures varying from 3 to 22 sugar units have been identified (German et al., 2008; Kobata, 2010; Kobata et al., 1978). The composition and content of oligosaccharides from different mothers can vary significantly, with milk from randomly

© Woodhead Publishing Limited, 2013

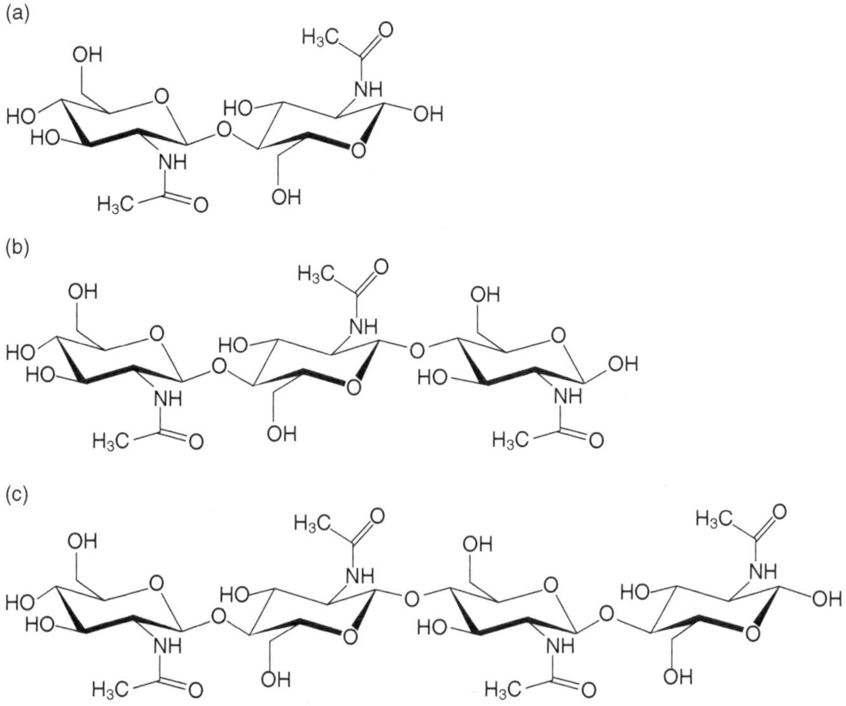

(a)

(b)

(c)

Fig. 18.7   Structures of some chito-oligosaccharides: (a) di-N-acetyl-chitobiose, (b) tri-N-acetyl-chitotriose and (c) tetra-N-acetyl-chitotetraose.

selected mothers containing between 23 to 130 different oligosaccharides. HMO are composed of the five monosaccharide building blocks D-glucose (Glc), D-galactose (Gal), N-acetylglucosamine (GlcNAc), L-fucose (Fuc) and sialic acid (N-acetylneuraminic acid), and thus they can be grouped into neutral and charged oligosaccharides, the latter being sialylated and comprising approximately 20% of all HMO (Wu *et al.*, 2010, 2011). The structures of HMO show typical patterns. Lactose (Gal-β-1,4-Glc) is found at the reducing end of HMO. This terminal lactose is typically elongated by lacto-N-biose units (LNB; Gal-β-1,3-GlcNAc) in type I or N-acetyl-lactosamine units (LacNAc; Gal-β-1,4-GlcNAc) in the rarer type II structures. Both LNB and LacNAc are attached via a β-1,3-linkage to the galactosyl moiety of the terminal lactose, with an additional β-1,6-linkage in branched HMO.

These LNB and LacNAc units can be repeated up to 25 times in larger HMO, forming the core region of these oligosaccharides. A further variation results from the attachment of fucosyl and sialic acid residues. Thus the simplest structures following this general scheme (apart from certain trisaccharides such as galactosyl-lactose, fucosyl-lactose and

© Woodhead Publishing Limited, 2013

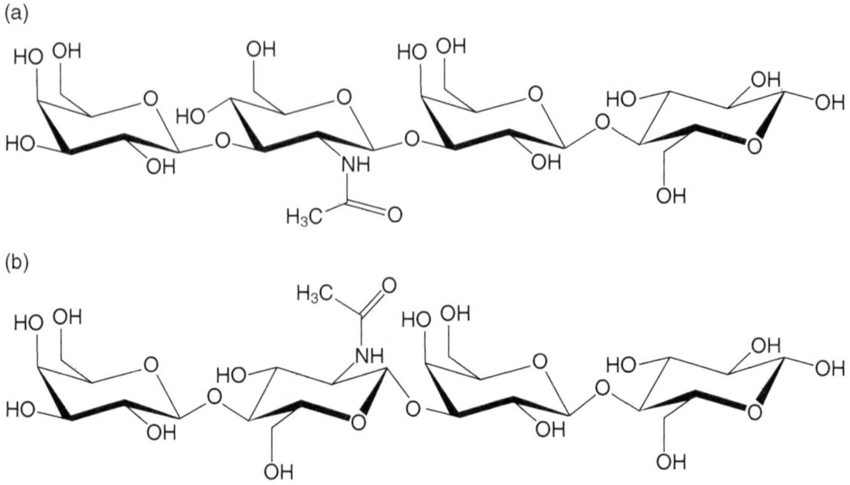

(a)

(b)

Fig. 18.8 Some simple structures of human milk oligosaccharides: tetrasaccharides (a) lacto-*N*-tetraose, Gal-β-1,3-GlcNAc-β-1,3-Gal-β-1,4-Glc and (b) lacto-*N-neo*-tetraose, Gal-β-1,4-GlcNAc-β-1,3-Gal-β-1,4-Glc.

sialyl-lactose) are the tetrasaccharides lacto-*N*-tetraose, Gal-β-1,3-GlcNAc-β-1,3-Gal-β-1,4-Glc (type I) and lacto-*N-neo*-tetraose, Gal-β-1,4-GlcNAc-β-1,3-Gal-β-1,4-Glc (type II) (Fig. 18.8) (Kobata, 2010; Kunz *et al.*, 2000; Wu *et al.*, 2010, 2011). As mentioned above, the composition and content of HMO can vary significantly between different mothers and is also affected by the time/length of lactation. Nevertheless, mono- and difucosyl-lactose, lacto-*N*-tetraose and its fucosylated derivatives as well as sialyl-lactose and the sialylated forms of lacto-*N*-tetraose are major HMO in human milk.

There is strong evidence that this complex mixture of HMO shows different beneficial effects for the breast-fed infants. Certain HMO show an anti-adhesive effect, that is, they prevent binding of potentially pathogenic microorganisms to oligosaccharides on the epithelial cell surface via specific glycan-binding proteins. This binding is essential for colonization and infection. These 'anti-adhesive' HMO thus serve as soluble ligand analogues and by acting as decoy binding sites for the pathogens they occupy the bacterial lectin site and thus prevent attachment of the pathogens. Furthermore, it has been proposed that some HMO may also have glycome-modifying effects, that is, they modulate the expression of intestinal epithelial cell surface glycans and thereby affect and modify the attachment sites of pathogens (Bode, 2009). In addition to preventing pathogens from colonizing infant intestines, HMO selectively support growth and activity of desired bacteria in the infant intestine, thus they have a prebiotic or bifidogenic effect since bifidobacteria, mainly *Bifidobacterium longum* biovar *infantis*, *B. longum* biovar *longum*, *B. breve* and *B. bifidum*, dominate in the gut of

© Woodhead Publishing Limited, 2013

breast-fed infants (Bode, 2009; Ventura *et al.*, 2001). The complex neutral HMO are especially linked to a prebiotic effect and metabolization/growth stimulation by bifidobacteria (LoCascio *et al.*, 2007), with *N*-acetylglucosamine-containing oligosaccharides being regarded as the most bifidogenic oligosaccharides in human milk (Kunz *et al.*, 2000). Recently, it was shown that among a number of gut-related bacteria studied only *Bifidobacterium* and *Bacteroides* species were able to consume HMO as a sole carbon source and reach high cell densities in these oligosaccharides (Marcobal *et al.*, 2010). Moreover, this ability to degrade and consume HMO such as lacto-*N*-tetraose or the LNB building block in type I HMO has been shown for the infant-type bifidobacteria, while adult-type bifidobacteria such as *B. adolescentis* or *B. animalis* cannot grow (Xiao *et al.*, 2010; Sela and Mills, 2010).

Different *Bifidobacterium* species appear to have developed different strategies for degrading HMO and thus utilizing them as a growth substrate. It was shown that *B. longum* subsp. *infantis* preferentially consumed HMO with a degree of polymerization of 7 or less (LoCascio *et al.*, 2007). This organism takes up these different oligosaccharides via a wide variety of transporters coupled with specific soluble binding proteins. Once taken up, a complement of intracellular glycosidases hydrolyse these different HMO to the respective monosaccharides, which enter the central metabolic pathways (Sela *et al.*, 2008; Sela and Mills, 2010; Zivkovic *et al.*, 2011). HMO consumption is quite different in *B. bifidum*, which secretes both fucosidases and sialidases so that the terminal fucosyl and sialyl groups are removed from HMO and the core structures are unmasked. Type I HMO are then further degraded extracellularly by lacto-*N*-biosidase, which releases LNB from lacto-*N*-tetraose or from the core of type I HMO. LNB is then imported via a specific ABC transporter system. Further utilization proceeds via an intracellular phosphorolytic cleavage of LNB by lacto-*N*-biose phosphorylase, yielding galactose-1-phosphate and GlcNAc (Fushinobu, 2010).

Because of the aforementioned health-related effects of HMO, considerable interest exists in the efficient production of these compounds, which is albeit hampered by the structural complexity as well as the complex mixture of HMO. Reported methods for HMO production include chemical and microbial/enzymatic synthesis as well as isolation from suitable natural sources. The milk of dairy animals (cow, sheep, goat) contains oligosaccharides that are identical or structurally related to HMO yet their concentrations are significantly lower than in human milk. Goat milk contains higher amounts of milk oligosaccharides (0.25–0.30 g l^{-1}) than milk from cows or sheep (0.02–0.06 g l^{-1}). Oligosaccharides from goat milk could be isolated in more than 80% yields by using a two-stage filtration process comprising ultrafiltration (50 kDa cut-off) and nanofiltration (1 kDa cut-off) with 3-sialyl-lactose, 6-sialyl-lactose, N-glycolyl-neuraminyl-lactose and 3-galactosyl-lactose as the main components (Martinez-Ferez *et al.*, 2006). Another promising source of milk oligosaccharides is permeate of bovine cheese

© Woodhead Publishing Limited, 2013

whey ultrafiltration. Even though the concentration of oligosaccharides is quite low in bovine milk/whey, whey permeate is a by-product from the production of whey protein concentrate and thus readily available in large amounts. Bovine whey contains a variety of neutral and sialylated oligosaccharides. A recent study identified 15 oligosaccharide structures in bovine whey permeate, half of which have the same composition as HMO (Barile *et al.*, 2009).

Classical chemical syntheses, often in combination with enzymatic methods, have been described for a number of HMO structures or their building blocks, for example for fucosylated LNB or LacNAc (La Ferla *et al.*, 2002). These reactions often include extensive protection/deprotection steps. These chemical approaches can provide access to small oligosaccharides such as 3′-and 6′-sialyl-lactose (Rencurosi *et al.*, 2002) but also more complex structures such as sialyl-lacto-*N*-tetraose (Schmidt and Thiem, 2010), albeit in different reaction yields. HMO structures such as lacto-*N*-neohexaose are also accessible via chemical solid-phase synthesis (Zhu and Schmidt, 2009).

The most promising approach for the production of defined HMO structures seems to be microbial, fermentative methods employing single, appropriately engineered microorganisms. This *in vivo* approach was introduced by the group of Eric Samain. They used a β-galactosidase-negative *Escherichia coli* strain overexpressing a β-1,3-*N*-acetylglucosaminyl-transferase gene from *Neisseria meningitidis*. When feeding lactose to this engineered strain, this disaccharide was taken up by the indigenous β-galactoside permease of *E. coli*. The recombinantly synthesized β-1,3-*N*-acetylglucosaminyl-transferase then utilized the intracellular pool of UDP-GlcNAc to transfer GlcNAc residues regiospecifically to lactose, resulting in the formation of the trisaccharide GlcNac-β-1,3-Gal-β-1,4-Glc. This compound was released into the extracellular medium in yields of 6 g l^{-1} (Priem *et al.*, 2002). In a similar manner, an engineered *E. coli* strain overexpressing an α-1,3-fucosyltransferase from *Helicobacter pylori*, genetically engineered to provide sufficient GDP-fucose as the intracellular substrate for the glycosyltransferase was used to produce various fucosylated HMO from lactose added to the medium (Dumon *et al.*, 2001; Drouillard *et al.*, 2006) An engineered *E. coli* strain over-expressing the α-2,3-sialyltransferase gene from *Neisseria meningitidis*, together with an engineered pathway to provide the activated sialic acid donor CMP-Neu5Ac as the substrate for the glycosyltransferase, produced 3-sialyl-lactose in concentrations of up to 25 g l^{-1} in high cell density cultivations with continuous lactose feed (Fierfort and Samain, 2008). 6-Sialyl-lactose was efficiently produced when employing α-2,6-sialyltransferase from *Photobacterium* sp. again in metabolically engineered *E. coli* (Drouillard *et al.*, 2010).

Lacto-*N*-biose was produced in a purely enzymatic approach, making use of the synthetic capacity of sugar phosphorylases. Lacto-*N*-biose phosphorylase, together with sucrose phosphorylase, UDP-glucose-

© Woodhead Publishing Limited, 2013

hexose-1-phosphate uridylyltransferase and UDP-glucose 4-epimerase produced LNB from sucrose and GlcNAc in the presence of phosphate and catalytic amounts of UDP-Glc in yields of 85% (Nishimoto and Kitaoka, 2007).

18.3.2   Mixed oligosaccharides

Some of the above-mentioned HMO structures or structurally related compounds can also be accessed through other approaches, albeit not in pure form. One such approach that has received some interest is based on β-galactosidase-catalysed transglycosylation with lactose as donor (thus transferring galactose onto suitable acceptors) and GlcNAc as acceptor, thus obtaining N-acetyl-lactosamine (LacNAc) and its regioisomers. Using this approach and a hyperthermophilic β-galactosidase from *Sulfolobus solfataricus*, Gal-β-1,6-GlcNAc together with an unidentified sugar were the main products starting from a mixture of 1 M lactose and 1 M GlcNAc, while LacNAc and Gal-β-1,3-GlcNAc were formed as well, yet in lower concentrations (Reuter *et al.*, 1999). This reaction was also optimized using β-galactosidase from *Bacillus circulans* as the biocatalyst. This enzyme is known for its propensity to synthesize β-1,4-linkages in its transgalactosylation mode and hence the main reaction product here was LacNAc together with smaller amounts of GlcNAc-containing higher oligosaccharides (one tri- and one tetrasaccharides) and Gal-β-1,6-GlcNAc. The total yield was 40% of these GlcNAc-containing oligosaccharides when starting from 0.5 M lactose and GlcNAc each (Li *et al.*, 2010). This reaction and the β-galactosidase from *B. circulans* were also compared to the enzyme from *Kl. lactis* with the reaction conditions optimized, and the latter enzyme was shown to form predominately Gal-β-1,6-GlcNAc. Again, both enzymes formed a mixture of various di- to tetrasaccharides (Bridiau and Maugard, 2011). Since these structures resemble the core of HMO they could be of interest as prebiotic compounds to be added to food.

Another approach to obtain mixed oligosaccharides composed with different sugar molecules with prebiotic potential is through enzymatic hydrolysis of suitable polysaccharides, mainly plant cell wall material. One study used soy arabinogalactan, sugar beet arabinan, wheat flour arabinoxylan, polygalacturonan and the rhamnogalacturonan fraction from apple as the starting material for enzymatic hydrolysis to yield different oligosaccharides, some of which were built from different sugar building blocks, such as the different rhamnogalacturono-oligosaccharides obtained from apple or unidentified arabinogalacto-oligosaccharides from soybeans. It was further shown that individual, relevant gut-related microorganisms such as bifidobacteria or *Bacteroides* could ferment some of these mixed oligosaccharides *in vitro*. These results do not allow prediction of the *in vivo* fermentation of these oligosaccharides and hence their prebiotic potential, yet they indicate that some of these oligosaccharides could be of potential

© Woodhead Publishing Limited, 2013

interest as non-digestible oligosaccharides specifically supporting growth and activity of desired microorganisms in the gut (van Laere *et al.*, 2000b). This approach of controlled hydrolysis of heteropolysaccharides could be of interest for the production of novel mixed oligosaccharides with prebiotic potential.

18.4   Conclusions

The relationship between certain oligosaccharide structures and prebiotic function is still poorly understood. Research, however, intensified in this field, and we are beginning to understand these structure/function relationships better now. This has already had an impact on novel prebiotic products that were brought to the market and will have a more important impact in the future. These insights into the structures of prebiotic oligosaccharides and their effect, together with advancement in the area of biotechnology, will certainly result in the production of functionally enhanced prebiotic oligosaccharides, and one can imagine that these will fulfil specific functions, for example one can imagine that these will target certain population groups such as infants or the elderly (Rastall and Maitin, 2002).

18.5   References

AACHARY, A. A. and PRAPULLA, S. G. (2008). 'Corncob-induced endo-1,4-β-D-xylanase of *Aspergillus oryzae* MTCC 5154: production and characterization of xylobiose from glucuronoxylan', *Journal of Agricultural and Food Chemistry*, **56**, 3981–8.

AACHARY, A. A. and PRAPULLA, S. G. (2011). 'Xylooligosaccharides (XOS) as an emerging prebiotic: microbial synthesis, utilization, structural characterization, bioactive properties, and applications'. *Comprehensive Reviews in Food Science and Food Safety*, **10**, 2–16.

AAM, B. B., HEGGSET, E. B., NORBERG, A. L., SØRLIE, M., VÅRUM, K. M. and EIJSINK, V. G. H. (2010). 'Production of chitooligosaccharides and their potential applications in medicine'. *Marine Drugs*, **8**, 1482–517.

ABRAMS, S. A., GRIFFIN, I. J., HAWTHORNE, K. M., LIANG, L., GUNN, S. K., DARLINGTON, G. and ELLIS, K. J. (2005). 'A combination of prebiotic short- and long-chain inulin-type fructans enhances calcium absorption and bone mineralization in young adolescents'. *American Journal of Clinical Nutrition*, **82**, 471–6.

ABRAMS, S. A., HAWTHORNE, K. M., ALIU, O., HICKS, P. D., CHEN, Z. and GRIFFIN, I. J. (2007). 'An inulin-type fructan enhances calcium absorption primarily via an effect on colonic absorption in humans'. *Journal of Nutrition*, **137**, 2208–12.

ADAM, A. C., RUBIO-TEXEIRA, M. and POLAINA, J. (2004). 'Lactose: The milk sugar from a biotechnological perspective'. *Critical Reviews in Food Science and Nutrition*, **44**, 5.

AKPINAR, O., AK, O., KAVAS, A., BAKIR, U. and YILMAZ, L. (2007). 'Enzymatic production of xylooligosaccharides from cotton stalks'. *Journal of Agricultural and Food Chemistry*, **55**, 5544–51.

AKPINAR, O., ERDOGAN, K. and BOSTANCI, S. (2009). 'Enzymatic production of xylooligosaccharide from selected agricultural wastes'. *Food and Bioproducts Processing*, **87**, 145–51.

© Woodhead Publishing Limited, 2013

ALBAYRAK, N. and YANG, S. T. (2002). 'Production of galacto-oligosaccharides from lactose by *Aspergillus oryzae* β-galactosidase immobilized on cotton cloth'. *Biotechnology and Bioengineering*, **77**, 8–19.

ALONSO, J. L., DOMINGUEZ, H., GARROTE, G., PARAJO, J. C., VAZQUEZ, M. J. (2003). 'Xylooligosaccharides: properties and production technologies'. *Electronic Journal of Environmental, Agricultural and Food Chemistry*, **2**, 230–2.

ÁLVARO-BENITO, M., DE ABREU, M., FERNÁNDEZ-ARROJO, L., PLOU, F. J., JIMÉNEZ-BARBERO, J., BALLESTEROS, A., POLAINA, J. and FERNÁNDEZ-LOBATO, M. (2007). 'Characterization of a β-fructofuranosidase from *Schwanniomyces occidentalis* with transfructosylating activity yielding the prebiotic 6-kestose'. *Journal of Biotechnology*, **132**, 75–81.

ASLAN, Y. and TANRISEVEN, A. (2007). 'Immobilization of *Penicillium lilacinum* dextranase to produce isomaltooligosaccharides from dextran'. *Biochemical Engineering Journal*, **34**, 8–12.

ASP, N. G., BURVALL, A. and DAHLQVIST, A. (1980). 'Oligosaccharide formation during hydrolysis of lactose with *Saccharomyces lactis* lactase (Maxilact®): Part 2. Oligosaccharide structures'. *Food Chemistry*, **5**, 147–53.

BARCLAY, T., GINIC-MARKOVIC, M., COOPER, P. and PETROVSKY, N. (2010). 'Inulin – A versatile polysaccharide with multiple pharmaceutical and food chemical uses'. *Journal of Excipients and Food Chemicals*, **1**, 27–50.

BARILE, D., TAO, N., LEBRILLA, C. B., COISSON, J. D., ARLORIO, M. and GERMAN, J. B. (2009). 'Permeate from cheese whey ultrafiltration is a source of milk oligosaccharides'. *International Dairy Journal*, **19**, 524–30.

BARRETEAU, H., DELATTRE, C. and MICHAUD, P. (2006). 'Production of oligosaccharides as promising new food additive generation'. *Food Technology and Biotechnology*, **44**, 323–33.

BLAUT, M. (2002). 'Relationship of prebiotics and food to intestinal microflora'. *European Journal of Nutrition*, **41**, 11–6.

BODE, L. (2009). 'Human milk oligosaccharides: prebiotics and beyond'. *Nutrition Reviews*, **67** Suppl 2, S183–91.

BOEHM, G. and STAHL, B. (2007). 'Oligosaccharides from milk'. *Journal of Nutrition*, **137**, 847S–9S.

BOON, M. A., JANSSEN, A. E. M. and VAN `T RIET, K. (2000). 'Effect of temperature and enzyme origin on the enzymatic synthesis of oligosaccharides'. *Enzyme and Microbial Technology*, **26**, 271–81.

BOUHNIK, Y., ACHOUR, L., PAINEAU, D., RIOTTOT, M., ATTAR, A. and BORNET, F. (2007). 'Four-week short chain fructo-oligosaccharides ingestion leads to increasing fecal bifidobacteria and cholesterol excretion in healthy elderly volunteers'. *Nutrition Journal*, **6**, art. no. 42.

BRIDIAU, N. and MAUGARD, T. (2011). 'A comparative study of the regioselectivity of the β-galactosidases from *Kluyveromyces lactis* and *Bacillus circulans* in the enzymatic synthesis of *N*-acetyl-lactosamine in aqueous media'. *Biotechnology Progress*, **27**, 386–94.

BROEKAERT, W. F., COURTIN, C. M., VERBEKE, K., VAN DE WIELE, T., VERSTRAETE, W. and DELCOUR, J. A. (2011). 'Prebiotic and other health-related effects of cereal-derived arabinoxylans, arabinoxylan-oligosaccharides, and xylooligosaccharides'. *Critical Reviews in Food Science and Nutrition*, **51**, 178–94.

BYSTRICKÝ, S., MEDOVARSKÁ, I. and MACHOVÁ, E. (2011). 'Separation of different types of monosaccharides by polyacrylamide column chromatography'. *Zeitschrift für Naturforschung – Section B Journal of Chemical Sciences*, **66**, 295–8.

CHI, Z., CHII, Z., ZHANG, T., LIU, G. and YUE, L. (2009). 'Inulinase-expressing microorganisms and applications of inulinases'. *Applied Microbiology and Biotechnology*, **82**, 211–20.

© Woodhead Publishing Limited, 2013

CHO, Y. J., SINHA, J., PARK, J. P. and YUN, J. W. (2001). 'Production of inulooligosaccharides from chicory extract by endoinulinase from *Xanthomonas oryzae* No. 5'. *Enzyme and Microbial Technology*, **28**, 439–45.

CHO, M. H., PARK, S. E., LEE, M. H., HA, S. J., KIM, H. Y., KIM, M. J., LEE, S. J., MADSEN, S. M. and PARK, C. S. (2007). 'Extracellular secretion of a maltogenic amylase from *Lactobacillus gasseri* ATCC33323 in *Lactococcus lactis* MG1363 and its application on the production of branched maltooligosaccharides'. *Journal of Microbiology and Biotechnology*, **17**, 1521–6.

COULIER, L., TIMMERMANS, J., RICHARD, B., VAN DEN DOOL, R., HAAKSMAN, I., KLARENBEEK, B., SLAGHEK, T. and VAN DONGEN, W. (2009). 'In-depth characterization of prebiotic galactooligosaccharides by a combination of analytical techniques'. *Journal of Agricultural and Food Chemistry*, **57**, 8488–95.

CRITTENDEN, R. G. and PLAYNE, M. J. (1996). 'Production, properties and applications of food-grade oligosaccharides'. *Trends in Food Science and Technology*, **7**, 353–61.

CUMMINGS, J. H., MACFARLANE, G. T. and ENGLYST, H. N. (2001). 'Prebiotic digestion and fermentation'. *American Journal of Clinical Nutrition*, **73**, 415S–20S.

DE ABREU, M. A., ÁLVARO-BENITO, M., PLOU, F. J., FERNÁNDEZ-LOBATO, M. and ALCALDE, M. (2011). 'Screening β-fructofuranosidases mutant libraries to enhance the transglycosylation rates of β-(2→6) fructooligosaccharides'. *Combinatorial Chemistry and High Throughput Screening*, **14**, 730–8.

DE ASSIS, C. F., ARAUÒIJO, N. K., PAGNONCELLI, M. G. B., DA SILVA PEDRINI, M. R., DE MACEDO, G. R. and DOS SANTOS, E. S. (2010). 'Chitooligosaccharides enzymatic production by *Metarhizium anisopliae*'. *Bioprocess and Biosystems Engineering*, **33**, 893–9.

DE PRETER, V., HAMER, H. M., WINDEY, K. and VERBEKE, K. (2011). 'The impact of pre- and/or probiotics on human colonic metabolism: Does it affect human health?' *Molecular Nutrition and Food Research*, **55**, 46–57.

DROUILLARD, S., DRIGUEZ, H. and SAMAIN, E. (2006). 'Large-scale synthesis of H-antigen oligosaccharides by expressing *Helicobacter pylori* α1,2-fucosyltransferase in metabolically engineered *Escherichia coli* cells'. *Angewandte Chemie – International Edition*, **45**, 1778–80.

DROUILLARD, S., MINE, T., KAJIWARA, H., YAMAMOTO, T. and SAMAIN, E. (2010). 'Efficient synthesis of 6'-sialyllactose, 6,6'-disialyllactose, and 6'-KDO-lactose by metabolically engineered *E. coli* expressing a multifunctional sialyltransferase from the *Photobacterium* sp. JT-ISH-224'. *Carbohydrate Research*, **345**, 1394–9.

DUEDAHL-OLESEN, L., MATTHIAS KRAGH, K. and ZIMMERMANN, W. (2000). 'Purification and characterisation of a malto-oligosaccharide-forming amylase active at high pH from *Bacillus clausii* BT-21'. *Carbohydrate Research*, **329**, 97–107.

DUMON, C., PRIEM, B., MARTIN, S. L., HEYRAUD, A., BOSSO, C. and SAMAIN, E. (2001). '*In vivo* fucosylation of lacto-*N*-neotetraose and lacto-*N*-neohexaose by heterologous expression of *Helicobacter pylori* α-1,3 fucosyltransferase in engineered *Escherichia coli*'. *Glycoconjugate Journal*, **18**, 465–74.

EIWEGGER, T., STAHL, B., HAIDL, P., SCHMITT, J., BOEHM, G., DEHLINK, E., URBANEK, R. and SZEPFALUSI, Z. (2010). 'Prebiotic oligosaccharides: *In vitro* evidence for gastrointestinal epithelial transfer and immunomodulatory properties'. *Pediatric Allergy and Immunology*, **21**, 1179–88.

ENGLYST, H. N. and HUDSON, G. J. (1996). 'The classification and measurement of dietary carbohydrates'. *Food Chemistry*, **57**, 15–21.

FENG, H. Y., DRONE, J., HOFFMANN, L., TRAN, V., TELLIER, C., RABILLER, C. and DION, M. (2005). 'Converting a β-glycosidase into a β-transglycosidase by directed evolution'. *Journal of Biological Chemistry*, **280**, 37088–97.

FERNÁNDEZ-ARROJO, L., MARÍN, D., GÓMEZ DE SEGURA, A., LINDE, D., ALCALDE, M., GUTIÉRREZ-ALONSO, P., GHAZI, I., PLOU, F. J., FERNÁNDEZ-LOBATO, M. and BALLESTEROS, A. (2007). 'Transformation of maltose into prebiotic isomaltooligosaccharides by

© Woodhead Publishing Limited, 2013

a novel α-glucosidase from *Xantophyllomyces dendrorhous'. Process Biochemistry*, **42**, 1530–6.

FIERFORT, N. and SAMAIN, E. (2008). 'Genetic engineering of *Escherichia coli* for the economical production of sialylated oligosaccharides'. *Journal of Biotechnology*, **134**, 261–5.

FIGUEROA-GONZÁLEZ, I., QUIJANO, G., RAMÍREZ, G. and CRUZ-GUERRERO, A. (2011). 'Probiotics and prebiotics-perspectives and challenges'. *Journal of the Science of Food and Agriculture*, **91**, 1341–8.

FOGARTY, W. M., BEALIN-KELLY, F., KELLY, C. T. and DOYLE, E. M. (1991). 'A novel maltohexose-forming α-amylase from *Bacillus caldovelox*: Patterns and mechanisms of action'. *Applied Microbiology and Biotechnology*, **36**, 184–9.

FOGARTY, W. M., BOURKE, A. C., KELLY, C. T. and DOYLE, E. M. (1994). 'A constitutive maltotetraose producing amylase from *Pseudomonas* sp. IMD 353'. *Applied Microbiology and Biotechnology*, **42**, 198–203.

FRANCK, A. and BOSSCHER, D. (2006). 'Inulin and oligofructose as prebiotics in infant feeding'. *Agro Food Industry Hi-Tech*, **17**, 53–5.

FUJIMOTO, Y., HATTORI, T., UNO, S., MURATA, T. and USUI, T. (2009). 'Enzymatic synthesis of gentiooligosaccharides by transglycosylation with β-glycosidases from *Penicillium multicolor'. Carbohydrate Research*, **344**, 972–8.

FUSHINOBU, S. (2010). 'Unique sugar metabolic pathways of bifidobacteria'. *Bioscience, Biotechnology and Biochemistry*, **74**, 2374–84.

GÄNZLE, M. G., HAASE, G. and JELEN, P. (2008). 'Lactose: Crystallization, hydrolysis and value-added derivatives'. *International Dairy Journal*, **18**, 685–94.

GERMAN, J. B., FREEMAN, S. L., LEBRILLA, C. B. and MILLS, D. A. (2008). 'Human milk oligosaccharides: evolution, structures and bioselectivity as substrates for intestinal bacteria'. *Nestle Nutrition Workshop Series Paediatric Programme*, **62**, 205–18; discussion 218–22.

GIACCO, R., CLEMENTE, G., LUONGO, D., LASORELLA, G., FIUME, I., BROUNS, F., BORNET, F., PATTI, L., CIPRIANO, P., RIVELLESE, A. A. and RICCARDI, G. (2004). 'Effects of short-chain fructo-oligosaccharides on glucose and lipid metabolism in mild hypercholesterolaemic individuals'. *Clinical Nutrition*, **23**, 331–40.

GIBSON, G. R. and ROBERFROID, M. B. (1995). 'Dietary modulation of the human colonic microbiota: Introducing the concept of prebiotics'. *Journal of Nutrition*, **125**, 1401–12.

GIBSON, G. R., PROBERT, H. M., VAN LOO, J., RASTALL, R. A. and ROBERFROID, M. B. (2004). 'Dietary modulation of the human colonic microbiota: Updating the concept of prebiotics'. *Nutrition Research Reviews*, **17**, 259–75.

GOPAL, P. K., SULLIVAN, P. A. and SMART, J. B. (2001). 'Utilisation of galacto-oligosaccharides as selective substrates for growth by lactic acid bateria including *Bifidobacterium lactis* DR10 and *Lactobacillus rhamnous* DR20'. *International Dairy Journal*, **11**, 19–25.

GOSLING, A., STEVENS, G. W., BARBER, A. R., KENTISH, S. E. and GRAS, S. L. (2010). 'Recent advances refining galactooligosaccharide production from lactose'. *Food Chemistry*, **121**, 307–18.

GOULAS, A. K., FISHER, D. A., GRIMBLE, G. K., GRANDISON, A. S. and RASTALL, R. A. (2004). 'Synthesis of isomaltooligosaccharides and oligodextrans by the combined use of dextransucrase and dextranase'. *Enzyme and Microbial Technology*, **35**, 327–38.

GOULAS, T. K., GOULAS, A. K., TZORTZIS, G. and GIBSON, G. R. (2007). 'Molecular cloning and comparative analysis of four β-galactosidase genes from *Bifidobacterium bifidum* NCIMB41171'. *Applied Microbiology and Biotechnology*, **76**, 1365–72.

GOULAS, T., GOULAS, A., TZORTZIS, G. and GIBSON, G. R. (2009). 'Comparative analysis of four β-galactosidases from *Bifidobacterium bifidum* NCIMB41171: Purification

© Woodhead Publishing Limited, 2013

and biochemical characterisation'. *Applied Microbiology and Biotechnology*, **82**, 1079–88.

GUTIÉRREZ-ALONSO, P., FERNÁNDEZ-ARROJO, L., PLOU, F. J. and FERNÁNDEZ-LOBATO, M. (2009). 'Biochemical characterization of a β-fructofuranosidase from *Rhodotorula dairenensis* with transfructosylating activity'. *FEMS Yeast Research*, **9**, 768–73.

HANCOCK, S. M., VAUGHAN, M. D. and WITHERS, S. G. (2006). 'Engineering of glycosidases and glycosyltransferases'. *Current Opinion in Chemical Biology*, **10**, 509–19.

HÉBERT, E. M., RAYA, R. R., TAILLIEZ, P. and DE GIORI, G. S. (2000). 'Characterization of natural isolates of *Lactobacillus* strains to be used as starter cultures in dairy fermentation'. *International Journal of Food Microbiology*, **59**, 19–27.

HINZ, S. W. A., VAN DEN BROEK, L. A. M., BELDMAN, G., VINCKEN, J. P. and VORAGEN, A. G. J. (2004). 'β-Galactosidase from *Bifidobacterium adolescentis* DSM20083 prefers β(1,4)-galactosides over lactose'. *Applied Microbiology and Biotechnology*, **66**, 276–84.

HOLZAPFEL, W. H. and SCHILLINGER, U. (2002). 'Introduction to pre- and probiotics'. *Food Research International*, **35**, 109–16.

HUNG, M.-N. and LEE, B. H. (2002). 'Purification and characterization of a recombinant β-galactosidase with transgalactosylation activity from *Bifidobacterium infantis* HL96'. *Applied Microbiology and Biotechnology*, **58**, 439–45.

HUNG, M. N., XIA, Z., HU, N. T. and LEE, B. H. (2001). 'Molecular and biochemical analysis of two β-galactosidases from *Bifidobacterium infantis* HL96'. *Applied and Environmental Microbiology*, **67**, 4256–63.

HUSAIN, Q. (2010). 'β-Galactosidases and their potential applications: A review'. *Critical Reviews in Biotechnology*, **30**, 41–62.

IQBAL, S., NGUYEN, T. H., NGUYEN, T. T., MAISCHBERGER, T. and HALTRICH, D. (2010). 'β-Galactosidase from *Lactobacillus plantarum* WCFS1: Biochemical characterization and formation of prebiotic galacto-oligosaccharides'. *Carbohydrate Research*, **345**, 1408–16.

IQBAL, S., NGUYEN, T. H., NGUYEN, H. A., NGUYEN, T. T., MAISCHBERGER, T., KITTL, R. and HALTRICH, D. (2011). 'Characterization of a heterodimeric GH2 β-galactosidase from *Lactobacillus sakei* Lb790 and formation of prebiotic galacto-oligosaccharides'. *Journal of Agricultural and Food Chemistry*, **59**, 3803–11.

ISAPP (2008). *6th Meeting of the International Scientifc Association of Probiotics and Prebiotics*, London, Ontario.

IUB-IUPAC (1982). 'Polysaccharide nomenclature. Recommendations 1980. IUB-IUPAC Joint Commission on Biochemical Nomenclature (JCBN)'. *Journal of Biological Chemistry*, **257**, 3352–4.

KANGO, N. and JAINS, S. C. (2011). 'Production and properties of microbial inulinases: Recent advances'. *Food Biotechnology*, **25**, 1532–4249.

KELLY, G. (2008). 'Inulin-type prebiotics – A review: Part 1'. *Alternative Medicine Review*, **13**, 315–29.

KILIAN, S., KRITZINGER, S., RYCROFT, C., GIBSON, G. and DU PREEZ, J. (2002). 'The effects of the novel bifidogenic trisaccharide, neokestose, on the human colonic microbiota'. *World Journal of Microbiology and Biotechnology*, **18**, 637–44.

KIM, C. S. and JI, E.-S. (2004). 'A new kinetic model of recombinant β-galactosidase from *Kluyveromyces lactis* for both hydrolysis and transgalactosylation reactions'. *Biochemical and Biophysical Research Communications*, **316**, 738–43.

KIM, S. K. and RAJAPAKSE, N. (2005). 'Enzymatic production and biological activities of chitosan oligosaccharides (COS): A review'. *Carbohydrate Polymers*, **62**, 357–68.

KIM, D. H., CHOI, Y. J., SONG, S. K. and YUN, J. W. (1997). 'Production of inulo-oligosaccharides using endo-inulinase from a *Pseudomonas* sp'. *Biotechnology Letters*, **19**, 369–71.

© Woodhead Publishing Limited, 2013

KIM, I. C., CHA, J. H., KIM, J. R., JANG, S. Y., SEO, B. C., CHEONG, T. K., DAE SIL, L., YANG DO, C. and PARK, K. H. (1992). 'Catalytic properties of the cloned amylase from *Bacillus licheniformis*'. *Journal of Biological Chemistry*, **267**, 22108–14.

KIM, S., KIM, W. and HWANG, I. K. (2003). 'Optimization of the extraction and purification of oligosaccharides from defatted soybean meal'. *International Journal of Food Science and Technology*, **38**, 337–42.

KITTUR, F. S., VISHU KUMAR, A. B., VARADARAJ, M. C. and THARANATHAN, R. N. (2005). 'Chitooligosaccharides – Preparation with the aid of pectinase isozyme from *Aspergillus niger* and their antibacterial activity'. *Carbohydrate Research*, **340**, 1239–45.

KOBATA, A. (2010). 'Structures and application of oligosaccharides in human milk'. *Proceedings of the Japan Academy – Series B: Physical & Biological Sciences*, **86**, 731–47.

KOBATA, A., YAMASHITA, K. and TACHIBANA, Y. (1978). 'Oligosaccharides from human milk'. *Methods in Enzymology*, **50**, 216–20.

KUNZ, C., RUDLOFF, S., BAIER, W., KLEIN, N. and STROBEL, S. (2000). 'Oligosaccharides in human milk: structural, functional, and metabolic aspects'. *Annual Review of Nutrition*, **20**, 699–722.

KURAKAKE, M., MASUMOTO, R. Y. O., MAGUMA, K., KAMATA, A., SAITO, E., UKITA, N. and KOMAKI, T. (2010). 'Production of fructooligosaccharides by β-fructofuranosidases from *Aspergillus oryzae* KB'. *Journal of Agricultural and Food Chemistry*, **58**, 488–92.

LA FERLA, B., PROSPERI, D., LAY, L., RUSSO, G. and PANZA, L. (2002). 'Synthesis of building blocks of human milk oligosaccharides. Fucosylated derivatives of the lacto- and neolacto-series'. *Carbohydrate Research*, **337**, 1333–42.

LEE, J. W., KWON, T. O. and MOON, I. S. (2003). 'Chromatographic separation of maltopentaose from maltooligosaccharides'. *Biotechnology and Bioprocess Engineering*, **8**, 47–53.

LEE, J. W., KWON, T. O. and MOON, I. S. (2004). 'Adsorption of monosaccharides, disaccharides, and maltooligosaccharides on activated carbon for separation of maltopentaose'. *Carbon*, **42**, 371–80.

LEE, D. X., XIA, W. S. and ZHANG, J. L. (2008a). 'Enzymatic preparation of chitooligosaccharides by commercial lipase'. *Food Chemistry*, **111**, 291–5.

LEE, M. S., CHO, S. K., EOM, H. J., KIM, S. Y., KIM, T. J. and HAN, N. S. (2008b). 'Optimized substrate concentrations for production of long-chain isomaltooligosaccharides using dextransucrase of *Leuconostoc mesenteroides* B-512F'. *Journal of Microbiology and Biotechnology*, **18**, 1141–5.

LI, Z., XIAO, M., LU, L. and LI, Y. (2008). 'Production of non-monosaccharide and high-purity galactooligosaccharides by immobilized enzyme catalysis and fermentation with immobilized yeast cells'. *Process Biochemistry*, **43**, 896–9.

LI, W., SUN, Y., YE, H. and ZENG, X. (2010). 'Synthesis of oligosaccharides with lactose and *N*-acetylglucosamine as substrates by using β-D-galactosidase from *Bacillus circulans*'. *European Food Research and Technology*, **231**, 55–63.

LIM, J. S., LEE, J. H., KANG, S. W., PARK, S. W. and KIM, S. W. (2007). 'Studies on production and physical properties of neo-FOS produced by co-immobilized *Penicillium citrinum* and neo-fructosyltransferase'. *European Food Research and Technology*, **225**, 457–62.

LIN, S. B., LIN, Y. C. and CHEN, H. H. (2009). 'Low molecular weight chitosan prepared with the aid of cellulase, lysozyme and chitinase: Characterisation and antibacterial activity'. *Food Chemistry*, **116**, 47–53.

LINDE, D., MACIAS, I., FERNÁNDEZ-ARROJO, L., PLOU, F. J., JIMÉNEZ, A. and FERNÁNDEZ-LOBATO, M. (2009). 'Molecular and biochemical characterization of a β-fructofuranosidase from *Xanthophyllomyces dendrorhous*'. *Applied and Environmental Microbiology*, **75**, 1065–73.

© Woodhead Publishing Limited, 2013

LOCASCIO, R. G., NINONUEVO, M. R., FREEMAN, S. L., SELA, D. A., GRIMM, R., LEBRILLA, C. B., MILLS, D. A. and GERMAN, J. B. (2007). 'Glycoprofiling of bifidobacterial consumption of human milk oligosaccharides demonstrates strain specific, preferential consumption of small chain glycans secreted in early human lactation'. *Journal of Agricultural and Food Chemistry*, **55**, 8914–9.

LV, Y., YANG, X., ZHAO, Y., RUAN, Y., YANG, Y. and WANG, Z. (2009). 'Separation and quantification of component monosaccharides of the tea polysaccharides from *Gynostemma pentaphyllum* by HPLC with indirect UV detection'. *Food Chemistry*, **112**, 742–6.

MACFARLANE, S., MACFARLANE, G. T. and CUMMINGS, J. H. (2006). 'Review article: Prebiotics in the gastrointestinal tract'. *Alimentary Pharmacology and Therapeutics*, **24**, 701–14.

MACFARLANE, G. T., STEED, H. and MACFARLANE, S. (2008). 'Bacterial metabolism and health-related effects of galacto-oligosaccharides and other prebiotics'. *Journal of Applied Microbiology*, **104**, 305–44.

MAHONEY, R. R. (1998). 'Galactosyl-oligosaccharide formation during lactose hydrolysis: a review'. *Food Chemistry*, **63**, 147–54.

MAISCHBERGER, T., LEITNER, E., NITISINPRASERT, S., JUAJUN, O., YAMABHAI, M., NGUYEN, T. H. and HALTRICH, D. (2010). 'β-Galactosidase from *Lactobacillus pentosus*: Purification, characterization and formation of galacto-oligosaccharides'. *Biotechnology Journal*, **5**, 838–47.

MAISCHBERGER, T., NGUYEN, T.-H., SUKYAI, P., KITTL, R., RIVA, S., LUDWIG, R. and HALTRICH, D. (2008). 'Production of lactose-free galacto-oligosaccharide mixtures: comparison of two cellobiose dehydrogenases for the selective oxidation of lactose to lactobionic acid'. *Carbohydrate Research*, **343**, 2140–7.

MARCOBAL, A., BARBOZA, M., FROEHLICH, J. W., BLOCK, D. E., GERMAN, J. B., LEBRILLA, C. B. and MILLS, D. A. (2010). 'Consumption of human milk oligosaccharides by gut-related microbes'. *Journal of Agricultural and Food Chemistry*, **58**, 5334–40.

MARTEAU, P. and BOUTRON-RUAULT, M. C. (2002). 'Nutritional advantages of probiotics and prebiotics'. *British Journal of Nutrition*, **87**, S153–7.

MARTINEZ-FEREZ, A., RUDLOFF, S., GUADIX, A., HENKEL, C. A., POHLENTZ, G., BOZA, J. J., GUADIX, E. M. and KUNZ, C. (2006). 'Goats' milk as a natural source of lactose-derived oligosaccharides: Isolation by membrane technology'. *International Dairy Journal*, **16**, 173–81.

MARX, S. P., WINKLER, S. and HARTMEIER, W. (2000). 'Metabolization of β-(2,6)-linked fructose-oligosaccharides by different bifidobacteria'. *FEMS Microbiology Letters*, **182**, 163–9.

MATHIESEN, G., NAMLØS, H. M., RISØEN, P. A., AXELSSON, L. and EIJSINK, V. G. H. (2004a). 'Use of bacteriocin promoters for gene expression in *Lactobacillus plantarum* C11'. *Journal of Applied Microbiology*, **96**, 819–27.

MATHIESEN, G., SØRVIG, E., BLATNY, J., NATERSTAD, K., AXELSSON, L. and EIJSINK, V. G. H. (2004b). 'High-level gene expression in *Lactobacillus plantarum* using a pheromone-regulated bacteriocin promoter'. *Letters in Applied Microbiology*, **39**, 137–43.

MAUGARD, T., GAUNT, D., LEGOY, M. D. and BESSON, T. (2003). 'Microwave-assisted synthesis of galacto-oligosaccharides from lactose with immobilized β-galactosidase from *Kluyveromyces lactis*'. *Biotechnology Letters*, **25**, 623–9.

MIN, B. C., YOON, S. H., KIM, J. W., LEE, Y. W., KIM, Y. B. and PARK, K. H. (1998). 'Cloning of novel maltooligosaccharide-producing amylases as antistaling agents for bread'. *Journal of Agricultural and Food Chemistry*, **46**, 779–82.

MOLLER, P. L., JORGENSEN, F., HANSEN, O. C., MADSEN, S. M. and STOUGAARD, P. (2001). 'Intra- and extracellular β-galactosidases from *Bifidobacterium bifidum* and *B. infantis*: molecular cloning, heterologous expression, and comparative characterization'. *Applied and Environmental Microbiology*, **67**, 2276–83.

© Woodhead Publishing Limited, 2013

MOURE, A., GULLÓN, P., DOMÍNGUEZ, H. and PARAJÓ, J. C. (2006). 'Advances in the manufacture, purification and applications of xylo-oligosaccharides as food additives and nutraceuticals'. *Process Biochemistry*, **41**, 1913–23.

MOURYA, V. K., INAMDAR, N. N. and CHOUDHARI, Y. M. (2011). 'Chitooligosaccharides: synthesis, characterization and applications'. *Polymer Science – Series A*, **53**, 583–612.

MUSSATTO, S. I. and MANCILHA, I. M. (2007). 'Non-digestible oligosaccharides: A review'. *Carbohydrate Polymers*, **68**, 587–97.

MUSSATTO, S. I., AGUILAR, C. N., RODRIGUES, L. R. and TEIXEIRA, J. A. (2009). 'Colonization of *Aspergillus japonicus* on synthetic materials and application to the production of fructooligosaccharides'. *Carbohydrate Research*, **344**, 795–800.

MUTANDA, T., WILHELMI, B. S. and WHITELEY, C. G. (2008). 'Response surface methodology: synthesis of inulooligosaccharides with an endoinulinase from *Aspergillus niger*'. *Enzyme and Microbial Technology*, **43**, 362–8.

NAIDOO, K., AYYACHAMY, M., PERMAUL, K. and SINGH, S. (2009). 'Enhanced fructooligosaccharides and inulinase production by a *Xanthomonas campestris* pv. *phaseoli* KM 24 mutant'. *Bioprocess and Biosystems Engineering*, **32**, 689–95.

NAKAKUKI, T. (2002). 'Present status and future of functional oligosaccharide development in Japan'. *Pure and Applied Chemistry*, **74**, 1245–51.

NAKAYAMA, T. and AMACHI, T. (1999). 'β-Galactosidase, enzymology'. In: *Encyclopedia of Bioprocess Technology: Fermentation, Biocatalysis, and Bioseparation.* FLICKINGER, M. C. and DREW, S. W. (eds) John Wiley, New York; pp 1291–305.

NGUYEN, T. H., SPLECHTNA, B., STEINBÖCK, M., KNEIFEL, W., LETTNER, H. P., KULBE, K. D. and HALTRICH, D. (2006). 'Purification and characterization of two novel β-galactosidases from *Lactobacillus reuteri*'. *Journal of Agricultural and Food Chemistry*, **54**, 4989–98.

NGUYEN, T. H., SPLECHTNA, B., KRASTEVA, S., KNEIFEL, W., KULBE, K. D., DIVNE, C. and HALTRICH, D. (2007). 'Characterization and molecular cloning of a heterodimeric β-galactosidase from the probiotic strain *Lactobacillus acidophilus* R22'. *FEMS Microbiology Letters*, **269**, 136–44.

NGUYEN, T. T., MATHIESEN, G., FREDRIKSEN, L., KITTL, R., NGUYEN, T. H., EIJSINK, V. G. H., HALTRICH, D. and PETERBAUER, C. K. (2011). 'A food-grade system for inducible gene expression in *Lactobacillus plantarum* using an alanine racemase-encoding selection marker'. *Journal of Agricultural and Food Chemistry*, **59**, 5617–24.

NING, Y., WANG, J., CHEN, J., YANG, N., JIN, Z. and XU, X. (2010). 'Production of neo-fructooligosaccharides using free-whole-cell biotransformation by *Xanthophyllomyces dendrorhous*'. *Bioresource Technology*, **101**, 7472–8.

NISHIMOTO, M. and KITAOKA, M. (2007). 'Identification of the putative proton donor residue of lacto-*N*-biose phosphorylase (EC 2.4.1.211)'. *Bioscience, Biotechnology and Biochemistry*, **71**, 1587–91.

OTIENO, D. O. (2010). 'Synthesis of β-galactooligosaccharides from lactose using microbial β-galactosidases'. *Comprehensive Reviews in Food Science and Food Safety*, **9**, 471–82.

PARK, A. R. and OH, D. K. (2010). 'Galacto-oligosaccharide production using microbial β-galactosidase: current state and perspectives'. *Applied Microbiology and Biotechnology*, **85**, 1279–86.

PARK, J. P., BAE, J. T., YOU, D. J., KIM, B. W. and YUN, J. W. (1999). 'Production of inulooligosaccharides from inulin by a novel endoinulinase from *Xanthomonas* sp'. *Biotechnology Letters*, **21**, 1043–6.

PATEL, S. and GOYAL, A. (2011). 'Functional oligosaccharides: production, properties and applications'. *World Journal of Microbiology and Biotechnology*, **27**, 1119–28.

PERUGINO, G., TRINCONE, A., ROSSI, M. and MORACCI, M. (2004). 'Oligosaccharide synthesis by glycosynthases'. *Trends in Biotechnology*, **22**, 31–7.

© Woodhead Publishing Limited, 2013

PETZELBAUER, I., ZELENY, R., REITER, A., KULBE, K. D. and NIDETZKY, B. (2000). 'Development of an ultra-high-temperature process for the enzymatic hydrolysis of lactose: II. Oligosaccharide formation by two thermostable β-glycosidases'. *Biotechnology and Bioengineering*, **69**, 140–9.

PIVARNIK, L. F., SENEGAL, A. G. and RAND, A. G. (1995). 'Hydrolytic and transgalactosylic activities of commercial β-galactosidase (lactase) in food processing'. In: *Advances in Food and Nutrition Research.* KINSELLA, J. E. and TAYLOR, S. L. (eds) Academic Press, Santiago, Volume 38, pp 1–102.

PLAYNE, M. J. and CRITTENDEN, R. G. (2009). 'Galacto-oligosaccharides and other products derived from lactose'. In: *Lactose, Water, Salts and Minor Constitutents*, 3rd edition. MCSWEENEY, P. L. H. and FOX, P. F. (ed.) Springer, New York; pp 121–201.

PRENOSIL, J. E., STUKER, E. and BOURNE, J. R. (1987). 'Formation of oligosaccharides during enzymatic lactose hydrolysis: Part I: State of art'. *Biotechnology and Bioengineering*, **30**, 1019–25.

PRIEM, B., GILBERT, M., WAKARCHUK, W. W., HEYRAUD, A. and SAMAIN, E. (2002). 'A new fermentation process allows large-scale production of human milk oligosaccharides by metabolically engineered bacteria'. *Glycobiology*, **12**, 235–40.

QIANG, X., YONGLIE, C. and QIANBING, W. (2009). 'Health benefit application of functional oligosaccharides'. *Carbohydrate Polymers*, **77**, 435–41.

RABIU, B. A., JAY, A. J., GIBSON, G. R. and RASTALL, R. A. (2001). 'Synthesis and fermentation properties of novel galacto-oligosaccharides by β-galactosidases from *Bifidobacterium* species'. *Applied and Environmental Microbiology*, **67**, 2526–30.

RASTALL, R. A. and MAITIN, V. (2002). 'Prebiotics and synbiotics: towards the next generation'. *Current Opinion in Biotechnology*, **13**, 490–6.

RASTALL, R. A., GIBSON, G. R., GILL, H. S., GUARNER, F., KLAENHAMMER, T. R., POT, B., REID, G., ROWLAND, I. R. and SANDERS, M. E. (2005). 'Modulation of the microbial ecology of the human colon by probiotics, prebiotics and synbiotics to enhance human health: An overview of enabling science and potential applications'. *FEMS Microbiology Ecology*, **52**, 145–52.

RENCUROSI, A., POLETTI, L., GUERRINI, M., RUSSO, G. and LAY, L. (2002). 'Human milk oligosaccharides: an enzymatic protection step simplifies the synthesis of 3′- and 6′-O-sialyllactose and their analogues'. *Carbohydrate Research*, **337**, 473–83.

REUTER, S., RUSBORG NYGAARD, A. and ZIMMERMANN, W. (1999). 'β-Galactooligosaccharide synthesis with β-galactosidases from *Sulfolobus solfataricus, Aspergillus oryzae*, and *Escherichia coli*'. *Enzyme and Microbial Technology*, **25**, 509–16.

RICCA, E., CALABRÒ, V., CURCIO, S. and IORIO, G. (2007). 'The state of the art in the production of fructose from inulin enzymatic hydrolysis'. *Critical Reviews in Biotechnology*, **27**, 129–45.

ROBERFROID, M. (2007a). 'Prebiotics: The concept revisited'. *Journal of Nutrition*, **137**, 830S–7S.

ROBERFROID, M. B. (2007b). 'Inulin-type Fructans: Functional food ingredients'. *Journal of Nutrition*, **137**, 2493S–502S.

ROBERFROID, M. and SLAVIN, J. (2000). 'Nondigestible oligosaccharides'. *Critical Reviews in Food Science and Nutrition*, **40**, 461–80.

ROBERFROID, M., GIBSON, G. R., HOYLES, L., MCCARTNEY, A. L., RASTALL, R., ROWLAND, I., WOLVERS, D., WATZL, B., SZAJEWSKA, H., STAHL, B., GUARNER, F., RESPONDEK, F., WHELAN, K., COXAM, V., DAVICCO, M. J., LÉOTOING, L., WITTRANT, Y., DELZENNE, N. M., CANI, P. D., NEYRINCK, A. M. and MEHEUST, A. (2010). 'Prebiotic effects: Metabolic and health benefits'. *British Journal of Nutrition*, **104**, S1–63.

RYCROFT, C. E., JONES, M. R., GIBSON, G. R. and RASTALL, R. A. (2001). 'Fermentation properties of gentio-oligosaccharides'. *Letters in Applied Microbiology*, **32**, 156–61.

© Woodhead Publishing Limited, 2013

SAKO, T., MATSUMOTO, K. and TANAKA, R. (1999). 'Recent progress on research and applications of non-digestible galacto-oligosaccharides'. *International Dairy Journal*, **9**, 69–80.

SANGEETHA, P. T., RAMESH, M. N. and PRAPULLA, S. G. (2005a). 'Fructooligosaccharide production using fructosyl transferase obtained from recycling culture of *Aspergillus oryzae* CFR 202'. *Process Biochemistry*, **40**, 1085–8.

SANGEETHA, P. T., RAMESH, M. N. and PRAPULLA, S. G. (2005b). 'Recent trends in the microbial production, analysis and application of fructooligosaccharides'. *Trends in Food Science and Technology*, **16**, 442–57.

SANGWAN, V., TOMAR, S. K., SINGH, R. R. B., SINGH, A. K. and ALI, B. (2011). 'Galactooligosaccharides: novel components of designer foods'. *Journal of Food Science*, **76**, R103–11.

SANZ, M. L., GIBSON, G. R. and RASTALL, R. A. (2005). 'Influence of disaccharide structure on prebiotic selectivity *in vitro*'. *Journal of Agricultural and Food Chemistry*, **53**, 5192–9.

SCHEPPACH, W., LUEHRS, H. and MENZEL, T. (2001). 'Beneficial health effects of low-digestible carbohydrate consumption'. *British Journal of Nutrition*, **85**, S23–30.

SCHMIDT, D. and THIEM, J. (2010). 'Chemical synthesis using enzymatically generated building units for construction of the human milk pentasaccharides sialyllacto-*N*-tetraose and sialyllacto-*N*-neotetraose epimer'. *Beilstein Journal of Organic Chemistry*, **6**, No 18.

SCHOLZ-AHRENS, K. E., SCHAAFSMA, G., VAN DEN HEUVEL, E. G. H. M. and SCHREZENMEIR, J. (2001). 'Effects of prebiotics on mineral metabolism'. *American Journal of Clinical Nutrition*, **73**, 459S–64S.

SCHUSTER-WOLFF-BÜHRING, R., FISCHER, L. and HINRICHS, J. (2010). 'Production and physiological action of the disaccharide lactulose'. *International Dairy Journal*, **20**, 731–41.

SELA, D. A. and MILLS, D. A. (2010). 'Nursing our microbiota: Molecular linkages between bifidobacteria and milk oligosaccharides'. *Trends in Microbiology*, **18**, 298–307.

SELA, D. A., CHAPMAN, J., ADEUYA, A., KIM, J. H., CHEN, F., WHITEHEAD, T. R., LAPIDUS, A., ROKHSAR, D. S., LEBRILLA, C. B., GERMAN, J. B., PRICE, N. P., RICHARDSON, P. M. and MILLS, D. A. (2008). 'The genome sequence of *Bifidobacterium longum* subsp. *infantis* reveals adaptations for milk utilization within the infant microbiome'. *Proceedings of the National Academy of Sciences of the United States of America*, **105**, 18964–9.

SHIDA, O., TAKANO, T., TAKAGI, H., KADOWAKI, K. and KOBAYASHI, S. (1992). 'Cloning and nucleotide sequence of the maltopentaose-forming amylase gene from *Pseudomonas* sp. KO-(8940)'. *Bioscience, Biotechnology, and Biochemistry*, **56**, 76–80.

SHIMODA, K., KUBOTA, N. and HAMADA, H. (2011). 'Synthesis of gentiooligosaccharides of genistein and glycitein and their radical scavenging and anti-allergic activity'. *Molecules*, **16**, 4740–7.

SINGH, R. S. and SINGH, R. P. (2010). 'Production of fructooligosaccharides from inulin by endoinulinases and their prebiotic potential'. *Food Technology and Biotechnology*, **48**, 435–50.

SOMKUTI, G. A., DOMINIECKI, M. E. and STEINBERG, D. H. (1998). 'Permeabilization of *Streptococcus thermophilus* and *Lactobacillus delbrueckii* subsp. *bulgaricus* with ethanol'. *Current Microbiology*, **36**, 202–6.

SØRVIG, E., GRÖNQVIST, S., NATERSTAD, K., MATHIESEN, G., EIJSINK, V. G. H. and AXELSSON, L. (2003). 'Construction of vectors for inducible gene expression in *Lactobacillus sakei* and *Lactobacillus plantarum*'. *FEMS Microbiology Letters*, **229**, 119–26.

© Woodhead Publishing Limited, 2013

SØRVIG, E., MATHIESEN, G., NATERSTAD, K., EIJSINK, V. G. H. and AXELSSON, L. (2005). 'High-level, inducible gene expression in *Lactobacillus sakei* and *Lactobacillus plantarum* using versatile expression vectors'. *Microbiology*, **151**, 2439–49.

SPLECHTNA, B., PETZELBAUER, I., BAMINGER, U., HALTRICH, D., KULBE, K. D. and NIDETZKY, B. (2001). 'Production of a lactose-free galacto-oligosaccharide mixture by using selective enzymatic oxidation of lactose into lactobionic acid'. *Enzyme and Microbial Technology*, **29**, 434–40.

SPLECHTNA, B., NGUYEN, T. H., STEINBÖCK, M., KULBE, K. D., LORENZ, W. and HALTRICH, D. (2006). 'Production of prebiotic galacto-oligosaccharides from lactose using β-galactosidases from *Lactobacillus reuteri*'. *Journal of Agricultural and Food Chemistry*, **54**, 4999–5006.

SPLECHTNA, B., NGUYEN, T. H. and HALTRICH, D. (2007a). 'Comparison between discontinuous and continuous lactose conversion processes for the production of prebiotic galacto-oligosaccharides using β-galactosidase from *Lactobacilius reuteri*'. *Journal of Agricultural and Food Chemistry*, **55**, 6772–7.

SPLECHTNA, B., NGUYEN, T. H., ZEHETNER, R., LETTNER, H. P., LORENZ, W. and HALTRICH, D. (2007b). 'Process development for the production of prebiotic galacto-oligosaccharides from lactose using β-galactosidase from *Lactobacillus* sp'. *Biotechnology Journal*, **2**, 480–5.

STEED, H., MACFARLANE, G. T. and MACFARLANE, S. (2008). 'Prebiotics, synbiotics and inflammatory bowel disease'. *Molecular Nutrition and Food Research*, **52**, 898–905.

SWENNEN, K., COURTIN, C. M. and DELCOUR, J. A. (2006). 'Non-digestible oligosaccharides with prebiotic properties'. *Critical Reviews in Food Science and Nutrition*, **46**, 459–71.

TAKASAKI, Y. (1982). 'Production of maltohexaose by α–amylase from *Bacillus circulans* G-6'. *Agricultural Biology and Chemistry*, **46**, 1539–47.

TAKASAKI, Y. (1985). 'An amylase producing maltotriose from *Bacillus subtilis*'. *Agricultural Biology and Chemistry*, **49**, 1091–7.

TAKASAKI, Y. (1989). 'Novel maltose-producing amylase from *Bacillus megaterium* G-2'. *Agricultural Biology and Chemistry*, **53**, 341–7.

TAKASAKI, Y., SHINOHARA, H., TSURUHISA, M., HAYASHI, S. and IMADA, K. (1991). 'Maltotetraose-producing amylase from *Bacillus* sp. MG-4'. *Agricultural Biology and Chemistry*, **55**, 1715–20.

TOBA, T., YOKOTA, A. and ADACHI, S. (1985). 'Oligosaccharide structures formed during the hydrolysis of lactose by *Aspergillus oryzae* β-galactosidase'. *Food Chemistry*, **16**, 147–62.

TORRES, D. P., GONÇALVES, M., TEIXEIRA, J. A. and RODRIGUES, L. R. (2010). 'Galacto-oligosaccharides: production, properties, applications, and significance as prebiotics'. *Comprehensive Reviews in Food Science and Food Safety*, **9**, 438–54.

TZORTZIS, G. and VULEVIC, J. (2009). 'Galacto-oligosaccharide prebiotics'. In: *Prebiotics and Probiotics Science and Technology*. CHARALAMPOPOULOS, D. and RASTALL, R. A. (ed.) Springer, New York; pp 207–44.

VAN LAERE, K. M. J., ABEE, T., SCHOLS, H. A., BELDMAN, G. and VORAGEN, A. G. J. (2000a). 'Characterization of a novel β-galactosidase from *Bifidobacterium adolescentis* DSM 20083 active towards transgalactooligosaccharides'. *Applied and Environmental Microbiology*, **66**, 1379–84.

VAN LAERE, K. M. J., HARTEMINK, R., BOSVELD, M., SCHOLS, H. A. and VORAGEN, A. G. J. (2000b). 'Fermentation of plant cell wall derived polysaccharides and their corresponding oligosaccharides by intestinal bacteria'. *Journal of Agricultural and Food Chemistry*, **48**, 1644–52.

VASILJEVIC, T. and JELEN, P. (2002). 'Lactose hydrolysis in milk as affected by neutralizers used for the preparation of crude β-galactosidase extracts from

© Woodhead Publishing Limited, 2013

Lactobacillus bulgaricus 11842'. *Innovative Food Science & Emerging Technologies*, **3**, 175–84.

VÁZQUEZ, M.J., ALONSO, J.L., DOMÍNGUEZ, H. and PARAJÓ, J.C. (2001). 'Xylooligosaccharides: manufacture and applications'. *Trends in Food Science and Technology*, **11**, 387–93.

VÁZQUEZ, M. J., ALONSO, J. L., DOMÍNGUEZ, H. and PARAJÓ, J. C. (2002). 'Enzymatic processing of crude xylooligomer solutions obtained by autohydrolysis of *Eucalyptus* wood'. *Food Biotechnology*, **16**, 91–105.

VENTURA, M., ELLI, M., RENIERO, R. and ZINK, R. (2001). 'Molecular microbial analysis of *Bifidobacterium* isolates from different environments by the species-specific amplified ribosomal DNA restriction analysis (ARDRA)'. *FEMS Microbiology Ecology*, **36**, 113–21.

VIJAYARAGHAVAN, K., YAMINI, D., AMBIKA, V. and SRAVYA SOWDAMINI, N. (2009). 'Trends in inulinase production–a review'. *Critical Reviews in Biotechnology*, **29**, 67–77.

VORAGEN, A. G. J. (1998). 'Technological aspects of functional food-related carbohydrates'. *Trends in Food Science and Technology*, **9**, 328–35.

WOLLOWSKI, I., RECHKEMMER, G. and POOL-ZOBEL, B. L. (2001). 'Protective role of probiotics and prebiotics in colon cancer'. *American Journal of Clinical Nutrition*, **73**, 451S–5S.

WU, S., TAO, N., GERMAN, J. B., GRIMM, R. and LEBRILLA, C. B. (2010). 'Development of an annotated library of neutral human milk oligosaccharides'. *Journal of Proteome Research*, **9**, 4138–51.

WU, S., GRIMM, R., GERMAN, J. B. and LEBRILLA, C. B. (2011). 'Annotation and structural analysis of sialylated human milk oligosaccharides'. *Journal of Proteome Research*, **10**, 856–68.

XIA, W., LIU, P., ZHANG, J. and CHEN, J. (2011). 'Biological activities of chitosan and chitooligosaccharides'. *Food Hydrocolloids*, **25**, 170–9.

XIAO, J. Z., TAKAHASHI, S., NISHIMOTO, M., ODAMAKI, T., YAESHIMA, T., IWATSUKI, K. and KITAOKA, M. (2010). 'Distribution of *in vitro* fermentation ability of lacto-*N*-biose I, a major building block of human milk oligosaccharides, in bifidobacterial strains'. *Applied and Environmental Microbiology*, **76**, 54–9.

YANAHIRA, S., KOBAYASHI, T., SUGURI, T., NAKAKOSHI, M., MIURA, S., ISHIKAWA, H. and NAKAJIMA, I. (1995). 'Formation of oligosaccharides from lactose by *Bacillus circulans* β-galactosidase'. *Bioscience, Biotechnology and Biochemistry*, **59**, 1021–6.

YI, S. H., ALLI, I., PARK, K. H. and LEE, B. (2011). 'Overexpression and characterization of a novel transgalactosylic and hydrolytic β-galactosidase from a human isolate *Bifidobacterium breve* B24'. *New Biotechnology*, **28**, 806–13.

YOKOTA, A., YAMAUCHI, O. and TOMITA, F. (1995). 'Production of inulotriose from inulin by inulin-degrading enzyme from *Streptomyces rochei* E87'. *Letters in Applied Microbiology*, **21**, 330–3.

YUN, J. W., KIM, D. H., KIM, B. W. and SONG, S. K. (1997). 'Production of inulo-oligosaccharides from inulin by immobilized endoinulinase from *Pseudomonas* sp'. *Journal of Fermentation and Bioengineering*, **84**, 369–71.

ZENG, J., ZHAO, X., LIU, B., LI, G., GAO, H. and LIANG, X. (2011). 'Properties and application of β-fructofuranosidase from *Aspergillus japonicus* 3. 3556'. *Journal of Food Biochemistry*, **35**, 1117–29.

ZHANG, L., JIANG, Y., JIANG, Z., SUN, X., SHI, J., CHENG, W. and SUN, Q. (2009). 'Immobilized transglucosidase in biomimetic polymer-inorganic hybrid capsules for efficient conversion of maltose to isomaltooligosaccharides'. *Biochemical Engineering Journal*, **46**, 186–92.

ZHANG, L., SU, Y., ZHENG, Y., JIANG, Z., SHI, J., ZHU, Y. and JIANG, Y. (2010). 'Sandwich-structured enzyme membrane reactor for efficient conversion of maltose into isomaltooligosaccharides'. *Bioresource Technology*, **101**, 9144–9.

© Woodhead Publishing Limited, 2013

ZHENGYU, J., JING, W., BO, J. and XUEMING, X. (2005). 'Production of inulooligosaccharides by endoinulinases from *Aspergillus ficuum*'. *Food Research International*, **38**, 301–8.

ZHOU, Q. Z. K. and CHEN, X. D. (2001). 'Effects of temperature and pH on the catalytic activity of the immobilized β-galactosidase from *Kluyveromyces lactis*'. *Biochemical Engineering Journal*, **9**, 33–40.

ZHOU, Q. Z. and CHEN, X. L. (2002). 'Kinetics of lactose hydrolysis by β-galactosidase of *Kluyveromyces lactis* immobilized on cotton fabric'. *Biotechnology and Bioengineering*, **81**, 127–33.

ZHOU, C., XUE, Y., ZHANG, Y., ZENG, Y. and MA, Y. (2009). 'Recombinant expression and characterization of *Thermoanaerobacter tengcongensis* thermostable α-glucosidase with regioselectivity for high-yield isomaltooligosaccharides synthesis'. *Journal of Microbiology and Biotechnology*, **19**, 1547–56.

ZHU, X. and SCHMIDT, R. R. (2009). 'New principles for glycoside-bond formation'. *Angewandte Chemie – International Edition*, **48**, 1900–34.

ZIEMER, C. J. and GIBSON, G. R. (1998). 'An overview of probiotics, prebiotics and synbiotics in the functional food concept: perspectives and future strategies'. *International Dairy Journal*, **8**, 473–9.

ZIVKOVIC, A. M., GERMAN, J. B., LEBRILLA, C. B. and MILLS, D. A. (2011). 'Human milk glycobiome and its impact on the infant gastrointestinal microbiota'. *Proceedings of the National Academy of Sciences of the United States of America*, **108**, 4653–8.

© Woodhead Publishing Limited, 2013

19

Microbial production of polyunsaturated fatty acids as nutraceuticals

C. Ratledge, University of Hull, UK

DOI: 10.1533/9780857093547.2.531

Abstract: The principal fatty acids that are, or have been, produced by microorganisms as nutraceuticals are: gamma-linolenic acid (GLA, 18:3 omega-6) derived from *Mucor circinelloides*, docosahexaenoic acid (DHA, 22:6 omega-3) from *Crypthecodinium cohnii* and *Schizochytrium/ Thraustochytrium* spp, arachidonic acid (ARA, 20:4 omega-6) from *Mortierella alpina* and eicosapentaenoic acid using genetically modified *Yarrowia lipolytica*. Their production by various companies and applications are reviewed. This is now a multi-billion dollar industry with the main uptake being the use of ARA + DHA mixtures for inclusion in infant formulas around the world. The background to the processes is reviewed as is the assessment of the safety of these materials. New processes that are under current evaluation for fatty acid production include the use of algae grown both photosynthetically and heterotrophically.

Key words: large-scale production, microbial oils, nutrition/nutraceuticals, polyunsaturated fatty acids, safety.

A note on lipid nomenclature: The standard nomenclature of fatty acids is to give the number of carbon atoms in the molecule followed by the number of double bonds. Thus, 18 : 1 indicates a molecule with 18 carbon atoms and one double bond. If the position of the bond has to be specified it is then usually numbered from the carboxylic acid group. Thus 18 : 1 (9) would indicate oleic acid. When there is more than one double bond present in a fatty acid, it is necessary to specify the position of these bonds. Normally, multiple bonds are separated by a single methylene group ($-CH_2-$) in a molecule. It is therefore expedient to give the position of only the final bond; once this is known then the positions of all the other bonds are self-evident. The bond position is given by counting from the terminal methyl group of the fatty acid and can be indicated, for example, as omega-3, omega-6 or omega-9; the alternative nomenclature, which is used in this chapter, is to give these respective positions as *n*-3, *n*-6 or *n*-9. Where individual bonds are not methylene-interrupted, the position of each bond has to be individually specified but, confusingly, these are specified by counting from the carboxylic acid end.

© Woodhead Publishing Limited, 2013

Although most double bonds in a fatty acid are of the *cis*-configuration, some bonds occur in the *trans*-configuration. These may also be written, respectively, as *c* and *t*. For the chemical purists who may be reading this chapter, these older designations have been superseded by the systematic names: *E* or *Z*, respectively. I will, however, use the older nomenclature as this is still favoured by most oil chemists and biochemists.

Finally, in spite of many (erroneous) reports to the contrary, fatty acids do not exist in cells as free entities as they are highly cytotoxic and membrane-disruptive. They usually occur as esters of glycerol: the triacylglycerols or triglycerides. In addition, all cells also contain glycerophospholipids. Unless stated otherwise, it is to be understood in this chapter that the microbial oils being described occur as triacylglycerols.

19.1   Introduction

The past 20 years have seen major developments in our understanding of the role of polyunsaturated fatty acids (PUFA) for the maintenance and improvement of our health and well being. These fatty acids are major components of the phospholipids that occur in all cell membranes. Their roles include regulation of membrane fluidity, attachment of specific enzymes to the membrane as well as being able to mediate signal transduction and other key metabolic processes (Li *et al.*, 2003; Gogus and Smith, 2010). Additionally, PUFAs are used for the biosynthesis of a range of bioactive molecules, such as various eicosanoids, leukotrienes, prostaglandins and resolvins (Serhan, 2005) which, in turn, have numerous functions including anti-inflammatory, anti-arrhythmic and anti-aggregatory effects (see Fig. 19.1). Many of these effects are involved in the maintenance and improvement of cardiovascular health (Lee *et al.*, 2009; Cottin *et al.*, 2011). There is also such strong evidence that certain PUFAs can improve the development of eye function and memory in newly born infants and adults (Innis, 2003; McCann and Ames, 2005; Sinclair and Jayasooriya, 2010) that two specific PUFAs, arachidonic acid (ARA; 20:4 *n*-6) and docosahexaenoic acid (DHA; 22:6 *n*-3), are now routinely added to infant formulas in most countries of the world (see Sections 19.4 and 19.5). Other research has suggested involvement of PUFAs in alleviating a variety of illnesses including Alzheimer's disease, chronic bowel disorder and some cancers (Janakiram and Rao, 2009).

This surge in appreciating the nutritional value of PUFAs and their applications in clinical medicine, has focussed attention on their possible sources. Fish, especially the oily fish, such as mackerel, trout, salmon, and sardines, represent a rich source of PUFAs but fish oils are always mixtures of DHA and eicosapentaenoic acid (EPA; 20:5 *n*-6) in various proportions. But for some applications, and particularly for incorporation into infant formulas, EPA is contra-indicated as it then leads to production of unwanted prostaglandins and hydroxy-PUFAs (see Fig. 19.1). Supplies of the major long-chain PUFAs, DHA and EPA, as single entities, are therefore extremely

© Woodhead Publishing Limited, 2013

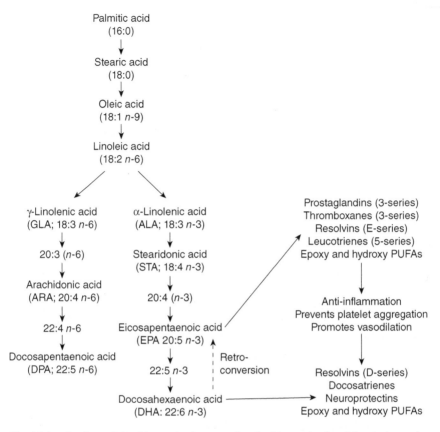

Fig. 19.1 Outline of the biosynthetic route for the biosynthesis of the major polyunsaturated fatty acids (PUFA) covered in this review (GLA, ARA, EPA and DHA) and their metabolites (adapted from Cottin *et al.*, 2011).

limited and, if these are produced from fish oils, they become prohibitively expensive as final purification of them relies on preparative-scale high performance liquid chromatography (HPLC). They cannot be produced by plants in spite of several decades of intensive research to develop appropriate transgenic plants for this purpose. Additionally, there is a problem in the supplementation of infant formulas with oils rich in DHA as the DHA is retro-converted into EPA (see Fig. 19.1) and this, for the reasons given above, is undesirable. To prevent this conversion, it has been found expedient to include arachidonic acid (ARA) along with the DHA in the dietary supplements for infants.

There is now a need to produce a variety of oils that will be rich in one or more of the various PUFAs, and particularly the ones known as the very long-chain PUFAs of C20 and C22. Where mixtures of EPA and DHA are

© Woodhead Publishing Limited, 2013

indicated, as for example as supplements to the adult diet, then these can largely be met by eating oily fish once or twice per week or by the intake of appropriate capsules containing EPA and DHA (Lee *et al.*, 2009). However, warnings about the possible danger of overeating oily fish have been given by various health authorities as the fish may contain undesirably high amounts of dioxins, heavy metals and other potentially toxic materials that have been ingested by the fish from the environment. Fish oil producers, though, are responding by manufacturing oils to higher specifications and also carefully monitoring the quality of the oils that they are offering for sale.

When oils rich in single PUFAs are required, as occurs in infant nutrition, one has to look elsewhere for a realistic provision of these materials. As already stated, they cannot be obtained from plants or indeed any other convenient source. It is therefore for this reason that considerable work has been carried out over the past 20 years into the production of individual PUFAs using microbial technology. Some microorganisms have turned out to be exceptionally good sources of these fatty acids with the result that commercial production of several PUFAs now takes place using large-scale fermentation technology. In addition to their sale as nutraceuticals and in infant formulas, microbial oils are also acceptable commodities for vegans and various religious groups who do not wish to consume fish or fish-derived products. This chapter now describes the various processes that are used for the production of various key PUFAs.

19.2   Production of microbial oils

Microbial oils destined for human consumption have been given the more marketable name of single cell oils (SCO) which then avoids any mention of fungal or mould oil although yeast oil and even algal oils would seem to be acceptable to some consumers. The ability of a microorganism, usually a eukaryotic cell such as a yeast, fungus or alga, to accumulate high levels of lipid in their cells is not ubiquitous and is found only in a minority of species. For example, although over 600 species of yeast have been identified, only about 30–40 can accumulate more than about 20% oil in their cells. These species are known as the oleaginous microorganisms. The oil is formed as a reserve storage material for subsequent metabolism should the organism become starved of a supply of carbon. The oil is in the form of triacylglycerols (= triglycerides) which is the same as occurs with plant seed oils and animals fats. The oils being produced by oleaginous yeasts, fungi and algae, therefore, are all edible materials (see below Section 19.8 on Safety) as their component fatty acids are also the same as are found in edible plant oils and animal fats (see Tables 19.1, 19.2 and 19.3).

© Woodhead Publishing Limited, 2013

Table 19.1   Oil contents and fatty acid profiles of various oleaginous yeasts[a]

	Maximum Lipid content (% w/w)	Major fatty acyl groups (relative % (w/w))						
		16:0	16:1	18:0	18:1	18:2	18:3 (n-3)	18:4 (n-3)
Candida diddensiae	37	19	3	5	45	17	5	1
Cryptococcus albidus	65	12	1	3	73	12	–	–
Cryptococcus curvatus	58	32	–	15	44	8	–	–
Lipomyces starkeyi	63	34	6	5	51	3	–	–
Rhodotorula glutinis	72	37	1	3	47	8	–	–
Rhodosporidium toruloides	66	18	3	3	66	–	–	–
Waltomyces lipofer[b]	64	37	4	7	48	3	–	–
Yarrowia lipolytica	36	11	6	1	28	51	1	–

[a]from Ratledge (1997), and Ratledge and Hopkins (2006a).
[b]formerly *Lipomyces lipofer*.

Oil accumulation in microorganisms has been long studied. It occurs when a microorganism is grown in a medium that has been formulated with a surfeit of carbon (usually glucose) which is always present but with another essential nutrient (nitrogen is the most frequently used) becoming exhausted part way through the cultivation. When this occurs, the cells remain viable but cannot continue to multiply as the nitrogen is essential for the biosynthesis of cellular materials such as proteins and nucleic acids. The cells, however, continue to assimilate the available glucose, convert it into fatty acids and thence into the triacylglycerols. As fatty acids are synthesized at all stages of growth – it being essential that phospholipids are continually synthesized for incorporation into cell membranes – there is no particular induction of any new set of enzymes for this activity. The existing enzymatic machinery for fatty acids biosynthesis is sufficient; this continues after the exhaustion of the limiting nutrient but the other activities of the cell involved in protein and nucleic acid biosynthesis stop. Thus, lipid accumulates, not because of increased fatty acid biosynthesis activities, but because other anabolic processes in the cell cease. Cells therefore become obese.

The process of oil accumulation in the oleaginous microorganism continues to some upper limit. Some yeasts and fungi can accumulate oil up to over 70% of the biomass (see Tables 19.1 and 19.2). However, in many oleaginous organisms, the highest level of oil that is produced is substantially less than this. The maximum oil content of a microbial cell is therefore

© Woodhead Publishing Limited, 2013

Table 19.2 Oil contents and fatty acid profiles of various fungi[a]

	Maximum Lipid content	Major fatty acyl groups (relative % (w/w))									
		16:0	16:1	18:0	18:1	18:2	18:3 (n-3)	18:3 (n-6)	20:3 (n-6)	20:4 (n-6)	20:5 (n-3)
Cunninghamella echinulata	24	13	1	2	46	16	–	19.5	–	–	–
Mortierella alpina	50	8	–	11	14	7	–	4	4	49	–
Mortierella isabellina	50	27	1	6	51	10	–	8	–	–	–
Mucor circinelloides	25	22	1	6	40	11	–	18	–	–	–
Pythion irregulare	>25	17	7	2	14	18	–	–	–	11	14
Pythium ultimum	>25	15	–	2	20	16	1	–	–	15	14
Syzgites megalocarpus	22	14	–	1	12	10	–	62	–	–	–

[a]From Ratledge (1997); Ratledge and Hopkins (2006a); Wynn and Ratledge (2006) and Ratledge (2006).

© Woodhead Publishing Limited, 2013

Table 19.3 Fatty acid profiles (relative percentage of total fatty acids) of various algae being considered for PUFA production[a]

	14:0	16:0	16:1	18:1	18:2	GLA 18:3 (n-6)	ARA 20:4 (n-6)	EPA 20:5 (n-3)	DHA22:6 (n-3)	Other major acids
Amphidinium carterae	–	12	1	2	1	3	20	–	24	18:4 (n-3) 19
Chlorella minutissima	12	13	21	1	2	–	3	45	–	–
Cylindrotheca fusiformis[b]	6	22	24	2	1	–	8	17	1	16:4 (n-1) 8
Isochrysis galbana	12	0	11	3	2	–	–	25	11	–
Monodus subterraneus	–	19	10	5	2	<1	14	34	–	–
Nannochloropsis oculata	4	15	22	3	1	–	4	38	–	–
Nitzschia laevis	12	22	33	3	3	1	5	15	–	–
Parietochloiris incisa[c]	–	10	2	12	17	–	43	1	–	–
Phaeodactylum tricornutum	–	10	21	1	4	1	1	33	–	–
Pinguiochrysis sp.[c]	30	1	1	0.6	1	–	3	55	11	–
Porphyridium cruentum	–	30	5	<1	5	1	16	–	–	–

[a]Data from Ratledge (1997); Ratledge and Hopkins (2006a), and Wynn and Ratledge (2005, 2006). Although the oil contents of many algae have been reported as being up to 35–36% (see for example Rodolfi *et al.*, 2009) or even higher, these have only been attained by using CO_2-enriched air in photobioreactors. Lipid yields in algae grown in outdoor opens and lagoons rarely exceeds 10% (w/w).
[b]Data from Liang *et al.* (2005).
[c]Data from Kawachi *et al.* (2002).

© Woodhead Publishing Limited, 2013

under genetic control in the same way as is the oil content of plant seeds. Not all cells continue to produce oil until the cell is physically replete and cannot store more without the danger of bursting. Some cells appear to be able to limit the amount of oil that they accumulate. Although we now have a shrewd idea about why some microbial species are oleaginous and others are not and can explain this on the basis of knowing the enzymatic make-up of the oleaginous cell (Ratledge and Wynn, 2002), it is less clear what mechanism is controlling the upper limit of oil accumulation. The current explanation suggests that it is the provision of reducing power (in the form of nicotinamide adenine dinucleotide phosphate, NADPH), which is essential for the fatty acid synthesizing system, that becomes rate limiting. As all fatty acids are synthesized from acetate units (CH_3CO-), each acetate has to be reduced to $-CH_2CH_2-$ in order to create a long-chain fatty acid. Thus from recent work carried out in the author's laboratory, it appears that the ability of a cell to accumulate oil is controlled, not so much by the supply of acetate units, but by the ability of the cell to produce sufficient NADPH to drive fatty acid biosynthesis. Once the supply of reducing power has stopped or slowed down, the cell ceases to accumulate further amounts of oil. However, as it is not the aim of this chapter to go into great detail about the biochemistry and molecular biology of oil accumulation in microorganisms, the interested reader is referred to the papers and reviews by the current author and his colleagues (Ratledge and Wynn, 2002; Ratledge, 2004; Zhang et al., 2009; Wang et al., 2011).

The following sections are designed to give the essential information regarding the various microbial oils that have been and are being produced as nutraceuticals for human consumption.

19.3   Gamma-linolenic acid (GLA, 18:3 n-6)

Although this fatty acid is no longer in commercial production, it is included in this review as it was the first microbial oil to be commercially produced and, as such, opened the way for all subsequent microbial oils to be given approval for human consumption.

GLA is found in the oil from seeds of evening primrose (Oenothera biennis), borage (Borago officinalis) and blackcurrant (Ribes nigrum). The first oil used to be considered useful for the treatment of multiple sclerosis although this claim is no longer made. Today, however, there is still a considerable market for evening primrose oil as a dietary supplement for the alleviation of premenstrual tension and also for improvement of various skin conditions. In the late 1970s, evening primrose oil commanded a very high price (at about £40–50/kg). This was sufficient incentive to search for alternative sources of GLA and the focus fell on the Mucorales fungi that were already known to produce this fatty acid. Work carried out in the author's laboratory was able to identify several potential species that could

© Woodhead Publishing Limited, 2013

be considered for commercial development and, in conjunction with the biotechnology company, J & E Sturge at Selby in North Yorkshire, UK, a collaborative venture was initiated for the production of a GLA-rich oil using *Mucor circinelloides* (formerly *M. javanicus*). A full description of the background to the work and the development of the process has been provided by the author (Ratledge, 2006). The specification of this fungal oil is given in Table 19.4.

Commercial production of GLA-SCO began in 1985 and finished in 1990. The process was carried out in stirred fermenters of 220 m^3 which normally were used by the company for the production of citric acid using *Aspergillus niger*. With *M. cirinelloides*, the fermentation was carried out with high carbon to nitrogen ratio in the medium to promote lipid accumulation (see Section 19.2). The process was complete in about 72–96 hours with a yield of cells at about 60 kg m^{-3} and an oil content of the cells at approx. 25%. Over the six years of operation, about 30 tonnes of the oil were produced and was sold, mainly in the UK, in an encapsulated form as

Table 19.4 Specifications of the first single cell oil: an oil rich in γ-gamma linoleic acid

GLA – SCO	
Production organism:	*Mucor circinelloides* (formerly *M. javanicus*)
Production company:	J. & E. Sturge, N. Yorks, UK (taken over by Rhone-Poulenc in 1990)
Cultivation system:	220 m^3 stirred fermenters. Duration: 90–96 h Biomass: 50–70 kg m^{-3} Oil content: 25% dry wt
Oil	
Trade name:	Oil of Javanicus (also sold as GLA-Forte)
Appearance:	Bright, clean yellow oil
Melting point:	12–14°C
Free fatty acids:	<0.1%
Triacylglycerol content of oil:	~97%
Added antioxidant:	Vitamin E
Stability:	No off-flavours noted with storage at 20°C over 10 years

Fatty acyl composition (rel % w/w)

14:0	1	18:1 (*n*-9)	38–40
16:0	22–24	18:2 (*n*-6)	10–11
16:1 (*n*-9)	1	18:3 (*n*-6) [GLA]	18–19
18:0	5–7	18:3 (*n*-3)	0.2

© Woodhead Publishing Limited, 2013

an over-the-counter nutraceutical. The oil, being the very first single cell oil, was scrutinized exceptionally carefully by the regulatory authorities for any possibly toxicities or deleterious effects but none were found (see Section 19.8).

The *Mucor* oil was expensive to produce. The high costs of fermentation, including the purchase of the glucose syrup as feedstock (some 5–5.5 tonnes of sugar are needed to produce 1 tonne of oil), plus the costs of oil extraction, refining and final deodorization, meant that there was an insufficient profit margin from the sales of the oil. The oil, known as Oil of Javanicus™, was competing with evening primrose oil itself and then subsequently with borage oil that came on the market in the late 1980s. Furthermore, the sales of the oil were not tremendous – about 8–10 tonnes/year at its height. This was principally because the consumers of the oil (almost entirely women taking the oil for the relief of premenstrual tension) did not readily appreciate that GLA was the active ingredient of evening primrose oil and that another oil with a higher GLA content and selling at about the same price would therefore be more effective. The consumers wished to buy evening primrose oil and, as the *Mucor* oil could not be so labelled, there was only a low take up of the oil. This then led to the termination of the process in 1990 as the new owners of the company (Rhone-Poulenc Co.) considered that its profitability was too low to be sustained. Nevertheless, this oil was the harbinger of many other microbial oils and its evaluation as a safe and reliable oil helped all subsequent SCOs to become equally accepted by regulatory authorities as suitable for human consumption.

19.4   Docosahexaenoic acid (DHA, 22:6 *n*-3)

19.4.1   DHA in infant formulas

Over the past two decades there has been increasing awareness of the importance of long-chain omega-3 fatty acids (mainly DHA and eicosapentaenoic acid, EPA; see Fig. 19.1) in the diet and the role that these fatty acids play in the brain, neural tissues and in the membranes of the eye. There was also an appreciation of the importance of these fatty acids to be included in the diet of premature and newly born infants (see Collins *et al.*, 2011; Birch *et al.*, 2010 for detailed references). But it was not recommended that fish oils should be used as a source of them since, although fish oils were good sources of both EPA and DHA, EPA was contraindicated for infants as it generated undesirable metabolites (see Fig. 19.1) for the developing brain and eye (Huang and Sinclair, 1998; Li *et al.* 2003; Sinclair and Jayasooriya, 2010). Then, in the late 1980s came the realization that it was possible to produce a DHA-only oil if microalgae were used and that such an oil, subject to safety checks, would then be the ideal oil for including in infant formulas. The person behind this realization was David Kyle who used his wide knowledge of algal lipids and his

© Woodhead Publishing Limited, 2013

Table 19.5 Commercially available nutraceutical oils containing decosahexaenoic acid (DHA): principal fatty acids (values are rel. % (w/w) total fatty acids)

	DHA-SCO[a]	Schizo-SCO[b]	Ulkenia-SCO[c]	Schizo-ONC[d]
14:0	20	8	3	13
16:0	18	22	30	27
16:1	2	<0.5	<0.5	2
18:0	<0.5	0.5	1	1
18:1	15	1	–	<1
22:5 (n-6)(DPA)	–	17	11	8
22:6 (n-3)(DHA)	40	41	43–46	40

[a]Oil from *Crypthecodinium colinii*; manufactured by Martek BioSciences Corp. (now DSM): trade name DHASCO. Also sold as life'sDHA™.
[b]Oil from *Schizochytrium* sp. manufactured by Martek Biosciences Corp. (now DSM): trade name DHASCO-S.
[c]Oil from *Ulkenia* sp., manufactured by Lonza (Switzerland), sold as DHA CL (Clear liquid) and DHAid™.
[d]Oil being produced by Ocean Nutrition Canada Ltd using *Thraustochytrium* sp. ONC-T8 (submission to UK Food Standards Agency, 10 Oct 2011; www.food.gov.uk/multimedia/pdfs/dharich); see also Burja *et al.* (2006).

appreciation of the crucial roles of the long-chain PUFAs in both infants and adults to launch the company, Martek Corporation. An account of the early days in the development of the DHA oil has been given by Kyle *et al.* (1992); the organism of choice was a non-photosynthetic dinoflagellate, *Crypthecodinium cohnii*.

Initially, most of the oil that was produced was donated for trials with premature babies and newly born infants. The results were extremely encouraging: a genuine demand for the oil quickly followed. In 2002, the oil, known as DHASCO™, was given GRAS (generally recognized as safe) status by the FDA in the USA and, from that date, the oil could then be added to infant formulas in the USA; other countries quickly followed. The fatty acid profile of the oil is given in Table 19.5. The recommended level of supplementation of DHA in infant formulas is between 0.32% and 0.64% of the total fatty acids (Birch *et al.*, 2010); higher amounts produce no added advantages regarding improvements in visual acuity.

Sales of the oil have increased year by year and the oil is now included in infant formulas that are sold in over 70 countries around the world. Martek Biosciences was sold in 2011 to DSM (Dutch State Mines) in the Netherlands for US$1.1 billion, a price that reflected the huge sales of the oil which was still the cornerstone of the company. Current sales of the oil are at about 2000 tonnes per year and are likely to increase still further in the years to come. Although prices of the oil are commercially sensitive, the declared financial results for Martek for 2010, their last year of trading before their purchase by DSM, gave a revenue of US$317 million for sales of oil for infant nutrition.

© Woodhead Publishing Limited, 2013

There was, though, one problem with the administration of a DHA-only oil to infants and that was its retro-conversion into the undesirable EPA (see Fig. 19.1). It was realized, however, that if another PUFA, arachidonic acid (ARA, 20:4 *n*-6), was administered simultaneously it would inhibit this undesirable reaction. In addition, and importantly, ARA itself is highly desirable as it is the most abundant PUFA in mammalian brain and neural tissues (Sinclair and Jaysooriya, 2010). A combination of ARA + DHA at 2 : 1 (v/v) was determined to be the optimal dosage. Therefore, two oils had to be produced for inclusion in infant formulas: an DHA-SCO and an ARA-SCO. An account of ARA production is given in Section 19.5 below.

The DHA oil from *C. cohnii* is used exclusively for incorporation into infant formulas and is sold under the trade name of *life's*DHA. It is currently purchased by 24 companies around the world, including Abbott Laboratories, Heinz, Materna Ltd, Mead Johnson, Nestle and Wyeth Ayest. These companies then cover 70% of the total market for infant formulas (Fichtali and Senananyake, 2010; this overview provides a complete list of the companies purchasing the DHA oil).

19.4.2   DHA for adult and animal nutrition

At more or less the same time as Martek was developing *Cryptbecodinium cohnii* for the production of DHA for infant nutrition, another company in the USA was investigating other marine microorganisms as possible sources of the same fatty acid. Omega-Tech Inc., led by Bill Barclay, found there was a large group of marine eukaryotic organisms, known as the thraustochytrids, that could produce high concentrations of lipid in their cells together with a high proportion of DHA in the fatty acids. However, the DHA was always accompanied by another long chain fatty acid – docosapentaenoic acid as the omega-6 isomer (DPA, 22:5 *n*-6). After screening many possible candidates, the final choice for production was a *Schizochytrium* sp. The profile of the fatty acids of this organism is given in Table 19.5.

Schizochytrium species are members of the thraustochyrid group of microorganisms. Originally, these were considered to be marine fungi but were subsequently placed, along with *Thraustochytrium* spp., into the heterokont algae group (Cavalier-Smith *et al.*, 1994) and were classified as being in the family of the Labyrinthulomycetes. Thraustochyrids are non-photosynthetic organisms and appear to be widely distributed in most coastal water of both tropical and temperate climes (Bowles *et al.*, 1999; Burja *et al.*, 2006). The taxonomy of this group of organisms is still somewhat confused but, at the moment, *Schizochytrium* is considered to consist of three genera: *Schizochytrium* itself, *Aurantiochytrium* and *Oblongichytrium* (Wong *et al.*, 2008). *Ulkenia* – also referred to as *Labyranathula* – is another related species; in common with thrautochytrids in general, it produces a significant amount of DPA (22:5 *n*-6) (see Table 19.5). Most species of these genera are of potential commercial interest because of their

© Woodhead Publishing Limited, 2013

abilities to produce DHA and also, in most cases, for their abilities to grow extremely rapidly and to attain very high cell densities in fermenters.

The presence of DPA (*n*-6) in the oils of *Schizochytrium* and related genera initially caused some consternation as this fatty acid had not been previously recognised as a 'typical' microbial fatty acid. However, when it was appreciated that DPA (*n*-6) was a component of human brain tissue and therefore could not be considered to be an undesirable fatty acid, opposition to its use for human and animal consumption was withdrawn and the oil was given full GRAS (generally recognized as safe) status by the FDA in 2004 (see Barclay *et al.*, 2005, 2010). Extensive feeding trials of the *Schizochytrium* oil to poultry, fish and other animals, including humans, has shown its complete safety (see Section 19.8). Trials of oil in healthy men and women have shown that the oil is well tolerated and does not adversely affect any cardiovascular risk (Sanders *et al.*, 2006).

The original producer of *Schizochytrium* oil, Omega-Tech Inc, was taken over by Martek BioSciences in 2002 and is now part of DSM. The process is run in large fermenters (~100–150 m^3) and the organism has the ability to attain cell densities of >200 kg m^{-3} in less than 72 h (Barclay *et al.*, 2010). The cells contain up to 60% (w/w) oil. At least 40% of the total fatty acids is DHA.

Besides being used directly as a source of DHA as a nutraceutical, *Schizochytrium* oil has also been incorporated into a wide variety of food products, including table spreads, mayonnaises, milk products and so on (Barclay *et al.*, 2010) so that the advantages of DHA can be realized without the need to take it in the form of capsules which is not to everyone's taste and particularly those of children or elderly people. The *Schizochytrium* SCO is also distributed and sold in capsules under a variety of names usually being advertised as an 'algal oil containing DHA' but without speci-fying the organism being used. Such oils may originate from DSM or may be purchased from other companies now entering this market.

Other companies who are active in the production of oils using similar organisms include Lonza Group AG in Switzerland using a species of *Ulkenia* (see Kiy *et al.*, 2005), New Horizons Global in the UK and Ocean Nutrition in Canada. The latter company has provided the fatty acid profile of their oil being produced by *Thraustochytrium* sp. ONC-T8 in a submis-sion to the Foods Standards Agency in the UK (October 2011) as a prelimi-nary to launching the product in the UK and probably in Europe as well (Food Standards Agency, 2011). The fatty acid profile of the oil is given in Table 19.5. The organism being used was originally described by Burja *et al.* (2006).

In May 2012, Ocean Nutrition was acquired by DSM to add to their portfolio of companies involved in the supply of vitamins and other nutraceuticals.

More recently, at least one Chinese biotechnology company (Jiangsu Tiankai Biotechnology Co. Ltd) has developed a process using a new strain of *Schizochytrium* for the production of a DHA-rich oil similar to that

© Woodhead Publishing Limited, 2013

described above. However, unlike the other *Schizochytrium* oils, this one is being directly incorporated, along with arachidonic acid (see below), into infant formulas. This company is based in Nanjing and is a spin-off company from Nanjing University of Technology.

19.5   Arachidonic acid (ARA, 20:4 *n*-6)

Japanese workers were the first to report that the filamentous fungus, *Mortierella alpina*, could produce an oil with a very high content of ARA (see Totani *et al.*, 1987). This, however, was using cultures grown on agar plates. Nevertheless, impressive amounts were still produced when the fungus was grown in submerged cultures with ARA representing >40% of the total fatty acids. The fatty acid profile of this oil is shown in Table 19.6. There are no realistic alternative sources of this fatty acid that could provide the amounts that are required and at a price that could compete with the biotechnological route of production.

Commercial production of the oil was developed in the 1990s by both Suntory Ltd (Japan) and Gist-brocades Co. (now Dutch State Mines, DSM) but the real interest in developing this oil for commercial purposes came from Martek Corp. wishing to use the oil in conjunction with their DHA oil (see Section 19.4) for incorporation into infant formulas. An agreement was then reached between Martek and Gist-brocades that resulted in ARA being produced by the latter company but which was sold exclusively to Martek for blending with DHA oil. The oil is blended with DHASCO at 2 : 1 (v/v) before being added to infant formulas.

Less information is available concerning the oil produced by Suntory Ltd but it is believed that this is sold exclusively in Japan as a high purity oil not blended with DHA or any other oil. The trade name for this oil is SUNTGA40S and a profile of its constituent fatty acids is given in Table 19.6.

The production processes for ARA-SCOs are carried out in large fermenters (100–150 m^3) using conventional technologies although the filamentous nature of the production organism can be troublesome in order to achieve the highest cell densities. The process must last for up to 9–10 days to give the highest yields of oil, which is nearly three times longer than the fermentation times achieved with *Mucor circinelloides* for the production of the GLA-SCO (see Section 19.3). The yield of oil per kg biomass is about 50% and the reasons why this is not higher parallel those found for *Mucor circinelloides* in the production of GLA (Section 19.3). Details of the process are somewhat limited but a brief account has been provided by Streekstra (2010).

A rival process to that of the DSM and Suntory processes has recently been indicated by Cargill Inc. Since the early 2000s, Cargill have had a joint venture with Wuhan Alking Bioengineering Co. Ltd in China for the

© Woodhead Publishing Limited, 2013

Table 19.6 Commercially available nutraceutical oils derived from *Mortierella alpina* containing arachidonic acid (ARA) (values are rel. % (w/w) total fatty acids)

	16:0	18:0	18:1	18:2	18:3 (n-6)	20:3 (n-6)	ARA 20:4 (n-6)	22:0	24:0
ARA-SCO[a]	8	11	14	7	4	4	49	–	1
CABIO oil[b]	7.5	6	9	6	2.5	4	43	3	9.5

[a]Oil produced by DSM (formerly Gist-brocades, Netherlands).
[b]Oil produced by Cargill Alking Bioengineering (Wuhan) Co., Ltd. (Hubei, China); from Casterton *et al.* (2009), and Kusumoto *et al.* (2007).

production of ARA-SCO using *Mort. alpina* and, in 2010, they opened a dedicated Chinese facility to boost production of ARA. Initially, because of existing Martek patents, this oil was restricted for sale only in China but in 2012, the European Union gave approval to Cargill to allow sales of the oil in Europe for infant nutrition (Cargill, 2012). The oil is similar in fatty acid profile to the oil produced by DSM/Martek and is given the (provisional) name of CABIO oil (see Table 19.6). This development may therefore herald the arrival of other companies intending to produce similar ARA-SCOs, as the original Martek patents of the early 1990s are now coming to an end. However, as with all oils and especially for those destined for the infant nutrition market, considerable emphasis will be placed on good manufacturing practice for all stages of the oil production process to ensure the highest possible safety of the oil (see Section 19.8).

19.6 Eicosapentaenoic acid (EPA, 20:5 *n*-3)

Some of the metabolic roles of EPA are given in Fig. 19.1. Production of a number of eicosanoids, including the 3-series prostaglandins, prostacyclin and thramboxane and the 5-series leukotrienes and lipoxins, arise from EPA and these fulfil numerous functions in human metabolism ranging from beneficial effects on blood pressure, platelet aggregation and inflammation. There is also considerable evidence that EPA together with DHA is useful in the secondary prevention of various cardiac problems (Li *et al.*, 2003); this would then indicate that consumption of fish oils that are rich in both DHA and EPA would be helpful in adult humans. However, the usefulness of EPA as a single PUFA has also been proposed as a treatment for a wide range of disorders.

Administration of EPA, as its ethyl ester, may have beneficial effects on some neuropsychiatric disorders, including manic-depression (bipolar disorder), depression and even schizophrenia (Peet and Stokes, 2005; Riediger *et al.*, 2009; Lin *et al.*, 2010), and also for attention deficit hyperactivity disorder in children (Li *et al.*, 2003). Zhu *et al.* (2010) listed several unique

© Woodhead Publishing Limited, 2013

and beneficial effects of EPA; these include a decrease of about 20% in major coronary events occurring in patients with a history of heart disease and its role in preventing and treating obesity, metabolic syndrome, non-alcoholic steatohepatitis and type-2 diabetes.

Whilst clinical administration of EPA for the treatment of any of the above disorders has still to take place on any significant scale, EPA is, however, now a prescribed treatment for people suffering from hypertri-glyceridemia, a high level of triacylglycerols in the blood (see Koski, 2008). The preparation that is used is purified from fish oils; it contains 465 mg EPA and 375 mg DHA per capsule as a daily dose. It is sold under the trade-name of Lovaza and is produced by GlaxoSmithKline. It is approved for use in most European countries and in the USA for the treatment of this disorder (Baker *et al.*, 2008). A product that is aimed at the same market is also produced by Amarin Corp plc in the USA under the initial name of AMR101. This, unlike Lovaza, is virtually pure ethyl EPA: >95% and has (presumably) been prepared from fish oils using the very expensive proce-dure of preparative-scale HPLC. This preparation, it is claimed (see Bays *et al.*, 2011), not only decreases the concentration of circulating triacylglyc-erols in the blood, and therefore is similar to Lovaza, but also decreases the levels of low-density lipoprotein (LDL) – the so-called 'bad' cholesterol – which is not a claim made for Lovaza.

In these respects, the EPA should be regarded as a 'medical food' rather than a nutraceutical. The difference between these two epithets for EPA is that a medical food is administered orally under the supervision of a physi-cian and is intended for the specific management of a disease or condition for which distinctive nutritional requirements have been established in appropriate clinical trials. Not unnaturally, the selling price of EPA as a medical food can then be much higher than it might attain if just sold as an over-the-counter (OTC) nutraceutical. This is shown by the costs of Lovaza compared to equivalent OTC capsules of EPA/DHA: the monthly cost of Lovaza therapy is over four times the cost of OTC-equivalent fish oil cap-sules (Baker *et al.*, 2008) although, in fairness, it is claimed that a patient has to take fewer capsules of Lovaza than fish oil capsules as the concentra-tion of the omega-3 fatty acids is considerably higher in the former.

Because of the very high costs of producing high-purity EPA from fish oils, considerable interest is now being shown in developing microbial sources of EPA mainly as a medical food and for clinical applications, rather than just an OTC nutraceutical. The aim is to develop an EPA-only oil, as clearly mixtures of EPA and DHA can be easily obtained in varying degrees of purity from fish oils, but the principal markets for EPA are as a 'medical food' and these require its purity to be over 90% and preferably over 95%. The choice of production organism is therefore between using existing species that are known to produce EPA, albeit in mixtures with other PUFAs including DHA and/or ARA, or genetically modifying an organism to produce EPA specifically.

© Woodhead Publishing Limited, 2013

Table 19.7   Productivities for EPA attained by various microalgae[a]

Organism	Culture system[b]	EPA yield (mg l^{-1})	EPA productivity (mg l^{-1} day^{-1})
Isochrysis galbana	Photobioreactor	–	24
Monodus subterraneus	Plate reactor	96	59
Nannochloropsis sp.	Photobioreactor	~100[c]	32
Nitzschia laevis	Stirred fermenter	1112	175
Phaeodactylum tricornutum	Glass tanks	130	34
Porphyridium cruentum	Glass flasks	–	4

[a]Data adapted from Wen and Chen (2003, 2010).
[b]All data derived from photoautotrophic cultivation systems except for *N. laevis* which was grown heterotrophically with glucose as principal carbon source.
[c]Calculated from EPA being 4% of biomass (Zittelli *et al.* 1999) and an assumed cell density of 2.5 g l^{-1}.

Microorganisms that produce EPA naturally are almost exclusively microalgae (see Table 19.3). Most of these species can only be grown photosynthetically and therefore their productivities for EPA are low. Final biomasses are rarely above 3–4 g l^{-1} even in photobioreactors. Table 19.7 indicates the low productivities of EPA in a number of microalgae that are themselves the best examples that can be identified for this process. Some attention has been given to the possibility of using *Nannochloropsis* spp. for this purpose (see Tables 19.3 and 19.7). Although lipid contents of this alga have been reported to be up to 36% of the cells, this is only in reactors to which CO_2-enriched air has been provided to ensure a surfeit of carbon is available (Rodolfi *et al.*, 2009). The lipid content of cells grown in outdoor lagoons or ponds, however, is not likely to exceed 10%. Moreover, the cells are extremely small, making harvesting expensive; they also have a hard cell wall making lipid extraction difficult. The main focus of work with this microalga would, though, seem to be for the production of fatty acids for biodiesel production (see also Section 19.7). The organism, however, has the advantage of now being genetically manipulatable (Kilian *et al.*, 2011).

An additional complication with algal lipids is that they are complex phospho- and glyco-lipids associated with the photosynthetic process; recovery of a fatty acid, such as EPA, therefore requires lipid extraction, hydrolysis and re-esterification of the fatty acid. This makes any process to produce an individual fatty acid extremely expensive. Only the highest value outlets for such fatty acids can therefore be envisaged and, for these reasons, EPA would need to be considered as a medical food rather than a nutraceutical.

The alternative to growing EPA-producing microalgae photosynthetically is to grow them heterotrophically; that is in closed fermenters with a fixed carbon source, usually glucose. However, only a small minority of algal species can be so cultivated. Several possible species have been highlighted

© Woodhead Publishing Limited, 2013

by Tan and Johns (1996) and by Wen and Chen (2003, 2010); these include *Cylindrotheca fusiformis, Navicula pelliculosa* and *Nitzschia laevis* (see also Wen *et al.*, 2002). The latter species appears the most attractive although, as can be seen from Table 19.7, its productivity is still relatively low even when cultivated heterotrophically. Companies that are using heterotrophic cultivation of microalgae to produce an EPA-only oil include Algisys LLC (Cleveland, OH) and Photonz (New Zealand). The lipid, however, that is produced by algae grown under heterotrophic conditions is still a mixture of complex entities rather than being mainly triacylglycerols, as found in yeasts and fungi. Thus, careful hydrolysis and re-esterification of the lipid will be needed to recover the EPA.

An alternative to using microalgae to produce EPA is to use a genetically modified microorganism. At least two companies are known to be using recombinant gene technology to achieve production of EPA (and perhaps other PUFAs as well) using yeasts. The pioneering work that has been done in this direction has been carried out by a team at DuPont, Wilmington, USA where the chosen route was to genetically engineer the oleaginous yeast, *Yarrowia lipolytica* (see Table 19.1). This yeast, at the time of commencing the work, was the only oleaginous yeast available whose genome profile has been elucidated. To achieve EPA biosynthesis, between 15 and 20 genes had to be individually introduced into *Y. lipolytica* (see Figure 19.2). By using an oleaginous yeast, no genes had to be introduced to promote high levels of lipid accumulation; the oil content of this yeast thus remained at about 35–40% of the cell biomass (Damude *et al.*, 2011). The process has been in commercial production since the beginning of 2010. The oil contains at least 50% of the total fatty acids as EPA and has received GRAS status from the FDA. It is marketed and sold through New Harvest™, a wholly owned subsidiary company of DuPont, and is intended as an OTC nutraceutical. A profile of the oil is given in Table 19.8.

Table 19.8   Lipid composition of a genetically modified strain of *Yarrowia lipolytica* for the production of eicosapentaenoic acid (EPA, 20:5 n-3)[a,b]

Rel. % (w/w) fatty acids			
16:0	2.5	20:3 (*n*-6) DGLA	1.7
16:1	0.8	20:4 (*n*-6) ARA	0.6
18:0	1.1	20:3 (*n*-3) EtrA	0.8
18:1	5.8	20:4 (5, 11, 14, 17)	0.7
18:2 (*n*-6) LA	18.3	20:4 (*n*-3) ETA	1.3
18:3 (*n*-3) GLA	2.6	20:5 (*n*-3) EPA	55.6
20:2 (*n*-6) EDA	2.6	Others	5.6

[a]Strain: *Y. lipolytica* Y4305; production company: E.I. du Pont de Nemours; growth conditions: 2 l stirred fermentor; duration of culture 148 h (~6 days); temperature: 30–32°C; cell dry wt: 45.5 g l^{-1}; total fatty acids: 21.7% of cell dry wt.
[b]from Xue *et al.* (2009), see also Fig. 19.2 for route of synthesis.

© Woodhead Publishing Limited, 2013

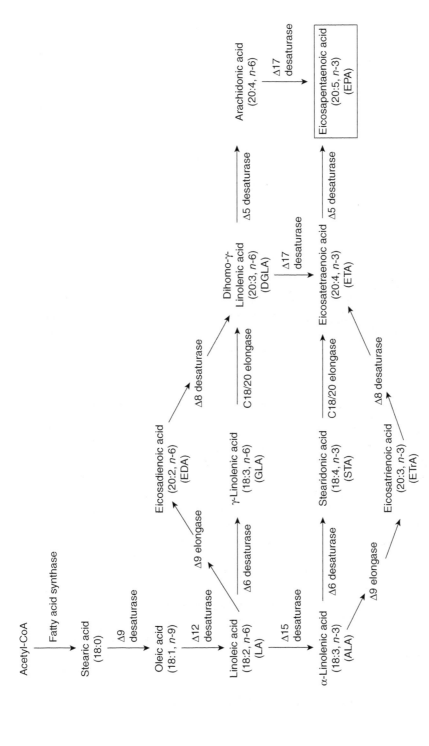

Fig. 19.2 Outline of biosynthetic pathways for the biosynthesis of EPA showing the enzymes for which genes had to be genetically engineered into *Yarrowia lipolytica* to achieve conversion of 18:2 *n*-6 (LA) to EPA. Adapted from Zhu *et al.* (2010).

© Woodhead Publishing Limited, 2013

A more recent development has been announced by a Danish company, Fluxome, which is developing a process using genetically modified baker's yeast (*Saccharomyces cerevisiae*) to produce an EPA-rich oil. No details of the process or the product are currently available [see, however, Gunnarsson *et al.* (2010) for information about the genetic strategy being used] but the task of producing a high content of EPA in this yeast will be more difficult than using *Y. lipolytica* as *S. cerevisiae* is not an oleaginous species and will probably have less than 10% of its biomass as a triacylglycerol oil. Therefore, to promote lipid accumulation, several additional genes will have to be introduced into the yeast besides those used by DuPont to genetically modify their yeast. This will then add a considerable genetic burden to the yeast.

However in October 2012, this company announced severe financial problems resulting in the (temporary?) abandonment of the EPA project using *S. cerevisiae* and with the company focussing on the production of the anti-oxidant, resveratrol.

The driving force behind these pioneering efforts is clearly the desire to produce a very high value oil that will command a premium price. The possible clinical applications of EPA-rich oils are, however, very likely to expand in the immediate future and therefore this may be a particularly desirable fatty acid to produce by whatever means are available.

19.7   PUFA oils from photosynthetically-grown microalgae

There has long been considerable interest in using photosynthetically grown microalgae for the production of polyunsaturated fatty acids. This interest has increased considerably over the past decade with the current interest in the nutraceutical value of such products. Major reviews on the subject include those by Adarme-Vega *et al.* (2012), Cohen and Khozin-Goldberg (2010), Guschina and Harwood (2006), Spolaore *et al.* (2006) and Ward and Singh (2005). The main problem with using microalgae is that most species produce a number of PUFAs, mainly ARA, EPA and/or DHA in varying proportions (see Table 19.3). In addition, and as mentioned in Section 19.6, algal lipids are complex entities. The highest lipid contents of cells, sometimes quoted as being up to 36% or higher in the algal cell (see for example Rodolfi *et al.*, 2009), can only be obtained in closed systems into which CO_2-enriched air is introduced. This then considerably increases the costs of production. If algae are grown outdoors, without CO_2 addition, the lipid contents of the cells rarely exceed 10% of the biomass. This low yield, coupled with the low cell densities, does not make this approach particularly attractive.

In spite of these difficulties, a number of companies that have developed large-scale algal technology for the production of algal oils to be used as biofuels have now begun to realize that these processes are unlikely to be

© Woodhead Publishing Limited, 2013

economical in the near future and that it would therefore make greater commercial sense if they focused on producing oils with a much greater value than biodiesel fatty acids. Algal production of EPA plus DHA or EPA plus ARA is therefore a very likely development in this field over the next few years. In all probability these new algal oils will be aimed at niche markets for those customers who do not wish to consume fish oils as sources of DHA/EPA; customers such as vegetarians and vegans as well as possibly some religious minority groups.

The organisms are being grown photosynthetically sometimes in photobioreactors (usually tubular bioreactors which appear the cheapest of these systems) or, more usually, in large outdoor lagoons and ponds. Both strategies have their problems. The cost of running photobioreactors is extremely expensive and, at present, only one product, astaxanthin, is produced commercially using this technology. The costs of producing an oil using photobioreactor technology, based on the costs of producing astaxanthin in this way using CO_2-enrichment, could be up to US $100,000 per tonne (Ratledge and Cohen, 2008). The main problem in using outdoor cultivation of algae in ponds or lagoons is that it is very difficult to achieve the conditions necessary to produce contents of lipid much greater than 10% of the biomass. This is because lipid accumulation requires a surfeit of carbon to be available to the cells (see Section 19.2) but this can only be achieved by using forced addition of CO_2 which needs to be both clean and plentiful. This difficulty has not yet been satisfactorily resolved for large-scale cultivation processes. Although outdoor cultivation is, in principle, cheap, the biomass productivity is low in comparison with heterotrophic cultivation methods and this means that lipid productivity is even lower (see Table 19.7 for some examples). Furthermore, the lipids are complex (see above) thereby requiring careful extraction and purification of the fatty acids.

A final problem comes in the acceptability of the final product: outdoor cultivation of algae means that the processes cannot be regarded as good manufacturing practice as there is always considerable danger of contamination. Thus, safety of the oil could be a significant issue and may necessitate each batch of oil having to be carefully checked. However, as the product is an extracted entity that has undergone considerable refinement and various post-harvesting treatments (as are all plant oils), the oil could be regarded as satisfying the regulatory requirements in much the same way as a plant oil is regarded as safe even though the plant itself will not have grown in an aseptic environment.

Companies that are currently active in this area include Aurora Algae (based in California and Australia), Qponics (in Queensland, Australia), AlgaeBio (a Canadian company with production facilities in Arizona), Bio-Process Algae (based in the USA), Algae Biotechnologies, Renewable Algal Energy and Solazyme who are using *Chlorella protothecoides* (based in San Francisco).

© Woodhead Publishing Limited, 2013

19.8   Safety

Microbial oils are a relatively recent development with the first one being only marketed in 1985 (see Section 19.3). As may be imagined, tremendous pressure was placed on the company producing the oil (J & E Sturge, North Yorks, UK) to have it extensively tested for any possible toxicity. As the oil was aimed as a substitute for an existing plant oil, evening primrose oil, there were strong objections to it being released for human consumption by the existing suppliers and distributors of the plant oil. Work that was carried out on the oil from *Mucor circinelloides* was therefore of crucial importance not only for the oil itself but, as it turned out, for all subsequent SCOs. Had the *Mucor* oil failed for any reason to have been declared fit for consumption, this would have inevitably delayed all subsequent launches of microbial oils. The initial toxicological trials were conducted with brine shrimps and then followed by animal feeding trials for 90 days with various rodents and other animals. All trials were completely successful and the data were then presented to the UK Advisory Committee on Novel Foods and Processes for evaluation and approval. No objections were raised to the oil being offered for sale in the UK. This decision was considerably influenced by the production organism itself having already being given GRAS status as it was a known constituent of several oriental fermented foods. Thus, as people had been consuming *Mucor circinelloides* for millennia, there could be no objection to a component of it, that is the oil, also being consumed by humans. Although the assessment of the oil was still continuing when production of it ceased in 1990, it was clear that the oil posed no danger in consumption; all evaluations had shown it to be equivalent or better than high purity plant oils. Indeed, the oil, in some respects, was purer than the conventional plant oils, as the levels of trace residues arising from the application of herbicides and pesticides during plant cultivation were virtually absent from the *Mucor* oil whereas these could be easily detected in the plant oils, including evening primrose oil. (It should be said, however, that the levels of the residual herbicides and pesticides in the plant oils were all well below the national and international safety limits.) Such considerations will therefore apply to all SCOs under current production.

In assessing the safety of any microbial oil, it is of considerable importance to evaluate the production organism itself. This should have no history of producing toxic metabolites or be capable of infecting animals or damaging the environment. Clearly, animal pathogens cannot be considered but equally neither can plant pathogens as there is always the danger that live organisms could escape from a production facility and infect nearby crops or plants. These issues are therefore relatively easy to check through the existing scientific literature. Equally important is that the organism itself should not produce allergic reactions: allergens may not be present in the final oil product but, if present in the production organism, would pose a

© Woodhead Publishing Limited, 2013

hazard to process operators. Therefore, diligent searching through the literature should provide sufficient support for the lack of any known allergic response. All current SCOs are considered by the FDA to be 'highly refined oils' that are not associated with allergic reactions (Ryan *et al.*, 2010). A major factor that helps considerably in establishing the safety of an SCO, is that the oil is just an oil: it has been extracted, refined and de-odorized (in common with all commercial plant oils) and can be analysed down to the last 0.1% of its content or even lower. Oils can therefore be shown to contain extremely low contents of non-lipid materials and cannot therefore contain any appreciable amount of any material that is likely to be a danger. However, the SCOs will, contain some non-triacylglycerol lipids such as carotenoids and perhaps a low amount of sterols or sterol esters. These can be identified and, if necessary, compared to similar materials that are commonly found in most commodity plant oils.

Another important factor to establish is that all the component fatty acyl groups within the oil are known to be safe: this is best achieved by showing that the fatty acids of the microbial oil also occur in plant oils that have a safe history of consumption. If, however, one of the component fatty acids in the oil is unusual, as happened with docosapentaenoic acid (DPA, *n*-6) when considering *Schizochytrium* sp. as a source of DHA, it was sufficient to show that this fatty acid was also a component of human brain lipid. Thus, there could be no objections to the consumption of the oil if one of its fatty acids was already known to be synthesized in humans.

Microbial oils have to be evaluated for human consumption using criteria that could be equally applied to commercial plant oils. Any new plant oil that is offered for sale, such as *Echium* oil which has recently started commercial production as a source of stearidonic acid (18:4 *n*-3), also needs to be evaluated and the criteria that are used here must equally apply to microbial oils intended for human consumption.

Testing of the oil is, however, still essential using appropriate feeding trials in animals. Detailed reviews of the methods used for the evaluation of the safety of SCOs has been provided by Zeller (2005) and Ryan *et al.* (2010) and these should be consulted for full details of the trials that the various current SCOs have undergone. Briefer information is available in the reviews of Wynn and Ratledge (2006) and Ratledge and Hopkins (2006b). The evidence provided for the safety of microbial oils, though, has to be finally evaluated by the various regulatory authorities around the world: the Food and Drugs Administration (FDA) in the USA, EU 'Regulations for novel foods and novel food ingredients' in Europe, Food and Drugs Regulations in Canada, Australia/New Zealand Food Standards Code and so on. It is these authorities that then give the necessary approval for the sale of a new food or nutraceutical.

When the oil is coming from a novel source, as was the case with the very first SCO from *M. circinelloides* (see Section 19.3) and also those from *C. cohnii, Schizochytrium* sp. (see Section 19.4) and *M. alpina* (see Section

© Woodhead Publishing Limited, 2013

19.5), all of which when first introduced represented 'new' sources of oil, the regulatory authorities required complete dossiers of the evidence for the safety of the oils. These may have included acute oral studies, sub-chronic feeding to rodents for 90 days, lifetime feeding to rodents, exposure of pregnant animals to the oil and short-term tests for carcinogenicity (Ryan *et al.*, 2010). For the oil from *M. alpina*, scrutiny was particularly intense to establish that there were no mycotoxins or other deleterious materials in the organism and over ten separate assessments of the oil were carried out before clearance was finally given (Hempenius *et al.* 1997, 2002; Streekstra, 1997; Zeller 2005; Ryan *et al.* 2010). More recent work has confirmed the safety of the oil when fed to albino rats at up to 30 g kg^{-1} body wt over 13 weeks (Nisha *et al.*, 2009).

'Substantial equivalence' can, however, be claimed for an oil if it is being produced by an organism that is already taxonomically related to an existing organism being used for SCO production. The European Food Safety Authority use a similar term, 'Qualified presumption of safety', to authorize use of microbial products in foods (EFSA, 2012). Thus, for example, changing from a *Schizochytrium* sp. to a *Thraustochytrium* sp. for the production of DHA should not require major evaluation and testing as the two organisms are within the same genetic family. However, this is not to say that no testing is needed but only that the minimum level of safety needs to be demonstrated. What then constitutes this 'minimum' level will depend on which regulatory authority is dealing with the request for the oil to be acceptable.

The quantity of oil that is likely to be consumed by a person is also a factor that is frequently taken into account. No one is likely to consume hundreds of grams of any oil per day on a regular basis let alone an expensive microbial oil. Nevertheless, trials have taken place in which 20 ml of ARA-SCO were fed to rats which corresponded to a 75 kg human ingesting 1.5 l of oil in a single dose. The only problem that was noted was some diarrhoea the day after administration (Wynn and Ratledge, 2006). Human studies, with intakes of 7.3 g DHASCO per day, also failed to detect any adverse effect; the most serious complaint was 'fishy burps'!

Microbial oils have now been consumed by infants, adults and a huge variety of animals including fish, poultry, pigs and so on for two decades. No substantiated report has been provided to indicate that there has been any problem with their consumption and the manufacturers and distributors of the SCOs have no reason to doubt the complete safety and reliability of their products. Microbial oils have been the most heavily scrutinized of all microbial products and have consistently been shown to pose no dangers to health or well being. Indeed, the advantages of including these oils in our diet and even that of our animals are now becoming increasingly well established and recognized. Few if any babies in western countries will have been raised on formula milk preparations that did not contain DHA and ARA derived from microorganisms. The benefits of PUFAs, especially the

© Woodhead Publishing Limited, 2013

long chain omega-3 fatty acids, EPA and DHA, are increasingly being appreciated by adults. Microbial oils therefore represent unique and valuable sources of these important nutraceuticals and are likely to be the principal renewable source of them for many years to come.

19.9   References

ADARME-VEGA, T.C., LIM, D.K.Y., TIMMINS, M., VERNEN, F., LI, Y. and SCHENK, P.M. (2012) 'Microalgal biofactories: a promising approach towards sustainable omega-3 fatty acid production.' *Microb. Cell Fact.*, **11**, 96–105.

BAKER, N.A., MURDOCK, N. and PUGMIRE, B. (2008) 'Lovaza: potent, pure and proven?' *Evidence-based Pract.*, **11**, 10–11.

BARCLAY, W., WEAVER, C. and METZ, J. (2005) 'Development of a docosahexaenoic acid production technology using *Schizochytrium*: a historical perspective', In *Single Cell Oils*, Cohen, Z. and Ratledge, C. (eds), AOCS Press, Champaign, IL, 36–52.

BARCLAY, W., WEAVER, C., METZ, J. and HANSEN, J. (2010) 'Development of a docosahexaenoic acid production technology using *Schizochytrium*: historical perspective and update', In *Single Cell Oils*, 2nd edition, Cohen, Z. and Ratledge, C. (eds), AOCS Press, Champaign, IL, 75–96.

BAYS, H.E., BALLANTYNE, C.M., KASTELEIN, J.J., ISAACSOHN, J.L., BRAECKMAN, R.A. and SONI, P.N. (2011) 'Eicosapentaenoic acid ethyl ester (AMR101) therapy in patients with very high triglyceride levels (from the Multi-center, plAcebo-controlled Randomized, double-bliNd, 12-week study with an open-lable Extension [MARINE] Trial),' *Am. J. Cardiol.*, **108**, 682–90.

BIRCH, E.E., CARLSON, S.E., HOFFMAN, R.R., FITZGERALD-GUSTAFSON, K.M. FU, V.L.N., DROVER, J.R., CASTAÑEDA, Y.S., MINNS, L., WHEATON, D.K.H., MUNDY, D., MARUNYCZ, J. and DIERSEN-SCHADE, D.A. (2010) 'The DIAMOND (DHA Intake And Measurement Of Neural Development) study: a double-masked, randomized controlled clinical trial of the maturation of infant visual acuity as a function of the dietary level of dococoshexaenoic acid,' *Am. Soc. Nutr.*, **91**, 848–59.

BOWLES, R.D., HUNT, A.E., BREMER, G.B., DUCHARS, M.G. and EATON, R.A. (1999) 'Long-chain *n*-3 polyunsaturated fatty acid production by members of the marine protistan group, the thraustochytrids: screening of isolates and optimization of docosahexaenoic acid production', *J. Biotechnol.*, **70**, 193–202.

BURJA, A.M., RADIANINGTYAS, H., WINDUST, A. and BARROW, C.J. (2006) 'Isolation and characterisation of polyunsaturated fatty acid producing Thraustochytrium species; screening of strains and optimization of omega-3 production', *Appl. Microbiol. Biotechnol.*, **72**, 1161–9.

CARGILL (2012) *Cargill's Arachidonic Acid-Rich Oil Authorized as a Novel Food by the European Commission*. Cargill.com/news/releases/2012/NA3053758.jsp.

CASTERTON, P.L., CURRY, L.L., LINA, B.A.R., WOLTERBEEK, A.P.M. and KRUGER, C.L. (2009) '90-Day feeding and genotoxicity studies on a refined arachidonic acid-rich oil', *Food Chem. Toxicol.*, **47**, 2407–18.

CAVALIER-SMITH, T., ALLSOPP, M.T.E.P. and CHAO, E.E. (1994) 'Thraustochytrids are chromists not fungi: 18sRNA signatures of heterokonta', *Phil. Trans. R. Soc. London B: Biol. Sci.*, **346**, 387–97.

COHEN, Z. and KHOZIN-GOLDBERG, I. (2010) 'Searching for polyunsaturated fatty acid-rich photosynthetic microalgae', In *Single Cell Oils*, 2nd edition, Cohen, Z. and Ratledge, C. (eds), AOCS Press, Champaign, IL, 201–24.

COLLINS, C.T., GIBSON, R.A. and MAKRIDES, M. (2011) 'The DINO trial – challenges for translation into clinical practice', *Lipid Technol.*, **23**, 200–2.

© Woodhead Publishing Limited, 2013

COTTIN, S.C., SANDERS, T.A. and HALL, W.L. (2011) 'The differential effects of EPA and DHA on cardiovascular risk factors', *Proc. Nutr. Soc.*, **70**, 215–31.

DAMUDE, H.G., GILLIES, P.J., MACOOL, D.J., PICATOAGGIO, S.K., POLLAK, D.M.W., RAGGHIANTI, J.J., XUE, Z., YADAV, N.S., ZHANG, H and ZHU, Q.Q. (2011) *High Eicosapentaenoic Acid Producing Strains of* Yarrowia lipolytica, US Patent, 7,932,077 B2 (Apr. 26, 2011).

EFSA (2012) *Qualified Presumption of Safety (QPS)*. www.efsa.europa.eu/en/topics/topic/qps.htm.

FICHTALI, J. and SENANAYAKE S.P.J.N. (2010) 'Development and commercialization of microalgae-based functional lipids', In *Functional Food Product Development*, Smith, E. and Charter, E. (eds)., Blackwell, Oxford, UK, 206–25.

FOOD STANDARDS AGENCY (2011) 'DHA-rich algal oil from *Schizochytrium* sp. ONC-Ti8', submission to the Foods Standards Agency in the UK (October 2011) in accordance with regulation EC 258/97 (www.food.gov.uk/multimedia/pdfs/dharich).

GOGUS, U. and SMITH, C. (2010) '*n*-3 Omega fatty acids: a review of current knowledge', *Inter. J Food Sci. Technol.*, **45**, 417–36.

GUNNARSSON, N.K., FORSTER, J. and NIELSEN, J.B. (2010) *Metabolically Engineered* Saccharomyces *Cells for the Production of Polyunsaturated Fatty Acids*, US Patent 7,736,884 B2 (Jun. 15, 2010).

GUSCHINA, I.A. and HARWOOD, J.L. (2006) 'Lipids and lipid metabolism in eukaryotic algae', *Prog. Lipid Res.*, **45**, 160–86.

HUANG, Y-S. and SINCLAIR, A.J. (eds) (1998) *Lipids in Infant Nutrition*, AOCS Press, Champaign, IL.

HEMPENIUS, R.A., VAN DELFT, J.M.H., PRINZEN, M. and LINA, B.A.R. (1997) 'Preliminary safety assessment of an arachidonic acid-enriched oil from *Mortierella alpina*: summary of toxicological data', *Food Chem. Toxicol.*, **38**, 573–81.

HEMPENIUS, R.A., LINA, B.A.R. and HAGGIT, R.C. (2000) 'Evaluation of a subchronic (13 week) oral toxicological study, preceded by an *in utero* exposure phase, with arachidonic acid oil derived from *Mortierella alpina* in rats', *Food Chem. Toxicol.*, **38**, 127–39.

INNIS, S.M. (2003) 'Perinatal biochemistry and physiology of long-chain polyunsaturated fatty acids', *J. Pediatr.*, **143**, 1–8.

JANAKIRAM, N.B. and RAO, C.V. (2009) 'Role of lipoxins and resolvins as anti-inflammatory and proresolving mediators in colon cancer', *Curr. Mol. Med.*, **9**, 565–79.

KAWACHI, M., INOUYE, I., HONDA, D., O'KELLY, C.J., BAILEY, J.C., BIDIGARE, R.R. and ANDERSEN, R.A. (2002) 'The *Pinguiophyceae classis nova*, a new class of photosynthetic stramenopiles whose members produce large amounts of omega-3 fatty acids', *Phycol. Res.*, **50**, 31–47.

KILIAN, O., BENEMANN, C.S.E., NIYOGI, K.K. and VICK, B. (2011) 'High-efficiency homologous recombination in the oil-producing alga *Nannochloropsis* sp.', *Proc. Natl. Acad. Sci. USA*, **108**, 21265–9.

KIY, T., RUSING, M. and FABRITIUS, D. (2005) 'Production of docosahexaenoic acid by the marine microalga, *Ulkenia* sp.', In *Single Cell Oils*, Cohen, Z. and Ratledge, C. (eds), AOCS Press, Champaign, IL, 99–106.

KOSKI, R.R. (2008) 'Omega-3-acid ethyl esters (Lovaza) for severe hypertriglcyeridemia', *Pharm. Therapeut.*, **33**, 271–303.

KUSUMOTO, A., ISHIKURA, Y., KAWASHIMA, H., KISO, Y., TAKAI, S. and MIYAZAKI, M. (2007) 'Effects of arachidonate-enriched triacylglycerol supplementation on serum fatty acids and platelet aggregation in healthy male subjects with a fish diet', *Brit. J. Nutr.*, **98**, 626–35.

KYLE, D.J., SICOTTE, V.J., SINGER, J.J. and REEB, S.E. (1992) 'Bioproduction of docosahexaenoic acid (DHA) by microalgae', In *Industrial Applications of Single*

© Woodhead Publishing Limited, 2013

Cell Oils, Kyle, D. J. and Ratledge, C., (eds), AOCS Press, Champaign, IL, 287–300.

LEE, J.H., O'KEEFE, J.H., LAVIE, C.J. and HARRIS, W.S. (2009) 'Omega-3 fatty acids: cardiovascular benefits, sources and sustainability', *Nat. Rev. Cardiol.*, **6**, 753–8.

LI, D., BODE, O., DRUMMOND, H. and SINCLAIR, A.J. (2003) 'Omega-3 (*n*-3) fatty acids'. In *Lipids for Functional Foods and Nutraceuticals*, Gunstone F.G. (ed). The Oily Press, Dundee, 225–62.

LIANG, Y, MAI, K. and SUN, S. (2005) 'Difference in growth, total lipid contents and fatty acid composition among 60 clones of *Cylindrotheca fusiformis*', *J. Appl. Phycol.*, **17**, 61–5.

LIN, P-Y., HUANG, S-Y. and SU, K-P. (2010) 'A meta-analytic review of polyunsaturated fatty acid compositions in patients with depression', *Biol. Psych.*, **68**, 140–7.

MCCANN, J.C. and AMES, B.N. (2005) 'Is docosahexaenoic acid, an *n*-3 long-chain polyunsaturated fatty acid, required for development of normal brain function? An overview of evidence from cognitive and behavioural tests in humans and animals', *Am. J. Clin. Nutr.*, **82**, 281–95.

NISHA, A., MUTHUKUMAR, S.P. and VENKATESWARAN, G. (2009) 'Safety evaluation of arachidonic acid rich *Mortierella alpina* in albino rats – a subchronic study', *Regul. Toxic. Pharm.*, **53**, 186–94.

PEET, M. and STOKES, C. (2005) 'Omega-3 fatty acids in the treatment of psychiatric disorders', *Drugs*, **65**, 1051–9.

RATLEDGE, C. (1997) 'Microbial lipids', In *Biotechnology*, 2nd edition, *Products of Secondary Metabolism*, Kleinkauf, H. and van Dohren, H. (eds), VCH, Weinheim, Germany, Vol. 7, 133–97.

RATLEDGE, C. (2004) 'Fatty acid biosynthesis in microorganisms being used for single cell oil production', *Biochimie*, **86**, 807–15.

RATLEDGE, C. (2006) 'Microbial production of gamma-linolenic acid', In *Handbook of Functional Lipids*, Akoh, A. A. (ed), Baco Raton, Taylor & Francis, 19–45.

RATLEDGE, C. and COHEN, Z. (2008) 'Microbial and algal oils: do they have a future for biodiesel or as commodity oils?' *Lipid Technol.*, **20**, 155–60.

RATLEDGE, C. and HOPKINS, S. (2006a) 'Lipids from microbial sources', In *Modifying Lipids for Use in Food*, Gunstone, F.G. (ed)., Woodhead Publishing, Abington, UK, 80–113.

RATLEDGE, C. and HOPKINS, S. (2006b) 'Applications and safety of microbial oils in food,' In *Modifying Lipids for Use in Food*, Gunstone, F.G. (ed)., Woodhead Publishing, Abington, UK, 567–86.

RATLEDGE, C. and WYNN, J.P. (2002) 'The biochemistry and molecular biology of lipid accumulation in oleaginous microorganisms', *Adv. Appl. Microbiol.*, **51**, 1–51.

RIEDIGER, N.D., OTHAMAN, R.A., SUH, M. and MOGHADASIAN, M.H. (2009) 'A systemic review of the roles of *n*-3 fatty acids in health and disease', *J. Am. Diet. Ass.*, **109**, 668–79.

RODOLFI, L., ZITTELLI, G.C., BASSI, N., PADOVANI, G., BIONDI, N., BONINI, G. and TREDICI, M.R. (2009) 'Microalgae for oil: strain selection, induction of lipid synthesis and outdoor mass cultivatin in a low-cost photobioreactor', *Biotech. Bioeng.*, **102**, 100–12.

RYAN, A., ZELLER, S. and NELSON, E.B. (2010) 'Safety evaluation of single cell oils and the regulatory requirements for use as food ingredients,' In *Single Cell Oils*, 2nd edition, Cohen, Z. and Ratledge, C. (eds), AOCS Press, Champaign, IL, 317–50.

SANDERS, T.A.B., GLEASON, K., GRIFFIN, B. and MILLER, G.J. (2006) 'Influence of an algal triacylglycerol containing docosahexaenoic acid (22:6*n*-3) and docosapentaenoic acid (22:5*n*-6) on cardiovascular risk factors in healthy men and women', *Brit. J. Nutr.*, **95**, 525–31.

SERHAN, C.N. (2005) 'Novel eicosanoid and docosaniod mediators: resolvins, docosatrienes, and neuroprotectins', *Curr. Op. Clin. Nutr. Metab. Care*, **8**, 115–21.

© Woodhead Publishing Limited, 2013

SINCLAIR, A.J. and JAYASOORIYA, A. (2010) 'Nutritional aspects of single cell oils: applications of arachidonic acid and docosahexaenoic acid oils', In *Single Cell Oils*, Cohen, Z. and Ratledge, C. (eds), AOCS Press, Champaign, IL, 351–68.

SPOLAORE, P., JOANNIS-CASSAN, C., DURAN, E. and ISAMBERT, A. (2006) 'Commercial applications of microalgae', *J. Biosci. Bioeng.*, **101**, 87–96.

STREEKSTRA, H. (1997) 'On the safety of *Mortierella alpina* for the production of food ingredients, such as arachidonic acid', *J. Biotechnol.*, **65**, 153–65.

STREEKSTRA, H. (2010) 'Arachidonic acid: fermentative production by *Mortierella* fungi', In *Single Cell Oils*, 2nd edition, Cohen, Z. and Ratledge, C. (eds)., AOCS Press, Champaign, IL, 97–114.

TAN, C.K. and JOHNS, M.R. (1996) 'Screening of diatoms for heterotrophic eicosapentaenoic acid production', *J. Appl. Phycol.*, **8**, 59–64.

TOTANI, N., WATANABE, A. and OBA, K. (1987) 'An improved method of arachidonic acid production by *Mortierella* sp. S-17', *J. Jpn. Oil Chem. Soc.*, **36**, 328–31.

WANG, L., CHEN, W., FENG, Y. REN, Y., GU, Z. CHEN, H., WANG, H., THOMAS, M.J., ZHANG, B., BERQUIN, I.M., LI, Y., WU, J., ZHANG, H., SONG, Y., LIU, X., NORRIS, J.S., WANG, S., DU, P., SHEN, J., WANG, N., YANG, Y., WANG, W., FENG, L., RATLEDGE, C., ZHANG, H. and CHEN, Y.Q. (2011) 'Genome characterization of the oleaginous fungus *Mortierella alpina*', PLoS ONE, **6**, e28319.

WARD, O.P. and SINGH, A. (2005) 'Omega-3/6 fatty acids: alternative sources of production', *Proc. Biochem.*, **4**, 3627–52.

WEN, Z.Y. and CHEN, F. (2003) 'Heterotrophic production of eicosapentaenoic acid by microalgae', *Biotechnol. Adv.*, **21**, 273–94.

WEN, Z.Y. and CHEN, F. (2010) 'Production of eicosapentaenoic acid using heterotrophically grown microalgae,' In *Single Cell Oils*, 2nd edition, Cohen, Z and Ratledge, C. (eds), AOCS Press, Champaign, IL, 151–77.

WEN, Z.Y., JIANG, Y. and CHEN, F. (2002) 'High cell density culture of the diatom *Nitzschia laevis* for eicosapentaenoic acid production: fed-batch development', *Proc. Biochem.*, **37**, 1447–53.

WONG, M.K.M., TSUI, C.K.M., AU, D.W.T. and VRIJMOED, L.L.P. (2008) 'Docosahexaenoic acid production and ultrastructure of the thraustochyrid *Aurantiochytrium mangrovei* MP under high glucose concentrations', *Mycoscience*, **49**, 266–70.

WYNN, J.P. and RATLEDGE, C. (2005) 'Oils from microorganisms', In *Bailey's Industrial Oil and Fat Products*, 6th editon, *Edible oil and fat products: specialty oils and oil products*, Shahidi, F. (ed)., Wiley-Interscience, New Jersey, Vol. 3, 121–53.

WYNN, J.P. and RATLEDGE, C. (2006) 'Microbial production of oils and fats,' In *Food Biotechnology*, 2nd edition, Shetty, K., Paliyath, G., Pometto, A. and Levin, R.E. (eds), Taylor & Francis, Boca Raton, 443–72.

XUE, Z., YADAV, N.S. and ZHU, Q.Q. (2009) *Optimized Strains of* Yarrowia lipolytica *for High Eicosapentaenoic Acid Production*, US Patent Application 2009/0093543 A1, Apr. 9 2009.

ZELLER, S. (2005) 'Safety evaluation of single cell oils and the regulatory requirements for use as food ingredients,' In *Single Cell Oils*, Cohen, Z and Ratledge, C. (eds), AOCS Press, Champaign, IL, 161–81.

ZHANG, Y, ADAMS, I.P. and RATLEGE, C. (2009) 'Malic enzyme: the controlling activity for lipid production? Overexpresssion of malic enzyme in *Mucor circinelloides* leads to a 2.5-fold increase in lipid accumulation', *Microbiology*, **153**, 2013–25.

ZHU, Q., XUE, Z., YADAV, N., DAMUDE, H. *et al.* (2010) 'Metabolic engineering of an oleaginous yeast for the production of omega-3 fatty acids', In *Single Cell Oils*, 2nd edition, Cohen, Z. and Ratledge, C. (eds), AOCS Press, Champaign, IL, 51–73.

ZITTELLI, G.C., LAVISTA, F., BASTIANINI, A., RODOLFI, L. VINCENZINI, M., TREDICI, M.R. (1999) 'Production of eicosapentaenoic acid by *Nannochloropsis* sp. cultures in outdoor tubular photobioreactors', *Prog. Indust. Microbiol.*, **35**, 299–312.

© Woodhead Publishing Limited, 2013

20

Microalgae as sources of food ingredients and nutraceuticals

B. Klein and R. Buchholz, Friedrich-Alexander-Universität
Erlangen-Nürnberg, Germany

DOI: 10.1533/9780857093547.2.559

Abstract: This chapter discusses the potential of microalgae and cyanobacteria to meet the challenging demands of the food industry regarding the development of new nutraceuticals. Substances such as coenzyme Q_{10}, α-tocopherol, and phycobilliproteins exert bioactive activity of high interest not only for the drug industry, but also for the food industry. Phototrophic microorganisms are the primary producers of polyunsaturated fatty acids such as docosahexaenoic acid, which makes these organisms even more valuable as nutraceuticals.

Key words: cyanobacteria, human nutrition, nutraceuticals, microalgae.

20.1 Introduction

Microalgae and cyanobacteria are a nearly untapped group of at least 50 000 species (Behrens and Delente, 1991). Up to the present day, only a small number of them have been isolated, cultivated, and thus analysed with regard to their metabolic potential (Wijffels, 2008). With this hitherto unexploited capacity, aquatic microorganisms have the potential to meet the challenging demands of both the food and the pharmaceutical industry for the development of new nutraceuticals or drugs. Utilization of phototrophic microorganisms can range from simple usage of biomass for food and animal feed to high value products used as pharmaceuticals or in diagnostics. Therefore, aquatic microorganisms have the ability to biosynthesize secondary metabolites like antioxidants, antivirals as well as anti-inflammatory or antibiotic agents (Chu *et al.*, 2004; Guzmán *et al.*, 2001; Matsukawa *et al.*, 2000; Rechter *et al.*, 2006). Substances such as coenzyme Q_{10}, α-tocopherol, sulfated polysaccharides, sulfoquinovosyldiacylglycerides, and phycobilliproteins exert bioactive activity of high interest not only for the drug industry, but also for the cosmetic and food industry (Arad *et al.*, 1985, 1988; Gantt, 1969; Guil-Guerrero *et al.*, 2000; Naumann, 2009; Klein *et al.*,

© Woodhead Publishing Limited, 2013

2011; Durmaz *et al.*, 2007). Microalgae and cyanobacteria are the primary producers of polyunsaturated fatty acids such as arachidonic acid, eicosapentaenoic acid or docosahexaenoic acid, which makes these organisms also valuable as nutraceuticals (Nuutila *et al.*, 1997; Fábregas *et al.*, 1998; Yongmanitchai and Ward, 1991; Aaronson and Bensky, 1967; De Swaaf *et al.*, 2003; Cohen, 1999; Otero *et al.*, 1997; Giménez Giménez *et al.*, 1998; Molina Grima *et al.*, 1996).

20.2   Microalgae and cyanobacteria and their potential as food supplements

Owing to the growing interest in functional foods, aquatic microorganisms such as microalgae and cyanobacteria have become more and more interesting for the food industry. High value products from microalgae can be applied either as health foods or food supplements. Microalgae and cyanobacteria biosynthesize several high value products like vitamins, ω-3 fatty acids (which will be discussed in a separate section), proteins, pigments, and antioxidants which makes them of high interest for application as food additives (Pulz and Gross, 2004). The availability of substances valuable for food industry varies within the group of microalgae. An overview of different microalgae products suitable for the food industry is given in Table 20.1.

Examples for the use of the biomass itself as functional food are *Dunaliella, Chlorella* or *Arthrospira. Dunaliella* powder is sold as a nutraceutical by Nutrimed Pty Ltd or as capsules by ICL Health. *Arthrospira* and *Chlorella* algae are widely distributed and sold as capsules or tablets by different companies including Earthrise Nutrition, Sanatur, Cyanotech or GSE Naturland for human nutrition.

Use of *Arthrospira* as food has long tradition going back about more than 700 years (Fig. 20.1). The Aztecs harvested *Arthrospira* from Lake Texcocco. Even before that time, the Kanembu tribe in the Lake Chad area

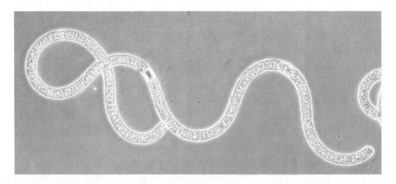

Fig. 20.1   Microscopic record of *Arthrospira platensis*.

© Woodhead Publishing Limited, 2013

Table 20.1 Selection of microalgal species and their products and application in human nutrition; modified from Carlsson et al. (2007)

Species	Product	Application areas	References
Spirulina (Arthrospira platensis)	Phycocyanin, biomass	Health food	Lee (2001); Costa et al., (2003)
Chlorella vulgaris	Biomass	Health food, food supplement	Lee (2001)
Dunaliella salina	Carotenoids, β-carotene	Health food, food supplement	Jin et al. (2003); Del Campo et al. (2007)
Haematococcus pluvialis	Carotenoids, astaxanthin	Health food	Del Campo et al. (2007)
Odontella aurita	Fatty acids	Baby food	Pulz and Gross (2004)
Porphyridium cruentum	Polysaccharides	Human nutrition	Rebolloso Fuentes et al. (1999)
Isochrysis galbana	Fatty acids	Nutrition (animal)	Molina Grima et al. (2003); Pulz and Gross (2004)
Phaedactylum tricornutum	Lipids, fatty acids	Human nutrition	Yongmanitchai and Ward (1991); Acién Fernández et al. (2003)
Lyngbya majuscule	Immune modulators	Human nutrition	Singh et al. (2005)

© Woodhead Publishing Limited, 2013

in Chad used *Arthrospira* as dried bread called 'dihe', which is eaten until nowadays (Abdulqader *et al.*, 2000). Several studies revealed the immunomodulating capacity of *Arthrospira* making this cyanobacteria an even more valuable nutraceutical. These features have been proven for mice, rats, chicken, prawns, catfish and humans (Belay, 2002). Corresponding investigations showed an increase in T-cell proliferation, enhanced activity of phagocytes, improved IFN-γ (interferon-γ) secretion activity, and NK-cell (natural killer-cell) cell damage activities (Belay, 2002).

20.2.1   Pigments

Pigments such as ß-carotene, astaxanthin, phycoerythrin or phycocyanin can be used as colouring agents, whereas ß-carotene and phycocyanin are also promising antioxidants with potential applications as food additives (Hirata *et al.*, 2000; Pulz and Gross, 2004; Palozza and Krinsky, 1991). Carotenoids from microalgae are valuable colorants for food (Denery *et al.*, 2004). Investigations of carotenoids have been performed intensively during the past decades. The total amount of ß-carotene (Fig. 20.2) can amount up to 10% dry weight from *Dunaliella salina* cultivated under light stress in a tubular outdoor photobioreactor (García-González *et al.*, 2005). The biosynthesis of astaxanthin (Fig 20.3) can account for up to 4% dry weight under nitrogen and phosphate starvation, resulting in commercialization of these products (Boussiba, 2000; Hosseini Tafreshi and Shariati, 2009; Guerin *et al.*, 2003).

Carotenoids represent a huge market of microalgae products including mainly ß-carotene and astaxanthin. ß-Carotene from *Dunaliella* is produced by several companies including Henkel-Cognis Nutrition and Health,

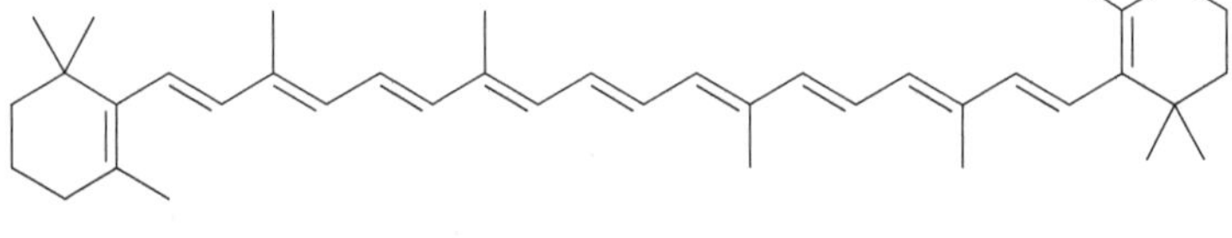

Fig. 20.2   β-carotene.

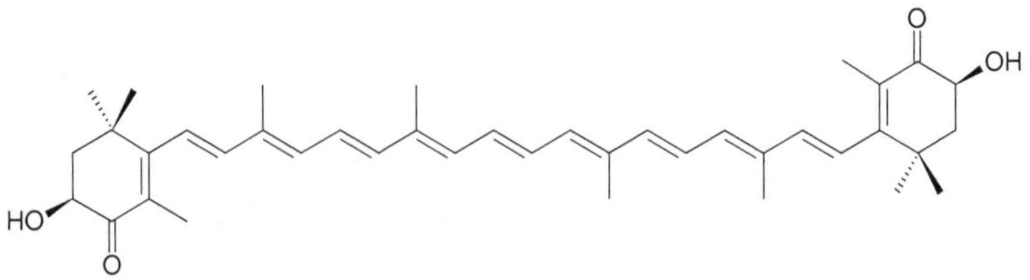

Fig. 20.3   Astaxanthin.

© Woodhead Publishing Limited, 2013

Cyanotech, Inner Mongolia Biological Eng., Nature Beta Technologies, Tianjin Lantai Biotechnology and Dutch State Mines (DSM; Food Specialities) (Raja *et al.*, 2007). Commercialization of astaxanthin produced by *Haematococcus pluvialis* has been carried out for some years by different companies such as Fuji Chemical Industry Co., Cyanotech, Algatech and Parry Nutraceuticals.

Phycobilliproteins serve as colorants for different applications. The phycocyanin concentration in *A. platensis* can amount 3.2 g l^{-1}. Phycocyanin, for example, replaced a synthetic blue dye, which is thought to cause attention deficit hyperactivity disorder and was used to give smarties a blue colour (Michel, 2011). Blue 'Gummy Bears' are now coloured with *Arthrospira* concentrates in order to substitute synthetic dyes (Trolli). In Japan chewing gums, ice sherberts, popsicles, candies, soft drinks, dairy products and wasabi are tinted with a colorant from *Arthrospira* which was developed by Dainippon Ink & Chemicals (Henrikson, 2010).

20.2.2   Vitamins

Some cyanobacteria such as *Arthrospira* (*Spirulina*), *Aphanizomenon* and *Nostoc* are produced, among others, for the food industry with a total annual output of up to 3000 tons (Watanabe, 2007). Since the 1990s, *Arthrospira* has been known to be a useful and superior source of vitamins. Vitamin A from *Arthrospira* showed more of a beneficial effect compared to synthetically produced vitamin A, since the uptake of the *Arthrospira* vitamin A was identified to be more effective (Annapurna *et al.*, 1991). *Arthrospira* is also known to be a source of other vitamins like vitamin B$_{12}$. Commercially available *Arthrospira* contains up to 244 µg vitamin B$_{12}$ per g dry weight (Watanabe *et al.*, 1999; Watanabe, 2007). Nevertheless, the major part of this vitamin is thought to be pseudovitamin B$_{12}$ (adeninly cobamide), which might barely be absorbed by the mammalian intestine, whereas more investigations have to be performed for ensured prediction (Watanabe, 2007). The predominant form of vitamin B$_{12}$ in *Aphanothece sacrum* is also pseudovitamin B$_{12}$ and its availability has therefore to be regarded as doubtful.

Other microalgae, such as *Tetraselmis suecica*, *Isochrysis galbana*, *Dunaliella tertiolecta* or *Chlorella stigmatophora* are a precious source of vitamins, containing higher or similar concentrations of vitamins (except ascorbic acid and pyridoxin) than oranges (see Table 20.2). A comparison with other foods like carrots, wheat flour, corn flour or soy flour showed these algae to be a valuable source of vitamins (Fabregas and Herrero, 1990).

The cyanobacterium *Aphanizomenon flos aquae* (AFA algae) has been identified to be another source of vitamin B$_{12}$, in which no pseudovitamin could be found. Nevertheless, no human study is known that deals with the vitamin B$_{12}$ uptake of this cyanobacterium in humans (Watanabe, 2007).

© Woodhead Publishing Limited, 2013

Table 20.2   Vitamin content of four different microalgal species in comparison with orange modified after: (Fabregas and Herrero, 1990). Vitamin A is given in IU/kg dry weight (1 IU = 0.6 µg of β-carotene). The other vitamins are given in mg/kg dry weight

	Tetraselmis suecica	*Isoehrysis galbana*	*Dunaliella tertiolecta*	*Chlorella stigmatophora*	Orange
Vitamin A	493,750	127,500	137,500	82,300	14,728
Tocopherol E	421.8	58.2	116.3	699.0	17.8
Thiamin (B1)	32.3	14.0	29.0	14.6	6.2
Riboflavin (B2)	19.1	30.0	31.2	19.6	2.3
Pyridoxin (B6)	2.8	1.8	2.2	1.9	9.3
Cobalamin (B12)	0.5	0.6	0.7	0.6	0.0
Folic acid	3.0	3.0	4.8	3.1	2.7
Nicotinic acid	89.3	77.7	79.3	82.5	15.5
Pantothenic acid	37.7	9.1	13.2	21.4	15.5
Biotin (H)	0.8	1.0	0.9	1.1	0.1
Ascorbic acid	191.0	119.0	163.2	100.2	3798.0

20.2.3   Proteins

Apart from lipids, vitamins or pigments, phototrophic microorganisms have the ability to biosynthesize large quantities of proteins, which can amount to up to 70% of dry weight under optimized cultivation conditions, making microalgae a superior source of proteins, for example compared to soy (Becker, 2004; Fabregas and Herrero, 1985; Paoletti *et al.*, 1980; Santillan, 1982). The protein concentrations of different microalgae range from 28% dry matter (*Porphyridium cruentum*) to 71% dry matter (*Spirulina maxima*). Comparing these values with human food sources which range from 8% (rice) to 43% (meat) microalgae can be considered as a valuable protein source in comparison to other foods (Becker, 2004).

The phototrophic cyanobacterium *A. platensis* is used as dietary supplement owing to its high protein content (Cyanotech Corporation, Hawaii, USA). The basic amino acid composition consists of isoleucine, leucine, lysine, methionine, phenylalanine, threonine, tryptophan, and valine (Santillan, 1982). These amino acids can be found as well in other microalgae like *Chlorella* spp., *I. galbana*, *Dunaliella* spp. and *T. suecica* and their levels in such algae meet World Health Organisation (WHO) requirements for human nutrition (Fabregas and Herrero, 1985).

20.2.4   Antioxidants

In the 1950s free radicals were identified to be responsible for ageing phenomena in humans (Harman, 1956). Damaged macromolecules caused by radicals can be prevented by antioxidants from microalgae that are consumed as nutraceuticals. Substances such as vitamin E, coenzyme Q_{10} and

© Woodhead Publishing Limited, 2013

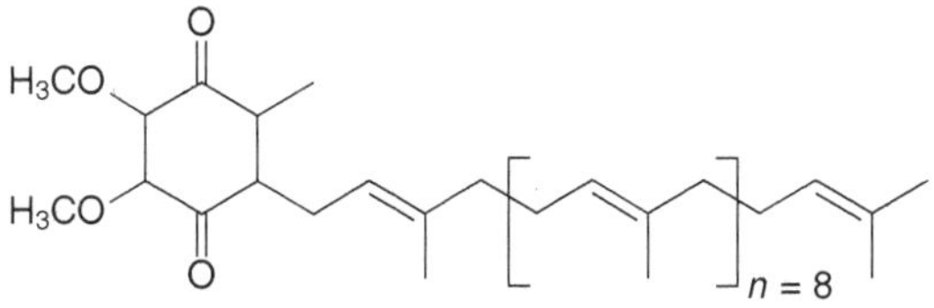

Fig. 20.4   Coenzyme Q_{10}.

astaxanthin as well as ß-carotene are biosynthesized by different micro-algae (Yaşar, 2007; Klein *et al.*, 2011).

Coenzyme Q_{10} is a component of the human mitochondrial respiratory chain and acts as an antioxidant, suppressing the formation of free radicals (Suzuki and King, 1983, Berg *et al.*, 2003). Coenzyme Q_{10}-concentration in the red microalga *Porphyridium purpureum* can amount up to 1.96 mg l^{-1} (Klein, 2010), making this alga a suitable source of antioxidative compounds and a useful nutraceutical.

Besides coenzyme Q_{10}, astaxanthin is a valuable antioxidant biosynthe-sized by microalgae. The production of astaxanthin by *Haematoccocus* has been commercialized by different companies such as Fuji Chemical Indus-try Co., Japan or Algatech, Israel. The main use of astaxanthin is still as an animal feed supplement, but astaxanthin as a supplement for human nutrition is a promising market as astaxanthin shows antioxidative activi-ties and can be used as a colouring agent (Kurashige *et al.*, 1990). Even though it is mainly used as feed colorant nowadays, the application as a natural colorant for human food is a possible application of this carotenoid owing to the fact that astaxanthin does not possess any health risk up to a dosage of about 20 mg per day (MeraPharmaceuticals, 1999; Manimeg-alai *et al.*, 2010).

In addition to astaxanthin phycocyanin which is, for example, produced by the cyanobacteria *A. platensis* also indicates antioxidative activity com-parable to the antioxidative capacity of α-tocopherol. Investigations of inhibition of the peroxidation of methyl linoleate and the oxidation of phosphatidylcholine liposomes demonstrated that phycocyanin is an effec-tive antioxidant which is capable of preventing initiation of radical chain reactions (Hirata *et al.*, 2000).

Another very important antioxidant produced by various microalgae is α-tocopherol, also known as vitamin E. α-Tocopherol producers are, among others, *D. tertiolecta*, *T. suecica*, *Euglena gracilis* and *Anabaena variabilis* (Carballo-Cárdenas *et al.*, 2003; Ogbonna *et al.*, 1999; Dasilva and Jensen, 1971). In humans α-tocopherol is supposed to be responsible for preventing light-induced pathologies of the skin and eyes, but it is also believed to obviate diseases such as atherosclerosis or cancer (Carballo-Cárdenas *et al.*, 2003). Therefore, α-tocopherol is crucial in human diets (Shintani and Del-laPenna, 1998).

© Woodhead Publishing Limited, 2013

20.3   Risks of microalgal products

Considering the use of microalgae as nutraceuticals, one issue nevertheless has to be taken into account: biomass of microalgae is mainly produced using open ponds which harbour the risk of contamination. Analysis of *Chlorella* products using 18s rDNA showed the existence of more than one species per product. Furthermore, many of the investigated algae samples included critical concentrations of bacteria. In addition, some of the *Chlorella* samples also contained cyanobacterial contaminations (Görs *et al.*, 2010).

20.4   Conclusion

Microalgae, as well as microalgal products, can be regarded as potential and valuable resources for human nutrition and nutraceuticals. Metabolites from phototrophic microorganisms exhibit a positive influence on human health, which makes those organisms precious. Besides using, for example, pigments as colourants, microalgal products are health-promoting agencies.

One important issue has to be considered when using microalgae and cyanobacteria and their products as nutraceuticals. These products have to be free from unhealthy by-products. This is especially important for consumption of microalgae on the whole. When microalgae are cultivated on a large scale, the biomass consists of concomitant organisms, which might make them less worthy than they actually appear.

20.5   References

AARONSON, S. and BENSKY, B. (1967). 'Effect of aging of a cell population in lipids and drug resistance in *Ochromonas danica*'. *Journal of Protozoologica*, **14**, 76–8.

ABDULQADER, G., BARSANTI, L. AND TREDICI, M.R. (2000). 'Harvest of *Arthrospira platensis* from Lake Kossorom (Chad) and its household usage among the Kanembu'. *Journal of Applied Phycology*, **12**, 493–8.

ACIÉN FERNÁNDEZ, F.G., HALL, D.O., CAÑIZARES GUERRERO, E., KRISHNA RAO, K. and MOLINA GRIMA, E. (2003). 'Outdoor production of *Phaeodactylum tricornutum* biomass in a helical reactor'. *Journal of Biotechnology*, **103**, 137–52.

ANNAPURNA, V., DEOSTHALE, Y. and BAMJI, M. (1991). 'Spirulina as a source of vitamin A'. *Plant Foods for Human Nutrition (formerly Qualitas Plantarum)*, **41**, 125–34.

ARAD, S., ADDA, M. and COHEN, E. (1985). 'The potential of production of sulfated polysaccharides from *Porphyridium*'. *Plant and Soil*, **89**, 117–27.

ARAD, S.M., FRIEDMAN, O.D. and ROTEM, A. (1988). 'Effect of nitrogen on polysaccharide production in a *Porphyridium* sp'. *Applied Environmental Microbiology*, 2411–4.

BECKER, W. (2004). 'Microalgae in human and animal nutrition'. In: *Handbook of Microalgal Culture*. Richmond, A. (ed.). Blackwell Publishing, Oxford.

© Woodhead Publishing Limited, 2013

BEHRENS, P.W. and DELENTE, J.J. (1991). 'Microalgae in the pharmaceutical industry'. *Biological and Pharmaceutical Bulletin*, **4**, 54–8.

BELAY, A. (2002). 'The potential application of spirulina (*Arthrospira*) as a nutritional and therapeutic supplement in health management'. *The Journal of the American Nutraceutical Association*, **5**, 27–49.

BERG, J.M., TYMOCZKO, J.L. and STRYER, L. (2003). '*Biochemie*, Spektrum Akademischer Verlag, Heidelberg - Berlin.

BOUSSIBA, S. (2000). 'Carotenogenesis in the green alga *Haematococcus pluvialis*: Cellular physiology and stress response'. *Physiologia Plantarum*, **108**, 111–7.

CARBALLO-CÁRDENAS, E.C., TUAN, P.M., JANSSEN, M. and WIJFFELS, R.H. (2003). 'Vitamin E (α-tocopherol) production by the marine microalgae *Dunaliella tertiolecta* and *Tetraselmis suecica* in batch cultivation'. *Biomolecular Engineering*, **20**, 139–47.

CARLSSON, A., BEILEN VAN, J., MÖLLER, R., CLAYTON, D. and BOWLES, D. (2007). 'Micro and macroalgae – utility for industrial applications Bioproducts, E.R.t.E.P.o.S.R. and Crops, f-N.-f., CNAP, University of York: 86.

CHU, C.Y., LIAO, W.R., HUANG, R. and LIN, L.P. (2004). 'Haemagglutinating and antibiotic activities of freshwater microalgae'. *World Journal Microbiology and Biotechnology*, **20**, 817–25.

COHEN, Z. (1999). '*Porphyridium cruentum*'. In: *Chemicals from Microalgae*. Cohen, Z. (ed.). Taylor and Francis, Philadelphia.

COSTA, J.A.V., COLLA, L.M. and PAULO DUARTE FILHO (2003). '*Spirulina platensis* growth in open raceway ponds using fresh water supplemented with carbon, nitrogen and metal ions'. *Zeitschrift fuer Naturforschung*, **58c**, 76–80.

DASILVA, E.J. and JENSEN, A. (1971). 'Content of α-tocopherol in some blue-green algae'. *Biochimica et Biophysica Acta (BBA) – Lipids and Lipid Metabolism*, **239**, 345–7.

DE SWAAF, M.E., SIJTSMA, L. and PRONK, J.T. (2003). 'High-cell-density fed-batch cultivation of the docosahexaenoic acid producing marine alga *Crypthecodinium cohnii*'. *Biotechnology and Bioengineering*, **81**, 666–72.

DEL CAMPO, J.A., MERCEDES GARCÍA-GONZÁLEZ, M. and GUERRERO, M.G. (2007) 'Outdoor cultivation of microalgae for carotenoid production: current state and perspectives'. *Applied Microbiology and Biotechnology*, **74**, 1163–74.

DENERY, J.R., DRAGULL, K., TANG, C.S. and LI, Q.X. (2004). 'Pressurized fluid extraction of carotenoids from *Haematococcus pluvialis* and *Dunaliella salina* and kavalactones from *Piper methysticum*'. *Analytica Chimica Acta*, **501**, 175–81.

DURMAZ, Y., MONTEIRO, M., BANDARRA, N., GÖKPINAR, Ş. and IŞIK, O. (2007). 'The effect of low temperature on fatty acid composition and tocopherols of the red microalga, *Porphyridium cruentum*'. *Journal of Applied Phycology*, **19**, 223–227.

FABREGAS, J. and HERRERO, C. (1985). 'Marine microalgae as a potential source of single cell protein (SCP)'. *Applied Microbiology and Biotechnology*, **23**, 110–3.

FABREGAS, J. and HERRERO, C. (1990). 'Vitamin content of four marine microalgae. Potential use as source of vitamins in nutrition'. *Journal of Industrial Microbiology and Biotechnology*, **5**, 259–63.

FÁBREGAS, J., GARCÍA, D., MORALES, E., DOMÍNGUEZ, A. and OTERO, A. (1998). 'Renewal rate of semicontinuous cultures of the microalga *Porphyridium cruentum* modifies phycoerythrin, exopolysaccharide and fatty acid productivity'. *Journal of Fermentation and Bioengineering*, **86**, 477–81.

GANTT, E. (1969). 'Properties and ultrastructure of phycoerythrin from *Porphyridium cruentum*'. *Plant Physiology*, **44**, 1629–38.

GARCÍA-GONZÁLEZ, M., MORENO, J., MANZANO, J.C., FLORENCIO, F.J. and GUERRERO, M.G. (2005). 'Production of *Dunaliella salina* biomass rich in 9-*cis*-β-carotene and lutein in a closed tubular photobioreactor'. *Journal of Biotechnology*, **115**, 81–90.

© Woodhead Publishing Limited, 2013

GIMÉNEZ GIMÉNEZ, A., IBÁÑEZ GONZÁLEZ, M.J., ROBLES MEDINA, A., MOLINA GRIMA, E., GARCÍA SALAS, S. and ESTEBAN CERDÁN, L. (1998). 'Downstream processing and purification of eicosapentaenoic (20:5n-3) and arachidonic acids (20:4n-6) from the microalga *Porphyridium cruentum*'. *Bioseparation*, **7**, 89–99.

GÖRS, M., SCHUMANN, R., HEPPERLE, D. and KARSTEN, U. (2010). 'Quality analysis of commercial Chorella products used as dietary supplement in human nutrition'. *Journal of Applied Phycology*, **22**, 265–76.

GUERIN, M., HUNTLEY, M.E. and OLAIZOLA, M. (2003). '*Haematococcus astaxanthin*: applications for human health and nutrition'. *Trends in Biotechnology*, **21**, 210–6.

GUIL-GUERRERO, J.L., BELARBI, E.-H. and REBOLLOSO-FUENTES, M.M. (2000). 'Eicosapentaenoic and arachidonic acids purification from the red microalga *Porphyridium cruentum*'. *Bioseparation*, **9**, 299–306.

GUZMÁN, S., GATO, A. and CALLEJA, J.M. (2001). 'Antiinflammatory, analgesic and free radical scavenging activities of the marine microalgae *Chlorella stigmatophora* and *Phaeodactylum tricornutum*'. *Phytotherapy Research*, **15**, 224–30.

HARMAN, D. (1956). 'Aging: a theory based on free radical and radiation chemistry'. *Journal of Gerontology*, **11**, 298–300.

HENRIKSON, R. (2010). '*Spirulina – World Food: How this Micro Algae can Transform your Health and Our Planet*'. CreateSpace Independent Publishing Platform (see also http://www.amazon.com/Spirulina-World-transform-health-planet/dp/1453766987)

HIRATA, T., TANAKA, M., OOIKE, M., TSUNOMURA, T. and SAKAGUCHI, M. (2000). 'Antioxidant activities of phycocyanobilin prepared from *Spirulina platensis*'. *Journal of Applied Phycology*, **12**, 435–9.

HOSSEINI TAFRESHI, A. and SHARIATI, M. (2009). 'Dunaliella biotechnology: methods and applications'. *Journal of Applied Microbiology*, **107**, 14–35.

JIN, E.S., FETH, B. and MELIS, A. (2003). 'A mutant of the green alga *Dunaliella salina* constitutively accumulates zeaxanthin under all growth conditions.' *Biotechnology and Bioengineering*, **81**, 115–24.

KLEIN, B.C. (2010). '*Strukturbasiertes Screening zur Gewinnung von Co-Enzmy Q10 aus phototrophen Mikroorganismen*'. Dissertation, Universität Erlangen-Nürnberg.

KLEIN, B.C., BARTEL, S.J., DARSOW, K.H., NAUMANN, I., WALTER, C., BUCHHOLZ, R. and LANGE, H.A. (2011). 'Identification of coenzyme Q_{10} from *Porphyridium purpureum* (Rhodophyta) by MALDI curve field reflectron mass spectrometry'. *Journal of Phycology*, **47**, 687–91.

KURASHIGE, M., OKIMASU, E., INOUE, M. and UTSUMI, K. (1990). 'Inhibition of oxidative injury of biological membranes by astaxanthin'. *Physiological Chemistry and Physics and Medical NMR*, **22**, 27–38.

LEE, Y.-K. (2001). 'Microalgal mass culture systems and methods: Their limitation and potential'. *Journal of Applied Phycology*, **13**, 307–15.

MANIMEGALAI, M., BUPESH, G., MIRUNALINI, M., VASANTH, S., KARTHIKEYINI, S. and SUBRAMANIAN, P. (2010). 'Color enhancement studies on *Etroplus maculatus* using astaxanthin and β-carotene'. *International Journal of Environmental Sciences*, **1**, 403–18.

MATSUKAWA, R., HOTTA, M., MASUDA, Y., CHIHARA, M. and KARUBE, I. (2000). 'Antioxidants from carbon dioxide fixing *Chlorella sorokiniana*'. *J. Appl. Phycol.*, **12**, 263–7.

MERAPHARMACEUTICALS (1999). 'Haematococcus pluvialis *and Astaxanthin Safety for Human Consumption*'. Technical Report TR.3005.001.

MICHEL, M. (2011). 'Natürliche Lebensmittel – Ein nachhaltiger Trend?' *Kooperationsforum – Natürliche Inhaltsstoffe*. Traunreut, Germany: Bayern Innovativ.

© Woodhead Publishing Limited, 2013

MOLINA GRIMA, E., ROBLES MEDINA, A., GIMÉNEZ GIMÉNEZ, A. and IBÁÑEZ GONZÁLEZ, M.J. (1996). 'Gram-scale purification of eicosapentaenoic acid (EPA, 20:5n-3) from wet *Phaeodactylum tricornutum* UTEX 640 biomass'. *Journal of Applied Phycology*, **8**, 359–67.

MOLINA GRIMA, E., BELARBI, E.H., ACIÉN FERNÁNDEZ, F.G., ROBLES MEDINA, A. and CHISTI, Y. (2003). 'Recovery of microalgal biomass and metabolites: process options and economics'. *Biotechnology Advances*, **20**, 491–515.

NAUMANN, I. (2009). '*Sulfoquinovosyldiacylglyceride – antiviral aktive Substanzen*. Dissertation. Universität Erlangen-Nürnberg.

NUUTILA, A.M., AURA, A.-M., KIESVAARA, M. and KAUPPINEN, V. (1997). 'The effect of salinity, nitrate concentration, pH and temperature on eicosapentaenoic acid (EPA) production by the red unicellular alga *Porphyridium purpureum*'. *Journal of Biotechnology*, **55**, 55–63.

OGBONNA, J.C., TOMIYAMA, S. and TANAKA, H. (1999). 'Production of α-tocopherol by sequential heterotrophic–photoautotrophic cultivation of *Euglena gracilis*'. *Journal of Biotechnology*, **70**, 213–21.

OTERO, A., GARCÍA, D. and FÁBREGAS, J. (1997). 'Factors controlling eicosapentaenoic acid production in semicontinuous cultures of marine microalgae'. *Journal of Applied Phycology*, **9**, 465–69.

PALOZZA, P. and KRINSKY, N.I. (1991). 'The inhibition of radical-initiated peroxidation of microsomal lipids by both α-tocopherol and β-carotene'. *Free Radical Biology and Medicine*, **11**, 407–14.

PAOLETTI, C., VINCENZINI, M., BOCCI, F. and MATERASSI, R. (1980). 'Composizione biochimica generale delle biomasse di *Spirulina platensis* e *S. maxima*'. In: *Prospettive della coltura di Spirulina in Italia*. Materassi, R. (ed.). Rome, Consiglio Nazionale delle Ricerche.

PULZ, O. and GROSS, W. (2004). 'Valuable products from biotechnology of microalgae'. *Applied Microbiology and Biotechnology*, **65**, 635–48.

RAJA, R., HEMAISWARYA, S. and RENGASAMY, R. (2007). 'Exploitation of Dunaliella for β-carotene production'. *Applied Microbiology and Biotechnology*, **74**, 517–23.

REBOLLOSO FUENTES, M.M., GARCIA SÁNCHEZ, J.L., FERNÁNDEZ SEVILLA, J.M., ACIÉN FERNÁNDEZ, F.G., SÁNCHEZ PÉREZ, J.A. and MOLINA GRIMA, E. (1999). 'Outdoor continuous culture of *Porphyridium cruentum* in a tubular photobioreactor: quantitative analysis of the daily cyclic variation of culture parameters'. *Marine Bioprocess Engineering*, **35**, 271–88.

RECHTER, S., KÖNIG, T., AUEROCHS, S., THULKE, S., WALTER, H., DÖRNENBURG, H., WALTER, C. and MARSCHALL, M. (2006). 'Antiviral activity of *Arthrospira*-derived spirulan-like substances'. *Antiviral Research*, **72**, 197–206.

SANTILLAN, C. (1982). 'Mass production of *Spirulina*'. *Cellular and Molecular Life Sciences*, **38**, 40–3.

SHINTANI, D. and DELLAPENNA, D. (1998). 'Elevating the vitamin E content of plants through metabolic engineering'. *Science*, **282**, 2098–100.

SINGH, S., BHUSHAN, N.K. and BANERJEE, U.C. (2005). 'Bioactive compounds from cyanobacteria and microalgae: an overview'. *Critical Reviews in Biotechnology*, **25**, 73–95.

SUZUKI, H. and KING, T.E. (1983). 'Evidence of an ubisemiquinone radical(s) from the NADH-ubiquinone reductase of the mitochondrial respiratory chain'. *Journal of Biological Chemistry*, **258**, 352–8.

WATANABE, F. (2007). 'Vitamin B12 sources and bioavailability'. *Exp Biol Med*, **232**, 1266–74.

WATANABE, F., KATSURA, H., TAKENAKA, S., FUJITA, T., ABE, K., TAMURA, Y., NAKATSUKA, T. and NAKANO, Y. (1999). 'Pseudovitamin B12 is the predominant cobamide of an algal health food, spirulina tablets'. *Journal of Agricultural and Food Chemistry*, **47**, 4736–41.

© Woodhead Publishing Limited, 2013

WIJFFELS, R.H. (2008). 'Potential of sponges and microalgae for marine biotechnology'. *TIBTECH*, **26**, 26–31.

YAŞAR, D. (2007). 'Vitamin E (α-tocopherol) production by the marine microalgae *Nannochloropsis oculata* (Eustigmatophyceae) in nitrogen limitation'. *Aquaculture*, **272**, 717–22.

YONGMANITCHAI, W. and WARD, O.P. (1991). 'Growth of and omega-3 fatty acid production by *Phaeodactylum tricornutum* under different culture conditions'. *Applied Environmental Microbiology*, **57**, 419–25.

© Woodhead Publishing Limited, 2013

21

Microbial production of vitamins

R. Ledesma-Amaro, M. A. Santos, A. Jiménez and J. L. Revuelta,
University of Salamanca, Spain

DOI: 10.1533/9780857093547.2.571

Abstract: This chapter is a short review of the production of vitamins with the main focus on the microbial fermentation processes that are currently being employed in the production of vitamins. The state of the art of the industry of vitamins is extensively analysed with a description of both chemical and biotechnological systems that have been developed for the production of each vitamin. Finally, future trends in the microbial production of vitamins are analysed and the recent advances in strain design for whole-cell production of vitamins are discussed.

Key words: biotechnology, metabolic engineering, microbial production, vitamin, vitamin production.

21.1 Introduction

Vitamins are defined as organic compounds that are essential for normal growth and nutrition and are required in small quantities in the diet because they cannot be synthesized by the body. According to their chemical nature, they can be divided into two groups: water-soluble vitamins and fat-soluble vitamins. These vitamins are naturally synthesized in microorganisms and plants and are essential for the metabolism of all living organisms.

The absence of sufficient amounts of these compounds in the diet leads to different health problems, which is especially important in cultures in which a varied diet is lacking, and this lack affects not only humans but also farm animals, with a huge economic impact. Thus, over the past decades, a vast industry relating to the production of vitamins has been successfully developed around the world and today vitamins are produced industrially and used widely not only as food and feed additives, but also as cosmetics, therapeutic agents and health and technical aids.

Traditionally, such vitamins have been produced by organic chemical synthesis, but this often requires a high number of reactions, using expensive devices as well as solvents that are usually undesirable pollutants harmful

© Woodhead Publishing Limited, 2013

to the environment. To overcome these drawbacks, biological production of vitamins has been developed, by identifying natural producer organisms, finding the most profitable culture conditions, scaling up production and optimizing downstream processes to extract the pure product. Nevertheless, the intrinsic limitations of most natural producers make some vitamin titers unable to compete with chemical synthesis. Thus, biotechnology emerges as an environmentally friendly way to increase vitamin production, enriching natural sources or creating new ones that are more suitable for industrial purposes. In this sense, 'green' biotechnology can produce crops with high vitamin contents, which can be extracted or used directly as a food source of vitamins, while 'white' biotechnology is able to modify microorganisms through genetic and metabolic engineering, turning them into vitamin producers (see Part I of this book).

Microbial production has several advantages since microbes grow fast, they do not have to rely on climatic conditions and seasons, they can be scaled up readily and they are not in competition with human food needs. Additionally, many disciplines and techniques are being developed by the scientific community to attempt to encourage the rapid improvement of 'white' biotechnology, such as metabolic engineering, bioinformatics, flux-omics, metabolomics, systems biology, synthetic biology, and so on. Most vitamins are still produced chemically, but the number of microbiologically produced products is increasing with time and, currently, lab-scale approaches to the modification of microorganisms with a view to producing large amounts of all kinds of vitamins are under way.

Throughout this chapter, we shall explore each vitamin within a group, depending on its chemical nature: water-soluble or fat-soluble. Brief basic notions about the vitamin molecule, its biological functions and dietary intake are discussed and large-scale production is addressed, followed by some notes from some lab-scale studies that have provided promising results in terms of the microbial production of vitamins. Some previous chapters and reviews addressing similar topics can be considered for further reading.[1-6]

21.2   Fat-soluble vitamins

21.2.1   Vitamin and pro-vitamin A

Vitamin A is a group formed by different retinoids, retinol, retinal, retinoic acid and retinyl esters. Pro-vitamin A is composed of various carotenoids, the most important one being beta-carotene, while others are alpha-carotene and beta-cryptoxanthin. In our bodies, these pro-vitamins can be converted to retinal and retinoic acid, which are the active forms of vitamin A (Fig. 21.1).

Vitamin A is involved in many functions such as in the immune system, vision, reproduction, cellular communication, cell growth and

© Woodhead Publishing Limited, 2013

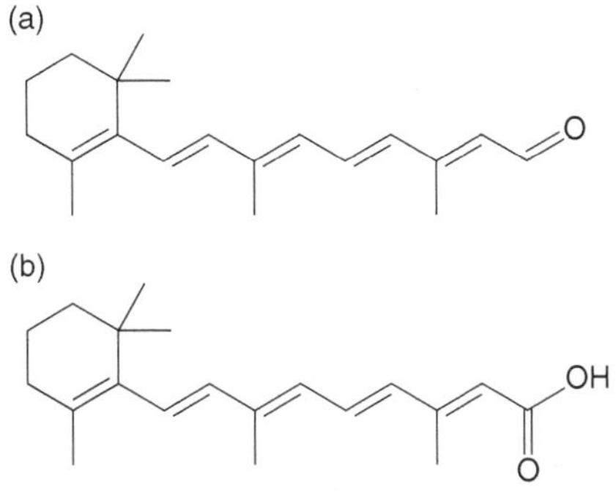

Fig. 21.1   Chemical structure of retinal (a) and retinoic acid (b).

differentiation, and it is critical for the normal development and maintenance of the heart, lungs, kidneys and other organs. It is essential for vision
since it is a key component of retinal receptors.

In the diet both vitamin A and pro-vitamin A can be obtained from food.
Two pre-formed vitamin forms are currently available in daily-consumed
products – retinol and retinyl ester – and these can be found in fish, meat,
milk and eggs, with higher concentrations in fish oil and liver. Pro-vitamins
are usually plant pigments found in leafy greens and orange and yellow
vegetables. Both pro-vitamin A and vitamin A must be metabolized to
retinal and retinoic acids.

At the industrial scale, vitamin A is mainly produced chemically (2700 t
per annum (t/a), Hoffmann-La Roche, BASF, Rhône-Poulenc), as is beta-
carotene (400 t/a, Hoffmann-La Roche, BASF).[2] Some microbial processes
have been put to efficient use, using the green microalga *Dunaliella* and the
fungus *Blakeslea trispora* to produce beta-carotene. Under certain conditions, *Dunaliella* can accumulate more than 0.1 g g^{-1} dry cells (20–30% salts,
nitrogen limited, 10,000 lux and 25–27°C for 3 months).[7] *B. trispora* fermentation can produce 0.2 g g^{-1} of dry cells after seven days and whole cells are
used directly as a feed additive; alternatively beta-carotene can be purified
from them.[8]

In recent years, several metabolic engineering approaches have been
developed in order to produce carotenoids in different organisms.[9] *Candida
utilis* and *Saccharomyces cerevisiae* have been modified to express bacterial
carotenoid biosynthesis genes, affording beta-carotene titres between 0.1
and 0.4 mg g^{-1} dry cells. *Escherichia coli* has also been bioengineered to
produce this carotenoid, using different strategies in which the expression
of genes from *Enterococcus faecalis* and *Streptococcus pneumoniae* is the
most productive, reaching titres of 460 mg l^{-1}.[10–12] *S. cerevisiae* has been
modified to express carotenogenic genes from *Xanthophyllomyces dendro-
rorhous*, producing beta-carotene levels that reach 6.3 mg g^{-1} of dry cells.[13,14]
Currently, other hosts are under development to produce carotenoids, such

© Woodhead Publishing Limited, 2013

as *Pichia pastoris* (from which, to date, only low amounts of beta-carotene have been produced: 0.34 mg g^{-1} dry cells) [15,16] and *Yarrowia lipolytica*.[17,18]

21.2.2   Vitamin D

Vitamin D is a fat-soluble compound derived from cholesterol and ergosterol. On the one hand, cholesterol is modified metabolically to produce 7-dehydrocholesterol, which can be cleaved by UV-radiation to form cholecalciferol (vitamin D3) (Fig. 21.2b). On the other hand, ergosterol can be transformed into ergocalciferol (vitamin D2) (Fig. 21.2a). Neither the D2 nor the D3 form is active and in humans they must undergo two hydroxylations. First, the liver converts vitamin D into 25-hydroxyvitamin D (calcidiol) and second the kidney transforms it into 1,25-dihydroxyvitamin D (calcitriol). Normally, vitamin D2 and D3 are used in food and feeds. This vitamin is essential as it allows calcium absorption and permits the normal mineralization of bone. It also modulates cell growth and has neuromuscular, immune and inflammation functions.

Vitamin D can be synthesized by most people through exposure to sunlight, but excessive radiation can be carcinogenic so it is prudent to counteract the exposure of the skin to sunlight with a complete diet including this vitamin. This vitamin can be found naturally mostly in the flesh of fatty fish and fish liver oils, but small amounts are also present in beef liver, cheese, egg yolks (D2) and some mushrooms (D3). Currently it is common to satisfy the need for this vitamin by supplementing its levels in food to create fortified foods. Vitamin D is synthesized chemically from sterols, ergosterol and cholesterol, using UV radiation (38 t/a; Solvay-Duphar (The Netherlands), Hoffmann-La Roche, BASF).[2]

Ergosterol is produced by bioprocesses including generally recognized as safe (GRAS) yeasts such as *S. cerevisiae, Saccharomyces uvarum* and *C.*

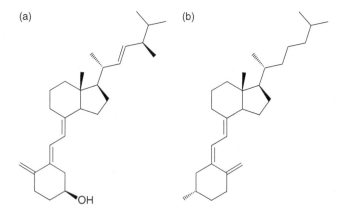

Fig. 21.2   Chemical structure of vitamin D2 (a) and vitamin D3 (b).

© Woodhead Publishing Limited, 2013

utilis (10–30 mg g^{-1} dry cells). For the preparation of vitamin concentrates, fish oils have been used directly to provide a natural source of vitamin D3. Fungi such as *Trichoderma, Cephalosporium* and *Fusarium* have also been investigated with regard to their capacity to accumulate ergosterol, but they afford lower production titres than *S. cerevisiae*.[6] *S. cerevisiae* has been bioengineered to increase the accumulation of ergosterol by overexpressing different enzymes involved in the biosynthetic pathway and using molasses as a cheap carbon source (52.6 mg g^{-1} dry weight).[19] In addition, strains producing cholesterol instead of ergosterol have been developed by metabolic engineering of *S. cerevisiae*.[20]

21.2.3   Vitamin E

The vitamin E group is formed by different molecules with antioxidant activities. There are eight chemical forms, four of tocopherol and four of tocotrienol (Fig. 21.3b). However, the only one able to meet human requirements is alpha-tocopherol (Fig. 21.3a).

Tocopherol is a fat-soluble antioxidant involved in the regulation of reactive oxygen species (ROS) produced during fat oxidation. Thus, its activity diminishes the damage caused by free radicals to cells, such free radicals being able to lead to the development of cardiovascular disease or cancer. In addition to this activity, vitamin E also has functions in the immune system, cell signalling and the regulation of gene expression. Vitamin E can be found naturally in many common foods and the highest amounts are present in nuts, seeds and vegetable oils.

This vitamin is currently used as a dietary supplement for humans, for food preservatives, in the manufacture of cosmetics and for the fortification of animal feed. Of the 40,000 t produced in 2002 (Eastman-Kodak (USA), Eizai (Japan), Hoffmann-La Roche) only 4000 t were extracted from natural sources, mainly soybean oil.[2] Owing to the high price of the vitamin when

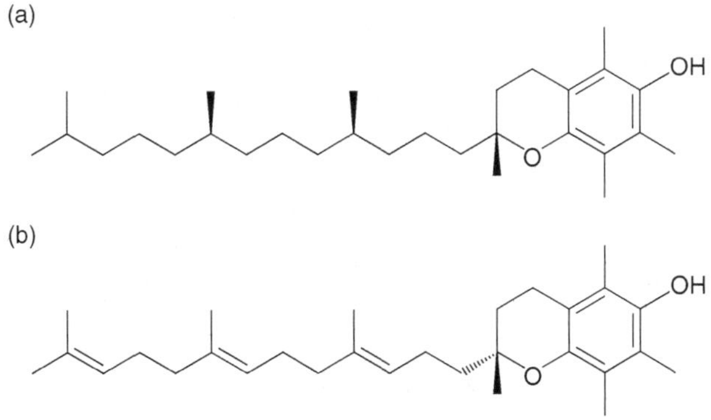

Fig. 21.3   Chemical structure of alpha-tocopherol (a) and tocotrienol (b).

© Woodhead Publishing Limited, 2013

obtained from natural sources (> US$ 20/kg), it is mainly used for human applications, while chemically synthesized alpha-tocopherol acetate produced from isophytol and trimethylhydroquinone only costs US$ 11/kg and is therefore widely used in animal feed.[21] This price may well increase in the future since the chemical precursors come from fossil fuels, whose cost is also expected to rise in the near future. Accordingly, many different approaches have been tested with a view to increasing tocopherol levels in natural sources, plants and microorganisms.

Natural sources from higher plants include *Vitis vinifera* seeds, soybean oil, rice bran, wheat germ and sunflower oil, and their vitamin E can be extracted by hexane or super-critical fluids. The main drawbacks of these sources are their low contents of tocopherol and generally low proportions of alpha-tocopherol. As an alternative to this approach, photosynthetic microorganisms are known to accumulate tocopherols and *Euglena gracilis* have been seen to be the best producer organisms, reaching 7.35 mg g^{-1} of dry cells, where 97% of tocopherols is in the alpha-isoform. Its major limitation is that in culture conditions it is easily contaminated; the proposed alternatives are *Dunaliella* and *Spirulina*.[22]

Most genetic and metabolic engineering aimed at producing alpha-tocopherol has been developed in *Synechocystis* sp., which has been used as a well-characterized model organism for this purpose, and strategies have been developed leading to a five-fold increase in tocopherol.[23] Unfortunately these approaches have not been transferred to the best producer, *E. gracilis*. Many other modifications have been made in higher plants, but they are aimed at increasing the proportion of alpha-tocopherol, which is not a problem in microorganisms because more than 90% is usually in the most active vitamin isoform.

21.2.4   Vitamin K

Vitamin K is a family of chemically related compounds: naphthoquinones. There are two kinds of naphthoquinones; phylloquinones (Fig. 21.4a), produced by plants and cyanobacteria, and menaquinones (Fig. 21.4b) produced by bacteria. The letter K for the name of this vitamin came from the word *koagulation* referring to its main role in healthy blood clotting. Apart from this well-known activity, it is also involved in bone health, in the prevention of blood vessel and heart valve calcification, and also in the protection of organisms against oxidative damage and inflammatory responses and in protecting the nervous system. All these features make it an effective agent for the prevention and treatment of different health conditions such as cystic fibrosis, liver and pancreatic cancer, osteoporosis, and so on.

More than 90% of our dietary vitamin K comes from plant foods (especially vegetables) but there is also a contribution by our intestinal bacteria, although this is very low. Certain fermented foods are enriched in this vitamin since some of the microorganisms involved in the fermentation

© Woodhead Publishing Limited, 2013

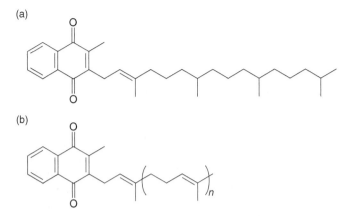

Fig. 21.4   Chemical structure of phylloquinone (a) and menaquinone (b).

process can produce and accumulate menaquinones. *Proprionibacterium* can produce cheese with a high content of the vitamin, while *Bacillus subtilis* can produce enriched fermented soy foods.

Phylloquinone is produced by a chemical process, around 3.5 t/a (Hoffmann-La Roche, Eizai),[2] and the chiral side-chain of the molecule can be obtained by biotransformation by *Geotrichum*. Phylloquinone is mainly used for clinical purposes. About 500 t/a of menadione are produced as an animal feed additive (Vanetta).[2] Menadione is not active as a vitamin, but in animals it can be converted to menaquinone after absorption. Some microorganisms are able to produce this vitamin. For example, *Flavobacterium* sp. has been found to be a potent producer. After extensive screening, a mutant able to produce 249 mg l^{-1} extracellularly and 40 mg l^{-1} or 2.7 mg g^{-1} of dry cells was obtained.[24] *B subtilis*, and *Propionibacterium freudenreichii*[25,26] have also been proposed as candidates for vitamin K production.

There is no metabolic engineering model for developing a microbial producer of vitamin K and hence *E. coli* has been selected and evaluated as a microbial platform for its production. Competitive side reactions for the substrate of the pathway were reduced by gene knockout and the genes involved in the formation of two precursors were overexpressed, leading to a five-fold increase in menaquinone production.[27]

21.3   Water-soluble vitamins

21.3.1   Vitamin B1

Vitamin B1, also called thiamine, is a water-soluble compound and there are five phosphate derivatives, including thiamine monophosphate (ThMP), thiamine diphosphate (ThDP), thiamine triphosphate (ThTP), adenosine

© Woodhead Publishing Limited, 2013

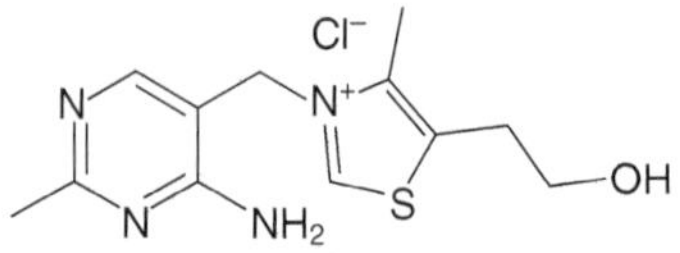

Fig. 21.5   Chemical structure of thiamine.

thiamine triphosphate (AThTP) and adenosine thiamine diphosphate (AThDP). The phosphorylated forms are thought to be the active forms of the vitamin, while thiamine (Fig. 21.5) is mainly the transportation form.

Thiamine has essential metabolic functions and deficiencies of this compound are associated with imbalances in carbohydrate status because it is involved in oxidative decarboxylation and transketolase reactions. It is also an active molecule of the nervous system. It can be found in the diet, especially in wheat germ, soy beans, dried beans and peas. Although it is widespread in foodstuffs, its concentration is often low because it is destroyed when food is cooked. Therefore, in developed countries rice and flour are usually fortified with this vitamin.

Vitamin B1 is produced chemically to supply human and animal needs (4200 t/a in 1996).[2] Over time, two methods of synthesis have been developed: (1) condensation of the pyrimidine and thiazole rings and (2) construction of the thiazole ring on a preformed pyrimidine portion. Recently, a patent protecting the metabolic engineering of bacteria to accumulate high amounts of thiamine in media has been developed using *B. subtilis*.[28] The patent refers to species of Bacillaceae, Lactobacillaceae, Streptococcaceae, Corynebacteriaceae and Brevibacteriaceae, in which a microorganism containing a mutation that deregulates thiamine production and causes thiamine products to be released from the cell is described. Several strategies of overexpression and deregulation of the genes involved in precursor synthesis and pathway engineering, among others, have also been described.

21.3.2   Vitamin B2

Vitamin B2 is also called riboflavin, which takes its name from its yellow colour (flavus). It is essential for the proper functioning of all the flavoproteins, since riboflavin is the central component of the FAD and FMN cofactors. These are involved in oxidation–reduction reactions, which are key activities in the energy metabolism of carbohydrates, fats, ketone bodies and proteins. Vitamin B2 is also involved in the metabolism of other vitamins such as B6, B3 and A, in glutathione recycling and homocysteine metabolism. The highest amounts of riboflavin in food can be found in crimini mushrooms and spinach, but also in asparagus, green beans, yogurt and cow's milk.

© Woodhead Publishing Limited, 2013

Industrial riboflavin production is a paradigm of how biotechnology can turn a chemical synthesis into a bioprocess with significant cost reductions by employing a genetic and metabolic bioengineering approach. Chemically, it is produced from D-glucose by three different processes. More than 9000 t/a of riboflavin were produced in 2010, around 75% being used for feed additive and the rest for human food and pharmaceuticals (Hoffmann-La Roche, BASF, ADM, Takeda).[2] This compound is naturally produced by several microorganisms such as ascomycete fungi (*Ashbya gossypii*, *Eremothecium ashbyii*), by yeasts such as *Candida flaeri* and *Candida famata*, and also by bacteria such as *B. subtilis* and *Corynebacterium ammoniagenes*.

Several metabolic approaches have been developed in *B. subtilis* by overexpression of the gene cluster involved in riboflavin synthesis[29] and including multiple copies of these genes.[30] Other approaches have attempted to express heterologous genes involved in riboflavin accumulation but only modest results have been achieved[31] and some modifications guided by transcriptional analysis have afforded a strain able to accumulate 15 g l^{-1} riboflavin.[32] However, most metabolic bioengineering strategies have been carried out in the main industrial producer *A. gossypii*.[33] All six genes of the riboflavin synthetic pathway have been overexpressed and its use to improve riboflavin production patented (Fig. 21.6).[34] Some other genes

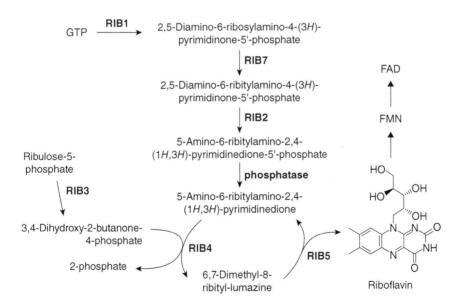

Fig. 21.6   Biosynthetic pathway of riboflavin in *Ashbya gossypii*. GTP = guanosine 5'-triphosphate; FAD = flavin adenine dinucleotide; FMN = flavin mononucleotide; RIB (1–5 and 7) = riboflavin biosynthesis gene(s).

© Woodhead Publishing Limited, 2013

have been reported to accumulate the vitamin when they are overexpressed, deregulated, or disrupted.[33] These genetic alterations lead to an
accumulation ranging from 1.4- to ten-fold increases relative to the wild-
type, affording strains able to produce more than 13 g l^{-1}. In recent years,
efforts in metabolic bioengineering have also been performed using *C.
famata* and strains accumulating 4.1-fold the wild-type amount of the
vitamin have been constructed.[35]

21.3.3   Vitamin B3

Vitamin B3 is a group formed of nicotinic acid, nicotinamide and other
compounds such as inositol hexanicotinate, that exhibit a related biological
activity. These compounds can be clustered together under the term niacin
(which sometimes is only used referring to nicotinic acid) (see chemical
structure in Fig. 21.7).

Like all B vitamins, niacin is involved in energy metabolism, in the use
of carbohydrates, fats and proteins, and it is therefore needed for healthy
skin, hair, eyes, liver and nervous system. Niacin also helps the body's stress-
and sex-related hormones. It is used, or is currently under investigation for
use, in therapies against high cholesterol, atherosclerosis and heart disease,
diabetes and osteoarthritis. This vitamin can be found in different foods but
can also be produced from the essential amino acid tryptophan, which can
be obtained from most sources of protein. To cite just some examples, niacin
is present in liver, chicken and beef, seeds, mushrooms and yeast and also
in some species of fish.

Nicotinamide and nicotinic acid are produced both chemically and biotechnologically (22,000 t/a, BASF, Lonza and Degussa).[2] The major use of
this production is for animal nutrition and the remaining 25% is used for
food enrichment and pharmaceutical applications. Chemically it is synthesized by the oxidation of 5-ethyl-2-methylpyridine or by total hydrolysis of
3-cyanopyridine. However, this latter transformation can also take place in
a bioprocess using nitrilase (to produce nicotinic acid) or nitrile hydratase
(to produce nicotinamide). Nitrilase has been overexpressed in *Rhodococ-
cus rhodochrous*, generating a strain able to convert almost all 3-cyanopy-
ridine into nicotinic acid.[36] Nitrile hydratase in *R. rhodochrous* exceeded
50% of the total cellular protein, permitting high production of nicotinamide from 3-cyanopyridine. Both reactions are almost stoichiometric, even

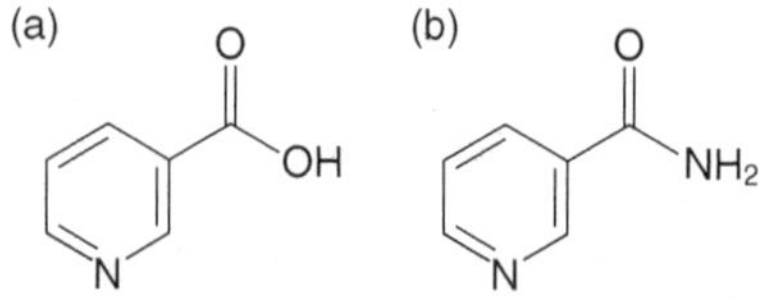

Fig. 21.7   Chemical structure of nicotinic acid (a) and nicotinamide (b).

© Woodhead Publishing Limited, 2013

when very high concentrations of the substrate are used, and no significant amounts of by-products are produced.

21.3.4   Vitamin B5

Vitamin B5 is also called pantothenic acid. Commercially it is available as D-pantothenic acid, as well as dexpanthenol and calcium pantothenate, which are chemicals made in the laboratory from D-pantothenic acid (Fig. 21.8).

Vitamin B5 has roles in the breakdown of carbohydrates and fats for energy, and this vitamin is also involved in the production of red blood cells, in stress- and sex-related hormones and in maintaining a healthy digestive tract, helping to obtain other vitamins. It is also essential for the synthesis of cholesterol and regulates fats in the blood. It has been suggested to help wound healing and rheumatoid arthritis.[37,38] Pantothenic acid can be found in a wide range of foods, especially unprocessed food, because most of the vitamin is lost after refining or freezing. To cite some examples, it is present in brewer's yeast, corn, tomatoes, beef (especially liver and kidney) and salmon.

Calcium D-pantothenic and D-pantothenyl alcohol is mainly produced (80%) for animal feed but also for human food and pharmaceuticals (6000 t/a calcium D-pantothenic and 1000 t/a D-pantothenyl alcohol, by Hoffmann-La Roche, Fuji, and BASF).[2] It has a large-scale production process involving a mixture of chemical and enzymatic reactions. These bioprocess steps are especially important for circumventing the expensive and troublesome optical resolution of racemic mixtures of D-L pantolactone. Chemically, pantolactone is obtained from isobutyraldehyde, formaldehyde and cyanide. D-pantolactone is condensed with beta-alanine to form pantothenic acid. The fungal enzyme lactohydrolase is the key protein for racemic resolution and D-pantolactone is its substrate but not L-pantolactone. It hydrolyses D-pantolactone into D-pantoic acid, which is easily separated from the L-enantiomer of pantolaconte. High activities of this enzyme have been reported in *Gibberella*, *Cylindocarpon* and *Fusarium*, and cells of *Fusarium oxysporum* have been immobilized in calcium alginate gels, affording high conversion rates (more than 90%).[39,40]

An alternative process that avoids optical resolution involves the chemical or biological (L-pantolactone dehydrogenase from *Nocardia asteroides*) synthesis of ketopantolactone, which is bioreduced to form D-pantolactone.

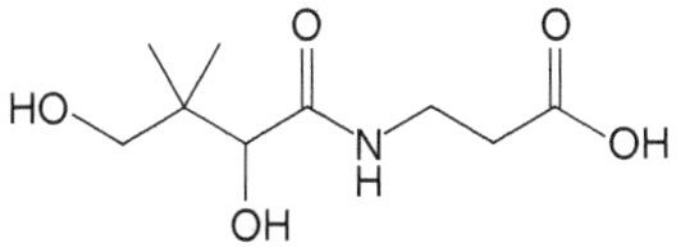

Fig. 21.8 Chemical structure of pantothenic acid.

© Woodhead Publishing Limited, 2013

Candida parapsilosis cells have been proposed for biotransformation owing to their high activity in carbonyl reductase.[41] A variation of this method is the hydrolysis of ketopantolactone to ketopanoic acid, which is reduced by *Agrobacterium* sp. into D-pantoic acid.[42]

Over the past decade some metabolic engineering approaches have been implemented in order to produce D-pantothenic acid or D-pantoic acid directly by fermentation. *E. coli* has been proposed and patented as a good producer of the vitamin after some genetic bioengineering processes, reaching 66 g l⁻¹ pantothenic acid.[43] Also, the characterization of the synthetic pathway in *Corynebacterium glutamicum* has permitted metabolic engineering aimed at producing this vitamin by biotechnological processes, although the amounts obtained are far from those obtained with bioengineered *E. coli*.[44]

21.3.5   Vitamin B6

Vitamin B6 is a water-soluble vitamin formed by different compounds; pyridoxine, pyridoxine 5′phosphate, pyridoxal, pyridoxal 5′phosphate (PLP), pyridoxamine and pyridoxamine 5′phosphate (PMP). PLP and PMP are the two active coenzyme forms of the vitamin. Figure 21.9 shows the chemical structure of these compounds.

These coenzymes have a wide variety of functions and are involved in more than 100 enzymatic reactions. This vitamin affects protein metabolism, amino acid metabolism, and it maintains homocysteine levels in the blood. It is also involved in the metabolism of one-carbon units, carbohydrates and lipids. It is essential for the synthesis of neurotransmitters and it is also involved in gluconeogenesis, glyconeogenesis, immune function and haemoglobin formation.

This vitamin can be obtained from the consumption of many different foods in the diet. It is especially found in liver and kidney, fish, starchy vegetables and fruits. After dephosphorylation in the jejunum, vitamin B6 is absorbed by passive diffusion.

Vitamin B6 is produced almost exclusively by chemical processes (2500 t/a, Takeda, Hoffman-La Roche, Fuji/Daiichi).[2] Pyridoxine hydrochloride is produced in large-scale batches via the Diels–Alder reaction of oxoazoles

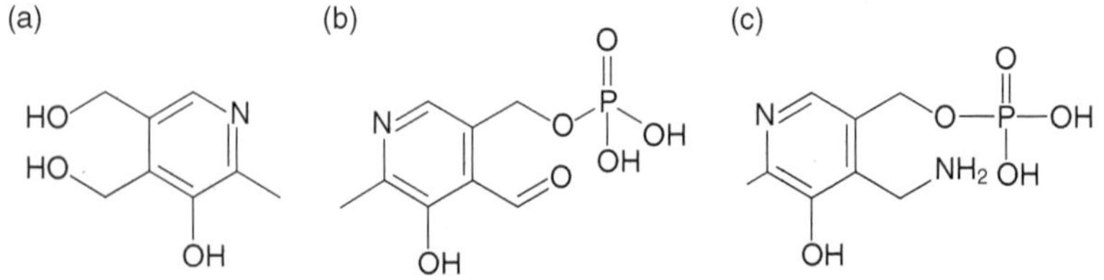

Fig. 21.9   Chemical structure of pyridoxine (a), pyridoxal phosphate, PLP (b) and pyridoxamine phosphate, PMP (c).

© Woodhead Publishing Limited, 2013

with maleic acid, among other methods. From this form, pyridoxal, pyridox-amine and their 5′phosphates can be synthesized. The pyridoxine biosyn-thetic pathway has been studied in both bacteria[45–47] and fungi.[48,49]

Some microorganisms have been screened for vitamin B6 production and *Flavobacterium* sp and *Rhizobium meliloti* (now *Sinorhizobium meliloti*) were identified as the best producers, with 20 mg l^{-1} and 84 mg l^{-1} respectively,[50] but they are still far from being useful for industrial fermen-tation. In order to increase its large-scale suitability, metabolic bioengineer-ing to increase vitamin B6 accumulation must be carried out. Currently this kind of approach has only been developed in plants, increasing vitamin amounts in plants and seeds.[51] The feasibility of the metabolic bioengineer-ing approach in both plants and fungi was revealed by the overexpression of two conserved genes of the synthesis pathway, but the increase in total vitamin production was low or undetected because of the strong regulation of the pathway.[52]

21.3.6   Vitamin B7

Vitamin B7 is also called biotin and is a molecule formed by a tetrahydroi-midizalone ring (Fig. 21.10) fused with a tetrahydrothiophene ring. It is synthesized by three enzymes from two precursors: alanine and pimeloyl-CoA.

Biotin is involved in a wide range of important cellular processes. Its function is important for fatty acid production, amino acid and fat metabo-lism, and the citric acid cycle. It also helps to transfer carbon dioxide and to regulate blood sugar levels. Biotin can readily be obtained from the diet since it is present in a wide range of foods such as egg, liver, soybeans, nuts, Swiss chard or whole wheat. Its deficiency is not frequent and most people can satisfy their needs for this vitamin from its synthesis by their own intes-tinal bacteria.

Vitamin B7 is mainly produced chemically (25 t/a, Hoffmann-La Roche, Tanable and Sumitomo).[2] However, several overproducing mutants have been isolated from *B. sphaericus* and *Serratia marcescens* which are resistant to biotin analogues. Nevertheless, the amounts of vitamin B7 obtained were still too low to compete with chemical synthesis.

The synthetic pathway to biotin has now been elucidated in several organisms such as *E. coli,*[53] *B. sphaericus*[54] and *B. subtilis.*[55] Several metabolic approaches have been developed in order to increase the

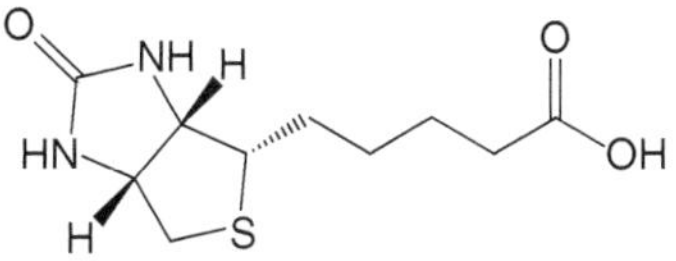

Fig. 21.10   Chemical structure of biotin.

© Woodhead Publishing Limited, 2013

biological production of biotin. Thus, *S. marcescens* harbouring a plasmid with an extra copy of the mutated biotin operon has been obtained, attaining a continuous production of 600 mg l^{-1}.[56] A recombinant *Sphingomonas* sp. has also been constructed with a plasmid with a part of the biotin operon.[57] Until biotechnology can overcome all the drawbacks of the biological production of biotin at higher levels, some approaches currently are attempting to combine traditional chemical synthesis with enzymatic steps using well characterized genes.[58]

21.3.7   Vitamin B9

Vitamin B9 is also known as folate, a generic name that includes a wide set of different molecules. Folic acid is the synthetic form of folate and it is found in supplements and in fortified food. It is important in periods of rapid cell growth and division such as pregnancy and infancy since it is essential for maintaining new cells and to make DNA and RNA. It is also important for the production of red blood cells and the prevention of anaemia and it is needed to maintain normal levels of homocysteine. Folate gets its name from the Latin word 'folium', for leaf, because of one of the main sources of folate in the diet are leafy green vegetables, although it can also be found in fruits, beans and peas, among others.

Folic acid is produced by chemical processes and there are several known methodologies. It is mainly used for animal feed, but also for human food and pharmaceuticals, reaching a production of 2000 t/a in 2008. The biosynthetic pathway (Fig. 21.11) is well known in bacteria and plants and hence metabolic engineering approaches have been carried out in both. Plants have been engineered by heterologous expression of genes involved in folate synthesis from mammals and *Arabidopsis thaliana* in tomato fruit and rice grain respectively[59,60] in order to avoid negative regulation. With these two approaches, 140-fold and 100-fold increases in folate level were achieved.

Crop biofortification is especially important in the fight against folic acid deficiency in developing countries. Microorganisms have been bioengineered to overproduce folates. *B. subtilis* was modified at three different levels previously predicted by computer-aided flux analysis and an eight-fold increase in folate levels was observed.[61] Metabolic engineering has also been developed in *Lactococcus lactis* leading to a more than three-fold increase.[62,63] Recently *Ketogulonigenium vulgare* has been engineered to overexpress the folate operon, eight-fold higher folate contents being reached.[64] A mixed fermentation/chemical process has been proposed and patented.[65]

21.3.8   Vitamin B12

Vitamin B12 is a group of water-soluble compounds that contain the element cobalt, which leads to them being called 'cobalamines'.

© Woodhead Publishing Limited, 2013

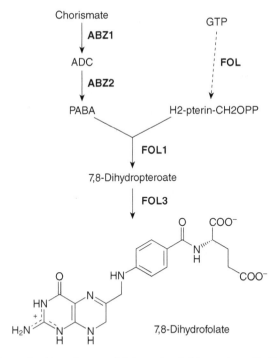

Fig. 21.11 Biosynthetic pathway of folates in *Ashbya gossypii*. ABZ (1–2) = *p*-aminobenzoic acid biosynthesis gene(s); ADC = 4-amino 4-deoxychorismate; PABA = *p*-aminobenzoic acid; FOL (1-3) = folic acid biosynthesis gene(s).

There are two active forms; methylcobalamine (Fig. 21.12) and 5-deoxyadenosylcobalamine.

This vitamin is involved in DNA synthesis, neurological function and red blood cell formation. It is a cofactor directly involved in the methylation of DNA, RNA, hormones, lipids and proteins, and also in protein and fat metabolism. In the diet it is found bound to proteins and it must be released in the stomach to be absorbed. It is mainly present in animal products, such as meat, fish, poultry, milk and eggs, and because of this vegetarians may have to obtain this vitamin from fortified foods.

Vitamin B12 is therefore produced industrially for pharmaceutical products, fortified foods and animal feed. It is produced biologically from *Pseudomonas* and *Propionibacterium* fermentation (10 t/a, Rhône-Poulenc, Aventis).[2] On the one hand, in the case of *Propionibacterium shermanii*, the vitamin is produced in two steps: (1) bacterial growth and the production of intermediates and (2) the production of vitamin B12 from corn steep liquor, glucose and $CoCl_2$, reaching production levels of 25–40 mg l^{-1}.[66] On the other hand, when made from *Pseudomonas denitrificans* in a medium with sugar beet molasses and 5,6-dimethylbenzimidazol, a production of 150 mg l^{-1} was obtained.[67]

© Woodhead Publishing Limited, 2013

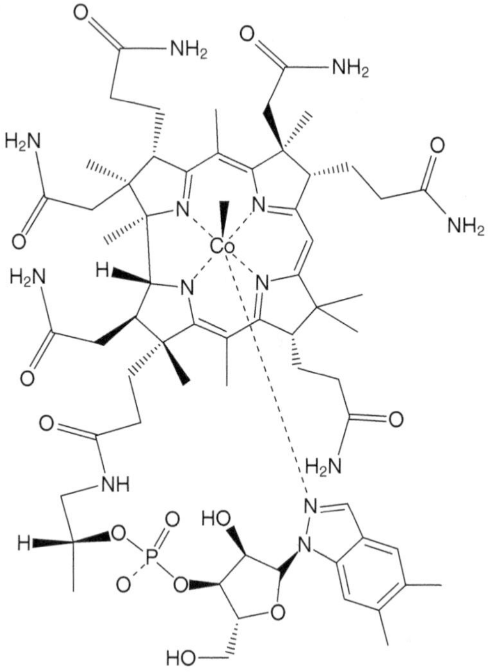

Fig. 21.12   Chemical structure of methylcobalamine.

These naturally overproducing strains have been modified in order to increase the vitamin production rate. Metabolic engineering has been carried out in *P. shermanii* by overexpression of a gene of the biosynthetic pathway, cobA[68] and recently genome shuffling has been carried out, obtaining a 61% improvement in cobalamine production.[69] Extensive metabolic engineering approaches have been developed in the model organism *Bacillus megaterium*, where biosynthetic and regulatory genes and operons were overexpressed, by-products from second pathway branches were silenced by antisense RNA, and three cobalamine-binding proteins were expressed heterologously to avoid feedback inhibition.[70]

The most important producer, *P. denitrificans*, has also been genetically engineered. The copy-number of one operon (*cobF-cobM*) encompassing the eight genes involved in the synthesis of the vitamin was increased (30%); two independent genes (*cobA* and *cobE*) were also amplified (20% increase) and, in addition, strong inducible promoters, highly efficient ribosomal binding sites and terminator sequences were applied to another important gene (*cobB*).[71] In order to avoid some regulatory steps, such as substrate inhibition, the heterologous expression of genes from *Methanobacterium ivanovii* and *Rhodobacter capsulatus* has been implemented.[71,72] The combination of both approaches – genetic engineering and random

© Woodhead Publishing Limited, 2013

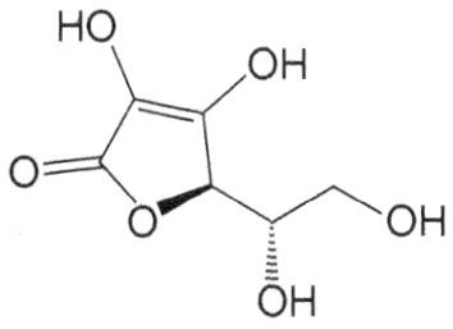

Fig. 21.13   Chemical structure of ascorbic acid.

mutagenesis – have led this process to being used to synthesize 80% of the world's production of vitamin B12.[73]

21.3.9   Vitamin C

Vitamin C is an essential dietary component that humans are unable to synthesize. It is also known as L-ascorbic acid (Fig. 21.13).

It is an important antioxidant which might prevent or delay certain cancers, cardiovascular and other diseases in which oxidative stress is a crucial factor. It is also involved in the biosynthesis of collagen, L-carnitine and certain neurotransmitters, and in protein metabolism. It also plays an important role in immune function and improves the absorption of non-haem iron. It is absorbed in the intestine via an active transporter and can be found in different food sources, mainly in fruits and vegetables. Citrus fruits, tomato and potatoes are major contributors of vitamin C.

It is used as a food and feed antioxidant and in pharmaceuticals (110,000 t/a, Hoffmann-La Roche, Dalry, Belvidere, Takeda, etc).[74] Normally, starch from corn or wheat is converted to glucose, which can be transformed chemically into sorbitol. From this sorbitol, and via a series of biotechnical, chemical processing and purification steps, vitamin C is produced. The so-called Reichstein method used to be the market-dominating process of synthesis but over the past two decades some bioconversions have simplified this method. The Reichstein process consists of seven steps. First, L-glucose is transformed into D-sorbitol. Second, *Gluconobacter oxydans* regiospecifically oxidizes D-sorbitol into L-sorbitol. L-sorbitol is then crystallized and condensed with acetone and sorbose-diacetone is formed, which is later oxidized to 2-keto-L-gluconic acid (2KLGA). Then, this is enolized and lactonized to form L-ascorbic acid, with a final yield of 50%. This process is still expensive owing to the high energy consumption of some steps, so an alternative is required.

Most microbial methodologies lead to the production of 2-KLGA and some single-strain and mixed cultures have been developed. The single-strain process includes the use of *Gluconobacter, Acetobacter, Ketogulonicigenium, Pseudomonas, Erwinia* and *Corynebacterium*.[75–78] The mixed culture processes include two fermentation steps: first to produce di-acetone-ketogluonic acid and second to produce 2-KLGA. One example of

© Woodhead Publishing Limited, 2013

this process is that carried out by *Erwinia* or *Acetobacter* (first step) and *Corynebacterium* (second step), although other microorganisms have been used in mixed methods, such as *Pseudomnoas striata*, *G. oxydans* and *B. megaterium*.[79,80]

Several genetic and metabolic bioengineering approaches have been developed to produce 2-KLGA from different microorganisms. *Erwinia herbicola* was bioengineered to express heterologous genes from *Corynebacterium*, reaching a production of 120 g l^{-1} of 2-KLGA.[81] *G. oxydans* was also genetically engineered to express genes from other strains and some promotors were exchanged to optimize their expression, producing 130 g l^{-1} 2-KLGA.[82] *Pseudomonas putida* has also been engineered to express genes from *G. oxydans*, reaching 16 g l^{-1} 2-KLGA, much lower than the production values of the earlier microorganisms.

After identification of the enzymes involved in direct conversion to L-ascorbic acid, these were expressed in *G. oxydans*, but only 4.2 g l^{-1} of vitamin C was produced.[83] The yeast *S. cerevisiae* has also been engineered since it is able to produce D-erythroascorbic acid. The biosynthetic pathway was modified by overexpression of some genes and heterologous expression of an *A. thaliana* gene, generating a yeast able to produce 100 mg l^{-1} of L-ascorbic acid.[84] Some microalgae are also under study to produce vitamin C directly, such as *Prototheca moriformis* or *Chlorella pyrenoidosa*, which were able to produce 2 g l^{-1} of ascorbic acid. The main drawback of microalgal cultures are their slow rates of growth and metabolic activities, which make them unprofitable.[85]

21.4   Future trends

The immediate conclusion of this chapter is that the large scale industrial production of vitamins is turning from chemical synthesis into microbial production. Biotechnology is the driving force of this change and this can be clearly seen not only in latest industrial achievements but also in recently published papers, as has been described case by case. Therefore all improvements in the biotechnical sciences are indirectly enhancing microbial production of vitamins. Among all this emerging science, methods most associated with vitamin production are genetic, protein and metabolic engineering, systems biology, fluxomics and synthetic biology. In the near future there will be no need to find a naturally producing organism as the ones most studied, such as *S. cerevisiae* or *E. coli*, could be synthetically engineered to become perfect vitamin factories. This will be allowed by the complete construction of new pathways through synthetic molecular biology and by the creation of enzymatic activities that are still unknown in nature using protein engineering. The *in silico* prediction of bottlenecks in metabolic fluxes and the most desired modifications to avoid them can be evaluated by fluxomics. To date, many bioinformatics programs have been

© Woodhead Publishing Limited, 2013

developed in order to analyse the huge amounts of data that these system sciences provide us with.

Currently there are some vitamins that are almost exclusively produced by microorganisms, such as vitamin B2 and B12. Vitamins C and A are produced both chemically and microbiologically and certain others such as Vitamin D, K, B3 and B5 have at least one or more microbial enzymatic step. The rest, which are mostly produced chemically, have been studied in depth in order to develop biological platforms to make their microbial production competitive enough to replace their chemical synthesis. This has been described previously, vitamin by vitamin, in the most recent publications and patents in microbial biotechnology. Thus these future developments in biotechnology should lead us to more sustainable, environmentally-friendly and economically competitive vitamin production systems using microorganisms as effective cell factories.

21.5   References

1. DE BAETS S and VANDEDRINCK S, EJ V (2000) 'Vitamins and related biofactors, microbial production'. In *Encyclopedia of Microbiology*, Lederberg J (ed.), Volume 4, 2nd edition, Academic Press, New York, 837–53.
2. EGGERSDORFER M, ADAM G, JOHN MWH and LABLER L (1996) 'Vitamins'. *Biotechnology*, Vol 4, Pape, H. and Rehm, H.-J., (eds), VCH, Weinheim. 114–58.
3. FLORENT J (1986) 'Vitamins'. In *Biotechnology – A Comprehensive Treatise*. Rehm, H.J. and Reed, G. (eds), VHC, Germany, Volume 1, 133–40.
4. DEMAIN A (2000) 'Microbial biotechnology'. *Trends Biotechnol*, **18**, 26–31.
5. VANDAMME EJ (1992) 'Production of vitamins, coenzymes and related biochemicals by biotechnological processes'. *J Chem Tech Biotechnol*, **53**, 313–27.
6. SHIMIZU S (2008) 'Vitamins and related compounds: microbial production'. *Biotechnology: Special Processes*, Volume 10, 2nd edition, H.-J. Rehm and G. Reed (eds), Wiley-VCH, Weinheim, Germany.
7. BOROWITZKA L and BOROWITZKA M (1988) *Microalgal Biotechnology*. Cambridge University Press, Cambridge.
8. BOROWITZKA L and BOROWITZKA M (1989) 'β-Carotene (provitamin A) production with algae'. *Biotechnology of Vitamins, Pigments and Growth Factors*, Vandamme, E. J., (ed.), Elsevier Applied Science, London, 15–26.
9. YE VM and BHATIA SK (2012) 'Pathway engineering strategies for production of beneficial carotenoids in microbial hosts'. *Biotechnol Lett*, **34**, 1405–14.
10. YOON SH, LEE SH, DAS A, RYU HK, JANG HJ, KIM JY, OH DK, KEASLING JD and KIM SW (2009) 'Combinatorial expression of bacterial whole mevalonate pathway for the production of beta-carotene in *E. coli*'. *J Biotechnol*, **140**, 218–26.
11. AJIKUMAR PK, XIAO WH, TYO KE, WANG Y, SIMEON F, *et al.* (2010) 'Isoprenoid pathway optimization for Taxol precursor overproduction in *Escherichia coli*'. *Science*, **330**, 70–4.
12. MA SM, GARCIA DE, REDDING-JOHANSON AM, FRIEDLAND GD, CHAN R, BATTH TS, HALIBURTON JR, CHIVIAN D, KEASLING JD, PETZOLD CJ, LEE TS and CHHABRA SR (2011) 'Optimization of a heterologous mevalonate pathway through the use of variant HMG-CoA reductases'. *Metab Eng*, **13**: 588–97.
13. VERWAAL R, WANG J, MEIJNEN JP, VISSER H, SANDMANN G, VAN DEN BERG JA and VAN OOYEN AJ (2007) 'High-level production of beta-carotene in *Saccharomyces*

cerevisiae by successive transformation with carotenogenic genes from *Xantho-phyllomyces dendrorhous*'. *Appl Environ Microbiol*, **73**, 4342–50.

14. YAN GL, WEN KR and DUAN CQ (2012) 'Enhancement of beta-carotene production by over-expression of HMG-CoA reductase coupled with addition of ergosterol biosynthesis inhibitors in recombinant *Saccharomyces cerevisiae*'. *Curr Micro-biol*, **64**, 159–63.

15. ARAYA-GARAY JM, FEIJOO-SIOTA L, ROSA-DOS-SANTOS F, VEIGA-CRESPO P and VILLA TG (2012) 'Construction of new *Pichia pastoris* X-33 strains for production of lyco-pene and beta-carotene'. *Appl Microbiol Biotechnol*, **93**, 2483–92.

16. BHATAYAA A, SCHMIDT-DANNERTA C and LEEB P (2009) 'Metabolic engineering of *Pichia pastoris* X-33 for lycopene production'. *Process Biochem*, 1095–102.

17. BAILEY R, MADDE K and TRUEHEAR J (2011) *Production of Carotenoids in Oleagi-nous Yeast and Fungi*. US Patent Pub No US2011/0021843 A1.

18. SABIROVA JS, HADDOUCHE R, VAN BOGAERT IN, MULAA F, VERSTRAETE W, TIMMIS KN, SCHMIDT-DANNERT C, NICAUD JM and SOETAERT W (2011) 'The "LipoYeasts" project: using the oleaginous yeast *Yarrowia lipolytica* in combination with specific bacterial genes for the bioconversion of lipids, fats and oils into high-value products'. *Microb Biotechnol*, **4**, 47–54.

19. HE X, GUO X, LIU N and ZHANG B (2007) 'Ergosterol production from molasses by genetically modified *Saccharomyces cerevisiae*'. *Appl Microbiol Biotechnol* **75**, 55–60.

20. SOUZA CM, SCHWABE TM, PICHLER H, PLOIER B, LEITNER E, GUAN XL, WENK MR, RIEZMAN I and RIEZMAN H (2011) 'A stable yeast strain efficiently producing cholesterol instead of ergosterol is functional for tryptophan uptake, but not weak organic acid resistance'. *Metab Eng*, **13**, 555–69.

21. VALENTIN HE and QI Q (2005) 'Biotechnological production and application of vitamin E: current state and prospects'. *Appl Microbiol Biotechnol*, **68**, 436–44.

22. OGBONNA JC (2009) 'Microbiological production of tocopherols: current state and prospects'. *Appl Microbiol Biotechnol*, **84**, 217–25.

23. QI Q, HAO M, NG WO, SLATER SC, BASZIS SR, WEISS JD and VALENTIN HE (2005) 'Appli-cation of the Synechococcus nirA promoter to establish an inducible expression system for engineering the Synechocystis tocopherol pathway'. *Appl Environ Microbiol*, **71**, 5678–84.

24. TAGHUCHI H, SHIBATA T, DUANGMANEE C and TANI Y (1989) 'Menaquinone-4 pro-duction by a sulfonamide-resistant mutant of *Favobacterium* sp 238–7'. *Agric Biol Chem*, **53**, 3017–23.

25. FURUICHI K, HOJO K, KATAKURA Y, NINOMIYA K and SHIOYA S (2006) 'Aerobic culture of *Propionibacterium freudenreichii* ET-3 can increase production ratio of 1,4-dihydroxy-2-naphthoic acid to menaquinone'. *J Biosci Bioeng*, **101**, 464–70.

26. SATO T, YAMADA Y, OHTANI Y, MITSUI N, MURASAWA H and ARAKI S (2001) 'Efficient production of menaquinone (vitamin K2) by a menadione-resistant mutant of *Bacillus subtilis*'. *J Ind Microbiol Biotechnol*, **26**, 115–20.

27. KONG MK and LEE PC (2011) 'Metabolic engineering of menaquinone-8 pathway of *Escherichia coli* as a microbial platform for vitamin K production'. *Biotechnol Bioeng*, **108**, 1997–2002.

28. GOESE MG, PERKINS P and SEHYNS G (2006) *Thiamin Production by Fermentation*. US Patent Pub No US2006/0127993 A1.

29. PERKINS JB, SLOMA A, HERMANN T, THERIAULT K, ZACHAGO E, ERDENBERGER T, HANNETT N, CHATTERJEE NP, WILLIAMSII V, RUFO GA JR, HATCH R and PERO J (1999) 'Genetic engineering of *Bacillus subtilis* for the commercial production of ribo-flavin'. *J Ind Microbiol Biotechnol*, **22**, 8–18.

30. HÜMBELIN M, GRIESSER V, KELLER T, SCHURTER W, HAIKER M, HOHMANN HP, RITZ H, RICHTER G, BACHER A and VAN LOON APGM (1999) 'GTP cyclohydrolase II and 3,4-dihydroxy-2-butanone 4-phosphate synthase are rate-limiting enzymes in

© Woodhead Publishing Limited, 2013

riboflavin synthesis of an industrial *Bacillus subtilis* strain used for riboflavin production'. *J Ind Microbiol Biotechnol*, **22**, 1–7.

31. WANG Z, CHEN T, MA X, SHEN Z and ZHAO X (2011) 'Enhancement of riboflavin production with *Bacillus subtilis* by expression and site-directed mutagenesis of *zwf* and *gnd* gene from *Corynebacterium glutamicum*'. *Bioresour Technol*, **102**, 3934–40.

32. SHI S, CHEN T, ZHANG Z, CHEN X and ZHAO X (2009) 'Transcriptome analysis guided metabolic engineering of *Bacillus subtilis* for riboflavin production'. *Metab Eng*, **11**, 243–52.

33. KATO T and PARK EY (2012) 'Riboflavin production by *Ashbya gossypii*'. *Biotechnol Lett*, **34**, 611–18.

34. REVUELTA D, BUITRAGO S and SANTOS G (1995) *Riboflavin Synthesis in Fungi*. Patent WO 9526406-A.

35. DMYTRUK KV, YATSYSHYN VY, SYBIRNA NO, FEDOROVYCH DV and SIBIRNY AA (2011) 'Metabolic engineering and classic selection of the yeast *Candida famata (Candida flareri)* for construction of strains with enhanced riboflavin production'. *Metab Eng*, **13**, 82–8.

36. NAGASAWA T and YAMADA H (1989) 'Microbial transformations of nitriles'. *Trends Biotechnol*, 153–8.

37. ELLINGER S and STEHLE P (2009) 'Efficacy of vitamin supplementation in situations with wound healing disorders: results from clinical intervention studies'. *Curr Opin Clin Nutr Metab Care*, **12**, 588–95.

38. PINS J and KEENAN JM (2006) 'Dietary and nutraceutical options for managing the hypertriglyceridemic patient'. *Prog Cardiovasc Nurs*, **21**, 89–93.

39. KATAOKA M, SHIMIZU K, SAKAMOTO K, YAMADA H and SHIMIZU S (1995) 'Optical resolution of racemic pantolactone with a novel fungal enzyme, lactonohydrolase'. *Appl Microbiol Biotechnol*, 2292–4.

40. KATAOKA M, SHIMIZU K, SAKAMOTO K, YAMADA H and SHIMIZU S (1995) 'Lactonohydrolase catalyzed optical resolution of pantonyl lactone: selection of potent enzyme producer and optimization of culture and reaction conditions for practical resolution'. *Appl Microbiol Biotechnol*, 333–8.

41. HATA H, SHIMIZU S and YAMADA H (1987) 'Enzymatic production of D-(–)pantonyl lactone from ketopantonyl lactone'. *Agric Biol Chem*, **51**, 3011–16.

42. KATAOKA M, SHIMIZU S and YAMADA H (1990) 'Steroespecific conversion of a racemic pantonyl lactone to D-(–)pantonyl lactone through microbial oxidation and reduction reactions'. *Recl Trav Chim Pays-Bas*, **110**, 155–7.

43. MORIYA T, HIKICHI Y, MORIYA Y and YAMAGUCHI T (1997) *Process for Producing D-Pantoic acid and D-Pantothenic acid or Salts Thereof*. WO1997010340.

44. SAHM H and EGGELING L (1999) 'D-pantothenate synthesis in *Corynebacterium glutamicum* and use of panBC and genes encoding L-valine synthesis for D-pantothenate overproduction'. *Appl Environ Microbiol*, **55**, 1973–9.

45. CANE DE, HSIUNG Y, CORNISH JA, ROBINSON JK and SPENSER ID (1998) 'Biosynthesis of vitamin B6: the oxidation of 4-(phosphohydroxy)-L-threonine by PdxA'. *J Am Chem Soc*, 1936–7.

46. CANE DV, DU SC, ROBINSON K, HSIUNG Y and SPENSER ID (1999) 'Biosynthesis of vitamin B6: enzymatic conversion of 1-deoxy-D-xylulose-5-phosphate to pyridoxol phosphate'. *J Am Chem Soc*, **121**, 7722–3.

47. LABER B, MAURER W, SCHARF S, STEPUSIN K and SCHMIDT FS (1999) 'Vitamin B6 biosynthesis: formation of pyridoxine 5_-phosphate from 4-(phosphohydroxy)-L-threonine and 1-deoxy-D-xylulose-5-phosphate by *PdxA* and *PdxJ* protein'. *FEBS Lett*, **449**, 45–8.

48. KONDO H, NAKAMURA Y, DONG YX, NIKAWA J and SUEDA S (2004) 'Pyridoxine biosynthesis in yeast: participation of ribose 5-phosphate ketol-isomerase'. *Biochem J*, **379**, 65–70.

© Woodhead Publishing Limited, 2013

49. WETZEL DK, EHRENSHAFT M, DENSLOW SA and DAU ME (2004) 'Functional comple-mentation between the *PDX1* vitamin B6 biosynthetic gene of *Cercospora nicotianae* and *pdxJ* of *Escherichia coli*'. *FEBS Lett*, **564**, 143–6.
50. TANI Y (1989) 'Microbial production of vitamin B6 derivatives'. *Biotechnology of Vitamins, Pigments and Growth Factors*, Vandamme E.J. (ed.), Elsevier Applied Science, London, 221–30.
51. CHEN H and XIONG L (2009) 'Enhancement of vitamin B(6) levels in seeds through metabolic engineering'. *Plant Biotechnol J*, **7**, 673–81.
52. HERRERO S and DAUB ME (2007) 'Genetic manipulation of vitamin B-6 biosynthe-sis in tobacco and fungi uncovers limitations to up-regulation of the pathway'. *Plant Sci*, **172**, 609–20.
53. SANYAL I, GIBSON KJ and FLINT DH (1996) 'Escherichia coli biotin synthase: An investigation into the factors required for its activity and its sulfur donor'. *Arch Biochem Biophys*, **326**, 48–56.
54. OHSHIRO T, YAMAMOTO M, IZUMI Y, BUI BT, FLORENTIN D and MARQUET A (1994) 'Enzymatic conversion of dethiobiotin to biotin in cell-free extracts of a *Bacillus sphaericus* bioB transformant'. *Biosci Biotechnol Biochem*, **58**, 1738–41.
55. BOWER S, PERKINS JB, YOCUM RR, HOWITT CL, RAHAIM P and PERO J (1996) 'Cloning, sequencing, and characterization of the *Bacillus subtilis* biotin biosynthetic operon'. *J Bacteriol*, **178**, 4122–30.
56. SAKURAI N, IMAI Y, MASUDA M, KOMATSUBARA S, and TOSA T (1995) 'Further improve-ment of D-biotin production by a recombinant strain of *Serratia marcescens*'. *Process Biochem*, **30**, 553–62.
57. SAITO II, HONDA H, KAWABE T, MUKUMOTO F, SHIMIZU M and KOBAYASHI T (2000) 'Comparison of biotin production by recombinant *Sphingomonas* sp. under various agitation conditions'. *Biochem Eng J*, **5**, 129–36.
58. HOSHINO TK, ASAKURA AF, KIYASU TF and NAGAHASHI YF (2002) *Production of Biotin*. United States Patent 6361978.
59. DIAZ DE LA GARZA RI, GREGORY JF, 3RD and HANSON AD (2007) 'Folate biofortifica-tion of tomato fruit'. *Proc Natl Acad Sci USA*, **104**, 4218–22.
60. STOROZHENKO S, DE BROUWER V, VOLCKAERT M, NAVARRETE O, BLANCQUAERT D, ZHANG GF, LAMBERT W and VAN DER STRAETEN D (2007) 'Folate fortification of rice by metabolic engineering'. *Nat Biotechnol*, **25**, 1277–9.
61. ZHU T, PAN Z, DOMAGALSKI N, KOEPSEL R, ATAAI MM and DOMACH MM (2005) 'Engi-neering of *Bacillus subtilis* for enhanced total synthesis of folic acid'. *Appl Environ Microbiol*, **71**, 7122–9.
62. SYBESMA W, STARRENBURG M, KLEEREBEZEM M, MIERAU I, DE VOS WM and HUGEN-HOLTZ J (2003) 'Increased production of folate by metabolic engineering of *Lactococcus lactis*'. *Appl Environ Microbiol*, **69**, 3069–76.
63. WEGKAMP A, VAN OORSCHOT W, DE VOS WM and SMID EJ (2007) 'Characterization of the role of *para*-aminobenzoic acid biosynthesis in folate production by *Lac-tococcus lactis*'. *Appl Environ Microbiol*, **73**, 2673–81.
64. CAI L, YUAN MQ, LI ZJ, CHEN JC and CHEN GQ (2012) 'Genetic engineering of *Ketogulonigenium vulgare* for enhanced production of 2-keto-L-gulonic acid'. *J Biotechnol*, **157**, 320–5.
65. MIYATA R and YONEHARA T (1999) *Method for Producing Folic Acid*. US Patent No 5,968,788.
66. FLORENT J (1986) 'Vitamins'. *Biotechnology*, Volume 4, Papa, H and Rehm, H.-J. (eds), VCH, Weinheim, 114–58.
67. SPALLA C, GREIN A, GAROFANO L and FERNI G (1989) 'Microbial production of vitamin B12'. *Biotechnology of Vitamins, Pigments and Growth Factors*, Van-damme, E.J. (ed.), Elsevier Applied Science, London, 257–84.

68. POUWELS P, VAN LUIJK N, JORE J and LUITEN R (1999) *Gist-Brocades*. World Patent 99/67356.

69. ZHANG Y, LIU JZ, HUANG JS and MAO ZW (2010) 'Genome shuffling of *Propionibacterium shermanii* for improving vitamin B12 production and comparative proteome analysis'. *J Biotechnol*, **148**, 139–43.

70. BIEDENDIECK R, MALTEN M, BARG H, BUNK B, MARTENS JH, DEERY E, LEECH H, WARREN MJ and JAHN D (2010) 'Metabolic engineering of cobalamin (vitamin B12) production in *Bacillus megaterium*. *Microb Biotechnol*, **3**, 24–37.

71. FRANCIS B, CAMERON B, CROUZET J, DEBUSSCHE L, LEVY-SCHIL S and THIBAUT D (1998) Rhône-Poulenc Biochimie. *Polypeptides Involved in the Biosynthesis of Cobalamines and/or Cobamides, DNA Sequences Coding for these Polypeptides, and their Preparation and Use*. Eur Patent 0516647 B1.

72. BLANCHE F, CAMERON B, CROUZET J, DEBUSSCHE L and THIBAUT D (1997) Rhône-Poulenc Rorer. *Biosynthesis Method Enabling Preparation of Cobalamins*. World Patent 97/43421.

73. MARTENS JH, BARG H, WARREN MJ and JAHN D (2002) 'Microbial production of vitamin B12'. *Appl Microbiol Biotechnol*, **58**, 275–85.

74. BREMUS C, HERRMANN U, BRINGER-MEYER S and SAHM H (2006) 'The use of microorganisms in L-ascorbic acid production'. *J Biotechnol*, **124**, 196–205.

75. URBANCE JW, BRATINA BJ, STODDARD SF and SCHMIDT TM (2001) 'Taxonomic characterization of *Ketogulonigenium vulgare* gen. nov., sp. nov. and *Ketogulonigenium robustum* sp. nov., which oxidize L-sorbose to 2-keto-L-gulonic acid'. *Int J Syst Evol Microbiol*, **51**, 1059–107.

76. SUGISAWA T, HOSHINO T, MASUDA S, NOMURA S, SETOGUCHI Y, TAZOE M, SHINJOH M, SOMEHA S and FUJIWARA A, (1990) 'Microbial production of 2-keto-L-gulonic acid from L-sorbose and D-sorbitol by *Gluconobacter melanogenus*'. *Agric Biol Chem*, **54**, 1201–9.

77. SONOYAMA T, TANI H, MATSUDA K, KAGEYAMA B, TANIMOTO M, KOBAYASHI K, YAGI S, KYOTANI H and MITSUSHIMA K (1982) 'Production of 2-keto-L-gulonic acid from D-glucose by two-stage fermentation'. *Appl Env Microbiol*, **43**, 1064–9.

78. ISONO M, NAKANISHI I, SASAJIMA K-I, MOTIZUKI K, KANZAKI I, OKAZAKI H and YOSHINO H (1968) '2-keto-L-gulonic acid fermentation. Part I. Paper chromatography characterization of metabolic products from sorbitol and sorbose by various bacteria'. *Biol Chem* **32**, 424–31.

79. AIGUO J and PEIJI G (1998) 'Synthesis of 2-keto-L-gulonic acid from gluconic acid by co-immobilized *Gluconobacter oxydans* and *Corynebacterium* sp'. *Biotechnol Lett*, **20**, 939–42.

80. ZINSHENG Y, ZENGXIN T, LONGHUA Y, GUANGHIN Y and WENZHU NC (1981) 'Studies on production of vitamin C precursor-2-keto-L-gulonic acid from L-sorbose by fermentation. II. Conditions for submerged fermentation of 2-keto-L-gulonic acid'. *Acta Microbiol Sin*, **21**, 185–91.

81. CHOTANI G, DODGE T, HSU A, KUMAR M, LADUCA R, TRIMBUR D, WEYLER W and SANFORD K (2000) 'The commercial production of chemicals using pathway engineering'. *Biochim Biophys Acta*, **1543**, 434–55.

82. SAITO Y, ISHII Y, HAYASHI H, IMAO Y, AKASHI T, YOSHIKAWA KNY, SOEDA S, YOSHIDA M, NIWA M, HOSODA J and SHIMOMURA K (1997) 'Cloning of genes coding for L-sorbose and L-sorbosone dehydrogenases from *Gluconobacter oxydans* and microbial production of 2-keto-L-gulonate, a precursor of lascorbic acid, in a recombinant *G. oxydans* strain'. *Appl Environ Microbiol*, **63**, 454–60.

83. BERRY A, LEE C, MAYER AF and SHINJOH M (2005) *Microbial Production of L-Ascorbic Acid*. World Intellectual Property Organisation. Patent WO 2005/017172.

© Woodhead Publishing Limited, 2013

84. SAUER M, BRANDUARDI P, VALLI M and PORRO D (2004) 'Production of L-ascorbic acid by metabolically engineered *Saccharomyces cerevisiae* and *Zygosaccharomyces bailii*'. *Appl Environ Microbiol*, **70**, 6086–91.
85. RUNNING JA, HUSS RJ and OLSON PT (1994) 'Heterotrophic production of ascorbic acid by microalgae'. *J Appl Phycol* **6**, 99–104.

© Woodhead Publishing Limited, 2013

Index

© Woodhead Publishing Limited, 2013

© Woodhead Publishing Limited, 2013

© Woodhead Publishing Limited, 2013

© Woodhead Publishing Limited, 2013

© Woodhead Publishing Limited, 2013

© Woodhead Publishing Limited, 2013

© Woodhead Publishing Limited, 2013

© Woodhead Publishing Limited, 2013

© Woodhead Publishing Limited, 2013

© Woodhead Publishing Limited, 2013

near infrared spectroscopy (NIRS),
134–8
advantages and limitations, 134–5
Neurospora crassa, 24
New Harvest, 548
niacin, 580
nicotinamide, 580
nicotinamide adenine dinucleotide
(NADH), 295
nicotinic acid, 580
Nisaplin, 354, 364, 370
nitrilase, 580
nitrogen limitation, 206
non-refined carbon sources, 218
Novozymes, 145
nutraceuticals
microalgae as source, 559–60
polyunsaturated fatty acids microbial
production, 531–55
systems and synthetic biology
application, 81–92
advantages, 84–6
definition and uses, 82–4
future trends, 91–2
omics approach for food grade
amino acids production, 86–9
systems approach used in food
production, 90–1

off gas analysis, 130–3
combination oxygen/carbon dioxide
off gas analyser integral pump,
132
magnetic sector mass spectrometer
and gas manifold, 133
Oil of Javanicus, 540
oligofructose, 497
Oligomate, 506
oligosaccharide, 495
one stage culture, 217–18
open cultures, 216
OptGene algorithm, 61
ordinary differential equations (ODE),
162
organic acids
citric acid cycle oxidative branch,
293–306
citric acid cycle reductive branch,
306–10
filamentous fungi, genetically
engineered bacteria and baker's
yeasts, 289
annual production of food-related
organic acids, 291

food-related organic acids
chemical structures, 290
future trends, 312–13
gluconic acid production, 289–93
kojic acid, 310–11
microbial production for food use,
288–313
oscillatory baffled bioreactor (OBR),
117–18
oxidative stress, 205–6
2-oxoglutamic acid, 396
oxygen demands, 104
oxygen transfer
function of scale, 159–61
scale down for two broth types,
160
oxygen transfer rates (OTR), 161,
168–9

paclitaxol, 237
pantothenic acid, 581
D-pantothenyl alcohol, 581
Paracoccus zeaxanthinifaciens, 196
pectinases, 267
pediocins, 357
Pediococcus acidilactici MCH14, 357
Pediococcus pentosaceus BCC3772, 357
Penicillium chrysogenum, 99
pentose-phosphate pathway (PPP), 87
peptidoglycan, 401
PGG, 429
nutritional and health effects, 451
2-phenylethanol, 182
phosphoglucomutase (PGM), 431
phospholipase, 269
photoregulation, 206–7
phycobilliproteins, 563
phycocyanin, 563, 565
phylloquinone, 577
Pichia anomala, 67–8
Pichia guillermondii, 63
pigments, 562–3
P7L, 506
Plantaricin C, 367
polyols, 469–89
biochemistry of sugar alcohol
metabolism, 475–80
erythritol, 475–6
hexose metabolism and mannitol
synthesis in heterofermentative
LAB, 479
mannitol, 479–80
metabolism of sorbitol production
in *Lactobacillus plantarum*, 478

© Woodhead Publishing Limited, 2013

© Woodhead Publishing Limited, 2013

© Woodhead Publishing Limited, 2013

© Woodhead Publishing Limited, 2013

water-soluble vitamins, 577–88
  ascorbic acid structure, 587
  biosynthetic pathway of riboflavin
    in *Ashbya gossypii*, 579
  biotin structure, 583
  folates biosynthetic pathway in
    Ashbya gossypii, 585
  methylcobalamine structure, 586
  nicotinic acid and nicotinamide
    structures, 580
  pantothenic acid structure, 581
  pyridoxine, pyridoxal phosphate,
    PLP and pyridoxamine
    phosphate structures, 582
  thiamine chemical structure, 578
Vivinal GOS, 505, 506

wastewater treatment, 7–8
whole genome shotgun (WGS), 25
wine production, 53–5

x-omics, 46, 50
xanthan, 415, 417
  biosynthesis, 432
  industrial production, 439, 440, 442
xanthan gum, 443
Xanthomonas campestris, 115
Xanthophyllomyces dendrorhous, 64,
    196, 197, 210
xylan, 507
xylanases, 267
xylinan *see* acetan
xylitol, 57–8, 480
  biochemistry of sugar alcohol
    metabolism, 476–7

biotechnological production
  strategies
  chemical dehydrogenation, 484–6
  microbiological production, 482
  patented production processes,
    485
history, 471
physiological effects, 474, 475
xylitol dehydrogenase (XDH), 476
xylitol phosphate dehydrogenase
  (XPDH), 477
xylo-oligosaccharides (XOS), 507–8
  general structure, 507
xylobiose, 507
xylose, 471
xylose reductase (XR), 476
xylotriose, 507
xylulokinase (XK), 476

Yarrowia lipolytica, 66–7, 182
  citric acid production, 298–9
yeast
  flavour production, 57–62
  science, 43–4
  systems biology and food and food
    ingredients production, 47, 50–2
  timeline, 51
yeast artificial chromosome (YAC),
    254–5

zeaxanthin, 221–2
ZerO$_2$, 339
Zymomonas mobilis, 197
Zymosan, 429
  nutritional and health effects, 451

© Woodhead Publishing Limited, 2013